Ecological Studies, Vol. 14

Analysis and Synthesis

Edited by

M.M. Caldwell, Logan, USA
G. Heldmaier, Marburg, Germany
O.L. Lange, Würzburg, Germany
H.A. Mooney, Stanford, USA
E.-D. Schulze, Jena, Germany
U. Sommer, Kiel, Germany

Ecological Studies

Volumes published since 1994 are listed at the end of this book.

Springer

Berlin
Heidelberg
New York
Barcelona
Hong Kong
London
Milan
Paris
Singapore
Tokyo

E.-D. Schulze (Ed.)

Carbon and Nitrogen Cycling in European Forest Ecosystems

With 183 Figures, 14 in Color, 106 Tables, and CD-ROM

Springer

Prof. Dr. Ernst-Detlef Schulze
Max-Planck-Institute for Biogeochemistry
P.O. Box 10 01 64
07701 Jena
Germany

Design of the cover illustration by Barbara Lühker

ISSN 0070-8356
ISBN 3-540-67025-4 Springer-Verlag Berlin Heidelberg New York (hardcover)
ISBN 3-540-67239-7 Springer-Verlag Berlin Heidelberg New York (softcover)

Library of Congress Cataloging-in-Publication Data
Carbon and nitrogen cycling in European forest ecosystems /E.-D. Schulze (ed.).
    p. cm. – (Ecological studies, ISSN 0070-8356; vol. 142)
    Includes bibliographical references.
    ISBN 3540670254 (hardcover : alk.paper) – ISBN 3540672397 (softcover : alk.paper)
    1. Carbon cycle (Biogeochemistry) – Europe.   2. Nitrogen cycle – Europe.   3. Forest
 ecology – Europe.   I. Schulze, E.-D. (Ernst-Detlef), 1941.   II. Ecological studies; v. 142.

This work is subject to copyright. All rights are reserved, whether the whole or part of the material is concerned, specifically the rights of translation, reprinting, reuse of illustrations, recitation, broadcasting, reproduction on microfilm or in any other way, and storage in data banks. Duplication of this publication or parts thereof is permitted only under the provisions of the German Copyright Law of September 9, 1965, in its current version, and permissions for use must always be obtained from Springer-Verlag. Violations are liable for prosecution under the German Copyright Law.

Springer-Verlag Berlin Heidelberg New York
a member of BertelsmannSpringer Science+Business Media GmbH

© Springer-Verlag Berlin Heidelberg 2000
Printed in Germany

The use of general descriptive names, registered names, trademarks, etc. in this publication does not imply, even in the absence of a specific statement, that such names are exempt from the relevant protective laws and regulations and therefore free for general use.

Production: PRO EDiT GmbH, Heidelberg, Germany
Cover design: Design & Production GmbH, Heidelberg, Germany
Typesetting: Best-set Typesetter Ltd., Hong Kong

SPIN 10698782 (hardcover) SPIN 10755267 (softcover)
31/3136/göh 5 4 3 2 1 0 – Printed on acid free paper

Preface

One might think that the principles of carbon and nitrogen cycles of forests are textbook knowledge. However, if we are asked to quantify the fluxes and to make firm estimates about relations to natural and human-induced disturbances, then our answers will rapidly become very vague. This ecological dilemma becomes even more obvious, when scientists are questioned by politicians to verifiably state whether a region is a source or a sink for carbon. The Kyoto protocol of the Climate Convention poses this question to the scientific community, and the answers are contradictory. Fan et al. [Science 282:442(1998)] claimed that the Northern Hemisphere sink for carbon is located solely along the continental US, while Ciais (J. Geophys. Res., 1999) located the Northern Hemisphere sink in Siberia, and Lloyd (Functional Ecology 1999) convincingly showed that the main sink is located in the tropics. Thus, we are almost lost and, as soon as we depart from natural vegetation into managed systems, the uncertainties are even greater.

Based on a 5-year investigation, this books, adopts the IGBP approach of studying continental transects, because only by comparison beyond our "backyard" we are able to identify general principles. The study of transects seems to be the most manageable approach to quantify ecosystem processes on a continental scale, which is needed if ecology wants to compete and co-operate with earth system science in the future. Thus, the demanding task was undertaken to study whole ecosystems in a comparative way, and under a joint protocol along a transect reaching from north Sweden to central Italy. It is quite clear that it remains impossible to study a geographic transect in a contiguous way. In the present work, the transect was used as a tool in order to maintain certain conditions as constant as possible, e.g. soils or type of vegetation. The focus was rather on other important variables (e.g. climate and deposition), and to determine processes that drive the ecosystem mass balance of carbon and nitrogen along these conditions. This task became further complicated by the insight (see Schulze and Mooney, Ecological Studies 99) that the biogeochemical cycles of terrestrial ecosystems can no longer be studied without considering the specific function of biodiversity. Obviously, European forests are not very diverse with respect to tree species. In fact, they are managed monocultures, but a wealth of species remains in the soil, and the subtle interaction between plants and mycorrhizae, and other soil organisms is not at all obvious. This book addresses this problem and investigates to what extent this diversity affects ecosystem functioning, and why.

In view of the present request by the public and to obtain answers that were posed by the Kyoto protocol, we would like to acknowledge the foreseeing actions of the EU to emphasise transect studies across Europe, and to foster basic understanding of processes. When the project started, the aim to support basic research was a consequence of the problems experienced with identifying the causes of forest decline. In view of the present demand for data on carbon sequestration, the early decision by the EU may demonstrate the importance of basic science for the future of a society. We would like to make this point at a time when scientists are increasingly asked to carry out research of public concern, or with direct applications.

Jena, June 2000 E.-D. Schulze

Contents

Part A Introduction to the European Transect

1	The Carbon and Nitrogen Cycle of Forest Ecosystems	
	E.-D. Schulze	3
1.1	Introduction	3
1.2	The Carbon and Nitrogen Cycles	3
1.3	The NIPHYS/CANIF Project	8
1.4	Experimental Design	10
1.5	Conclusions	11
References		11

2	Experimental Sites in the NIPHYS/CANIF Project	
	T. Persson, H. van Oene, A.F. Harrison, P.S. Karlsson, G.A. Bauer, J. Cerny, M.-M. Coûteaux, E. Dambrine, P. Högberg, A. Kjøller, G. Matteucci, A. Rudebeck, E.-D. Schulze, and T. Paces	14
2.1	Site Description. The NIPHYS/CANIF Transect	14
2.2	Soil Characteristics	26
2.3	Ecosystem C and N Pools	36
2.4	Database	39
2.5	Conclusions	42
References		46

Part B Plant-Related Processes

3	Tree Biomass, Growth and Nutrient Pools	
	G. Scarascia-Mugnozza, G.A. Bauer, H. Persson, G. Matteucci, and A. Masci	49
3.1	Introduction	49
3.2	Experimental Background	50
3.3	Biomass	52
3.4	Forest Productivity	54
3.5	Carbon and Nutrient Pools	56
3.6	Allometric and Functional Relations	58
3.7	Conclusions	60
References		62

4	**Linking Plant Nutrition and Ecosystem Processes** G.A. Bauer, H. Persson, T. Persson, M. Mund, M. Hein, E. Kummetz, G. Matteucci, H. van Oene, G. Scarascia-Mugnozza, and E.-D. Schulze	63
4.1	Introduction	63
4.2	Experimental Approach	64
4.3	Nutrient Concentrations	65
4.4	Nutrient Contents	73
4.5	Nitrogen Partitioning in Different Tree Compartments	82
4.6	Ecosystem C and N Pools	88
4.7	Conclusions	95
	References	95
5	**Root Growth and Response to Nitrogen** C. Stober, E. George, and H. Persson	99
5.1	Introduction	99
5.2	Approaches to the Study of Root Growth	100
5.3	Root Growth Measurements Obtained by Soil Coring	105
5.4	Root Growth Measurements Obtained by Root Windows	107
5.5	Root Growth Measurements Obtained by In-Growth Cores	113
5.6	Root Growth at Different European Forest Sites	116
5.7	Conclusions	118
	References	119
6	**Nitrogen Uptake Processes in Roots and Mycorrhizas** T. Wallenda, C. Stober, L. Högbom, H. Schinkel, E. George, P. Högberg, and D.J. Read	122
6.1	Introduction	122
6.2	Approaches to Study Different Aspects of the N Uptake Process	123
6.3	Studies with Excised Roots and Mycorrhizas	125
6.4	Field-Based Experiments	135
6.5	Conclusions	139
	References	141
7	**The Fate of ^{15}N-Labelled Nitrogen Inputs to Coniferous and Broadleaf Forests** G. Gebauer, B. Zeller, G. Schmidt, C. May, N. Buchmann, M. Colin-Belgrand, E. Dambrine, F. Martin, E.-D. Schulze, and P. Bottner	144
7.1	Introduction	144
7.2	Sites of Investigation	145
7.3	Approaches to Study the Fate of ^{15}N-Labelled Nitrogen Inputs	147
7.4	N Release and Tree Uptake from ^{15}N-Labelled Decomposing Litter in a Beech Forest in Aubure	150
7.5	Ecosystem Partitioning of ^{15}N-Labelled Ammonium and Nitrate on the Sites in the Fichtelgebirge and Steigerwald	157
7.6	Conclusions	166
	References	168

Contents

8 Canopy Uptake and Utilization of Atmospheric Pollutant Nitrogen
A.F. Harrison, E.-D. Schulze, G. Gebauer, and G. Bruckner 171

8.1 Introduction 171
8.2 Atmospheric Nitrogen Pollutants 172
8.3 Pathways for Canopy Uptake of Nitrogen 174
8.4 Approaches to the Determination of Canopy Uptake of Nitrogen 175
8.5 Review of Research 177
8.6 Role in the Critical Load 182
8.7 Ecophysiological Consequences of Canopy N Uptake 182
8.8 Conclusions 183
8.9 Way Forward 183
8.10 Policy Implications 184
References 184

9 Biotic and Abiotic Controls Over Ecosystem Cycling of Stable Natural Nitrogen, Carbon and Sulphur Isotopes
G.A. Bauer, G. Gebauer, A.F. Harrison, P. Högberg, L. Högbom,
H. Schinkel, A.F.S. Taylor, M. Novak, F. Buzek, D. Harkness, T. Persson,
and E.-D. Schulze 189

9.1 Introduction 189
9.2 Approaches to the Study of Stable Isotopes in the Field 189
9.3 $\delta^{15}N$ of Ammonium and Nitrate in Wet Deposition 191
9.4 Stable Isotope Signatures in Different Ecosystem Compartments 197
9.5 $\delta^{15}N$ Signatures as Indicators
    of N Saturation in Forest Ecosystems 207
9.6 Conclusions 211
References 213

Part C Heterotrophic Processes

10 Soil Respiration in Beech and Spruce Forests in Europe: Trends, Controlling Factors, Annual Budgets and Implications for the Ecosystem Carbon Balance
G. Matteucci, S. Dore, S. Stivanello, C. Rebmann, and N. Buchmann ... 217

10.1 Introduction 217
10.2 Approaches to Measuring Soil Respiration 218
10.3 Daily and Seasonal Trends in Soil Respiration
     and Climatic Variables 221
10.4 Factors Controlling Soil Respiration 226
10.5 Comparison of Chamber Measurements
     with the Eddy Covariance Measurements Below the Canopy 230
10.6 Annual Budgets of Soil Respiration 231
10.7 Conclusions 233
References 234

11	**Annual Carbon and Nitrogen Fluxes in Soils Along the European Forest Transect, Determined Using the ^{14}C-Bomb**	
	A.F. Harrison, D.D. Harkness, A.P. Rowland, J.S. Garnett, and P.J. Bacon	237
11.1	Introduction	237
11.2	Forests, Sampling Procedure and Analysis	239
11.3	Model Description	241
11.4	Estimations of C and N Pools and Fluxes	242
11.5	Pools and Distribution of Carbon and Nitrogen in Soil Profiles	243
11.6	Variations in the Carbon Age and Mean Residence Times (MRTs)	246
11.7	Annual Carbon and Nitrogen Fluxes	248
11.8	General Discussion	252
11.9	Conclusions	253
References		255
12	**Carbon Mineralisation in European Forest Soils**	
	T. Persson, P.S. Karlsson, U. Seyferth, R.M. Sjöberg, and A. Rudebeck	257
12.1	Introduction	257
12.2	Experimental Background	257
12.3	C Mineralisation in the North-South Transect	260
12.4	Long-Term Fertilisation Experiments	269
12.5	Mean Residence Time	271
12.6	Comparison of Intact and Sieved Soil Cores	271
12.7	Conclusions	274
References		275
13	**Litter Decomposition**	
	M.F. Cotrufo, M. Miller, and B. Zeller	276
13.1	Introduction	276
13.2	Factors Affecting the Decomposition Process	276
13.3	Enzymatic Activity	278
13.4	Nitrogen Dynamics in Decomposing Litter	279
13.5	Decomposition Studies in Europe: from DECO, VAMOS, MICS to CANIF	280
13.6	Decomposition Studies Within a Latitudinal Transect of European Beech Forests	281
13.7	Conclusions	292
References		293
14	**Soil Nitrogen Turnover – Mineralisation, Nitrification and Denitrification in European Forest Soils**	
	T. Persson, A. Rudebeck, J.H. Jussy, M. Colin-Belgrand, A. Priemé, E. Dambrine, P.S. Karlsson, and R.M. Sjöberg	297
14.1	Background and Aim of the Study	297
14.2	Methods Used to Study N Turnover	299

14.3	Net N Mineralisation Based on Laboratory Studies	304
14.4	Net Nitrification Based on Laboratory Studies	307
14.5	Manipulation of pH, N Availability and Nitrifier Density in the Laboratory	312
14.6	Autotrophic Versus Heterotrophic Nitrification	314
14.7	Net N Mineralisation and Nitrification in N-Fertilisation Experiments	316
14.8	Comparison of N Turnover in Similar Soils at Different Climate	318
14.9	Comparison of N Turnover in Sieved and Intact Soil Cores	319
14.10	In Situ Mineralisation Studies at Aubure	321
14.11	Comparison of in Situ and Laboratory-Based Mineralisation Studies	323
14.12	Denitrification	324
14.13	Final Discussion	326
14.14	Conclusions	328
References		329

15	**Nitrogen and Carbon Interactions of Forest Soil Water** B.R. Andersen and P. Gundersen	332
15.1	Introduction	332
15.2	Approaches to Studying the Forest Soil Waters	333
15.3	Soil Water Concentrations of Nitrogen and Carbon	334
15.4	Correlation Between Dissolved Organic Nitrogen and Carbon	337
15.5	Conclusions	340
References		340

Part D Diversity-Related Processes

16	**Fungal Diversity in Ectomyccorhizal Communities of Norway Spruce [*Picea abies* (L.) Karst.] and Beech (*Fagus sylvatica* L.) Along North-South Transects in Europe** A.F.S. Taylor, F. Martin, and D.J. Read	343
16.1	Introduction	343
16.2	Analysis of Ectomycorrhizal Community Structure and Diversity	344
16.3	ECM Communities of Spruce Forests	347
16.4	ECM Communities of Beech Forests	353
16.5	Genetic Diversity Within a Population of *Laccarina amethystina*	353
16.6	Isolation and Growth of ECM Fungal Isolates on an Organic N Source	357
16.7	Comparative Evaluation of Ectomycorrhizal Diversity	358
16.8	Conclusions	362
References		363

17	**Diversity and Role of the Decomposer Food Web** V. Wolters, A. Pflug, A.R. Taylor, and D. Schroeter	366

17.1	Introduction	366
17.2	Approaches to Investigating Decomposer Communities	368
17.3	The Microflora	369
17.4	The Soil Fauna	372
17.5	Contribution of the Decomposer Food Web to C and N Flows	376
17.6	Conclusions	378
References		379

18 Diversity and Role of Microorganisms
A. Kjøller, M. Miller, S. Struwe, V. Wolters, and A. Pflug 382

18.1	Introduction and Background	382
18.2	Experimental Background	383
18.3	Community of Microfungi in Beech Forests	385
18.4	Functional Diversity of Bacteria in the Litter of Coniferous Forests	393
18.5	Conclusions	399
References		400

Part E Integration

19 Spatial Variability and Long-Term Trends in Mass Balance of N and S in Central European Forested Catchments
E. Dambrine, A. Probst, D. Viville, P. Biron, T. Paces, M. Novak, F. Buzek, J. Cerny, M. C. Belgrand, and H. Groscheova 405

19.1	Introduction	405
19.2	Approaches to Studying Long-Term Changes in Watersheds	405
19.3	Temporal Variations and Trends	407
19.4	Budgets	415
19.5	Biological Cycling of Sulphur	416
19.6	Conclusions	417
References		418

20 Model Analysis of Carbon and Nitrogen Cycling in *Picea* and *Fagus* Forests
H. van Oene, F. Berendse, T. Persson, A.F. Harrison, E.-D. Schulze, B.R. Andersen, G.A. Bauer, E. Dambrine, P. Högberg, G. Matteucci, and T. Paces 419

20.1	Introduction	419
20.2	Model Description	420
20.3	Input Data and Parameter Values	427
20.4	Model Calibration and Comparison with Measured Data	431
20.5	Model Analysis	442
20.6	Conclusions	462
References		463

21	Interactions Between the Carbon and Nitrogen Cycle and the Role of Biodiversity: A Synopsis of a Study Along a North-South Transect Through Europe E.-D. Schulze, P. Högberg, H. van Oene, T. Persson, A.F. Harrison, D. Read, A. Kjøller, and G. Matteucci	468
21.1	Introduction	468
21.2	Change of Ecosystem Processes Along the European Transect	468
21.3	What Limits the C and N Fluxes in These Forest Ecosystems?	472
21.4	What Are Net Ecosystem Productivity (NEP) and Net Biome Productivity (NBP) and How Do They Relate to Ecosystem Parameters?	477
21.5	Are There Thresholds and Non-Linearities?	482
21.6	What Role Does Biodiversity Play in Ecosystem Processes?	483
21.7	Conclusions	487
References		488

Subject Index	493
Species Index	499

CD-ROM containing the database of this volume enclosed at the end of the book

Contributors

(bold letters indicate principal authors of chapters and principal investigators of project)

ANDERSEN B.

Danish Forest and Landscape Research Institute, Hoersholm Kongevej 11, 2970 Hoersholm, Denmark, Tel.: 0045 45 76 3200, 0045 4517 8276, Fax: 0045 45 76 3233, e-mail: bra@fsl.dk

BACON P.J.

Banchory Research Station, Centre of Ecology and Hydrology, Hill of Brathens, Glassel, Banchory, Kincardineshire, AB31 4BY, UK

BAUER G.A.

Harvard University, Dept. of Organismic & Evolutionary Biology, 16 Divinity Ave., Cambridge, MA 02138, USA, Tel.: +1 617 495 8791, Fax: +1 617 495 9300, e-mail: gbauer@oeb.harvard.edu

BELGRAND M.C.

Cycles Biogéochimiques, INRA-Nancy, 54280 Seichamps, France

BERENDSE F.

Nature Conservation and Plant Ecology Group, Wageningen University, Bornsesteeg 69, 6708 PD Wageningen, The Netherlands, Tel.: +31 3174 84973, Fax: +31 3174 84845, e-mail: Frank.Berendse@staf.ton.wau.nl

BIRON T.

CEREC – ULP, 3 rue de l'Argonne, 67083 Strasbourg, France

BOTTNER P.

CGS – CNRS, 1 rue de Blessig, 67083 Strasbourg, France

BRUCKNER G.

Universität Bayreuth, Universitätsstr. 30, 95440 Bayreuth, Germany

BUCHMANN N.

Max-Planck-Institut für Biogeochemie, PO Box 10 01 64, 07701 Jena, Germany

BUZEK F.

Czech Geological Survey, Klarov 3, 11821 Prague, Czech Republic

CERNY J.

Czech Geological Survey, Klarov 3, 11821 Prague, Czech Republic

COTRUFO M.F.

Dipartimento di Scienze Ambientali, Seconda Universitá di Napoli, Via Vivaldi 43, 81100 Caserta, Italy, Tel.: +39 0823 274647, Fax: +39 0823 274605, e-mail: fcotrufo@tin.it

COÛTEAUX M.M.

CGS – CNRS, 1 rue de Blessig, 67083 Strasbourg, France

DAMBRINE E.

INRA-Nancy, Cycles Biogéochimiques, 54280 Seichamps, France, Tel.: +33 383 394071, Fax: +33 383 394069, e-mail: dambrine@nancy.inra.fr

DORE S.

University of Tuscia, Department of Forest Environment and Resources, Via San Camillo de Lellis, 01100 Viterbo, Italy

GARNETT J.S.

Natural Environment Research Council, Centre of Ecology and Hydrology, Merlewood Research Station, Grange-over-Sands, Cumbria LA11 6JU, UK

GEBAUER G.

Universität Bayreuth, Universitätsstr. 30, 95440 Bayreuth, Germany, Tel.: +49 921 / 55 27 48, 2798, 25 67, Fax: +49 921 / 55 25 64, e-mail: Gerhard.Gebauer@uni-bayreuth.de

GEORGE E.

Universität Hohenheim, Institut für Pflanzenernährung, Postfach 70 05 62, 70593 Stuttgart, Germany, Tel.: +49 711 459 3664, Fax: +49 711 459 3295, e-mail: gorge@uni-hohenheim.de

GROSCHEOVA H.

Czech Geological Survey, Klarov 3, 11821 Prague, Czech Republic

GUNDERSEN P.

Danish Forest and Landscape Research Institute, Hoersholm Kongevej 11, 2970 Hoersholm, Denmark

Contributors

HARKNESS D.D.

NERC Radiocarbon Laboratory, Scottish Enterprise Technology Park, Rankine Ave, East Kilbride, Glasgow G75 0QF, UK

HARRISON A.F.

Centre of Ecology and Hydrology, Natural Environment Research Council Merlewood Research Station, Grange-over-Sands, Cumbria, LA11 6JU, UK, Tel.: 00441 5395 32264, Fax: 00441 5395 34705, e-mail: T.Harrison@CEH.AC.UK

HEIN M.

Max-Planck-Institut für Biogeochemie, P.O. Box 10 01 64, 07701 Jena, Germany

HÖGBERG P.

Swedish University of Agricultural Sciences (SLU), Dept. of Forest Ecology, 90183 Umeå, Sweden, Tel.: 0046 90786 5007, Fax: 0046-90786 7750, e-mail: Peter.Hogberg@sek.slu.se

HÖGBOM L.

Swedish University of Agricultural Sciences (SLU), Dept. of Forest Ecology, 90183 Umeå, Sweden
Present address: Skog Forsk, Uppsala Science Park, 75183 Uppsala, Sweden

JUSSY J.H.

INRA-Nancy, Cycles Biogéochimiques, 54280 Seichamps, France

KARLSSON P.S.

Swedish University of Agricultural Sciences (SLU), Dept. of Ecology and Environmental Res., P.O. Box 7072, 75007 Uppsala, Sweden

KJØLLER A.

University of Copenhagen, Department of General Microbiology, Sølvgade 83 H, 1307 Copenhagen K, Denmark, Tel.: +45 35 322045, Fax: +45 35 322040, e-mail: ak@mermaid.molbio.ku.dk

KUMMETZ E.

Universität Bayreuth, Universitätsstr. 30, 95440 Bayreuth, Germany

MARTIN F.

INRA-Nancy, Cycles Biogéochimiques, 54280 Seichamps, France, Tel.: +33 83 39 40 80, Fax: +33 83 39 40 69, e-mail: fmartin@nancy.inra.fr

MATTEUCCI G.

DISAFRI – University of Tuscia, Dept. of Forest Environm. And Resources, Via San Camilo de Lellis, 01100 Viterbo, Italy, Tel.: +39 0761 357292, Fax: +39 0761 357389, e-mail: macchia@unitus.it

MAY C.

Universität Bayreuth, Universitätsstr. 30, 95440 Bayreuth, Germany

MASCI A.

University of Tuscia, Department of Forest Environment and Resources, Via San Camillo de Lellis, 01100 Viterbo, Italy

MILLER M.

University of Copenhagen, Department of General Microbiology, Sølvgade 83 H, 1307 Copenhagen K, Denmark

MUND M.

Max-Planck-Institut für Biogeochemie, PO Box 10 01 64, 07701 Jena, Germany

NOVAK M.

Czech Geological Survey, Klarov 3, 11821 Prague, Czech Republic

PACES T.

Czech Geological Survey, Klarov 3, 11821 Prague, Czech Republic, Tel.: +420-2-6816945, Fax: +420-2-581 6945, +420-2-581 8748 or 581 6748, e-mail: paces@cgu.cz

PERSSON H.

Swedish University of Agricultural Sciences, Dept. of Ecology and Environmental Res., P.O. Box 70 72, 750 07 Uppsala, Sweden, Tel.: +46 18 67 2426, Fax: +46 18 67 3430, e-mail: Hans.Persson@eom.slu.se

PERSSON T.

Swedish University of Agricultural Sciences, Dept. of Ecology and Environmental Res., P.O. Box 70 72, 750 07 Uppsala, Sweden, Tel.: +46 18 67 2448, Fax: +46 18 67 3430, e-mail: Tryggve.Persson@eom.slu.se

PFLUG A.

Justus-Liebig-Universität, Stephanstr. 24, 35390 Giessen, Germany

PRIEMÉ A.

University of Copenhagen, Department of General Microbiology, Sølvgade 83 H, 1307 Copenhagen K, Denmark

PROBST A.

CGS – CNRS, 1 rue de Blessig, 67083 Strasbourg, France

READ D.J.

The University of Sheffield, Dept. of Animal and Plant Sciences, Sheffield, S10 2UQ, UK, Tel.: 0044 1 14 222 4318, Fax: 0044 1 14 276 0159, e-mail: D.J.Read@sheffield.ac.uk

REBMANN C.

Max-Planck-Institut für Biogeochemie, P.O. Box 10 01 64, 07701 Jena, Germany

ROWLAND A.P.

Natural Environment Research Council, Centre of Ecology and Hydrology, Merlewood Research Station, Grange-over-Sands, Cumbria, LA11 6JU, UK

RUDEBECK A.

Swedish University of Agricultural Sciences (SLU), Dept. of Ecology and Environmental Res., P.O. Box 7072, 75007 Uppsala, Sweden

SCARASCIA-MUGNOZZA G.

DISAFRI-University of Tuscia, Dept. of Forest Environment and Resources, Via San Camillo de Lellis, 01100 Viterbo, Italy, Tel.: 0039 0761 357395, Fax: 0039 0761 357389, e-mail: gscaras@unitus.it

SCHINKEL H.

Swedish University of Agricultural Sciences (SLU), Dept. of Forest Ecology, 90183 Umeå, Sweden

SCHMIDT G.

Universität Bayreuth, Universitätsstr. 30, 95440 Bayreuth, Germany

SCHROETER D.

Justus-Liebig-Universität, Stephanstr. 24, 35390 Giessen, Germany

SCHULZE E.-D.

Max-Planck-Institut für Biogeochemie, P.O. Box 10 01 64, 07701 Jena, Germany, Tel.: +49 3641 643702, Fax: +49 3641 643794, e-mail: detlef.schulze@bgc-jena.mpg.de

SEYFERTH U.

Swedish University of Agricultural Sciences, Dept. of Ecology and Environmental Res., P.O. Box 70 72, 750 07 Uppsala, Sweden

SJÖBERG R.M.

Swedish University of Agricultural Sciences, Dept. of Ecology and Environmental Res., P.O. Box 70 72, 750 07 Uppsala, Sweden

STIVANELLO S.

DISAFRI – University of Tuscia, Dept. of Forest Environment and Resources, Via San Camilo de Lellis, 01100 Viterbo, Italy

STOBER C.

Universität Hohenheim, Institut für Pflanzenernährung, Postfach 70 05 62, 70593 Stuttgart, Germany, Tel.: +49 711 459 2344, Fax: +49 711 459 3295, e-mail: george@uni-hohenheim.de

STRUWE S.

University of Copenhagen, Dept. of Microbiology, Sølvegade 83H, 1307 Copenhagen, Denmark, Tel.: +45 35 32 20 44, Fax: +45 35 32 20 40, e-mail: struwe@mermaid.molbio.ku.dk

TAYLOR A.F.S.

The University of Sheffield, Dept. of Animal and Plant Sciences, Sheffield, S10 2UQ, UK, Tel.: +46-18-672797 (direct), Fax: +46-18-309245, e-mail: Andy.Taylor@mykopat.slu.se
Present address: Swedish University of Agricultural Sciences, Dept. of Forest Mycology and Pathology, P.O. Box 7076, 75007 Uppsala, Sweden

TAYLOR A.R.

Justus-Liebig-Universität, Stephanstr. 24, 35390 Giessen, Germany

VAN OENE H.

Wageningen University, Nature Conservation and Plant Ecology Group, Bornesteeg 69, 6708 PD Wageningen, The Netherlands, Tel.: +31 317 484167, Fax: +31 317 484845, e-mail: Harmke.vanOene@staf.ton.wau.nl

VIVILLE D.

CEREC – ULP, 3 rue de l'Argonne, 67083 Strasbourg, France

WALLENDA T.

The University of Sheffield, Dept. of Animal and Plant Sciences, Sheffield, S10 2UQ, UK, Tel.: 0044 1 14 222 4318, Fax: 0044 1 14 276 0159, e-mail: D.J.Read@sheffield.ac.uk

WOLTERS V.

Justus-Liebig-Universität, Stephanstr. 24, 35390 Giessen, Germany, Tel.: +49 641 / 9 93 56 20, Office: +49 641 / 9 93 56 21, Fax: +49 641 / 9 93 56 29, e-mail: Volkmar.Wolters@allzool.bio.uni-giessen.de

ZELLER B.

INRA-Nancy, Cycles Biogéochimiques, 54280 Seichamps, France

Part A
Introduction to the European Transect

1 The Carbon and Nitrogen Cycle of Forest Ecosystems

E.-D. Schulze

1.1 Introduction

Our understanding of the biology of major biogeochemical cycles came initially from, and is still based upon, field observations (Bolin et al. 1979; Clark and Rosswall 1981; Apps and Price 1996). This is in contrast to very advanced models, which explore the physics of the climate system and are based on laws of physics or chemistry with a mechanistic understanding of the underlying processes (Houghton et al. 1996; Bengtsson 1999). For the biologist, the responses of organisms reach far beyond physicochemical reactions, and they include genetically regulated changes in physiological pathways or activation of enzyme systems as part of acclimations and adaptations that are coupled with climate and species composition changes. Generic predictions thus remain elusive because there are too many species and pathways. Although climate greatly influences the biogeochemical cycles, models that include biology thus remain at a correlative level. Moreover, the cycling of elements like carbon (C) cannot readily be separated from the abundance, state and cycles of other elements, especially nitrogen (N) (Schulze et al. 1994) which, in turn, is tied to the cycling of other elements (Ulrich 1987). Nevertheless, detailed knowledge of the biology of C cycling and that of other major and minor elements is urgently needed because the Kyoto Protocol demands strategies to balance industrial emissions by biological C fixation (WBGU 1998; IGBP 1998). By this protocol, mankind is taking a first step to deliberately engineer the biology of the global C cycle; but without full understanding of the underlying processes, there is a risk of serious deleterious side effects (Schellnhuber and Wenzel 1998; Schellnhuber 1999). Forests are a major focus in this new endeavour of global engineering because of their high C storage capacity coupled with a relatively low demand for N (Melillo et al. 1996).

This book has the timely goal of elucidating the regulation of C and N pools and fluxes in forest ecosystems. We use European coniferous and deciduous forests along a transect of climate and pollution deposition as models, and include analyses of the role of biodiversity at plant, soil faunal and microbial levels in the determination of C and N cycles. Thus, we hope to understand limitations of resource use and supply in view of the new global engineering endeavour.

1.2 The Carbon and Nitrogen Cycles

Ecosystem C and N cycles are strongly coupled by several nested loops driven by the activity of specific organisms with specialised abilities to carry out a very diverse set of functions that transform and store C and N products.

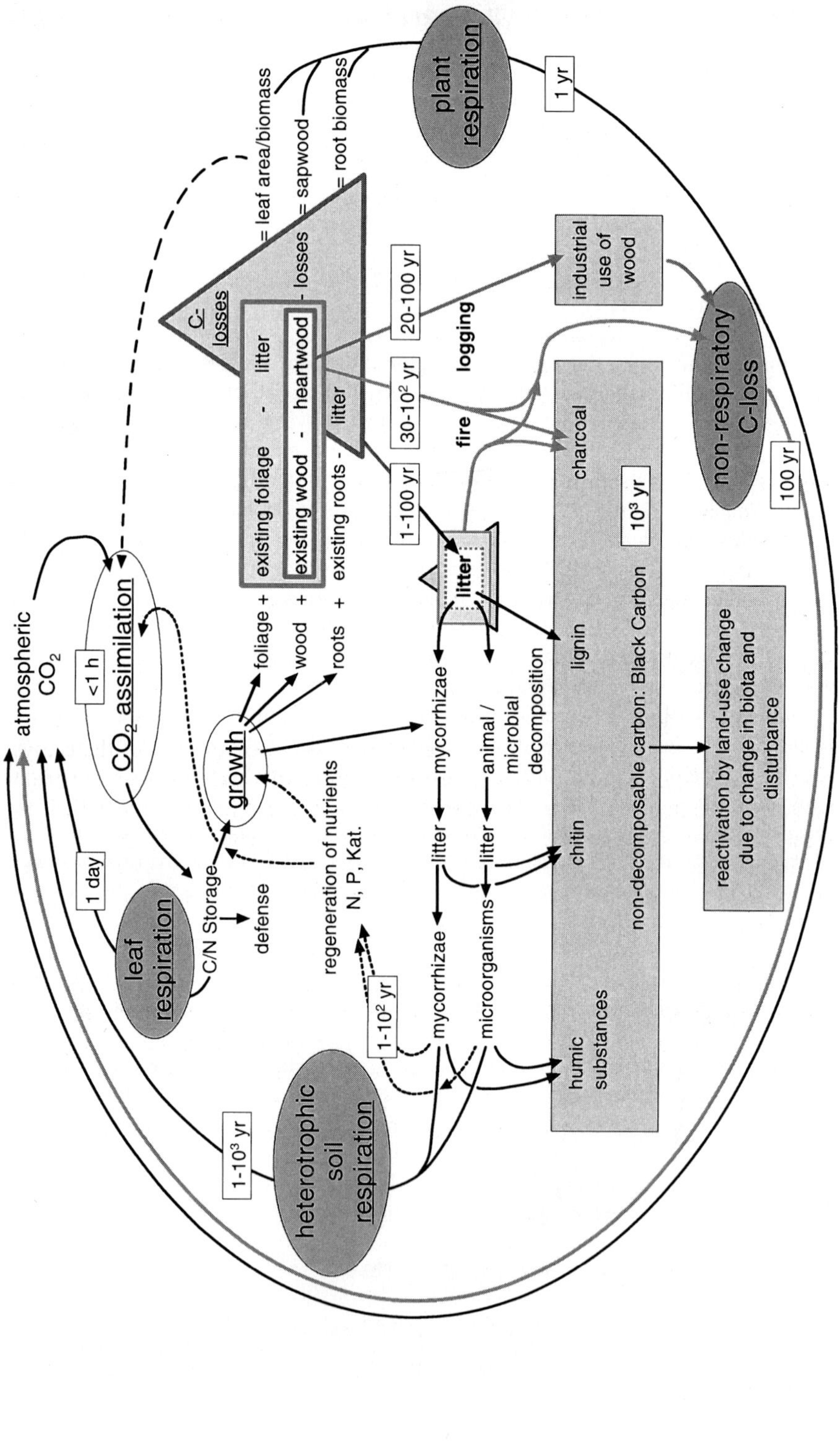
atmospheric CO2
<1 h
CO2 assimilation
1 day
leaf respiration
C/N Storage
defense
growth
foliage +
wood +
roots +
existing foliage - litter
existing wood - heartwood
existing roots - litter
C-losses
- losses
= leaf area/biomass
= sapwood
= root biomass
plant respiration
1 yr
regeneration of nutrients
N, P, Kat.
1-10² yr
mycorrhizae
litter
microorganisms
animal / microbial decomposition
litter
1-100 yr
30-10² yr
20-100 yr
fire
logging
industrial use of wood
non-respiratory C-loss
100 yr
humic substances
chitin
lignin
charcoal
10³ yr
non-decomposable carbon: Black Carbon
reactivation by land-use change due to change in biota and disturbance
heterotrophic soil respiration
1-10³ yr

The Carbon and Nitrogen Cycle of Forest Ecosystems

The C cycle (Fig. 1.1) begins with the process of CO_2 assimilation by plants. This is virtually an instantaneous process, depending on light energy. CO_2 assimilation determines the delivery of assimilates to the plant internal store, which may then be used for growth, reserve or defense (Chapin et al. 1990). In trees, growth adds biomass as foliage, wood and roots (Schulze and Chapin 1987; Stitt and Schulze 1994). However, this process is not simple. For example, the allocation of C to leaf growth leads to a positive feedback on the C-assimilation capacity of the tree, but there is an upper limit in tree canopies for leaf development, because there is a maximum leaf area index, depending on the availability of light and on leaf orientation, i.e. the distribution of foliage in three-dimensional space (Monsi and Saeki 1953). Growth is essential in plants because they operate as open systems for which the formation of new tissue offsets the internal cellular clock of cell senescence in ageing tissues. There is an upper limit for growth determined by the balance of assimilation and respiration, as growth with its associated respiration cannot exceed the cost of maintenance. Plants solve this dilemma by aptosis (Havel and Durzan 1996), in which some living tissue dies and is either shed as litter or, as in the case of heartwood in trees (Leopold 1980; Pennell and Lamb 1997), is retained to serve structural purposes. The processes involved have different time constants (or turnover rates), such as seconds or days for assimilation, and perhaps from growing seasons to years for senescence and abscission of plant parts. This further complicates the common practice of integrating growth processes over annual periods. Nevertheless, the difference between assimilation and respiration, which is equivalent to growth above and below the ground is commonly used as resulting quantity, known as net primary productivity (NPP).

The annual cycle of plant part losses, arising in the form of the litterfall derived from above- and below-ground parts, feeds back to the heterotrophs of the ecosystem which use the energy stored in the organic matter and recycle nutrients as a major resource for further plant growth. However, the decomposer chain is by no means a single pathway. Several groups of organisms competing for discarded plant organic matter include bacteria, fungi and soil animals. In this context, symbiotic mycorrhizal associations between tree roots and fungi are especially important in forests because, while supported by carbohydrates received directly from the living plant, the fungi produce enzymes enabling nutrients such as N to be recycled from plant litter without mineralisation (Smith and Read 1997). At the same time, more complex residues like cellulose and lignin are comminuted by soil animals, degraded by saprophytic fungi, and recycled by bacteria, and new compounds are synthesised by the decomposing organisms. Despite these very specialised and diverse microbial activities, some particularly recalcitrant fractions of the litter or products of the decomposers are not degradable for the decomposer community involved. These left-over products accu-

---

Fig. 1.1. The ecosystezm carbon cycle. This figure illustrates the main fluxes of carbon in forest ecosystems, neglecting minor pathways through flowers and fruits and bark. Herbivory is assumed to be part of litter. The times that are depicted at the different cycles illustrate the usual integration time of measurements or the mean residence time. *Solid arrows* depict C fluxes, *gray arrows* depict nutrient fluxes, the *broken line* between leaf biomass and assimilation indicates a major feedback of C allocation. Black carbon encompasses the recalcitrant, non-decomposable fraction of humus, which to a large proportion consists of charcoal

mulate as humus (Meyer 1993; Zech and Kögel-Knabner 1994). This pool of residues serves a valuable ecosystem function because it acts as a temporary and reversible store of nutrients, including N, which can be exchanged against equivalent charges of other ions. However, under certain conditions, for example following a change in species composition in the flora, the humus itself can also be remobilised if the edaphic conditions or the decomposing organisms change. Decomposition of litter or humus may also lead to losses of dissolved C to groundwater.

The heterotrophic respiration associated with decomposition leads to a release of CO_2 from the ecosystem. This inclusion of heterotrophic respiration with the NPP term at the ecosystem level yields net ecosystem productivity (NEP = assimilation minus autotrophic plant respiration and minus heterotrophic respiration). The NEP term is what is measured using the eddy covariance technique, though it is then commonly called net ecosystem exchange (NEE), because these measurements generally are carried out over relatively short periods of time and do not include periodic or long-term disturbances (Schulze and Heimann 1998). NEP quantifies the change in C storage of the ecosystem, including growth of woody biomass and changes in soil C.

The decomposer chain is only one possible pathway for the breakdown of organic C. The above-ground biomass and the organic layer of the soil may also be consumed by fires, which produce charcoal on one hand but release CO_2, bypassing heterotrophic respiration, on the other. Harvest of trees is the other main process disrupting the forest C cycle. Harvest generates fresh litter on one hand, but extracts a major fraction of organic material and the associated elements (nutrients) from the ecosystem on the other, and this may affect soils in the long term (Schulze and Ulrich 1991). The C balance, including disturbances that export C from the ecosystem bypassing respiration, was termed net biome productivity, NBP (Schulze and Heimann 1998). Thus, NBP includes changes in soil C, excluding changes in woody biomass. Soil C may include a short-term component (changes during stand development) and a long-term component considering the effect of disturbances, e.g. during harvest. The recalcitrant component of soil C, which may originate from fire or from soil processes, is termed black carbon.

In the following text we will not deal with fire or logging, but we will try to unravel the interaction between plants and decomposers, which covers time constants between hours and millenia of years, and where the pool of soil C acts as an "ecological memory" that is regulated by parameters and organisms other than the assimilatory process, and which may significantly determine the actual rate and the interannual variability of NEE. The time lag between assimilation and dissimilation may serve as an indicator of the mobilisation of deposited organic material. In the forest ecosystems under investigation we may neglect N_2-fixation.

While nutrients appear as an environmental parameter in the C cycle where decomposers release nutrients that are required for new growth, the emphasis changes when inspecting the N cycle. In this case, C is substantially involved only for a fairly short part of the pathway, and a shuttle between organic an inorganic forms of N play a major role at all time scales.

Nitrogen (Fig. 1.2) enters into the biological cycle through N_2 assimilation or through oxidation by electrical discharge or combustion processes. From then on, N shuttles between reduced and oxidized inorganic N and organic N, which is mainly associated with amino acids, nucleic acid and wall materials such as chitin. The plant

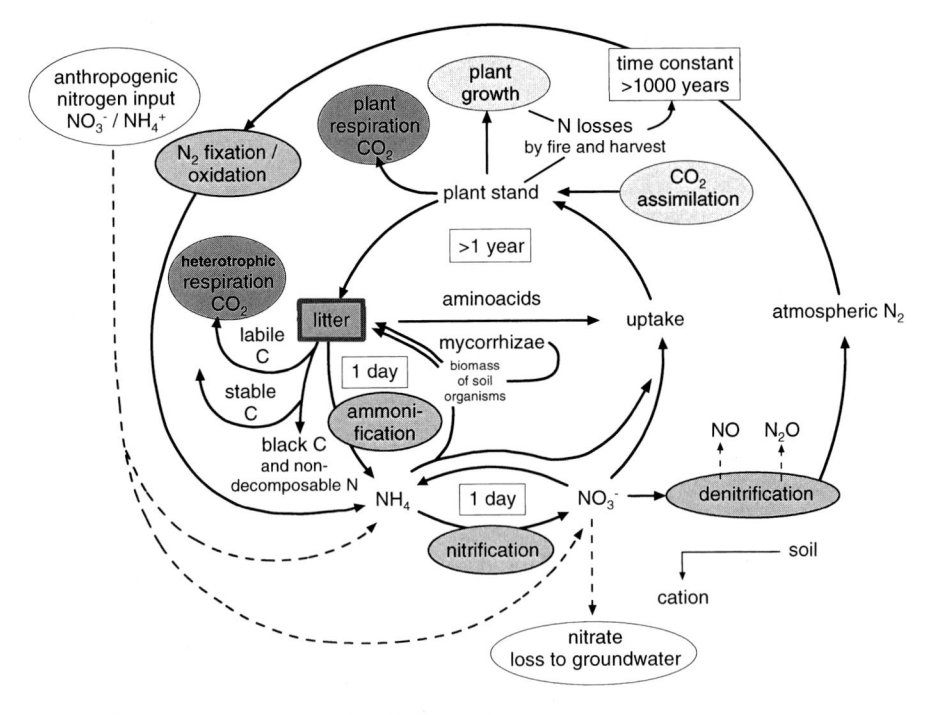

The Carbon and Nitrogen Cycle of Forest Ecosystems

Fig. 1.2. The ecosystem nitrogen cycle. This figure illustrates the major pathways of N in forest ecosystems. The times that are depicted at the different cycles illustrate the usual integration time of measurements or the mean residence time. *Solid arrows* depict C fluxes, *broken arrows* illustrate the path of anthropogenically added N

cover can potentially use all forms of N, organic as well as inorganic. Generally, however, it assimilates inorganic N and releases organic N in form of litter. Depending on conditions, the N cycle shows a number of nested cycles during decomposition. Mycorrhizae are capable of breaking down proteins (Abuzinadah and Read 1986) and most likely contribute to the capacity of plants to take up amino acids from soils (Naesholm et al. 1998; Wallenda and Read 1999). Under nutrient-deficient conditions this process is very efficient, i.e. it may, in fact, compete successfully with other soil organisms. When more N is available, bacteria, decomposer fungi and soil fauna will mineralise organic N and release ammonium, which may again be utilised by other microorganisms for their own metabolism or structure, or be taken up by plant roots. If excess ammonium is available, and soil conditions are appropriate, a part of the ammonium may be converted to nitrate by nitrifying microorganisms. Nitrate may, in turn, be assimilated by microorganisms for their own metabolism, be taken up by plants or be denitrified, if conditions are suitable (mostly anoxic conditions). Excess nitrate may be leached to the groundwater. This latter process is associated with a loss of cations and soil acidification.

The N cycle, with all its shortcuts, is thought to be a closed system under natural conditions, especially if N is limiting; but the cycle can be disrupted by anthropogenic N deposition from industrial and agricultural sources. The input of ammonium and nitrate initially stimulates the N cycle in the soil and in the plant cover (more N is circulated faster) until a limit is reached, where input exceeds the uptake capacity of the system, and nitrate is leached to groundwater. This situation is termed N saturation in hydrological studies. The associated flux of cations increases in the case of leaching leading to serious changes in the soil chemical properties. There is also evidence, that at high N inputs the storage of humus may increase (Berg and Matzner 1997; Nadelhoffer et al. 1999). N is also lost from ecosystems by denitrification as well as by harvest or fire.

1.3 The NIPHYS/CANIF Project

The present book is based on two projects of the European Community (NIPHYS: NItrogen PHYSiology of Forest Plants and Soils and CANIF: Carbon and Nitrogen Cycling in Forest Ecosystems) aimed at studying key processes of the C and N cycle in coniferous (*Picea abies*) and deciduous (*Fagus sylvatica*) forest ecosystems along a north-south transect through Europe. *Picea abies* and *Fagus sylvatica* dominate about 60% of the European forests (Stanners and Bourdeau 1995).

NIPHYS and CANIF were experimental investigations of the processes involved in the contribution of C and N to the overall biogeochemical cycles and of the functional significance of biodiversity in the biogeochemical cycles in forest ecosystems. Thus, CANIF is an investigation of the present effects of climate, and soil-borne and deposited N on C and N assimilation as well as on forest organism functioning along a climatic transect through Europe. The study was based on the idea that acid and N deposition has changed forest ecosystems dynamics in Central Europe (Schulze and Ulrich 1991), but it remained difficult to quantify these effects because all habitats and regions of central Europe are affected (Teller et al. 1992). Therefore, this study had the following major objectives:

1. To investigate effects of N deposition on ecosystem processes, particularly the C cycle, by extending the range of study to the northern and southern limit of spruce and beech forests where N deposition was reported by the European Environmental Agency to be lower (Fig. 1.3; Stanners and Bourdeau 1995). Since climate and N deposition changed over the geographic range of this study, the specific effects of N deposition were expected to become apparent by studying low deposition in the north and in the south. In order to separate effects of climate and N deposition, the attempt was made to maintain soil conditions as constant as possible by selecting acidic bedrock and soils for as many sites as possible. We hypothesised that a number of ecosystem parameters would reach a specific maximum or minimum in central Europe when compared with the south or the north. Thus, a main emphasis of this study was to identify the interactions between the forest C and N cycle and their effect on forests being C sources or sinks.
2. In contrast to studies of the biogeochemical cycles per se, it was the specific aim of this study to investigate the significance of soil organisms and their associated C and N transformations and C and N fluxes. There was an evident lack in knowledge to what extent soil organisms regulate the biogeochemical cycles and affect the soil conditions, and to what extent the decomposer community responds to the climatic and edaphic conditions. In this case, ecosystem processes may be maintained constant due to a compensatory effect of a change in the organism

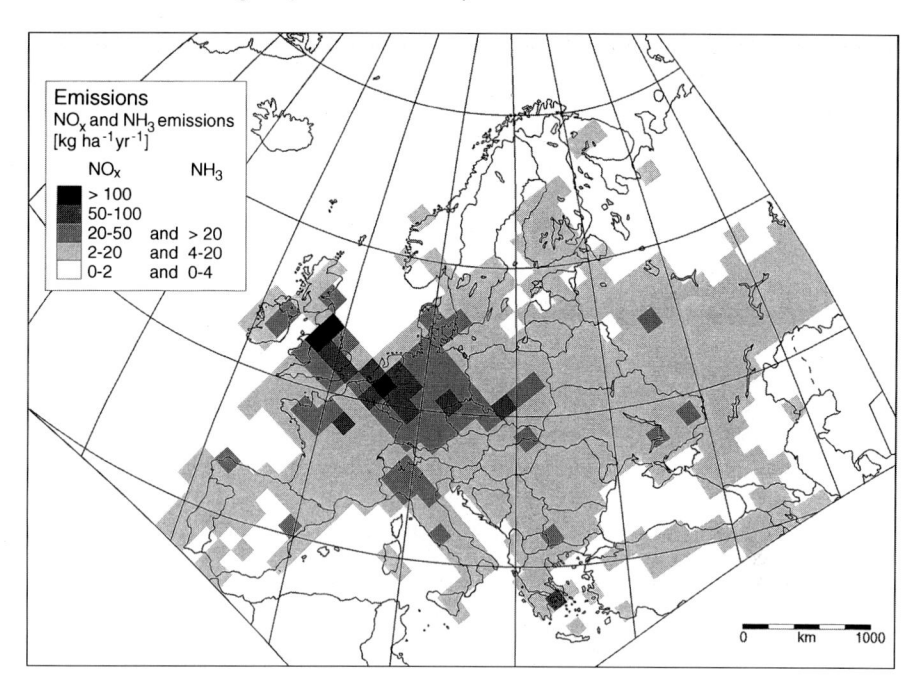

Fig. 1.3. Emissions of NO$_x$ and NH$_3$ across Europe. (After Stanners and Bourdeau 1995)

assembly (Schulze 1984). The role of biodiversity would be to contain a reservoir of organisms that is able to maintain ecosystem function under conditions of variable climate and N deposition. It is important to know if soil organisms adjust to or change their environment and maintain their specific function in the process of element cycling. It also remains unclear if ecosystem processes depend on keystone species, a balance of many species or a variable minimum number of species, and what is the role of the heterogeneity in soil processes. We hypothesise that certain products (e.g. soil organic matter) would accumulate if the capacity for compensatory response of the soil biota were overstressed.

3. Given the fact that there are a number of nested cycles in the pathway of N through the ecosystem, it was unclear if certain thresholds exist, where one or the other pathway dominates in the transformation of N, and if such thresholds would relate to N deposition. If the decomposer chain was C-limited and delivers N, and C assimilation of plants was N-limited but delivers C for the decomposer community, then thresholds should occur under conditions of N deposition, such as the loss of nitrate to groundwater. We hypothesised that the soil solution chemistry and the output of dissolved organic C and N, as well as the loss of nitrate, would be indicators for such thresholds. However, before such interpretation is possible, a careful assessment of the pathway in the soil is necessary.

4. It is quite clear that soil processes determine NEP in a time-lapse manner (Melillo et al. 1996), i.e. pools of C and N compounds may serve as sources for CO_2 asynchronously to the plant production process. It is very difficult to decide if soils are C sinks or sources because their net balance is determined by litter inputs and by

respiration, and both processes may not be in phase, resulting in SOM accumulation or losses. Since changes in C and N pools are difficult to quantify due to small fluxes and large pools in soils, this study intended to quantify the underlying processes and predict SOM turnover using a whole range of approaches. We hypothesised that a hierarchical approach of predictions (extrapolations) from the lower level of C transformations (bottom-up prediction) and verifications from whole ecosystem measurements (top-down verifications) would make it possible to quantify changes in SOM.

1.4 Experimental Design

In order to test these hypotheses, data were collected and experiments were carried out by the same scientist, being specialist for certain processes, i.e. samples were taken and analysed by the same laboratory along the whole transect. This seemed the only way to obtain comparable data, but also put a limitation on the number of sites that could be studied for certain parameters. This is the reason why some processes were studied only in Central Europe (France or Germany) and compared with few sites in Sweden or Italy. Key processes studied were:

- Plant-related processes: NPP, nutrient relations, root growth and nutrient uptake, dry deposition.
- Soil processes: decomposition, mineralisation and C and N leaching.
- Biodiversity-related processes: function of mycorrhizal species and of functional groups of invertebrates and microorganisms.

The study was carried out on plots representing a European transect ranging from North Sweden to Central Italy (see Persson et al. Chap. 2, this Vol.: Fig. 2.1 and Plates 2.1 to 2.12). The study site in North Sweden was a typical mixed boreal forest with dominance of *Picea*. *Fagus* does not reach as far north, but *Betula* served as deciduous comparison. In western Denmark, the Klosterhede Forest was initially used as a study site because of the ongoing roof experiment (EXMAN, Teller et al. 1992), but since there was no comparable *Fagus* forest in this area, the preferred *Picea* site was moved to Skogaby, which was closer to Gribskov, the Danish *Fagus* site. In Germany, the Waldstein *Picea* site was the main study area of a local ecosystem project, while Schacht was a newly established site for *Fagus* at the same elevation and climate as Waldstein. In that region, forest management has almost totally replaced *Fagus* by *Picea*. Therefore, some of the stable isotope labelling work was carried out at Steigerwald, which is a large deciduous forest region further west, but at lower elevation. In France, the Aubure sites were associated with a former catchment experiment (ENCORE, Teller et al. 1992) where *Picea* and *Fagus* were growing almost adjacent to each other. Also in the Czech Republic the study sites were associated with long-term catchment observations. It was difficult to find comparable sites in southern Europe. Initially, it was decided to use a pine and an oak forest near Thezan (southwest of Montpellier) due to the presence of acid sandstone; but these sites carried Mediterranean shrubs as ground cover, which made comparisons with the ground vegetation in central European difficult. Therefore, this site was moved to the EUROFLUX site at Collelongo, Central Italy, for *Fagus* (Valentini et al. 1999), although this forest grows on loamy soil derived from limestone as bedrock. It also remained difficult to find *Picea* in Central Italy; the forest at Monte di Mezzo was about the only plantation that could be found in that region, and so it was used as a study site although it was an afforestation on former agricultural land.

1.5 Conclusions

The IGBP core project GCTE has emphasised the value of continental transects as research tools while recognising both the inherent heterogeneity of the European landscape and the transformations brought about in it by human activity (Koch et al. 1995). The present project employed the transect approach but attempted to minimise the impacts of heterogeneity by selecting sites, where possible, which were of similar acidic soil conditions and land-use type. The study was designed to evaluate the interrelationships between ecosystem processes along a transect in which the primary variables were those of climate and pollutant deposition.

Acknowledgements. The CANIF consortium would like to acknowledge the support by the EEC (NIPHYS: ENV5V-CT92-0143, and CANIF: ENV4-CT95-0053 with subcontract ERB IC20CT960024) which made this study possible. We would also like to acknowledge the financial contributions and scientific support by universities and research institutes in each nation. The idea of using continental transects to study large-scale ecosystem processes originated from the IGBP project; the NIPHYS/CANIF study was a core project of IGBP-GCTE. It was mainly the vision of both Philipe Bourdeau and Pierre Mathy that established ecosystem research and the transect idea, respectively, in the EEC's Framework 4 program. The Terrestrial Ecosystem Initiative (TERI) was the platform on which several transect studies and networks were established, CANIF being one of them. We are especially grateful to Pierre Mathy, who had always been a mentor for ecological research, and who followed this project with insight and understanding. We should like to acknowledge the work of Dr. Guntram Bauer, Michaela Hein and Dr. Harmke van Oene, who started and maintained the data bank on behalf of all participants. We thank Barbara Lühker for editing all figures and artwork, and Kerstin Seume for final editorial revisions. Last but not least, we thank Margit Barrera and Werner Potzel for many years of administrative support, without which this work would not have been possible.

References

Abuzinadah RA, Read DJ (1986) The role of proteins in the nutrition of ectomycorrhizal plants. I. Utilisation of peptides and proteins by ectomycorrhizal fungi. New Phytol 103:481–493
Apps MJ, Price DT (1996) Forest ecosystems, forest management and the global carbon cycle. NATO ASI Series I: Global environmental change. Springer, Berlin Heidelberg New York, 452 pp
Bengtsson L (1999) From short-range barometric modelling to extended-range global weather predictions: a 40-year perspective. Tellus 51A–B:13–32
Berg B, Matzner E (1997) Effect of N deposition on decomposition of plant litter and soil organic matter in forest systems. Environ Rev 5:1–25
Bolin B, Degens ET, Kempe S, Ketner P (1979) The global carbon cycle. SCOPE 13, 491 pp
Chapin FS III, Schulze E-D, Mooney HA (1990) The ecology and economics of storage in plants. Annu Rev Ecol Syst 21:423–447
Clark FE, Rosswall T (1981) Terrestrial nitrogen cycles. Ecol Bull (Stockholm) 33, 714 pp
Havel L, Durzan DJ (1996) Aptosis in plants. Bot Acta 109:268–277

Houghton JT, Filho LGM, Callander BA, Harris N, Kattenberg A, Maskell K (1996) Climate change 1995. The science of climate change. Cambridge University Press, Cambridge, 572 pp
IGBP Terrestrial Carbon Working Group (1998) The terrestrial carbon cycle: implications for the Kyoto protocol. Science 280:1393–1394
Koch GW, Scholes RJ, Steffen WL, Vitousek PM, Walker BH (1995) The IGBP terrestrial transects: science Plan. IGBP Report No 36, IGBP Secretariat, Stockholm
Leopold AC (1980) Aging and senescence in plant development. In: Thimann KV (ed) Senescence in plants. CRC Press, Boca Raton, pp 1–13
Melillo J, Prentice C, Schulze E-D, Farquahr G, Sala O (1996) Terrestrial ecosystems: Respiration to global environmental change and feedbacks to climate. IPCC Chap 9:445–482
Meyer O (1993) Functional groups of microorganisms. In: Schulze E-D, Mooney HA (eds) Biodiversity and ecosystem function. Ecological studies 99. Springer, Berlin Heidelberg New York, pp 67–96
Monsi M, Saeki T (1953) Über den Lichtfaktor in den Pflanzengesellschaften und seine Bedeutung für die Stoffproduktion. Jpn J Bot 14:22–52
Nadelhoffer KJ, Emmett BA, Gundersen P, Kjönaas OJ, Koopmans CJ, Schleppi P, Tietema A, Wright RF (1999) Nitrogen deposition makes a minor contribution to carbon sequestration in temperate forests. Nature 398:145–147
Naesholm T, Ekblad A, Nordin A, Giesler R, Hoegberg M, Hoegberg P (1998) Boreal forest plants take up organic nitrogen. Nature 392:914–916
Pennell RI, Lamb C (1997) Programmed cell death in plants. Plant Cell 9:1157–1168
Schellnhuber HJ (1999) Globales Umweltmanagement oder Dr. Lovelock übernimmt Dr. Frankensteins Praxis. In: Jahrbuch Ökologie. CH Beck, München, pp 168–187
Schellnhuber HJ, Wenzel V (1998) Earth system analysis. Springer, Berlin Heidelberg New York, 530 pp
Schlesinger WH (1997) Biogeochemistry: an analysis of global change. Academic Press, San Diego, 588 pp
Schulze E-D (1994) Flux control at the ecosystem level. TREE 10:40–43
Schulze E-D, Chapin FS III (1987) Plant specialization to environments of different resource availability. In: Schulze E-D, Zwölfer H (eds) Potentials and limitations of ecosystem analysis. Ecological studies 61. Springer, Berlin Heidelberg New York, pp 120–148
Schulze E-D, Heimann M (1998) Carbon and water exchange of terrestrial systems. In: Galloway J, Melillo J (eds) Asian change in the context of global climate change. Cambridge University Press, Cambridge, IGBP Book Ser 4:145–161
Schulze E-D, Ulrich B (1991) Acid rain – a large-scale, unwanted experiment in forest ecosystems. SCOPE 45:89–106
Schulze E-D, Kelliher FM, Körner Ch, Lloyd J, Leuning R (1994) Relationships between plant nitrogen nutrition, carbon assimilation rate, and maximum stomatal and ecosystem surface conductance for evaporation. A global ecological scaling exercise. Annu Rev Ecol Syst 25:629–660
Smith SE, Read DJ (1997) Mycorrhizal symbiosis. Academic Press, San Diego, 605 pp
Stanners D, Bourdeau (1995) Europe's environment. The Dobris assessment. European Environmental Agency, Copenhagen, 676 pp
Stitt M, Schulze E-D (1994) Plant growth, storage, and resource allocation: from flux control in a metabolic chain to the whole-plant level. In: Schulze E-D (ed) Flux control in biological systems. Academic Press, San Diego, pp 57–118
Teller A, Mathy P, Jeffers JNR (1992) Responses of forest ecosystems to environmental change. Elsevier Applied Science, London, 1009 pp
Ulrich B (1987) Stability, elasticity, and resilience of terrestrial ecosystems with respect to matter balance. In: Schulze E-D, Zwölfer H (eds) Potentials and limitations of ecosystem analysis. Ecological Studies 61. Springer, Berlin Heidelberg New York, pp 11–49
Valentini R, Matteucci G, Dolman AJ, Schulze E-D, Rebmann C, Moors EJ, Granier A, Gross P, Jensen NO, Pilgaard K, Lindroth A, Grelle A, Bernhofer Ch, Grünwald T, Aubinet M, Ceulemans R, Kowalski AS, Vesala T, Rannik Ü, Berbigier P, Lousteau D, Gudmundsson J,

Thorgeirsson H, Ibrom A, Morgenstern K, Clement R, Montcrieff J, Montagnani L, Minerbi S, Jarvis PG (1999) Respiration is the main determinant of European forest Carbon balance. Nature (in press)

Wallenda T, Read DJ (1999) Kinetics of amino acid uptake by ectomycorrhizal roots. Plant Cell Environ 22:179-187

WBGU (1998) The accounting of biological sinks and sources under the Kyoto protocol: a step forward or backward for global environmental protection? German Advisory Council on Global Change. Special Report 1998, 75 pp WBGU Secretariat, Bremerhaven

Zech W, Kögel-Knabner I (1994) Patterns and regulation of organic matter transformation in soils: litter decomposition and humification. In: Schulze E-D (ed) Flux control in biological systems. Academic Press, San Diego, pp 303-334

2 Experimental Sites in the NIPHYS/CANIF Project

T. Persson, H. van Oene, A.F. Harrison, P.S. Karlsson, G.A. Bauer, J. Cerny, M.-M. Coûteaux, E. Dambrine, P. Högberg, A. Kjøller, G. Matteucci, A. Rudebeck, E.-D. Schulze, and T. Paces

2.1 Site Description. The NIPHYS/CANIF Transect

The NIPHYS/CANIF sites (Fig. 2.1) were chosen to represent a latitudinal, climatic and deposition transect through Europe containing conifer and broadleaf stands. The main species are Norway spruce, *Picea abies* L. (Karst.) and European beech, *Fagus sylvatica* L. In southern France, *Picea* and *Fagus* are lacking at suitable sites and were replaced by maritime pine, *Pinus pinaster* Ait. (*P. maritima* Lam.), Aleppo pine, *P. halepensis* Mill. and white oak, *Quercus pubescens* Willd. for some of the studies. *F. sylvatica* has its most northern distribution in southern Scandinavia. Therefore, in northern Sweden, *F. sylvatica* was replaced with birch, *Betula pubescens* Ehrh.

The main characteristics of the NIPHYS/CANIF sites are given in Table 2.1. In this table the sites are ordered following the gradient from high to low latitudes. Figures 2.4–2.16 (colour photos at end of chapter) give a visual impression of the stands.

2.1.1 Climatic Conditions

The climatic conditions along the gradient vary from boreal in North Sweden, humic oceanic conditions at the southern Scandinavian sites, humic continental at the Mid-European sites to montane Mediterranean conditions at the Italian sites. Only the southern French site Thezan has an explicit Mediterranean climate. Mean annual temperature is low at Åheden in the north (1 °C) and high at the Mediterranean site Thezan (15 °C), but does not seem to vary much between the other sites (5.4–8.5 °C; Table 2.2). The south Scandinavian sites with their oceanic influence have a higher mean annual temperature than the Mid-European sites with a continental influence. The latter sites in Germany and the Czech Republic are all located at high altitudes (700–1050 m above sea level) as well as the north French site with a mountainous oceanic climate and the Italian sites (900–1600 m asl). Annual precipitation does not show a clear gradient, but is lowest in the north. Mean monthly temperature and precipitation are given in Table 2.2. The North Swedish site is most extreme with a temperature far below zero during 5 months and has consequently a short growing season. At the mountainous sites, the early start of winter and the later onset of spring limit the length of the growing season. The *P. abies* stand in Italy (Monte di Mezzo) deviates in having a less cold winter. The Mediterranean sites are characterised by summer drought. The other sites have rainfall rather evenly distributed throughout the year with perhaps a less rainy period during spring.

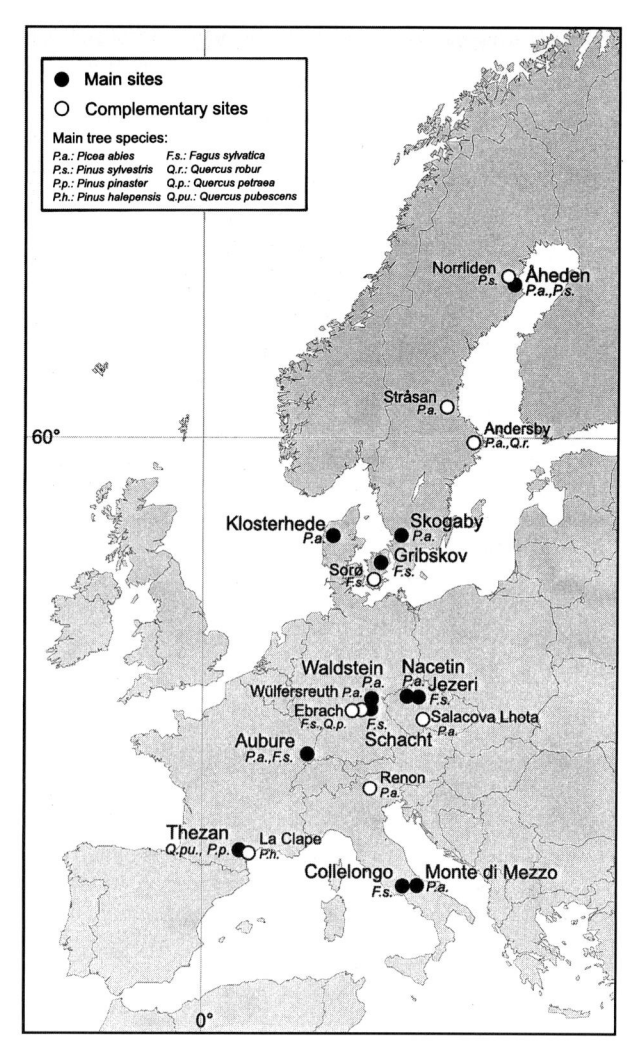

Fig. 2.1. Location of the NIPHYS/CANIF sites in Europe representing a transect from North Sweden to South France and Central Italy. Main sites (in NIPHYS and/or CANIF) are indicated by *filled circles* and complementary sites by *open circles*

2.1.2 Deposition

Total N deposition is highest for the Mid-European (the German and Czech) sites and some of the South Scandinavian sites (Klosterhede and Skogaby) and decreases towards the north and the south (Fig. 2.2). The most northern site, Åheden, is virtually without any deposition and a virgin undisturbed site. Total S deposition shows a picture similar to total N deposition with the highest deposition levels in Mid-Europe. However, the Czech sites distinguish themselves by a very high dry S deposition.

Table 2.1. Characteristics of the sites in the NIPHYS/CANIF project. Wet deposition values are means for 1993–1997. Dry deposition data are from EMEP (Barrett and Berge 1996; Berge 1997) and means for 1993–1996

Country	Sweden		Denmark		Czech Republic	
Site	Åheden	Skogaby	Klosterhede	Gribskov	Nacetin	Jezeri
Site abbreviation	Åhe	Sko	Klo	Gri	Nac	Jez
Location	64°13′N 19°30′E	56°33′N 13°13′E	56°29′N 08°24′E	55°58′N 12°15′E	50°35′N 13°15′E	50°33′N 13°28′E
Elevation (m asl)	175	95–115	27	45	775	750
Species	*Pinus, Picea, Betula*	*Picea*	*Picea*	*Fagus*	*Picea*	*Fagus*
Type of stand	Natural, unmanaged	Planted	Planted	Natural regeneration	Planted residual	Planted
Stand age (in 1995)	180	31	76	118	58	79
Field layer[a]	Moderate Vm, Vv, moss	Patches moss, Df	Patches moss, Df	Sparse Oa, Go	Dense Cv, Df, Vm	None
Climate[b]	B	ho	ho	ho	hc	hc
Bud break	Early June	Mid-May	Mid-May	Early May	Mid-May	Late April
N wet/dry/ total deposition (kg N ha^{-1} a^{-1})	1.1/0.6 1.7	12.8/3.6 16.4	7.1/13.5 20.6	8.0/3.6 11.6	11.9/6.7 18.6	14.2/6.7 20.9
S wet/dry/ total deposition (kg S ha^{-1} a^{-1})	5.0/0.8 5.8	7.7/4.9 12.6	12/18 30	5.8/4.9 10.8	12.7/29.1 41.8	16.9/29.1 45.9
Soil type	Regosol	Haplic podzol	Haplic podzol	Arenosol	Spodo-dystric cambisol	Dystric cambisol
Soil pH (H$_2$O) FH layer	3.93	4.06	3.88	4.29	3.62	4.11
Soil pH (H$_2$O) (0–10 cm)	4.45	4.01	3.94	4.07	3.47	3.93
Soil texture (0–10 cm)	Sand	Sandy loam	Sand	Fine sand	Sandy loam	Sandy silt loam
Stoniness (0–10 cm) (%)	0	0	0	0	25	30

Experimental Sites in the NIPHYS/CANIF Project

Germany		France			Italy	
Waldstein	Schacht	Aubure	Aubure	Thezan	Collelongo	Monte di Mezzo
Wal	Sch	AuP	AuF	The	Col	MdM
50°12′N 11°53′E	50°04′N 11°50′E	48°12′N 07°11′E	48°12′N 07°11′E	43°07′N 02°45′E	41°52′N 13°38′E	41°45′N 14°53′E
700	850	1050	1000	170	1560	905
Picea	*Fagus*	*Picea*	*Fagus*	*Pinus pinaster*	*Fagus*	*Picea*
Planted	Natural, slightly managed	Planted	Natural regeneration	Natural regeneration	Natural regeneration	Planted on former pasture
142	120	92	161	>100	104	37
Dense Vm, Cv, Df	Sparce, Dfm, Vm	Moderate Df, Dfm, Ru	Sparse Po, Dfm	Dense Qi, Es, Jo	Sparse Go, Gr	Very sparse Hh
hc	hc	ho	ho	m	mm	mm
Late April	Early May	Late April	Early May	Late April	Early May	Mid-Late April
13.5/6.6	13.5/6.6	9.1/5.5	9.1/5.5	?	6.4/4.4	?/4.0
20.1	20.1	14.7	14.7		10.8	?
10.0/7.0	10.0/7.0	6.8/5.0	6.8/5.0	?	6.3/3.4	?/4.4
17.0	17.0	11.8	11.8		9.7	?
Cambic podzol	Dystric cambisol	Dystric cambisol	Haplic podzol	Chromic luvisol	Humic alfisol	?
3.69	4.34	3.48	4.01	–	5.2	–
3.52	3.99	3.71	3.57	5.74	5.69	6.92
Loamy sand	Clay silt – sandy silt	Sandy loam	Loamy sand	Sandy silt, loamy sand	Silty loam – silty clay	Clay
3	20	46	40	5–10	0–10	0

Table 2.1. *Continued*

Country	Sweden		Denmark		Czech Republic	
Site	Åheden	Skogaby	Klosterhede	Gribskov	Nacetin	Jezeri
Site abbreviation	Åhe	Sko	Klo	Gri	Nac	Jez
LAI (1994 or 1995)	2.5	7	5.7	4.9	5.8	6.9
Tree density[c] (trees ha^{-1})	433 (Ps) 434 (Pa) 51 (Bp)	<2285	735	400	616	568
Mean tree height (m)	18.0 (Ps) 14.3 (Pa) 11.7 (Bp)	15	20	26.3	20.7	24
DBH[c] (cm)	24.1 (Ps) 16.0 (Pa) 12.1 (Bp)	14.5	22.9	40.5	28.5	30
Basal area[c] (m^2 ha^{-1})	22.7 (Ps) 10.4 (Pa) 0.7 (Bp)	34	30	24.6	37.6	40.4
Annual volume increment[c] (m^3 ha^{-1} a^{-1})	3.9 (Ps)	15	8.6	8.9–9.6	11.8	11

[a] Bp = *Betula pubescens*, Cv = *Calamagrostis villosa*, Df = *Deschampsia flexuosa*, Dfm = *Dryopteris filix-mas*, Es = *Erica scoparia*, Go = *Galium odoratum*, Gr = Geranium robertianum, Hh = *Hedera helix*, Jo = *Juniperus oxycedrus*, Oa = *Oxalis acetosella*, Pa = *Picea abies*, Ps = *Pinus sylvestris*, Po = *Polystichum spinulosum*, Qi = *Quercus ilex*, Ru = *Rubus* sp., Up = *Ulex parviflorus*, Vm = *Vaccinium myrtillus*, Vv = *Vaccinium vitis-idaea*.
[b] b Boreal, ho: humic oceanic, hc: humic continental, m: Mediterranean, mm: montane Mediterranean.
[c] Data Åhe 1998; AuP: 1988, AuF: 1994; other sites 1995 or 1996.

These dry deposition data are EMEP data (Barrett and Berge 1996; Berge 1997) and represent average dry deposition data for a grid cell of 150 × 150 km. Any effects of scavenging of air-borne pollutants by forest are not included in these data.

2.1.3 Stand Characteristics and Site History

The site in North Sweden, Åheden, is a 180-year-old unmanaged *P. sylvestris* L. forest, mixed with *P. abies* and *B. pubescens*. *P. sylvestris* is the dominant species. The area is almost undisturbed, although forest fires have probably occurred at times and reindeer grazing is a normal feature in this region of Sweden. The site is situated at 175-m altitude on level and thick delta/lacustrine sediments, which form a recharge area for groundwater (Högbom and Högberg 1991). This boreal site is characterised by a

Experimental Sites in the NIPHYS/CANIF Project

Germany		France			Italy	
Waldstein	Schacht	Aubure	Aubure	Thezan	Collelongo	Monte di Mezzo
Wal	Sch	AuP	AuF	The	Col	MdM
6.2	4.1	5.4	5.8	?	4.5	5.5
363	372	568	352	700	885	1197
26.7	24.0	27.9	22.4	8–10	18.0	20.5
36.5	33.6	40.7	34.9	30–35	21.2	19.5
39.4	37.3	72	?	?	32.1	38.4
10.3	?	7.7	2.4	?	4.7	3.5

bottom layer of forest mosses. The dwarf shrubs *Vaccinium myrtillus* and *Vaccinium vitis-idaea* dominate the field layer.

The Swedish site, Skogaby, is a young homogeneous *P. abies* plantation lying about 20 km from the sea (Kattegat) in the southwestern part of Sweden. Formerly, the area was a grazed *Calluna* heathland that was planted with *P. sylvestris* in 1913. The *P. abies* stand is the second-generation forest and replaced the pine forest in 1966. The bedrock belongs to the Swedish gneiss area covered by a more than 2-m-thick sandy loamy till layer. Mosses make up the bottom layer for about 50% and *Deschampsia flexuosa* occurs in gaps. Beside the control plots that have been used in the NIPHYS/CANIF project, the site also consists of plots with experimental drought, irrigation, fertilisation and acidification manipulations (the Skogaby experiment, Nilsson and Wiklund 1992).

The Danish *P. abies* site Klosterhede is located in western Jutland, 15 km from the North Sea. Throughfall and soil water chemistry are dominated by high sea salt inputs (Beier et al. 1995). The topography is flat and the parent material consists of glacifluvial deposits. The soil is well-drained, nutrient-poor and consists mainly of coarse sand with low clay content (<4%) (Beier et al. 1995; Gundersen 1998). The soil was ploughed prior to the planting of the trees. The present forest stand is the second after afforesting a former *Calluna* heathland about 120 years ago (Gundersen and

Table 2.2. Mean monthly and annual air temperature and precipitation for the NIPHYS/CANIF sites during the period 1993–1997. For some sites, data were not available for all years. Data for Collelongo are based on 1996–1998. Annual data might not correspond completely with the given monthly data due to incompleteness of the climatic datasets. Only complete years are used for calculating annual means, whereas monthly data are based on availability of monthly data

Mean air temperature (°C)

Site[a]	Jan	Feb	Mar	Apr	May	Jun	Jul	Aug	Sep	Oct	Nov	Dec	Annual
Åhe	−10.5	−11.9	−4.8	0.0	5.1	12.4	14.9	13.6	7.1	1.4	−6.5	−9.8	1.0
Sko	0.2	−0.3	2.0	7.1	11.8	14.1	17.3	16.3	11.7	7.7	2.5	0.1	7.6
Klo	0.3	0.7	2.3	6.4	10.0	13.0	15.8	16.6	11.7	8.4	3.9	0.9	7.5
Gri	−2.1	−0.5	1.5	6.1	9.1	14.5	16.3	19.2	12.5	8.5	4.5	0.6	7.5
Nac, Jez	−2.7	−1.9	0.9	5.5	10.3	12.7	15.6	15.4	10.3	6.6	0.5	−2.7	5.9
Wal, Sch	−3.1	−1.7	1.1	5.5	10.3	12.9	15.1	15.1	9.6	5.7	−0.4	−3.1	5.5
AuP, AuF	−1.7	−0.8	0.8	4.1	8.6	11.3	15.0	13.5	8.2	5.0	1.4	−0.6	5.4
The	8.8	8.5	11.0	13.0	17.3	21.1	24.2	23.8	18.8	16.1	11.6	9.4	15.4
Col	−0.1	0.9	−0.7	1.9	8.4	13.3	15.3	15.8	10.8	6.4	4.2	−1.1	6.4
MdM	2.1	2.4	3.1	5.7	11.1	14.3	16.7	16.3	12.4	8.9	5.2	3.1	8.5

Precipitation (mm)

	Jan	Feb	Mar	Apr	May	Jun	Jul	Aug	Sep	Oct	Nov	Dec	Annual
Åhe	34	34	33	51	19	54	73	35	46	21	54	34	588
Sko	132	88	130	62	43	110	115	115	151	83	76	132	1237
Klo	97	67	68	55	28	41	56	100	113	101	70	107	903
Gri	5	47	17	30	87	78	76	39	52	86	67	47	632
Nac, Jez	62	48	80	68	96	113	90	96	115	41	74	54	935
Wal, Sch	74	62	66	54	59	76	121	98	73	62	46	99	890
AuP, AuF	169	69	99	73	116	74	118	72	122	113	42	171	1192
The	66	65	31	77	44	34	20	35	92	110	93	120	787
Col	72	46	64	150	83	30	19	20	123	118	285	76	1109
MdM	67	74	72	105	61	22	64	58	120	90	152	147	1032

[a] The sites Nac and Jez have the same meteorological station as well as AuP and AuF. Sch had too few data to be separately treated and is joined with Wal (located 10 km away).

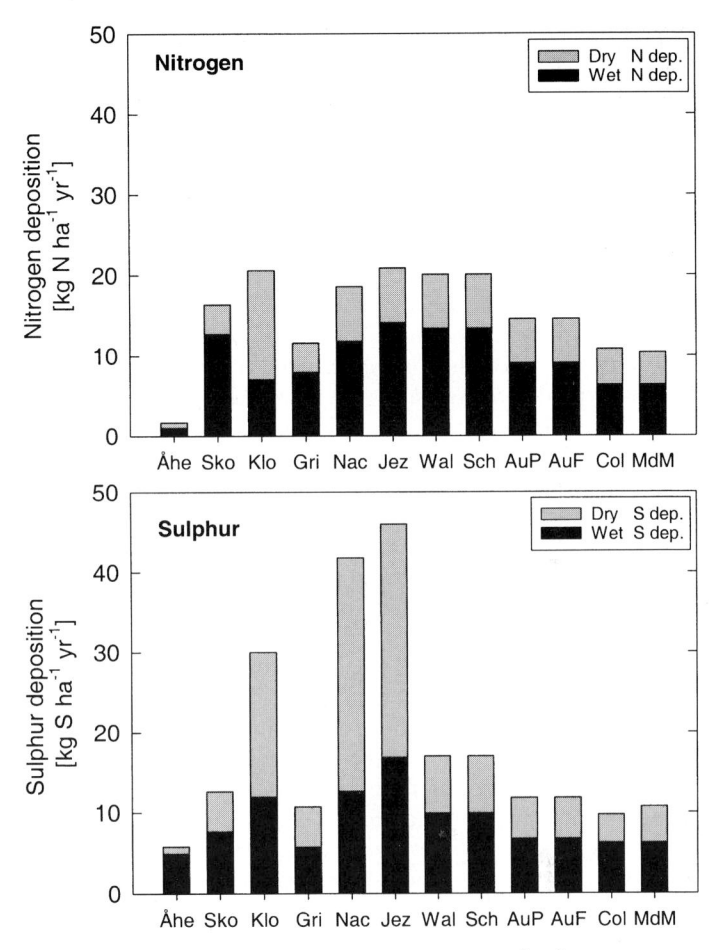

Experimental Sites in the NIPHYS/CANIF Project

Fig. 2.2. Mean wet and dry nitrogen (kg N ha^{-1} yr^{-1}) and sulphur (kg S ha^{-1} yr^{-1}) deposition at the main CANIF sites. Mean wet deposition refers to 1993–1997 and dry deposition to 1993–1996. Wet deposition is measured at individual sites. Wet deposition at MdM is assumed equal to wet deposition at Col. Dry deposition data are from EMEP (Barrett and Berge 1996; Berge 1997), except for Klosterhede. Values are given in Table 2.1

Rasmussen 1995). The plots used for NIPHYS are situated close to the EXMAN (Rasmussen et al. 1990) and NITREX (Gundersen 1998) plots.

The Danish *F. sylvatica* site, Gribskov, is a 118-year-old forest on northern Sealand, situated about 20 km from Kattegat and 40 km north of Copenhagen. Beech forests have probably covered this area for several centuries. There is no distinct field layer. The distance to Klosterhede is about 250 km and to Skogaby 90 km. Climatically, Gribskov is probably more similar to the Swedish site Skogaby than to the more oceanic site Klosterhede.

The Czech sites, Nacetin and Jezeri, are located in an area with spruce forests dating from the 16th century. The 58-year-old *P. abies* stand at Nacetin is planted, while the

79-year-old *F. sylvatica* stand at Jezeri is formed by natural regeneration. The sites are situated 17 km apart in the Ore Mountains (Krušné Hory) at the northwestern side of the heavily industrialised North Bohemian lignite basin fairly close to the city of Most. The sites have experienced high levels of atmospheric S deposition for many decades, but the S deposition has decreased during recent years. Jezeri is situated on a south-directed slope, facing the lignite-burning power stations, while Nacetin is on the northern slope and somewhat protected from major point sources of pollution (Novak et al. 1995). The stand at Nacetin was one of the spruce stands that were left and not completely defoliated by a combination of air pollution and a severe temperature inversion in early 1979. The *P. abies* site has a dense field layer vegetation (cover 90%) consisting of mainly *Deschampsia flexuosa, Calamagrostis villosa* and *Vaccinium myrtillus*. The *F. sylvatica* stand is almost without a field layer.

The German sites, Waldstein and Schacht, are located at the northwestern border of the Fichtel Mountains (Fichtelgebirge, NE Bavaria), 9 km apart. The area mainly consists of planted *P. abies* forests; the first forests were already planted in the 16th century. The *P. abies* stand at Waldstein, probably the second generation of the same tree species, is a 142-year-old plantation (1995) with a dense field layer vegetation dominated by *Vaccinium myrtillus, Calamagrostis villosa*, and *Deschampsia flexuosa*. The 120-year-old *F. sylvatica* forest at Schacht, however, is a relict stand that has been managed only to a slight extent. It has a heterogeneous field layer vegetation with grasses and ericaceous plants.

The French sites at Aubure are located in the Strengbach catchment at the northeastern side of the Vosges Mountains. The old *P. abies* stand is situated at a south-facing slope whereas the *F. sylvatica* stand is on the north-facing slope, ca. 500 m from each other. The old *P. abies* stand is a 92-year-old forest planted after an old (100–200 year-old) grazed declining fir forest. Since 1983 the canopy has been partly (about 30%) defoliated and some needles are yellow and deficient in magnesium. Patches of *Dryopteris filix-mas* and *Deschampsia flexuosa* make up the field layer. The *F. sylvatica* stand is 161 years old and is formed by natural regeneration from a mixed beech and fir forest. It has a field layer with a sparse cover of *Polystichum spinulosum* and *Athyrium filix-femina*.

The site Thezan is located about 20 km WSW the city of Narbonne in southern France at an elevation of 170 m above mean sea level. The conifer stand consists of 100-year-old *Pinus pinaster* with a dense understorey of *Erica scoparia, Juniperus oxycedrus, Ulex parviflorus, Quercus coccifera* and *Calluna vulgaris*. The stand is growing on Permian sandstone, partly with an organic layer and partly without. The NIPHYS plots had almost no organic layer and the soil was extensively influenced by earthworm activities, whereas the adjacent area used for the VAMOS project had a thick organic layer. A *Quercus pubescens* stand, situated about 500 m apart from the *P. pinaster* stand, was used for some comparative studies within NIPHYS.

The Italian sites Collelongo and Monte di Mezzo are located in the Apennines (Appenino Abruzzese) about 50 km from each other. The *F. sylvatica* stand, close to the village of Collelongo, is situated NW of the Abruzzo National Park at an elevation of 1500–1600 m above mean sea level. It is a 104-year-old stand formed by natural regeneration and growing on a slope. The *Fagus* stand has a sparse field layer of herbs, mainly *Galium odoratum*. The *P. abies* stand at Monte di Mezzo is a 37-year-old stand planted on a former pasture. Before that, the area was arable land that was used for cereal crops. The stand has not been thinned until now and has therefore a high tree density. It has a very sparse field layer. *Hedera helix* and some deciduous tree seedlings can be found, especially at the periphery of the stand. Both Italian sites deviate from the other sites in having a soil with calcareous parent material.

2.1.4 Complementary Sites

Sites other than those mentioned in Table 2.1 were used by some of the research groups within NIPHYS and CANIF to elucidate specific questions (see individual chapters, this Vol.). Data from these sites are summarised in Table 2.3.

Norrliden is located in the same area as Åheden. The stand is a 44-year-old *P. sylvestris* stand that was planted after prescribed burning. From 1971 onwards, the area was used for long-term fertilisation experiments with annual additions of different doses of N fertilisers.

Stråsan is located in central Sweden, about 200 km NW of Uppsala. In 1956–1957 the former coniferous forest was clear-cut and slash-burned. One year later the area was planted with *P. abies*. Also this area has been used for long-term fertilisation experiments since 1967.

Andersby is a nature reserve about 40 km north of Uppsala situated in an area that has been affected by man since about 1500–2000 years, when the land was lifted up from the Baltic Sea. The dominant tree species are 100–200-year-old *Quercus robur* together with *Betula pubescens* and *B. pendula* (Hytteborn 1975). The shrub layer is dominated by *Corylus avellana*. In some areas *P. abies* has formed isolated stands. The ground vegetation is rich and variable, with a high frequency of *Anemone nemorosa* and *Hepatica nobilis*. The area that was included in the present study had been subjected to moderate cattle grazing before the grazing ceased in 1918. The soil is a cambisol with a high content of limestone and high pH. This site was used for a comparison with the *Quercus* stand at Thezan and the *Pinus* stand at La Clape (see below) as regards soil nitrogen turnover (see Persson et al. Chap. 14, this Vol.).

Sorø is an 80-year-old *F. sylvatica* forest situated about 50 km SW of Copenhagen on central Sealand in Denmark. The forest covers about 1 km^2 lying as an island in the flat agricultural landscape. The forest has been dominated by beech since at least 1770 and probably before. The soil is a luvisol with a sparse cover of *Anemone nemorosa* and *Mercurialis perennis*.

Salacova Lhota is a catchment in the eastern part of the Bohemian Moravian uplands, Czech Republic, situated on the southern slope of the hill Straziste near the town of Pacov. It is facing a rural countryside without major point sources of industrial pollution. The main stand consists of 87-year-old *P. abies*. The distance between this site and the heavily polluted site Nacetin (see above) is about 160 km.

At Ebrach in Germany (Steigerwald, NE Bavaria) a 30-year-old mixed *Fagus sylvatica/ Quercus petraea* stand is situated within a large broadleaf forest area. Due to a high tree density and a closed canopy, no understorey vegetation is present.

The Wülfersreuth site in Germany is located in the Fichtelgebirge (NE Bavaria). The site consists of a 15-year-old *Picea abies* plantation. Due to an open canopy, a dense field layer vegetation with ericaceous shrubs and *Deschampsia flexuosa* is present.

At Aubure in NE France, there is a 45-year-old *P. abies* stand (AuP45) situated on the same south-facing slope as the old *P. abies* stand (AuP90). The younger stand is

Table 2.3. Characteristics of the complementary sites of the NIPHYS/CANIF project

Country	Sweden				Denmark
Site	Norrliden	Stråsan	Andersby	Andersby	Sorø
Site abbreviation	Nor	Str	AnP	AnQ	Sor
Location	64°21′N 19°46′E	60°55′N 16°01′E	60°09′N 17°49′E	60°09′N 17°49′E	55°29′N 11°38′E
Elevation [m a.s.l.]	260	350	30	30	40
Species	*Pinus sylvestris*	*Picea abies*	*Picea abies*	*Quercus robur*	*Fagus sylvatica*
Type of stand	Planted	Planted	Natural regener.	Natural regener.	Natural regener.
Stand age (in 1995)	44	42	80	150	80
Field layer[a]	Moderate Vv, Vm, mosses	Moderate Vv, Vm, mosses	Sparse Oa, mosses	Moderate Hn, An	Sparse An, Mp, grasses
Climate[b]	b	b	bn	bn	ho
Mean air temp.	1.2	3.2	5.5	5.5	8
Annual mean precip.	595	740	570	570	630
Bud break	Early June	Early June	Late May	Mid-May	Apr-May
Soil type	Orthic podzol	Orthic podzol	Dystric cambisol	Eutric cambisol	Luvisol
Soil pH (H$_2$O) FH layer	3.98	4.12	5.04	[c]	[c]
Soil pH (H$_2$O) (0–10 cm)	4.61	4.40	no data	5.84	4.62

[a] Aa = *Athyrium alpestre*, An = *Anemone nemorosa*, Ca = *Calluna vulgaris*, Cv = *Calamagrostis villosa*, Df = *Deschampsia flexuosa*, Es = *Erica scoparia*, Ga = *Galium* sp., Ha = *Homogyne alpina*, Hn = *Hepatica nobilis*, Hs = *Hieracium sylvaticum*, Jo = *Juniperus oxycedrus*, Jp = *Juniperus phoenica*, Mp = *Mercurialis perennis*, Ms = *Melampyrum sylvatica*, Oa = *Oxalis acetosella*, Po = *Polystichum spinulosum*, Qi = *Quercus ilex*, Sa = *Sorbus aucuparia*, Vm = *Vaccinium myrtillus*, Vv = *Vaccinium vitis-idaea*.
[b] Boreal, bn: Boreo-nemoral, ho: Humic oceanic, hc: humic continental, m: Mediterranean, mm: montane Mediterranean, mo: montane.
[c] No data or soil layer missing.

denser and the current growth is high, despite the fact that the needles are yellowing and deficient in magnesium. Although the young stand is growing on the same slope as the old one, some soil processes are markedly different (see Persson et al. Chap. 14, this Vol.).

Experimental Sites in the NIPHYS/CANIF Project

Czech Republic	Germany		France			Italy
Salacova Lhota	Ebrach	Wülfersreuth	Aubure	La Clape	Thezan	Renon
SLP	Ebr	Wül	AuP45	LaC	ThQ	Ren
49°31′N 14°59′E	49°52′N 10°29′E	50°04′N 11°46′E	48°12′N 07°11′E	43°09′N 03°08′E	43°07′N 02°45′E	46°36′N 11°28′E
557–744	440	670	1020	90	170	1720
Picea abies	*Fagus sylv./ Q. petraea*	*Picea abies*	*Picea abies*	*Pinus halepensis*	*Quercus pubesc.*	*Picea abies*
Planted	Planted	Planted	Planted	Natural regener.	Natural regener.	Natural regener.
87	30	15	45	100	>100	80
Sa, Df, Cv, Vm, Aa, Ga, Po	None	Dense Vm, Ca, Df	None	Jo, Jp, Qi	Es, Jo	Moderate Vm, Vv, Ms, Ha, Hs
hc	hc	hc	ho	m	m	mo
6	7.5	5.9	5.4	14.8	14.4	4
685	750	1072	1192	587	578	1010
Early May	Early May	Late April	Late April	?	?	?
Dystric/eutr. cambisol	Cambisol	Spododyst. cambisol	Dystric cambisol	Chromic luvisol	Chromic luvisol	?
3.98	4.5	3.8	3.77	4.71	–	?
4.21	4.5	4.0	3.45	7.3	6.47	?

At Thezan in southern France a *Quercus pubescens* stand (ThQ) is situated about 500 m apart from the *P. pinaster* stand (The) (see above). The stand is growing on a west-facing slope on a chromic luvisol.

La Clape site in southern France is located in a hilly area (Massif de la Clape) close to the coastal line near the city of Narbonne. The site consists of a 100-year-old *Pinus halepensis* stand growing on hard calcareous bedrock (karst) forming an organic horizon classified as neutral or acidic xeromoder. The understorey is very variable with, for example, *Juniperus oxycedrus, J. phoenica* and *Quercus ilex*.

Renon, in Northern Italy, is located north of Bolzano in the Alps close to the Rittner Horn. It has a southwest aspect with a mean slope of 35%. The site has an 80-year-old *Picea abies* stand of natural origin and managed through selective cuttings, less intensively during recent years. The stand is mixed with *Pinus cembra* (13%) and *Larix decidua* (4%). The understorey consists of dwarf shrubs and herb species. This

Table 2.4. Depths (cm) and texture (% cl = clay < 0.002 mm, si = silt 0.002–0.06 mm, sa = sand 0.06–2 mm) of soil horizons (Hor.) at main sites in the NIPHYS/CANIF transect.

Åhe (1)

Hor.	Depth (cm)	Texture sa/si/cl (%)
L	5–4	
FH	4–0	
E	0–1	82/15/3
Bs	1–30	82/15/3
C	>30	82/15/3

Sko (2)

Hor.	Depth (cm)	Texture sa/si/cl (%)
L	7–5	
FH	5–0	
A/E	0–11	68/26/6
Bhs	11–23	68/27/5
Bs	23–37	70/26/4
BC	37–58	70/25/5
C	>58	70/25/5

Klo (3, 4)

Hor.	Depth (cm)	Texture sa/si/cl (%)
L	9–8	
FH	8–0	
A	0–7	91/5/4
E	7–11	91/5/4
Bh	11–15	91/5/4
Bhs	15–35	93/5/2
Bs	35–54	98/2/0

Wal (7)

Hor.	Depth (cm)	Texture sa/si/cl (%)
L	9–8	
F	8–4	
H	4–0	
Ahe	0–9	?/?/12
Bhs	9–24	?/?/12
Bsv	24–38	?/?/15
BvCv	38–54	Loam
Cv	>54	Sa loam

Sch (3)

Hor.	Depth (cm)	Texture sa/si/cl (%)
L	8–6	
FH	6–0	
Ah	0–5	14/78/8
Bsh	5–9	14/78/8
Bhsw	9–40	14/78/8
BwCw	40–55	31/65/4

AuP90 (8)

Hor.	Depth (cm)	Texture sa/si/cl (%)
L	3–1	
FH	1–0	
A	0–15	?/?/20
B	15–60	?/?/20
BC	60–120	
C	>120	

Data on horizons from (1) Giesler (1996), (2) J. Bergholm (pers. comm.), (3) Matschonat and Matzner (1996), (4) Beier et al. (1991), (5) J. Bille-Hansen (pers. comm.), (6) Šefrna

site is used for complementary soil respiration measurements (see Matteucci et al. Chap. 10, this Vol.).

2.2 Soil Characteristics

2.2.1 Soil Profile Description

Extensive data on soil profiles in the NIPHYS/CANIF transect are given in the data base on CD-ROM. Selected data on depths and texture (% of fine soil <2 mm) of the soil horizons are shown in Table 2.4. It can be noted that different research groups within the project delimited the soil layers differently. Most groups defined the 0-level to be the borderline between the organic and mineral soil horizons (as in Table 2.4). However, ITE had to include the H layer within the 0–5 cm layer because of the

Litter layer (L) is often intermixed with mosses (Ahe, Sko, Klo) or stem bases of grasses (Nac, Wal, Sch)

Gri (5)			Nac (6)			Jez (6)		
Hor.	Depth (cm)	Texture sa/si/cl (%)	Hor.	Depth (cm)	Texture sa/si/cl (%)	Hor.	Depth (cm)	Texture sa/si/cl (%)
L	5–4		LF	7–4		LF	8–2	
FH	4–0		H	4–0		H	2–0	
A	0–5	Fine sa	A	0–10	54/33/13	A(e)	0–8	?/?/16
Bsv	5–36	Fine sa	Bvs	10–30	58/35/7	Bvs	8–38	?/?/10
Bwv	36–94	Fine sa	Bv	30–70	62/30/8	Bv	38–85	?/?/?
Bw	94–168	Med sa	C	>70	72/20/8	CB	85–110	?/?/9
BC	>168	Coar sa						

AuF (8)			The (9)			Col (10)		
Hor.	Depth (cm)	Texture sa/si/cl (%)	Hor.	Depth[a] (cm)	Texture sa/si/cl (%)	Hor.	Depth (cm)	Texture sa/si/cl (%)
L	3–2		L	10–5		L	2–1	
FH	2–0		FH	5–0		F	1–0	
Ah	0–10	?/?/7	A/E	0–60	?	A	0–6	Si + loam
A/E	10–20	?/?/8	Bt	60–110	?	E	6–45	Si + loam
Bh	20–30	?/?/8	BC	>110	?	BC	45–80	Si + cl
Bs	30–110	?/?/8				CC	> 80	Si + cl
C	>110							

(1994), (7) Göttlein (1995), (8) Lefèvre (1988), (9) P. Bottner and M.-M. Coûteaux (pers. comm.), (10) anonymous and members of the NIPHYS/CANIF staff.
[a] L and FH layers at Thezan very variable (see text on p. 22).

limitations in the number of ^{14}C analyses which could be carried out (see Harrison et al. Chap. 11, this Vol.). This must be taken into consideration when results from different chapters are compared.

2.2.2 Field Sampling

Soil samples including the litter and humus layers were taken from all main sites in the NIPHYS and CANIF transect (see Table 2.1) for determination of chemical characteristics and C and N pools. In addition, samples were also taken from some of the complementary sites to elucidate certain questions. At each site, the area was divided into four plots varying from 20 × 20 m at some sites to 45 × 45 at others.

Two research groups, the Uppsala group (T. Persson and coworkers) and the ITE group (A.F. Harrison and coworkers), performed soil sampling on most sites. The Uppsala group sampled litter (L) and humus (FH) layers quantitatively with a 250-cm^2 frame, five units per plot. Just below the humus layer, the upper mineral soil was sampled with an 83-cm^2 corer (three to four units per plot), and the soil was divided into 0–10 and 10–20 cm layers independent of generic horizons. Below 20 cm, the 20–30 and 30–50 cm soil layers were collected with a 16-cm^2 corer (three cores per plot). The 30–50 cm layer was not sampled at Schacht because of extreme stoniness. Samples from the same soil layer were pooled for each plot. At sites without a visible humus layer, the 0–10 cm mineral soil layer was subdivided into 0–5 cm (sampled by the 250-cm^2 frame) and 5–10 cm (sampled by the 83-cm^2 corer).

The ITE group took at random nine replicate cores to 20-cm depth in the mineral soil from each site, at different sampling positions from the four plots sampled by the Uppsala group. The cores were divided into an organic (L+F) layer and 0–5 (including the H layer), 5–10 and 10–20 cm soil layers. In addition, nine replicate L+F-layer samples were taken from random 0.25-m^2 areas to compare these estimates of litter mass with those taken with the soil core samples; L+F present on the cores was considered too small to give an accurate estimate of L+F mass.

2.2.3 Laboratory Treatment

The samples were immediately transported to the laboratory, where they were treated without being dried or frozen. The procedure for the Uppsala group was that litter samples were separated from living plants and twig fragments, meaning that the remaining litter material mainly consisted of leaf or needle litter. Humus samples were passed through a 5-mm sieve, and the mineral soil through a 2-mm sieve to remove roots and stones. The pH was determined with a glass electrode in the supernatant after shaking for 2 h and sedimentation in an open flask for 24 h in both distilled water and in 1 M KCl. The proportions between fresh soil and extractant were 1:1 by volume (about 1:10 by dry matter to water for litter and humus and 1:2.5 for mineral soil). Fresh weight/dry weight ratios were determined after the samples were dried at 105 °C for 24 h. Loss-on-ignition (LOI) was determined after combustion at 550 °C for 3 h. Total C and N concentrations were analysed in a Carlo-Erba NA 1500 Analyzer. Soluble aluminium was determined after extraction with 1 M KCl solution. Exchangeable base cations were extracted with a 1 M NH$_4$OAc solution adjusted to pH 7. Cation exchange capacity (CEC) and base saturation (BS) were calculated from the data on base cations (Ca^{2+}, Mg^{2+}, K$^+$ and Na$^+$) determined on an inductively coupled plasma (ICP) analyser and titratable acidity in the buffer solution. Inorganic N was extracted with 1 M KCl solution for 1 h (see details in Persson et al. Chap. 14, this Vol.).

The procedure for the ITE group included sorting of L+F and soil materials to remove all roots and living plant material. The roots were washed and weighed. The soil samples were sieved through a 2-mm sieve to separate the fine soil from the coarser material. The fine soil was air-dried and weighed. The stones were washed and weighed and their volume was estimated by water displacement. The bulk density of the <2-mm fraction of the soil was calculated, i.e. omitting the masses and volumes of stones and roots. Subsamples of each layer from all cores were analysed for C and N. Carbon was analysed as organically bound C by the modified Tinsley method

(Kalembasa and Jenkinson 1973) to avoid the inclusion of inorganic carbon forms, such as charcoal particles derived from past fires on the site. N was analysed as total N by sulphuric acid-peroxide digestion (Rowland and Grimshaw 1985) and continuous-flow colorimetric method using indophenol blue (SKALAR).

2.2.4 Bulk Density and Soil Pools

Data on bulk density can be found in the database on CD-ROM. The estimated amounts of litter, sifted humus layer (<5 mm) and fine soil (<2 mm) to a depth of 50 cm in the mineral soil are also given in Table 2.5. The soil pool of sieved dry matter varied considerably between sites due to the presence of organic matter and gravel/stones. The sandy soils at Klosterhede, Gribskov and Åheden almost lacked stones and had a soil pool of 590–680 kg fine soil m^{-2} to a depth of 50 cm. The sites at Jezeri, Schacht and Aubure had a high proportion of stone and gravel components. At these sites, only 180–250 kg fine soil m^{-2} was estimated in the top 50 cm. Figures considerably lower than 90–100 kg m^{-2} for each 10-cm depth in the mineral soil mostly indicate presence of stones. However, at Collelongo, the low dry mass (and bulk density) in the 0–50 cm layer is a consequence of high amount of organic matter (see also Table 2.8).

2.2.5 Acid/Base Conditions

The sites differed with regard to the pH value. High or moderately high pH values were observed at sites with a high content of carbonate (Monte di Mezzo, Collelongo and Sorø) or sandstone (Thezan) in the parent material (Table 2.5). At most other sites, pH had a minimum in the FH and 0–10 cm mineral soil layers and higher values in both the litter layer and in the deeper mineral soil. Very low pH values (pH 3.45–3.80) were found in the 0–10 and 10–20 cm mineral soil layers in the central European sites Nacetin, Waldstein and Aubure, all sites subjected to heavy deposition of acidifying nitrogen and sulphur compounds during decades. A comparison of spruce and beech stands showed that spruce litter had lower pH than beech litter for comparable sites (Sko/Gri, Nac/Jez, Wal/Sch and AuP/AuF). Spruce stands had generally lower pH than beech stands also in the FH layer, but at greater depths there was no clear difference between dominating tree species. A comparison between Col and MdM was not meaningful because of contrasting land use.

KCl-extractable aluminium (Al) mostly occurred in high concentrations where the pH value was low and in low concentrations where the pH value was high (Table 2.5). Very high concentrations of Al were found in the mineral soil at the Czech, German and French (Aubure) sites (Table 2.5). High Al levels in relation to pH were found in the deep mull soil at Collelongo, where the CEC was comparatively high (see below).

The concentrations of extractable base cations (mmol$_c$ kg^{-1}) were high in the litter layer and mostly much lower at greater depths (Table 2.6). Extractable Ca^{2+} was higher in the beech than in the spruce litter for comparable stands (Sko < Gri, Nac < Jez, Wal < Sch, AuP45 < AuF), and the same tendency was found for the humus layer, where such a layer was developed. At greater depths (10–50 cm), very low Ca concentrations were found at Skogaby and Klosterhede, indicating base-poor minerals. High concentrations were found at Monte di Mezzo, Collelongo, Thezan, Sorø and Schacht.

Table 2.5. Mean dry mass (±SE), pH(H$_2$O) (SE mostly 0.05–0.15) and KCl-extractable Al (SE mostly 5–15% of the mean) in the litter (L), humus (FH) and sieved soil layers (roots

	Soil layer	Åhe	Sko	Klo	Gri	(Sor)	Nac
Dry mass (kg m^{-2})	L	0.7 ± 0.1	0.6 ± 0.0	1.0 ± 0.2	1.2 ± 0.2	1.2 ± 0.2	1.4 ± 0.1
	FH	2.8 ± 0.6	6.4 ± 0.5	11.7 ± 0.9	41 ± 2.9[a]	23 ± 6.5[a]	13.8 ± 0.8
	0–10 cm	89 ± 2.9	93 ± 5.4	108 ± 5.5	55 ± 2.7[b]	57 ± 3.4[b]	50 ± 10
	10–20 cm	123 ± 2.3	96 ± 5.9	104 ± 4.8	122 ± 4.8	105 ± 15	66 ± 9.4
	20–30 cm	128 ± 4.7	95 ± 4.7	124 ± 2.0	140 ± 3.3	141 ± 3.1	70 ± 3.1
	30–50 cm	247 ± 13	195 ± 6.4	327 ± 5.7	248 ± 11	232 ± 22	167 ± 8.6
	Total	591 ± 11	485 ± 15	676 ± 10	606 ± 16	559 ± 31	369 ± 23
pH(H$_2$O)	L	4.28	4.04	4.14	5.15	5.40	4.06
	FH	3.93	4.06	3.88	4.29[a]	4.65[a]	3.62
	0–10 cm	4.45	4.01	3.94	4.07[b]	4.58[b]	3.47
	10–20 cm	5.09	4.48	4.22	4.15	5.12	3.76
	20–30 cm	5.18	4.52	4.47	4.40	6.05	3.97
	30–50 cm	5.27	4.57	4.60	4.65	7.24	4.28
KCl-extr. Al (mmol$_c$ kg^{-1})	L	6.5	4.9	1.6	0.6	1.1	6.7
	FH	24.6	35.0	27.8	7.0[a]	13.6[a]	78.4
	0–10 cm	47.6	30.2	16.2	3.7[b]	34.1[b]	86.4
	10–20 cm	14.5	21.3	35.7	18.1	23.8	86.5
	20–30 cm	11.9	14.6	13.3	14.2	14.1	60.8
	30–50 cm	3.5	7.8	3.2	10.0	0.5	33.3

Where no FH layer was found, the 0–10 cm layer was subdivided into 0–5[a] and 5–10 cm[b]. At AuP90 the L and FH layers were pooled (Incl FH). At some sites, the figure for dry mass

Relatively high concentrations of extractable K$^+$ were found in the FH layer at Åheden and at greater depths at Thezan, Collelongo and Monte di Mezzo, but K$^+$ showed less difference between sites than Ca^{2+} (Table 2.6). Na$^+$ was especially high in the upper soil layers at Klosterhede, showing an influence of sea salt. Similar to Ca^{2+}, extractable Mg^{2+} was higher in the beech than in the spruce litter for comparable stands (Sko < Gri, Nac < Jez, Wal < Sch, AuP45 < AuF). In the FH layer, Skogaby and Klosterhede had high Mg^{2+} concentrations, indicating influence of sea salt, whereas the concentrations at 10–50-cm depth were lower (Skogaby) or as low as those at Nacetin, Waldstein and Aubure, where needle yellowing because of Mg deficiency was observed.

Cation exchange capacity (CEC) was calculated to be high (400–800 mmol$_c$ kg^{-1}) in most organic (L and FH) layers (Table 2.7). The only exception was the litter layer from Thezan (270 mmol$_c$ kg^{-1}) consisting of fairly stiff needle litter of *P. pinaster*. High CEC values were also found in the fine-textured mineral soil at Monte di Mezzo, in the mull/mineral soil at Collelongo and in the upper mineral soil at Schacht and Waldstein.

Base saturation (BS) was high in the soils with high pH and low in the soils with low pH. BS was especially low in the mineral soil layers at Nacetin and Skogaby but also, for example, at Waldstein, Aubure and Klosterhede. For comparable sites (Sko/Gri, Nac/Jez, Wal/Sch and AuP/AuF), the spruce stands had considerably lower

and stones >2 mm removed) in the NIPHYS/CANIF transect. Complementary sites given within parentheses. For abbreviations see Tables 2.1 and 2.3

Jez	Wal	Sch	AuP90	(AuP45)	AuF	The	Col	MdM
1.2 ± 0.2	0.9 ± 0.1	1.5 ± 0.1	Incl FH	0.8 ± 0.1	0.7 ± 0.1	0.8 ± 0.1	2.1 ± 0.2	3.9 ± 0.7
13.4 ± 3.0	14.8 ± 1.4	6.7 ± 0.6	4.5 ± 1.1	3.3 ± 0.5	3.0 ± 0.3	34 ± 0.9[a]	19 ± 1.7[a]	34 ± 2.2[a]
34 ± 2.1	80 ± 9.2	43 ± 4.0	44 ± 5.0	45 (ap)	36 ± 5.7	46 ± 1.1[b]	32 ± 1.6[b]	52 ± 3.7[b]
34 ± 6.8	62 ± 8.8	46 ± 5.6	45 ± 3.2	45 (ap)	47 ± 3.3	95 ± 3.2	79 ± 2.5	108 ± 2.1
33 ± 5.7	60 ± 5.9	51 ± 6.8	50 (ap)	50 (ap)	54 ± 2.2	95 (ap)	58 ± 7.5	104 ± 7.3
66 ± 11	181 ± 17	–	100 (ap)	110 (ap)	110 (ap)	190 (ap)	175 ± 25	209 ± 15
182 ± 24	398 ± 29	148 ± 8.7	244 (ap)	254 (ap)	251 (ap)	460 (ap)	365 ± 18	511 ± 25
4.86	4.72	5.03	Incl FH	4.15	4.61	4.57	5.89	6.62
4.11	3.69	4.34	3.96	3.77	4.01	5.58[a]	5.92[a]	6.85[a]
3.93	3.52	3.99	3.60	3.45	3.57	5.90[b]	5.45[b]	6.98[b]
4.12	3.80	4.32	3.93	3.78	3.80	5.88	5.45	7.32
4.18	4.26	4.49	4.27	4.01	4.09	5.90	5.50	7.37
4.27	4.44	–	4.48	4.24	4.43	5.75	5.60	7.61
8.8	3.8	1.3	Incl FH	5.5	3.1	0.6	0.5	0.2
90.7	96.1	30.0	35.7	25.6	29.0	0.2[a]	0.4[a]	0.0[a]
81.1	63.8	60.0	76.6	81.7	61.1	0.1[b]	27.6[b]	0.0[b]
72.8	108.4	79.6	84.9	91.5	68.4	0.2	43.9	0.0
65.2	58.0	69.0	81.1	92.0	76.0	0.4	44.8	0.0
48.8	24.9	–	72.5	61.8	67.0	3.9	39.2	0.0

is an approximation (ap) based on few replicates. At Schacht the 30–50 cm layer was not sampled because of stones.

BS than the beech stands in the 0–10 and 10–20 cm soil layers (Table 2.7). At greater depths the differences were less evident.

2.2.6 Soil C and N Pools

Carbon concentrations (% of sieved dry matter) were high (36–52%) in the organic horizons but decreased markedly with increasing depth and were especially low (0.2%) at 30–50 cm depth at Åheden (Table 2.8). In the mineral soil, high C concentrations were found in the 0–10 cm layer at Schacht and Nacetin. At the latter site, the 0–10 and 10–20 cm layer had considerable amounts of charcoal in one of the plots. As a whole, the Czech, German and Italian sites had high C concentrations in the mineral soil. No clear differences were found between spruce and beech sites.

Nitrogen concentrations (percent of sieved dry matter) generally decreased with soil depth. Moderately high N concentrations (1.4–2.1%) were found in most L layers studied, but at Thezan and Åheden the concentrations were considerably lower (0.6 and 1.1%, respectively; Table 2.8). At greater depths, the decrease in N concentration was a consequence of increased mineral materials.

The C:N ratios often decreased with increasing depth, but there were many exceptions (Table 2.8). Low C:N ratios were found at sites with calcareous subsoil (Monte di Mezzo and Collelongo). These sites had also high pH (Table 2.5). However, the possible correlation between low C:N ratio and high pH does not fit for Thezan, where the C:N ratio was fairly high throughout the soil profile. Other sites with high C:N ratios were Klosterhede and Åheden. Åheden has a very low N input (Table 2.1) and probably a very tight N cycle. Klosterhede has experienced much higher inputs of

Table 2.6. NH$_4$OAc-extractable base cations (Ca, K, Na and Mg) (SE mostly 5–15% of the mean) in litter (L), humus (FH) and sieved soil layers (roots and stones >2 mm removed) in the NIPHYS/CANIF transect. For abbreviations and symbols see Tables 2.1, 2.3 and 2.5

	Soil layer	Åhe	Sko	Klo	Gri	(Sor)	Nac
Ca (mmol$_c$ kg^{-1})	L	127.3	53.4	57.8	232.6	367.1	77.4
	FH	98.5	28.0	30.3	14.4[a]	68.0[a]	22.9
	0–10 cm	2.8	0.4	0.7	3.0[b]	23.6[b]	2.5
	10–20 cm	1.1	0.2	0.3	0.9	35.5	0.9
	20–30 cm	1.2	0.1	0.1	0.6	66.1	0.4
	30–50 cm	0.7	0.1	0.0	0.2	180.3	0.3
K (mmol$_c$ kg^{-1})	L	18.7	12.1	17.4	24.5	20.4	12.4
	FH	19.7	9.4	8.2	2.4[a]	2.6[a]	5.1
	0–10 cm	1.6	0.9	0.5	1.0[b]	0.7[b]	1.4
	10–20 cm	1.4	0.3	0.5	0.6	0.5	0.7
	20–30 cm	1.3	0.2	0.2	0.3	0.3	0.5
	30–50 cm	0.6	0.2	0.1	0.1	0.4	0.4
Na (mmol$_c$ kg^{-1})	L	1.0	4.0	15.8	5.6	6.2	1.9
	FH	0.9	6.7	19.6	0.8[a]	1.6[a]	1.4
	0–10 cm	0.2	0.6	1.0	0.3[b]	0.8[b]	0.4
	10–20 cm	0.2	0.4	1.1	0.2	0.8	0.2
	20–30 cm	0.2	0.3	0.5	0.2	0.8	0.2
	30–50 cm	0.1	0.2	0.2	0.2	1.2	0.1
Mg (mmol$_c$ kg^{-1})	L	29.1	25.0	43.5	64.4	80.9	13.3
	FH	21.1	23.8	57.8	4.0[a]	12.0[a]	5.8
	0–10 cm	1.0	0.8	1.8	1.0[b]	3.1[b]	1.0
	10–20 cm	0.6	0.2	0.9	0.4	2.3	0.5
	20–30 cm	0.7	0.1	0.2	0.2	2.8	0.3
	30–50 cm	0.5	0.1	0.0	0.1	5.7	0.1
Ca + K + Na + Mg (mmol$_c$ kg^{-1})	L	176.1	94.4	134.5	327.0	474.5	104.9
	FH	140.3	67.9	115.8	21.6[a]	84.3[a]	35.2
	0–10 cm	5.7	2.7	4.1	5.3[b]	28.3[b]	5.4
	10–20 cm	3.2	1.1	2.8	2.1	39.1	2.3
	20–30 cm	3.4	0.8	1.0	1.3	70.1	1.4
	30–50 cm	1.9	0.5	0.3	0.7	187.7	1.0

inorganic N due to the vicinity of agricultural areas. Still, the organic layers that developed both under the former *Calluna* heathland and the present coniferous forest have high C:N ratios. The Czech and German sites with a high input of N had moderate C:N ratios, not very different from the South-Swedish site Skogaby and the Danish site Gribskov. The high C:N ratio in the 0–20-cm mineral soil at Nacetin may reflect the presence of charcoal. For comparable sites (Sko/Gri, Nac/Jez, Wal/Sch and AuP/AuF) spruce seemed to have higher C:N ratios than beech, with Aubure as an exception.

Loss-on-ignition (LOI) determined at 550 °C provides a means of estimating organic matter (OM) in the organic horizons, but hard-bound water (present at

Table 2.6. *Continued*

Jez	Wal	Sch	AuP90	(AuP45)	AuF	The	Col	MdM
265.9	114.0	289.4	Incl FH	121.0	194.6	82.7	458.0	673.9
62.7	27.9	127.9	53.8	41.4	58.0	128.3[a]	246.2[a]	356.4[a]
5.6	9.7	43.1	3.0	0.9	11.7	83.3[b]	110.9[b]	343.7[b]
2.4	1.7	18.7	1.0	0.4	1.5	55.0	58.9	350.5
1.9	0.5	2.0	0.6	0.4	0.7	49.8	44.9	351.4
1.3	0.2	–	0.5	0.4	0.4	56.7	37.6	332.9
20.4	17.4	17.5	Incl FH	18.2	21.3	21.8	16.7	14.0
4.4	4.9	6.6	9.0	11.6	7.6	9.8[a]	13.8[a]	9.2[a]
1.4	3.2	3.2	1.9	2.1	3.7	7.9[b]	7.3[b]	6.7[b]
1.1	1.3	1.4	1.3	1.4	2.1	6.8	4.0	6.2
1.0	0.9	0.5	1.2	1.4	1.3	8.3	3.2	5.9
0.8	1.1	–	1.1	1.2	0.9	9.4	3.1	4.9
2.1	1.6	1.7	Incl FH	1.1	2.1	4.3	1.8	2.0
0.9	0.8	1.2	0.9	1.0	0.8	2.3[a]	1.1[a]	1.1[a]
0.3	0.4	0.7	0.3	0.2	0.5	1.8[b]	1.1[b]	1.1[b]
0.2	0.4	0.4	0.2	0.1	0.3	1.5	1.2	1.0
0.2	0.1	0.3	0.1	0.1	0.2	1.7	1.4	1.0
0.2	0.1	–	0.2	0.1	0.2	2.3	1.3	1.0
49.0	21.0	40.2	Incl FH	10.3	28.7	33.7	67.6	43.9
16.7	5.0	15.0	6.9	5.5	8.5	19.8[a]	32.0[a]	28.5[a]
2.7	2.2	5.6	0.8	0.6	2.6	13.8[b]	15.9[b]	25.9[b]
1.2	0.8	2.0	0.3	0.3	0.5	9.6	9.7	24.3
0.8	0.2	0.4	0.2	0.2	0.2	10.5	8.0	22.7
0.5	0.1	–	0.2	0.2	0.2	13.7	6.8	19.3
337.4	154.0	348.8	Incl FH	150.6	246.7	142.4	544.2	733.8
84.8	38.6	150.6	70.6	59.6	74.9	160.2[a]	293.1[a]	395.1[a]
10.0	15.5	52.6	5.9	3.9	18.5	106.6[b]	135.1[b]	377.4[b]
4.9	4.1	22.5	2.9	2.2	4.3	72.8	73.9	382.0
3.9	1.7	3.2	2.1	2.1	2.4	70.4	57.6	381.0
2.9	1.5	–	1.9	1.8	1.6	82.1	48.8	358.1

105 °C) is also included in the estimates for the mineral soil. The LOI content to a depth of 50 cm was lowest (12–14 kg m^{-2}) at Thezan and in the AuP90 stand at Aubure and highest (40–60 kg m^{-2}) at Collelongo, Waldstein and Monte di Mezzo (details not shown).

Total carbon pools (estimated by the Uppsala group) varied between 39 (Thezan) and 228 (Collelongo) ×10^3 kg C ha^{-1} in the European transect (Table 2.8). The contribution of the organic soil layer (LFH) to total C in the soil profile studied varied between 3 and 44%. At sites where a humus layer had developed, 12–44% of the total C pool was found in the LFH layer, but without a humus layer, only 3–7% of the total C pool belonged to the organic layer. Thus, most of the C pool was found in the

Table 2.7. CEC (at pH 7) and % BS (at pH 7) (SE mostly 5–15% of the mean) in the litter (L), humus (FH) and sieved soil layers (roots and stones >2mm removed) in the NIPHYS/CANIF transect. For abbreviations and symbols see Tables 2.1, 2.3 and 2.5

	Soil layer	Åhe	Sko	Klo	Gri	(Sor)	Nac	Jez	Wal	Sch	AuP90	(AuP45)	AuF	The	Col	MdM
CEC	L	453	478	453	586	659	443	658	503	746	Incl FH	535	611	271	672	791
(mmol$_c$ kg^{-1})	FH	495	657	693	114[a]	176[a]	572	419	606	742	493	474	410	74[a]	405[a]	410[a]
	0–10cm	75	88	63	61[b]	83[b]	196	130	299	402	116	117	170	44[b]	233[b]	381[b]
	10–20cm	36	55	84	47	76	114	91	176	202	91	80	88	30	190	372
	20–30cm	32	48	39	40	92	85	82	101	114	85	81	77	28	170	368
	30–50cm	13	29	12	27	178	56	69	57	–	75	71	84	36	148	334
BS (%)	L	39	20	30	56	72	23	51	31	47	Incl FH	28	40	45	81	92
	FH	28	10	17	20[a]	49[a]	6	20	6	20	15	13	18	65[a]	72[a]	96[a]
	0–10cm	7	3	7	9[b]	34[b]	3	8	5	13	5	3	9	70[b]	62[b]	99[b]
	10–20cm	9	2	3	4	47	2	5	2	11	3	3	5	69	38	100
	20–30cm	11	2	3	3	64	2	5	2	3	2	3	3	69	34	100
	30–50cm	14	2	3	2	100	2	4	3	–	3	3	2	66	33	100

mineral soil, and a considerable fraction was often found at 0–10-cm depth. Typically, the amount of C decreased with increasing depth. However, at Klosterhede the 10–20-cm layer was affected by a former humus layer that was ploughed in when the former *Calluna* heathland was changed into a tree plantation about 100 years ago. At Monte di Mezzo the upper 0–30 cm of the mineral soil was affected by tilling of the former pasture land. At this site, pH in the subsoil was higher than at any other site, indicating presence of carbonate. Because the Carlo-Erba method included all types of C in the determinations, the C pool estimated for the 30–50-cm layer probably included a considerable amount of carbonate C. This is indicated by the C:LOI relations at Monte di Mezzo, that deviated from all other sites. Taking this relation into account, the total organic C pool at Monte di Mezzo should be about 175×10^3 kg ha^{-1}. The data indicates that the warm-temperate site, Thezan, and the boreal site, Åheden, had low C pools. At the other sites, the pools were generally higher but quite variable. Permanently forested areas like Fichtelgebirge (Waldstein), Krusne Hory (Nacetin) and the Abruzzo Mountains (Collelongo) had all fairly high C pools. The Aubure sites with a more variable land use and, in addition, coarse-textured soils had lower C pools. A comparison between spruce and beech stands showed no consistent pattern, probably because of different land use (Sko/Gri), different slope (Nac/Jez) or different water regimes (AuP/AuF). Because the C pools have developed for centuries, the continuity of the same tree species is a prerequisite for a proper comparison.

The soil carbon pools estimated by the ITE group to a depth of 20 cm were generally only about 80% of those estimated by the Uppsala group (see above). One explanation is that the two groups divided the upper soil layers differently. The Uppsala group included the H layer in the organic layer, whereas the ITE group included the H layer in the 0–20 cm soil layer. This meant that the 20-cm depth was shallower for the ITE than for the Uppsala group when an H layer was present. Another difference between the two groups was that the ITE group had generally slightly lower estimates of bulk density. This can also be a result of the different soil depths considered. A further difference is that the method of the Uppsala group included charcoal, which could have contributed to slightly higher values at some sites. In the following chapters, the estimates of the Uppsala group form a basis for C and N mineralisation calculations (see Persson et al. Chaps. 12 and 14, this Vol.) and those of the ITE group form a basis for C turnover estimates in Harrison et al. (Chap. 11, this Vol.).

Total nitrogen pools varied between 1.4 (Thezan) and 18.1 (Collelongo) $\times 10^3$ kg N ha^{-1} in the European transect (Table 2.8). The Italian sites Collelongo and Monte di Mezzo had high N pools (about 18×10^3 kg ha^{-1}), the German and Czech sites had intermediate (5–8.5×10^3 kg ha^{-1}) and the south-Scandinavian and north-French sites had slightly lower N pools (3–5.5×10^3 kg ha^{-1}). The warm-temperate Thezan and the boreal Åheden had low N pools (about 1–2×10^3 kg ha^{-1}). At most sites the N pools were large where the C pools were large. However, at the Italian sites the C:N ratios were much lower than at the other sites and, consequently, the N pools were proportionally higher than expected from the C pools (Fig. 2.3).

KCl-extractable NH_4-N and NO_3-N vary considerably during the year, depending on N deposition, N mineralisation, nitrification, plant N uptake and N leaching. The data in Table 2.9 thus provide a snapshot of possibilities. Some conclusions can, however, be drawn. Åheden is special in having very low pools of inorganic N, of both ammonium and nitrate. Skogaby and Klosterhede had almost no nitrate, but also Nacetin, AuP45 and Thezan had fairly low nitrate levels. High ammonium and nitrate

Table 2.8. Mean C and N concentrations (mg g^{-1} sieved dry matter), mean C and N pools (10^3 kg ha^{-1}) and mean C:N ratios ±SE in different soil layers in the NIPHYS/CANIF

	Soil layer	Åhe	Sko	Klo	Gri	(Sor)	Nac	Jez
C	L	491 ± 8	511 ± 2	516 ± 2	445 ± 21	363 ± 70	458 ± 16	456 ± 7
(mg g^{-1})	FH	403 ± 23	432 ± 15	430 ± 10	47 ± 4[a]	94 ± 40[a]	370 ± 16	289 ± 9
	0–10 cm	24 ± 2	42 ± 4	26 ± 2	24 ± 4[b]	25 ± 3[b]	120 ± 42	76 ± 6
	10–20 cm	8 ± 1	23 ± 3	29 ± 2	16 ± 3	18 ± 2	44 ± 13	32 ± 4
	20–30 cm	5 ± 1	19 ± 2	15 ± 2	12 ± 2	9 ± 1	25 ± 2	25 ± 3
	30–50 cm	2 ± 0	12 ± 2	4 ± 1	8 ± 1	11 ± 2	13 ± 1	19 ± 2
N	L	10.8 ± 0.3	14.2 ± 0.5	18.8 ± 0.3	17.1 ± 0.9	15.0 ± 2.5	17.0 ± 1.4	17.9 ± 0.7
(mg g^{-1})	FH	10.9 ± 1.0	15.4 ± 0.7	14.0 ± 0.4	2.5 ± 0.2[a]	5.3 ± 1.9[a]	15.2 ± 0.6	14.6 ± 0.4
	0–10 cm	0.7 ± 0.1	1.7 ± 0.2	0.7 ± 0.0	1.2 ± 0.3[b]	1.7 ± 0.2[b]	3.9 ± 0.8	3.8 ± 0.3
	10–20 cm	0.3 ± 0.0	1.0 ± 0.1	0.9 ± 0.1	0.7 ± 0.1	1.1 ± 0.1	1.6 ± 0.3	1.6 ± 0.2
	20–30 cm	0.2 ± 0.0	0.9 ± 0.1	0.5 ± 0.1	0.5 ± 0.1	0.6 ± 0.1	1.1 ± 0.1	1.2 ± 0.1
	30–50 cm	0.1 ± 0.0	0.6 ± 0.1	0.1 ± 0.0	0.4 ± 0.0	0.6 ± 0.1	0.6 ± 0.0	1.0 ± 0.1
C	L	3.4 ± 0.6	3.3 ± 0.2	4.9 ± 1.0	4.4 ± 1.0	4.5 ± 1.0	6.5 ± 0.7	5.5 ± 0.8
(10^3 kg ha^{-1})	FH	11.5 ± 2.5	27.8 ± 2.4	50.7 ± 4.7	19.3 ± 2.6[a]	15.2 ± 1.8[a]	50.8 ± 3.9	38.9 ± 8.6
0.4[b]	0–10 cm	21.2 ± 1.6	38.4 ± 2.5	27.5 ± 0.9	12.7 ± 1.4[b]	14.0 ± 1.1[b]	47.5 ± 6.8	25.5 ± 1.8
	10–20 cm	9.3 ± 1.5	21.6 ± 1.4	29.9 ± 1.9	18.9 ± 3.1	18.0 ± 2.4	25.7 ± 2.9	10.2 ± 1.3
	20–30 cm	5.7 ± 0.5	18.2 ± 2.5	17.9 ± 1.7	17.1 ± 2.6	13.3 ± 1.9	17.4 ± 1.9	7.7 ± 0.5
	30–50 cm	4.0 ± 0.3	22.5 ± 3.2	13.1 ± 1.5	20.0 ± 3.1	23.1 ± 3.8	21.8 ± 0.3	12.1 ± 1.4
	Total	55 ± 4	132 ± 9	144 ± 5	92 ± 12	88 ± 6	170 ± 10	100 ± 10
	% in LFH	27	24	39	5	5	34	44
N	L	0.1 ± 0.0	0.1 ± 0.0	0.2 ± 0.0	0.2 ± 0.0	0.2 ± 0.0	0.2 ± 0.0	0.2 ± 0.0
(10^3 kg ha^{-1})	FH	0.3 ± 0.1	1.0 ± 0.1	1.6 ± 0.2	1.0 ± 0.1[a]	0.9 ± 0.1[a]	2.1 ± 0.2	1.9 ± 0.4
	0–10 cm	0.7 ± 0.1	1.5 ± 0.1	0.7 ± 0.0	0.6 ± 0.1[b]	0.9 ± 0.1[b]	1.7 ± 0.3	1.3 ± 0.1
	10–20 cm	0.4 ± 0.0	0.9 ± 0.0	0.9 ± 0.0	0.8 ± 0.2	1.1 ± 0.1	1.0 ± 0.0	0.5 ± 0.1
	20–30 cm	0.3 ± 0.0	0.8 ± 0.1	0.6 ± 0.1	0.7 ± 0.1	0.8 ± 0.1	0.8 ± 0.1	0.4 ± 0.0
	30–50 cm	0.3 ± 0.0	1.1 ± 0.1	0.5 ± 0.1	0.9 ± 0.1	1.3 ± 0.1	1.0 ± 0.0	0.6 ± 0.1
	Total	2.1 ± 0.2	5.4 ± 0.3	4.6 ± 0.2	4.3 ± 0.6	5.2 ± 0.2	6.8 ± 0.3	5.0 ± 0.4
	% in LFH	19	20	39	5	4	34	42
C:N	L	46 ± 1.4	36 ± 1.3	27 ± 0.4	26 ± 1.2	24 ± 0.8	27 ± 2.2	25 ± 1.3
	FH	37 ± 1.6	28 ± 0.7	31 ± 0.2	19 ± 0.4[a]	17 ± 1.0[a]	24 ± 0.8	20 ± 0.4
	0–10 cm	32 ± 1.9	25 ± 0.6	39 ± 0.4	21 ± 1.2[b]	15 ± 1.1[b]	29 ± 4.0	20 ± 0.5
	10–20 cm	24 ± 1.4	24 ± 0.7	33 ± 0.9	23 ± 1.3	17 ± 1.4	27 ± 3.0	20 ± 1.0
	20–30 cm	18 ± 1.3	22 ± 1.6	30 ± 1.9	23 ± 1.2	17 ± 1.9	22 ± 0.4	20 ± 0.9
	30–50 cm	12 ± 0.5	21 ± 0.7	27 ± 0.7	22 ± 0.7	17 ± 2.6	21 ± 0.5	20 ± 0.6
	Total	26 ± 0.3	24 ± 0.6	32 ± 0.4	21 ± 0.9	17 ± 1.0	25 ± 1.5	20 ± 0.4

concentrations were found at Gribskov, Sorø, Jezeri, Waldstein, Schacht, Aubure (excepting AuP45) and Collelongo. At Monte di Mezzo, the ammonium levels were low in comparison with the nitrate levels. Except for Monte di Mezzo, that had a distribution of inorganic N similar to arable fields, and the AuP90 stand at Aubure, there was a tendency for the beech stands to have higher nitrate concentrations than comparable spruce stands (Sko/Gri, Nac/Jez and possibly Wal/Sch).

2.3 Ecosystem C and N Pools

The ecosystem pools of organic C and N were calculated from data on tree biomass (see Scarascia-Mugnozza et al. Chap. 3 and Stober et al. Chap. 5, this Vol.) with fine-root biomass (0–2 mm diam) to a rooting depth of 30 cm (see heading of Table 2.10),

transect through Europe. Complementary sites within parentheses. Further explanations in Tables 2.1 and 2.3. SE = 0 indicates a value <0.5 and 0.0 < 0.05

Wal	Sch	AuP90	(AuP45)	AuF	The	Col	MdM
497 ± 4	473 ± 9	Incl FH	496 ± 10	488 ± 3	513 ± 4	374 ± 14	339 ± 39
377 ± 25	431 ± 23	389 ± 23	367 ± 45	259 ± 26	21 ± 1[a]	146 ± 15[a]	65 ± 5[a]
59 ± 11	155 ± 31	38 ± 2	49 ± 6	92 ± 34	10 ± 1[b]	90 ± 5[b]	41 ± 3[b]
58 ± 9	72 ± 2	15 ± 0	23 ± 3	32 ± 4	6 ± 1	67 ± 2	37 ± 2
52 ± 8	63 ± 4	11 ± 1	21 ± 5	22 ± 41	7 ± 1	60 ± 2	35 ± 2
20 ± 4	-	9 ± 1	16 ± 2	26 ± 5	6 ± 1	49 ± 13	26 ± 3
20.6 ± 0.5	19.6 ± 0.5	Incl FH	17.9 ± 0.3	19.3 ± 0.7	5.6 ± 0.5	18.1 ± 0.8	16.1 ± 1.3
17.0 ± 1.4	21.9 ± 0.9	15.0 ± 0.4	14.7 ± 1.3	12.1 ± 1.1	0.7 ± 0.0[a]	11.0 ± 0.8[a]	5.6 ± 0.3[a]
2.4 ± 0.6	7.6 ± 1.2	2.3 ± 0.2	2.8 ± 0.3	5.2 ± 1.8	0.4 ± 0.0[b]	7.3 ± 0.3[b]	4.2 ± 0.2[b]
2.3 ± 0.4	3.4 ± 0.2	1.1 ± 0.1	1.3 ± 0.2	1.8 ± 0.2	0.2 ± 0.0	5.4 ± 0.2	4.1 ± 0.1
2.0 ± 0.3	2.8 ± 0.3	0.9 ± 0.0	1.1 ± 0.3	1.1 ± 0.2	0.3 ± 0.0	4.8 ± 0.3	3.8 ± 0.2
0.9 ± 0.2	-	0.7 ± 0.1	0.8 ± 0.1	1.2 ± 0.2	0.2 ± 0.1	4.0 ± 1.1	2.3 ± 0.3
4.3 ± 0.5	7.0 ± 0.5	Incl FH	4.0 ± 0.6	3.5 ± 0.4	4.3 ± 0.6	7.7 ± 1.0	13.2 ± 3.2
54.7 ± 2.2	28.9 ± 3.4	17.5 ± 4.6	12.7 ± 3.5	7.6 ± 0.4	7.1 ± 0.6[a]	26.8 ± 1.7[a]	22.0 ± 1.5[a]
44.8 ± 4.6	63.7 ± 5.3	16.0 ± 1.9	22.1 ± 2.6	27.7 ± 3.7	4.4 ± 0.3[b]	28.4 ± 1.7[b]	21.0 ±
33.4 ± 0.5	33.0 ± 3.3	6.4 ± 0.3	10.5 ± 1.4	14.9 ± 1.8	5.5 ± 0.8	52.9 ± 2.1	39.8 ± 1.7
30.0 ± 2.8	31.1 ± 2.4	5.7 ± 0.7	10.5 ± 2.5	12.0 ± 2.3	6.2 ± 0.7	35.1 ± 4.7	35.9 ± 2.5
34.5 ± 3.1	-	8.8 ± 0.5	17.2 ± 2.2	28.4 ± 6.0	11.3 ± 2.7	77.1 ± 12	54.3 ± 5.9
202 ± 5	164 ± 12	54 ± 6	77 ± 4	94 ± 4	39 ± 4	228 ± 18	186 ± 7
29	22	32	22	12	11	3	7
0.2 ± 0.0	0.3 ± 0.0	Incl FH	0.1 ± 0.0	0.1 ± 0.0	0.05 ± 0.0	0.4 ± 0.1	0.6 ± 0.1
2.5 ± 0.1	1.5 ± 0.2	0.7 ± 0.2	0.5 ± 0.1	0.4 ± 0.0	0.3 ± 0.0[a]	2.0 ± 0.1[a]	1.9 ± 0.2[a]
1.8 ± 0.2	3.1 ± 0.2	1.0 ± 0.1	1.3 ± 0.1	1.6 ± 0.2	0.2 ± 0.0[b]	2.3 ± 0.2[b]	2.2 ± 0.1[b]
1.3 ± 0.1	1.5 ± 0.1	0.5 ± 0.0	0.6 ± 0.1	0.8 ± 0.1	0.2 ± 0.0	4.2 ± 0.1	4.4 ± 0.1
1.1 ± 0.1	1.4 ± 0.1	0.5 ± 0.0	0.6 ± 0.1	0.6 ± 0.1	0.3 ± 0.0	2.8 ± 0.4	3.9 ± 0.2
1.5 ± 0.1	-	0.7 ± 0.0	0.9 ± 0.1	1.3 ± 0.2	0.5 ± 0.1	6.3 ± 1.1	4.6 ± 0.4
8.4 ± 0.2	7.8 ± 0.4	3.3 ± 0.3	4.0 ± 0.3	4.9 ± 0.2	1.4 ± 0.1	18.1 ± 1.7	17.7 ± 0.7
32	23	21	15	10	4	2	3
24 ± 0.7	24 ± 0.3	Incl FH	28 ± 0.8	25 ± 1.0	94 ± 7.2	21 ± 0.2	21 ± 0.8
22 ± 0.4	20 ± 0.5	26 ± 0.9	25 ± 0.9	21 ± 0.3	28 ± 0.8[a]	13 ± 0.3[a]	12 ± 0.5[a]
26 ± 1.4	20 ± 0.7	16 ± 0.3	18 ± 0.9	17 ± 0.5	25 ± 0.4[b]	12 ± 0.5[b]	10 ± 0.4[b]
26 ± 0.8	22 ± 0.6	13 ± 0.7	19 ± 1.2	18 ± 0.2	25 ± 1.3	13 ± 0.4	9 ± 0.2
26 ± 0.5	23 ± 0.9	12 ± 0.6	19 ± 0.3	19 ± 0.7	23 ± 0.4	13 ± 0.4	9 ± 0.3
23 ± 0.3	-	12 ± 0.7	18 ± 0.8	21 ± 1.2	25 ± 1.1	12 ± 0.7	12 ± 1.4
24 ± 0.5	21 ± 0.6	16 ± 0.9	19 ± 0.5	19 ± 0.3	27 ± 0.6	13 ± 0.4	11 ± 0.1

C and N concentrations (see Bauer et al. Chap. 4, this Vol.) and soil data (Table 2.8). The total pools varied between about 150 (Åheden) and 360 × 10³ kg C ha⁻¹ (Collelongo and Waldstein) and 2.6 (Åheden) and 18.5 × 10³ kg N ha⁻¹ (Collelongo), respectively (Table 2.10). Thus, the N pools varied much more between sites than the C pools.

The biomass C pool was lowest in the 30-year-old spruce stand at Skogaby (74 × 10³ kg C ha⁻¹) and highest in the old beech stand at Gribskov (192 × 10³ kg C ha⁻¹). On average, the C pool in the tree biomass amounted to 40–50% of that in the whole ecosystem (to a depth of 50 cm in the soil), but the variation was high. For example, the biomass in the young Skogaby stand contained only 36% of the ecosystem C pool, whereas the old spruce stand at Aubure contained as much as 73%. For mature stands,

Table 2.9. KCl-extractable NH$_4$-N and NO$_3$-N (µg N g^{-1} LOI) ±SE in different soil layers in the NIPHYS/CANIF transect through Europe on different sampling occasions.

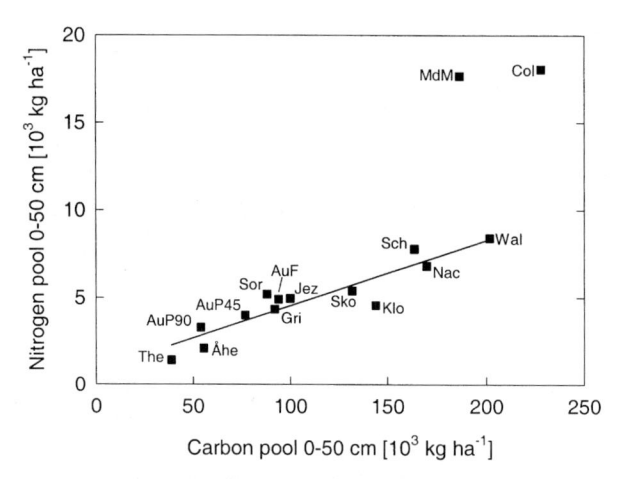

	Sampling	Åhe Sep 94	Sko May 97	Klo May 94	Gri May 94	(Sor) Oct 97	Nac Oct 97	Jez Oct 97
NH$_4$-N	L	4 ± 2	15 ± 2	88 ± 5	187 ± 37	182 ± 36	129 ± 16	180 ± 60
	FH	3 ± 2	16 ± 2	35 ± 5	165 ± 13[a]	69 ± 8[a]	27 ± 13	63 ± 17
	0–10 cm	1 ± 0	49 ± 11	31 ± 9	47 ± 16[b]	45 ± 11[b]	11 ± 5	31 ± 5
	10–20 cm	3 ± 0	48 ± 9	22 ± 7	56 ± 11	62 ± 15	12 ± 3	34 ± 5
	20–30 cm	7 ± 2	13 ± 7	15 ± 4	40 ± 3	48 ± 9	12 ± 3	27 ± 2
	30–50 cm	11 ± 2	11 ± 5	25 ± 10	30 ± 2	40 ± 3	3 ± 1	18 ± 5
NO$_3$-N	L	0 ± 0	0 ± 0	2 ± 2	15 ± 4	93 ± 45	5 ± 2	155 ± 96
	FH	0 ± 0	0 ± 0	0 ± 0	47 ± 14[a]	71 ± 40[a]	14 ± 7	115 ± 49
	0–10 cm	0 ± 0	0 ± 0	0 ± 0	29 ± 10[b]	39 ± 20[b]	5 ± 3	37 ± 21
	10–20 cm	0 ± 0	1 ± 0	0 ± 0	16 ± 5	20 ± 3	4 ± 2	13 ± 4
	20–30 cm	0 ± 0	2 ± 1	0 ± 0	10 ± 3	12 ± 2	4 ± 1	10 ± 4
	30–50 cm	1 ± 1	1 ± 1	2 ± 1	6 ± 1	8 ± 4	14 ± 6	7 ± 2

Fig. 2.3. Relation between carbon and nitrogen pools at the NIPHYS/CANIF sites to a depth of 50 cm in the mineral soil. The regression (R^2 = 0.86) based on data from all sites excluding Collelongo and Monte di Mezzo, two sites with a calcareous subsoil and very high N pools

there was no difference in biomass C between spruce and beech stands (Nac/Jez, Wal/Sch, and AuP/AuF).

The biomass N pool varied between 2% (Collelongo and Monte di Mezzo) and 19% (Åheden) of the total ecosystem N pool. The low percentage for the Italian sites depends on extraordinarily high soil N storage that must have accumulated long

Experimental Sites in the NIPHYS/CANIF Project 39

Determination made after sieving in the laboratory. Further explanations in Tables 2.1 and 2.3. 0 indicates a value less than 0.5. The samples at Wal and Sch were not properly replicated (SE not given)

Wal Sep 98	Sch Sep 98	AuP90 Aug 94	(AuP45) Aug 94	AuF Aug 94	The Apr 93	Col Apr 96	MdM Apr 96
339	198	Incl FH	57 ± 9	205 ± 16	5 ± 2	79 ± 33	33 ± 16
59	119	74 ± 35	164 ± 25	86 ± 9	105 ± 29[a]	222 ± 76[a]	27 ± 12[a]
31	35	81 ± 10	89 ± 5	57 ± 10	112 ± 46[b]	89 ± 17[b]	26 ± 12[b]
43	7	52 ± 7	52 ± 15	78 ± 27	64 ± 15	78 ± 8	14 ± 5
10	14	40 ± 19	40 ± 6	30 ± 12	77 ± 8	53 ± 5	12 ± 4
14	–	45 ± 16	37 ± 13	23 ± 10	53 ± 9	22 ± 7	7 ± 1
87	166	Incl FH	0 ± 0	18 ± 11	2 ± 1	71 ± 30	129 ± 32
53	67	51 ± 12	1 ± 1	98 ± 21	1 ± 1[a]	114 ± 21[a]	133 ± 40[a]
46	84	67 ± 17	19 ± 10	83 ± 24	4 ± 1[b]	66 ± 21[b]	102 ± 18[b]
13	64	79 ± 10	20 ± 8	65 ± 7	6 ± 1	30 ± 11	125 ± 12
38	26	51 ± 13	16 ± 7	40 ± 3	7 ± 1	24 ± 10	122 ± 10
21	–	47 ± 10	16 ± 8	25 ± 4	7 ± 1	15 ± 4	78 ± 21

before the present stands. Except for the Italian sites, the biomass N pools were, on average, 10–11% of the total ecosystem N pool. This proportion is similar to the values found by Cole and Rapp (1981) for mature broadleaf and conifer stands. For comparable sites, spruce stands contained more N in their needles than the beech stands in their leaves. On the other hand, the beech stands had more N than the spruce stands in the branch biomass. For the sites investigated in this study, there was no clear difference in N pools between spruce and beech biomass (Table 2.10).

2.4 Database

For the main sites of the CANIF project (Thezan and Klosterhede were sites in the NIPHYS project), data are collected in a database. This database is available on CD-ROM as a part of this book. The data in this database are free to use for anyone when crediting data products and data holders by correct citation. The collected data are meant to represent the total ecosystem and include data on tree species, field layer vegetation, soil organic matter, soil data, soil chemistry data, hydrology as well as climatic conditions and deposition data. The database covers in principle the years 1993–1998, although also information from earlier years is included when available. An overview of the collected data is given in Table 2.11. The data included are primary data from the sites. Measured tree biomass and tree growth were, however, not available for all sites. For those sites, an estimate of biomass and growth was made based on inventories of the site for tree height and diameter distributions. Since all sites were not similarly investigated, the data availability varies between the sites. This is indicated in Table 2.11 by minus or plus signs. More plus signs indicate more detailed information on a subject or a longer recording period than normal.

Table 2.10. Mean ecosystem C and N pools in the NIPHYS/CANIF transect through Europe. C concentration taken as 48% of dry wt in plant matter. At Åhe data include *P abies*, *P sylvestris* and *B pubescens*. Branch biomass for Gri and Sch estimated from a regression of branch biomass against stem biomass of the other *Fagus* sites. Coarse roots include stumps. Stump biomass for AuF, Gri and Sko estimated from a regression of stump biomass against stem biomass for the other *Fagus* and *Picea* sites. Fine roots are here defined as <2 mm diam. in the upper 30 cm of the soil profile. Data were extrapolated for some sites from fine-root data of other diameter classes or depths. For AuP and AuF the available fine root biomass for <1 mm was extrapolated to a fine root biomass for <2 mm using the biomass ratio of <1 to <2 mm at Nac, Sch and Wal. The fine-root biomass <2 mm at Nac for 0–15-cm depth was recalculated to 0–30-cm depth using the ratio of fine-root biomass of 0–20-cm depth to that at 0–30-cm depth at AuP, AuF, Sch, Sko and Wal. The fine root biomass <5 mm at Col and MdM at 80-cm depth was recalculated using the ratio of fine roots <2 mm to fine roots <5 mm at Wal assuming all fine roots <2 mm to be in the upper 30 cm of the soil profile. For Gri, Jez and Åhe no fine-root data were available. For these sites the average of the other sites, per tree species, were used. Further explanations in Table 2.1. Data of NPP$_C$ and NPP$_N$ are listed in Tables 3.3 and 3.4 of Chapter 3, this Volume

Category		Åhe	Sko	Gri	Nac	Jez	Wal	Sch	AuP	AuF	Col	MdM
C (10^3 kg ha^{-1})	Leaves	4	7	2	6	2	8	2	4	2	1	8
	Branches	14	11	29	11	19	15	22	9	23	15	8
	Stem	56	41	128	56	77	93	107	112	98	84	63
	Coarse roots	20	13	34	29	23	36	26	21	26	24	25
	Fine roots	2.7	2.6	1.4	0.6	1.4	0.9	1.2	0.4	0.7	2.4	1.4
	Total plant	97	74	194	103	122	152	158	147	150	125	106
	Soil OM	55	132	92	170	100	202	164	54	94	228	175[a]
	Total	152	206	284	273	222	355	322	201	244	355	281
	% in plants	64	36	68	38	55	43	49	73	61	35	38
N (10^3 kg ha^{-1})	Leaves	0.09	0.17	0.13	0.20	0.04	0.24	0.09	0.13	0.08	0.07	0.15
	Branches	0.14	0.11	0.24	0.18	0.28	0.14	0.18	0.06	0.14	0.07	0.04
	Stem	0.13	0.10	0.30	0.17	0.20	0.22	0.25	0.28	0.28	0.13	0.08
	Coarse roots	0.05	0.04	0.11	0.10	0.08	0.12	0.09	0.07	0.09	0.08	0.08
	Fine roots	0.08	0.05	0.04	0.02	0.03	0.02	0.03	0.01	0.02	0.06	0.04
	Total plant	0.48	0.48	0.82	0.67	0.62	0.74	0.64	0.55	0.61	0.41	0.39
	Soil OM	2.08	5.40	4.35	6.83	4.98	8.40	7.80	3.33	4.90	18.05	17.68
	Total	2.56	5.88	5.04	7.50	5.60	9.14	8.44	3.88	5.52	18.46	18.07
	% in plants	19	8	16	9	11	8	8	14	11	2	2

[a] Denotes a value at MdM corrected from Table 2.8 (see text).

Table 2.11. Overview of the data and sites included in the CANIF database on CD-ROM. A minus sign means that no data are available. A plus sign means that data are available. More plusses indicate more information. See also text

	Sweden		Denm.	Czech Rep.		Germany		France		Italy	
	Åhe	Sko	Gri	Nac	Jez	Wal	Sch	AuP	AuF	Col	MdM
Tree biomass (measured)	–	++	–	+	–	+	–	+	+	++	+
Tree biomass (estimated)	+	–	+	+	+	–	+	–	–	–	–
Tree growth (measured)	–	++	+	+	–	+	–	+	–	++	+
Tree growth (estimated)	+	–	–	+	+	–	–	–	+	–	–
Tree nutrient data	+	++++	+	++	++	+	++	+++	++	+	++
Tree root biomass	–	+++	–	+	+	++	–	++	++	++	++
Tree root nutrient data	–	+++	–	+	+	+	+	+++	+++	++	++
Tree mortality	–	++	–	+	–	+	–	–	++	–	–
Understorey vegetation	–	–	nf[a]	+	nf	+	–	–	–	–	–
Soil organic matter data	+++	++++	+++	+++	+++	+++	+++	+++	+++	+++	+++
Soil characteristics	+	+++	+	+++	++	++	+	+++	+++	+++	+++
Soil solution data	+	+++	++	+	–	+	–	+++	+++	–	–
Leaching data	+	++	–	+	–	–	–	+++	+++	–	–
Wet deposition	++	+++	++	++	+	++	+	++	++	+	–
Throughfall	–	+++	–	++	+	+	+	+++	++	+	–
Dry deposition	+++	+++	+++	+++	++	+++	+++	+++	+++	+++	+++
Climate data	+++	++++	+	++	+++	+++	+	++	++	++	+++
Soil Mycorrhizal fungi[b]	+	–	+	–	–	+	–	+	+	+	–
Soil fauna[b]	+	+	–	–	–	+	–	+	–	–	–

[a] nf: no field layer present at that site.
[b] Data on soil micro-fungi and soil fauna are also available for Sorø (Denmark) and Klosterhede (Denmark), respectively.

Data on abundance and diversity of soil fauna and soil fungi are also part of the database. These data are given in separate files containing lists of species present and abundances of these species. The sites that were investigated are indicated in Table 2.11.

Additionally, the Czech site Salacova Lhota is included in the database. This site is discussed in Chapter 19, this Vol., on long-term trends in whole catchments. The site has a long recording period.

2.5 Conclusions

The climatic gradient along which the sites were to be chosen at the start of the project has one representative main site for a cold climate (northern Sweden) and one for a Mediterranean climate (southern France). The other sites have reasonably similar climatic conditions due to the low altitudes of the South-Swedish and Danish sites and the high altitudes of the northeastern French, German, Czech and Italian sites. The deposition gradient shows a rather small range of values, except for the most northern site, that is almost unaffected by deposition. The Czech sites deviate in having very high S deposition levels. At a local scale, a pairwise comparison of "closely" situated *Picea* and *Fagus* was not always possible to perform properly because of differences in land use (Sko/Gri), slopes (Nac/Jez) or water regimes and health status of the trees (AuP/AuF).

Low pH and base saturation (BS) and high levels of soluble aluminium were found at the central European sites Waldstein, Nacetin and Aubure, indicating high deposition (historical or recent) of acidifying sulphur and nitrogen compounds. However, also the Danish site Klosterhede and the South-Swedish site Skogaby had fairly low pH and BS values in some of the soil layers. The latter sites had very low exchangeable calcium levels, whereas the magnesium and sodium concentrations were higher, especially for Klosterhede, indicating the influence of sea salt. In the upper soil layers, beech stands had generally higher pH, Ca and Mg values than the spruce stands, but at greater depths there was mostly no clear difference between tree species.

N concentration in the litter layer was low in northern Sweden and southern France ($<11\,\mathrm{mg\,g^{-1}}$) and high in the central European sites ($>17\,\mathrm{mg\,g^{-1}}$) including NE France, indicating great differences in litterfall concentrations. Total C and N pools followed a similar pattern, with low amounts in northern Sweden and southern France and high amounts at the German, Czech and Italian sites. The NE French sites deviated from the other central European sites in having relatively low C and N pools, probably depending on site history but also on the coarse soil texture. At most sites, there was a reasonably good correlation between C and N soil pools. However, the Italian sites differed from this pattern in having much higher N pools than expected from the C pools. These sites were the only ones having clay or clayish soils, which indicates that clay soils can immobilise N more than coarse-textured soils. KCl-extractable inorganic N was very low at the northern Swedish site. At all other sites, there was an abundance of inorganic N (at least of ammonium). Beech stands had a tendency to have higher nitrate concentrations than comparable spruce stands. The N storage in the tree biomass was, on average, 10–11% of the total ecosystem N pool with a range from 2 to 19%.

Acknowledgments. The study was supported by grants from EEC (ENV5V-CT92-1043 and ENV-CT95-0053) and the Swedish Environmental Protection Agency.

Experimental Sites in the NIPHYS/CANIF Project

Fig. 2.4. The Åheden site in N Sweden with a mixed 180-year-old stand of mainly *Picea abies* and *Pinus sylvestris* with an understorey of forest mosses and *Vaccinium* species. (Photo: October 1993, E.-D. Schulze)

Fig. 2.5. The Skogaby site in SW Sweden. The 33-year-old stand of *Picea abies* (in 1997) has a sparse understorey of forest mosses. In the Skogaby experiment, control plots and, for certain studies, N-fertilisation plots, were used within the NIPHYS/CANIF project. (Photo: May 1997, T. Persson)

Fig. 2.6. The Klosterhede site in W Denmark with a 73-year-old stand of *Picea abies* (in 1992). The understorey consists of forest mosses and *Deschampsia flexuosa*. (Photo: October 1992, T. Persson)

Fig. 2.7. The Gribskov site in E Denmark with a 119-year-old stand of *Fagus sylvatica* (in 1996). The understorey is almost missing except for *Galium odoratum* and *Oxalis acetosella*. (Photo: June 1996, A. Taylor)

Fig. 2.8. The Nacetin site in the Ore Mountains in NW Bohemia in the Czech Republic. The stand consists of 57-year-old *Picea abies* (in 1994) with a patchy understorey of *Calamagrostis villosa* and *Deschampsia flexuosa*. (Photo: June 1994, T. Persson)

Fig. 2.9. The Jezeri site in the Ore Mountains in NW Bohemia in the Czech Republic. The stand consists of 81-year-old *Fagus sylvatica* (in 1997) without any understorey. (Photo: October 1997, T. Persson)

Fig. 2.10. The Waldstein site in Fichtelgebirge in NE Bavaria, Germany. The stand consists of 143-year-old *Picea abies* (in 1996) with an understorey of *Calamagrostis villosa*, *Deschampsia flexuosa* and *Vaccinium myrtillus*. (Photo: June 1996 during biomass harvest, E.-D. Schulze)

Experimental Sites in the NIPHYS/CANIF Project 45

Fig. 2.13. The 159-year-old *Fagus sylvatica* stand (in 1993) at Aubure in the Vosges Mountains, northeastern France. The stand has a sparse understorey of *Athyrium filix-femina* and *Polystichum spinulosum*. (Photo: March 1993, T. Persson)

Fig. 2.14. The 100-year-old *Pinus pinaster* stand at Thezan, southern France. The stand has a dense understorey of *Erica scoparia*, *Juniperus oxycedrus*, *Quercus ilex* and *Ulex parviflorus*. (Photo: April 1993, E.-D. Schulze)

Fig. 2.15. The 105-year-old *Fagus sylvatica* stand (in 1996) at Collelongo in the Abruzzo Mountains, Central Italy. The stand has a sparse understorey of *Galium odoratum* and *Geranium robertianum*. (Photo: April 1996, T. Persson)

Fig. 2.16. The 37-year old *Picea abies* stand at Monte di Mezzo. The stand is the first generation of a reforestation of agricultural land. (Photo: G. Scarascia-Mugnozza)

◄────────────────────────────────────

Fig. 2.11. The Schacht site in Fichtelgebirge in NE Bavaria, Germany. The stand consists of 123-year-old *Fagus sylvatica* (in 1998) with a sparse understorey of *Dryopteris filix-mas* and *Vaccinium myrtillus*. (Photo: September 1998, T. Persson)

Fig. 2.12. The 90-year-old *Picea abies* stand (in 1993) at Aubure in the Vosges Mountains, northeastern France. The stand has a sparse understorey of *Deschampsia flexuosa* and *Dryopteris filix-mas*. (Photo: May 1993, A. Taylor)

References

Barrett K, Berge E (eds) (1996) Transboundary air pollution in Europe. MSC-W status report 1996. EMEP/MSC-W, Report 1/96, Norwegian Meteorological Institute, Oslo

Beier C, Rasmussen R, Hansen K (1991) The Klosterhede project. Lab of Environmental Sciences and Ecology, Technical University of Denmark, Lyngby

Beier C, Gundersen P, Hansen K, Rasmussen L (1995) Experimental manipulation of water and nutrient input to a Norway spruce plantation at Klosterhede, Denmark. Plant Soil 168-169:613-622

Berge E (ed) (1997) Transboundary air pollution in Europe. MSC-W status report 1997. EMEP/MSC-W, Report 1/97, Norwegian Meteorological Institute, Oslo

Cole DW, Rapp M (1981) Elemental cycling in forest ecosystems. In: Reichle DE (ed) Dynamic properties of forest ecosystems. Int Biol Programme 23:341-409

Giesler R (1996) Chemistry of soil solution extracted by centrifugation – methodology and field applications. Swedish University of Agricultural Sciences, Department of Forest Ecology, Dissertation, Umeå

Göttlein A (1995) Bodenkundliche Charakterisierung der Intensiv-Meßfläche Coulissenhieb. In: Manderscheid B, Göttlein A (eds) Wassereinzugsgebiet Lehstenbach – das BITÖK-Untersuchungsgebiet am Waldstein (Fichtelgebirge, NO-Bayern), Bayreuther Forum Ökologie 18

Gundersen P (1998) Effects of enhanced nitrogen deposition in a spruce forest at Klosterhede, Denmark, examined by moderate NH_4NO_3 addition. For Ecol Manage 101:251-268

Gundersen P, Rasmussen L (1995) Nitrogen mobility in a nitrogen limited forest at Klosterhede, Denmark, examined by NH_4NO_3 addition. For Ecol Manage 71:75-88

Högbom L, Högberg P (1991) Dynamics of soil nitrate after forest fertilization as monitored by the plant nitrate reductase assay. For Ecol Manage 44:223-238

Hytteborn H (1975) Deciduous woodland at Andersby, eastern Sweden. Above-ground tree and shrub production. Acta Phytogeogr Suec 61, 96 pp

Kalembasa SJ, Jenkinson DS (1973) A comparative study of titrimetric and gravimetric methods for determining organic carbon in soils. J Sci Food Agric 24:1085-1090

Lefèvre Y (1988) Les sols du bassin versant d'Aubure (Haut-Rhin): caractérisation et facteur de répartition. Ann Sci For 45:417-422

Matschonat G, Matzner E (1996) Soil chemical properties affecting NH_4^+ sorption in forest soils. Z Pflanzenernähr Bodenkd 159:505-511

Nilsson L-O, Wiklund K (1992) Influence of nutrient and water stress on Norway spruce production in south Sweden – the role of air pollutants. Plant Soil 147:251-265

Novak M, Bottrell SH, Groscheová H, Buzek F, Cerný J (1995) Sulphur isotope characteristics of two northern Bohemian forest catchments. Water Air Soil Pollut 85:1641-1646

Rasmussen L, Beier C, Van Breemen N, De Visser P, Kreutzer K, Schierl R, Matzner E, Farrell EP (1990) Study on acid depositon effects by manipulating forest ecosystems. New title: EXMAN-Experimental MANipulation of forest ecosystems in Europe. Air Pollut Res Rep 24, Commission of the European Communities, Brussels, 42 pp

Rowland AP, Grimshaw HM (1985) A wet oxidation procedure suitable for total nitrogen and phosphorus in soil. Comm Plant Soil Anal 16:551-560

Part B
Plant-Related Processes

3 Tree Biomass, Growth and Nutrient Pools

G. Scarascia-Mugnozza, G.A. Bauer, H. Persson, G. Matteucci, and A. Masci

3.1 Introduction

The role of forest ecosystems in the carbon cycle is a central problem of global ecology and has important, practical implications from the points of view of global change and production of woody raw material. Furthermore, in forest ecosystems the biogeochemical cycles of carbon and mineral nutrients, particularly nitrogen, are intimately related because foliage nutrient content strongly controls carbon assimilation by forests and, therefore, their primary productivity (Waring and Running 1998).

Net primary production is the amount of carbon accumulated by a forest over 1 year; it corresponds to the difference between gross primary production and the autotrophic respiration, but commonly it is calculated by repeated measurements of biomass accumulation, litter production and herbivory losses in forest stands (Landsberg and Gower 1997). However, the quantification of carbon allocation to roots is problematic because of the practical difficulties and labour requirements of this type of studies (Santantonio et al. 1977; Persson 1983); this is the reason why many of the studies reported in the literature do not include these data. Biomass assessment is also ecologically relevant per se because it is related to the amount of carbon and nutrients immobilised in the living trees.

Primary production and its distribution between above-ground and below-ground components may change in response to ecological gradients of nutrients availability, soil properties and climatic conditions. In boreal coniferous forests primary production is generally lower and biomass allocation to roots higher than in temperate forest ecosystems because of reduced nutrient availability and extreme environmental conditions (VanCleve et al. 1983). Human influences can also modify the nutrient balance of forest ecosystems, causing impacts that are modulated by soil type and bedrock characteristics. Water limitations, often experienced by forests growing in southern Europe, may also interact with nutrient levels, thus controlling biomass production and its distribution into the soil compartment relative to the aboveground portions. However, the relationships between nitrogen cycling and primary productivity and controls over above-ground to below-ground allocation remain largely unresolved.

The European transect, spanning from northern Scandinavia to the Mediterranean region, includes coniferous (Norway spruce) and broadleaf (beech) forest ecosystems, thus providing a unique opportunity to study net primary production, biomass accumulation and carbon allocation in response to gradients of climate, soil characteristics and nitrogen deposition levels.

In this chapter we report on the results of the investigations conducted on biomass and net primary production as well as on carbon and nutrient pools of forest ecosystems of beech and Norway spruce, along the European ecological transect.

3.2 Experimental Background

Primary production is usually estimated through repeated biomass sampling or, retrospectively, by the measure of increments of diameter, height and, therefore, volume of tree stems; measurements should also include branches and leaf production and possibly the below-ground components. A mean increment value can also be obtained for even-aged stands by dividing the total biomass, estimated through sampling, by the age of the stand.

All these methods are based on the felling of an adequate number of sampling trees, the larger the more variable the stand. Sampling trees should be representative of the average growth conditions of the trees within the study stand or within the tree diameter classes that compose the stand. The selection of sampling trees implies a preliminary inventory of the structural and dendrometrical characteristics of the stand that should include the diameter at breast height distribution of trees, the mean height of trees within each diameter class, and the relationship between height and diameter; generally, diameter classes are 5 cm wide, starting from the class of 5 cm onwards. Where yield and biomass tables are available, estimates of standing biomass and production can be derived simply from the dendrometrical characteristics of the stand.

The quantification of biomass per unit of land surface can be obtained in different ways. The first consists of multiplying the biomass of the average tree, per each diameter class, by the number of trees of that class. In the second case, if a sufficient number of trees has been measured in different diameter and height classes, allometric regressions can be derived between the tree biomass and the different tree mensurational variables; the biomass of the entire stand can be derived by summing up the biomass of individual trees calculated by the regression applied to the size variable (height, diameter, height × diameter, etc.) measured for each tree.

Along the European transect, measurements of above-ground biomass components and of fine root biomass, and measurements of the above-ground primary production were conducted in 7 out of 12 forest sites (Skogaby, Nacetin, Waldstein, Aubure-*Fagus*, Aubure-*Picea*, Collelongo and Monte di Mezzo), whereas in the remaining five sites (Åheden, Klosterhede, Gribskov, Jezeri and Schacht) estimates of above-ground biomass and production were derived from volume and yield tables, after conversion of volumetric data into dry weights. In these latter sites, fine root biomass was estimated as mean values of other *Fagus* and *Picea* stands except for the site of Schacht and Åheden, where fine roots biomass was directly measured. For some of the sites, the values of fine root biomass reported in this chapter are to some extent different from those reported in Stober et al. (Chap. 5, this Vol.), as there the definition of fine roots is mainly functional (higher-order lateral roots and primary roots), while here it is merely dimensional (<2 mm). Furthermore, the years of sampling are sometimes different.

In only four of the directly measured stands (Aubure-*Fagus*, Aubure-*Picea*, Collelongo and Monte di Mezzo) were stump and coarse roots biomass and production also investigated by excavating the below-ground parts of some of the sampling trees. In these sites fine root production was assessed by repeated soil core extraction. Also for this category, some differences can be present (see Stober et al., Chap. 5, this Vol.), basically due to the different methods applied (sequential coring vs. root windows). In the remaining sites, fine root production was estimated assuming the annual production equal to their biomass (Jackson et al. 1997).

Above-ground biomass was analysed on at least 20 to 30 sampling trees per stand, according to the within-stand variability (Masci et al. 1998). Sampling trees, two to five per each diameter class, were selected to be representative of the mean tree height and form for that class. After trees were felled, the volume of the stem was measured and the volume-to-dry-weight conversion factor was assessed on stem sections. At regular height intervals, every 1 to 4 m depending on tree size, cross sections were cut from the stems, and the increment of stem diameter at the various tree heights was determined to reconstruct the increment of the stem volume over the course of the life of the sample tree (stem analysis method).

Branches were cut and separated in size classes, and samples of the branches were measured for dry weight. At the Collelongo and Monte di Mezzo sites, branch growth was reconstructed on a selected number of branches measuring the increments of diameter on cross sections cut at the base of branches.

After tree felling, all the leaves were weighed and some subsamples of the foliage were dried and weighed again to determine the fresh to dry leaf weight conversion factor. Leaf production was assessed differently in beech vs. spruce: in deciduous trees foliage biomass corresponds to the annual leaf production provided that trees are felled in the period of fully expanded foliage, while in evergreen trees it is necessary to separate needles in the different age classes and then calculate the amount of annual leaf production as the needle biomass of the last age class.

The estimate of below-ground biomass and productivity is more complex; this is why far fewer studies have included this component in their determinations. Biomass of the stump and main root system, including large and small roots, with the lower limit of 5 to 2 mm, was measured using the method of the sampling trees, as for the above-ground biomass components. In fact, on a subsample of the harvested trees, below-ground biomass was analysed by stump and coarse root excavation within holes 2 m in radius and 1 m deep; roots were washed and a subsample collected for dry weight. Growth of the stump and the coarse roots was estimated, adapting the method of stem analysis to the root material.

A different approach was followed for the fine root component, corresponding to roots with a diameter smaller than 2 mm. Fine root biomass was determined on soil cores collected at each site, only at one sampling date. The number of extracted cores should adequately represent the variability of the forest site, whereas the depth of extraction should include all the soil layers where fine roots are present; at each site, 15 soil cores were extracted from randomly located points. After sieving and sorting live from dead roots, the dry weight of live fine roots was determined.

At some sites, fine root production was assessed by repeated extractions of soil cores throughout the year, especially during spring and summer, paying attention that the time interval between two successive extractions be sufficiently short to avoid fine roots being born, and dying and decomposing, within the period thus not being available for calculation of below-ground productivity (Persson 1983). In fact, root turnover can be particularly fast in forest ecosystems, particularly under conditions of low nutrient availability and water limitations. Therefore, throughout the growing season, five to ten successive corings were performed. Annual fine root production was thus obtained from the difference between maximum and minimum live fine root biomass within the growing season (max-min method); additional estimates (Fogel 1983) were also obtained from changes, statistically significant, in monthly values of live and dead fine root biomass (monthly increment method).

The tree biomass components, once dried, were also sampled and analysed for nutrients concentration to calculate nutrient pools and nitrogen uptake for primary production. Nitrogen uptake was estimated by multiplying the increments of the various biomass components by their nitrogen concentration; however, N concentrations of foliage and fine roots were corrected to account for the process of nitrogen recycling inside the trees (Bonneau 1981).

3.3 Biomass

Total biomass of the Norway spruce stands (Table 3.1) studied within the European transect amounted on average to 237 t ha^{-1} (SD 58.1). The most northern site of the transect, Åheden in North Sweden, had one of the lowest values of total biomass (198 t ha^{-1}), even though it was the oldest Norway spruce stand in the study (180 years). The other spruce stand in Sweden, Skogaby, had the lowest value of total biomass (154 t ha^{-1}) but was also the youngest stand along the European transect (33 years). Moving southward, the Danish stand at Klosterhede and the Czech stand at Nacetin had similar total biomass ranging from 242 to 215 t ha^{-1}; however, Nacetin was a much younger stand (58 years) than Klosterhede (120 years). The highest biomass values were measured at the German (Waldstein) and French (Aubure) sites (respectively 317 and 307 t ha^{-1}), while in the most southern spruce site, Monte di Mezzo, total biomass decreased to 223 t ha^{-1}; the latter stand, with about the same age as Skogaby, had a 35% greater total biomass.

Along the transect, notwithstanding the differences of age and fertility among sites, total biomass of spruce forests first increased from northern to central Europe, then decreased slightly to southern Europe.

The average total biomass of the beech stands was 313 t ha^{-1} (SD 59.2), thus about 30% greater than for the spruce forests (Table 3.1). The Danish site, Gribskov, had the highest biomass value of the transect (405 t ha^{-1}), although it was not the oldest beech stand. Again, the German and French stands, at Schacht and Aubure, had a high total biomass of more than 300 t ha^{-1}, whereas the biomass was smaller at Jezeri and at Collelongo (255 and 268 t ha^{-1}, respectively).

The biomass of the understorey vegetation was measured at only one site, in the spruce stand of Waldstein (G. Schmidt, pers. comm.); it was composed of herbs (i.e. *Deschampsia villosa*, *Calamagrostis villosa* and *Oxalis acetosella*) and shrubs (i.e. *Vaccinium myrtillus*) and amounted to 12.6 t ha^{-1} that represented 4% of the trees' total biomass. The below-ground portion made up 40% of the understorey biomass.

Needle biomass was generally relevant in the spruce stands (3 to 9% of the total biomass) but varied among sites: highest values were reached at Skogaby and Monte di Mezzo, whereas the lowest were found at Aubure and Åheden. In spruce stands, branches represented about 7 to 10% of the total biomass for most sites but reached a maximum for the Swedish and Danish spruce stands, where the branch component was 14–15%. The stem represented most of the total biomass, ranging from 50 to 60% in all sites except at Aubure, where it amounted to 76%. Total root biomass ranged from 33 t ha^{-1} at Skogaby to 76 t ha^{-1} at Waldstein; the root fraction of the total biomass varied around 20%, with the highest value (29%) for Nacetin and the lowest (14%) for Aubure. This range is in agreement with previous findings reported by Harris et al. (1980). The fraction of fine roots varied from 0.3 to 3.6%; however, different

Table 3.1. Total biomass and its components (t dw ha^{-1}) of the CANIF forest stands; data for Åheden, a mixed forest, include biomass of *P. abies*, *P. sylvestris* and *B. pendula*

	Beech stands					Spruce stands						
	Col	AuF	Sch	Jez	Gri	MdM	AuP	Wal	Nac	Klo	Sko	Åhe
Total	267.7	310.5	328.8	255.4	404.7	223.0	307.2	316.5	215.7	241.6	154.3	198.2
Foliage	2.8	2.8	3.5[a]	3.7[a]	4.7	16.8	9.2	16.6	13.3	15.0	14.0	9.1[a]
Stems	174.2	203.7	222.9[a]	160.7[a]	265.9	131.3	234.1	193.2	117.0	153.0	85.5	116.4[a]
Branches and twigs	31.8	48.3	45.8[a]	40.2[a]	60.4[a]	17.2	19.5	30.2	23.4	26.1	22.2	29.3[a]
Fine roots	3.8	1.5	2.6	2.9[a]	2.9[a]	2.9	0.8	2.3	1.2	3.5[a]	5.5	1.8
Coarse roots and stump	55.2	54.2[a]	54.0	47.9[a]	70.8[a]	54.8	43.7	74.2	60.8[a]	44.0	27.1[a]	41.7[a]
Total roots	59.0	55.7	56.6	50.8	73.7	57.7	44.5	76.5	62.0	47.5	32.6	43.5

[a] Estimated data.

Table 3.2. Allometric ratios for the CANIF forest stands

	Beech stands					Spruce stands						
	Col	AuF	Sch	Jez	Gri	MdM	AuP	Wal	Nac	Klo	Sko	Åhe
Root/shoot	0.28	0.22	0.21	0.25	0.22	0.35	0.17	0.32	0.40	0.24	0.27	0.28
Root/tot. biom	0.22	0.18	0.17	0.20	0.18	0.26	0.14	0.24	0.29	0.20	0.21	0.22
Harvest index[a]	0.65	0.66	0.68	0.63	0.66	0.59	0.76	0.61	0.54	0.63	0.55	0.59
Epig. harvest index[b]	0.83	0.80	0.82	0.79	0.80	0.79	0.89	0.81	0.76	0.79	0.70	0.75

[a] Harvest index = stem biomass/total biomass.
[b] Epigeal harvest index = stem biomass/above-ground biomass.

dates of root samples collection are known to contribute significantly to the variability of fine root biomass data (Santantonio et al. 1977).

In the beech stands, the foliage component represented a much lower fraction of total biomass (about 1%) than in the spruce forests. Conversely, in beech, the branch component made up a greater proportion of the total biomass, 10 to 16%, and the stem fraction was generally slightly higher than for spruce (63 to 68%). Total root biomass varied from 51 t ha^{-1}, in Jezeri, to 74 t ha^{-1}, in Gribskov; the fraction of roots represented about 20% of the total biomass and was less variable than in spruce stands, with the highest value (22%) in Collelongo and the lowest (18%) in Aubure. Live fine root biomass accounted for between 0.5 and 1.4% of total stand biomass for beech sites.

Allometric ratios (Table 3.2) provide further insights to understand strategies of resource investment into the different tree components. The root-to-shoot ratio (R/S) varied between 0.2 and 0.4 considering all stands of both species; in the spruce stands the range was from 0.17 in Aubure to 0.4 in Nacetin, with an average of 0.29 (SD 0.08). In the beech stands R/S ratio was generally lower than for the spruce sites with an average of 0.24 (SD 0.03) and a range between 0.2 in Aubure to 0.28 in Collelongo. The southern European sites of the transect had a high R/S ratio for both forest species.

The stem harvest index, calculated either over total biomass or only over the above-ground part, was slightly greater for beech (0.6 to 0.7) than for spruce stands (0.5 to 0.7). Nevertheless, the maximum harvest index was reached in the spruce stand of Aubure.

3.4 Forest Productivity

Net primary productivity is presented in two ways (Table 3.3): one takes into consideration all the components, including also the roots (NPP), while the other considers only the above-ground part (ANPP). We decided to compute both parameters because fine root and coarse root productivities were measured in a few study sites, Aubure, Collelongo and Monte di Mezzo.

In the spruce stands, ANPP ranged from 4 to 12 t ha^{-1} a^{-1}, with the lowest value for Åheden and the highest for Skogaby. In the beech forests, above-ground productivity ranged from 7 to 12.5 t ha^{-1} a^{-1}, with the lowest value measured in Aubure and the highest in Jezeri. The ratio of ANPP to the above-ground biomass was 5.4% (SD 2.5) in the spruce stands and 3.7% (SD 1.5) in the beech sites; the higher value of the spruce stands is due to a combined effect of greater above-ground productivity and smaller standing biomass.

Net primary productivity, that includes also the below-ground components, could be significantly greater than ANPP; it reached 18.8 t ha^{-1} a^{-1} for spruce, in Skogaby, and 16.6 t ha^{-1} a^{-1} for beech, in Jezeri. The mean NPP values were 13.4 t ha^{-1} a^{-1} (SD 4.5) and 12.4 t ha^{-1} a^{-1} (SD 2.9), respectively, for the spruce and beech stands. The ratio of NPP to the total stand biomass was 4 and 6% for beech and spruce sites, respectively.

In the spruce stands, a larger amount of ANPP was allocated to the stem (3.9 ± 1.5 t ha^{-1} a^{-1}) than to the foliage (2.5 ± 1.4 t ha^{-1} a^{-1}) or the branches (2.9 ± 1.1 t ha^{-1} a^{-1}); on the contrary, in the beech forests, particularly in Collelongo, the ANPP allocated to the stem (2 to 5 t ha^{-1} a^{-1}) was similar to the productivity allocated to branches and foliage, in agreement also with a number of studies reported by Cannell (1982).

Table 3.3. Growth (tdwha^{-1}a^{-1}), nitrogen and carbon uptake (tha^{-1}a^{-1}) of forest tree stands in the CANIF sites[a]

	Beech stands					Spruce stands							
	Col	AuF	Sch	Jez	Gri	MdM	AuP	Wal	Nac	Klo	Sko	Åhe
Foliage	2.8	2.8	3.5	3.7	4.7	2.8	2.1	4.7	3.6	1.0	2.5	0.9
Stem	2.8	2.0[d]	2.0[d]	4.2	2.0[d]	4.5	3.0	3.3	3.6	6.0	5.2	1.7
Branches and twigs	3.1	2.2[d]	2.2[d]	4.6[d]	2.2[d]	3.6	2.2[d]	2.3	2.7[d]	4.5[d]	3.9[d]	1.3[d]
ANPP[b]	8.7	7.0	7.7	12.5	8.9	10.9	7.4	10.3	9.9	11.5	11.6	3.9
Coarse roots and stump	1.0	0.5[d]	0.5[d]	1.2[d]	0.5[d]	2.9	0.6	1.3[d]	1.9[d]	1.7[d]	1.7[d]	0.6[d]
Fine roots	3.8	1.5[d]	2.6[d]	2.9	2.9[d]	2.9[d]	0.8[d]	2.3[d]	1.2[d]	3.5[d]	5.5[d]	1.8[d]
NPP	13.5	9.0	10.8	16.6	12.3	16.7	8.8	13.9	13.0	16.7	18.8	6.3
N-NPP[c]	0.085	0.061	0.079	0.098	0.099	0.072	0.042	0.096	0.074	0.072	0.098	0.031
C-NPP	6.5	4.4	5.3	8.0	6.0	8.1	4.3	6.8	6.3	8.0	9.1	3.0

[a] In Sch, Jez, Gri, Wal, Nac, Klo, Sko and Åhe, data for coarse roots were estimated assuming the same percent increment as the stem component; in beech, branches production is calculated assuming that nearly the same proportion of above-ground NPP is allocated to branches (35%) and stems (32%) according to Nihlgard and Lindgreen, 1977 (southern Sweden) and Holm and Jensen, 1981 in Cannell, 1982 (Denmark). In Coll, branches production was estimated on the basis of branches basal area increment for branches older than 1 year and on the basis of the ratio of the weight of 1-year old twigs to leaf weight in 76 sample branches of different basal diameter and different position in the crowns. We summed branches increment and twigs increment; the results obtained are in agreement with the above references.
Branches production for the spruce sites has been estimated on the basis of the mean ratio of branch increment to stem increment as measured in MdiM and Wal.
[b] Above-ground net primary production.
[c] Nitrogen uptake has been calculated taking into account N retranslocation for the foliage and the fine roots.
[d] Estimated data.

Coarse and fine root production accounted for a significant amount of the net primary productivity of the studied stands. Below-ground productivity ranged from 1.4 to 7.2 t ha^{-1} a^{-1} in the spruce sites, with the highest values measured in the Swedish stand of Skogaby; for the boreal sites (Skogaby and Åheden) most of the below-ground productivity (75%) was allocated to fine roots, whereas in the temperate spruce forests fine root production was 40 to 60% of below-ground productivity. In the beech forests, the NPP allocated below ground was between 2 and 5 t ha^{-1} a^{-1}, with the highest value measured in Collelongo; in the beech forests fine root productivity represented 70 to 85% of below-ground NPP. Fine root contribution to total stand production ranged from 17 to 28% in beech stands and between 9 and 30% in the spruce forests. The highest contribution of fine root production to NPP in beech forests was measured in Collelongo, the most Southern site of the European transect, while for spruce, the highest values were recorded in the northern European stands, confirming that boreal forests have a higher proportion of total ecosystem fixed C allocated below ground (Ruess et al. 1996).

Leaf area index (Table 3.4) varied from 2.5 (Åheden) to 7 m^2 m^{-2} (Skogaby and Waldstein) in the spruce stands and from 4 (Schacht) to 7 m^2 m^{-2} (Jezeri) in the beech stands. The specific leaf area (SLA, Table 3.4) was around 3 to 5 m^2 kg^{-1} in spruce forests and varied from 16 to 24 m^2 kg^{-1} in the beech stands. The SLA, which can be used as a sensitive measure of growing conditions (Bauer et al. 1997), showed the lowest values at the northern and southern extremes of the European transect, northern Sweden and central Italy, in parallel with the trend of N depositions.

3.5 Carbon and Nutrient Pools

The carbon pools of the forest biomass (Table 3.5) showed the same geographic pattern as the distribution of the individual biomass components.

In spruce forests, the highest C pool was measured at Waldstein (152 t ha^{-1}) and the lowest at Skogaby (74 t ha^{-1}); the average value for the spruce trees was 114 t ha^{-1}. In the beech forests the total C pool measured in the trees amounted, on average, to 150 t ha^{-1}, with values ranging from 120 to 190 t ha^{-1}. In both species, most of the carbon was bound in the stem (65%) whereas about 20% of the total carbon was found in the roots.

Nutrient pools were determined for all the sites and for the different biomass components (Table 3.6). On average, the spruce stands contained about 560 kg ha^{-1} of N in the tree biomass with large variations among sites (SD 133 kg ha^{-1}); Waldstein had the highest value (745 kg ha^{-1}) and Monte di Mezzo the lowest (393 kg ha^{-1}). Calcium content was also very variable, with relevant values in Waldstein and Monte di Mezzo (730 kg ha^{-1}) and the lowest in Skogaby (280 kg ha^{-1}). Potassium and phosphorus pools were relatively more uniform among sites. Generally, sulphur pools varied between 40 and 80 kg ha^{-1}, except for the Aubure site, where the sulphur content of the biomass reached an unusually high value (230 kg ha^{-1}).

In the beech stands the average N pool was 620 kg ha^{-1}, with values ranging from 398 kg ha^{-1}, in Collelongo, to 819 kg ha^{-1}, in Gribskov. The highest Ca pool was found in Gribskov (1110 kg ha^{-1}) and the smallest in Aubure (720 kg ha^{-1}). More uniform values were measured for the pools of K (510 kg ha^{-1} on average), P (66 kg ha^{-1}) and S (74 kg ha^{-1}).

Table 3.4. Foliage parameters in the CANIF sites

	Beech stands					Spruce stands						
	Col	AuF	Sch	Jez	Gri	MdM	AuP	Wal	Nac	Klo	Sko	Åhe
LAI ($m^2 m^{-2}$)	4.5	5.8	4.1	6.9	4.9	5.5	5.4	6.2	5.8	5.7	7.0	2.5
SLA ($m^2 kg^{-1}$)	16.0	24.0	18.5	18.6	19.4	3.4	4.7	4.7	4.4	3.2	5.0	3.4

Table 3.5. Carbon pools (t ha^{-1}) in the CANIF sites

	Beech stands					Spruce stands						
	Col	AuF	Sch	Jez	Gri	MdM	AuP	Wal	Nac	Klo	Sko	Åhe
Foliage	1.4	1.4	1.8	1.8	2.4	8.4	4.6	8.3	6.7	7.5	7.0	4.5
Stem	83.6	97.8	107.0	77.1	127.6	63.0	112.4	92.7	56.2	73.4	41.0	55.9
Coarse roots and stump	26.5	26.0	25.9	23.0	34.0	26.3	21.0	35.6	29.2	21.1	13.0	20.0
Fine roots	1.8	0.7	1.2	1.4	1.4	1.4	0.4	1.1	0.6	1.7	2.7	0.9
Branches and twigs	15.2	23.2	22.0	19.3	29.0	8.2	9.4	14.5	11.2	12.5	10.6	14.0
Stand total	128.6	149.1	157.9	122.6	194.4	107.4	147.7	152.2	103.8	116.3	74.3	95.3

Table 3.6. Nutrient pools in the CANIF sites (kg ha^{-1})

	Beech stands					Spruce stands					
	Col	AuF	Sch	Jez	Gri	MdM	AuP	Wal	Nac	Klo	Sko	Åhe
Foliage												
N	66.1	74.9	91.6	40.4	130.3	147.3	128.7	241.7	202.8	188.4	168.8	69.3
Ca	25.8	8.3	14.7	19.8	33.7	225.2	34.0	103.4	63.7	65.6	28.3	44.6
K	31.4	26.0	12.8	25.0	32.9	151.3	61.0	115.5	70.2	111.9	52.8	53.3
P	2.1	5.1	3.9	3.0	6.4	20.3	17.9	27.9	16.9	19.4	11.3	14.5
S	4.6	21.7	5.1	2.9	7.8	17.0	9.3	21.6	33.5	18.6	10.5	6.2
Mg	4.5	1.9	2.1	1.6	7.5	16.8	6.8	9.3	9.2	23.4	8.7	9.9
Stems												
N	130.0	279.1	251.6	204.1	300.1	79.7	276.2	218.0	173.2	189.6	106.6	131.4
Ca	498.2	344.3	454.0	250.7	541.6	260.4	311.3	258.9	108.4	205.0	95.7	156.0
K	235.5	266.8	292.9	205.7	349.4	101.5	145.1	137.7	65.3	109.1	76.9	83.0
P	23.6	21.0	24.7	14.5	29.4	24.0	11.7	21.8	13.2	17.2	10.7	13.1
S	27.8	32.5	35.6	25.6	42.4	39.2	173.2	57.7	6.0	45.7	9.0	34.8
Mg	30.7	46.9	39.3	43.4	46.9	18.2	32.8	26.8	19.0	21.2	18.4	16.1
Branches												
N	66.9	141.4	181.6	275.6	239.5	42.1	60.4	137.8	178.1	118.2	112.7	133.5
Ca	141.1	122.1	160.0	141.0	211.1	40.2	72.1	98.5	67.7	85.2	62.3	95.5
K	63.1	85.4	92.5	92.4	121.9	32.7	37.0	64.1	73.0	55.4	46.0	62.1
P	15.7	9.2	21.9	30.1	28.9	7.3	7.8	15.6	18.8	13.5	12.0	15.1
S	10.9	16.5	15.7	13.8	20.7	9.5	10.8	16.8	13.8	14.5	11.5	16.3
Mg	11.7	14.0	10.7	14.9	14.1	6.6	5.8	11.6	14.5	10.1	13.3	11.3
Roots												
N	135.6	109.8	118.4	104.3	149.4	124.2	81.7	147.5	112.2	158.0	99.3	89.3
Ca	129.0	246.9	252.0	343.6	327.9	200.7	76.7	275.6	206.8	141.8	90.4	127.9
K	94.3	116.2	117.0	134.4	154.7	73.8	66.2	161.0	143.2	82.8	49.6	75.9
P	10.8	17.9	19.4	22.2	23.4	17.9	10.2	6.7	33.0	14.9	10.2	12.5
S	19.3	15.7	19.4	13.8	21.9	27.2	36.3	4.8	31.7	22.9	16.1	20.3
Mg	42.9	38.6	43.6	35.5	52.8	17.4	9.1	13.5	32.9	14.8	10.9	13.3
Stand total												
N	398.6	605.1	643.3	624.3	819.3	393.2	546.9	745.0	666.3	654.1	487.4	423.5
Ca	794.1	721.6	880.7	755.1	1114.3	726.5	494.1	736.5	446.6	497.5	276.7	424.0
K	424.3	494.5	515.1	457.5	659.0	359.3	309.3	478.4	351.7	359.2	225.4	274.3
P	52.2	53.2	69.8	69.8	88.1	69.5	47.6	71.9	81.9	65.0	44.2	55.2
S	62.6	86.4	75.7	56.1	92.9	93.0	229.7	100.9	85.1	101.7	47.1	77.6
Mg	89.9	101.4	95.7	95.4	121.3	59.0	54.5	61.2	75.6	69.4	51.3	50.6

3.6 Allometric and Functional Relations

Several relationships were tested to verify which structural and ecological parameters could be functionally related to biomass pools and primary productivity.

Total standing biomass of all spruce and beech sites, pooled together, was significantly and positively correlated with the mean diameter (D) ($r^2 = 0.76, p < 0.05$) and with the stem biomass of each stand ($r^2 = 0.95, p < 0.05$). Considering single species, spruce standing biomass was more significantly correlated with mean height

(H, $r^2 = 0.92, p < 0.05$) than with the diameter ($r^2 = 0.84, p < 0.05$), while the contrary was found for beech (D: $r^2 = 0.69, p < 0.05$; H: $r^2 = 0.44$). This difference can be related to the fact that spruce stands were generally more regular and uniform, resulting in a lower variability of height within the stand.

The volume index, calculated as the product of the squared diameter and height (D^2H), is often used as a predictor of the tree volume and biomass. Total stand biomass was significantly ($p < 0.05$) correlated with the mean volume index of single stands (Fig. 3.1) with r^2 of 0.82 and 0.79, for beech and spruce, respectively. The volume index proved also to be related to single biomass components of the studied forests. As expected, the volume index proved to be a good predictor for stem biomass in both species (r^2 of 0.87 and 0.74, $p < 0.05$, data not shown). In beech stands (Fig. 3.2), mean D^2H was significantly related to the biomass of branches ($r^2 = 0.98$, $p < 0.001$) but not so with total below-ground biomass ($r^2 = 0.42$; Fig. 3.2), coarse roots and stump ($r^2 = 0.49$).

Considering some silvicultural characteristics of the study stands, a significant negative relationship was observed between total stand biomass and stand density ($r^2 = 0.77, p < 0.05$; Fig. 3.3); this relationship cannot be explained by an age effect on the stand biomass because this correlation was not significant.

Concerning productivity, the beech stands showed a negative and significant relationship of NPP and ANPP versus stand age ($r^2 = 0.92$ and 0.76, respectively). As for the climatic parameters interesting relationships were found between NPP and global radiation in beech ($r^2 = 0.93$) while in spruce NPP was significantly correlated with the mean air temperature of the May–September period ($r^2 = 0.85, p < 0.05$; Fig. 3.4) and with the number of months with mean air temperature above 5 °C ($r^2 = 0.53$). Nitrogen depositions were related to stand productivity only for the spruce sites; in these stands ANPP was linearly and significantly correlated to total N depositions with $r^2 = 0.91$.

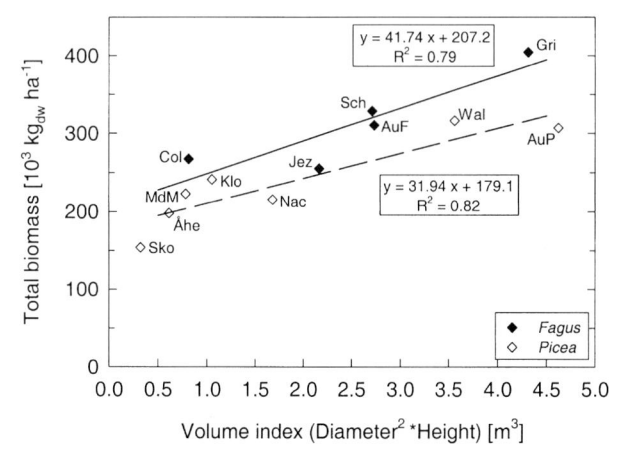

Fig. 3.1. Relationship of total biomass of the stand to the volume index (diameter squared × height). The diameter is the mean dbh and the height is the mean height of the stand

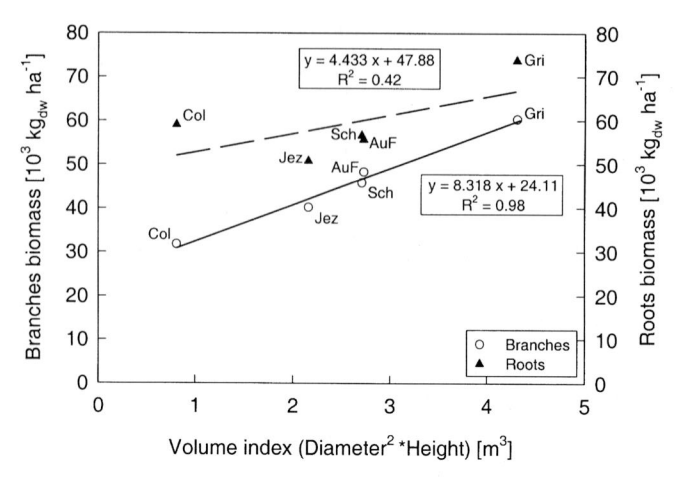

Fig. 3.2. Relationship of branch and root biomass to the volume index (diameter squared × height) of the stand at the beech sites. The diameter is the mean dbh and the height is the mean height of the stand

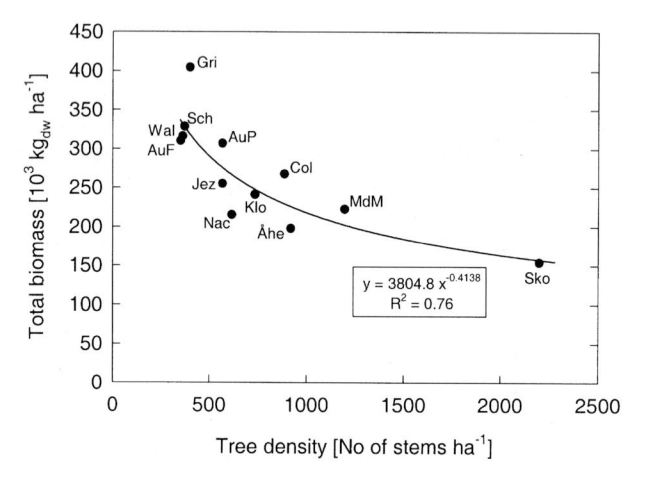

Fig. 3.3. Stand biomass in relation to the stand density of all study sites

3.7 Conclusions

A preliminary consideration pertains to the silvicultural history of the study stands. With the exception of Åheden, all the spruce forests along the European transect originated from plantations, whereas the beech stands were naturally regenerated although they were all subjected to silvicultural management. However, the natural beech forests had, on average, greater biomass and greater carbon and nutrient pools than the spruce forests. Also the harvest index was slightly greater in the beech than

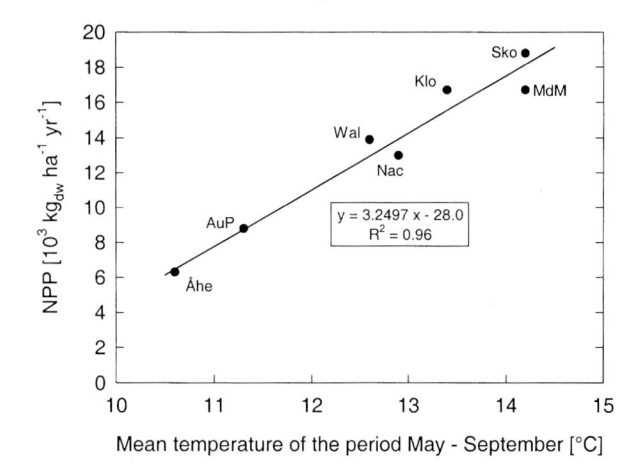

Fig. 3.4. Net primary productivity in relation to the mean air temperature of the period May–September, in the Norway spruce sites

in the spruce stands. Nevertheless, the productivity and the root-to-shoot ratio was greater in the spruce forests, which showed, however, also a larger data variability.

Below-ground components may significantly contribute to the total stand biomass and productivity in forest ecosystems (Harris et al. 1980). In our study, coarse roots and stumps represented a relevant fraction of total stand biomass in both types of stands, particularly in spruce. Furthermore, fine roots contributed significantly to the NPP of the stands of both species. The contribution of fine roots to NPP increased towards the geographic extremes of the European transect, in southern Europe and, particularly, in northern Europe, as observed also in other forests of the boreal region (Ruess et al. 1996).

In a number of studies all over the world, net productivity was found to decrease with the age of the stand (Landsberg and Gower 1997). A similar trend was also found in the European transect for both total and above-ground NPP, but only in beech forest sites and not for spruce. More firm conclusions can certainly be drawn only through more detailed studies conducted on chronosequences of different forest ecosystems. In Norway spruce stands, primary productivity was also directly related with nitrogen depositions, indicating that additional nutrients may sometimes increase productivity in some of the European transect sites, as was also confirmed by the measured values of SLA, that were lowest for beech and spruce where N depositions reached the minimum values in the transect (Bauer et al. 1997).

Acknowledgements. The authors wish to thank Dr. Giovanni Potena and Dr. Domenico Tascione of the Italian Forest Service (CFS) together with their collaborators and the Collelongo township administration for the support provided during the data collection campaign at Collelongo and Monte di Mezzo forest sites. The skillful technical assistance of Mr. R. Zompanti, Mr. Tullio Oro (DISAFRI, University of Tuscia) and of Mr. R. Bimbi (IFA-CNR, Rome) is also gratefully acknowledged. The authors are also grateful to Dr. Gisela Schmidt for kindly providing the data on the understorey vegetation and to Dr. L. Portoghesi for his useful comments during the review of the chapter.

References

Bauer G, Schulze ED, Mund M (1997) Nutrient contents and concentrations in relation to growth of *Picea abies* and *Fagus sylvatica* along a European transect. Tree Physiol 17:777-786

Bonneau M (1981) Le Hêtre et le maintien de l'équilibre naturel. In: Teissier du Cros E (ed) Le hêtre. INRA, Paris, pp 118-135

Cannell MGR (1982) World forest biomass and primary production data. Academic Press, London

Fogel R (1983) Root turnover and productivity of coniferous forests. Plant Soil 71:75-85

Harris WF, Santantonio D, Mc Ginty D (1980) The dynamic belowground ecosystem. In: Waring RH (ed) Forest: fresh perspective from ecosystem analysis. Oregon State University, Corvallis, pp 119-129

Jackson RB, Mooney HA, Schulze ED (1997) A global budget for fine root biomass, surface area, and nutrient contents. Proc Natl Acad Sci USA 94:7362-7366

Landsberg JJ, Gower ST (1997) Applications of physiological ecology to forest production. Academic Press, San Diego

Masci A, Napoli G, Dore S, Matteucci G, Scarascia Mugnozza G (1998) Aboveground and belowground biomass production in a beech forest and in a norway spruce plantation growing on the Italian Apennines. In: Borghetti M (ed) Proc 1st Congr of the Italian Sylviculture and Forest Ecology Society 1, Padova, Italy, pp 225-232

Nihlgard B, Lindgreen L (1977) Plant biomass, primary production and bioelements of three mature beech forests in South Sweden. Oikos 28:95-104

Persson H (1983) The distribution and productivity of fine roots in boreal forests. Plant Soil 1:87-101

Ruess RW, Van Cleve K, Yarie J, Viereck LA (1996) Contribution of fine root production and turnover to the carbon and nitrogen cycling in taiga forests of the Alaskan interior. Can J For Res 26:1326-1336

Santantonio D, Hermann RK, Overton WS (1977) Root biomass studies in forest ecosystems. Pedobiologia 17:1-31

Van Cleve K, Oliver L, Schlentner R, Viereck LA, Dyrness CT (1983) Productivity and nutrient cycling in taiga forest ecosystems. Can J For Res 13:747-766

Waring RH, Running SW (1998) Forest ecosystems: analysis at multiple scales. Academic Press, San Diego

4 Linking Plant Nutrition and Ecosystem Processes

G.A. BAUER, H. PERSSON, T. PERSSON, M. MUND, M. HEIN, E. KUMMETZ,
G. MATTEUCCI, H. VAN OENE, G. SCARASCIA-MUGNOZZA, and E.-D. SCHULZE

4.1 Introduction

Mineral nutrients are a major part of all the physiological and biogeochemical processes in forest ecosystems. This is especially true for forests across Europe, which were deprived of nutrients due to intensive wood and litter use, and which experienced deposition of acids, nitrogen and sulphur over the second half of this century, resulting in significant nutrient imbalances for growth (Schulze 1989). Decreased nutrient availability can lead to a reduction of leaf size (Linder 1987), resulting in an almost instantaneous decrease in current year growth. In this way, the nutrient status of long-lived conifer needles might influence net primary production (NPP) long after a transient nutrient shortage, caused, e.g. by one dry season, has occurred. In natural forest ecosystems nutrient uptake from soil solution and nutrient release through litterfall and fine root turnover should balance each other such that the turnover time of nutrients within the system meets the requirements for stand growth (Gorham et al. 1979; Miller 1986; Attiwill and Adams 1993) and keeps the ecosystem nutrient cycle tight. Any deviation from this cycle due to anthropogenic influence (e.g. Vitousek et al. 1997) or natural disturbance (e.g. Foster et al. 1997) could alter one or more processes within the nutrient cycle with long-lasting effects on forest functioning.

The forests of the present study are, with two exceptions, more than 80 years of age. Therefore we can assume that they are in a steady-state stage of development, in which the canopy is fully expanded and nutrient demands are lower because the peak in annual biomass increment has been passed (Ryan et al. 1997). Also, the forests under study are characterised by a diverse forest floor vegetation, which ranges from a completely dense lichen and moss layer in northern Sweden via large and dense patches of ericaceous and grass species in France and Germany to a very sparse occurrence of calcareous annuals in Italy (see Persson et al. Chap. 2, this Vol.). The presence of a persistent forest floor vegetation in addition to spring ephemerals (Zak et al. 1990) adds another – biological – component to the nutrient cycle of a forest ecosystem. Ericaceous shrubs and grasses generally do not produce large amounts of supporting structure; but plants like *Deschampsia flexuosa* or *Vaccinium myrtillus* are physiologically active throughout most of the growing season. Due to a continuous uptake of nutrients along with their transpiration water, they can therefore accumulate nutrients in their foliage (Bormann and Likens 1979; Ellenberg et al. 1986). From this perspective, it seems reasonable to consider a dense understorey vegetation as a competitive but transient sink for available resources (Buchmann et al. 1996).

In the present chapter we will analyse how mineral nutrition influences growth of Norway spruce (*Picea abies*) and European beech (*Fagus sylvatica*) forests across a transcontinental transect through Europe. Specifically, we ask how changing environmental conditions across this transect affect the mineral nutrition of two important forest species and what the consequences for above-ground net primary productivity (ANPP) of these forest ecosystems are. Apart from climatic differences throughout this transect, we assumed that nitrogen availability is a key parameter with direct influence on productivity. In annual plants and crops, photosynthesis and, consequently, primary production are directly connected to leaf nitrogen concentration (Field and Mooney 1986; Evans 1989, 1993). Therefore, we hypothesised that increased atmospheric N deposition would cause leaf nitrogen concentrations across the European transect to increase, leading to increased tree growth.

There is evidence that increased N availability affects the composition and the total amount of nitrogen in leaves (van Dijk and Roelofs 1988; Perez-Soba et al. 1994; Weber et al. 1998) and also affects soil N availability. We assumed that growth across the European transect would correlate with the amount of N in leaves. In the same way, we also expected that soil N pools would be a good predictor for forest growth, assuming that soil uptake dominates over canopy uptake of nitrogen.

Changes in nitrogen availability mediated through atmospheric inputs or through altered soil N mineralisation as a result of these inputs are likely to increase the amount of N accessible for plant uptake. This has given rise to the debate about competition for N between forest plants (in connection with mycorrhizae) and soil microorganisms (Zak et al. 1990; Kaye and Hart 1997). Atmospheric nitrogen deposition is known to stimulate canopy uptake of N, which could counteract root N uptake and disrupt the plant internal C/N balance. On the other hand, N-tracer studies revealed that N immobilisation by soil microorganisms can be a stronger sink for soil N than plant/mycorrhizal uptake (Buchmann et al. 1996; but see Aber et al. 1998). Increased competition for available N or other nutrients therefore could offset the relationship between nutrient uptake rate and productivity (Agren 1988; Ingestad and Agren 1995). Therefore, we used this transect to investigate if forest growth is a function of soil nutrient supply.

In this chapter we will analyse nutrient concentrations in different parts of the forest ecosystems, followed by comparisons of nutrient contents in relation to leaf growth and nitrogen partitioning in leaves. The results on organ and individuum level will then be used to infer the response of ANPP across the European transect.

4.2 Experimental Approach

Tree foliage was collected from four neighbouring trees ($n = 3$ at Nacetin) at each site after termination of foliage growth in 1994 (beech in July, spruce in September/October). Only foliage from the upper third of the crown was used for analysis. Sampling of *Deschampsia flexuosa* leaves was carried out during leaf unfolding in spring 1993 and once again after leaf unfolding in summer 1994. Patches of *Deschampsia* in the vicinity to the canopy trees used in this study were chosen for leaf sampling.

Root sampling was performed by collecting soil cores with the help of a cylindrical steel corer, 4.5 cm in diameter (cf. Vogt and Persson 1991). Sets of two soil cores (depth about 40 cm in the mineral soil) were taken on five sampling occasions during 1995–1997 at distances of 1.5, 2.5, 3.5 and 4.5 m from the chosen sampling tree with a root window. The distribution pattern of fine roots indicated no dependency on the distance to the nearest tree. Therefore the root data are presented for the total stand, disregarding the importance of small-scale distribution patterns in relation to the sampling trees. The core samples were divided into five layers, starting from the top of the organic soil horizon and extending to a maximum depth of 40 cm into the mineral soil. The soil samples were transported to the laboratory and stored in a freezer at −4 °C in plastic bags until root separation could take place.

Immediately after thawing, roots were picked out from the plastic bags from each soil layer and sorted. Fine roots < 1 mm in diameter were separated into live and dead categories, based on morphological characteristics (Vogt and Persson 1991). Live fine roots were to a varying degree brownish/suberised and often well-branched, with the main part of the root tips light and turgid. The stele was white to slightly brown and elastic. In dead roots the stele was brownish and showed a reduced elasticity. The roots were dried at 70 °C for 48 h, weighed to the nearest mg and finally ground to a fine powder. Elemental analysis was carried out after acid digestion according to Schramel et al. (1980), using an ICP-AES (Model XMP, GBC, Australia). Carbon and nitrogen concentrations were determined with an C/N analyser (Model 1500, Carlo ERBA, Italy).

Amino acid analysis was carried out from aliquots of frozen plant material. Samples were ground in 250 µl a 1:1(vol) mixture of acetone and Na-acetate (50 mM) under liquid nitrogen. The powderised plant material was dissolved in 1750 µl of the extraction medium and was placed for 45 min at 3 °C in the dark. After the incubation, the crude extract was centrifuged at 15000 g for 15 min. The supernatant was placed underneath a hood to evaporate the acetone and the remaining liquid was used for amino acid analysis. The detection of single amino acids was carried out via precolumn derivatisation of the primary amino group with 0-phtalaciddialdehyd (OPA). For the derivatisation 40 µl of the extracted plant material were mixed with a mixture of 5% (w/v) OPA/methanol, 0.8 M borate buffer (pH 10.4, KOH) and 3-mercaptoethanol in a ratio of 10:90:1. After an incubation time of 2 min at room temperature, 40 µl of the reaction mixture were injected onto a Hypersil ODSII column. During the passage of the column the fluorescense was detected with a data processing system (Kontron D450, Germany) and continuously recorded. Further details on the chromatographic separation are given in Bauer (1997).

4.3 Nutrient Concentrations

4.3.1 Overview for Different Forest Compartments

Across the transect, European beech (*Fagus sylvatica*) and Norway spruce (*Picea abies*) showed significant differences in above-ground net primary production (ANPP, kg m^{-2} a^{-1}; Table 4.1). But in contrast to what has been reported for other old-growth forests (e.g. Gower et al. 1996; Ryan et al. 1997), ANPP for spruce and beech along the transect did not correlate with stand age. Although differences in specific leaf area (SLA) between spruce and beech are evident (see Scarascia-Mugnozza et al., Chap. 3, this Vol.), species-specific differences emerged to a greater extent only for total nitrogen. Leaf and needle N concentrations remained very conservative, except for lower N concentrations in northern Scandinavia. Leaf nitrogen, together with carbon concentrations, remained rather constant, whereas differences in other element concentrations were much larger, probably reflecting rather variation in soil supply than differences in plant demand.

Table 4.1. Above-ground net primary production (ANPP, kg m^{-2} a^{-1}) and element concentrations (mg g^{-1}) for major ecosystem components across the European transect. Where no FH layer was present, the 0–10 cm layer was subdivided into 0–5 (*) and 5–10 cm (**)

	Collelongo	Aubure	Jezeri	Schacht	Gribskov	MdiM	Aubure	Nacetin	Waldstein	Klosterh.	Skogaby	Åheden
Stand age (years)	90	150	78	120	120	38	100	60	145	80	30	145
ANPP (kg m^{-2} a^{-1})	0.87	0.70	1.25	0.77	0.89	1.09	0.73	0.99	1.03	1.14	1.16	0.37
Leaves, needles (mg g^{-1} dw)												
C	478.30	498.50		506.20	476.70	499.20	495.60		506.30	496.10	468.35	497.30
N	24.22	26.74		26.18	27.72	13.26	12.44	15.25	14.84	12.56	10.70	7.66
P	1.23	1.82	0.81	1.11	1.37	1.57	1.95	1.27	1.68	1.29	0.81	1.60
S	1.64	7.74		1.46	1.67	1.01	1.01	2.52	1.30	1.24	0.75	0.69
Ca	7.56	2.97	5.36	4.21	7.18	10.62	3.70	4.79	6.23	4.37	2.02	4.93
K	6.39	9.28	6.78	3.66	7.00	6.67	6.64	5.28	6.96	7.46	3.77	5.89
Mg	1.59	0.69	0.43	0.59	1.59	0.75	0.74	0.70	0.56	1.56	0.62	1.09
Twigs (mg g^{-1} dw)												
C	464.40	480.00		477.60	474.00	498.00	486.00		515.00	489.60		490.80
N	9.64	6.83	6.86	7.18	7.53	5.21	3.92	7.61	8.40	2.62	4.63	5.78
P	0.74	1.24	0.75	1.24	1.48	0.78	0.41	0.80	1.66	0.38	0.50	
S	0.39	0.36	0.34	0.38	0.42		0.78	0.59	0.91	0.24	0.54	
Ca	5.66	0.70	3.51	1.96	1.81	6.59	3.37	2.89	3.79	0.66	2.42	
K	5.13	3.41	2.30	2.24	3.16	4.13	1.87	3.12	5.81	2.00	1.85	
Mg	0.37	0.22	0.37	0.36	0.41	0.06	0.32	0.62	0.98	0.45	0.53	
Branches (mg g^{-1} dw)												
C	475.79					481.10			515.00			
N	1.29	2.93				0.30		7.61	4.48		4.63	
P	0.50	0.19				0.22		0.01	0.67		0.50	
S								0.59	0.68		0.54	
Ca	2.96	2.53				1.15		0.93	6.04		2.42	
K	1.77	1.77				0.88		0.56	1.54		1.85	
Mg	0.04	0.29				0.04		0.16	0.63		0.53	

	Sapwood (mg g⁻¹ dw)										
C	465.60	481.20	485.50	474.00	470.40	489.60		497.20	480.20	483.60	483.60
N	0.67	0.42	1.26	1.40	0.14	1.18	1.48	0.84	0.98	0.73	0.98
P	0.13	0.08	0.10		0.12	0.05	0.01	0.04		0.12	
S		0.08				0.74	0.05	0.05		0.11	
Ca	1.41	0.45	0.27		1.11	0.23	0.93	0.67		1.06	
K	1.29	0.98	1.04		0.45	0.62	0.56	0.21		0.91	
Mg	0.12	0.12	0.29		0.03	0.14	0.16	0.08		0.22	

	Stembark (mg g⁻¹ dw)			
C	468.04	486.20	494.40	
N	4.97	4.63	4.62	5.30
P	0.28	0.76	0.51	
S			0.55	
Ca	19.61	9.56	12.12	
K	2.06	3.58	1.69	
Mg	0.04	0.09	0.81	

	Coarse roots (mg g⁻¹ dw)					
C	465.60	470.40			496.56
N	0.04	0.02	1.60	5.97	2.80
P	0.00	0.01	0.21	0.52	0.06
S			0.81	0.51	0.03
Ca	0.03	0.01	1.70	3.34	3.61
K	0.02	0.02	1.50	2.30	2.15
Mg	0.00	0.00	0.20	0.53	0.17

Table 4.1. *Continued*

	Collelongo	Aubure	Jezeri	Schacht	Gribskov	MdiM	Aubure	Nacetin	Waldstein	Klosterh.	Skogaby	Åheden
Stand age (years)	90	150	78	120	120	38	100	60	145	80	30	145
ANPP (kg m^{-2} a^{-1})	0.87	0.70	1.25	0.77	0.89	1.09	0.73	0.99	1.03	1.14	1.16	0.37
Fine roots (mg g^{-1} dw)												
C	494.50	527.00		500.70		452.45	509.00		484.30			
N	4.42	17.50	9.53	17.30		6.65	16.10	12.96	16.90		9.70	
P	0.29	1.47	0.73	1.45		1.14	1.31	1.20	1.05		0.60	
S		1.44	0.63	2.28			1.18	0.37	1.28		0.70	
Ca	2.80	2.19	4.68	3.58		9.58	2.96	3.09	3.40		2.20	
K	1.84	1.25	3.08	1.18		3.67	0.77	2.82	0.65		0.40	
Mg	0.04	0.44	0.69	2.27		0.33	0.43	0.56	0.40		0.50	
Carbon (mg g^{-1} dw)												
L	373.50	488.20	456.10	472.50	445.20	337.80	Incl. FH	458.00	496.60	515.50	510.80	490.50
FH	146.2*	258.80	289.10	430.80	47*	64.8*	388.80	369.70	377.10	430.10	432.00	403.30
A 0–10	89.7**	92.40	75.60	164.70	23.8**	41.2**	36.10	119.50	59.10	25.80	41.80	24.00
A 10–20	67.10	31.80	32.00	72.20	15.70	36.80	14.40	43.90	58.00	28.90	23.00	7.60
A 20–30	60.30	21.90	25.00	62.60	12.20	34.60	11.40	24.60	51.70	14.50	19.10	4.50
A 30–50	48.90	25.80	19.10		8.10	26.30	8.80	13.20	20.10	4.00	11.80	1.60
Nitrogen (mg g^{-1} dw)												
L	18.05	19.30	17.90	19.60	17.20	16.10	Incl. FH	17.00	20.60	18.79	14.20	10.78
FH	11.00	12.10	14.60	21.90	2.5*	5.61*	15.00	15.20	16.96	13.97	15.40	10.93
A 0–10	7.29	5.24	3.80	7.56	1.16**	4.21**	2.18	3.90	2.40	0.67	1.70	0.74
A 10–20	5.36	1.80	1.60	3.36	0.70	4.08	1.10	1.60	2.31	0.87	1.00	0.32
A 20–30	4.82	1.15	1.20	2.80	0.53	3.78	0.94	1.10	1.99	0.52	0.90	0.25
A 30–50	4.05	1.22	1.00		0.37	2.25	0.72	0.60	0.88	0.15	0.60	0.14

In all woody compartments nutrient concentrations remained below the levels of the leaves, indicating the importance of wood for structure and transport rather than for growth processes. There was also no evidence for a general difference between beech and spruce in the nutrient concentrations of woody structure. In twigs, branches, sapwood and coarse roots nutrient concentrations in beech were of the same order as in spruce. Across the transect we found a trend to somewhat higher C and N concentrations at sites in France (Aubure), Czech Republic (Jezeri, Nacetin) and Germany (Schacht, Waldstein; Table 4.1). Only fine root nutrient concentrations were much higher than in other woody structure and showed significantly higher N concentrations at sites in Central Europe than elsewhere along the transect. Soil C and N concentrations closely matched the pattern of fine roots, with highest N concentrations in France, the Czech Republic and Germany, whereas soil C concentrations were highest in south and mid-Scandinavia. Also, soil C and N concentrations differed much more strongly across the transect than leaf concentrations.

The distribution of nutrients among forest components and the lack of correlation with ANPP and stand age does not support the general view that nutrient limitation of old-growth forests is a key factor inducing an age-related decline in productivity. Rather, it seems that the constant supply of N from air pollution extends the phase of active growth in mature forests.

4.3.2 Foliar Analysis

In current-year needles of spruce, harvested in September/October 1994, variation in nutrient concentrations is large (Table 4.2), probably reflecting incompleted needle maturity (Bauer et al. 1997). Given the differences in productivity among the experimental sites (see Scarascia-Mugnozza, Chap. 3, Table 3.1, this Vol.), it is striking that low nutrient concentrations, for example, do not correspond with low productivity. In northern Sweden N and S concentrations reached a minimum, but P was lowest at Nacetin and Mg was lowest in Skogaby. In 1-year-old needles nitrogen concentrations did not change significantly between Italy and Danemark (Table 4.2), despite differences in stand growth. Only trees in mid- and northern Sweden had significantly lower needle N concentration. For both needle classes S concentrations showed a geographic pattern similar to N, but with significantly higher values at the sites in Central Europe. The concentrations of phosphorus and the alkali ions K and Ca were variable, mainly reflecting differences in local soil conditions. Magnesium showed an inverse relation to N, with higher values in the south and north of the transect. Nutrient concentrations in the needles of *Pinus pinaster* and *Pinus sylvestris* matched those of the geographically corresponding spruce stand. Despite different leaf morphology in pine, the nutrient concentrations followed the trend in Norway spruce of generally higher N, P and K values in 0-year needles. In beech, differences in leaf N concentration among the sites were even smaller than in spruce. With the exception of the beech stand in Aubure, this holds true also for S and P. Only the cations K, Ca and Mg varied significantly across the transect, with again an inverse relationship of Mg to N, as in spruce.

Desampsia flexuosa was present in seven of the stands, but the percentage of ground cover differed largely among those stands. Irrespective of the stand type, there was a trend to lower N, S, P and K concentrations in summer than in spring (Table 4.3), whereas Ca and Mg tended to be lower in spring than in summer, while the tree

Table 4.2. Element concentrations (mmol g^{-1} for nitrogen; μmol g^{-1} for S, P, K, Ca and Mg) in needles and leaves of *Picea abies*, *Pinus pinaster*, *Pinus sylvestris*, *Fagus sylvatica* and *Betula pendula* on sites along the European transect (Bauer et al. 1997)

	Nitrogen (mmol g^{-1})	Sulphur (μmol g^{-1})	Phosphorus (μmol g^{-1})	Potassium (μmol g^{-1})	Calcium (μmol g^{-1})	Magnesium (μmol g^{-1})
Picea abies, 0-year-old needles						
M di Mezzo	1.05 ± 0.04bcd	30.6 ± 1.6ab	57.0 ± 7.9abc	206.7 ± 24.4ab	140.5 ± 39.8b	32.9 ± 8.3a
Aubure	0.87 ± 0.07bc	32.0 ± 3.1bc	77.0 ± 9.4bc	216.2 ± 33.6ab	72.3 ± 15.8ab	45.4 ± 12.2ab
Nacetin	1.20 ± 0.03d	63.6 ± 3.0d	50.1 ± 2.2ab	140.0 ± 24.9a	75.5 ± 18.9ab	33.6 ± 1.7a
Schacht	1.27 ± 0.12d	38.4 ± 2.5bc	82.9 ± 10.3c	220.3 ± 17.9ab	70.8 ± 17.6a	37.0 ± 2.7a
Waldstein	1.15 ± 0.22cd	36.7 ± 4.2bc	73.3 ± 16.8bc	249.0 ± 6.2b	76.5 ± 27.7ab	33.9 ± 8.6a
Klosterhede	0.99 ± 0.04bcd	39.5 ± 2.5c	50.8 ± 6.7ab	247.2 ± 39.7b	68.5 ± 9.7a	70.8 ± 12.8b
Skogaby	0.77 ± 0.07ab	22.2 ± 3.1a	31.4 ± 10.0a	138.7 ± 36.3a	37.6 ± 10.2a	28.0 ± 5.8a
Åheden	0.54 ± 0.14a	22.0 ± 2.9a	61.2 ± 1.9bc	204.5 ± 25.8ab	64.7 ± 26.8a	47.2 ± 10.8ab
Picea abies, 1-year-old needles						
M di Mezzo	1.00 ± 0.21b	32.0 ± 4.7bc	52.8 ± 14.6bc	172.1 ± 28.3b	249.3 ± 25.3c	31.3 ± 12.1b
Aubure	0.91 ± 0.11ab	32.0 ± 2.7bc	62.4 ± 4.5c	173.8 ± 27.3b	92.4 ± 14.5ab	29.8 ± 9.2b
Nacetin	1.00 ± 0.13b	72.5 ± 4.0d	37.3 ± 5.8ab	128.4 ± 15.1ab	112.7 ± 21.0ab	28.2 ± 1.2b
Schacht	1.10 ± 0.12b	40.1 ± 1.7c	64.3 ± 6.5c	168.4 ± 9.2b	104.4 ± 33.7ab	31.4 ± 3.7b
Waldstein	1.06 ± 0.13b	37.6 ± 4.3c	49.1 ± 7.5bc	138.7 ± 12.2ab	125.3 ± 8.9b	21.4 ± 3.5b
Klosterhede	0.88 ± 0.11ab	37.4 ± 4.2c	41.9 ± 9.2abc	175.2 ± 13.2b	134.5 ± 29.8b	74.9 ± 15.5a
Skogaby	0.76 ± 0.21ab	22.1 ± 1.44ab	21.1 ± 2.7a	85.0 ± 5.6a	47.1 ± 6.4a	23.6 ± 3.3b
Åheden	0.53 ± 0.10a	21.2 ± 3.1a	48.1 ± 3.8bc	165.3 ± 19.7b	123.5 ± 39.0ab	45.4 ± 12.3b

Pinus pinaster, Thezan						
0-year-old	0.62 ± 0.03	40.2 ± 2.2	22.6 ± 1.5	161.8 ± 19.1	72.6 ± 28.1	81.5 ± 17.0
1-year-old	0.53 ± 0.07	38.5 ± 2.1	18.2 ± 0.8	114.1 ± 33.4	134.3 ± 35.5	114.3 ± 32.3
Pinus sylvestris, Åheden						
0-year-old	0.94 ± 0.12	31.2 ± 3.3	57.1 ± 3.8	190.9 ± 10.4	40.0 ± 18.6	46.1 ± 12.1
1-year-old	0.79 ± 0.06	27.4 ± 2.9	44.3 ± 3.3	143.7 ± 6.5	65.2 ± 28.8	39.3 ± 9.8
Fagus sylvatica						
Collelongo	1.73 ± 0.11a	51.3 ± 8.3a	41.4 ± 12.0a	163.5 ± 25.1abc	188.7 ± 39.7b	65.5 ± 5.1b
Aubure	1.91 ± 0.03ab	54.2 ± 13.8a	60.1 ± 27.0a	237.3 ± 99.7c	74.1 ± 23.2a	28.4 ± 4.5a
Schacht	1.88 ± 0.16ab	45.7 ± 5.1a	41.6 ± 17.8a	93.5 ± 26.1a	104.9 ± 17.5a	24.2 ± 5.4a
Waldstein	1.87 ± 0.16ab	55.2 ± 8.0a	44.5 ± 18.8a	130.3 ± 40.6ab	155.4 ± 16.2b	51.0 ± 15.4b
Gribskov	1.98 ± 0.17b	52.2 ± 4.6a	49.0 ± 16.6a	179.0 ± 11.7bc	179.1 ± 28.0b	65.5 ± 20.4b
Betula pubescens						
Åheden	1.32 ± 0.14	39.2 ± 4.9	65.2 ± 9.3	223.5 ± 28.9	172.0 ± 19.4	72.4 ± 7.9

Values are the means of $n = 4$ repetitions ± SD. Numbers in a single column followed by different letters are significantly different at the $p = 0.05$ level.

Table 4.3. Element concentrations (mmol g^{-1} for nitrogen; µmol g^{-1} for S, P, K, Ca and Mg) in leaves of *Deschampsia flexuosa* in the understorey vegetation of spruce and beech stands along the European transect during leaf unfolding of the canopy (1993) and underneath a fully developed canopy (1994). (Bauer 1997)

Picea stands	Nitrogen (mmol g^{-1})	Sulphur (µmol g^{-1})	Phosphorus (µmol g^{-1})	Potassium (µmol g^{-1})	Calcium (µmol g^{-1})	Magnesium (µmol g^{-1})
Leaf unfolding						
Aubure	2.12 ± 0.16a	67.2 ± 0.9a	100.1 ± 19.0b	860.3 ± 146.0a	11.6 ± 3.7a	46.5 ± 6.6a
Waldstein	2.41 ± 0.18a	77.2 ± 4.5a	84.4 ± 20.0ab	661.3 ± 96.2a	7.0 ± 1.6a	40.7 ± 5.1a
Klosterhede	1.85 ± 0.09b	45.3 ± 5.5b	56.2 ± 7.0a	722.0 ± 40.3a	12.9 ± 5.8a	49.8 ± 9.8a
Åheden	1.88 ± 0.09b	46.7 ± 6.7b	93.4 ± 12.0b	702.4 ± 71.3a	14.5 ± 5.1a	39.5 ± 5.7a
Full canopy						
Aubure	2.17 ± 0.07a	61.2 ± 1.8a	80.7 ± 6.3a	669.5 ± 45.0a	15.0 ± 3.7a	54.9 ± 8.0a
Waldstein	1.80 ± 0.23a	62.3 ± 7.1a	63.2 ± 16.2a	565.1 ± 70.5a	21.9 ± 8.5b	48.4 ± 8.7a
Klosterhede	1.40 ± 0.14b	n.d.	n.d.	n.d.	n.d.	n.d.
Åheden	0.73 ± 0.36b	22.3 ± 3.1a	59.3 ± 15.9a	527.8 ± 52.0a	14.8 ± 5.0ab	23.0 ± 3.3b

Fagus stands	Nitrogen (mmol g^{-1})	Sulphur (µmol g^{-1})	Phosphorus (µmol g^{-1})	Potassium (µmol g^{-1})	Calcium (µmol g^{-1})	Magnesium (µmol g^{-1})
Leaf unfolding						
Aubure	2.25 ± 0.49a	61.5 ± 20.6a	111.9 ± 34.5a	770.7 ± 138.6a	13.7 ± 0.9a	43.9 ± 12.9a
Schacht	2.21 ± 0.16a	67.7 ± 6.2a	69.3 ± 4.3a	832.3 ± 101.9a	21.5 ± 5.2a	47.5 ± 8.7a
Gribskov	2.38 ± 0.34a	63.8 ± 9.2a	129.5 ± 37.9a	1083.4 ± 91.3b	20.7 ± 5.7a	49.9 ± 9.2a
Full canopy						
Aubure	2.07 ± 0.13a	54.2 ± 3.2a	95.3 ± 23.8a	952.9 ± 133.0a	28.1 ± 3.0a	57.5 ± 3.6ab
Schacht	2.11 ± 0.22a	61.7 ± 4.9a	83.4 ± 2.9a	805.3 ± 184.7a	55.0 ± 15.0b	66.5 ± 10.8b
Gribskov	1.26 ± 0.08b	32.2 ± 3.1a	70.4 ± 20.1a	737.9 ± 49.8a	40.6 ± 5.3ab	44.6 ± 4.5a

Values are the means of $n = 4$ repetitions ± SD. Numbers in a single column followed by different letters are significantly different at the $p = 0.05$ level.

canopy had not reached full closure. Within the spruce stands N, S and partially P concentration in the leaves of *Deschampsia* were higher at the sites in Aubure and Waldstein compared to the other sites, whereas at the two northern stands Mg, and to a certain extent Ca, tended to be higher than elsewhere along the transect. The nutrient concentrations in the *Deschampsia* leaves underneath the beech canopies hardly differed from each other. Only a very dry summer at the site in Gribskov probably caused the low nutrient concentrations in *Deschampsia* leaves at that site.

The nutrient concentrations in needles and leaves across the European transect showed that significant differences between the sites for most of the nutrients were lacking. In fact, this finding is contrary to what was expected for a comparison of forest ecosystems ranging from central Italy to northern Sweden which significantly differ in N mineralisation, cation exchange capacity and atmospheric deposition of N and S compounds. Therefore, we further explored the nutrient concentrations at different spatial scales (Table 4.4). For nitrogen, variation in spruce needles in southern Sweden, or in leaves of *Fagus* stands across northern Bavaria is the same as the range of observations across the entire European transect. Although Norway spruce needles at the site in northern Sweden had the lowest N concentrations in our study (Table 4.1), their variation at the local scale was in the same range as along the transect. Needle sulphur concentrations in forests surrounding the Waldstein site and across Bavaria varied more strongly than those in our survey for the European transect. However, S concentrations in Norway spruce in southwest and north Sweden showed a smaller range. Also for P, Ca, K and Mg there was no clear evidence that the larger-scale study covered a wider range of nutrient availability than small-scale studies. Only for the understorey grass *Deschampsia* did we clearly find a much wider range of nutrient concentrations along the transect than in comparison to local studies in Sweden (Högbom and Högberg 1991).

4.3.3 Fine Roots

In living fine roots of 1–2 mm in diameter nutrient concentrations generally decreased along the soil profile from 0–40 cm of depth (Table 4.5). This pattern changed for P and K at the spruce stand Waldstein, where P and K concentrations increased, and at Aubure, where both elements remained at almost constant concentration over the soil profile. However, despite the general decrease with soil depth, element concentrations in living fine roots even in the mineral soil remained at or even above the concentrations in above-ground parts of the trees (see Table 4.1). This pattern seems to be the result of a complicated allocation pattern. On the one hand, nutrients have to be supplied to the canopy in order to meet the growth demands. On the other, root growth in acid forest soils might have increased due to the fact that roots have to grow deeper in order to forage for base cations like Mg, which are subjected to leaching. Under these conditions the trees might have to maintain a high root vigour over a larger profile in order to maximise the uptake of limiting resources.

4.4 Nutrient Contents

The forests which were investigated in the present study are assumed to be in a steady-state stage of canopy development, as indicated by the generally high leaf area index. We can furthermore assume that differences in foliage nutrition are not caused by

Table 4.4. Continental, regional and local means, standard deviations and maximum and minimum observations of element concentrations for *Picea abies*, *Fagus sylvatica* and *Deschampsia flexuosa*. The regional data for Sweden were recalculated from Ericsson et al. (1993, 1995; for *Picea* and Högbom and Högberg (1991; for *Deschampsia*). The data for Bavaria are taken from Kaupenjohann et al. (1989) and Rehfuess and Rodenkirchen (1984). (Bauer et al. 1997)

Picea abies		Nitrogen (mmol g^{-1})	Phosphorus (μmol g^{-1})	Sulphur (μmol g^{-1})	Calcium (μmol g^{-1})	Postassium (μmol g^{-1})	Magnesium (μmol g^{-1})
Southwest Sweden ($n = 60$)	Av ± S.D.	0.78 ± 0.09	34.8 ± 8.0	28.8 ± 3.0	174 ± 54.5	96.4 ± 16.6	37.9 ± 9.6
	Min	0.54	19.4			46.2	
	Max	1.36	61.3			138.5	
North Sweden ($n = 15$)	Av ± S.D.	0.68 ± 0.11	40.9 ± 9.0	25.6 ± 3.0	190 ± 14	101 ± 14	37.9 ± 6.0
Germany Waldstein ($n = 5$)	Av ± S.D.	0.94 ± 0.3	27.9 ± 10.0	33.5 ± 19.2	107.6 ± 97.6	87.2 ± 34.3	19.5 ± 12.9
	Min	0.56	6.3	6.3	22	26.2	1.8
	Max	2.12	51.7	90.6	399	156	45.6
Germany Bavaria ($n = 22$)	Av ± S.D.	0.98 ± 0.1	53.2 ± 11.9	43.6 ± 11.6	86.3 ± 43.3	140 ± 43.3	34.9 ± 14.2
	Min	0.70	32.3	25.9	35.0	79.5	10.8
	Max	1.21	87.1	68.8	250	351	83.3
Transect ($n = 7$)	Av ± S.D.	0.88 ± 0.23	48.5 ± 15.6	31.6 ± 7.7	125.2 ± 63.8	154.6 ± 35.5	37.3 ± 19.5
	Min	0.40	18.9	17.3	40.7	79.9	16.7
	Max	1.21	73.9	42.7	282.3	207.4	92.7

Fagus sylvatica		Nitrogen (mmol g⁻¹)	Phosphorus (μmol g⁻¹)	Sulphur (μmol g⁻¹)	Calcium (μmol g⁻¹)	Potassium (μmol g⁻¹)	Magnesium (μmol g⁻¹)
Germany Bavaria (n = 4)	Av ± S.D.	1.66 ± 0.18	60.8 ± 9.9	59.9 ± 9.0	271.9 ± 138.9	207.8 ± 63.3	64.7 ± 31.5
	Min	1.29	45.2	36.9	142.9	114.8	14.1
	Max	1.96	82.8	74.4	649.8	353.4	132.4
Transect (n = 5)	Av ± S.D.	1.87 ± 0.14	47.3 ± 17.9	51.7 ± 8.2	140.4 ± 49.6	160.7 ± 65.5	46.9 ± 20.6
	Min	1.58	25.7	37.2	52.6	64.5	17.4
	Max	2.18	94.0	70.1	242.3	376.5	95.9

Deschampsia flexuosa		Nitrogen (mmol g⁻¹)	Phosphorital (μmol g⁻¹)	Sulphur (μmol g⁻¹)	Calcium (μmol g⁻¹)	Potassium (μmol g⁻¹)	Magnesium (μmol g⁻¹)
South Sweden (n = 12)	Av ± S.D.	1.64 ± 0.38	71.6 ± 16.5		24.5 ± 7.8	636.2 ± 86.8	46.0 ± 7.8
	Min	1.15	42.0		17.5	511.5	32.9
	Max	2.20	93.6		42.4	800.5	57.6
North Sweden (n = 3)	Av ± S.D.	1.22 ± 0.07	67.8 ± 21.2		29.9 ± 2.5	574.6 ± 62.9	31.5 ± 4.8
	Min	1.16	48.4		27.5	503.8	28.8
	Max	1.29	90.4		32.4	624.0	37.0
Transect (n = 7)	Av ± S.D.	1.90 ± 0.50	84.4 ± 27.2	55.6 ± 16.3	22.0 ± 14.5	760.8 ± 172.3	47.2 ± 12.1
	Min	0.32	40.1	19.0	5.3	469.3	18.7
	Max	2.98	185.1	89.9	77.2	1178.2	81.3

Table 4.5. Nutrient concentrations in living fine roots (1–2 mm) over the soil depth profile (0–40 cm) for two beech stands (Aubure, Schacht) and three spruce stands (Aubure, Waldstein, Skogaby) across the European transect. (Data for Skogaby H. Persson pers. comm.)

Nutrient	Depth (cm)	Aubure (*Fagus*)	Aubure (*Picea*)	Schacht (*Fagus*)	Waldstein (*Picea*)	Skogaby (*Picea*)
N (mg g^{-1})	0–2.5	18.3 ± 1.13	16.2 ± 0.14	17.3	16.9	9.7
	2.5–5	15.7 ± 0.71	15.6 ± 0	15.8	16.6	9.0
	5–10	16.9 ± 0.57	14.45 ± 0.78	15.3	16.8	8.0
	10–20	13.6 ± 0.28	13.3 ± 1.7	13.1	15.8	7.0
	20–40	10.45 ± 0.07	12.5 ± 0.57	11.5	15.3	
S (mg g^{-1})	0–2.5	1.36 ± 0.11	1.17 ± 0.02	2.28	1.28	0.7
	2.5–5	1.18 ± 0.1	1.19 ± 0.06	2.74	1.25	0.7
	5–10	1.29 ± 0.05	1.19 ± 0.02	0.95	1.12	0.7
	10–20	0.99 ± 0.05	1.0 ± 0.09	0.84	1.05	0.6
	20–40	0.77 ± 0.02	0.84 ± 0.13	1.51	1.09	
P (mg g^{-1})	0–2.5	1.33 ± 0.21	1.3 ± 0.01	1.45	1.05	0.6
	2.5–5	1.06 ± 0.13	1.44 ± 0.13	1.54	1.04	0.8
	5–10	1.27 ± 0.08	1.38 ± 0.01	0.64	1.04	0.8
	10–20	1.13 ± 0.08	1.26 ± 0.16	0.56	1.16	0.7
	20–40	1.0 ± 0.05	1.17 ± 0.29	0.91	1.34	
Ca (mg g^{-1})	0–2.5	2.43 ± 0.34	2.94 ± 0.04	3.58	3.40	2.2
	2.5–5	1.69 ± 0.27	2.82 ± 0.01	4.86	2.86	1.3
	5–10	1.71 ± 0.47	2.87 ± 0.54	3.14	2.71	1.0
	10–20	1.32 ± 0.33	2.46 ± 0.22	2.28	2.29	1.1
	20–40	1.15 ± 0.01	2.09 ± 0.66	3.01	2.08	
K (mg g^{-1})	0–2.5	0.91 ± 0.48	0.75 ± 0.03	1.18	0.65	0.4
	2.5–5	0.76 ± 0.4	0.80 ± 0.17	1.44	0.60	0.3
	5–10	1.0 ± 0.45	0.75 ± 0.23	0.77	0.71	0.2
	10–20	0.86 ± 0.32	0.71 ± 0.05	0.60	0.90	0.2
	20–40	1.03 ± 0.17	0.74 ± 0.33	0.84	0.86	
Mg (mg g^{-1})	0–2.5	0.45 ± 0.01	0.41 ± 0.03	0.32	0.40	0.50
	2.5–5	0.34 ± 0.07	0.39 ± 0.05	0.41	0.31	0.40
	5–10	0.43 ± 0.13	0.39 ± 0.1	0.42	0.33	0.40
	10–20	0.37 ± 0.08	0.37 ± 0.08	0.29	0.33	0.50
	20–40	0.39 ± 0.01	0.35 ± 0.11	0.31	0.34	

changes in canopy structure during stand development, and can thus be interpreted as indicators for nutrient supply of closed canopies. Nutrient content, which is the amount of a nutrient per leaf (mass based concentrations times leaf dry weight), and leaf dry weight are the result of a seasonal development, and if the nutrient supply is ample during the growing season, both should increase proportionally to each other in order to keep the balance between structure and function of the leaf (Oren and Schulze 1989; Luo et al. 1994). On this basis, one can expect that each mole fraction of a nutrient taken up is invested in new biomass. In the case of supply in excess of demand, for example (sensu Chapin et al. 1990), the amount of a given nutrient should increase independently of leaf dry weight (Timmer and Armstrong 1987).

Across all tree species nutrient contents showed a much stronger geographic variation (Table 4.6) than the nutrient concentrations. Generally, there was at least a twofold difference for any given species and/or needle age class. Nitrogen contents (μmol leaf^{-1} or μmol needle^{-1} for deciduous and coniferous trees, respectively) varied significantly across the entire transect and reached highest values at Klosterhede and Nacetin for spruce and at Gribskov and Aubure for beech. Sulphur closely followed that pattern, but showed a clear accumulation at Nacetin (Ore Mountains, Czech Republic), where atmospheric sulphur pollution obviously is still a major concern (Paces 1985). Potassium, calcium, magnesium and, in part, phosphorus showed an inverse geographic pattern to N.

Nutrient contents of fully mature leaves might be taken as integrative measures for the nutrient supply throughout the growing season. By using this approach, it might not be possible to distinguish between actual uptake and the amount of nutrients remobilised from storage pools. However, assuming a functional balance between leaf size and nutrients necessary for the production of a leaf, a comparison of nutrient contents has the advantage of detecting imbalances between accumulation of dry weight and nutrient supply. Therefore, we used the nutrient contents in relation to foliage dry weight from all the sites to further analyse the nutrient supply. For Norway spruce, N content of 1-year-old needles increased linearly with needle dry weight (Fig. 4.1), except for northern Sweden, where needles reached higher dry weight at very low N content. This seems to be the result of needle growth under N limitation. In contrast to this, Norway spruce at Klosterhede, Waldstein and Nacetin reached higher N contents at the same range of needle dry weight than the other sites. At these sites a continuous supply of N resulted in a proportional increase in needle dry weight, whereas in northern Sweden needles accumulated carbon, which is typical for N shortage. Sulphur content was also positively related to needle dry weight, but at the site in Nacetin a strong S accumulation occurred, in contrast to lower S contents at Skogaby and Åheden, despite higher needle dry weight. The relationship for P was similar for all the sites except for lower P contents in Skogaby. At this site also Ca and K were in low supply, whereas Ca accumulated in needles of Norway spruce on the calcareous soils in Italy (Monte di Mezzo). The relationship between Mg content and needle dry weight clearly indicates an Mg accumulation in Norway spruce at Klosterhede, where the stand is very close to the sea shore. This is contrary to the stands in the Ore Mountains and the Fichtelgebirge, which stock on acid soils and have experienced acid rain history and therefore are in short supply of magnesium.

For the deciduous trees, leaf dry weight increased linearly with N content (Fig. 4.2). This means that, despite high atmospheric N inputs over most of Europe, each mole fraction of nitrogen taken up is proportionally invested in new biomass and that there is no imbalance between N content and leaf dry weight. However, this would indicate an N limitation for the beech stands. Sulphur contents were also linearly related to leaf dry weight, as were phosphorus and potassium, but at a higher variation among the sites. For Mg and Ca we found a clear tradeoff between nutrient content and leaf dry weight for the sites at Aubure and Schacht. Both stands growing on acid soils showed similar values for leaf dry weight at much lower nutrient contents, as compared to the other sites.

The variable growth conditions across the transect resulted in site specific changes in leaf morphology within each species (see Scarascia-Mugnozza et al., Chap. 3, this

Table 4.6. Element contents (μmol needle⁻¹ or μmol leaf⁻¹) in needles and leaves of *Picea abies*, *Pinus pinaster*, *Pinus sylvestris*, *Fagus sylvatica* and *Betula pendula* on sites along the European transect. (Bauer et al. 1997; Bauer 1997)

	Nitrogen (μmol leaf⁻¹)	Sulphur (μmol leaf⁻¹)	Phosphorus (μmol leaf⁻¹)	Potassium (μmol leaf⁻¹)	Calcium (μmol leaf⁻¹)	Magnesium (μmol leaf⁻¹)
Picea abies, 0-year-old needles						
M di Mezzo	3.24 ± 0.37ab	0.10 ± 0.02ab	0.18 ± 0.04a	0.64 ± 0.11a	0.44 ± 0.12a	0.10 ± 0.02b
Aubure	2.01 ± 0.50a	0.08 ± 0.02a	0.18 ± 0.05a	0.50 ± 0.13a	0.17 ± 0.05a	0.11 ± 0.03b
Nacetin	5.91 ± 0.69c	0.31 ± 0.06c	0.25 ± 0.03a	0.70 ± 0.19a	0.37 ± 0.05a	0.16 ± 0.02b
Schacht	3.77 ± 0.91abc	0.12 ± 0.03ab	0.24 ± 0.05a	0.65 ± 0.13a	0.22 ± 0.10a	0.11 ± 0.02b
Waldstein	4.39 ± 0.15abc	0.14 ± 0.04ab	0.29 ± 0.10a	0.97 ± 0.17a	0.30 ± 0.13a	0.14 ± 0.06b
Klosterhede	5.78 ± 1.62bc	0.23 ± 0.06bc	0.30 ± 0.10a	1.46 ± 0.51a	0.40 ± 0.12a	0.42 ± 0.15a
Skogaby	3.78 ± 0.88abc	0.11 ± 0.04ab	0.15 ± 0.03a	0.71 ± 0.38a	0.19 ± 0.10a	0.14 ± 0.06b
Åheden	2.92 ± 0.54a	0.13 ± 0.07ab	0.37 ± 0.19a	1.28 ± 0.85a	0.36 ± 0.13a	0.27 ± 0.09ab
Picea abies, 1-year-old needles						
M di Mezzo	3.89 ± 1.38ab	0.13 ± 0.04ab	0.22 ± 0.13a	0.69 ± 0.31a	0.97 ± 0.27b	0.12 ± 0.04a
Aubure	2.82 ± 0.60a	0.10 ± 0.02a	0.19 ± 0.03a	0.54 ± 0.15a	0.28 ± 0.01a	0.09 ± 0.02a
Nacetin	4.59 ± 0.34bc	0.33 ± 0.04c	0.17 ± 0.02a	0.59 ± 0.10a	0.51 ± 0.06ab	0.13 ± 0.01a
Schacht	4.16 ± 1.44bc	0.16 ± 0.06ab	0.24 ± 0.04a	0.64 ± 0.21a	0.40 ± 0.18a	0.12 ± 0.02a
Waldstein	4.53 ± 1.16bc	0.16 ± 0.06ab	0.21 ± 0.06a	0.62 ± 0.25a	0.54 ± 0.16ab	0.09 ± 0.04a
Klosterhede	5.01 ± 0.83bc	0.22 ± 0.03b	0.24 ± 0.05a	1.01 ± 0.19a	0.79 ± 0.26ab	0.43 ± 0.12b
Skogaby	4.56 ± 1.72bc	0.13 ± 0.02ab	0.12 ± 0.03a	0.50 ± 0.07a	0.28 ± 0.05a	0.14 ± 0.03a
Åheden	2.42 ± 0.47a	0.10 ± 0.02a	0.23 ± 0.07a	0.77 ± 0.18a	0.56 ± 0.14ab	0.21 ± 0.05a

Pinus pinaster, Thezan					
0-year-old	46.8 ± 11.3	3.06 ± 0.82	1.73 ± 0.52	12.2 ± 3.0	5.66 ± 3.02	6.31 ± 2.47
1-year-old	49.3 ± 23.2	3.55 ± 1.23	1.68 ± 0.57	10.3 ± 3.4	12.19 ± 4.06	9.94 ± 1.76
Pinus sylvestris, Åheden					
0-year-old	9.06 ± 1.90	0.30 ± 0.08	0.55 ± 0.12	1.84 ± 0.32	0.38 ± 0.18	0.45 ± 0.15
1-year-old	5.54 ± 0.57	0.19 ± 0.01	0.31 ± 0.03	1.00 ± 0.10	0.45 ± 0.20	0.28 ± 0.08
Fagus sylvatica					
Collelongo	164.7 ± 27.4a	4.78 ± 0.09a	3.95 ± 1.23b	15.4 ± 2.4b	17.6 ± 2.4b	6.2 ± 0.8b
Aubure	216.0 ± 46.8a	5.94 ± 0.77a	6.54 ± 2.31c	25.6 ± 7.6c	8.2 ± 2.2a	3.2 ± 0.5a
Schacht	142.4 ± 26.2a	3.45 ± 0.65b	3.18 ± 1.56ab	7.0 ± 1.9a	7.8 ± 1.0a	1.8 ± 0.4a
Waldstein	73.5 ± 4.2b	2.17 ± 0.23c	1.81 ± 0.92a	5.0 ± 1.1a	6.1 ± 0.6a	2.0 ± 0.6a
Gribskov	242.2 ± 34.4a	6.37 ± 0.80a	5.86 ± 1.46bc	21.9 ± 3.2c	22.2 ± 6.0b	8.2 ± 3.4b
Betula pubescens					
Åheden	86.5 ± 5.7	2.57 ± 0.21	4.27 ± 0.39	14.7 ± 2.3	11.3 ± 0.7	4.8 ± 0.6

Values are the means of $n = 4$ repetitions ± SD. Numbers in a single column followed by different letters are significantly different at the $p = 0.05$ level.

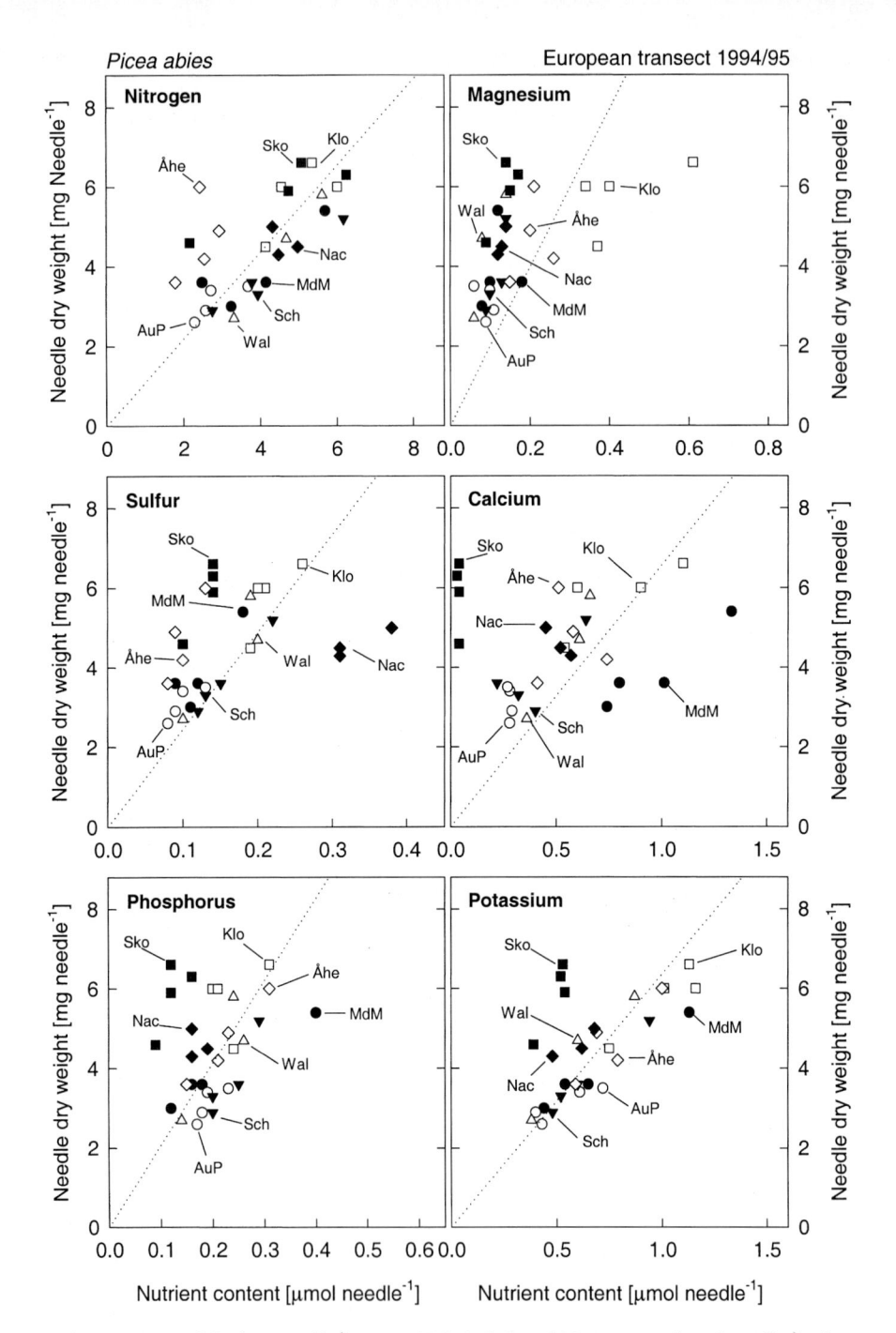

Fig. 4.1. Dry weight (mg needle^{-1}) versus N, P, S, C, P and Mg contents (μmol needle^{-1}) of 1-year-old *Picea abies* needles. The *dotted line* represents the correlation between the two parameters for fully developed needles. *Data points above this line* indicate accumulation of carbon whereas *data points below this line* indicate accumulation of each element. Each data point represents a measurement of one individual tree (n = 4 per site), at eight Norway spruce stands along the European transect: *It* Monte di Mezzo; *F* Aubure; *Ger$_S$* Schacht; *Ger$_W$* Waldstein; *Dk* Klosterhede; *S-Sw* Skogaby; *N-Sw* Åheden; *Cz* Nacetin

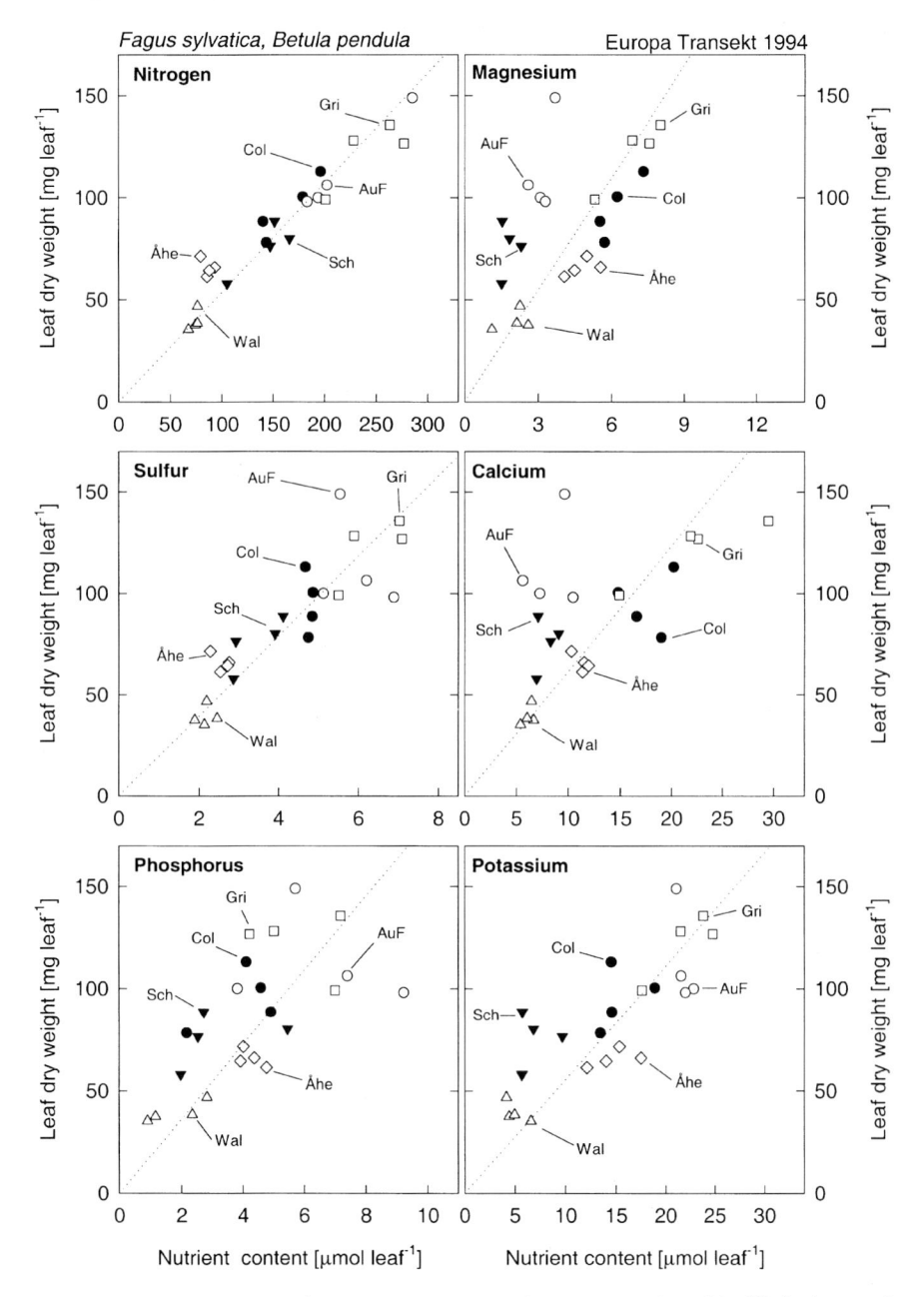

Linking Plant Nutrition and Ecosystem Processes 81

Fig. 4.2. Dry weight (mg leaf^{-1}) versus N, P, S, C, P and Mg contents (µmol leaf^{-1}) for leaves of *Fagus sylvatica* and *Betula pendula* (northern Sweden, Åheden). The *dotted line* represents the correlation between the two parameters for fully developed needles. *Data points above this line* indicate accumulation of carbon whereas *data points below this line* indicate accumulation of each element. *Each data point* represents a measurement of one individual tree (n = 4 per site), at six stands and along the European transect: *It* Collelonge; *F* Aubure; *Ger$_S$* Schacht; *Ger$_W$* Waldstein; *Dk* Gribskov; *N-Sw* Åheden

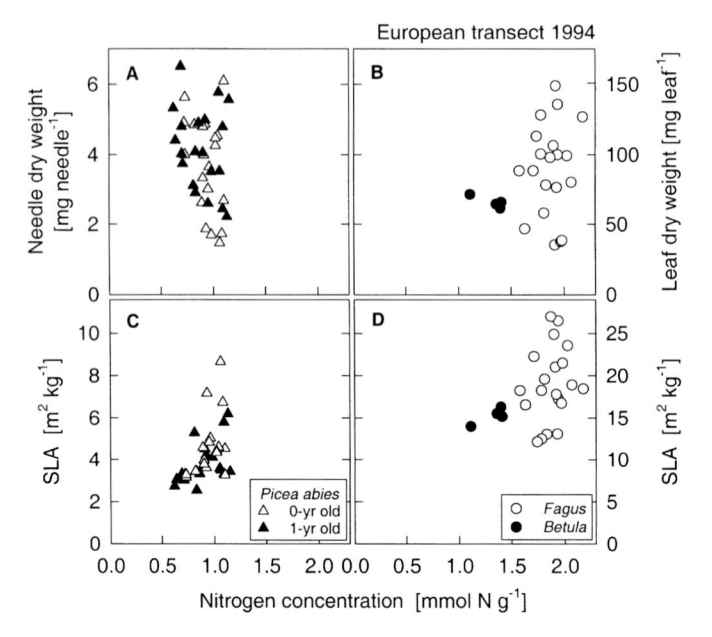

Fig. 4.3A–D. Needle and leaf dry weight (mg needle^{-1} and mg leaf^{-1}, respectively), and specific leaf area (SLA, m^2 kg^{-1}) versus nitrogen concentration (mmol N g^{-1}) of 0- and 1-year-old needles of *Picea abies* and leaves of *Fagus sylvatica* and *Betula pendula*

Vol.). Additionally to N limitation, the climatic stress of cold winters in northern Sweden and summer drought in Italy caused needles and leaves at these two sites to be scleromorphic. This can be seen in the much higher specific leaf area (SLA, m^2 kg^{-1}) for Norway spruce and beech in Central Europe (lower needle and leaf dry weight), whereas needles and leaves in Italy and northern Sweden become heavier due to accumulation of carbon. However, despite these consequences for leaf morphology due to environmental conditions, there is no significant correlation of leaf dry weight or SLA with foliage N concentrations for Norway spruce and European beech (Fig. 4.3). Specific leaf area and leaf dry weight changed independently of N concentration. This indicates for both species that the photosynthetically active leaf area is not regulated via the N concentration of needles and leaves. Instead, the strong proportionality between N content and leaf dry weight (Figs. 4.1 and 4.2) resulted in a concomittant increase in leaf area (Fig. 4.4). Thus, the much higher amount of N per leaf at Nacetin, Schacht, Waldstein and Klosterhede is likely to support a higher C assimilation and thus a higher overall stand growth (Stitt and Schulze 1994).

4.5 Nitrogen Partitioning in Different Tree Compartments

4.5.1 Nitrate

The N contents per needle and leaf indicated that nitrogen accumulated in the leaves at those sites which receive high inputs of atmospheric N without a proportional

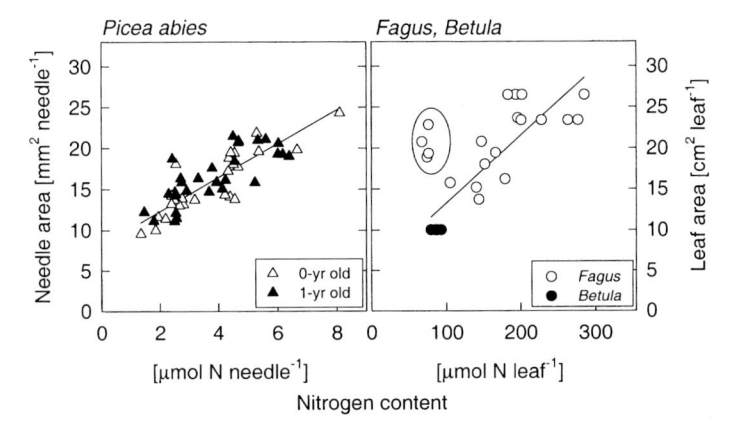

Fig. 4.4. Needle and leaf area (mm² needle⁻¹ and cm² leaf⁻¹, respectively) versus nitrogen content (μmol N needle⁻¹ μmol N leaf⁻¹) of 0- and 1-year-old needles of *Picea abies* and leaves of *Fagus sylvatica* and *Betula pendula*

increase in other elements. Although there was a generally linear trend between N and dry weight, there were sites which deviated from this trend, with North Sweden above and Waldstein below this line. Therefore, it is not certain whether the measured amount of N is essential for leaf growth at the particular sites, because in Sweden and Italy, for example, leaves were in the same range of dry weight at much lower N contents. We may interpret this result to mean that the stands at Schacht, Waldstein, Nacetin and Aubure accumulate N in excess of the demand for leaf growth. The question remains as to the form and physiological function of this N surplus in the foliage. Nitrate is known not to be a major storage form of N in trees (Pate 1983). Also in the present study we found only traces of nitrate in plant tissues, which physiologically were insignificant. Although soil nitrate availability and atmospheric N deposition varied strongly across the transect (see Persson et al., Chap. 2, this Vol.), no accumulation of nitrate for any given species and plant compartment was observed. Tissue nitrate concentrations remained below 1% of total nitrogen (Bauer 1997), which supports results from other studies stating that in trees the amount and composition of the amino acid pool is much more responsive and ultimately is physiologically more important than accumulated nitrate. If nitrate is being taken up, it appears to enter immediately into the nitrate reductase pathway.

4.5.2 Amino Acids

Amino acid analysis has proven to be a very indicative measure of nutritional disturbance across a wide selection of tree species (e.g. van Dijk and Roelofs 1988; Gezelius and Naesholm 1993; Huhn and Schulz 1996). The primary response of the trees in these investigations was that the concentration of the free amino acid (FAA) pool increased, either as a result of excess N availability (Perez-Soba and Van der Eerden 1993; Schneider et al. 1996) or due to higher supply of other major nutrients (Perez-Soba and de Visser 1994). Also in our study we found an increase in FAA concentration, but in contrast to other investigations, the major changes across the

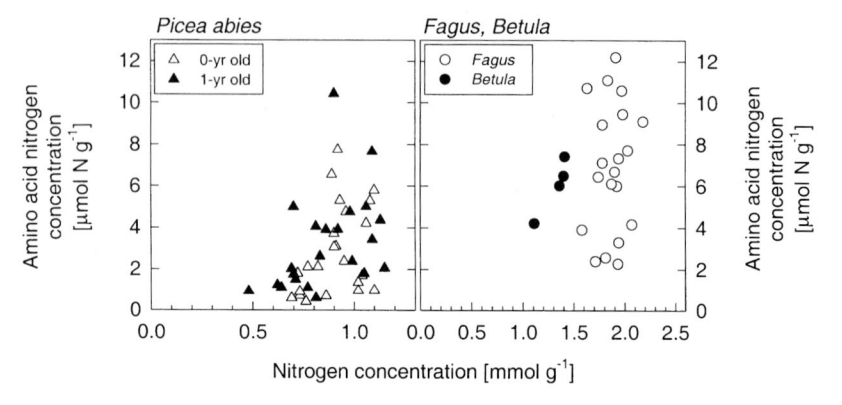

Fig. 4.5. Nitrogen concentration (mmol N g^{-1}) versus amino acid nitrogen (µmol N g^{-1}) in 0- and 1-year-old needles of *Picea abies* and leaves of *Fagus sylvatica* and *Betula pendula*

transect occurred within the composition of the leaf N pool without any affect on leaf N concentrations. Given the low contribution of amino-nitrogen to total N concentration of less than 1% in leaves and needles (Bauer 1997), this result confirms the role of amino acids as primary physiological signals rather than being the prevailing form of N storage. For Norway spruce as well as for European beech, the amino acid-nitrogen changed across the transect at constant N concentration (Fig. 4.5), and thus seems to be a much more sensitive indicator for plant N status than does dry-weight-based N concentration.

In needles of Norway spruce the concentration of total free amino acids (FAA) at Åheden and Monte di Mezzo was significantly lower than at sites in mid-Europe (Fig. 4.6). The arginine (Arg) concentrations partly mirrored this pattern for the 0- and 1-year-old needles, but generally Arg concentrations in needles were lower by a factor of 10 than those of the FAA (note different scales for needle FAA and Arg in Fig. 4.6). Apart from the needles, the FAA concentrations in twigs and bark increased by a factor of more than 10, for arginine they increased by a factor of more than 100. However, large variation in amino acid concentrations in the twigs obscured differences between the sites. In beech and birch the differences in the FAA concentration between leaves and twigs were smaller than in Norway spruce (Fig. 4.7). Arginine increased by a factor of 100 from leaves to twigs and dropped again by a factor of about 10 in the bark of 1-year-old twigs. Also, the differences between the sites were smaller. In part, these results support the hypothesis that increased arginine concentrations in trees indicate growth under excess input of N (Perez-Soba and Van der Eerden 1993; Perez-Soba et al. 1994; Naesholm et al. 1994), but given the large difference between needles and twigs, it seems much more likely that Arg concentrations in the foliage are more indicative of excess N at the lower range of supply, whereas at high supply Arg becomes more important as a cycling N form throughout the tree.

Among the other amino acids glutamine (Gln), glutamate (Glu) and aspartate (Asp) were the dominant forms in spruce, beech and birch (Figs. 4.8 and 4.9), but, in contrast to the FAA and arginine, their concentrations in foliage and twigs remained in the same order of magnitude. In Norway spruce needles Gln and Glu were highest

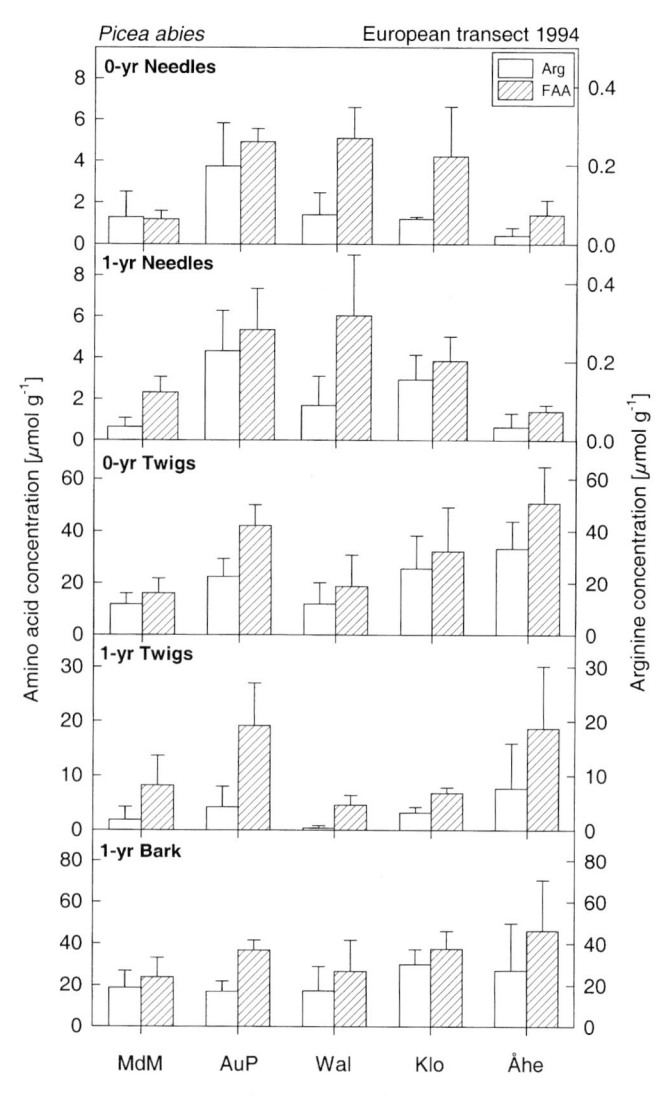

Fig. 4.6. Concentration of total amino acids (free amino acid pool, FAA) and arginine (µmol g⁻¹) in needles and twigs of *Picea abies* on five stands along the European transect. Values are the means of $n = 4$ repetitions ± SD. Note different scales for the y-axis for FAA and arginine in needles

at Aubure, Waldstein and Klosterhede with no apparent change in Asp across the transect (Fig. 4.8). In twigs and bark, glutamate concentration was much lower than in the needles. Also, the geographical pattern changed. In 1-year-old and partly in 0-year-old twigs Gln, Glu and Asp were higher at Monte di Mezzo and Åheden. This change in the predominant form from Glu to Gln, along with the change in the geographic pattern, indicates that in Italy and northern Sweden glutamine, the primary

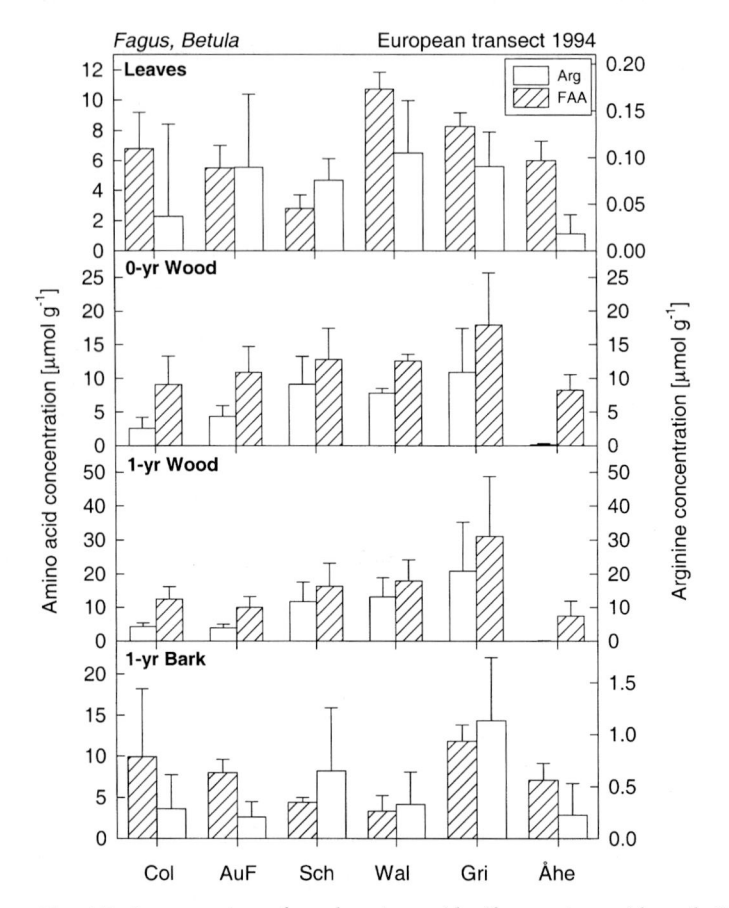

Fig. 4.7. Concentration of total amino acids (free amino acid pool, FAA) and arginine (µmol g⁻¹) in leaves and twigs of *Fagus sylvatica* and *Betula pendula* on six stands along the European transect. Values are the means of $n = 4$ repetitions ± SD. Note different scales for the y-axis for FAA and arginine in leaves and 1-year-old bark

product of ammonium synthesis and main transport form (Pate 1983; Oaks 1994), is transported to the canopy in order to supply needle growth with C and N. Glutamate, the primary acceptor of ammonium, accumulated in the needles in France, Germany and Denmark, indicating a high assimilatory activity in the canopy.

In the deciduous trees glutamate was the prevailing amino acid in the leaves, followed by glutamine and aspartate (Fig. 4.9). In contrast to Norway spruce, there was a significant difference between the low Glu concentration at the Schacht and the other European beech stands. Also, Glu concentrations decreased, whereas Gln levels increased in the twigs. In the bark of 1-year-old twigs, Glu increased again at the sites in Italy, Denmark and northern Sweden. Thus, following the interpretation of the amino acid pattern in Norway spruce, this indicates that Gln, Glu and Asp are transported to the crown, where they are used for further synthesis of organic compounds.

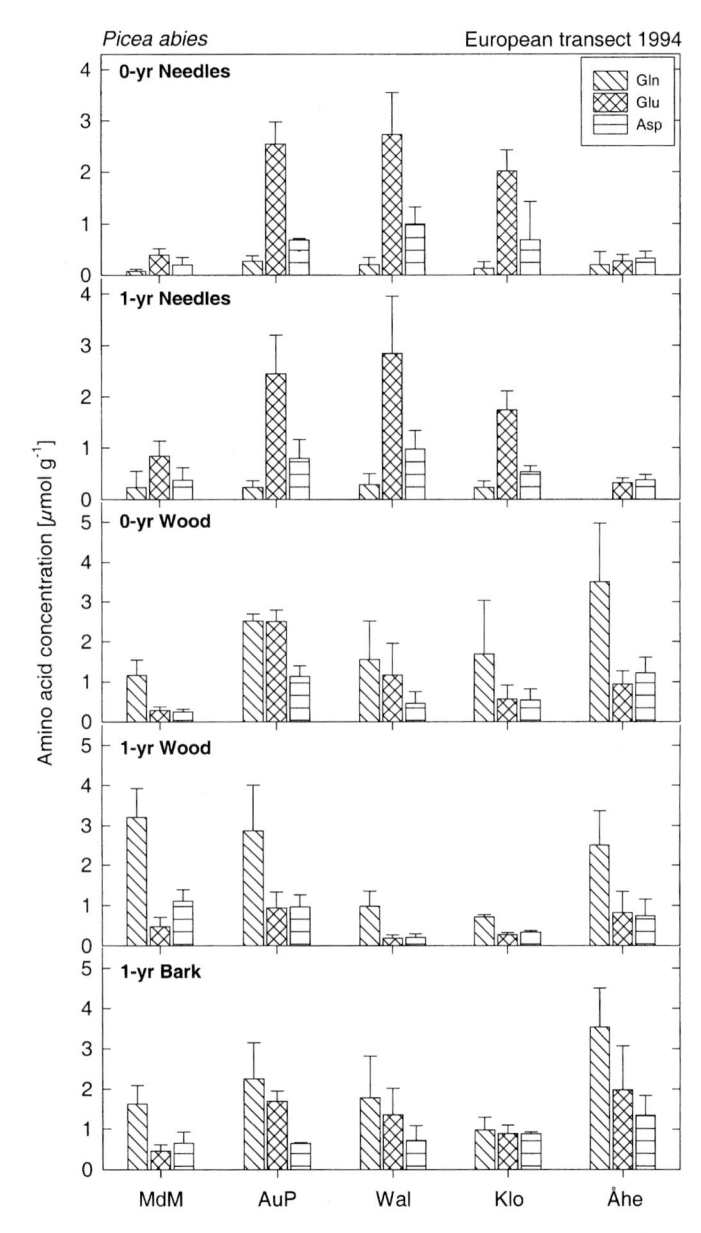

Fig. 4.8. Concentrations of glutamine (*Gln*), glutamate (*Glu*) and aspartate (*Asp*, µmol g^{-1}) in needles and twigs of *Picea abies* on five stands along the European transect. Values are the means of $n = 4$ repetitions ± SD

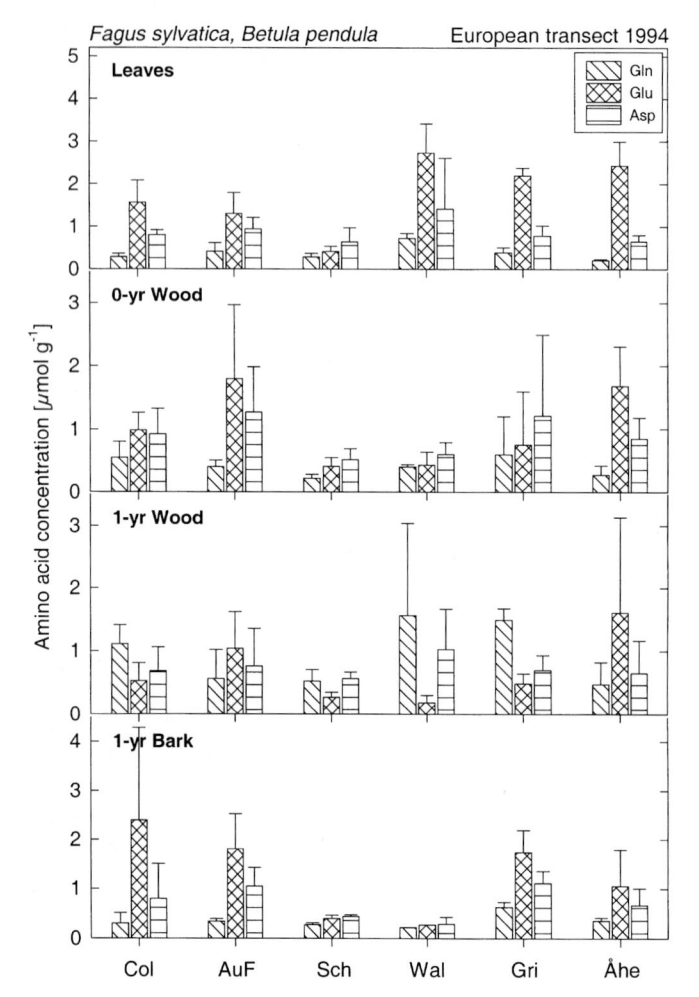

Fig. 4.9. Concentrations of glutamine (*Gln*), glutamate (*Glu*) and aspartate (*Asp*, µmol g^{-1}) in leaves and twigs of *Fagus sylvatica* and *Betula pendula* on six stands along the European transect. Values are the means of $n = 4$ repetitions ± SD

4.6 Ecosystem C and N Pools

Stand net primary productivity and biomass as determined by destructive harvest showed in part an opposite pattern across the European transect (Table 4.7). Aboveground net primary production (ANPP; kg m^{-2} a^{-1}) was highest at Jezeri for beech and at Skogaby for spruce, whereas stand biomass for beech was highest at Hilleroed and for spruce at Waldstein. Most of the forest stands which were investigated in this study were old-growth forests. Thus, biomass as well as C and N pools probably reflect past management regimes and soil conditions. However, from the analysis of leaf nutrient contents (Figs. 4.1 and 4.2) it emerges that measured rates of ANPP are ultimately linked to ample supply of major plant nutrients under present conditions.

As a consequence of the biomass distribution, the nitrogen (N) pools of the tree layer showed the same pattern, with highest N pools for beech in Gribskov (1.142 t N ha^{-1}) and for spruce at Nacetin (0.917 t N ha^{-1}). Nitrogen pools in forest biomass generally were much lower than in the humus layer (LFH layer, Table 4.7). However, tree N pools at Aubure and Gribskov were at or even above the amount of N in the forest floor. This probably reflects litter removal, which has been a traditional management regime in that region for decades. Soils in Italy had the highest C and N content on an area basis (Table 4.7, Fig. 4.10). The large soil pools at Monte di Mezzo appear to result from the fact that this stand was planted on a former pasture and still is surrounded by actively used pasture. The very high C and N pools at the beech stand in Collelongo are more likely related to the high altitude (elevation of 1500 m asl), which results in a much slower decomposition at this remote site.

In contrast to nitrogen, carbon pools in the soils and forest biomass (Table 4.7) displayed a pattern opposite to what has been reported for other forest ecosystems (Waring and Schlesinger 1985; Turner et al. 1995; Schlesinger 1997). Carbon bound in the forest biomass of all the stands clearly exceeded the amount of C in the humus layer. More striking is the fact that at Aubure (beech and spruce), Jezeri and Gribskov (beech only), carbon pools in the forest biomass are even higher when compared to the soil C pool down to 50 cm of depth. Additionally, there are sites where the C pool in the forest biomass was close to the amount of C bound in the soil (Fig. 4.10). The boreal climate in northern Sweden largely explains a slow soil C turnover. On the other hand, it emerged that high amounts of C and N in the forest biomass occurred at sites with a good nutrient supply, e.g. at Gribskov, or in combination with high N inputs from the atmosphere, e.g. at Jezeri or Schacht in central Europe.

In most cases where forest growth has been studied, primary production ceased after the stand had reached a peak in leaf area index (Gower et al. 1996; Ryan et al. 1997). However, ANPP of the oldest stands across the European transect was equal to

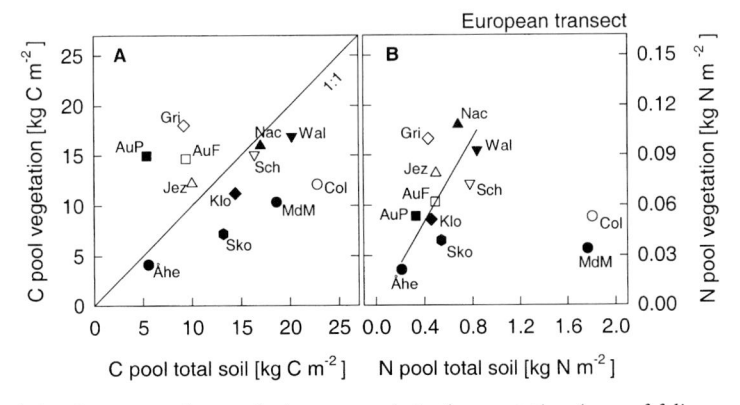

Fig. 4.10A,B. Relation between carbon and nitrogen pools in the vegetation (sum of foliage, branches, stems and coarse roots) and in the soil (whole depth profile LFH-50 cm) for spruce and beech stands across the European transect

Table 4.7. Annual dry matter production (kg m^{-2} a^{-1}), tree biomass (kg m^{-2}) and vegetation and soil C and N pools (kg m^{-2} and to ha^{-1}, respectively) for *Picea abies* and *Fagus sylvatica* stands across the European transect

	Beech stands				
	Collelongo (90)	Aubure (150)	Jezeri (78)	Schacht (120)	Gribskov (120)
Productivity (kg m^{-2} a^{-1})					
Foliage	0.28	0.28	0.37	0.35	0.47
Stem	0.28	0.20[b]	0.42	0.20[b]	0.20[b]
Branches + twigs	0.31	0.22	0.46	0.22	0.22
ANPP	0.87	0.70	1.25	0.77	0.89
Standing biomass (kg m^{-2})					
Foliage	0.28	0.28[c]	0.37	0.35	0.47
Stem	17.42	20.37	16.07	22.29	26.59
Branches + twigs	3.18	4.83	4.02	4.58	6.04
Coarse roots + stump	5.52	5.42	4.79	5.40[b]	7.08
Total	26.40	30.62	25.25	32.62	40.18
Nitrogen pools (t ha^{-1})					
Foliage	0.068	0.075	0.098	0.092	0.130
Stem	0.117	0.086	0.204	0.281	0.372
Branches + twigs	0.307	0.330	0.276	0.329	0.455
Coarse roots + stump	0.144	0.141	0.217	0.140	0.184
Total (trees)	0.636	0.631	0.794	0.842	1.142
L + F + H	2.390	0.500	2.170	1.760	1.230
Total (L-50 cm)	18.060	4.910	4.960	7.780	4.340
Carbon pools (kg m^{-2})					
Foliage	0.13	0.14	0.19	0.18	0.22
Stem	8.11	9.80	7.81	10.82	12.60
Branches + twigs	1.50	2.32	1.92	2.19	2.87
Coarse roots + stump	2.57	2.61	2.33	2.63	3.36
Total (trees)	12.31	14.87	12.24	15.82	19.06
L + F + H	3.45	1.11	4.44	3.59	2.45
Total (L-50 cm)	22.79	9.40	10.00	16.36	9.31

[a] Data given in the table account for spruce only. [b] Assumption based on literature comparison for stands in the same age and growing in the same geographic region. [c] Value represents leaf and twig biomass. [d] According Nilsson and Wiklund (1992). [e] Nitrogen pool for the stem at Monte di Mezzo was calculated based on the average N concentration in sapwood across the transect (1.03 mg N g^{-1} dw).

or even higher than the productivity of the younger stands (Table 4.7). The 145-year-old spruce stand at Waldstein had a much higher ANPP despite lower needle cation concentrations compared, for example, to the well nutrient-supplied and younger spruce stands in Monte di Mezzo or Skogaby. Except for a significantly lower needle

Linking Plant Nutrition and Ecosystem Processes

Table 4.7. *Continued*

Spruce stands						
MdiM (38)	Aubure (100)	Nacetin (60)	Waldstein (145)	Klosterhede (80)	Skogaby (30)	Åheden[a] (145)
0.28	0.21	0.36	0.47	0.10	0.25[b]	0.09
0.45	0.30	0.36	0.33	0.60	0.52	0.17
0.36	0.22	0.27	0.23	0.44	0.39	0.11
1.09	0.73	0.99	1.03	1.14	1.16	0.37
1.68	0.92	1.33	1.66	1.50	1.40[d]	0.61
13.10	23.41	11.70	19.32	15.30	8.55	5.20
1.72	1.95	2.34	3.02	2.61	2.22	1.39
5.48	4.37	6.08	7.42	4.40	2.71	1.32
21.98	30.64	21.45	31.42	23.81	14.88	8.52
0.223	0.114	0.203	0.246	0.188	0.150	0.047
0.135[e]	0.276	0.173	0.162	0.150	0.062	0.051
0.089	0.076	0.178	0.254	0.068	0.103	0.080
0.153	0.069	0.363	0.208	0.123	0.076	0.037
0.601	0.535	0.917	0.870	0.530	0.390	0.215
2.550	0.620	2.330	2.640	1.830	1.080	0.400
17.670	3.290	6.810	8.390	4.570	5.400	2.100
0.84	0.46	0.68	0.84	0.74	0.66	0.30
6.16	11.46	5.69	9.61	7.35	4.13	2.52
0.86	0.95	1.15	1.56	1.28	1.09	0.68
2.58	2.14	2.96	3.68	2.14	1.31	0.64
10.44	15.00	10.48	15.69	11.51	7.18	4.14
3.52	1.60	5.73	5.90	5.56	3.11	1.49
18.63	5.32	16.98	21.07	14.40	13.18	5.54

N concentration in northern Sweden, which seems to limit forest growth, ANPP for spruce across the European transect changed independently of needle N concentration (Fig. 4.11A; $r^2 = 0.439$, $p = 0.105$), which is the opposite to what is known for annual or crop plants, where photosynthesis and, in turn, growth change in relation to leaf N concentration (Field and Mooney 1986; Evans 1993). Stand productivity for these forests appears to be under the control of yet another factor other than leaf nitrogen concentration. For both spruce and beech, ANPP also was only weakly related to canopy N content (g N m^{-2} leaf area; $r^2 = 0.676$, $p = 0.023$; Fig. 4.11B). However, ANPP showed a stronger relationship to the N content of suncrown needles

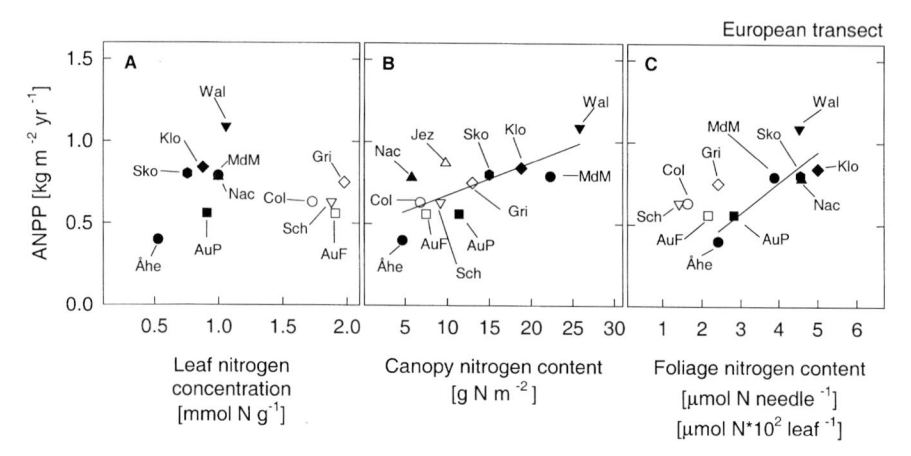

Fig. 4.11A–C. Above-ground net primary production (ANPP) of *Picea abies* and *Fagus sylvatica* in relation to **A** leaf N concentration (mmol N g^{-1}), **B** total foliage N pool (g N m^{-2}) and **C** leaf N content (µmol N needle^{-1} and (µmol N × 10^2 leaf^{-1} for spruce and beech, respectively). (Bauer et al. 1997)

(Fig. 4.11C; $r^2 = 0.818$, $p = 0.005$), indicating the higher contribution of the upper canopy to the overall carbon gain (Schulze et al. 1977). Needles and leaves showed a proportional increase in dry weight over a wide range of N contents (Figs. 4.1 and 4.2). In the same way, the capacity of the leaves to assimilate carbon might have increased, as indicated by the concomitant increase in leaf area (Fig. 4.4). These findings indicate that even old-growth forests such as those in the Ore Mountains (Czech Republic) or the Fichtelgebirge (Germany), which suffer from cation deficiency, maintain their level of productivity at high supply of nitrogen. In this situation it becomes clear that under excess supply of N, other nutrients like Mg or K, which are limiting, can constrain carbon assimilation at the plant and ecosystem level (Schulze 1989; Bazzaz 1997).

Across the European transect, roots of spruce and beech showed a lower uptake capacity for ammonium in central Europe than compared to northern Sweden (Högberg et al. 1998). Despite a constant and much lower uptake capacity for nitrate than for ammonium, root nitrate reduction was higher in central Europe than elsewhere along the transect. Thus, a high root uptake potential in central Europe correlates with high growth rates. However, ANPP of both species did not correlate with soil N pool (Fig. 4.12A) and reached a limit at high values for soil nitrogen. It is unclear whether the constant ANPP at high soil N is a result of limited uptake capacity of the roots (Högberg et al. 1998), a limitation due to lack of major cations or whether it is due to N losses to the groundwater table (Durka et al. 1994) resulting in a lower supply for uptake. In contrast to soil N pools, ANPP was more linearly related to soil carbon pools (Fig. 4.12B). In fact, increased N immobilisation by soil microorganisms (Zak et al. 1990), N uptake by the understorey vegetation or abiotic N immobilisation (Aber et al. 1998) could constrain the growth of trees. A comparison between N pools in the foliage of trees and *Deschampsia flexusoa* (Fig. 4.13) shows that at some of the sites the grass immobilised almost as much nitrogen in its leaves

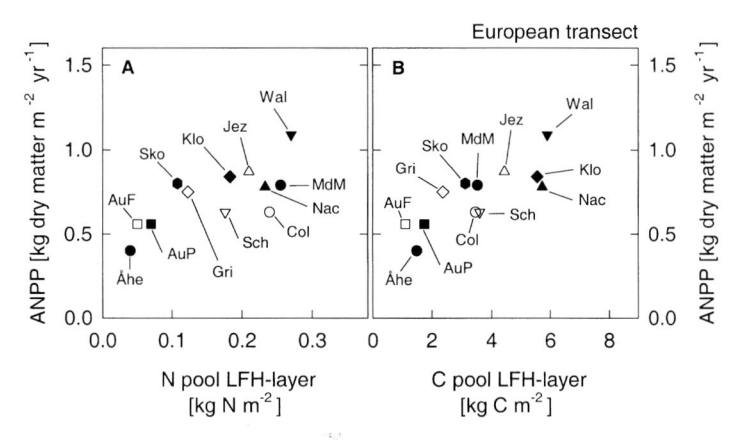

Linking Plant Nutrition and Ecosystem Processes

Fig. 4.12A,B. Above-ground net primary production (ANPP) of *Picea abies* and *Fagus sylvatica* forests across the European transect in relation to N and C pools in the humus layer (LFH layer only)

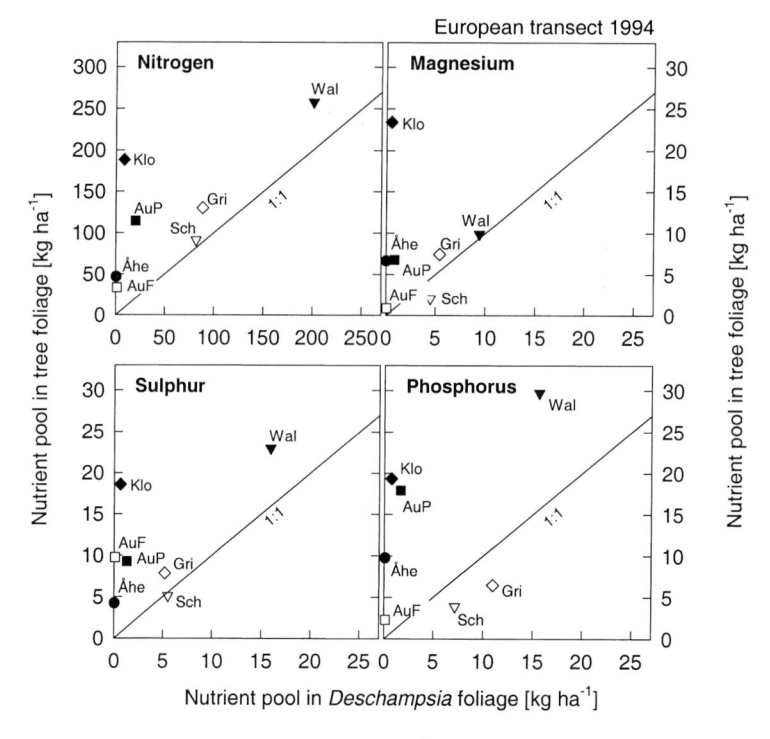

Fig. 4.13. Nitrogen, S, P and Mg pools in tree foliage (kg ha^{-1}) versus nutrient pools in the foliage of *Deschampsia flexuosa* (kg ha^{-1}) on three beech stands and four spruce stands along the European transect

as the trees. This was also evident in the case of Mg, S and P. Thus, nitrogen supply and, in turn, photosynthetic capacity for leaves and needles can partly be limited by strong competition for soil N with leaves of understorey plants.

The annual N use in above-ground biomass of the beech stands and two of the spruce stands (Nacetin and Waldstein) changed independently of mineralisation, whereas the low annual N use for most of the spruce stands appeared to result from low rates of net N mineralisation (Fig. 4.14A). The beech stands at Aubure, Jezeri and Gribskov had an annual N demand for growth well above the soil supply, indicating the importance of aerial N uptake. Also there were spruce stands which had an annual N use in the same range as the beech stands. Nitrogen productivity (g dry matter $g^{-1}N a^{-1}$), as the product of ANPP and N pools in the vegetation, for those sites with a high annual N demand is much lower than in the N-limited stands with proportional response to mineralisation rates (Fig. 4.14B). A low nitrogen productivity in this case means that N is used less efficiently for growth (Lambers et al. 1998), which in the case of the beech stands and some of the spruce stands points to the fact that there is probably more N present in the plant than is needed to support growth. Although ANPP of spruce was linearly related to leaf N contents (Fig. 4.11C), the saturating behaviour of ANPP at increasing soil N pool (Fig. 4.12A) clearly points toward a limitation of some other element. The above-ground N use in Fig. 4.14 represents a gross uptake, which does not consider the amount of remobilised N used for growth. Estimates for the amount of stored N used to support leaf growth can reach up to 90% of the total leaf N demand in broadleaf trees (Millard 1996). Thus, the relationship between above-ground N use, especially in beech and net N mineralisation, is likely to change in a net nitrogen balance. In contrast to spruce, the ANPP of beech seemed to be quite unresponsive to both leaf N content and net N mineralisation. Their similar age and the comparable leaf nutrient supply therefore are in line with smaller variations in ANPP (Table 4.7).

Across the European transect, sustained productivity of old-growth forests was in line with N accumulation, suggesting that atmospheric N input has to be regarded as

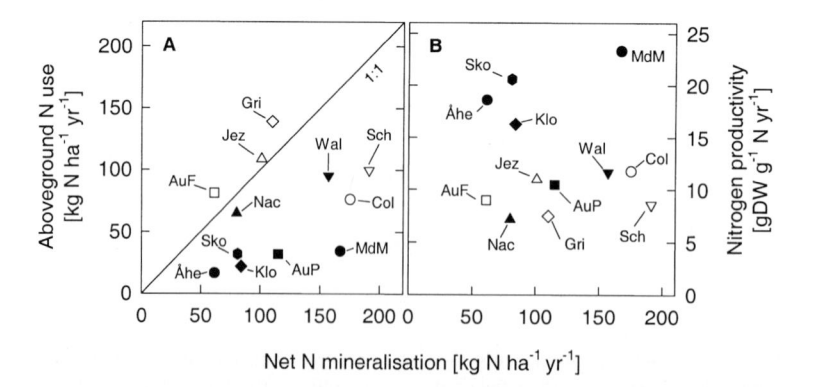

Fig. 4.14A,B. Above-ground nitrogen requirement (g N m-2 a^{-1}) and nitrogen productivity (NP, g $g^{-1} a^{-1}$) for spruce and beech stands across the European transect in relation to soil net N mineralisation rates (g N ha^{-1} day^{-1}) over the whole soil profile. Nitrogen productivity was calculated as the product of ANPP and N pool in the vegetation according Table 4.6

one key parameter leading to the enhanced growth. The close connection between the ecosystem N cycle and the C balance thus forced a stronger "withdrawal" of C into the system, which resulted in a close match of soil and vegetation C pools. However, in a long-term perspective, atmospheric N pollution does not seem to act as a cure against rising CO_2 concentrations (Nadelhoffer et al. 1999). From a nutritional point of view, it rather seems likely that regional and local variablity in nutrient supply has a strong impact on productivity. Moreover, competition between trees, soil microorganisms and understorey vegetation for limited nutrients other than N are likely to limit the forest's capacity for C sequestration.

4.7 Conclusions

- Nutrient concentrations in spruce and beech forests across the European transect were surprisingly constant and largely obscured a strong variability on a local scale. However, nutrient pools in the standing biomass and in the soil were highest in Central Europe. In addition to climatic conditions, there was evidence for a strong impact of past land use strategies in combination with atmospheric N deposition on plant and soil nutrient status.
- Major shifts in the amino acid pattern indicated more clearly that the plant internal N allocation is much more indicative of environmental disturbance and that accumulation of N at the stand level is responsible for a high ANPP in Central Europe.
- Despite nutritional imbalances, dry matter production and nutrient uptake proportionally increased in response to N deposition, with the effect that nutrient concentrations were kept at a constant level, which allowed the plants to maintain a constant C/N balance. As a result, old-growth spruce and beech forests maintained a high ANPP, probably due to aerial uptake of N.
- Increased competition for available soil resources between trees/mycorrhiza, understorey vegetation and soil microorganisms could be a possible reason for a limitation of forest ANPP. Thus, despite excess supply of nitrogen, strong resource competition among different life forms, in combination with low supply of other essential nutrients, is not likely to promote a long-term increase in forest carbon pools.

References

Aber JD, McDowell W, Nadelhoffer KJ, Magill A, Berntson GM, Kamakea M, McNulty S, Currie W, Rustad L, Fernandez I (1998) Nitrogen saturation in temperate forest ecosystems – hypothesis revisited. BioScience 48(11):921–934

Agren GI (1988) Ideal nutrient productivities and nutrient proportions in plant growth. Plant Cell Environ 11:613

Attiwill PM, Adams MA (1993) Nutrient cycling in forests. New Phytol 124:561–582

Bauer G (1997) Stickstoffhaushalt und Wachstum von Fichten- und Buchenwäldern entlang eines Europäischen Nord-Süd-Transektes. Bayreuther Forum Ökologie, Bd 53, Univ Bayreuth, Bitök, 176 pp

Bauer G, Mund M, Schulze E-D (1997) Nutrient contents and concentrations in relation to growth of *Picea abies* and *Fagus sylvatica* along a European transect. Tree Physiology 17:777–786

Bazzaz FA (1997) Allocation of resources in plants: state of the science and critical questions. In: Bazzaz FA, Grace J (eds) Plant resource allocation. Academic Press, London

Bormann FH, Likens GE (1979) Paltern and process in a forested ecosystem. Springer Berlin Heidelberg New York, 253 pp

Buchmann N, Gebauer G, Schulze E-D (1996) Partitioning of ^{15}N-labelled ammonium and nitrate among soil, litter, below- and above-ground biomass of trees and understory in a 15-year-old *Picea abies* plantation. Biogeochemistry 33:1–33

Chapin I FS, Schulze E-D, Mooney HA (1990) The ecology and economics of storage in plants. Annu Rev Ecol Syst 21:423–447

Durka W, Schulze E-D, Gebauer G, Voerkelius S (1994) Effects of forest decline on uptake and leaching of deposited nitrate determined from ^{15}N and ^{18}O measurements. Nature 372:765–767

Ellenberg H, Mayer R, Schauermann J (1986) Ökosystemforschung. Ergebnisse des Sollingprojektes 1966–1986. Eugen Ulmer, Stuttgart, 507 pp

Ericsson A, Norden L-G, Näsholm T, Walheim M (1993) Mineral Nutrient imbalances and arginine concentrations in needles of *Picea abies* (L.) Karst. from two areas with different levels of airborne deposition. Trees 8:67–74

Ericsson A, Walheim M, Norden L-G, Näsholm T (1995) Concentrations of Mineral Nutrients and orginine in needles of *Picea abies* trees from different areas in southern Sweden in relation to nitrogen deposition and humus form. Ecol Bull 44:147–157

Evans JR (1989) Photosynthesis and nitrogen relationships in leaves of C3 plants. Oecologia 78(1):9–19

Evans JR (1993) Photosynthetic acclimation and nitrogen partitioning within a Lucerne canopy. I. Canopy characteristics. Aust J Plant Physiol 20:55–67

Field C, Mooney HA (1986) The photosynthesis-nitrogen relationship in wild plants. In: Givinish TJ (ed) On the economy of plant form and function. Cambridge University Press, Cambridge, pp 25–55

Foster DR, Aber JD, Melillo JM, Bowden RD, Bazzaz FA (1997) Forest response to disturbance and anthropogenic stress. BioScience 47:437–445

Gezelius K, Naesholm T (1993) Free amino acids and protein in Scots pine seedlings cultivated at different nutrient availabilities. Tree Physiol 13:71–86

Gorham E, Vitousek PM, Reiners WA (1979) The regulation of chemical budgets over the course of terrestrial ecosystem succession. Ann Rev Ecol Syst 10:53–84

Gower ST, McMurtrie RE, Murty D (1996) Aboveground net primary production decline with stand age: potential causes. TREE 11:378–382

Högberg P, Högbom L, Schinkel H (1998) Nitrogen-related root variables of trees along an N-deposition gradient in Europe. Tree Physiol 18:823–828

Högbom L, Högberg P (1991) Nitrate nutrition of *Deschampia flexuosa* (L.) Trin. in relation to nitrogen deposition in Sweden. Oecologia 97:488–494

Huhn G, Schulz H (1996) Contents of free amino acids in Scots pine needles from field sites with different levels of nitrogen deposition. New Phytol 134:95–101

Ingestad T, Agren GI (1995) Plant nutrition and growth: basic principles. Plant Soil 168–169:15–20

Kaupenjohann M, Zech W, Hanschel R, Horn R, Schueider Bu (1989) Mineral nutrition of forest trees: a regional survey. In: Schulze E-D, Lange OL, Oren R (eds) Air pollution and forest decline. A study of spruce (*Picea abies*) on acid soils. Ecological studies, vol 77. Springer, Berlin Heidelberg New York, 475 pp

Kaye JP, Hart SC (1997) Competition for nitrogen between plants and soil microorganisms. TREE 12:139–143

Lambers H, Chapin FS III, Pons TL (1998) Plant physiological ecology. Springer, Berlin Heidelberg New York, 540 pp

Linder S (1987) Response to water and nutrients in coniferous ecosystems. In: Schulze E-D, Zwoelfer H (eds) Potentials and limitations of ecosystem analysis. Springer, Berlin Heidelberg New York, pp 180–202

Luo Y, Field CB, Mooney HA (1994) Predicting responses of photosynthesis and root fraction to elevated [CO2]a: interactions among carbon, nitrogen, and growth. Plant Cell Environ 17:1195-1204

Millard P (1996) Ecophysiology of the internal cycling of nitrogen for tree growth. Z Pflanzenernaehr Bodenkd 159:1-10

Miller HG (1986) Carbon × nutrient interactions – the limitations to productivity. Tree Physiol 2:373-385

Nadelhoffer KJ, Emmett BA, Gundersen P, Kjonaas OJ, Koopmans CJ, Schleppi P, Tietema A, Wright RF (1999) Nitrogen deposition makes a minor contribution to carbon sequestration in temperate forests. Nature 398:145-148

Näsholm T, Edfast AB, Ericsson A, Norden LG (1994) Accumulation of amino acids in some boreal forest plants in response to increased nitrogen availability. New Phytol 126:137-143

Nilsson L-O, Wiklund K (1992) Influence of nutrient and water stress on Norway spruce production in south Sweden – the role of air pollutants. Tree Physiol 9:185-207

Oaks A (1994) Primary nitrogen assimilation in higher plants and its regulation. Can J Bot – Rev Can Bot 72:739-750

Oren R, Schulze E-D (1989) Nutritional disharmony and forest decline: a conceptual model. In: Schulze E-D, Lange OL, Oren R (eds) Forest decline and air pollution. A study of spruce (*Picea abies*) on acid soils. Ecological studies, vol 77. Springer, Berlin Heidelberg New York, 475 pp

Paces T (1985) Sources of acidification in central europe estimated from elemental budgets in small basins. Nature 315:31-36

Pate JS (1983) Distribution of metabolites. In: Steward FC (ed) Plant physiology – a treatise. Academic Press, London, 335 pp

Perez-Soba M, de Visser PHB (1994) Nitrogen metabolism of Douglas fir and Scots pine as affected by optimal nutrition and water supply under conditions of relatively high atmospheric deposition. Trees 9:19-25

Perez-Soba M, van der Eerden LJM (1993) Nitrogen uptake in needles of Scots pine (*Pinus sylvestris* L.) when exposed to gaseous ammonia and ammonium fertiliser in the soil. Plant Soil 153:231-242

Perez-Soba M, Stulen I, van der Eerden LJM (1994) Effect of atmospheric ammonia on the nitrogen metabolism of Scots pine (*Pinus sylvestris*) Needles. Physiol Plant 90:629-636

Rehfuess KE, Rodenkirchen H (1984) Über die Nadelröte der Fichte (*Picea abies* Karst.) in Süddeutschland. Forstwiss Centralbl 103:245-262

Ryan MG, Binkley D, Fownes JH (1997) Age-related decline in forest productivity: patterns and process. In: Begon M, Fitter AH (eds) Advances in ecological research. Academic Press, London, pp 213-262

Schlesiner WH (1997) Biogeochemistry. An analysis of global change. Academic Press, London, 588 pp

Schneider S, Gessler A, Weber P, von Sengbusch D, Hanemann U, Rennenberg H (1996) Soluble N compounds in trees exposed to high loads of N: A comparison of spruce (*Picea abies*) and beech (*Fagus sylvatica*) grown under field conditions. New Phytol 134:103-114

Schramel P, Wolf A, Seif R, Klose BJ (1980) Eine neue Apparatur zur Druckveraschung von biologischem Material. Fresenius Z Anal Chem 302:62-64

Schulze E-D (1989) Air pollution and forest decline in a spruce (*Picea abies*) forest. Science 244:776-783

Schulze E-D, Fuchs M, Fuchs M (1977) Spacial distribution of photosynthetic capacity and performance in a mountain spruce forest of northern Germany. Oecologia 30:239-248

Stitt M, Schulze E-D (1994) Plant growth, storage and resource allocation: from flux control in a metabolic chain to the whole-plant level. In: Schulze E-D (ed) Flux control in biological systems. From enzymes to populations and ecosystems. Academic Press, London, 494 pp

Timmer VR, Armstrong G (1987) Diagnosing nutritional status of containerised tree seedlings: comparative plant analyses. Soil Sci Soc Am J 51:1082-1086

Turner DP, Koerper GJ, Harmon ME, Lee JL (1995) A carbon budget for forests of the conterminous United States. Ecol Appl 5(2):421–436

van Dijk HFG, Roelofs JGM (1988) Effects of exessive ammonium deposition on the nutritional status and condition of pine needles Physiol Plant 73:494–501

Vitousek PM, Aber JD, Howarth RW, Likens GE, Matson PA, Schindler DW, Schlesinger WH, Tilman DG (1997) Human alteration of the global nitrogen cycle: sources and consequences. Ecol Appl 7:737–750

Vogt KA, Persson H (1991) Root methods. In: Lassoie JP, Hinckley TM (eds) Techniques and Approaches in Forest Tree Physiology. CRC Press, Boca Raton, pp 477–502

Waring RH, Schlesinger WH (1985) Forest ecosystems. Academic Press, Orlando

Weber P, Stoermer H, Gessler A, Schneider S, von Sengbusch D, Hanemann U, Rennenberg H (1998) Metabolic responses of Norway spruce (*Picea abies*) trees to long-term forest management practices and acute $(NH_4)_2SO_4$ fertilisation: transport of soluble non-protein nitrogen compounds in xylem and phloem. New Phytol 140:401–485

Zak DR, Groffman PM, Pregitzer KS, Christensen S, Tiedje JM (1990) The vernal dam: plant-microbe competition for nitrogen in northern hardwood forests. Ecology 71:651–656

5 Root Growth and Response to Nitrogen

C. STOBER, E. GEORGE, and H. PERSSON

5.1 Introduction

Environmental conditions and nutrient supply do not only affect the above-ground growth of forest trees, but also the below-ground growth. In model experiments, N additions to soil can result in both increases and decreases in root dry weight or root length. More consistently, in most cases a decrease in the root/shoot biomass ratio is observed as a result of increased N supply in soil (George and Seith 1998; see also Chapin III et al. 1987). Some nitrogen fertilisation experiments suggest that much of the increased above-ground production may be due to a carbon translocation from below-ground to above-ground parts (Linder and Axelsson 1982). Although this change in carbon allocation is not necessarily harmful to the tree, on low-nutrient soils a nutrient imbalance may occur in the tree as a consequence of the decreased root/shoot ratio, and this may be one of the factors causing forest decline symptoms. Such effects can also be studied in the field in N-fertilisation experiments (Ahlström et al. 1988; Persson et al. 1995a) or N-addition and removal experiments (Clemensson-Lindell and Persson 1995; Persson et al. 1998). In the present experiment, we compared root growth of Norway spruce (*Picea abies*) and European beech (*Fagus sylvatica*) at different sites with contrasting climate and N deposition.

Depth distribution of roots in soil can best be determined with the soil coring method (Vogt and Persson 1991). Due to a high spatial variability of established fine roots, many replicated samples must be analysed to obtain representative values for a site, so that comparative observations can often be made at a small number of sites and at few instances during the growing season only. Therefore, other methods are also used to determine root growth and biomass in forests. For example, in the in-growth core method, root growth into initially root-free soil is quantified. This results in a measure of fine-root growth potential at a certain forest site or in treatment areas (Persson and Ahlström 1994). Other methods quantify root growth by observing roots along transparent interfaces in soil, for example in the minirhizotron method (Joslin and Wolfe 1999).

A spatially restricted nutrient supply in soil differs in its effect on root growth and activity from the effects of a homogeneous supply of the same nutrients (George et al. 1997). A locally enhanced supply of N in soil can lead to increased proliferation of tree roots in the N-rich soil patch. In this respect, a local N supply has the same effect on root growth as has been shown for a local P supply in soil (Jackson and Caldwell 1996). Model experiments indicate that this increased root proliferation in the N-rich soil patches occurs only when the tree is deficient in N (Friend et al. 1990). In the present experiment, the substrate in the in-growth cores was either not fertilised or fertilised with N in a slow-release form. It was hypothesised that root proliferation in the N-rich soil patch occurs in low-N forests (e.g. northern Sweden), but not in the central European forests with higher N deposition and higher soil N supply (see Persson et al., Chap. 2, this Vol.).

Fine-root growth of forest trees is determined not only by the nutrient supply or toxic elements in soil, but also by soil temperature (Steele et al. 1997; Hahn and Marschner 1998) and soil water conditions during the growing season (Santantonio and Hermann 1985). Seasonal changes in growth dynamics of fine roots generally reflect changes in environmental conditions and may be followed by non-destructive methods such as root windows or rhizotrons. Minirhizotrons have been used frequently for that purpose in the past two decades (e.g. Majdi and Persson 1995). In the present study, root windows were installed in different European forest sites. In these root windows, a Plexiglas plate is pressed against a soil profile. This allows repeated observations of root growth along the plate. Sequential soil coring adjacent to the root windows was carried out in order to obtain additional data on fine-root dry weight and length from the same sampling period. The data from soil coring were also used in order to describe the depth distribution of the fine roots.

The aims of the present study were:

1. to compare fine-root growth and root length at several European forest sites differing in climatic conditions, N deposition and dominating tree species;
2. to describe differences in seasonal root growth pattern between the different sites;
3. to test the effect of a local supply of N on fine-root growth at the different sites; and
4. to compare several methods to estimate root growth activity.

5.2 Approaches to the Study of Root Growth

5.2.1 Soil Coring

Destructive root sampling was performed by collecting soil cores with the help of a cylindrical steel corer, 4.5 cm in diameter (Vogt and Persson 1991). In Aubure (European beech and Norway spruce), sets of two soil cores (depth about 40 cm in the mineral soil) were taken on five sampling occasions during 1996–1997 at distances of 1.5, 2.5, 3.5 and 4.5 m from the chosen sampling tree with a root window (see below). The two soil cores were taken at 90° from each other on a radius around the tree and were pooled to give one sample. The horizontal distribution pattern of fine roots indicated no dependency on the distance to the nearest tree. Therefore the root data are presented for the total stand, disregarding the importance of small-scale distribution patterns in relation to the sampling trees. The same sampling pattern was carried out in Schacht (European beech) and Skogaby and Waldstein (Norway spruce), but on these sites only one sampling was completed (in 1997). The sampling was carried out in the same way as described for Aubure.

The core samples were divided into 0–2.5-, 2.5–5-, 5–10-, 10–20-, 20–30- and 30–40-cm segments from the top of the organic soil horizon and penetrating into the mineral-soil horizons. Thus, the litter layer was excluded, but O_f and O_h horizons (see Persson et al., Chap. 2, this Vol.) were included in the measurements. The soil samples were transported as soon as possible to the laboratory and stored in plastic bags in a deep freeze at −4 °C (a temperature that did not damage the live root tissue and caused no changes in ion concentrations; cf. Clemensson-Lindell and Persson 1992) until sorting could take place.

Immediately after thawing, roots were picked out from the plastic bags from each soil layer and sorted. To distinguish biomass (i.e. living biomass) from necromass, fine roots <1 mm in diameter were separated into live and dead categories, based on morphological characteristics (Vogt and Persson 1991). Roots >1 mm in diameter were discarded. Live fine roots were defined as roots that were to a varying degree brownish/suberised and often well branched, with the main part of the root tips light and turgid or changed from mycorrhizal clusters (Agerer 1987–93). The stele was white to slightly brown and elastic. In roots considered as dead, the stele was brownish and easily broken, and the elasticity was reduced. The roots were dried to constant weight at 70 °C for 48 h and weighed to the nearest milligram. Root length was measured using a Comair root-length scanner on air-dried fine-root samples (<1 mm in diameter).

Fine root biomass from soil coring was calculated on an area basis using the volume of the core and a soil depth of 40 cm. Net primary production of fine roots per year was calculated from the positive differences in total fine root mass (living and dead) between the sampling dates. Only significant differences were included in this calculation. Necromass accumulation was calculated by adding the positive differences in dead biomass (necromass) between the sampling dates. The fine root turnover index was calculated by dividing fine root production (as the sum of fine root net primary production and the positive difference of fine root mass between the last and the first sampling date) by the mean (living) fine root biomass of the sampling period. For some of these calculations, only roots in 0–30-cm soil depth were considered, to reduce data variation.

5.2.2 Root Window Observations

The time course of root growth was studied in situ by continuous root window observations (Vogt and Persson 1991). Root windows are soil profiles covered with a Plexiglas plate. These windows were installed in summer 1993 in Åheden, Skogaby, Klosterhede, Waldstein and Aubure in Norway spruce stands, in Gribskov, Schacht and Aubure in European beech stands, and in Åheden in a birch (*Betula pendula*) stand (for site characteristics see Persson, Chap. 2, this Vol.). To prepare the Plexiglas plates, $50 \times 40 \times 0.6$ cm plates were put in the laboratory into an oven. The oven was heated to approximately 60 °C and until the plate was shaped slightly concave, in order to obtain a better direct window-soil contact in the field. The root windows were then installed by first hammering a straight and heavy sharp-edged steel plate (approximately 54 cm wide) without deflections into the soil at the angle of 60° from the vertical, to obtain a smooth soil profile.

A soil space of approximately 55×55 cm was then excavated in front of the steel plate to the depth of installation (Fig. 5.1). Resulting from the angle of the soil profile and the few centimetres of Plexiglas covering the litter layer, the effective soil depth of root windows was approximately 40 cm. The excavated soil was carried away completely from the site to minimise local disturbances of soil chemistry. The steel plate was then carefully removed from the soil profile. The concave-shaped Plexiglas plate was tightly placed with the convex side against the soil profile. At the left and right side of the plates, wooden sticks were hammered deep into the soil in perpendicular

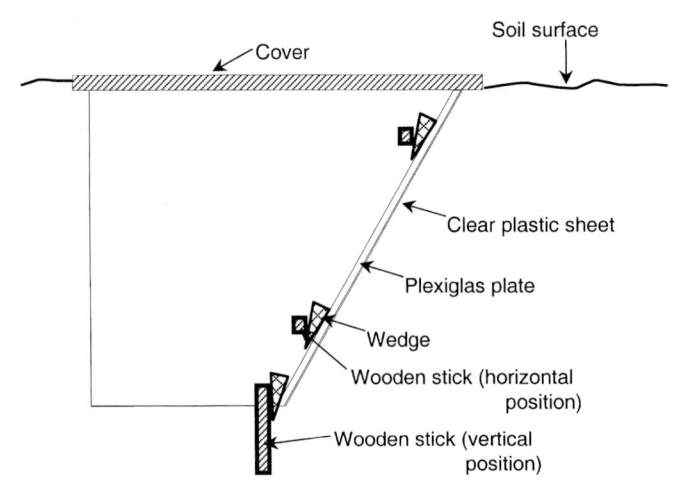

Fig. 5.1. Schematic drawing of an installed root window (side view)

direction to the plate, and wedges were pressed between the Plexiglas plate and the wooden sticks (Fig. 5.1). With the help of the wedges, the sides of the plate were tightly pressed against the soil profile. Because of the concave shape, the Plexiglas was also in tight connection with the soil in the centre of the plate.

A total of 32 root windows were installed per site. At stony sites or where the homogeneous stand area was small, the number of windows was reduced to 10, 16 or 24 (Table 5.1). The root windows were installed approximately in 1.5-m distance from the trunk of dominating or codominating trees. Two windows were installed per tree at an angle of 90° on a radius around the tree, and data from these two windows were pooled for further analysis. Thus, 16 independent replications were obtained at the sites with the full number of root windows, and 5, 8 or 12 replications at the other sites. After installation, the soil profile and the pit were insulated against light and against temperature fluctuations different from the bulk soil by covering the pit with a block of commercially available insulation material (Styrodur, BASF, Ludwigshafen, Germany). At most sites, new roots appeared on the window surface soon after installation, so that growth measurements began a few months after installation (Table 5.1). At Åheden, root growth was slower and measurements began approximately 1 year after installation.

Root observations started in December 1993 or in spring 1994. Further growth was monitored throughout three growing seasons by repeated observations. At the first observation, and also at the first registration of each season, all visible roots were copied to transparent sheets placed on the Plexiglas. At subsequent registrations during the season, root growth increments (elongation of existing roots, appearance of new roots, lateral branching) were recorded. By using different-coloured pens for each registration, a time course of new root growth during the growing season was obtained.

Tree roots can be separated into three categories: coarse supportive roots, small-diameter roots acting as conduits, and fine or mycorrhizal roots active in nutrient

Table 5.1. Location, tree species, and recording details of root window experiments

Site	Tree species	No. of windows	Installation of windows	Start of observation	No. of recordings			
					94	95	96	97
Åheden	Betula pendula	10	08/93	06/94	4	4	3	1
Gribskov	Fagus sylvatica	32	08/93	10/93	10	4	4	1
Schacht	Fagus sylvatica	24	07/93	10/93	5	4	3	1
Aubure	Fagus sylvatica	24	07/93	10/93	5	4	6	1
Åheden	Picea abies	32	08/93	08/94	3	4	3	1
Skogaby	Picea abies	16	10/93	05/94	3	5	4	1
Klosterhede	Picea abies	32	10/93	04/94	4	4	3	1
Waldstein	Picea abies	32	07/93	10/93	5	4	3	1
Aubure	Picea abies	24	07/93	10/93	5	4	6	1

uptake (Vogt and Bloomfield 1991). Coarse supportive roots started to appear at some windows 3 years after installation. They have a low turnover and were not recorded in the present investigation. In the present recording, primary roots were distinguished from higher-order lateral roots (fine roots). This morphological distinction of two parts of the root system with different function (transport of nutrients and elongation of the root system in primary roots, uptake of nutrients in higher-order roots) was possible due to the relatively large viewing surface of the root windows. Primary roots usually had a diameter of 1.0–3.0 mm in Norway spruce and 0.5–1.5 mm in beech, while higher-order lateral roots had a diameter of approximately 0.3–1.3 mm in Norway spruce and 0.3–0.5 mm in beech. Repeatedly, a reduction in root diameter of primary roots was observed with ageing of the roots. Typical mycorrhizal root clusters appeared at the root windows usually 2 years after installation. Length of roots in mycorrhizal clusters was difficult to determine due to overlapping of individual root tips, and is not included in the present records. However, root length in mycorrhizal clusters was always small and was estimated to be less than 10% of the total length of all roots in most windows.

The transparent sheets were analysed by the line intersect method (Tennant 1975) to quantify total root length and length increment for each observation period (in cm root length m^{-2} root window surface). All windows at each site were included in the analysis except those where large animals such as moles had destroyed the soil profile (not more than two windows at each site) and those where no root growth was observed. At least 1 year after installation, root growth was observed at all windows at all sites, except at Åheden, where no roots occurred at approximately half of the total number of windows installed. Windows without root growth were not included in the analysis because the lack of growth may be an artefact of installing the windows. However, root growth at Åheden may have been overestimated in the present calculations due to the exclusion of windows without roots from calculation. Total length growth for one growing season was calculated by adding all length increments measured during that growing season and was expressed as cm root length m^{-2} root window surface $year^{-1}$. From these data, net primary production of the different root classes (primary roots and higher-order lateral roots) was estimated using a depth of view on the root window of 2 mm to calculate root length on an area basis (Taylor et al. 1970; Merrill and Upchurch 1994) and a conversion factor of 15 m g^{-1} for Norway spruce and 20 m g^{-1} for European beech (unpubl. data, see also Table 5.6 and for spruce Steele et al. 1997) to convert data from length to dry matter basis. Carbon investment for net primary production of roots was estimated using a mean carbon concentration of roots of 500 mg C g^{-1} dry matter (Ågren et al. 1980). The turnover index (for nomenclature see Steele et al. 1997) of roots can be estimated not only from root mortality measurements, but also by dividing annual root production by the standing root biomass (for example, Aber et al. 1985; Ruess et al. 1996). Using the root window data, the annual turnover index of roots was estimated by dividing the sum of total root length increments during 1 year by the total root length at a window at the beginning of that growing season. The total root length at the beginning of the growing season was used for this calculation, as this was the only total length estimate available from the root windows for each season.

5.2.3 In-Growth Cores

In-growth cores were used to measure activity of root growth into a defined volume of soil (Vogt and Persson 1991). For installation of the in-growth cores, soil cores (8 cm in diameter) were taken next to dominant or codominant trees at a distance of 140 cm from the trunk in 10-cm intervals down to 40 cm. Soil material was sieved through a 5-mm mesh to remove large organic matter particles and stones. At each site, 300 mg N (kg soil)$^{-1}$ was added to half of the in-growth cores. Nitrogen was added as urea in a first experiment, and as a mix of urea and NH_4NO_3 in a second experiment. A nylon mesh bag (mesh size 2 mm) was inserted into the hole with the help of a plastic tube before the holes were refilled according to the original depth layers.

The first in-growth core experiment was set up in autumn 1993 in four Norway spruce forests (Åheden, Klosterhede, Aubure, Waldstein) and harvested after two growing seasons in autumn 1995. At Åheden, additional cores were also harvested after three growing seasons. A second in-growth core experiment was set up in autumn 1995 in three Norway spruce forests (Åheden, Klosterhede, Aubure) and two European beech forests (Gribskov, Aubure) and harvested after 1 year. In all instances, four replicate in-growth cores were harvested per treatment and site.

For sampling, roots which had grown into the in-growth cores were cut at the outer side of the meshbag with a long knife before the meshbag was pulled out of the soil. Soil cores were kept at 4 °C during transportation and storage. Roots were carefully washed free of soil and then visually separated into four root classes: living and dead fine roots (<2 mm in diameter), coarse roots (2–5 mm in diameter) and woody roots (>5 mm in diameter). Each sample was photocopied on paper for subsequent root length measurement. The sample was then dried for 48 h at 80 °C and dry matter was determined. In the present report, data for total root mass and length only are presented. The N concentration of living fine roots was measured by a Macro-N analysing system (Heraeus, Hanau, Germany). In cases when the sample size was too small for analysis, replicated samples were pooled together.

The photocopies were scanned with a videocamera controlled by an image analysis system (Image C, Imtronic, Berlin, Germany). Images were analysed for total root length using the Rootedge software (Kaspar and Ewing 1997). Root biomass data from in-growth cores were calculated on an area basis using the volume of the cores and a soil depth of 40 cm. Hyphal lengths densities in soil were also determined in in-growth cores of selected treatments using the method described by Li et al. (1991). These cores were sampled 3 (Aubure) or 12 (Åheden) months after installation in summer 1995.

5.2.4 Statistics

Statistical analyses were carried out with the SAS and SigmaStat statistical programmes. Data from soil coring were corrected for missing values with the SAS ANOVA and GLM procedures. In all calculations, for convenience, site was treated as one experimental factor in the analyses. Differences between means were evaluated by the Duncan or Tukey tests at an alpha value of 0.05. In many investigations on below-ground processes in forests, an alpha value of 0.10 is used to overcome the problem of large data variation (see, for example, Steele et al. 1997).

5.3 Root Growth Measurements Obtained by Soil Coring

Soil coring was used in order to obtain estimates of the spatial distribution of living (biomass) and dead (necromass) fine roots, both in relation to the distance to the trees and with depth in the soil profile. The highest density of fine roots was found in the upper parts of the soil profile (Tables 5.2 and 5.3). In the top 2.5 cm of soil, 48 and 30% of the total amount of fine root biomass were recorded in the European beech stands at Schacht and Aubure and 61, 49 and 13% in the Norway spruce stands at Skogaby, Waldstein and Aubure, respectively. The stands at Aubure differed from other sites at Skogaby, Waldstein and Schacht because of their thin raw-humus layers (see Persson et al., Chap. 2, this Vol.). The bulk of fine root biomass of all sites was found in the top 10 cm of the soil. Thus, 76 and 67% of the fine root biomass was found in the 0–10-cm depth in the European beech stands at Schacht and Aubure and 93, 75 and 78% at the Norway spruce stands at Skogaby, Waldstein and Aubure (Table 5.2).

Table 5.2. Live fine-root mass (biomass; <1 mm in diameter, g dry weight m^{-2}) in different depths at different European forest sites. Data (mean and SE) are averages of different sampling dates for Aubure (May, July, September, October 1996, May 1997; $n = 5 \times 20$) and are from one sampling date only for Schacht and Waldstein ($n = 24$; May 1997) and Skogaby ($n = 32$: October 1997)

Depth (cm)	Fagus sylvatica		Picea abies		
	Schacht	Aubure	Skogaby	Waldstein	Aubure
0–2.5	82 (12)	31 (4)	89 (9)	52 (6)	9 (4)
2.5–5	45 (6)	21 (4)	32 (4)	16 (3)	17 (5)
5–10	31 (9)	17 (3)	15 (4)	11 (2)	26 (5)
10–20	20 (3)	16 (3)	6 (2)	21 (7)	9 (2)
20–30	20 (2)	9 (2)	2 (1)	6 (3)	2 (1)
30–40	10 (2)	9 (2)	2 (1)	1 (1)	4 (1)

Table 5.3. Total (live and dead) fine root mass (<1 mm in diameter, g dw m^{-2}) at different European forest sites. Data (mean and SE) are averages of different sampling dates for Aubure (May, July, September, October 1996, May 1997; $n = 5 \times 20$) and are from one sampling date only for Schacht and Waldstein ($n = 24$; May 1997) and Skogaby ($n = 32$: October 1997)

Depth (cm)	Fagus sylvatica		Picea abies		
	Schacht	Aubure	Skogaby	Waldstein	Aubure
0–2.5	107 (16)	40 (5)	145 (10)	62 (6)	11 (4)
2.5–5	83 (10)	30 (5)	94 (9)	23 (3)	21 (6)
5–10	77 (17)	32 (4)	68 (8)	20 (4)	38 (8)
10–20	64 (6)	39 (5)	52 (4)	41 (9)	16 (4)
20–30	49 (6)	28 (4)	34 (3)	22 (3)	8 (2)
30–40	21 (4)	19 (2)	29 (4)	3 (1)	9 (2)

Table 5.4. Seasonal fluctuations in the amount of live and dead fine roots (<1 mm in diameter) and live/dead ratio of roots of beech and Norway spruce at Aubure in 0–30-cm soil depth. Means ($n = 20$) followed by the same letter are not significantly different by Duncan's multiple range test

	Live roots (g m^{-2})	Dead roots (g m^{-2})	Total roots (g m^{-2})	Live/dead ratio (g g^{-1})
Fagus sylvatica				
May 1996	94 cde	75 cd	169 b	2.13 bcd
July 1996	43 a	139 f	183 b	0.37 a
September 1996	120 e	111 e	231 a	1.30 ab
October 1996	79 bc	74 cd	154 bd	1.10 ab
May 1997	77 bd	151 f	229 a	0.63 a
Picea abies				
May 1996	70 abc	45 ab	115 a	3.01 c
July 1996	63 abc	64 ac	127 ab	1.13 ab
September 1996	56 ab	38 b	93 ac	1.76 abc
October 1996	40 a	37 b	77 c	2.46 bc
May 1997	57 ab	97 de	155 e	0.91 ad

Of the total amount of fine roots (live and dead), 27 and 54% were detected in the top 10 cm in the European beech stands at Schacht and Aubure and 73, 61 and 68% at the Norway spruce stands at Skogaby, Waldstein and Aubure (Table 5.3). Most fine roots in the top 10 cm of the soil profile consisted of living tissue, while in the deeper soil profile fine roots were rated dead to a great extent.

Significant seasonal fluctuations in the amount of live and dead fine roots were found at the Aubure site for both European beech and Norway spruce (Table 5.4). High seasonal growth dynamics may be inferred from those fluctuations. A similar general growth pattern at the Aubure sites in live/dead ratios for the two tree species

Table 5.5. Average live (biomass) and dead (necromass) fine root dry weight (g m^{-2}) in 0–30 cm soil depth from five sampling occasions 1996–1997, calculated fine root net primary production and necromass accumulation (g m^{-2} a^{-1}; minimum estimates see Vogt and Persson 1991) and turnover index (g g^{-1} a^{-1}) at Aubure during the period May 1996 to May 1997. For data see Table 5.4, for calculations see text

Species	Biomass (g m^{-2})	Necromass (g m^{-2})	Net primary production (g m^{-2} a^{-1})	Necromass accumulation (g m^{-2} a^{-1})	Turnover index (g g^{-1} a^{-1})
Fagus sylvatica	83	110	137	141	2.4
Picea abies	57	56	78	60	2.1

in May, July and September 1996 suggested that specific climatic factors should have caused the fluctuations during this period. Soil moisture stress during the summer and low temperature in winter seem to be factors which limit fine root growth most (Persson 2000). Fluctuations in the amounts of live and dead fine roots were used to calculate the total fine-root production at the Aubure sites (Table 5.5; see Chap. 5.2, this Vol., and Persson 1978). These calculations resulted in estimations of fine-root net primary production, necromass accumulation, and in a turnover index for fine-root biomass of more than 2 for both tree species (Table 5.5).

Both the average level of fine roots (live and dead) and the turnover rates at the Aubure site were higher in the European beech than in the Norway spruce stand. The fine roots in the European beech stand were distributed more deeply and more evenly in the total soil profile (Tables 5.2 and 5.3). The live/dead ratio in the European beech stand in Aubure was lower in all horizons than in the Norway spruce stand (Table 5.4). This was to a great extent a result of a higher dead mass (necromass) accumulation in the Norway spruce stand (Table 5.5). Specific root length of live fine roots in Aubure was higher in European beech than in Norway spruce (Table 5.6).

When fine-root biomass and length in Aubure as obtained by the soil coring method were compared to measurements of root length increments or biomass production rates of higher-order roots as obtained by the root window method (Table 5.6), no correlation between root growth rates observed at the windows and root biomass fluctuations as measured by soil coring were apparent.

5.4 Root Growth Measurements Obtained by Root Windows

After the windows had been installed, root windows allowed non-destructive observations of root growth during more than three growing seasons. Differences in root growth between sites and tree species and the interaction of species and site were statistically significant for most observation periods. As an example, significance values for selected sampling dates are shown in Table 5.7. A vigorous increase in root density early in the season occurred particularly at the sites with mild winter conditions (see Persson et al., Chap. 2, this Vol. for characterisation of sites). Root growth at the root windows in the first season (1994) was dominated by primary roots (Figs. 5.2 and 5.3; Tables 5.8 and 5.9). Vigorous growth of higher-order roots started later. Both primary and higher-order roots of Norway spruce continued to grow until late autumn, and

Table 5.6. Amount of live fine roots (biomass; <1 mm in diameter; g m^{-2}), length (m m^{-2}) and specific root length (m g^{-1}) at different European forest sites as estimated in the total soil cores in 0–40 cm soil depth. For Aubure, where soil cores were taken at different sampling dates, data from root windows are given for comparison. Data from root windows (with SE in parentheses) are growth increments of higher-order lateral roots at the root window surface in the time period between the last and the current sampling [mm (m window surface)$^{-2}$ day^{-1}] and the calculated biomass production rate per day (g m^{-2} day^{-1}; for calculations see text)

	Data from soil coring			Data from root windows	
	Biomass (g m^{-2})	Root length (m m^{-2})	Specific root length (m g^{-1})	Root length increments [mm (m window surface)$^{-2}$ day^{-1}]	Biomass production rate (g m^{-2} day^{-1})
Aubure – *Fagus sylvatica*					
May 1996	103	1256	15.0		
Jul. 1996	43	909	20.9	10.1 (2.6)	5.0 (1.3)
Sep. 1996	126	1410	11.6	16.1 (4.9)	8.0 (2.4)
Oct. 1996	83	1724	21.4	7.7 (2.0)	3.8 (1.0)
May 1997	84	1991	25.9		
Aubure – *Picea abies*					
May 1996	70	672	10.3		
July 1996	63	560	9.3	2.1 (0.4)	1.4 (0.3)
Sep. 1996	56	558	11.4	3.1 (0.1)	2.1 (0.6)
Oct. 1996	48	703	15.2	1.7 (0.1)	1.1 (0.4)
May 1997	57	689	14.0		
Schacht – *Fagus sylvatica*					
May 1997	207	5064	25.0		
Waldstein – *Picea abies*					
May 1997	106	1869	18.9		

Table 5.7. Probability values for ANOVAs for data on root length of beech and Norway spruce at the root windows at different sites for three sampling dates

Source of variation	Primary roots			Higher-order roots		
	04/95	10/95	04/96	04/95	10/95	04/96
Sites (gradient)	<0.001	<0.001	<0.001	<0.001	<0.001	<0.001
Tree species	0.108	<0.001	<0.001	<0.001	0.206	0.755
Site × tree species	0.012	0.002	<0.001	0.003	<0.001	<0.001

in some cases (for example, Åheden 1995, Skogaby 1996), higher-order roots showed highest growth rates at the end of the season (Fig. 5.3). From the second season (1995), increase in root length throughout the growing season and decrease during winter were within the same range, i.e. root length at the windows was in a steady state.

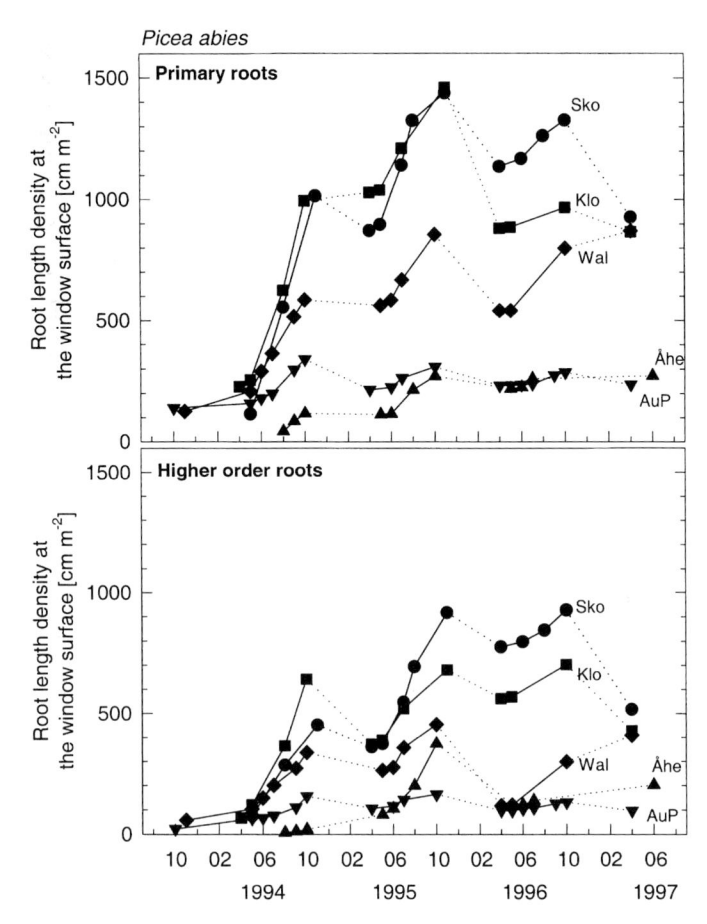

Fig. 5.2. Root length density of *Fagus sylvatica* (European beech) primary roots (*top*) and higher-order roots (*bottom*) at the root window surface at Åheden (birch), Gribskov, Schacht and Aubure. At the first observation date of each growing season the total root length at each window was recorded and for subsequent observation dates the length of newly grown roots since the previous observation date was recorded. Measurements were taken from October 1993 to June 1997

In a comparison of years, root growth was very high in 1994 (Tables 5.8 and 5.9). This may be due to seasonal climatic effects, but is most likely also an effect of window installation in the previous year. For example, root growth of European beech at the windows was exceptionally high at Gribskov in 1994, the first season of observation (Fig. 5.2). In the following seasons, root length growth was higher at Aubure than at Gribskov or Schacht (Tables 5.8 and 5.9). In all years, Schacht had lower root length growth than the other two sites. Birch roots observed at Åheden grew with particularly high rates during 1995. In contrast to Norway spruce, decreasing growth rates of primary and higher order roots were usually observed in European beech in autumn. Higher-order roots contributed relatively more to total root length in beech (Fig. 5.2) compared to Norway spruce (Fig. 5.3, see also Tables 5.8 and 5.9).

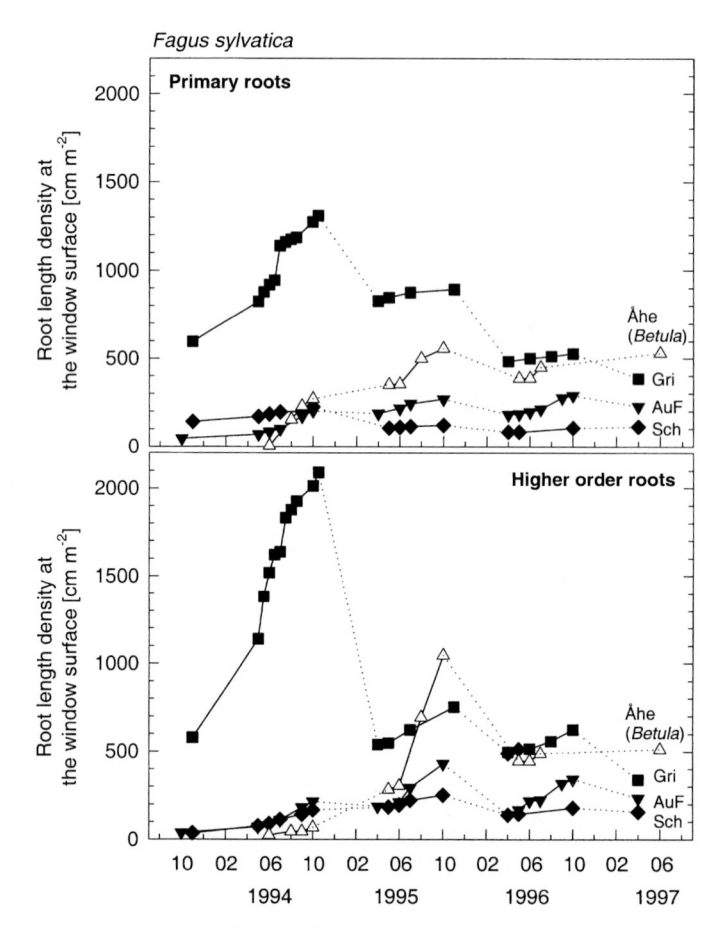

Fig. 5.3. Root length density of *Picea abies* (Norway spruce) primary roots (*top*) and higher-order roots (*bottom*) at the root window surface at Åheden, Skogaby, Klosterhede, Waldstein and Aubure. At the first observation date of each growing season the total root length at each window was recorded and for subsequent observation dates the length of newly grown roots since the previous observation date was recorded. Measurements were taken from October 1993 to June 1997

Also in Norway spruce, the intensity and the temporal dynamics of root growth differed markedly between the sites. Growth of both primary and higher order roots was low at the northern Swedish site at Åheden and also in the Vosges mountains at Aubure (Fig. 5.3, Tables 5.8 and 5.9). In contrast, higher root growth was observed at the south Scandinavian sites Klosterhede and Skogaby, in particular in 1994 and 1995. At Skogaby, in 1995 growth of higher order lateral roots was especially high. Root growth at Waldstein was intermediate in 1994 and 1995, but was relatively high in 1996 (Tables 5.8 and 5.9).

Measurements at the root windows were also used to calculate root net primary production, the amount of carbon used for this biomass production, and the root

Table 5.8. Length growth [cm (m^2 root window area)$^{-1}$ a^{-1}], calculated biomass production (g dw m^{-2} a^{-1}), calculated carbon investment for biomass production (g C m^{-2} a^{-1}), and turnover index (root length) (a^{-1}) of primary roots of different tree species at different European forest sites. Data are from root window observations, for calculations see text. Means and SE in parenthesis (n = 5–16). Data within one row and tree species are not significantly different between sites when followed by the same letter (Tukey test). If necessary, data were log$_{10}$ or square root-transformed before analysis to obtain normality

	Betula	Fagus sylvatica			Picea abies				
	Åheden	Gribskov	Schacht	Aubure	Åheden	Skogaby	Klosterhede	Waldstein	Aubure
Length growth [cm (m^2 root window area)$^{-1}$ a^{-1}]									
1994	n.a.[1]	587.4[b] (95.1)	87.4[a] (27.5)	167.0[a] (42.6)	128.9[a] (36.5)	827.0[b] (108.6)	847.3[b] (59.5)	686.1[b] (157.8)	253.0[a] (66.8)
1995	291.1 (45.3)	73.6[b] (12.6)	31.0[a] (12.5)	131.0[a] (53.3)	191.6[ab] (23.3)	565.7[c] (69.2)	518.5[c] (70.0)	378.0[bc] (80.0)	116.2[a] (26.2)
1996	173.8 (73.0)	43.9[a] (14.0)	50.3[ab] (11.6)	137.9[a] (44.7)	111.1[a] (27.1)	189.6[a] (24.9)	115.3[a] (25.4)	407.8[b] (59.2)	65.5[a] (16.6)
Mean 1995–1996	232.4	58.8	40.6	134.4	151.4	377.6	316.9	392.9	90.8
Net primary production (g dry weight m^{-2} a^{-1})									
1994	n.a.	293.7 (47.5)	43.7 (13.8)	83.5 (21.3)	85.9 (24.3)	551.3 (72.4)	564.9 (39.6)	457.4 (105.2)	162.0 (44.6)
1995	145.5 (22.6)	36.8 (6.3)	15.5 (6.2)	65.5 (26.6)	127.7 (15.6)	377.1 (46.1)	345.7 (46.5)	252.0 (53.1)	77.5 (17.5)
1996	86.9 (36.5)	21.9 (7.0)	25.1 (5.8)	68.9 (22.3)	74.1 (18.1)	126.4 (16.6)	76.8 (16.9)	271.9 (39.5)	43.7 (11.1)
Mean 1995–1996	116.2	29.4	20.3	67.2	100.9	251.8	211.25	262.0	60.6
Carbon investment for biomass production (g C m^{-2} a^{-1})									
Mean 1995–1996	58.1	14.7	10.2	33.6	50.4	125.9	105.6	131.0	30.3
Turnover index [root length] (a^{-1})									
1994	n.a.	0.84[b] (0.06)	0.37[a] (0.07)	2.59[c] (0.91)	1.31[a] (0.30)	9.00[c] (4.29)	5.17[bc] (1.21)	4.25[abc] (1.23)	2.36[ab] (0.69)
1995	1.41 (0.44)	0.11[a] (0.03)	0.29[ab] (0.11)	0.77[b] (0.27)	1.96[b] (0.30)	0.70[a] (0.10)	0.56[a] (0.06)	0.83[a] (0.13)	0.54[a] (0.15)
1996	0.45 (0.18)	0.11[a] (0.04)	0.61[ab] (0.13)	0.72[b] (0.17)	0.63[bc] (0.18)	0.23[ab] (0.08)	0.13[a] (0.03)	1.14[c] (0.28)	0.28[a] (0.07)
Mean 1995–1996	0.93	0.11	0.45	0.74	1.30	0.46	0.34	0.98	0.41

[1] Data not available (no root growth at window surface).

Table 5.9. Length growth [cm (m² root window area)⁻¹ a⁻¹], calculated biomass production (g dw m⁻² a⁻¹), calculated carbon investment for biomass production (g C m⁻² a⁻¹), and turnover index (root length) (a⁻¹) of higher-order lateral roots of different tree species at different European forest sites. Data are from root window observations, for calculations see text. Means and SE in parenthesis (n = 5–16). Data within one row and tree species are not significantly different between sites when followed by the same letter (Tukey test). If necessary, data were log₁₀ or square root-transformed before analysis to obtain normality

	Betula	Fagus sylvatica			Picea abies				
	Åheden	Gribskov	Schacht	Aubure	Åheden	Skogaby	Klosterhede	Waldstein	Aubure
Length growth [cm (m² root window area)⁻¹ a⁻¹]									
1994	n.a.ᵃ	1491.0b (122.5)	n.a.	243.6a (23.0)	n.a.	450.0ab (84.1)	615.4b (76.5)	238.7a (56.6)	n.a.
1995	1179.2 (426.4)	213.0a (15.6)	117.7a (27.5)	278.5a (93.2)	n.a.	670.7d (41.3)	324.5c (47.0)	200.0b (25.3)	94.8a (21.5)
1996	130.8 (35.0)	127.9ab (20.2)	57.4a (11.8)	202.6b (33.6)	88.5a (19.4)	183.5a (37.5)	152.2a (20.3)	337.5b (54.1)	59.9a (13.4)
Mean 1995–1996	655.0	170.4	87.6	240.6	88.5	427.1	238.4	268.8	77.4
Net primary production (g dry weight m⁻² a⁻¹)									
1994	n.a.	745.5 (61.2)	n.a.	121.8 (11.5)	n.a.	300.0 (56.0)	410.2 (51.0)	159.1 (37.7)	n.a.
1995	589.6 (213.2)	106.5 (7.8)	58.9 (13.7)	139.2 (46.6)	n.a.	447.1 (27.5)	216.4 (31.3)	133.3 (16.9)	63.2 (14.4)
1996	65.4 (17.5)	64.0 (10.1)	28.7 (5.9)	101.3 (16.8)	59.0 (12.9)	122.4 (25.0)	101.5 (13.5)	225.0 (36.1)	39.9 (8.9)
Mean 1995–1996	327.5	85.2	43.8	120.2	59.0	284.8	159.0	179.2	51.6
Carbon investment for biomass production (g C m⁻² a⁻¹)									
Mean 1995–1996	163.8	42.6	21.9	60.1	29.5	142.4	79.5	89.6	25.8
Turnover index [root length] (a⁻¹)									
1994	n.a.	3.31a (0.46)	n.a.	3.15a (0.48)	n.a.	5.04a (1.12)	7.44b (1.08)	2.36a (0.72)	n.a.
1995	13.52 (5.40)	0.49a (0.08)	0.77ab (0.16)	2.05b (0.79)	n.a.	2.68b (0.39)	0.94a (0.14)	1.00a (0.17)	1.42a (0.46)
1996	0.41 (0.18)	0.35a (0.08)	0.48a (0.11)	1.70b (0.33)	0.75a (0.17)	0.25a (0.05)	0.28a (0.04)	3.14b (0.84)	0.68a (0.17)
Mean 1995–1996	6.96	0.42	0.62	1.88	0.75	1.46	0.61	2.07	1.05

ᵃ Data not available (root growth at window surface in less than three windows per site).

length turnover index. It was estimated that trees invested between 32 (European beech at Schacht) and 268 (Norway spruce at Skogaby) $gCm^{-2}a^{-1}$ for biomass formation of primary and higher-order lateral roots. The high value at Skogaby is probably related to the relatively young age of this stand (see Persson et al., Chap. 2, this Vol.). Net primary production of higher-order lateral roots at Aubure, for example, was estimated to be 101 (European beech) and 40 (Norway spruce) $g\,dw\,m^{-2}a^{-1}$ in 1996 (Table 5.9). This is somewhat lower than the estimation of fine root net primary production by the soil coring method of 137 (European beech) and 78 (Norway spruce) $g\,dw\,m^{-2}a^{-1}$ for approximately the same period (Table 5.5). However, when values obtained at the root windows for primary roots (Table 5.8) and for higher-order lateral roots (Table 5.9) are added, the root window method resulted in an estimation of root net primary production of 170 (European beech) and 84 (Norway spruce) $g\,dw\,m^{-2}a^{-1}$ and thus in values somewhat higher than estimated for fine roots by the soil coring method.

High root turnover indices measured in 1994 in many of the sites are possibly a response to a stimulation of root growth in 1993 after installation of the windows. This probably also explains relatively high turnover indices for Åheden in 1995, a site where disturbance of existing roots may have longer-lasting effects on the following root growth than at other sites. In 1995 (except Åheden) and 1996, turnover indices varied from 0.11 to 3.14 $year^{-1}$ and were affected by root part (primary root and higher order lateral root), site and year of observation (Tables 5.8 and 5.9). As can be expected, turnover indices were higher for higher order lateral roots (Table 5.9) as for primary roots (Table 5.8). In higher-order roots, turnover indices in European beech were distinctly high at Aubure (Table 5.9) and in Norway spruce were distinctly high at Skogaby (1995) and Waldstein (1996).

Root turnover values in European beech were broadly within the same range as measured in Norway spruce. However, at Aubure, where Norway spruce and European beech stands were located at the same site, turnover indices were higher in European beech than in Norway spruce (Tables 5.8 and 5.9).

A comparison of turnover indices obtained from the root windows and those obtained by the soil coring method is difficult because sequential soil coring was restricted to one site and vegetation period (Aubure in 1996). For this site and year, turnover indices estimated for higher order lateral roots from the root windows were lower (1.7 for European beech and 0.7 for Norway spruce; Table 5.9) than those estimated from the sequential coring method (2.4 for European beech and 2.1 for Norway spruce; Table 5.5). One reason for this may be that only roots with a diameter of <1 mm were considered in the sequential coring method, while higher order lateral roots were sometimes slightly larger in diameter.

5.5 Root Growth Measurements Obtained by In-Growth Cores

The in-growth core investigations tested site-specific differences both in root growth capacity and in response to a locally N-enriched soil zone. In the first experiment, two growing seasons (1994 and 1995) after installation, root dry matter of Norway spruce in non-fertilised in-growth cores reached high values at Klosterhede, intermediate values at Waldstein and Aubure, and only low values at Åheden (Table 5.10). Thus, the relatively higher root growth at Klosterhede compared to Waldstein and

Table 5.10. Total biomass (g dw m^{-2}) of roots grown into in-growth cores (0–40 cm depth) in two separate experiments (one or two growing seasons). Means and SE ($n = 4$)

Site	Tree species	Non-fertilised soil	Fertilised soil	p
First experiment (harvested after two growing seasons, 1994 and 1995)				
Åheden	Picea abies	12.2 (1.4)	105.6 (28.8)	0.029
Klosterhede	Picea abies	789.1 (105.2)	280.6 (33.5)	0.004
Waldstein	Picea abies	183.0 (44.2)	171.4 (89.7)	0.486
Aubure	Picea abies	182.7 (70.0)	44.7 (20.5)	0.107
Second experiment (harvested after one growing season, 1996)				
Gribskov	Fagus sylvatica	175.2 (49.3)	177.7 (19.4)	0.963
Aubure	Fagus sylvatica	14.1 (6.2)	22.6 (6.4)	0.375
Åheden	Picea abies	18.4 (2.8)	20.5 (2.8)	0.570
Klosterhede	Picea abies	119.2 (25.1)	153.2 (72.7)	0.675
Aubure	Picea abies	4.5 (2.6)	3.5 (2.9)	0.686

Aubure observed in 1994 and 1995 at the root windows (Fig. 5.3 and Tables 5.8 and 5.9) was confirmed with the in-growth core method. The high root growth in Klosterhede may be related to a former humus layer buried at this site in 10–20-cm soil depth (see Persson et al., Chap. 2, this Vol.). Low root growth at Åheden was apparent both at the root windows (Fig. 5.3 and Tables 5.8 and 5.9) and in the in-growth cores. Root mass estimated from the in-growth cores (Table 5.10) was higher than root mass estimated by the coring method (Table 5.6) for both the Aubure and the Waldstein sites. However, only fine roots <1 mm were evaluated with the coring method, while all roots newly grown into the in-growth cores within 2 years were evaluated in the in-growth core method. Root lengths measured in the in-growth cores in non-fertilised soil were equivalent to 127 (Åheden), 2418 (Klosterhede), 1056 (Waldstein) and 1489 (Aubure) m m^{-2}. Thus, at Aubure Norway spruce root length in the in-growth cores was higher than root lengths measured at the same site by the coring method (Table 5.6). In contrast, at the Waldstein site higher root lengths were measured by the coring method than by in-growth cores.

Nitrogen supply to the cores had a significant effect on root growth at Åheden and at Klosterhede (Table 5.10). At Åheden, the localised soil N supply caused a distinct increase of root growth in the N-fertilised soil patch. In contrast, at Klosterhede root growth was distinctly decreased in the N-rich soil (Table 5.10). A similar effect was observed at Aubure, but this effect was not significant as variation of data within the treatments was particularly large.

Another sampling was carried out at Åheden 36 months after in-growth core installation. At this sampling, non-fertilised and N-fertilised cores had on average an equivalent of 70 and 293 g root dry weight m^{-2}, respectively. The difference between treatments was significant ($p = 0.029$) and confirmed the earlier observation.

During the short-term (one growing season, 1996) second experiment, no significant differences in root growth were observed between non-fertilised and N-fertilised soil (Table 5.10). However, differences between sites were distinct: in European beech, root production was much higher at Gribskov than at Aubure. Both

European beech stands had a higher root dry weight production than the corresponding Norway spruce stands (Table 5.10). In Norway spruce, root growth was intense at Klosterhede and low at Åheden and Aubure. Low root growth in Norway spruce at Aubure measured with the in-growth cores confirmed the exceptionally low root growth at this site for Norway spruce as measured with the root windows in 1996 (Tables 5.8 and 5.9). Thus, 1996 was very likely a year with particularly unfavourable root growth conditions in the Norway spruce stand at Aubure. In the in-growth cores of this experiment, fine root length of European beech in the non-fertilised soil was equivalent to 148 m (Aubure) and 2557 m (Gribskov) m^{-2}. Fine root length of Norway spruce in the non-fertilised soil was equivalent to 169 m m^{-2} at Åheden, 83 m m^{-2} at Aubure, and 730 m m^{-2} at Klosterhede.

In the first in-growth core experiment (two growing seasons, 1994 and 1995), root N concentrations varied between 11 and 14 mg N g^{-1} dry matter and were slightly higher at Waldstein and Aubure than at Åheden (Fig. 5.4). Nitrogen fertilisation caused a small but significant increase in N concentrations in roots at Åheden and at Klosterhede and a relatively larger increase at Waldstein and Aubure (Fig. 5.4). Thus, as predicted from model experiments, a local increase in soil N supply increased local root growth at an N-deficient site (Åheden) and increased root N concentration but not growth at N-sufficient sites (Waldstein, Aubure).

Nutrients can be taken up not only by roots but also by mycorrhizal hyphae. Because elements taken up by mycorrhizal hyphae can later be delivered to the host tree, mycorrhizal hyphae are, as well as roots, important for the tree nutrient uptake (see also Wallenda et al., Chap. 6, this Vol.). Therefore, hyphal lengths were measured in in-growth cores, with the assumption that a large part of these hyphae originated from mycorrhizal fungi. Hyphal length densities in non-fertilised soil were higher in Aubure than in Åheden (Fig. 5.5). Both sites also differed in their response to local N supply. At Åheden, N supply increased hyphal length density, whereas at Aubure hyphal length density was not significantly affected by local soil N supply (Fig. 5.5).

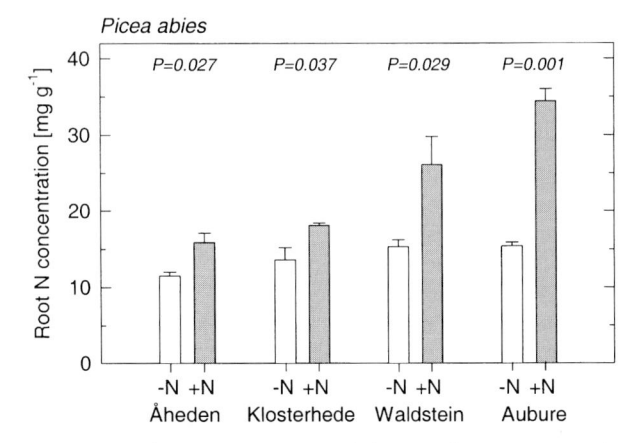

Fig. 5.4. Nitrogen concentrations in roots of Norway spruce roots in in-growth cores. In-growth cores were installed autumn 1993 and sampled in autumn 1995. The soil in the in-growth cores was either not fertilised (–N) or supplied with 300 mg N (kg soil)$^{-1}$ as urea (+N)

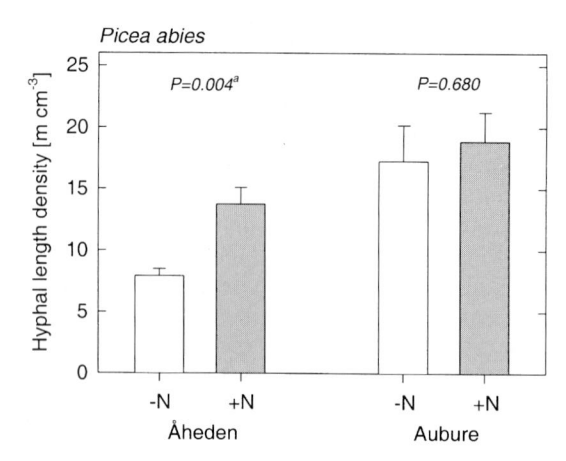

Fig. 5.5. Length density of hyphae in the mineral soil (0–10 cm soil depth) under Norway spruce in in-growth cores. In-growth cores were installed 3 (Aubure) or 12 (Åheden) months before sampling, and soil in the in-growth cores was either not fertilised (−N) or fertilised with 300 mg N (kg soil)$^{-1}$ (+N)

5.6 Root Growth at Different European Forest Sites

In the present investigations, three methods of studying fine-root growth dynamics were used, i.e. the soil coring method, the root window method and the in-growth core method. All methods have both benefits and limitations in determining root distribution and growth, root net primary production and root turnover. The problems of all presently existing methods have been discussed in detail elsewhere (Nadelhoffer and Raich 1992; Bloomfield et al. 1996; Hendrick and Pregitzer 1996; Vogt and Persson 1991; Vogt et al. 1996, 1998). The determination of below-ground biomass formation is much more difficult than the determination of above-ground growth, and is particularly difficult in forests due to (1) the existence of different root parts with different functions (see Sect. 5.2), (2) a high spatial heterogeneity of biomass pools and growth processes, and (3) large differences in growth and turnover at different times of the year. It is possible to use indirect methods, budget assessments (Hendricks et al. 1993) and model calculations (Mäkelä and Vanninen 2000) to estimate root growth and turnover, but also these budget calculations and models have to be equilibrated against a background of field data.

The depth distribution of roots has been determined in the present study with conventional soil cores only. All methods which require a prior disturbance of the soil, for example due to installation of instruments, very likely affect root depth distribution. Also, when roots grow against an obstacle such as root windows or minirhizotrons, their growth direction is affected. Correction factors have been used for calculations of root depth distribution from data obtained from minirhizotrons, but these correction factors are certainly not universally applicable. The coring data from the present study confirm earlier observations that length density of living fine roots (<1 mm diameter) is highest in the top 10 cm of the soil. Approximately 75% of all living fine roots were observed in this soil depth. Root density will also be affected by the presence of organic layers on top of the mineral soil, but this was not investigated in the present study.

The coring method is a very time-consuming method. In the present project, repeated samples at different times through the season were taken only in 1 year

(1996–1997) and at one site (the Norway spruce and European beech stands at Aubure). Thus, with the present soil coring data no site differences in root growth or turnover can be described. In addition, a large number of soil cores have to be taken in each forest site to obtain statistically meaningful results. If the number of cores is restricted and random spatial differences may influence the repeated measurements, sometimes large differences in the amount of fine roots will appear between two consecutive sampling dates (see Table 5.6). These large short-term leaps in biomass are very unlikely to occur in the soil. Root growth rates were observed at the root windows to be relatively steady at Aubure, with highest growth rates from July to early September and lower growth rates from September to October (Table 5.6). This predictable behaviour of roots has been observed in many other studies in boreal or temperate forests: root growth is positively correlated to soil temperature (see, for example, Steele et al. 1997; Hahn and Marschner 1998), except when soils become very dry in summer. Although root death or decay rates in soil are more difficult to follow than root growth rates, no reason is apparent for any catastrophic events in soil leading to sporadic large root death events during the vegetation period.

When comparing coring data from different studies, it is also a problem that fine roots are defined by different diameter classes in different studies. Also, although a distinction of live and dead roots is possible with the coring method and (with some additional problems) also with observation methods, the criteria used to distinguish live and dead roots are often not the same in different studies. In addition, dead roots very likely have no more function in nutrient uptake, but may act as transport organs for nutrients taken up by younger root parts. Moreover, it is not an unusual observation that new roots grow out of roots previously defined as dead (Eissenstat and Yanai 1997). Thus, any mortality estimates obtained from these classifications of dead and live have an additional uncertainty, and must be distinguished from turnover measurements based on disappearance of roots.

Transparent interfaces such as minirhizotrons or root windows are an accepted tool to study length growth increments of roots. Erroneous results may be obtained when the soil-observation plane contact is not complete and roots can grow in the resulting air space. Inflatable minirhizotrons (Gijsman et al. 1991) may be used in future to avoid this problem. Root windows allow a direct soil-interface contact when properly handled. However, even in this case the data are difficult to convert directly to a dry weight basis, and estimations of net primary production or carbon expenditure for root growth obtained from mini-rhizotrons or root windows (Tables 5.8 and 5.9) are based on assumptions on depth of view on these windows and a constant specific root length. However, these sources of error should not affect a relative comparison of site, tree species or year effect on root growth from root window data.

The present data show that root window and in-growth core method corresponded well in estimating differences in root growth between the sites investigated in this study. For example, for Norway spruce, Klosterhede was a site with high root growth in 1994 and 1995, while Aubure and Åheden showed low root growth, in particular in 1996 (Tables 5.8, 5.9 and 5.10). The root window method indicated that root primary production during 1995 and 1996 in European beech was particularly high at Aubure compared to Gribskov or Schacht (Tables 5.8 and 5.9). In Norway spruce, root primary production during this period was high at Skogaby, Klosterhede and Waldstein, with highest values for Skogaby in 1995. Thus, at Waldstein Norway spruce did not produce less roots than at other sites although aluminium levels in soil were highest at Waldstein among all sites tested (Table 2.4). Åheden and Aubure were sites with relatively low root primary production in Norway spruce (Tables 5.8 and 5.9).

The root turnover indices estimated from the root windows were usually lower in 1996 than in 1995, except at Waldstein and Schacht (primary roots only; Tables 5.8 and 5.9). For European beech, root turnover indices in 1995 and 1996 were higher at Aubure than at Gribskov or Schacht. For Norway spruce higher-order lateral roots, Skogaby (1995) and Waldstein (1996) were sites with particularly high turnover indices. The calculated root turnover indices compare well with the limited data published from other field studies. For example, Fahey and Hughes (1994) estimated the lifespan of early season roots at Hubbard Brook Experimental Forest as 8–10 months. At the hardwood Huntington Forest, annual turnover rates ranged from 0.8 to 1.2 (Burke and Raynal 1994). At two different sites in Canada, estimations for yearly turnover rates of spruce fine roots were between 1.3 and 3.2 (Steele et al. 1997).

In the present study, the current level of N deposition at each site did not have a clear effect on root turnover indices. For root turnover as well as for root growth dynamics (Figs. 5.2 and 5.3; see also Côté et al. 1998), many factors other than soil N supply, such as soil temperature and supply of other nutrients, also have an effect. Thus, a consistent effect of increased N deposition on root growth and turnover cannot be expected in a comparison of a few sites only. However, the present data suggest that current levels of high N deposition do not lead to a significant decrease in root growth at the sites investigated in this study.

The response of root growth to a local supply of N in soil was highly dependent on the N nutritional status of the trees and thus also on local N deposition (Table 5.10). Thus, in-growth cores with differential fertilisation are a convenient method to study nutrient limitations in forest ecosystems (Raich et al. 1994). The in-growth core study showed that Åheden was the only site under study where trees were severely N-limited (Table 5.10). This conclusion is confirmed by the low total N contents in needles of Norway spruce at Åheden (see Bauer et al., Chap. 4, this Vol.) and by studies of the N uptake capacity of roots from the different sites (see Wallenda et al., Chap. 6, this Vol.).

In addition to the roots, hyphae may take up large amounts of N at all sites and deliver it to the host plant (see Wallenda et al., Chap. 6, this Vol.). This hyphal N uptake is likely to be even more efficient than root N uptake, in particular because hyphal length densities in soil (Fig. 5.5) are much higher than root length densities.

5.7 Conclusions

Studies on root biomass, growth and turnover in forest ecosystems are notoriously more difficult than studies of above-ground growth processes. In the present investigation, several methods were used to quantify roots, but only with the root window method were sufficient data obtained to compare different sites and to extend the study to several years. The other methods, soil coring and in-growth cores, were used to corroborate data obtained at the root windows.

Large differences in root growth and turnover between sites were measured in this study, but large differences also occurred between years. In general, Norway spruce produced more root biomass than European beech, but at Aubure this trend was reversed due to exceptionally low root growth of the Norway spruce trees at this site, in particular in 1996. Of all Norway spruce sites, Skogaby showed the highest root growth, probably due to the relatively mild climatic conditions and the relatively young age of the stand. At all sites with relatively mild climatic conditions, Skogaby (Norway spruce), Klosterhede (Norway spruce) and Gribskov (European beech), root growth started early in the season and roots reacted to a disturbance quickly with new root growth (in-growth cores and root windows in the first year after installation). However, in the following years annual root growth was also high at Waldstein, a site with colder climate. The measurements showed that Norway spruce root growth was relatively low at Åheden, the site with the coldest climatic conditions. Similarly to root growth, also indices for annual root turnover were highly variable between sites and years. Turnover values were much higher for higher-order, lateral fine roots than for primary extension roots, and tended to be lowest in the Danish sites at Gribskov (European beech) and Klosterhede (Norway spruce). The present data indicated that the trees invested between 22 (European beech at Schacht) and 142 (Norway spruce at Skogaby) $gCm^{-2}a^{-1}$ for growth of (non-woody) extension and higher-order lateral roots.

Among these variations between sites at least partly related to climate and soil properties, it is not possible to extract the effect of one possible cause of differential root growth, i.e. the N deposition. Nevertheless, by comparison of the central European sites of this study with the site in northern Sweden, it is not apparent that current levels of N deposition in central Europe have caused a dramatic decline in root growth activity or a drastic change in root turnover indices. This supports the results of studies on N uptake by roots (see Wallenda et al., Chap. 6, this Vol.) which also show the root systems in central European forest sites to be fully functional. It is possible that a chronic deposition of relatively small annual amounts of N into the ecosystem has different effects from high amounts of N often applied in short-term fertiliser experiments. However, subtle changes in below-ground processes, for example due to higher levels of nitrate supply in soil, will be difficult to detect with the presently available methods but can have long-term consequences for uptake by the tree of scarce nutrients or water.

References

Aber JD, Melillo JM, Nadelhoffer KJ, McClaugherty CA, Pastor J (1985) Fine root turnover in forest ecosystems in relation to quantity and form of nitrogen availability: a comparison of two methods. Oecologia 66:317–321

Agerer R (ed) (1987–1993) Colour atlas of ectomycorrhizae. Einhorn-Verlag Eduard Dietenberger, Schwäbisch-Gmünd, Germany

Ågren GI, Axelsson B, Flower-Ellis JGK, Linder S, Persson H, Staaf H, Troeng E (1980) Annual carbon budget for a young Scots pine. Ecol Bull 32:307–313

Ahlström K, Persson H, Börjesson I (1988) Fertilisation in a mature Scots pine (*Pinus sylvestris* L.) stand – effects on fine roots. Plant Soil 106:179–190

Bloomfield J, Vogt K, Wargo PM (1996) Tree root turnover and senescence. In: Waisel Y, Eshel A, Kafkafi U (eds) Plant roots – the hidden half, 2nd edn. Marcel Dekker, New York, pp 363–381

Burke MK, Raynal DJ (1994) Fine root growth phenology, production, and turnover in a northern hardwood forest ecosystem. Plant Soil 162:135–146

Chapin FS III, Bloom AJ, Field CB, Waring RH (1987) Plant responses to multiple environmental factors. BioScience 37:49–57

Clemensson-Lindell A, Persson H (1992) Effects of freezing on rhizosphere and root nutrient content using two soil sampling methods. Plant Soil 139:39–45

Clemensson-Lindell A, Persson H (1995) The effects of nitrogen addition and removal on Norway spruce fine-root vitality and distribution in three catchment areas Gårdsjön. For Ecol Manage 71:123-131

Côté B, Hendershot WH, Fyles JW, Roy AG, Bradley R, Biron PM, Courchesne F (1998) The phenology of fine root growth in maple-dominated ecosystem: relationships with some soil properties. Plant Soil 201:59-69

Eissenstat DM, Yanai RD (1997) The ecology of root lifespan. Adv Ecol Res 27:1-59

Fahey TJ, Hughes JW (1994) Fine root dynamics in a northern hardwood forest ecosystem, Hubbard Brook Experimental Forest, NH. J Ecol 82:533-548

Friend AL, Eide MR, Hinckley TM (1990) Nitrogen stress alters root proliferation in Douglas fir seedlings. Can J For Res 20:1524-1529

George E, Seith B (1998) Long-term effects of a high nitrogen supply to soil on growth and nutritional status of young Norway spruce trees. Environ Pollut 101:1-6

George E, Seith B, Schaeffer C, Marschner H (1997) Responses of *Picea*, *Pinus* and *Pseudotsuga* roots to heterogeneous nutrient distribution in soil. Tree Physiol 17:39-45

Gijsman AJ, Floris J, Noordwijk MV, Brouwer G, Van-Noordwijk M (1991) An inflatable minirhizotron system for root observations with improved soil/tube contact. Plant Soil 134:261-269

Hahn G, Marschner H (1998) Effect of acid irrigation and liming on root growth of Norway spruce. Plant Soil 199:11-22

Hendrick RL, Pregitzer KS (1996) Applications of minirhizotrons to understand root function in forests and other natural ecosystems. Plant Soil 185:293-304

Hendricks JJ, Nadelhoffer KJ, Aber JD (1993) Assessing the role of fine roots in carbon and nutrient cycling. Trends Ecol Evol 8:174-178

Jackson RB, Caldwell MM (1996) Integrating resource heterogeneity and plant plasticity: modelling nitrate and phosphate uptake in a patchy soil environment. J Ecol 84:891-903

Joslin JD, Wolfe MH (1999) Disturbances during minirhizotron installation can affect root observation data. Soil Sci Soc Am J 63:218-221

Kaspar TC, Ewing RP (1997) Rootedge – software for measuring root length from desktop scanner images. Agron J 89:932-940

Li X-L, Marschner H, George E (1991) Acquisition of phosphorus and copper by VA-mycorrhizal hyphae and root-to-shoot transport in white clover. Plant Soil 136:49-57

Linder S, Axelsson B (1982) Changes in carbon uptake and allocation patterns as a result of irrigation and fertilisation in a young *Pinus sylvestris* stand. In: Waring RH (ed) Carbon uptake and allocation in subalpine ecosystems as a key to management. For Res Lab, Oregon State University, Corvallis, pp 38-44

Majdi H, Persson H (1995) A study on fine-root dynamics in response to nutrient applications in a Norway spruce stand using the minirhizotron technique. Z Pflanzenernähr Bodenkd 158:429-433

Mäkela A, Vanninen P (2000) Estimation of fine root mortality and growth: a method based on system dynamics. Trees (in press)

Merrill SD, Upchurch DR (1994) Converting root numbers observed at minirhizotrons to equivalent root length density. Soil Sci Soc Am J 58:1061-1067

Nadelhoffer KJ, Raich JW (1992) Fine root production estimates and belowground carbon allocation in forest ecosystems. Ecology 73:1139-1147

Persson H (1978) Root dynamics in a young Scots pine stand in central Sweden. Oikos 30:508-519

Persson H (2000) Adaptive tactics and characteristics of tree fine roots. Plant Soil (in press)

Persson H, Ahlström K (1994) The effects of alkalizing compounds on fine-root growth in a Norway spruce stand in SW Sweden. J Environ Sci Health 29:803-820

Persson H, Ahlström K (1999) Effect of nitrogen deposition on tree roots in boreal forests. In: Rastin N, Bauhus J (eds). Going underground – ecological studies in forest soils. Research Signpost, Trivandrum, India, pp 221-238

Persson H, von Firks Y, Majdi H, Nilsson LO (1995a) Root distribution in a Norway spruce [*Picea abies* (L.) Karst.] stand subjected to drought and ammonium-sulphate application. Plant Soil 168-169:161-165

Persson H, Majdi H, Clemensson-Lindell A (1995b) Effects of acid deposition on tree roots. Ecol Bull 44:158–167

Persson H, Ahlström K, Clemensson-Lindell A (1998) Nitrogen addition and removal at Gardsjon – effects on fine-root growth and fine-root chemistry. For Ecol Manage 101:199–205

Raich JW, Riley RH, Vitousek PM (1994) Use of root-ingrowth cores to assess nutrient limitations in forest ecosystems. Can J For Res 24:2135–2138

Ruess RW, Van Cleve K, Yarie J, Viereck LA (1996) Contributions of fine root production and turnover to the carbon and nitrogen cycling in taiga forests of the Alaskan interior. Can J For Res 26:1326–1336

Santantonio D, Hermann RK (1985) Standing crop, production and turnover of fine roots on dry, moderate, and wet sites of mature Douglas-fir in western Oregon. Ann Sci For 42:113–142

Steele SJ, Gower ST, Vogel JG, Norman JM (1997) Root mass, net primary production and turnover in aspen, jack pine and black spruce forests in Saskatchewan and Manitoba, Canada. Tree Physiol 17:577–587

Taylor HM, Huck MG, Klepper B, Lund ZF (1970) Measurement of soil-grown roots in a rhizotron. Agron J 62:807–809

Tennant D (1975) A test of a modified line intersect method of estimating root length. J Ecol 63:995–1001

Vogt KA, Bloomfield J (1991) Tree root turnover and senescence. In: Waisel Y, Eshel A, Kafkafi U (eds) Plant roots – the hidden half. Marcel Dekker, New York, pp 287–306

Vogt KA, Persson H (1991) Measuring growth and development of roots. In: Lassoie JP, Hinckley TM (eds) Techniques and approaches in forest tree ecophysiology. CRC Press, Boca Raton, pp 477–501

Vogt KA, Vogt DJ, Palmiotto PA, Boon P, O'Hara J, Asbjornsen H (1996) Review of root dynamics in forest ecosystems grouped by climate, climatic forest type and species. Plant Soil 187:159–219

Vogt KA, Vogt DJ, Bloomfield J (1998) Analysis of some direct and indirect methods for estimating root biomass and production of forests at an ecosystem level. Plant Soil 200:71–89

6 Nitrogen Uptake Processes in Roots and Mycorrhizas

T. Wallenda, C. Stober, L. Högbom, H. Schinkel, E. George, P. Högberg, and D. J. Read

Dedicated to the memory of Horst Marschner

6.1 Introduction

At each site along the European north–south transect studied, a predominant pool of nitrogen (N) is in the organic form. The processes of mineralisation followed by nitrification and anthropogenic deposition will, in addition, combine and provide the inorganic forms ammonium (NH_4^+) and nitrate (NO_3^-) in amounts and ratios which will depend upon local conditions of soil, climate and pollutant deposition. While Åheden, the most northerly site of the transect, supports little net mineralisation (see Cotrufo et al., Chap. 13, this Vol.) and receives virtually no pollutant N (see Persson et al., Chap. 2, this Vol.), the southern Scandinavian and central European sites have experienced considerable inputs of NH_4^+ and NO_3^- deposition over recent years.

In this study we aimed to evaluate the influence of site-specific N supply upon the biology of N assimilation of spruce and beech and on dominant and codominant trees at various points along the transect.

We tested the hypothesis that the ability of roots and mycorrhizas of these trees to use the different N sources varies according to their position on the gradient of increasing N deposition. Laboratory-based experiments with excised roots and mycorrhizas as well as field experiments with intact root systems were employed to test this hypothesis. The approach was multifaceted, sets of experiments being designed to examine different aspects of the N uptake process:

6.1.1 Studies with Excised Roots and Mycorrhizas

1. The N concentration of fine roots, their rates of uptake of $^{15}NH_4^+$ and $^{15}NO_3^-$ and nitrate reductase activity were measured using material collected from sites along the transect (experimental set 1).
2. The extent and the kinetics of amino acid uptake were determined in excised mycorrhizal roots of known morphotypes collected along the transect (experimental set 2).
3. The impact of the presence of inorganic N on amino acid uptake was examined (experimental set 3).

6.1.2 Field-Based Experiments

4. Uptake of $^{15}NH_4^+$ and $^{15}NO_3^-$ by non-mycorrhizal primary roots was determined in situ using the root window technique (experimental set 4, see also Stober et al., Chap. 5, this Vol.).

5. The relative abilities of ectomycorrhizal mycelium of ectomycorrhizal roots to take up both inorganic and organic N sources from soil was assessed (experimental set 5).

6.2 Approaches to Study Different Aspects of the N Uptake Process

In this study, roots of all the major tree species of the transect were examined. These were *Picea abies, Fagus sylvatica, Pinus sylvestris* and *Betula* spp. For site descriptions, soil type and pH see Persson et al. (Chap. 2, this Vol.).

6.2.1 Root N Concentration, Fine Root Nitrate Reductase Activity (NRA) and Inorganic N Uptake (Experimental Set 1)

Fine root samples (Ø < 2mm) were collected from the organic layers at Åheden, Klosterhede, Aubure, Schacht and Waldstein during the summers of 1993, 1994 and 1995 and analysed.

Cleaned subsamples of 150 mg root material were used for in vivo NRA assays by the anaerobic incubation method described by Gebauer et al. (1988). Uptake of $^{15}NH_4^+$ and $^{15}NO_3^-$ (applied as NH_4Cl or $NaNO_3$) by fine root samples, collected during the summer of 1995, at a concentration of $1\,mg\,N\,l^{-1}$ (corresponding to $70\,\mu M$) was studied according to Jones et al. (1991). For calculation of uptake rates see Högberg et al. (1998).

Portions of all material from all species and sites were dried (70°C, 48h) and their N concentrations and ^{15}N natural abundance were determined as described by Högberg et al. (1996).

6.2.2 Kinetic Properties of Amino Acid Uptake by Mycorrhizal and Non-Mycorrhizal Roots (Experimental Sets 2 + 3)

Ectomycorrhizal roots were collected from the organic layers at the CANIF sites Åheden, Gribskov, Schacht and Waldstein over the period August–October 1997. In the case of *Fagus sylvatica* mycorrhizal roots were also periodically collected throughout 1997 and 1998 from a pure stand of this species at Ridgeway Side Wood, Hathersage, Derbyshire, UK (NGR SK 227830). The soil type at this site was a Cambisol (brown earth), pH 4.0, with an annual N deposition of $25-30\,kg\,N\,ha^{-1}\,a^{-1}$ at the site (Department of the Environment 1994).

Because non-mycorrhizal short roots were virtually absent in the field (see Taylor et al., Chap. 16, this Vol.), it was necessary in order to obtain such material for comparative purposes to produce them on seedlings grown in inoculum free peat (M3, Levington Horticulture Ltd., Ipswich, UK).

Seeds of beech, birch, Norway spruce and Scots pine were surface-sterilised. In the case of beech, the seeds were placed on moist filter paper to germinate, then grown initially in small pots before being transferred to transparent observation chambers. After 7 months growth at $100\,\mu mol$ photons $m^{-2}\,s^{-1}$ and 20°C, they had produced sufficient non-mycorrhizal roots to satisfy the requirements of the experimental design. Seedlings of birch, Norway spruce and Scots pine were also grown in peat and watered with distilled water, non-mycorrhizal fine roots being harvested after 6 months. Healthy, non-suberised, non-mycorrhizal roots (Ø < 1 mm) were collected and treated as described below for ectomycorrhizas.

After cleaning (see Wallenda and Read 1999 for more details), mycorrhizal roots were detached from the root system using forceps and scalpel and sorted according to types, referred to as morphotypes, based upon their ramification patterns, the colour and texture of their mantles and the presence or absence of rhizomorphs. Where possible, the identity of the fungal species forming these morphotypes was established using published descriptions (Brand and Agerer 1986; Agerer 1987–1998; Brand 1989, 1991). Types dominating particular sites were selected for this study.

Amino acids, in all cases labelled with ^{14}C, were used as tracers for the uptake experiments (Chalot et al. 1995). After the incubation period, mycorrhizal roots were freeze-dried and apparent uptake rates were calculated using the specific activity [total activity/(amount ^{14}C + amount ^{12}C)] of the uptake solution and the total radioactivity per unit dry weight.

Five to 12 substrate concentrations of amino acids were used in the range of 0.001 to 10 mM. V_{max} and K_M values were calculated by non-linear curve fitting of the experimental data to the Michaelis-Menten equation $[v = (V_{max} \cdot [S])/(K_M + [S])]$. Standard errors of the fitted parameters have been included ($n = 16$–28). For more details on the uptake experiments, see Wallenda and Read (1999).

6.2.3 Field-Based Studies of Plant N Uptake (Experimental Sets 4 + 5)

Field studies were carried out to test whether the contrasting N deposition and soil N supply resulted in characteristic differences in plant N uptake patterns. Two different methodologies were used.

In the first, intact tips of primary roots (Ø 1–2 mm) of Norway spruce and beech were carefully excavated from soil profiles. The roots were taken from a depth of 0–20 cm in June 1996. The soil profiles had been excavated several years prior to this study, to allow growth of new roots along a Plexiglas plane covering the soil profile. For details of these root windows see Stober et al. (Chap. 5, this Vol.). Roots selected for this study had a diameter of 1–2 mm and were healthy in appearance. The mycorrhizal status of the roots was not assessed, but root tips of primary roots were not covered by mycorrhizal mantles. The excavated root tips were exposed to agar blocks containing ^{15}N-labelled NH_4^+ or NO_3^- (95% ^{15}N). A release of N from the agar blocks to the soil was avoided by separating agar blocks and soil by an aluminium foil. For this purpose, the root tips were carefully lifted a few millimetres from the soil and the aluminium foil was inserted between the root tip and the soil surface. The N in the agar blocks was supplied over a range of different NH_4^+/NO_3^- ratios (1:9, 1:1, 9:1) with a constant N concentration of 1 mM. While soil solution N concentrations will be much lower than those used in the agar blocks, N supply to the root surface in the blocks will be decreased compared to those in soil solution due to reduced diffusion of ions in agar.

After application for 24 h, the roots, including their supporting basal parts, which were usually 8–12 cm in length, were excavated from the soil profile. The excavated roots were cut into the labelled root tip, a 1-cm buffer segment and the basal part of the system. The ^{15}N concentrations were measured by mass spectrometry (Europa Scientific, Crewe, UK) after drying, weighing and milling the samples. Nitrogen uptake in this experiment was defined as the amount of ^{15}N in the basal root parts.

After the labelling period of 24 h, no ^{15}N label was found in root parts more than 12 cm away from the root tip. Thus, the measured amount was used as an indicator of the N translocated from the root tip to the rest of the plant during the uptake period. The buffer segment was excluded from analysis because some ^{15}N may have diffused over the root surface from the agar block to this portion of the system.

In the second approach, mesh bags were used to determine the relative abilities of roots and mycorrhizal hyphae to utilise different N forms. The bags, which were of pore sizes 2 mm, 30 µm and 0.45 µm, were buried in the soil next to 3- to 5-year-old Norway spruce or beech trees in August 1995 (Åheden, Norway spruce only) or April 1996 (other sites). They contained material (1 kg) from the organic litter layer and the top 10 cm of soil and were penetrable to roots and hyphae (2 mm), hyphae but not roots (30 µm) or were not penetrable to roots or hyphae (0.45 µm = control). In July, at all sites a solution with a total amount of 700 (high supply) or 70 µmol N (with 95% ^{15}N) was supplied to the bags, the N being in the form of NH_4^+, NO_3^- or the amino acid, glutamic acid. Six weeks later, plant shoots were covered with air-tight plastic bags and exposed to ^{13}C-labelled CO_2 for 8 h. After another 6 days, shoots, roots and mesh bags were harvested. Plants were separated into several components before drying, weighing, grinding and mass spectrometry. Nitrogen uptake in this experiment was defined as the amount of ^{15}N in the above-ground plant parts. Possible dilution of the different N forms in the soil solution was not considered in the uptake calculations because application levels were relatively high compared to the amount of N originally in the mesh bag.

Data presented are means of 3 to 12 independent observations for each N treatment per site. Data were subjected to one-way analyses of variance or t-tests, in which, for convenience, site was treated as one experimental factor.

6.3 Studies with Excised Roots and Mycorrhizas

6.3.1 Fine Root N Concentration and Nitrate Reductase Activity (Experimental Set 1)

Fine root N concentration was highly variable, both within and between sites (Fig. 6.1). In ectomycorrhizal fine roots of *P. abies* there was a considerable variation in N concentration, ranging from 11 mg g^{-1} in northern Sweden to 21 mg g^{-1} at Aubure (Fig. 6.1).

The increase in N concentration of fine roots of *P. abies* from Åheden to Aubure was paralleled by the increase in N concentration in needles of the same species at the same sites (Bauer et al. 1997). In a study at a site close to Åheden, Giesler et al. (1998) found a similar range of N concentrations in roots of *P. abies* (from 12 to 20 mg g^{-1}) along a 90-m-long natural hydrochemical N supply gradient. In their study the highest value was associated with a groundwater discharge area. All sites investigated in our study were groundwater recharge rather than discharge areas. However, their data show that in comparative studies hillslope hydrology has to be taken into account because local patterns of hydrochemical variation, determined by topography, can obviously be superimposed upon regional patterns (Giesler et al. 1998).

Root in vivo NRA was lowest for *P. abies* at Åheden, slightly higher at Klosterhede and Aubure and highest at Waldstein (Fig. 6.2). There were no differences between values of NRA obtained in *F. sylvatica* and *P. abies* at sites in Denmark, France and

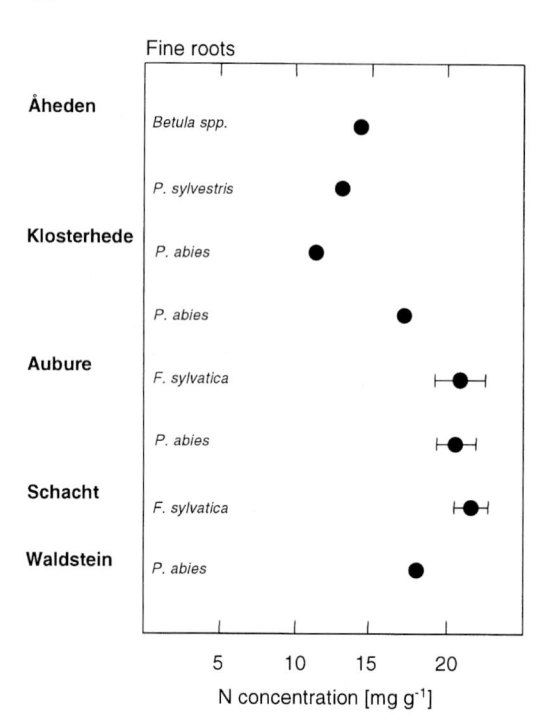

Fig. 6.1. Concentrations of N in fine roots collected at the different European sites in 1995. Data are means ± SE ($n = 18–25$)

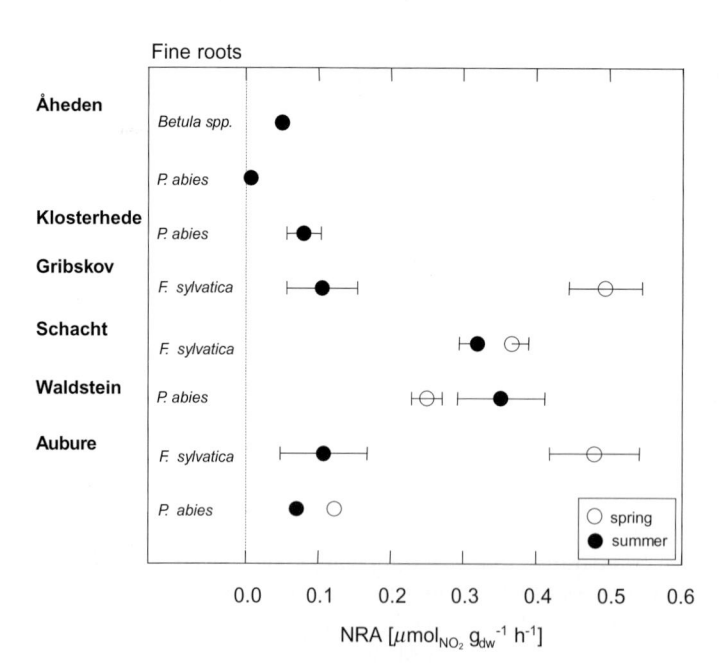

Fig. 6.2. In vivo nitrate reductase activity (NRA) in the spring (○) and summer (●) in fine roots at the different sites. Data are means ± SE ($n = 18–25$)

Germany. However, *Betula* spp. had higher NRA than *P. abies* at Åheden. In May 1993, NRA of roots at Gribskov and Aubure was four times higher than those measured at the same sites during July–August 1994, a pattern observed for neither *P. abies* nor *F. sylvatica* at the German sites (Fig. 6.2).

The in vivo NRA assay used in this study reflects the amount of active enzyme rather than the activity in situ. More precisely, it relates to the balance between synthesis/degradation and activation/inactivation of the enzyme (e.g. Srivastava 1980). Synthesis of NR is positively correlated with the supply of NO_3^- (e.g. Larsson 1994). The turnover of the enzyme is fast and in vivo NRA can therefore be used to monitor temporal changes in NO_3^- supply (e.g. Högberg et al. 1986), thus suggesting that NO_3^- is an important N source, especially at the German sites Schacht and Waldstein.

6.3.2 Uptake of NO_3^- vs. NH_4^+ by Excised Fine Roots (Experimental Set 1)

Uptake rates for NO_3^- were low with no differences being observed between species or sampling sites (Fig. 6.3). In contrast, uptake rates of NH_4^+ showed large variations between species and sites, values ranging from 0.14 to 0.6 $\mu mol\, g^{-1}\, dw\, h^{-1}$. At the N concentration used in the bioassay, uptake rates for NH_4^+ were on average 21 ± 5 (n = 6) times higher than those for NO_3^-. In the case of *P. abies* there was a decline up to a factor 4 in NH_4^+ uptake rates in central Europe compared to those in northern Sweden (one-way ANOVA, $p < 0.001$; Fig. 6.3). A similar pattern was found for broad-

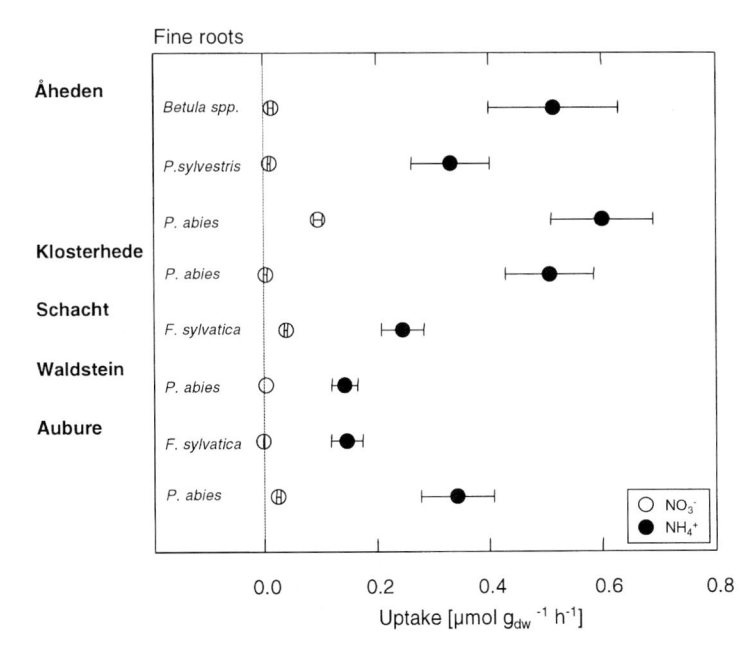

Fig. 6.3. Uptake of NO_3^- (○) and NH_4^+ (●) by excised fine roots at a single substrate concentration (70 µM) at the different sites. Means ± SE (n = 8–15)

leaved trees: the NH_4^+ uptake rates of *Betula* spp. at Åheden were two to three times higher than those obtained for *Fagus sylvatica* at the central European sites (Fig. 6.3).

Although highest NRA was observed at the sites in central Europe, NO_3^- uptake by fine roots collected at these sites did not differ from that found in northern Europe. It has frequently been shown that NH_4^+ and amino acids can repress NO_3^- uptake in both herbaceous (Lee and Drew 1989; Padgett and Leonard 1993) and tree species (Kreuzwieser et al. 1997; Fig. 6.9, this Chap.) when supplied at similar concentrations. A similar repression may have taken place at the investigated sites if NH_4^+-N concentrations reached appropriate levels in the soil solution.

Jones et al. (1991) developed the root N uptake bioassay employed as a test for the N status of plants. They showed that NH_4^+ uptake was inversely related to internal N concentrations, which may reflect a downregulation of NH_4^+ uptake systems in plants of high N status. It was also demonstrated experimentally (Jones et al. 1991, 1994) that N-fertilised trees had a lower potential uptake than N-limited trees from solutions with similar N concentration. Correspondingly, we found the highest uptake rates for NH_4^+ at Åheden, where fine roots had the lowest N concentration (Fig. 6.1). Wang et al. (1993) demonstrated the involvement of a negative-feedback regulation of NH_4^+ availability on high-affinity NH_4^+ uptake at both levels, affinity and capacity of the uptake system in rice. Similarly, Kronzucker et al. (1996a) observed a negative-feedback control for NH_4^+ uptake by *Picea glauca* roots. In addition, it is necessary to recognise that differences in N availability also have effects upon root morphology (Seith et al. 1996; Wallenda and Kottke 1998). In our study of inorganic N uptake by excised roots, a size category of less than 2 mm was selectively used. Under the conditions of increased N deposition experienced by the central European sites, larger average root diameters within this category are to be expected, with a corresponding reduction of uptake per unit of dry weight.

6.3.3 Kinetics of Amino Acid Uptake by Ectomycorrhizas and Non-Mycorrhizal Roots (Experimental Sets 2 + 3)

Increasing substrate concentration resulted in increased apparent uptake rates with no saturation observed over a range of 0.001 to 10 mM (Fig. 6.4a). Correspondingly, typical Eadie-Hofstee plots (of apparent uptake rate against the ratio of this rate to substrate concentration) showed a biphasic kinetic (for example Fig. 6.4c). This curvilinear and upwardly concave nature of the Eadie-Hofstee plots for amino acid uptake by ectomycorrhizal roots necessitates a dual or multiphasic interpretation of the systems responsible for uptake. Similar curves have been observed for a wide range of substrates and non-mycorrhizal species (see Reinhold and Kaplan 1984; Nissen 1991 for review). The biphasic nature of the curves indicates that both *high* and *low affinity transport systems* (HATS and LATS, respectively, Wang et al. 1993) are functioning.

In accordance with these results, the addition of the metabolic inhibitor 2,4-dinitrophenol at a concentration of 0.5 mM to equilibration and labelled uptake solution resolved a linear, non-inhibitable low-affinity component in the total apparent amino acid uptake, which contributed substantially to total amino acid uptake at concentrations >0.5 mM (Fig. 6.4a). Similarly, Chalot et al. (1996), working with *Paxillus involutus*, also observed a linear amino acid uptake component which was independent of that following Michaelis-Menten kinetics and which was not inhibited by the

Nitrogen Uptake Processes in Roots and Mycorrhizas

Fig. 6.4a–c. Uptake of ^{14}C-labelled glutamine into excised *Lactarius subdulcis*/*Fagus sylvatica* mycorrhizas as a function of glutamine concentration: glutamine uptake in the range of 0 to 10 mM (**a**) or 0 to 0.5 mM (**b**) in the absence (——) or presence (·····) of additional 0.5 mM 2,4-dinitrophenol. Active uptake (-----) calculated as difference of means. Eadie-Hofstee plot of glutamine uptake in the absence of DNP in a substrate range of 0.001–10 mM (**c**). Data are means ± SE ($n = 4$). Sampling site was Hathersage

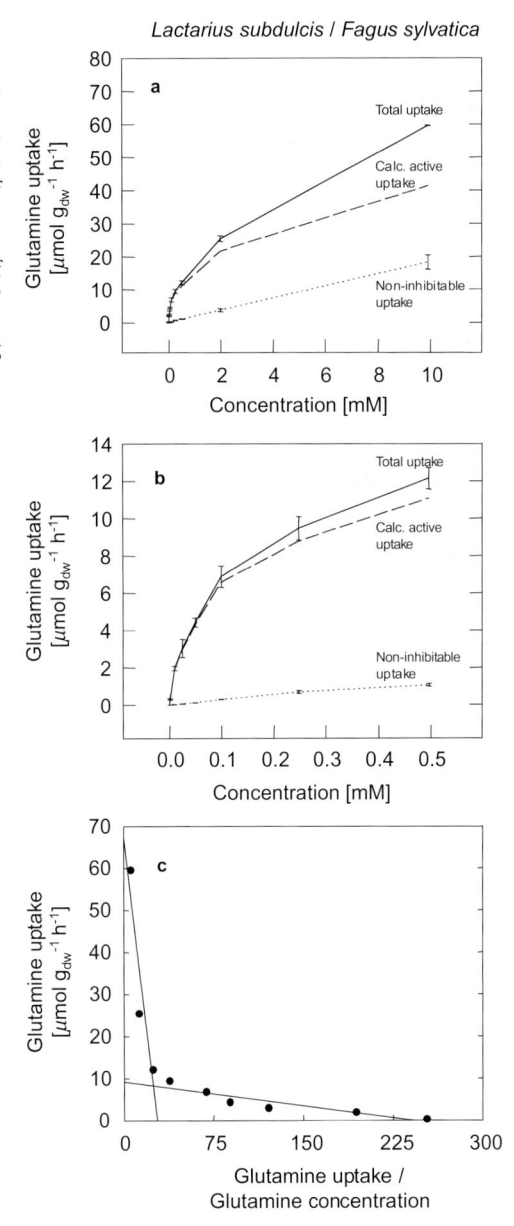

uncoupler 2,4-dinitrophenol (DNP). In our study, apparent uptake followed Michaelis-Menten kinetics up to concentrations of 0.5 mM (Fig. 6.4b). However, for several samples apparent uptake rates at a substrate concentration of 0.5 mM diverged from simple Michaelis-Menten kinetics (see, e.g. Fig. 6.6). Therefore, K_M values were calculated for a substrate concentration of 0.001 to 0.25 mM (mycorrhizal roots) and 0.001 to 0.35 mM (non-mycorrhizal roots), respectively. In this concentration range the contribution by the linear, non-inhibitable uptake component to the total apparent uptake was negligible and amino acid uptake took place via high-affinity transport systems. Correspondingly, calculation of apparent V_{max} and K_M values resulted in similar kinetic parameters regardless of whether uptake rates in the presence of DNP were taken into account or not (Wallenda and Read 1999). As the low-affinity uptake systems contribute to total uptake only at higher substrate concentrations, high affinity systems are likely to be by far the most important over the µmolar range of amino acid concentrations found in soil solutions of tundra and forest systems (Chapin et al. 1993; Kielland 1994).

Uptake of ^{14}C labelled glutamine by the often predominant mycorrhizal type in *Fagus*, *Lactarius subdulcis*, could be competitively inhibited by the addition of unlabelled glycine to equilibration and uptake solutions (Fig. 6.5a). The inhibition constant (K_i) for glycine was 0.175 mM. The presence of inorganic N sources (NH_4^+ or NO_3^-) in equilibration and uptake solutions had no effect on the uptake of glutamine (Fig. 6.5b) thus suggesting that these ions will not affect in situ uptake rates of amino acids. In a related study, Chalot et al. (1995) also showed that neither NH_4^+ nor NO_3^- had any effect on amino acid uptake by *Paxillus involutus* when applied at concentrations usually found in forest soil solutions (0.05–0.5 mM). These results therefore indicate that the processes involved in uptake of organic and inorganic N sources operate independently. Inhibition of ^{14}C-glutamine uptake by *L. subdulcis*/*F. sylvatica* mycorrhizas by non-labelled glycine indicates a rather unspecific transport system which could resemble the general amino acid permease reported for the ectomycorrhizal fungus *P. involutus* (Chalot et al. 1996) and yeasts (Sophianopoulou and Diallinas 1995). Other factors, which might be involved in regulation of amino acid uptake, include inducible transport systems and feedback control. Effects of these

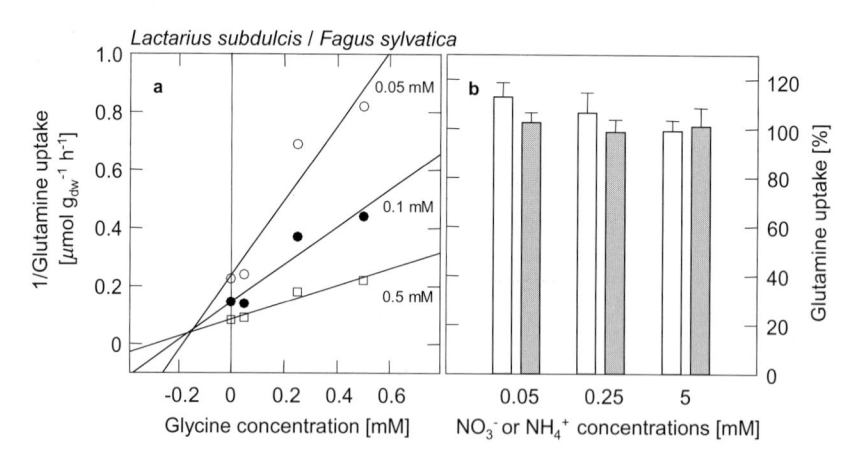

Fig. 6.5a,b. Effect of organic and inorganic N sources added to preincubation and uptake medium on the uptake of ^{14}C-glutamine by excised *Lactarius subdulcis*/*Fagus sylvatica* mycorrhizas. **a** Dixon plot of the inhibition by non-labelled glycine. Glutamine concentrations were 0.05 mM (○), 0.1 mM (●) and 0.5 mM (□). **b** Relative ^{14}C-glutamine uptake in the presence of NH_4^+ (applied as NH_4Cl, *empty bars*) or NO_3^- (applied as KNO_3, *filled bars*); 100% represents uptake in the absence of these ions and values are averages for three (0.05, 0.1, 0.5 mM) glutamine concentrations ($n = 10$–12)

kinds have so far only been characterised for inorganic N uptake in plants (von Wirén et al. 1997 and see above) and amino acid transporters of non-mycorrhizal lower eukaryotes (Sophianopoulou and Diallinas 1995).

6.3.4 Kinetic Parameters of Amino Acid Uptake Along a European North-South Transect (Experimental Set 2)

Lactarius subdulcis/Fagus sylvatica mycorrhizas showed almost identical uptake kinetics for glutamine at the three beech sampling sites (Fig. 6.6, Table 6.1). Higher apparent V_{max} and K_M values were determined for glycine, with greater variability among the sites (Wallenda and Read 1999). The mycorrhizas formed by *Russula ochroleuca* had the lowest apparent V_{max} values of all morphotypes independent of the sampling site, the host plant, or the type of amino acid supplied. Their apparent K_M values for glutamine were in the same range as those calculated for other morphotypes. The K_M values for glycine in *Russula ochroleuca* roots were, however, more variable, as they were, also in the case of *Lactarius subdulcis*, mycorrhizas. At Gribskov, *Russula ochroleuca* mycorrhizas collected from two different plant species showed similar glycine uptake kinetics independent of the nature of the host plant. A number of unidentified morphotypes showed apparent V_{max} values ranging from 7.7 up to 24.7 µmol g^{-1} dw h^{-1} and K_M values between 19 and 130 µM (glutamine). The corresponding values for glycine were 8.3–34.1 µmol g^{-1} dw h^{-1} and 21–113 µM (Table 6.1).

The kinetic data show that all the mycorrhizal types investigated were able to take up amino acids at rates which were significantly higher than those observed for NH_4^+ and NO_3^-. K_M values obtained for the uptake of amino acids were generally in the same range (9–377 µM) as those observed by Kielland (1994) for mycorrhizal fine roots of a number of arctic plant species. The lowest recorded K_M was seen in an unidentified morphotype of *F. sylvatica*, which produced rates of uptake similar to those calculated for pure cultures of the fungus *Paxillus involutus* by Chalot et al. (1996).

With the exception of ectomycorrhizas formed by *Russula ochroleuca*, V_{max} values were higher than those reported by Kielland (1994). This, however, may be attributable to the selection, in our study, of an incubation temperature of 20 °C, which was 6 °C higher than that used by Kielland. Clearly, in vivo uptake rates will be lower at naturally occurring soil temperatures but the relative rates of assimilation of organic and inorganic N sources are unlikely to be changed. Our data clearly show that mycorrhizal roots possess high-affinity transport systems which should enable them effectively to take up amino acids from soil solutions where concentrations of these organic sources, which are reported to be in the range 10–100 µM, can often exceed those of inorganic N (Abuarghub and Read 1988a, b; Németh et al. 1988; Kielland 1994; Näsholm et al. 1998; see Andersen and Gundersen, Chap. 15, this Vol.). Uptake capacities (V_{max}) and substrate affinity (K_M) of the investigated systems indicate the potential for a substantial uptake of organic N from the soil solution. In so doing, they will enable their host plant to be, at least partly, independent of the activities of separate populations of decomposers for ammonium release by mineralisation. The validity of this conclusion has recently been shown by Näsholm et al. (1998), who injected dual-labelled (^{13}C, ^{15}N) and ^{15}N-labelled NH_4^+ into the mor layer of a boreal forest. Ratios of $^{13}C:^{15}N$ in the roots showed that at least 40% of the N from the

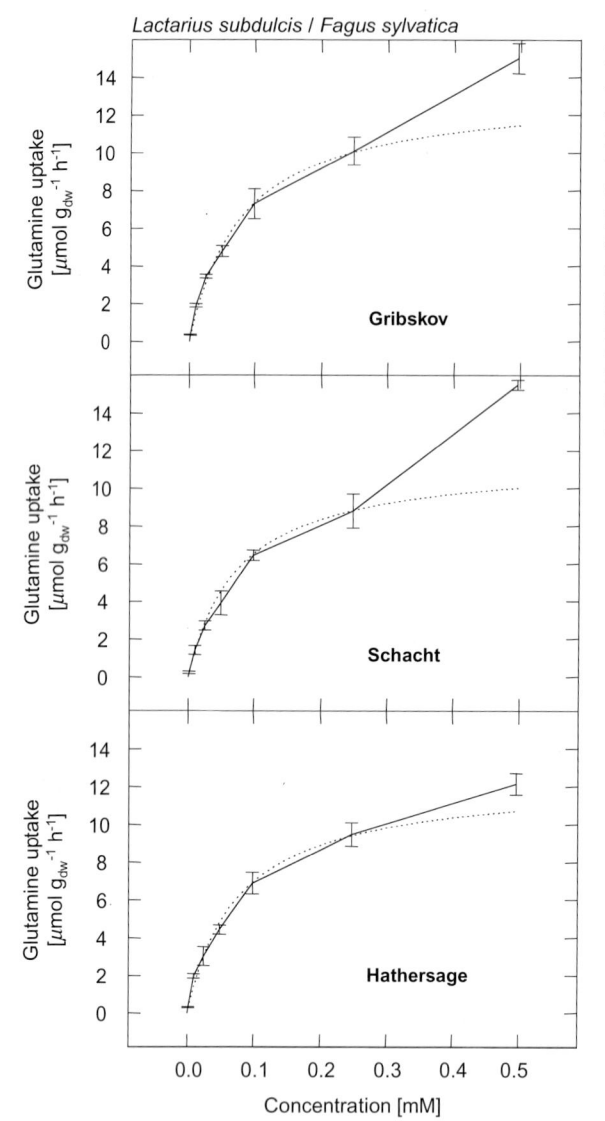

Fig. 6.6. Uptake of ^{14}C-labelled glutamine by excised *Lactarius subdulcis*/*Fagus sylvatica* mycorrhizas as a function of glutamine concentration in the range of 0 to 0.5 mM at three different sampling sites. Curves representing measured data (——) or calculated data using kinetic parameters obtained by non-linear regression of the Michaelis-Menten equation (·····) in a substrate concentration range of 0.001–0.25 mM. Data are means ± SE ($n = 4$)

absorbed glycine was taken up intact. In addition, the relative amount of glycine taken up was similar to that of ^{15}N-ammonium. Together, these data indicate that organic N is important for mycorrhizal plants, even when they are competing with each other and with non-symbiotic microorganisms.

Non-mycorrhizal fine roots (Ø < 1 mm) of four tree species had V_{max} and K_M values in the same range as those found in the different mycorrhizal types. However, most of the mycorrhizal morphotypes of *F. sylvatica* had higher V_{max} values than non-mycorrhizal roots. Only *R. ochroleuca-F. sylvatica* mycorrhizas had V_{max} values for uptake of glutamine and glycine that were similar to those of non-mycorrhizal roots.

Table 6.1. Kinetic parameters for amino acid uptake by mycorrhizal roots and fine roots collected along a European north-south transect and fine roots collected from non-mycorrhizal seedlings grown in peat in the greenhouse. V_{max} and K_M values (±standard errors of the fitted parameters) were calculated by non-linear curve fitting of the experimental data to the Michaelis-Menten equation [$v = (V_{max} \cdot [S])/(K_M + [S])$] in a substrate concentration range of 0.001–0.25 mM (mycorrhizal roots) or 0.001–0.35 mM (non-mycorrhizal roots)

Location	Tree species	Glutamine V_{max} (µmol g$_{dw}^{-1}$ h^{-1})	K_M (µM)	Glycine V_{max} (µmol g$_{dw}^{-1}$ h^{-1})	K_M (µM)	Glutamic acid V_{max} (µmol g$_{dw}^{-1}$ h^{-1})	K_M (µM)
Lactarius subdulcis							
Gribskov	Fagus sylvatica	13.3 ± 1.0	81 ± 14	18.1 ± 1.5	105 ± 18	n.a.	n.a.
Schacht	Fagus sylvatica	11.6 ± 1.2	77 ± 18	14.5 ± 1.1	102 ± 16	n.a.	n.a.
Hathersage	Fagus sylvatica	12.4 ± 0.9	79 ± 13	22.8 ± 4.2	197 ± 67	10.1 ± 1.3	115 ± 33
Russula ochroleuca							
Gribskov[a]	Fagus sylvatica	5.3 ± 0.6	79 ± 21	4.4 ± 0.9	127 ± 49	n.a.	n.a.
	Picea abies	n.a.[b]	n.a.	4.8 ± 0.6	135 ± 30	n.a.	n.a.
Hathersage	Fagus sylvatica	1.4 ± 0.1	86 ± 20	3.1 ± 0.3	233 ± 35	n.a.	n.a.
Unidentified morphotypes							
Åheden	Pinus sylvestris	11.6 ± 1.6	130 ± 38	8.3 ± 1.0	50 ± 17	n.a.	n.a.
	Picea abies	7.7 ± 1.4	36 ± 22	10.7 ± 1.8	42 ± 20	n.a.	n.a.
Gribskov[a]	Picea abies	24.7 ± 3.1	104 ± 29	n.a.	n.a.	n.a.	n.a.
Waldstein	Picea abies	24.0 ± 6.7	111 ± 59	34.1 ± 5.6	113 ± 38	n.a.	n.a.
Schacht	Fagus sylvatica	14.7 ± 1.6	19 ± 8	17.9 ± 1.7	21 ± 7	n.a.	n.a.
Non-mycorrhizal roots							
Greenhouse,	Pinus sylvestris	28.1 ± 3.6	136 ± 43	31.1 ± 5.2	89 ± 41	22.3 ± 2.5	86 ± 28
Sheffield	Picea abies	13.4 ± 1.4	33 ± 13	28.1 ± 2.9	107 ± 28	33.3 ± 6.2	107 ± 51
	Betula pubescens	8.8 ± 3.6	219 ± 179	5.3 ± 0.8	83 ± 34	6.0 ± 0.9	92 ± 37
	Fagus sylvatica	4.0 ± 0.7	105 ± 47	3.2 ± 0.5	64 ± 30	3.8 ± 0.7	88 ± 48

[a] Mycorrhizas were collected in a Norway spruce forest adjacent to the beech site at Gribskov.
[b] n.a.: not available.

V_{max} values of non-mycorrhizal roots of the broad-leaved tree species *B. pubescens* and *F. sylvatica* were significantly lower ($p = 0.002$) than those found in the coniferous species. For all amino acids tested, *B. pubescens* showed higher V_{max} values than *F. sylvatica*. It therefore appears that mycorrhizal colonisation is particularly important for uptake of soluble organic N sources by broadleaf tree species. Besides this enhanced uptake of soluble organic N sources especially for the broadleaf species, mycorrhizal colonisation will provide access to insoluble organic N sources (see Taylor et al., Chap. 16, this Vol.) and mobilisation of nutrients far from host roots by a vigorous vegetative mycelium (Read 1991 and see above).

When examining identical morphotypes, we observed no obvious relationships between the apparent V_{max} and K_M values and the location of the sampling site on the N deposition gradient (Table 6.1). Similarly, no apparent correlation existed between sample site location and apparent uptake rates of various ectomycorrhizal morphotypes at a single (40 µM) amino acid concentration (Fig. 6.7). This suggests that increasing availability of inorganic N will have no effect on the capabilities of these ectomycorrhizal types to take up amino acids. In accordance with the low V_{max} values observed in non-mycorrhizal roots of *B. pubescens* and *F. sylvatica*, these also had the lowest uptake rates for all amino acids tested. For both species all mycorrhizal roots had higher uptake rates at a substrate concentration of 40 µM.

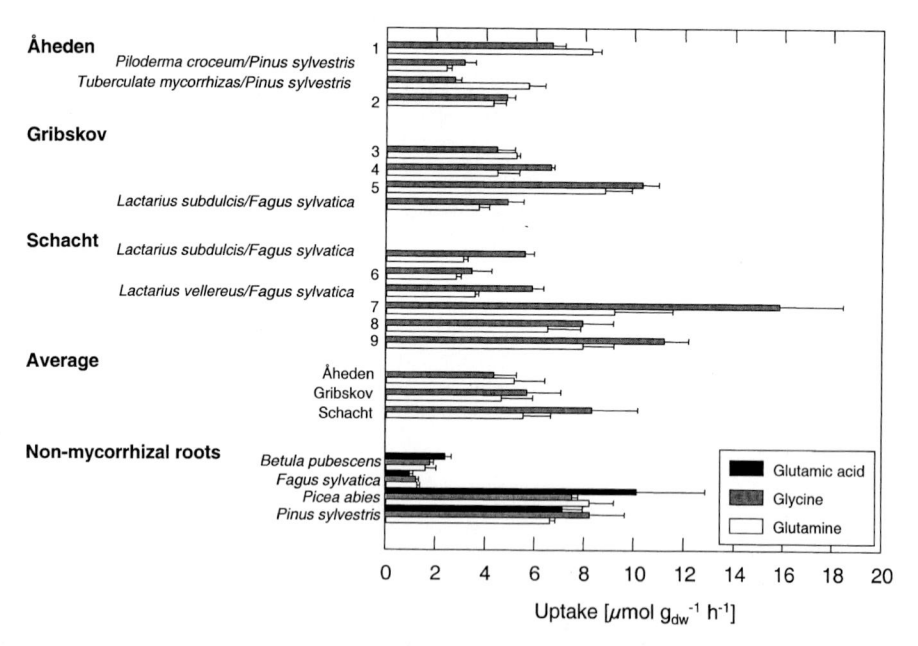

Fig. 6.7. Uptake of amino acids by mycorrhizal roots collected along a European north-south transect and non-mycorrhizal fine roots (Ø < 1 mm) at a single substrate concentration (40 µM). Average uptake rates for glycine or glutamine were not significantly different between different sampling sites (ANOVA, $p = 0.05$). FIGURES ON THE ORDINATE represent unidentified mycorrhizal morphotypes: *1 P. sylvestris* morphotype; *2 B. pubescens* morphotype; *3–9 F. sylvatica* morphotypes

6.4 Field-Based Experiments

6.4.1 Nitrogen Uptake by Single, Intact Roots in the Field (Experimental Set 4)

In this set of experiments, uptake of N was defined as the amount of ^{15}N exported from the root part supplied with label to the rest of the plant. Intact root tips of Norway spruce (Figs. 6.8 and 6.9) and beech (data not shown) took up both NH_4^+ and NO_3^-. Uptake rates were usually higher for NH_4^+ than for NO_3^-. This confirms several previous studies at other forest sites (Marschner et al. 1991; Buchmann et al. 1995; Gessler et al. 1998) and under controlled conditions (Kronzucker et al. 1995, 1996b; Gessler et al. 1998). Ammonium uptake rates were high in particular at Åheden, as was also shown in a test with excised roots (Fig. 6.3). This suggests that roots at Åheden, the site with lowest concentrations of mineral N in the soil solution (see Andersen and Gundersen, Chap. 15, this Vol.), in addition to having low N concentrations (Fig. 6.1 and Stober et al., Chap. 5, this Vol.), also have a particularly high potential ability to take up NH_4^+ when supplied with freely available amounts of this N form. In contrast, NO_3^- uptake rates were low in Åheden and highest in Waldstein, a site where the solution in the mineral soil is dominated by high NO_3^- concentrations (see Andersen and Gundersen, Chap. 15, this Vol.). Nitrogen uptake rates of beech (data not shown) were in most cases not significantly different from those of Norway spruce. The ability to take up organic N was not examined in this experiment.

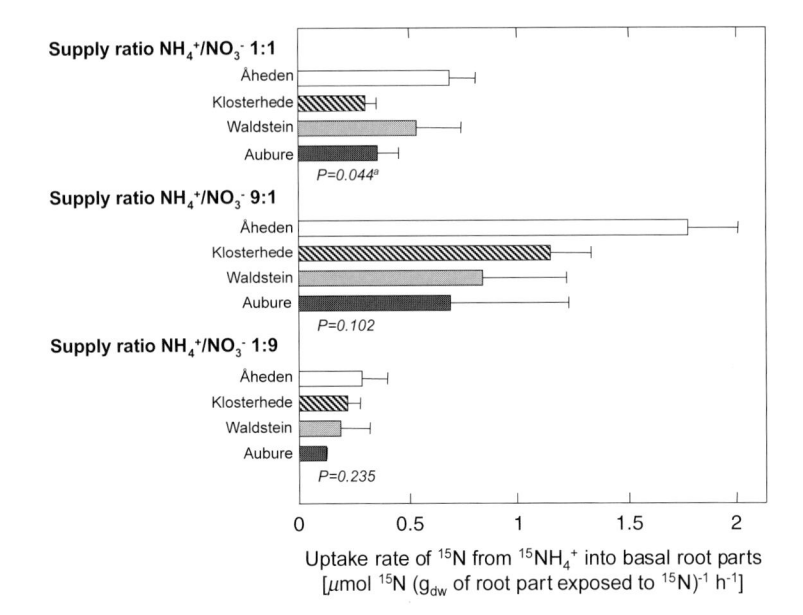

Fig. 6.8. Uptake of ^{15}N into basal parts of intact roots of Norway spruce after supply of $^{15}NH_4^+$ in agar blocks to excavated root tips at different European forest sites. Nitrogen was supplied in different NH_4^+/NO_3^- ratios. Means ± SE (n = 3–12); p probability of incorrectly rejecting the null hypothesis; [a] data were square root-transformed before ANOVA

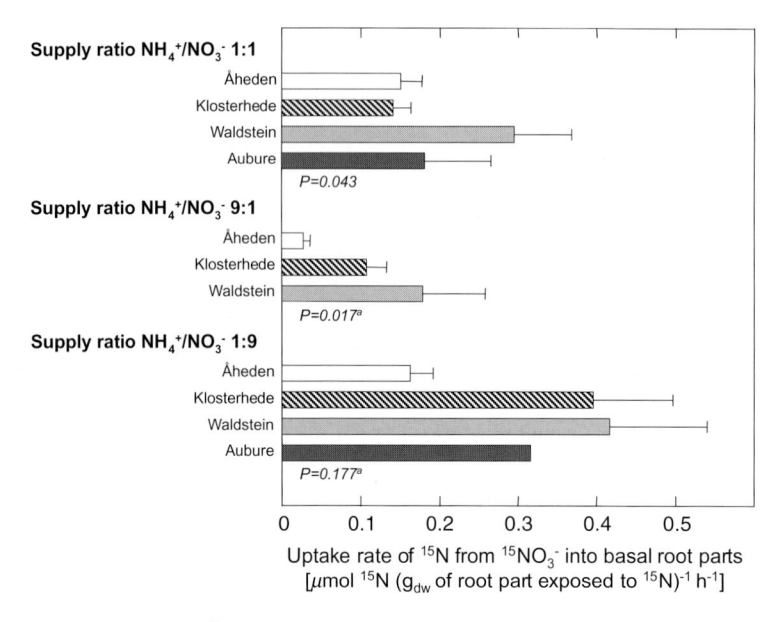

Fig. 6.9. Uptake of ^{15}N into basal parts of intact roots of Norway spruce after supply of ^{15}NO$_3^-$ in agar blocks to excavated root tips at different European forest sites. Nitrogen was supplied in different NH$_4^+$/NO$_3^-$ ratios. Means ± SE (n = 3–8); p probability of incorrectly rejecting the null hypothesis; [a] data were log-transformed before ANOVA

Intact primary roots of Norway spruce (Figs. 6.8 and 6.9) and of beech (data not shown) took up substantial amounts of both NH$_4^+$ and NO$_3^-$. Norway spruce at all sites examined (Åheden, Klosterhede, Waldstein and Aubure) took up more NH$_4^+$ than NO$_3^-$. In contrast to excised roots, intact roots also took up considerable amounts of NO$_3^-$ (Fig. 6.9), in particular when NH$_4^+$/NO$_3^-$ ratios in the supply solution were reduced to levels often found in solutions of the mineral forest soil (see Andersen and Gundersen, Chap. 15, this Vol.; Mitchell et al. 1992; Daldoum and Ranger 1994; Manderscheid and Matzner 1995; Ingerslev 1997). The contrasting behaviour of intact vs. excised roots in NO$_3^-$ uptake is probably related to different water fluxes in the two systems and to the absence of shoot signals in excised roots which are governing the nutrient uptake of intact roots (Marschner et al. 1997).

At a NH$_4^+$/NO$_3^-$ supply ratio of 1:9, not uncommon in the solution of the mineral soil at sites in Central Europe (see Andersen and Gundersen, Chap. 15, this Vol.), NO$_3^-$ uptake was at least as high as NH$_4^+$ uptake (Figs. 6.8 and 6.9). As experiments were carried out at one level of total N supply, a change in NH$_4^+$/NO$_3^-$ ratio was equivalent also to a change in NH$_4^+$ or NO$_3^-$ supply concentration. Thus, in agreement with model studies (George and Seith 1998; George et al. 1999a,b), the present results show that, although roots of Norway spruce roots prefer NH$_4^+$ over NO$_3^-$, they can also utilise NO$_3^-$ and amino acids, and actual uptake rates are largely determined by the supply rates and concentrations of the mineral N forms. In future field studies, the effect of the different organic N constituents of the soil solution on uptake of mineral N by primary roots and mycorrhizas should be of main interest.

6.4.2 In Situ N Uptake by Roots and Mycorrhizal Mycelium (Experimental Set 5)

When ^{15}N was supplied in different amounts and N forms to mesh bags in soil, penetrable to roots or hyphae close to young beech (at Aubure, Fig. 6.10) or Norway spruce trees, ^{15}N was translocated to the shoots of these trees. No ^{15}N signal was found in trees situated adjacent to mesh bags not penetrable by roots or hyphae, indicating that diffusion of N out of the mesh bags did not take place during the experimental period. The ^{15}N enrichment in the plants was higher when a higher amount of NH_4^+ was supplied to the mesh bags (Fig. 6.10). However, even at a lower rate of supply (70 µmol N tree^{-1}), detectable amounts of ^{15}N tracer were observed in the plants.

Nitrogen uptake rates from higher NH_4^+ supply were not consistently different for beech and Norway spruce (Fig. 6.11). However, NH_4^+ uptake rates in Norway spruce were clearly decreased at Åheden compared to those at Skogaby. The highest uptake rates of this ion were obtained at Aubure.

Nitrogen uptake rates for Norway spruce at the lower rate of N supply were in most cases also much lower at Åheden than at the other sites (Fig. 6.12). As an exception, uptake of NH_4^+ and NO_3^- by hyphae (but not by roots) was considerable at Åheden. Across all sites there was no clear dominance in uptake of any of the N forms tested (NH_4^+, NO_3^-, and glutamic acid). Nitrogen uptake from glutamic acid was at most sites at least as high as N uptake from NH_4^+ or NO_3^-. The results very clearly showed that at all sites and for all N forms supplied (NH_4^+, NO_3^- or glutamic acid), hyphae alone were as effective in N uptake as roots and hyphae together (Figs. 6.11 and 6.12). This suggests that under field conditions, plant N uptake takes place to a large extent by ectomycorrhizal mycelium and not by the (mycorrhizal or non-mycorrhizal) root itself. This is probably related to a high affinity of hyphae for different N forms, but even more to a much higher hyphal length density in soil compared to root length density. Along the north-south transect, hyphal densities of 5–20 m cm^{-3} of soil were measured in in-growth cores in 0–10 cm soil depth, and many of these hyphae were probably from ectomycorrhizal fungi. In comparison, root length densities in the uppermost soil horizons were, as an average of all the sites, approximately

Fig. 6.10. Enrichment of ^{15}N in young leaves of 3- to 5-year-old beech plants at Aubure, 6 weeks after supply of different ^{15}N-labelled N forms to mesh bags in soil penetrable either to roots and hyphae (*Roots*), to mycorrhizal hyphae only (*Hyphae*) or not accessible to roots or hyphae (*Control*). Means ± SE (*n* = 5)

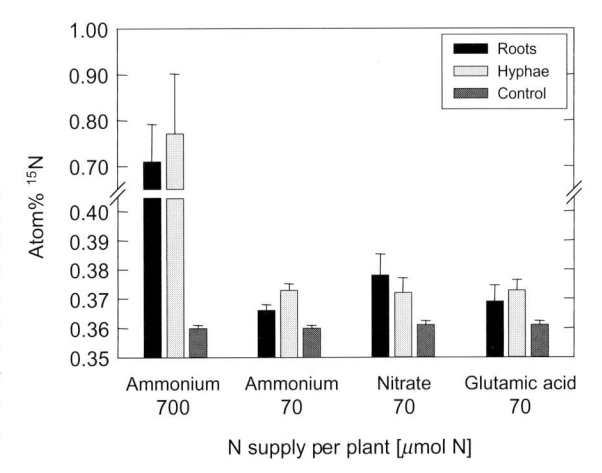

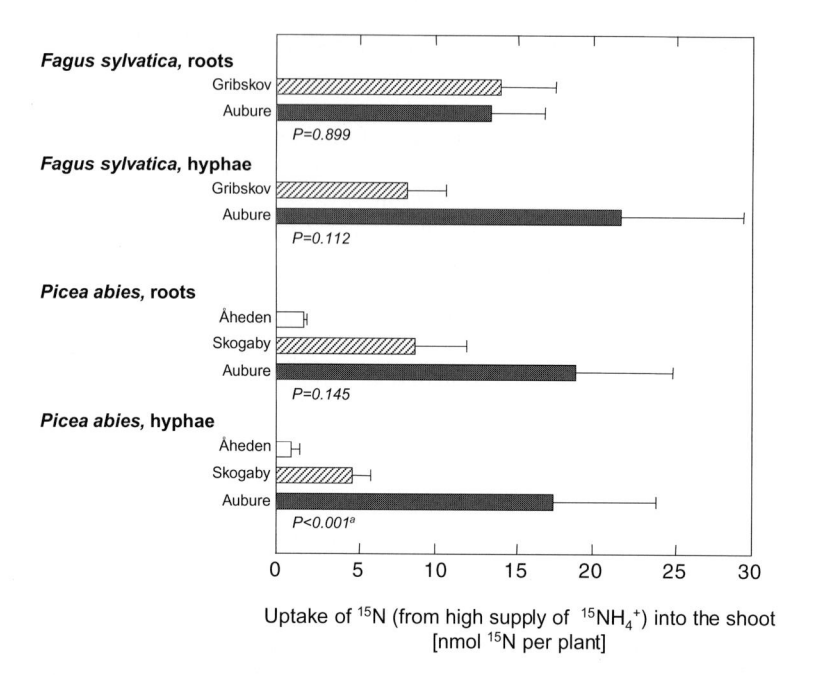

Fig. 6.11. Uptake of ^{15}N into the shoots of 3- to 5-year-old beech and Norway spruce plants at different European forest sites. Nitrogen was supplied at a rate of 700 µmol N in the form of ^{15}NH$_4^+$ to mesh bags in soil penetrable either to roots and hyphae (*Roots*) or to mycorrhizal hyphae only (*Hyphae*). Means ± SE (*n* = 3–10); *p* probability of incorrectly rejecting the null hypothesis; [a] data were log-transformed before ANOVA

5 mm cm^{-3} of soil. Hyphal densities were also high at Aubure, a site with relatively high N deposition. The current results demonstrate that the mycelium of ectomycorrhizal fungi is a vitally important organ of N absorption of Norway spruce and beech even on forest sites with currently high levels of N deposition.

In the mesh bag experiment, plant uptake of N was clearly higher at the central European sites than at Åheden in north Sweden (Figs. 6.11 and 6.12). This reflects higher root and hyphal growth rates at the central European sites (see Stober et al., Chap. 5, this Vol.) and a possible immobilisation of N in the low-N soil at Åheden. These aspects are not considered where N is applied to single roots.

Significant amounts of ^{13}C were found in different plant parts after the ^{13}CO$_2$-labelling period. At Åheden, compared to Skogaby or Aubure, a significantly higher amount of ^{13}C was accumulated in old needles (Fig. 6.13). In contrast, more ^{13}C was accumulated in roots in Aubure compared to Åheden. Thus, at the site with a low root N uptake activity during the study period (Åheden), the plants also allocated little C into the roots, while at the site with high root N uptake activity (Aubure), the plants also allocated large amounts of C into the roots. Thus, C and N cycles may be closely linked within the plants. This hypothesis should be verified in further time-course studies, as measurements in the present study were carried out only once during the growth period.

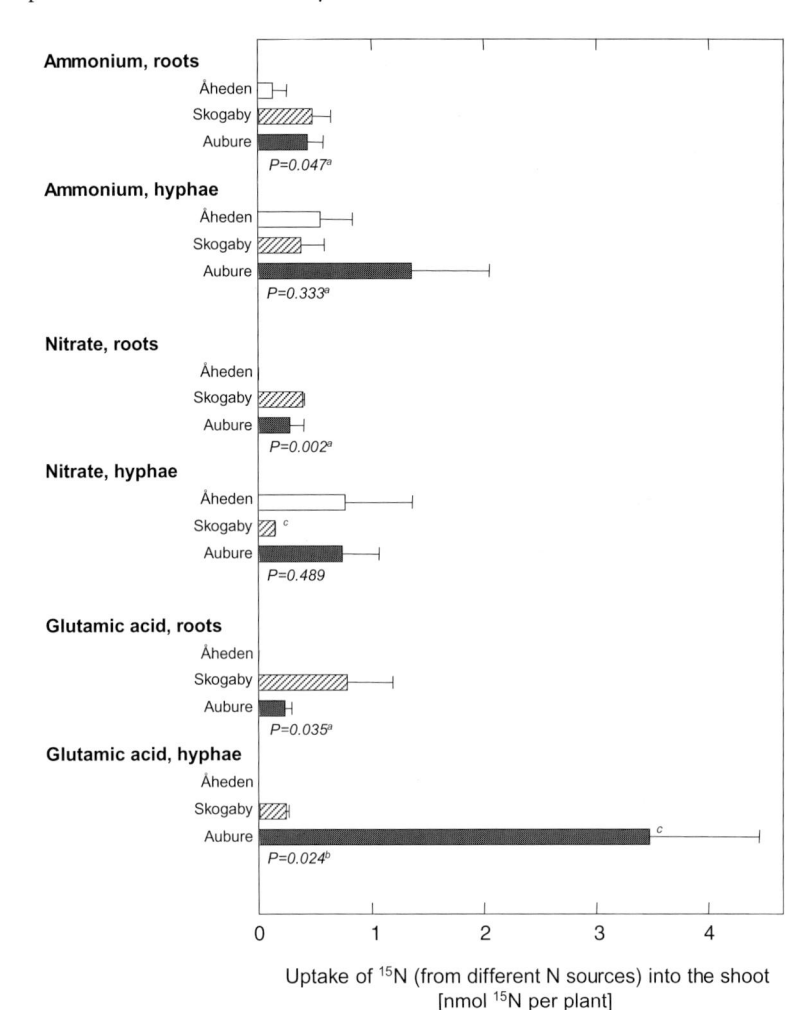

Nitrogen Uptake Processes in Roots and Mycorrhizas

Fig. 6.12. Uptake of ^{15}N into the shoots of 3- to 5-year-old Norway spruce plants at different European forest sites. Nitrogen was supplied at a rate of 70 µmol N in the form of ^{15}NH$_4^+$, ^{15}NO$_3^-$ or ^{15}N-glutamic acid to mesh bags in soil penetrable either to roots and hyphae (*Roots*) or to mycorrhizal hyphae only (*Hyphae*). Means ± SE (n = 3–10); p probability of incorrectly rejecting the null hypothesis; [a] data were square root-transformed before ANOVA; [b] Kruskal-Wallis ANOVA on ranks; [c] significant ($p < 0.01$) difference between roots and hyphae within the same site, data were square root-transformed before ANOVA

6.5 Conclusions

- Our data clearly indicate that within a single host species there can be considerable differences in the kinetic properties of amino acid uptake by ectomycorrhizal systems which are determined by the fungal partner. Therefore, the actual composition of the below-ground ectomycorrhizal community appears to determine the

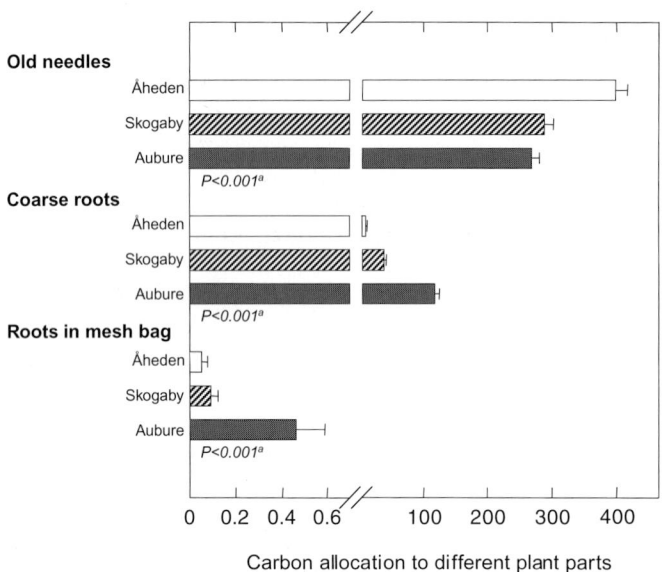

Fig. 6.13. Allocation of ^{13}C into different parts of 3- to 5-year-old Norway spruce plants at different European forest sites, 6 days after labelling the shoot with $^{13}CO_2$. Data are averages for different N supply treatments. Means ± SE ($n = 50–68$); p probability of incorrectly rejecting the null hypothesis; [a] data were square root-transformed before ANOVA

extent to which ectomycorrhizal plants can supplement their acquisition of N through the uptake of amino acids.
- Our studies along the gradient of N deposition suggest that increasing availability of inorganic N will have no effect on the capabilities of these ectomycorrhizal types to take up amino acids. However, the possibility that long-term exposure to inorganic N deposition could change the structure of ectomycorrhizal fungal communities and thereby affect patterns of amino acid uptake should not be ignored (Wallenda and Kottke 1998; Taylor et al., Chap. 16, this Vol.).
- High-affinity transport systems for amino acids described in this study are also a prerequisite for the ability of ectomycorrhizal fungi and roots to use polymeric peptides and proteins as N sources (Abuzinadah and Read 1986a,b) after cleavage into amino acids by extracellular proteases (Zhu et al. 1990 and see Taylor et al., Chap. 16, this Vol.).
- The present studies did not indicate that N uptake patterns were fundamentally different between Norway spruce and beech. Rather, tree N uptake was characterised by a high diversity, where all N forms available to the plants appeared to be used. In addition, recalcitrant N forms can be made available by plant or fungal exudates or ectoenzymes.
- This diversity is very likely related to an N demand of plants on all sites, a high species diversity of ectomycorrhizal fungi, but also to the ability of these trees to use different forms of N, depending on their availability in soil. The contribution of

organic N forms to plant N uptake has been underestimated in many previous studies and is not restricted to sites with extreme deficiency of soil mineral N forms.
- Further research on the diversity of ectomycorrhizal species in forest ecosystems and their physiological properties is necessary to evaluate the possible connection between the close association of organic substrates and ectomycorrhizal mycelium (Harley 1978; Bending and Read 1995) to gain more insight into the complex processes involved in the acquisition of organic and inorganic N sources in N-limited forest ecosystems and their interaction with an increased availability of inorganic N sources.

References

Abuarghub SM, Read DJ (1988a) The biology of mycorrhizas in the Ericaceae. XI. The distribution of nitrogen in soil of a typical upland *Callunetum* with special reference to the "free" amino acids. New Phytol 108:425–431

Abuarghub SM, Read DJ (1988b) The biology of mycorrhizas in the Ericaceae. XII. Quantitative analysis of individual "free" amino acids in relation to time and depth in the soil profile. New Phytol 108:433–441

Abuzinadah RA, Read DJ (1986a) The role of proteins in the nitrogen nutrition of ectomycorrhizal plants. I. Utilisation of peptides and proteins by ectomycorrhizal fungi. New Phytol 103:481–493

Abuzinadah RA, Read DJ (1986b) The role of proteins in the nitrogen nutrition of ectomycorrhizal plants. III. Protein utilisation by *Betula*, *Picea* and *Pinus* in mycorrhizal association with *Hebeloma crustuliniforme*. New Phytol 103:507–514

Agerer R (1987–1998) Colour atlas of ectomycorrhizae. Einhorn-Verlag Schwäbisch-Gmünd

Bauer G, Schulze E-D, Mund M (1997) Nutrient contents and concentrations in relation to growth of *Picea abies* and *Fagus sylvatica* along a European transect. Tree Physiol 17:777–786

Bending GD, Read D (1995) The structure and function of the vegetative mycelium of ectomycorrhizal plants. VI. Activities of nutrient mobilising enzymes in birch litter colonised by *Paxillus involutus* (Fr.) Fr. New Phytol 130:411–417

Brand F (1989) Studies on ectomycorrhizae XXI. Beech ectomycorrhizae and rhizomorphs of *Xerocomus chrysenteron* (Boletales). Nova Hedwigia 48:469–483

Brand F (1991) Ektomykorrhizen an *Fagus sylvatica*. Charakterisierung und Identifizierung, ökologische Kennzeichnung und unsterile Kultivierung. Libri Bot 2:1–229

Brand F, Agerer R (1986) Studien an Ektomykorrhizen VIII. Die Mykorrhizen von *Lactarius subdulcis*, *Lactarius vellereus* und *Laccaria amethystina* an Buche. Z Mykol 52:287–320

Buchmann N, Schulze E-D, Gebauer G (1995) ^{15}N-ammonium and ^{15}N-nitrate uptake of a 15-year-old *Picea abies* plantation. Oecologia 102:361–370

Chalot M, Kytöviita M-M, Brun A, Finlay RD, Söderström B (1995) Factors affecting amino acid uptake by the ectomycorrhizal fungus *Paxillus involutus*. Mycol Res 99:1131–1138

Chalot M, Brun A, Botton B, Söderström B (1996) Kinetics, energetics and specificity of a general amino acid transporter from the ectomycorrhizal fungus *Paxillus involutus*. Microbiology 142:1749–1756

Chapin FS III, Moilanen L, Kielland K (1993) Preferential use of organic nitrogen for growth by a non-mycorrhizal arctic sedge. Nature 361:150–153

Daldoum MA, Ranger J (1994) The biogeochemical cycle in a healthy and highly productive Norway spruce (*Picea abies*) ecosystem in the Vosges, France. Can J For Res 24:839–849

Department of the Environment (1994) Impacts of nitrogen deposition in terrestrial ecosystems. Report of the United Kingdom Review Group on Impacts of Atmospheric Nitrogen. Department of the Environment, London

Gebauer G, Rehder H, Wollenweber B (1988) Nitrate, nitrate reduction and organic nitrogen in plants from different ecological and taxonomical groups of Central Europe. Oecologia 75:371–385

George E, Seith B (1998) Long-term effects of a high nitrogen supply to soil on growth and nutritional status of young Norway spruce trees. Environ Pollut 102:301–306

George E, Kircher S, Schwarz P, Tesar A, Seith B (1999a) Effect of high soil nitrogen supply on growth and nutrient uptake of young Norway spruce plants grown in a shaded environment. J Plant Nutr Soil Sci 162:301–307

George E, Stober C, Seith B (1999b) The use of different soil nitrogen sources by young Norway spruce plants. Trees 13:199–205

Gessler A, Schneider S, von Sengbusch D, Weber P, Hanemann U, Huber C, Rothe A, Kreutzer K, Rennenberg H (1998) Field and laboratory experiments on net uptake of nitrate and ammonium by the roots of spruce (*Picea abies*) and beech (*Fagus sylvatica*) trees. New Phytol 138:275–285

Giesler R, Högberg M, Högberg P (1998) Soil chemistry and plants in Fennoscandian boreal forests as exemplified by a local gradient. Ecology 79:119–137

Harley JL (1978) Ectomycorrhizas as nutrient-absorbing organs. Proc R Soc Lond 203B:1–21

Högberg P, Granström A, Johansson T, Lundmark-Thelin A, Näsholm T (1986) Plant nitrate reductase activity as an indicator of availability of nitrate in forest soils. Can J For Res 16:1165–1169

Högberg P, Högbom L, Schinkel H, Högberg M, Johannison C, Wallmark H (1996) ^{15}N abundance of surface soils, roots and mycorrhizas in profiles of European forest soils. Oecologia 108:207–214

Högberg P, Högbom L, Schinkel H (1998) Nitrogen-related root variables of trees along an N-deposition gradient in Europe. Tree Physiol 18:823–828

Ingerslev M (1997) Effects of liming and fertilisation on growth, soil chemistry and soil water chemistry in a Norway spruce plantation on a nutrient-poor soil in Denmark. For Ecol Manage 92:55–66

Jones HE, Quarmby C, Harrison AF (1991) A root bioassay test for nitrogen deficiency in forest trees. For Ecol Manage 42:267–282

Jones HE, Högberg P, Ohlsson H (1994) Nutritional assessment of a forest fertilisation experiment in northern Sweden by root bioassays. For Ecol Manage 64:59–69

Kielland K (1994) Amino acid absorption by arctic plants: implications for plant nutrition and nitrogen cycling. Ecology 75:2373–2383

Kreuzwieser J, Herschbach C, Stulen I, Wiersema P, Vaalburg W, Rennenberg H (1997) Interactions of NH_4^+ and L-glutamate with NO_3^- transport processes of non-mycorrhizal *Fagus sylvatica* roots. J Exp Bot 48:1431–1438

Kronzucker HJ, Siddiqi MY, Glass ADM (1995) Compartmentation and flux characteristics of ammonium in spruce. Planta 196:691–698

Kronzucker HJ, Siddiqi MY, Glass ADM (1996a) Kinetics of NH_4^+ influx in spruce. Plant Physiol 110:773–779

Kronzucker HJ, Siddiqi MY, Glass ADM (1996b) Conifer root discrimination against soil nitrate and the ecology of forest succession. Nature 385:59–61

Larsson C-M (1994) Responses of the nitrate uptake system to external nitrate availability: a whole plant perspective. In: Roy J, Garnier A (eds) A whole-plant perspective on carbon-nitrogen interactions. SPB Academic Publishing, The Hague, pp 31–45.

Lee RB, Drew MC (1989) Rapid, reversible inhibition of nitrate influx in barley by ammonium. J Exp Bot 40:741–752

Manderscheid B, Matzner E (1995) Spatial and temporal variation of soil solution chemistry and ion fluxes through the soil in a mature Norway Spruce [*Picea abies* (L.) Karst.] stand. Biogeochemistry 30:99–114

Marschner H, Häussling M, George E (1991) Ammonium and nitrate uptake rates and rhizosphere pH in non-mycorrhizal roots of Norway spruce [*Picea abies* (L.) Karst.]. Trees 5:14–21

Marschner H, Kirkby EA, Engels C (1997) Importance of cycling and recycling of mineral nutrients within plants for growth and development. Bot Acta 110:265–273

Mitchell MJ, Burke MK, Shepard JP (1992) Seasonal and spatial patterns of S, Ca, and N dynamics of a northern hardwood forest ecosystem. Biogeochemistry 17:165–189

Näsholm T, Ekblad A, Nordin A, Giesler R, Högberg M, Högberg P (1998) Boreal forest plants take up organic nitrogen. Nature 392:914–916

Németh K, Bartels H, Vogel M, Mengel K (1988) Organic nitrogen compounds extracted from arable and forest soils by electro-ultrafiltration and recovery rates of amino acids. Biol and Fertil Soils 5:271–275

Nissen P (1991) Multiphasic uptake mechanisms in plants. Int Rev Cytol 126:89–134

Padgett PE, Leonard RT (1993) Regulation of nitrate uptake by amino acids in maize cell-suspension culture and intact roots. Plant Soil 156:156–162

Read DJ (1991) Mycorrhizas in ecosystems. Experientia 47:376–391.

Reinhold L, Kaplan A (1984) Membrane transport of sugars and amino acids. Annu Rev Plant Physiol 35:45–83

Seith B, George E, Marschner H, Wallenda T, Schaeffer C, Einig W, Wingler A, Hampp R (1996) Effects of high soil nutrient supply on Norway spruce (*Picea abies* [L.] Karst.). I. Shoot and root growth and nutrient uptake. Plant Soil 184:291–298

Sophianopoulou V, Diallinas G (1995) Amino acid transporters of lower eucaryotes: regulation, structure and topogenesis. FEMS Microbiol Rev 16:53–75

Srivastava HS (1980) Regulation of nitrate reductase activity in higher plants. Phytochemistry 19:725–733

Von Wirén N, Gazzarrini S, Frommer WB (1997) Regulation of mineral nitrogen uptake in plants. Plant Soil 196:191–199

Wallenda T, Kottke I (1998) Nitrogen deposition and ectomycorrhizas. New Phytol 139:169–187

Wallenda T, Read DJ (1999) Kinetics of amino acid uptake by ectomycorrhizal roots. Plant Cell Environ 22:179–187

Wang MY, Siddiqi MY, Ruth TJ, Glass ADM (1993) Ammonium uptake by rice roots. II. Kinetics of $^{13}NH_4^+$ influx across the plasmalemma. Plant Physiol 103:1259–1267

Zhu H, Guo D-C, Dancik BP (1990) Purification and characterisation of an extracellular acid proteinase from the ectomycorrhizal fungus *Hebeloma crustuliniforme*. Appl Environ Microbiol 56:837–843

7 The Fate of ¹⁵N-Labelled Nitrogen Inputs to Coniferous and Broadleaf Forests

G. Gebauer, B. Zeller, G. Schmidt, C. May, N. Buchmann, M. Colin-Belgrand, E. Dambrine, F. Martin, E.-D. Schulze, and P. Bottner

7.1 Introduction

Nitrogen in forest soils is mainly composed of organic N compounds originating from litterfall. During leaf senescence of the forest vegetation, N compounds are either allocated to perennial tissues or remain in the leaf litter, mainly as polyphenol-protein condensates. For example, senescent beech leaves are composed of 45% cellulose and hemicellulose, 5 to 10% lignin and 25 to 35% brown polyphenol condensates which contain about 70% of the litter N (Berthelin et al. 1994). Beech litter has a C/N mass ratio of 50–70 and evolves into soil organic matter with a C/N ratio ranging from 10 to 30 depending on the humus type. These organic N compounds in forest soils are highly protected from major N losses due to their high chemical stability and low mobility.

These organic N compounds of high molecular weight have to be decomposed into compounds of smaller molecular weight by soil animals, fungi and microorganisms before they can be taken up by forest plants. Mineral N compounds (ammonium and nitrate) are the end products of this slow decomposition process. Net N mineralisation along the European transect ranges from about $70\,kg\,N\,ha^{-1}\,a^{-1}$ (*Picea* stand in Aubure, Vosges Mountains, France) to almost $200\,kg\,N\,ha^{-1}\,a^{-1}$ (*Fagus* site Schacht, Fichtelgebirge, Germany). Net N mineralisation and net nitrification along the European transect are higher in broadleaf than in coniferous forests (see Persson et al., Chap. 14, this Vol.). Exceptional is the boreal forest site at Åheden (N Sweden) with almost no net N mineralisation (see Persson et al., Chap. 14, this Vol.). In boreal forests, organic N compounds of low molecular weight (i.e. amino acids) play an important role as N source for the forest vegetation (Näsholm et al. 1998; see Wallenda et al., Chap. 6, this Vol.).

Mineral N compounds (ammonium and nitrate) from atmospheric deposition are additional N sources available to the forest vegetation. Chronically elevated N inputs from atmospheric deposition can lead to changes in tree growth, understorey species composition, occurrence of forest decline symptoms and nitrate leaching to the groundwater (see e.g. Aber et al. 1989; Schulze 1989; Durka et al. 1994). N deposition is partially taken up by the forest canopy (see Harrison et al., Chap. 8, this Vol.). However, a major part of the N deposition reaches the soil via throughfall deposition. Mineral N from throughfall adds to the soil-borne mineral N as N source available for uptake to plant roots, fungi and soil microbia. Deposition-derived mineral N may, furthermore, be immobilised by pure physicochemical reactions in the soil (Matschonat and Matzner 1996). Throughfall N deposition along the European transect is highest for the central European and some of the southern Scandinavian sites (about $20\,kg\,N\,ha^{-1}\,a^{-1}$) and decreases towards the south (Collelongo, Italy:

Ecological Studies, Vol. 142
E.-D. Schulze (ed.) Carbon and Nitrogen Cycling in European Forest Ecosystems
© Springer-Verlag Berlin Heidelberg 2000

10.8 kg N ha^{-1} a^{-1}) and the north (Åheden, N Sweden: 1.7 kg N ha^{-1} a^{-1}; see Persson et al., Chap. 2, this Vol.). In wide areas of Europe, the molar ammonium-to-nitrate ratio in the throughfall deposition is close to 1:1 (see Harrison et al., Chap. 8, this Vol.).

This chapter summarises results from several reports about studies tracing the fate of mineral N compounds in forest ecosystems, i.e. the partitioning between the soil inorganic N pool and soil organic matter as well as between trees and understorey vegetation. In particular, two major aspects are addressed:

1. The rate of N release from decomposing litter, its fate in the soil and its contribution to the N uptake by adult trees is studied in a 50-year-old beech forest in the Vosges Mountains (France). Many of the previous decomposition experiments in forests were based on litterbag studies. From these studies, rates of mass loss as well as N dynamics in the litter are already well known (see, e.g. Staaf 1980; Blair 1988). However, as external N may accumulate in litter, little is known about the rate and the forms of litter N released into the soil and redistributed within the ecosystem. Information on litter decomposition and litter N mineralisation has been obtained using ^{15}N-labelled crop residues (Bottner et al. 1998) or needle litter (Berg 1988; Preston and Mead 1994a). Up to now, no attempt has been made to trace the fate of litter N on a forest ecosystem scale.

2. The response of forest trees, understorey vegetation and soil microbia to mineral N inputs (ammonium and nitrate) in wet deposition is compared for three forest stands in NE Bavaria (Germany): (1) a young *Picea abies* plantation (15 years old) with a still-open canopy and thus, a dense understorey vegetation, (2) a mature *Picea abies* stand (140 years old) with a dense understorey vegetation and (3) a young mixed broadleaf forest stand (30 years old) composed of *Fagus sylvatica* and *Quercus petraea* with a closed canopy and thus lacking an understorey vegetation. Previous labelling experiments on N inputs to forests were focussed on studying either N fertilisation effects in N-limited forests (e.g. Nambiar and Bowen 1986; Nômmik and Larsson 1992; Preston and Mead 1994b) or simulated chronic N inputs from atmospheric deposition by repeated tracer applications to the same plots (Nadelhoffer et al. 1995, 1999a; Koopmans et al. 1996; Tietema et al. 1998). In contrast, the approach of our studies is focussed on studying in a mechanistic manner the competition among vegetation for ammonium and nitrate from atmospheric inputs and from soil-borne processes after a single N deposition event.

Both approaches use ^{15}N-labelled compounds to trace the fate of mineral N, originating from either decomposing litter or atmospheric deposition, and both approaches focus on tracer mass balances on complete forest ecosystem levels.

7.2 Sites of Investigation

7.2.1 Aubure Site

The site for this experiment (0.1 ha) is located in a dense 50-year-old beech stand in the communal forest of Aubure (Strengbach catchment, Vosges Mountains, NE France) at an altitude of 1080 m (see Persson et al., Chap. 2, this Vol.). Site characteristics are given by Zeller (1998). The soil type varies between dystric cambisol and cambic podzol (FAO 1988) with a poorly developed moder-type humus. The humus layer is composed of an L layer (2–3 cm), an F layer (1 cm) and a discontinuous H layer (1 cm). The soil is acidic, sandy and rich in gravel with a low base saturation

Table 7.1. Soil characteristics at the Aubure beech forest experimental site. Sampling was carried out in 1994. Average mean (± SD) of 5 combined samples consisting of 20 single samples

Horizon (depth)	pH (KCl)	C (%)	N (%)	C/N	SOM (mg g^{-1})	Bulk density (g cm^{-3})	N (kg N ha^{-1})
O$_{f\text{-}h}$	2.85 (0.01)	25.1 (3.1)	1.20 (0.1)	21	430.7 (52.7)	0.018	59.1
0–10 cm	3.07 (0.40)	8.5 (3.2)	0.35 (0.05)	24	147.6 (55.6)	0.60 (0.12)	2100

(Table 7.1). Average height of the trees is 12 m, mean DBH (diameter at breast height) is 11 cm (±4 cm) and tree density is 2700 trees ha^{-1}. Five codominant trees of approximately the same DBH (8 cm) were selected for the experiment. The diameter class of the selected trees (6 to 10 cm) represented 32% of the stand. Herbaceous plants were present only in gaps.

7.2.2 Fichtelgebirge and Steigerwald Sites

Three forest stands differing in tree age, stand structure and species composition were selected.

1. A 15-year-old *Picea abies* plantation located near the village of Wülfersreuth in the Fichtelgebirge (NE Bavaria, Germany) at 670 m altitude: the soil type of this stand is a spodo-dystric cambisol (FAO 1988) developed from phyllite and characterised by a high stone content, low pH values (pH$_{H2O}$ 3.8 to 4.0) and C/N ratios of about 30 in the organic layer and in the A horizon. The plantation (1.5 × 1.5 m planting distance, 4000 trees ha^{-1}) was established on a windfall area with 14% slope facing southwest. At the time of this experiment (1991 to 1992), the average height of the trees was 3.6 m, and the average above-ground biomass (dry weight) of the trees 1.86 kg m^{-2}. Needles contributed 35%, twigs 27% and stems 38% to the above-ground tree biomass. The stand had not reached canopy closure; therefore, the ground was completely covered with understorey vegetation, dominated by *Vaccinium myrtillus* (0.33 kg m^{-2}), *Calluna vulgaris* (0.14 kg m^{-2}) and *Deschampsia flexuosa* (0.02 kg m^{-2}). The total root biomass was 0.26 kg m^{-2}. Fifteen plots (two treatments and a reference with five replicates each) of 40–70 m^2, and randomly distributed in the study area were selected for the labelling experiment. For more details on site characteristics and experimental design see Buchmann et al. (1995, 1996).
2. A 140-year-old *Picea abies* stand located in the northwestern part of the Fichtelgebirge (NE Bavaria, Germany) at 760 m altitude (Waldstein site): The stand has a 5° slope facing west. The relief of the stand is influenced by large granite rocks originating from weathering and Pleistocene solifluction. The soil type of this stand is a cambic podzol (FAO 1988) developed from granite, and it is characterised by low pH values (pH$_{H2O}$ 3.7 to 3.9) and C/N ratios of about 24. The average height of the trees was 24 m (1994). The mean tree density was 363 trees ha^{-1}, and

the average total biomass (dry weight) of the trees 37.1 kg m^{-2}. Needles contributed 4%, twigs 10%, stems 60% and roots 26% to the tree biomass. Most of the soil surface was covered with understorey vegetation dominated by *Deschampsia flexuosa* (0.35 kg m^{-2}) and *Vaccinium myrtillus* (0.06 kg m^{-2}). For a detailed description of the characteristics of this NIPHYS/CANIF site see Persson et al., Chapter 2, this Volume. For the ^{15}N-labelling experiment reported here (tracer application in spring), 15 trees (two treatments and a reference with five replicates) of similar size and apparent health status were selected. Each tree with its surrounding area was defined as a plot of the size of the mean crown area (approx. 30 m^2), thus including soil and understorey vegetation. For more details of the experimental design see May et al. (1996) and Schmidt et al. (1996).
3. A 30-year-old mixed broadleaf tree plantation composed of *Fagus sylvatica* and *Quercus petraea* located near the village of Ebrach in the Steigerwald (NE Bavaria, Germany, about 150 km west of the Fichtelgebirge sites) at 440 m altitude: the soil type of this stand is a cambisol (FAO 1988) developed from sandstone (Keuper). Its organic layer and A horizon is characterised by pH values of pH$_{H2O}$ 4.5 and C/N ratios of about 20. The average height of the trees was 12 m, and the mean tree density is 800 beech trees ha^{-1} and 1733 oak trees ha^{-1}. The total mean tree biomass (dry weight) in the year of the tracer experiment (1996) was 30.9 kg m^{-2} for beech and 20.1 kg m^{-2} for oak. The contribution of the various compartments to the total tree biomass was 11% (beech) or 6% (oak) for the leaves, 12% (beech and oak) for the twigs, 38% (beech) or 62% (oak) for the stems and 39% (beech) or 20% (oak) for the roots. Due to the closed canopy of the stand, no understorey vegetation was present. For the ^{15}N-labelling experiment reported here (tracer application in spring), nine plots (two treatments and a reference with three replicates) were selected randomly. Each plot comprised one pair of dominant and closely neighbouring beech and oak trees. The definition of the plot size (7.5 m^{-2}) was based on the mean crown area of each pair of trees. For more details of site characteristics and the experimental design, see May (1999).

7.3 Approaches to Study the Fate of ^{15}N-Labelled Nitrogen Inputs

7.3.1 ^{15}N-Labelled Beech Litter Application in Aubure

A circular plot of 4 m^2 around each of the five selected beech trees was isolated from surrounding trees by trenching roots and insertion of a plastic ring down to 0.3 m soil depth (Fig. 7.1). In November 1994, the L layer was carefully removed from the plot surface, to avoid confusion of partly decomposed litter with the added ^{15}N-labelled litter during the course of the experiment. The annual litterfall was replaced by an equivalent amount of ^{15}N-labelled beech litter (0.9% N, 3.2% ^{15}N excess), produced by spraying ^{15}N-labelled urea on the foliage of another beech stand (Zeller et al. 1998). The N, K and P concentrations of the labelled litter were lower, whereas the Ca concentration was higher than in the Aubure beech litter. In autumn 1995 and 1996, the non-labelled original litter was removed and similar amounts of labelled litter (200 g m^{-2}) were spread on the five experimental plots. A litte rbag experiment was set up in parallel (see Cotrufo et al., Chap. 13, this Vol.) to assess litter mass loss rates.

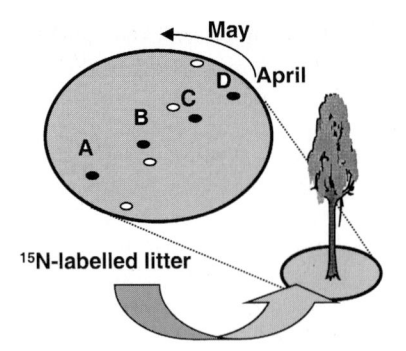

Fig. 7.1. Location of the site and the experimental plots in the 50-year-old beech stand at Aubure (Strengbach catchment, Vosges Mountains, NE France)

Decomposing labelled litter samples were collected and analysed five times from April to October 1995, seven times from April to November 1996, and four times from April to August 1997. Four soil cores (0–10 cm soil depth) per experimental plot were sampled with an auger exactly where the litter had been sampled. At the last sampling occasion of each year (November), four additional soil cores per plot were taken down to 30 cm soil depth. In 1995 to 1997, two soil cores (0–10 cm) per experimental plot and sampling time were cut in approximately 2-cm layers. About 10 g of the undisturbed soil was taken from each layer, dried at 65 °C, roots and gravels were removed, and soil samples were ground prior to ^{15}N analysis. After the collection of all roots from each soil core, the layers from each soil core were combined, and soil samples ($n = 20$) were sieved (mesh size 4 mm) prior to the extraction of mineral N.

Nitrate and ammonium concentrations (1 M KCl extractable) were measured in all soil samples. Microbial biomass was measured using the fumigation-extraction technique (Vance et al. 1987). The ^{15}N concentrations of fumigated and non-fumigated extracts were measured after mineralisation of 100 ml of the extract (Kjeldahl method), followed by steam distillation and by evaporation of the $(NH_4)_2SO_4$ distillates, adjusted to pH 4 (Bottner et al. 1998). The ^{15}N concentrations of nitrate and ammonium were measured after steam distillation of 100 ml of the extract. Microbial N was calculated from the difference between fumigated and non-fumigated extracts, using a correction factor ($k_N = 0.68$; Vance et al. 1987). Roots collected from the soil samples were divided into mycorrhizal root tips and lignified fine roots (1–3 mm). These two root classes represented approximately 80% of the roots present in the soil samples. Samples were carefully cleaned from adhering soil and organic matter, dried at 65 °C and stored for further analysis. All mycorrhizal root tips collected on a plot were combined to one composite sample for ^{15}N analysis.

Beech leaves ($n = 20$–30) from the upper crown of each tree in the five plots were collected during the growing season from June to August, twice in 1995 and four times in 1996 and 1997. At the beginning of September, each tree crown was completely wrapped with a large nylon net (mesh size 10 mm) to collect the annual litterfall. After the end of the autumnal litterfall, all leaf litter was collected, separated from dead branches and weighted. In February 1995 to 1997, two branches from each tree were cut and divided into buds, wood and bark of the twigs and remaining branches. In December 1997, five bark and wood samples were cored from the stem of each tree. The biomass and N content of three representative control trees was calculated by weighting trunk, branches, shoots and buds and analysing for total N concentrations.

Ground samples (<40 mesh) were analysed for ^{15}N and total N using an elemental analyser (Carlo Erba NA 1500) coupled to a mass spectrometer (Finnigan MAT) at the Service Central d'Analyse of CNRS (Casabianca 1995).

7.3.2 ^{15}N-Ammonium and ^{15}N-Nitrate Applications Simulating Wet Deposition in the Fichtelgebirge and Steigerwald

The ^{15}N-labelling experiments reported here were distributed over a period of 6 years from 1991 (labelling of the 15-year-old *Picea* stand) to 1996 (labelling of the *Fagus/Quercus* stand). However, the experimental designs were similar, thus allowing a comparison of the experiments. The tracers were always applied in spring before bud break of the trees. Five (*Picea* stands) or three plots (*Fagus/Quercus* stand) per site were labelled with ^{15}N-ammonium and five (*Picea* stands) or three plots (*Fagus/Quercus* stand) per site were labelled with ^{15}N-nitrate. The remaining five (*Picea* stands) or three plots (*Fagus/Quercus* stand) per site served as controls. Each plot was treated with 4.1 mmol ^{15}N m^{-2} (=0.62 kg ^{15}N ha^{-1}) sprayed homogeneously on the ground surface of the entire plot area. Highly enriched ^{15}N compounds (95 to 99% ^{15}N enrichment) were used in low amounts to minimise N fertilisation effects. The tracers were applied as solutions of either ^{15}NH$_4$Cl or K^{15}NO$_3$, simulating a single wet deposition event on the soil surface as a minor rain event (1 mm). The tracer amount applied is equivalent to about 3% of the annual throughfall N deposition in the area of investigation. Immediately after the tracer application, the plots were sprayed with additional water to minimise above-ground tracer uptake by the understorey vegetation. There is good evidence to assume that above-ground tracer uptake by the vegetation was negligible (see Buchmann et al. 1995). The control plots were treated similarly, but with unlabelled NH$_4$NO$_3$.

The temporal development of the ammonium and nitrate tracer distributions between the soil inorganic and organic N pool in different soil horizons, the different compartments of the trees (needles or leaves, twigs, stems and roots) and the understorey vegetation was studied throughout the growing season (beginning 12 days after the tracer application and ending up to one and a half years after tracer application). Subsamples of the respective ecosystem compartments were taken from ^{15}N-labelled and control plots, in addition to measurements of the total N pool sizes and the relative ^{15}N abundance in each compartment. The ^{15}N enrichment was calculated from the difference between ^{15}N abundances sampled from the ^{15}N-labelled and the control plots. The total uptake of tracer into the respective compartments was calculated from the difference between ^{15}N pool sizes sampled from the ^{15}N-labelled and the control plots. Soil mineral N was separated from soil total N by soil water (both of the *Picea* stands) or KCl (1 M) extraction (140-year-old *Picea* stand and 30-year-old *Fagus/Quercus* stand) and subsequent fractionation of the ammonium and nitrate by steam distillation (see May et al. 1996). Mineral N obtained from the water extractions was assumed to represent the minimum concentration of plant-available ammonium and nitrate. Mineral N obtained from the KCl extractions contains additional ammonium adsorbed to the surface of clay particles and organic matter, and thus, was assumed to represent the maximum concentration of plant available ammonium and nitrate. Ammonium and nitrate concentrations in the soil extracts were analysed using a Flow Injection Analyser (QuickChem AE, Lachat). Total N concentrations and relative ^{15}N abundances of distillates or dried and ground soil and plant samples were measured using an on-line system combining an elemental analyser (CARLO ERBA NA 1500) for Dumas combustion and a FINNIGAN MAT delta E gas isotope ratio mass spectrometer. For more details of the sampling, sample preparation and analysis and data calculation procedure for each stand see Buchmann et al. (1995, 1996), May et al. (1996), Schmidt et al. (1996) and May (1999).

7.4 N Release and Tree Uptake from ^{15}N-Labelled Decomposing Litter in a Beech Forest in Aubure

7.4.1 N and ^{15}N Dynamics in Decomposing Beech Litter

Litter N concentrations increased sharply during the first year, slightly during the second year and levelled off in the third year, approximately following a logarithmic function ($r^2 = 0.94$). ^{15}N excess of the litter decreased following a logarithmic function ($r^2 = 0.91$) (see Cotrufo et al., Chap. 13, this Vol.). During the decomposition of the ^{15}N-labelled beech litter, a simultaneous release of litter original N and an incorporation of N from different sources (fungal, throughfall, soil N) occurred. This ^{15}N release was linearly correlated to the time of decomposition ($r^2 = 0.97, p < 0.001$) and mass loss (Fig. 7.2). About 16, 25 and 64% of the initial amount of litter N had been released after 4 months, 1 year and 3 years, respectively. While N was released from the litter, external N was simultaneously incorporated into the decomposing litter. Almost all external N was incorporated during the first year. Over a 2 year period, the incorporation of external N balanced the amount of N released from the litter. A net release of N from the litter occurred from the third year. After a rapid and early release of soluble N during the first months, the release rate of ^{15}N from the labelled beech litter remained nearly constant and equivalent to mass loss, as suggested by results of Berg (1988). In parallel, external N was incorporated in the decomposing litter, mainly during the first year. Ergosterol and chitin assays indicated that 35% of incor-

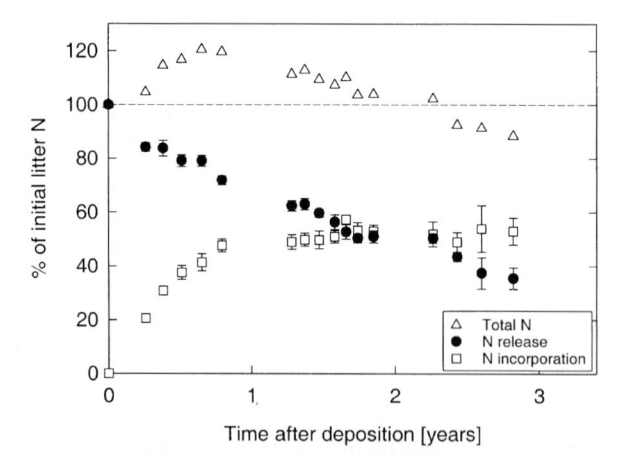

Fig. 7.2. Temporal variation of total N content in the decomposing litter at Aubure. Incorporation and release of N in the ^{15}N-labelled litter during litter decomposition. *Error bars* indicate standard errors of the mean ($n = 10$)

porated N was fungal N, the remaining being most likely provided by throughfall, bacteria and faecal pellets of the fauna (Zeller et al. 1998).

7.4.2 ^{15}N in Soil Compartments

The content of microbial N increased almost twofold from 45 to 90 mg N kg^{-1} soil over the 3 years (Fig. 7.3A). Collection of the L layer before the deposition of the ^{15}N-labelled litter affected soil microbia, which resulted in lower microbial N contents during 2 years. Three years after that disturbance, microbial N reached nearly constant values, as confirmed by measurements during 1998 (data not shown). Maybe the high Ca content in the labelled litter had a positive influence on the establishment of the soil microbial biomass. Extractable nitrate and ammonium increased slightly during the first year and then decreased to lower levels. Ammonium concentration was always higher than nitrate. The content of total extractable N (NO$_3^-$ + NH$_4^+$ + organic N) in the top soil, which was not measured during the first year, remained stable during the 2 following years (Fig. 7.3B). Organic N accounted for about 30% of total extractable N.

The difference between the ^{15}N abundance measured in labelled and control plant-soil compartments is mentioned as Δ^{15}N. Δ^{15}N of extractable N in the top soil (0–10 cm) increased rapidly after deposition of the ^{15}N-labelled litter, peaked during the second year, and decreased sharply during the third (Fig. 7.3C). Both nitrate and ammonium fractions showed quite similar ^{15}N enrichments. Δ^{15}N values for extractable nitrate and ammonium were highly variable, ranging from 20 to 360‰ (mean value = 130‰) during the first 2 years. Highest Δ^{15}N values were observed in summer. No systematic difference was noted between the Δ^{15}N before and after fumigation. Approximately 1 to 2% of the extractable ^{15}N and microbial biomass ^{15}N originated from the deposited litter ^{15}N assuming that the ^{15}N enrichment of the soluble N was similar to that of the litter.

In the control soil, δ^{15}N increased from −3‰ in the upper organic layer (0–2 cm) to slightly positive values of +2‰ in the mineral soil at 10 cm depth (Fig. 7.4). Six months after the deposition of ^{15}N-labelled litter, δ^{15}N increased slightly in the upper 2 cm of the soil. δ^{15}N values increased progressively during the 3 years, especially in the upper organic layer, but also deeper in the mineral soil profile. After 3 years, ^{15}N enrichment reached +50‰ in the upper 2 cm of the soil profile and almost +8‰ down at 10 cm soil depth. The variation in soil δ^{15}N values was linearly correlated to the time of litter decomposition (soil$_{0-2cm}$, $r^2 = 0.97$ and soil$_{2-10cm}$, $r^2 = 0.54$). Most of the litter N was released with litter fragments and faecal pellets, which mostly accumulated at the soil surface (0–2 cm) (Table 7.2). During the third year almost 4% of litter released N was transported below 2-cm soil depth by soil fauna, probably by earthworms. The form in which N is released from the litter, as ammonium, nitrate or organic N is important for the further utilisation by different organisms. The accumulation of litter-released N at the soil surface suggested that this N remained in an organic form, and became therefore part of the large soil N pool. Large amounts of litter-released organic N were associated with humic substances, i.e. protein–tannin complexes, only little organic N as amino acids or proteins (Northup et al. 1995). Possible mechanisms for the release of ^{15}N from the litter were: leaching of soluble litter N (Joergensen and Meyer 1989; Tietema and Wessel 1994), biochemical degradation by bacteria, saprophytic and symbiotic fungi and fragmentation and consumption by

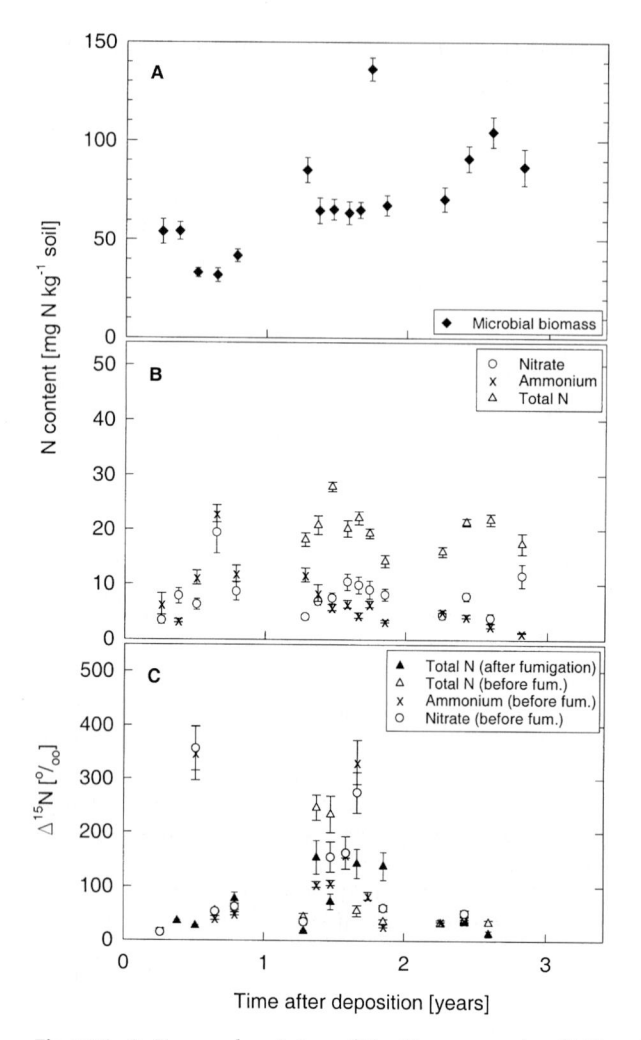

Fig. 7.3A–C. Temporal variation of the N concentration (**A,B**) and ^{15}N abundance (**C**) in the extractable soil N fraction (NO_3^-, NH_4^+ and organic N) and in the microbial biomass N in the soil (0–10 cm) at Aubure. Δ^{15}N in K_2SO_4-extractable nitrate, ammonium and total N before fumigation and total N after fumigation. *Error bars* indicate standard errors of the mean ($n = 20$)

soil fauna (Setälä et al. 1996). Observation of beech leaves in the second year of decomposition showed they were partly digested by enchytraeid activity (F. Toutain, pers. comm.). The soil fauna at the Aubure experimental site is largely dominated by microbial feeders belonging to the mesofauna (enchytraeids, Collembola and Acarina) (see Wolters et al., Chap. 17, this Vol.), but the role of the mesofauna in the decomposition of such litter was most likely limited during the first year. Leaching of soluble N, mediated or not by microbial degradation, certainly played a major role. These N compounds leached from the litter were extractable from the soil and incorporated in soil microbial biomass as early as 9 months after ^{15}N-labelled litter deposition. Soluble N released by microbes from easily degradable N compounds of the

Fig. 7.4. Temporal variation of $\delta^{15}N$ in the soil profile in relation to soil depth over the time course of the labelling experiment at Aubure. ^{15}N-labelled litter was deposited in November 1994. *Error bars* indicate standard errors of the mean ($n = 40$)

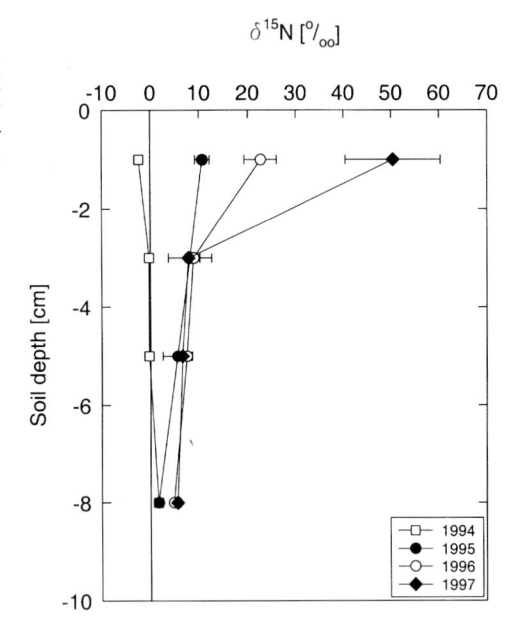

Table 7.2. Recovery of litter-derived ^{15}N in the following compartments after three years of litter decomposition at the Aubure site: ^{15}N-labelled litter, soil (0–2 and 2–30 cm depth), mycorrhizal root tips, fine roots, leaves, aerial biomass (without leaves) and total recovery for each experimental plot

	^{15}N in the litter	Labelled litter	Soil (0–2 cm)	Soil (2–30 cm)	Mycorrhizal root tips	Fine roots	Leaf litter	Aerial biomass	Total recovery
	mg ^{15}N plot^{-1}	Applied ^{15}N (%)							
Plot 1	239.94	35.2	40.7	8.4	0.58	0.44	0.13	1.21	86.7
Plot 2	256.63	37.6	33.5	8.5	0.31	0.35	0.34	1.09	81.7
Plot 3	217.19	35.2	37.0	8.7	0.13	0.32	0.39	1.46	83.2
Plot 4	197.12	34.4	41.2	9.0	0.19	0.57	0.18	1.05	86.6
Plot 5	267.14	35.6	37.1	8.3	0.18	0.63	0.01	0.59	82.4

litter (or litter fragments) was the most likely source for the peak in soil extractable ^{15}N during the first and second year. The decrease in soil-extractable ^{15}N concentration during the third year indicates that mineralisation of litter-derived soil N was low compared to the flush of soluble N originating from the litter.

7.4.3 Incorporation of Litter ^{15}N in Trees

7.4.3.1 Fine Roots and Mycorrhizal Root Tips

^{15}N originating from the labelled litter was detected in mycorrhizal root tips and fine roots as early as 6 months after ^{15}N-labelled litter deposition (Fig. 7.5) and increased

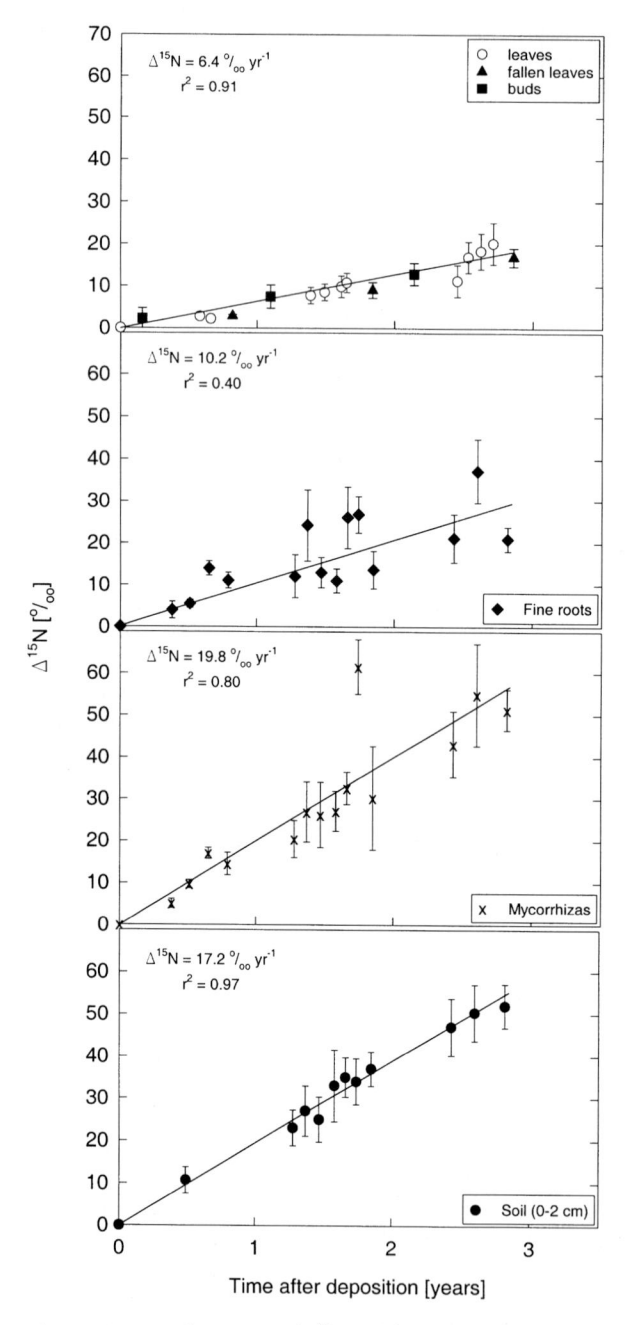

Fig. 7.5. Temporal variation of $\Delta^{15}N$ in the soil (0–2 cm), mycorrhizal root tips, fine roots, leaves, leaf litter and buds during the 3-year period at Aubure. *Error bars* indicate standard errors of the mean ($n = 5$)

linearly throughout the 3-year decomposition period, at a rate of about 20 and 11‰ a^{-1}, respectively. In autumn (end of October) the Δ^{15}N of mycorrhizal root tips and roots decreased drastically, suggesting a massive input of non-labelled N in these tissues. The time course of ^{15}N enrichment of mycorrhizal root tips and the upper 2 cm of the soil was similar throughout the 3 years. Mycorrhizal root tips, which were mainly located here, were in apparent isotopic equilibrium with this soil layer. This identity suggested that soil organic N was the main source of mycorrhizal N, as proposed by Read (1991). The Δ^{15}N of mycorrhizas decreased sharply in autumn each year, especially after the second year, indicating a dilution of soil-derived N by tree N. This sink function of mycorrhiza for glutamate has been described by Martin and Lorillou (1997). Cyclic variations of fine root Δ^{15}N during the second year suggested that fine root Δ^{15}N was strongly affected by N allocation from the tree N pool. This is in agreement with the relatively high Δ^{15}N values of soluble N and microbial biomass N measured in the soil, especially during the second year. This is also supported by Jussy (1998) and Bauer et al. (1997), who showed in an adjacent 150-year-old beech stand that mineral N was the main N source taken up.

7.4.3.2 Leaves

A change in the ^{15}N content of leaves from the upper crown was measured 9 months after ^{15}N-labelled litter deposition, indicating that litter-released ^{15}N was already allocated to the shoots. During the first year, Δ^{15}N increased in leaves, senescent leaves and buds at a rate of about 6‰ a^{-1}. After 3 years, average Δ^{15}N values of leaves was +20‰, a value much lower than root Δ^{15}N. In the autumn of the second and third year, Δ^{15}N of fallen leaves was about 20% lower than mature leaf Δ^{15}N, indicating a preferential allocation of ^{15}N-labelled compounds (i.e. amino acids) to buds and other perennial storage tissues. Whereas the Δ^{15}N of buds was similar to that of the leaves after bud break, the contribution of litter-derived N to total beech N increased linearly during litter decomposition.

7.4.4 Total Recovery of ^{15}N

After 3 years of litter decomposition, 42 to 49% of litter-released ^{15}N were found in the soil, about 36% remained in the ^{15}N-labelled beech litter and 1.4 to 2.4% had accumulated in the trees (Table 7.2, Fig. 7.6). ^{15}N incorporated in mycorrhizal root tips (0.1 to 0.6%), fine roots (0.3 to 0.6%) and leaf litter (0.01 to 0.4%) varied strongly among the five experimental plots. The highest amounts of ^{15}N incorporated in fine roots were found in trees from plots (4 and 5) in which accumulation in above-ground biomass was the lowest, varying between 0.6 and 1.5% of the initial litter ^{15}N. Most of the ^{15}N was stored in buds, twigs and branches, whereas only little ^{15}N was detected in stem bark and wood. We calculated the average Δ^{15}N of the soil N immobilised in the above-ground part of trees during the whole decomposition period as the ratio between the amount of ^{15}N and total N incorporated. An excess of 1.16 µg ^{15}N tree^{-1} was incorporated in 3 years' litterfall (0.51 µg ^{15}N) plus the final content in stem (0.23 µg ^{15}N) and branches (0.42 µg ^{15}N). An annual N incorporation of 6.28 g N tree^{-1} was computed from the mean annual increment and N concentration in stem and branches plus the annual litterfall (Table 7.2). From these values, we calculated a 3 years' average δ^{15}N of 19‰ for the N taken up and allocated in above-ground parts.

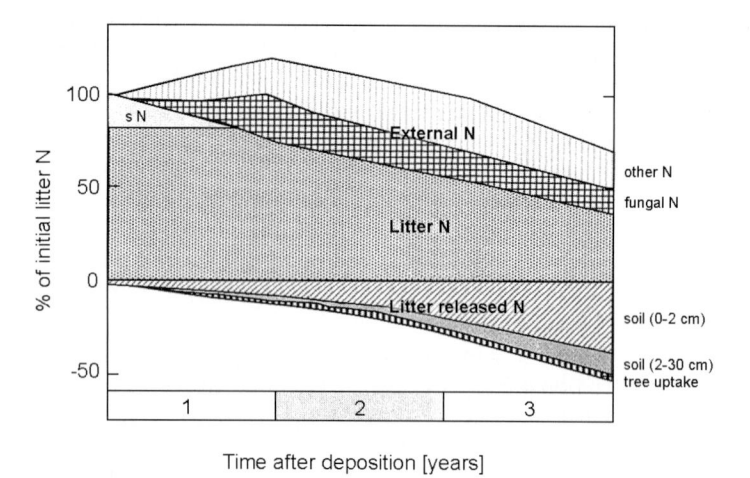

Fig. 7.6. Fate of beech litter N during decomposition at Aubure. *sN* Soluble N fraction

This value is intermediate between that of roots (3 years' average = 15‰) and mycorrhizae (3 years' average = 27‰). It is probably an underestimate of the soil source, as large roots, which have large N reserves, have certainly diluted the soil N isotopic signal. During the whole experiment, mycorrhizal and root ^{15}N enrichment remained higher than that of leaves, because of the dilution of soil derived N by isotopically lighter tree N.

In order to calculate the contribution of soil N and tree N to leaf N, we assumed that the mean annual Δ^{15}N of the N taken up from the upper 2 cm of the soil during 1 year was equal to the higher (September) value of mycorrhizal Δ^{15}N collected in this layer. Spring bud Δ^{15}N was considered as the tree N source for leaves. We calculated that between 6 and 17% of leaf N in August would be supplied each year from the soil if the upper 2 cm of soil was the only source of N. Using root maximum Δ^{15}N as the N source in the soil upper 10 cm, we computed that between 15 and 27% (depending on the year) of soil N would be directly incorporated in leaves. These higher percentages simply reflect the lower Δ^{15}N of roots in comparison to mycorrhizas. Compared to similar measurements using trees grown in pots (Preston and Mead 1994a; Setälä et al. 1996), these relatively low contributions of soil N to tree N may be explained by the poor growth of the beech stand, caused by the climatic conditions, and by the relative richness in N of the stand.

As a general conclusion, it appeared that N released from decomposing litter contributed very little to tree N nutrition during the first years of decomposition. This reflects the large size of the pool of organic N in the first 10 cm of the soil compared to the annual deposition in litterfall and also the high mineralisation potential of these soils as measured by Jussy at an adjacent beech site (B150, see Persson et al., Chap. 14, this Vol.). The accumulation of most of the N taken up in below-ground parts probably reflects the weak demand for N of poorly growing stands. Canopy uptake of N may be an alternative explanation for the low demand of soil-borne N of this beech stand. Highly ^{15}N-enriched mineral N contributed to tree N uptake, but organic N compounds are additional N sources for mycorrhizae. Due to the design of this experiment, the relative contributions of mineral N and organic N to total N uptake could not be quantified.

7.5 Ecosystem Partitioning of ^{15}N-Labelled Ammonium and Nitrate on the Sites in the Fichtelgebirge and Steigerwald

7.5.1 Ammonium and Nitrate Concentrations in the Soil

Ammonium was in almost all cases the dominant mineral N form in the soil solution, irrespective of the soil horizon, time of sampling, forest stand or soil extraction technique (Figs. 7.7–7.9). The vertical profile of ammonium concentrations showed a distinct pattern for all forest sites and sampling times: concentrations were always highest in the litter layer and decreased with increasing soil depth. A comparison of the ammonium concentrations obtained from soil water extractions (minimum of plant available ammonium) and soil KCl extractions (maximum of plant available ammonium) at the 140-year-old *Picea abies* stand indicated about threefold higher ammonium concentrations in the KCl extracts of the litter layer (Fig. 7.8). Nitrate concentrations were also highest in the litter layer. However, the vertical concentration profiles of nitrate were less pronounced (Figs. 7.7–7.9). Soil nitrate concentrations were unaffected by the type of soil extraction technique (Fig. 7.8). No major differences in soil ammonium and nitrate concentrations were found between the two age classes of the coniferous stands or between the 140-year-old coniferous and the broadleaf stand when data using the same soil extraction technique are compared. Ammonium and nitrate concentrations as well as depth gradients on the respective stands remained rather constant during the growing season for all three stands.

Ammonium and nitrate from atmospheric deposition are mixed with soil-borne mineral N compounds when entering the litter layer. Ammonium-to-nitrate ratios in the litter layer thus provide information on the relative mixing ratios of deposition-derived and soil-borne N species. In this study, the ammonium-to-nitrate ratios in

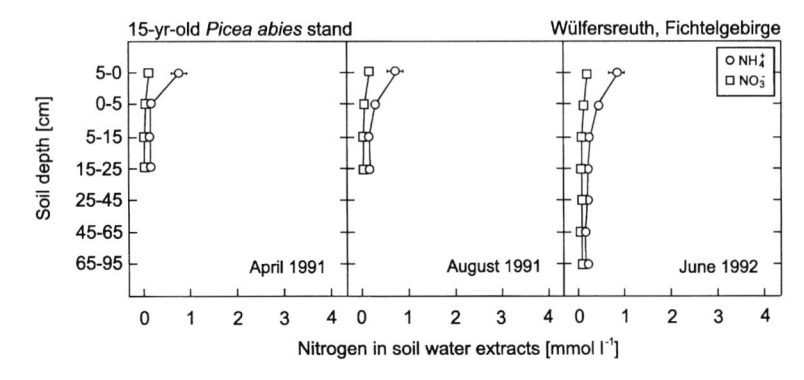

Fig. 7.7. Means and standard errors (n = 15) of ammonium and nitrate concentrations within the soil profile of a 15-year-old *Picea abies* stand in the Fichtelgebirge (Wülfersreuth, NE Bavaria, Germany). The concentrations are based on soil water extracts. The data indicate a mean NH_4^+-to-NO_3^- ratio of 8.9:1 in the soil organic layer. (Buchmann et al. 1995)

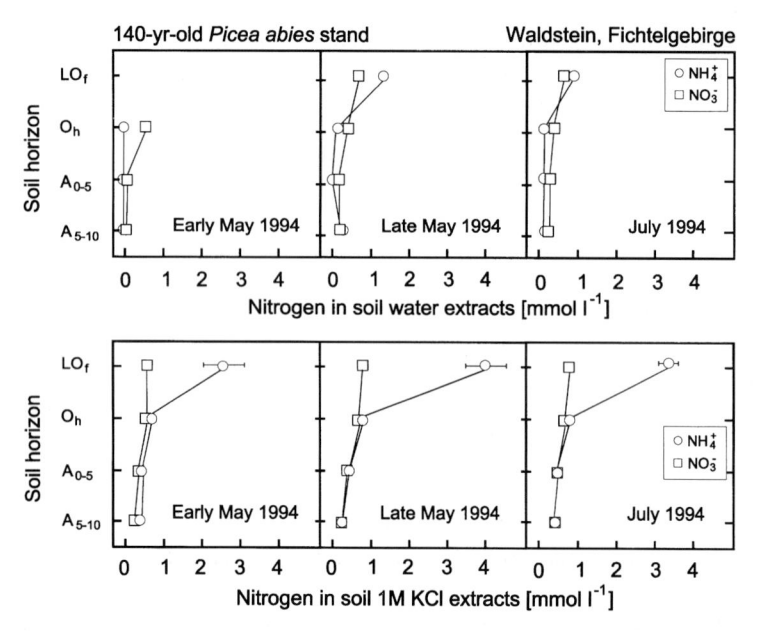

Fig. 7.8. Means and standard errors ($n = 15$) of ammonium and nitrate concentrations within the soil profile of a 140-year-old *Picea abies* stand in the Fichtelgebirge (Waldstein, NE Bavaria, Germany). The concentrations are based either on soil water extracts (*top*) or on soil 1 M KCl extracts (*bottom*). The data obtained from water extracts indicate a mean NH_4^+-to-NO_3^- ratio of 2.6:1 in the LO_f horizon of the soil organic layer (C. May, G. Schmidt, G. Gebauer and E.-D. Schulze, unpubl. data). The data obtained from 1 M KCl extracts indicate a mean NH_4^+-to-NO_3^- ratio of 7.6:1 in the LO_f layer. (May et al. 1996)

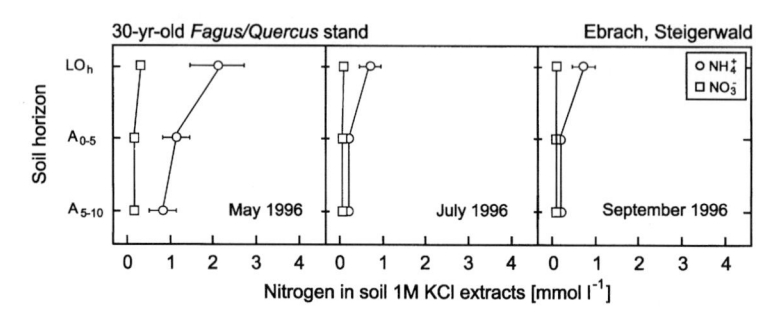

Fig. 7.9. Means and standard errors ($n = 6$) of ammonium and nitrate concentrations within the soil profile of a 30-year-old mixed *Fagus sylvatica/Quercus petraea* stand in the Steigerwald (Ebrach, NE Bavaria, Germany). The concentrations are based on soil 1 M KCl extracts. The data indicate a mean NH_4^+-to-NO_3^- ratio of 9.3:1 in the LO_f layer. (C. May and G. Gebauer, unpubl. data)

the litter layer varied, dependent on the forest stand and the soil extraction technique. The mean ammonium-to-nitrate ratio in the litter layer obtained from soil water extractions was 8.9:1 in the 15-year-old *Picea abies* stand and 2.6:1 in the 140-year-old *Picea abies* stand. Based on KCl extractions, ammonium-to-nitrate ratios were

The Fate of ^{15}N-Labelled Nitrogen Inputs to Coniferous and Broadleaf Forests 159

estimated to range from 7.6:1 in the 140-year-old *Picea abies* stand to 9.3:1 in the mixed broadleaf forest stand. All ratios indicate a considerably higher dilution of ammonium from atmospheric deposition with soil-borne ammonium compared to nitrate irrespective of the soil extraction technique, after pulse-labelling with an ammonium-to-nitrate ratio of 1:1. Thus, the different dilutions of ammonium and nitrate from deposition have to be considered in further uptake calculations.

7.5.2 ^{15}N Partitioning in the Soil Ammonium and Nitrate Fractions

Ammonium and nitrate from atmospheric or simulated deposition are not only diluted by soil-borne N species when entering the forest floor. In addition, both of them may be rapidly immobilised and/or metabolised by soil microbia, fungi or by physicochemical reactions before being taken up by the roots of forest plants. Especially two processes have to be considered: (1) nitrification of deposition-derived ammonium and (2) immobilisation of both of the deposition-derived mineral N compounds. The temporal development of the dilution process and the immobilisation and transformation of ammonium and nitrate from deposition becomes visible from the analysis of ^{15}N enrichment in the soil ammonium and nitrate fractions after ^{15}N-ammonium or ^{15}N-nitrate applications (May et al. 1996).

Two weeks after the tracer application to the 140-year-old *Picea* stand and the 30-year-old mixed broadleaf forest (early May), both the ^{15}N-ammonium and the ^{15}N-nitrate treatments, showed ^{15}N enrichments in those soil mineral N fractions that correspond to the tracer form applied (Figs. 7.10 and 7.11). The highest ^{15}N enrich-

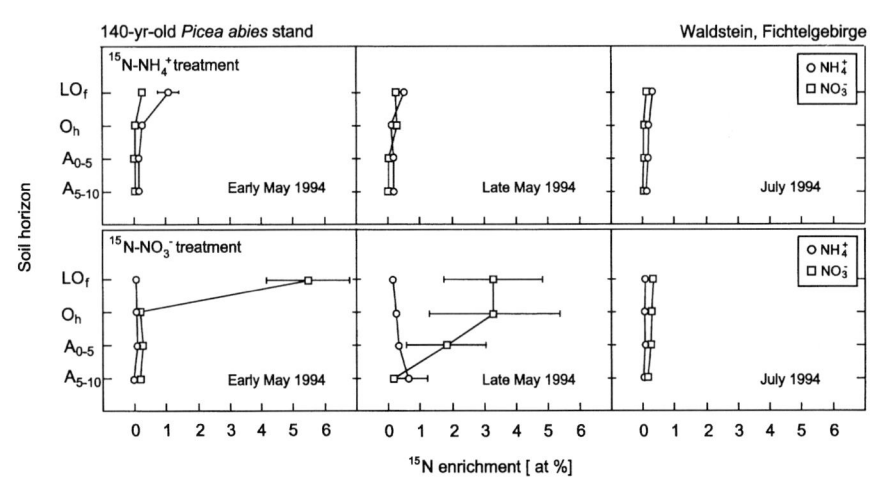

Fig. 7.10. Means and standard errors ($n = 5$) of the relative ^{15}N frequencies of ammonium and nitrate within the soil profile of a 140-year-old *Picea abies* stand in the Fichtelgebirge (Waldstein, NE Bavaria, Germany) 12 days, 1 month and 3 months after the ^{15}N tracer application. *Top* ^{15}N-ammonium application; *bottom* ^{15}N-nitrate application. The tracers were applied in April 1994 as a single wet deposition of 4.1 mmol m^{-2} ^{15}NH$_4^+$ or ^{15}NO$_3^-$, respectively. The relative ^{15}N frequencies are based on soil 1 M KCl extracts. The relative ^{15}N frequencies in the soil ammonium and nitrate fractions in early May indicate a low nitrification rate and a total NH$_4^+$-to-NO$_3^-$ ratio of 4.3:1 in the LO$_f$ layer. (May et al. 1996)

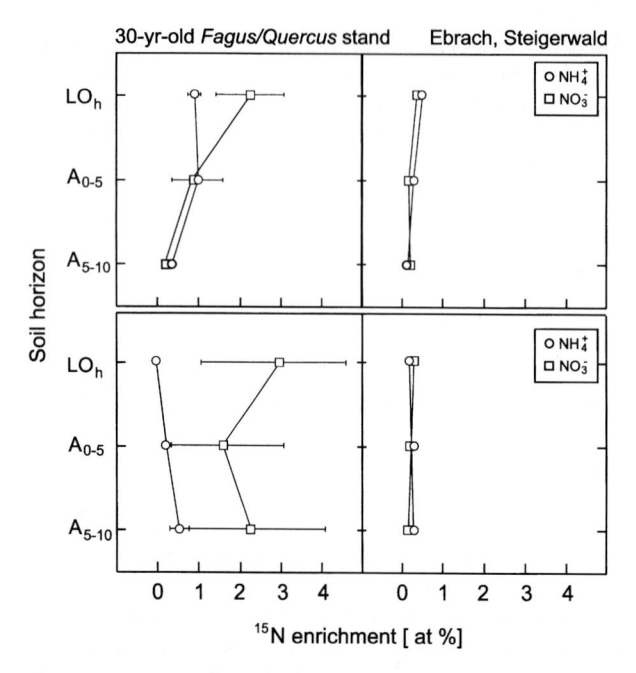

Fig. 7.11. Means and standard errors ($n = 3$) of the relative ^{15}N frequencies of ammonium and nitrate within the soil profile of a 30-year-old mixed *Fagus sylvatica/Quercus petraea* stand in the Steigerwald (Ebrach, NE-Bavaria, Germany) 12 days and 1 month after the ^{15}N tracer application. *Top* ^{15}N-ammonium application; *bottom* ^{15}N-nitrate application. The tracers were applied in April 1996 as a single wet deposition of 4.1 mmol m^{-2} ^{15}NH$_4^+$ or ^{15}NO$_3^-$, respectively. The relative ^{15}N frequencies are based on soil 1 M KCl extracts. The relative ^{15}N frequencies in the soil ammonium and nitrate fractions indicate a high nitrification rate. In this case the relative ^{15}N frequencies in the soil ammonium and nitrate fractions are less suitable to indicate NH$_4^+$-to-NO$_3^-$ ratios available for plant uptake. (C. May and G. Gebauer, unpubl. data)

ments were found in the litter layer, although ^{15}N-nitrate had a higher tendency to be transported to the mineral soil. This transport to greater depths was expressed stronger in the broadleaf stand (Fig. 7.11), in which the litter and the organic layers were much shallower than those in the coniferous stand (data not shown). The ^{15}N-ammonium tracer in both stands was more diluted by soil-borne ammonium than the ^{15}N-nitrate tracer was diluted by soil-borne nitrate. Thus, the ratio of ^{15}N enrichment in the nitrate and ammonium pool (i.e. the ratio of ^{15}N-nitrate to ^{15}N-ammonium) should, in principle, provide information similar to the ammonium-to-nitrate concentration ratios in the soil solution, provided no major transformations between both pools occur. For the 140-year-old *Picea abies* stand, 2 weeks after the tracer application this ratio is 4.3 and lies exactly between the maximum and minimum ratios obtained from the ammonium and nitrate concentrations in the soil solution (see above).

Two weeks after the ammonium tracer application (in early May), significant ^{15}N enrichments were found in the soil nitrate fraction (Figs. 7.10 and 7.11). For the coniferous site, this enrichment was low compared to the ^{15}N enrichment in the ammonium fraction, indicating a rather slow rate of net nitrification (Fig. 7.10). In contrast, 2 weeks after the ammonium tracer application, the ^{15}N enrichment in the soil nitrate fraction of the broadleaf stand was even higher than the ^{15}N enrichment in the ammonium fraction, indicating a much higher net nitrification rate (Fig. 7.11). These findings for the broadleaf forest soil correspond (1) to a lower C/N ratio of the soil organic matter, (2) to a higher soil pH value and (3) to the general tendency of higher incubation-based nitrification rates in broadleaf than in coniferous forests along the European transect (see Persson et al., Chap. 14, this Vol.). Both ratios, ammonium-to-nitrate concentrations and ^{15}N enrichments in the soil nitrate and ammonium fractions, allow estimations of the relative uptake of both mineral N compounds from the soil under conditions with low net nitrification rates, as in the coniferous forest. Under conditions with high nitrification rates, as in the broadleaf forest, both parameters are obviously less suitable to estimate relative uptake of mineral N compounds from the soil.

Immobilisation of deposition-derived ammonium and nitrate tracers was rather fast for both forest stands (see also Nadelhoffer et al. 1999a). Two weeks after the tracer application, the ^{15}N remaining in the soil mineral N pool represents only a few percent of the total amount of tracer applied (see May et al. 1996; May 1999). Four weeks (broadleaf forest) or 12 weeks after the tracer application (coniferous forest), hardly any ^{15}N label could be detected in either of the two mineral N fractions throughout the soil profile (Figs. 7.10 and 7.11). Thus, most of the ^{15}N label is to found in the soil bound in a non-plant-available form or has been taken up by the vegetation.

7.5.3 Ammonium and Nitrate Uptake by the Trees

Trees in forest ecosystems have to compete for N compounds with soil microbia, fungi and understorey vegetation. Trees represent the major part of the living biomass in forests (see Sect. 7.2.2) and their growth has been described as being severely N-limited (Tamm 1991). The highest fine root density of forest trees is usually found in the organic layer (Schneider et al. 1989; Gebauer and Dietrich 1993). Thus, the spatial fine root distribution coincides with the occurrence of the highest mineral N concentrations throughout the soil profile and with the location of throughfall N input into the soil. Based on these findings, the organic soil layer is assumed to be the site of major mineral N uptake by trees. This assumption is confirmed by analyses of natural N isotope abundance, indicating that a major part of the N incorporated into trees originates from the organic layer (Gebauer and Schulze 1991; Gebauer and Dietrich 1993; Nadelhoffer and Fry 1994). N uptake for most of the forest trees in temperate and boreal climates is, furthermore, improved by an increase of the surface available for nutrient uptake, e.g. due to the formation of ectomycorrhizae (Smith and Read 1997). Thus, in the first instance, we might assume that trees are the superior competitor for soil mineral N compounds, irrespective of whether the compounds originate from soil-borne mineralisation or from atmospheric deposition.

The ^{15}N tracer retained by the trees after a pulse deposition of ammonium or nitrate indicates a very different pattern for coniferous and broadleaf forest stands (Tables 7.3–7.5). Irrespective of stand age and considerable differences in tree biomass per ground area, conifers of both stands retained less than 10% of the applied tracers (Tables 7.3 and 7.4). On the other hand, the two broadleaf tree species in the

Table 7.3. Partitioning and recovery of ^{15}N tracer retained by 15-year-old *Picea abies* in the Fichtelgebirge (Wülfersreuth, NE Bavaria, Germany) after one growing season. The tracers were applied in March 1991 as a single wet deposition of 4.1 mmol m^{-2} $^{15}NH_4^+$ or $^{15}NO_3^-$, respectively. The recovery of the tracers was measured in needles, twigs, stems and roots of *Picea abies* in November 1991 ($n = 5$). The relative contribution of ammonium and nitrate to total mineral nitrogen uptake from the soil was estimated based on the assumption that the tracers were diluted proportionally to the NH_4^+-to-NO_3^- ratio in water extracts from the soil organic layer. (Recalculated from Buchmann et al. 1995)

Compartment	Absolute ^{15}N tracer uptake (µmol ^{15}N m^{-2})		Estimated contribution to total N uptake from the soil based on a NH_4^+-to-NO_3^- ratio of 8.9:1 in soil water extracts (%)	
	^{15}N-NH_4^+	^{15}N-NO_3^-	NH_4^+	NO_3^-
Needles	85	178	43	10
Twigs	46	76	23	4
Stem	8	26	4	2
Roots	20	68	10	4
Total	159	349	80	20
Recovery (%)	3.9	8.5		

Table 7.4. Partitioning and recovery of ^{15}N tracer retained by 140-year-old *Picea abies* in the Fichtelgebirge (Waldstein, NE Bavaria, Germany) 4 weeks after tracer application. The tracers were applied in April 1994 as a single wet deposition of 4.1 mmol m^{-2} $^{15}NH_4^+$ or $^{15}NO_3^-$, respectively. The recovery of the tracers was measured in needles, twigs, stems and roots of *Picea abies* in May 1994 ($n = 5$). The relative contribution of ammonium and nitrate to total mineral nitrogen uptake from the soil was estimated based on two assumptions: (1) the tracers were diluted proportionally to the NH_4^+-to-NO_3^- ratio in water extracts from the soil organic layer, (2) the tracers were diluted proportionally to the NH_4^+-to-NO_3^- ratio in 1M KCl extracts from the soil organic layer. (G. Schmidt, C. May, G. Gebauer and E.-D. Schulze, unpubl. data)

Compartment	Absolute ^{15}N tracer uptake (µmol ^{15}N m^{-2})		Estimated contribution to total N uptake from the soil based on a NH_4^+-to-NO_3^- ratio of:			
	^{15}N-NH_4^+	^{15}N-NO_3^-	2.6:1 in soil water extracts (%)		7.6:1 in soil 1M KCl extracts (%)	
			NH_4^+	NO_3^-	NH_4^+	NO_3^-
Needles	47	76	17	11	24	5
Twigs	28	39	10	6	14	3
Stem	29	64	11	9	15	4
Roots	53	122	19	17	27	8
Total	157	301	57	43	80	20
Recovery (%)	3.8	7.3				

Table 7.5. Partitioning and recovery of ^{15}N tracer retained by *Fagus sylvatica* and *Quercus petraea* in a 30-year-old mixed stand in the Steigerwald (Ebrach, NE Bavaria, Germany) 4 weeks after tracer application. The tracers were applied in April 1996 as a single wet deposition of 4.1 mmol m^{-2} ^{15}NH$_4^+$ or ^{15}NO$_3^-$, respectively. The recovery of the tracers was measured in leaves, twigs, stems and roots of the trees in May 1996 ($n = 3$). (C. May and G. Gebauer, unpubl. data)

Compartment		Absolute ^{15}N tracer uptake (µmol ^{15}N m^{-2})	
		^{15}N-NH$_4^+$	^{15}N-NO$_3^-$
Fagus	Leaves	580	329
	Twigs	549	537
	Stem	139	115
	Roots	309	328
Subtotal		1577	1319
Quercus	Leaves	619	715
	Twigs	595	430
	Stem	200	395
	Roots	399	156
Subtotal		1813	1696
Total		3390	3015
Recovery (%)		82.7	73.5

mixed stand retained more than 70% of the applied tracers (Table 7.5). Thus, broadleaf trees utilize a greater fraction of the throughfall N deposition than conifers. For both coniferous forest stands, the recoveries of ^{15}N in the trees (in % applied tracer) were twice as high after ^{15}N-nitrate than after ^{15}N-ammonium addition. Thus, nitrate N deposition to the forest floor affects N nutrition of the conifers more directly than ammonium N deposition. In contrast, for the broadleaf stand, the ^{15}N recovery after ^{15}N-ammonium addition was higher than after ^{15}N-nitrate addition. Due to low net nitrification rates in both coniferous stands, a major part of the ^{15}N label was presumably taken up in the form in which it was applied. In the broadleaf stand, a considerable part of the ^{15}N-ammonium deposition was obviously nitrified before it was taken up by the trees. The amount of ^{15}N-ammonium input that was taken up by the broadleaf trees as nitrate, however, remains difficult to estimate.

Nitrification in the broadleaf stand makes it difficult to estimate the relative usage of ammonium versus nitrate, based on data from this ^{15}N-labelling experiment. At the present stage we can only presume that nitrate plays a more important role as mineral N source for broadleaf trees than for conifers. The low nitrification rate in the coniferous stands, however, provides a good opportunity to estimate the relative contribution of ammonium and nitrate to the total mineral N uptake. This estimate is based on the amount of ^{15}N tracer taken up by the trees and on the ammonium-to-nitrate ratio in the soil solution obtained either from soil water or soil KCl extraction. Based on this estimate, both coniferous stands showed a preference for ammonium over nitrate as mineral N source (Tables 7.3 and 7.4). The relative contribution of ammonium and nitrate used by the conifers as mineral N source ranged between 80:20% and 57:43%. The preference of *Picea abies* for ammonium as mineral N source is less pronounced than might be expected from the ammonium-to-nitrate concentration ratios in the soil. A higher mobility of nitrate in the soil and thus, a faster flow of nitrate to the root surface or to the surface of ectomycorrhizal hyphae, may be the reason for this finding (see Marschner et al. 1991). The difference in preference for ammonium or nitrate as mineral N source found for the conifers and the broadleaf trees corresponds to a considerably lower nitrate than ammonium uptake capacity of conifers (Kronzucker et al. 1997) and to a considerably higher nitrate assimilation capacity of many broadleaf tree species common to the temperate European forest vegetation (Gebauer and Schulze 1997 and references therein).

The ^{15}N tracer uptake by the trees in the 15-year-old spruce stand was almost completely terminated within the first 2 weeks after tracer application (Buchmann et al. 1995). A similar situation could be expected for the 140-year-old coniferous and for the broadleaf stand due to the fast immobilisation of the ^{15}N-ammonium and ^{15}N-nitrate into soil organic matter on both of these stands (see Figs. 7.10 and 7.11). The distribution of the ^{15}N label among different compartments of the trees was, however, time-dependent. One month after the tracer application to the 140-year-old *Picea abies* stand, 34 to 41% of the ^{15}N label was incorporated in the roots, 34 to 36% in the stems and twigs and 25 to 30% in the needles (Table 7.4). Eight months after the tracer application to the 15-year-old *Picea abies* stand, only 13 to 19% of the incorporated ^{15}N label was found in the roots, while the proportion of ^{15}N label transported to the needles had increased to more than 50% (Table 7.3); 34 to 39% of the incorporated ^{15}N label was still found in stems and twigs. The proportion of ^{15}N label transported to above-ground compartments was between 75 and 90% within 4 weeks after the tracer application in the 30-year-old mixed broadleaf stand (Table 7.5). Only 10 to 25% of the incorporated ^{15}N label was left in the roots of both broadleaf tree species. This finding might indicate a faster transport of the ^{15}N label to above-ground compartments for broadleaf trees than for conifers. The distribution of ^{15}N label between roots and above-ground compartments of broadleaf trees, however, is also influenced by the type of ^{15}N-labelled nitrogen input (^{15}N-labelled ammonium or nitrate versus ^{15}N-labelled litter). After application of ^{15}N-labelled ammonium or nitrate the proportion of ^{15}N label transported to above-ground compartments was higher than after application of ^{15}N-labelled litter (cf. Sect. 7.4.4).

7.5.4 Total Ecosystem Recovery of ^{15}N

Based on the low recovery of throughfall N deposition found in trees for both coniferous stands, the question arises as to what extent other ecosystem compartments function as sinks for the throughfall N input. ^{15}N mass balances on a total ecosystem level provide information for this question. This aspect also relates to the role of N deposition as stimulating agent for carbon sequestration in forests (see Nadelhoffer et al. 1999b).

The soil was the most important sink for the ammonium and nitrate deposition to the forest ground of the 15-year-old *Picea abies* stand (Fig. 7.12). Eight months after

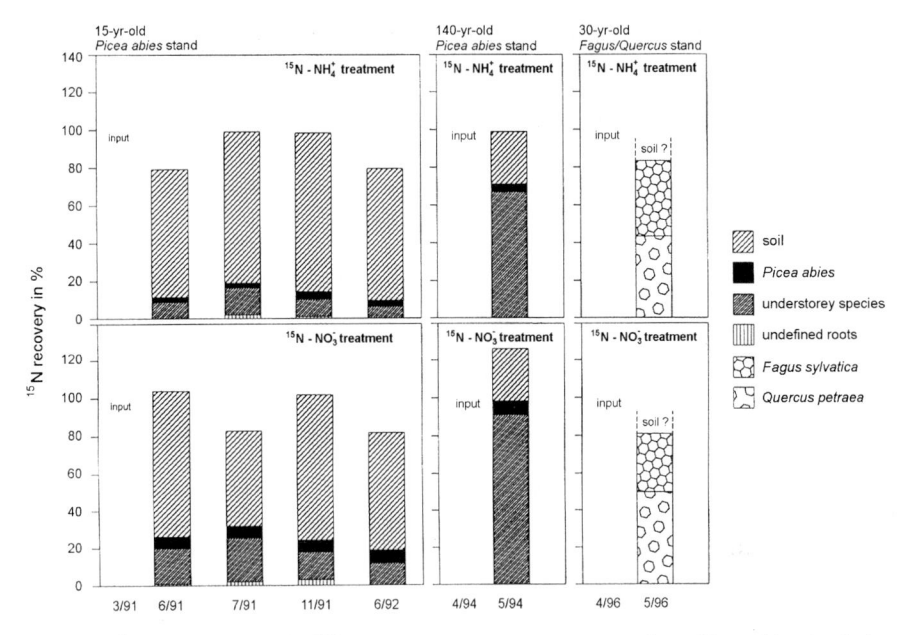

Fig. 7.12. ^{15}N recovery rates in different ecosystem compartments of two *Picea abies* stands in the Fichtelgebirge, NE Bavaria, Germany (15-year-old stand: Wülfersreuth; 140-year-old stand: Waldstein) and one 30-year-old mixed *Fagus sylvatica/Quercus petraea* stand in the Steigerwald (Ebrach, NE Bavaria, Germany) after ^{15}N-ammonium (*top*) or ^{15}N-nitrate treatments (*bottom*), respectively. Soil includes organic layers and mineral horizons. All tissues of the trees and the understorey species (if present) are summed up and presented in the individual bars. (Buchmann et al. 1996; G. Schmidt, C. May, G. Gebauer and E.-D. Schulze, unpubl. data; C. May and G. Gebauer, unpubl. data)

the ^{15}N tracer application to the 15-year-old *Picea abies* stand (in November), 79% (^{15}N-nitrate treatment) or 87% (^{15}N-ammonium treatment) of the recovered tracer were found in the soil. Although the major part of the ^{15}N label was still found in the organic layer (Buchmann et al. 1996), it was obviously no longer available for plant N uptake. The ^{15}N found here was either immobilised in organic N compounds due to incorporation by soil microbia or fungi or it was bound due to physicochemical reactions (Matschonat and Matzner 1996). The release of this soil-bound ^{15}N label into new plant-available N compounds obviously has a rather long time constant (more than 1 year). Thus, the soil of young coniferous stands functions as a temporal sink for N from atmospheric deposition, at least in the short term.

The understorey vegetation of the 15-year-old *Picea abies* stand turned out to be a major competitor for ammonium and nitrate from throughfall deposition compared to the spruce trees (Fig. 7.12). The incorporation of ^{15}N tracer by the forest-ground vegetation reached a maximum with 15% (^{15}N-ammonium treatment) to 23% (^{15}N-nitrate treatment) of the applied tracer at the peak of the growing season in July (4 months after the tracer application). The ^{15}N retention in the living understorey biomass decreased until November due to litter fall. Nevertheless, at the end of the vegetation period, still 9% (^{15}N-ammonium treatment) to 15% (^{15}N-nitrate treatment) of the applied tracer were incorporated in the understorey biomass. Thus, even after litterfall at the end of the growing season, more than twice as much N from throughfall deposition was bound by the understorey vegetation than was incorporated in tree biomass, despite the much lower understorey biomass per unit ground area.

The sink strength of the understorey vegetation for N from throughfall input was even more pronounced in the 140-year-old *Picea abies* stand (Fig. 7.12). One month after the tracer application to this mature coniferous stand, more than two thirds of the ^{15}N-ammonium and ^{15}N-nitrate inputs were sequestered by the understorey vegetation, although it comprised only 1% of the total plant biomass in this stand. As detailed above, less than 10% of the tracers was taken up by the trees, despite the fact that the trees represented 99% of the vegetation biomass. Almost 30% of the applied tracers was found in the soil, mainly bound in a non-plant-available form. The sink function of the soil for N from throughfall deposition was less pronounced in the 140-year-old *Picea abies* stand than in the 15-year-old stand. The importance of the understorey vegetation as competitor for mineral N compounds from throughfall deposition is presumably based on a higher capacity of understorey species to assimilate mineral N compounds. For example, the capacity per unit of biomass of herbaceous understorey plants to assimilate nitrate is higher by up to 2 orders of magnitude than the capacity of *Picea abies* trees (Gebauer and Schulze 1997; Gebauer et al. 1998).

The broadleaf forest stand was completely free from understorey vegetation. Therefore, the question remains open whether the high sink strength of both broadleaf tree species for mineral N (Fig. 7.12) is based solely on a higher capacity of deciduous broadleaf trees in temperate climate to assimilate mineral N compounds compared to that of conifers (see Gebauer and Schulze 1997 and references therein) or if simply the understorey vegetation is missing as strong competitor for mineral N uptake. A higher sink strength for N from throughfall deposition was recently also found by Nadelhoffer et al. (1999a) comparing broadleaf trees and conifers. The difference in the sink strength between both forest types, however, was not as pronounced as in the present investigation. Unfortunately, the role of understorey vegetation in the study of Nadelhoffer et al. (1999a) is not clear; either there was no understorey or it was not investigated.

If the strong short-term sink strength of deciduous broadleaf trees and the understorey vegetation for N from throughfall deposition also remains valid in the long term, the above findings have implications for the N and C cycles. In comparison to evergreen conifers, forest understorey species (e.g. grasses) and deciduous broadleaf trees have shorter foliage life spans. In addition, their litter is of better quality (lower C/N ratio) and therefore results in faster N mineralisation. This might speed up the circulation of N from throughfall deposition within deciduous broadleaf forests or in forests with a dense cover of understorey vegetation. Consequently, the N fertilisation effect that generally leads to increased biomass production might be greater for broadleaf tree and understorey species than for conifers. On the other hand, immobilisation of N from throughfall deposition in the soil profile seems to be higher in evergreen coniferous than in deciduous forests.

7.6 Conclusions

This chapter summarises the present knowledge about the fate of mineral N compounds in the soil of coniferous and broadleaf forests in temperate climate. The fate is represented by the partitioning between soil mineral N pools, soil organic matter, trees and understorey vegetation. In particular, two major aspects are addressed: (1) the rate of N release from decomposing litter, its fate in the soil and its contribution to the N uptake by adult beech trees and (2) the response of trees, forest understorey vegetation and soils to ammonium and nitrate inputs from throughfall N deposition in young and mature Norway spruce stands and in a mixed broadleaf forest stand. Both of the approaches are based on ^{15}N-tracer experiments. The following conclusions are derived from the data:

- Decomposing beech litter acts as a sink for external N from different sources (fungi, microbes, throughfall) and as a source of N due to a continuous release of original litter N. This N release is related to litter mass loss. The incorporation of external N into the litter balances the N release for almost 2 years, in the third year a net N loss occurs.
- Litter-released N is rapidly incorporated into different soil N pools, taken up by mycorrhiza and transferred into the host plant. During the first year following labelled litter deposition, early-released N is preferentially mineralised, creating a pulse of labelled nitrate and ammonium. Later, the ^{15}N enrichment of soil nitrate and ammonium remains low, indicating that the turnover rate of litter N is the same as that for soil organic matter. The contribution of litter N to the overall N supply to adult trees is very low. Most of the litter N is incorporated in the large soil N pool, almost exclusively in organic form.
- In the most densely rooted organic soil layer of coniferous and broadleaf forests on acid soils, the ammonium fraction dominates over the nitrate fraction throughout the growing season. Therefore, ammonium from throughfall N deposition is diluted more strongly by soil-borne ammonium than is deposition-derived nitrate diluted by soil-borne nitrate.
- Most of the N from throughfall deposition to coniferous and broadleaf forests is immobilised within less than 2 weeks. Within that short period of time, the microbial net nitrification in the coniferous forests is too slow to transform major amounts of ^{15}N-ammonium input into ^{15}N-nitrate. Thus, conifers take up N from throughfall in the form in which it was deposited. In contrast, considerable amounts of ^{15}N-ammonium inputs are nitrified in broadleaf forests before they are immobilised. Broadleaf trees, therefore, take up parts of the ammonium deposition in the form of nitrate.
- Ammonium dominates over nitrate as mineral N source for *Picea abies* under the growing conditions in temperate climate on acid soils. Nitrate is more important as mineral N source for broadleaf trees than for conifers.
- During the study period, conifers from young and mature stands retained less than 10% of the throughfall N deposition. Broadleaf trees retained 70% and more from throughfall N input. Understorey vegetation in coniferous forests proved to be a strong competitor for N from throughfall input. Despite their much lower biomass, understorey species immobilised more N from throughfall than conifers. The importance of the understorey vegetation as a temporary N sink has been underestimated in the past and must be considered with more emphasis in the future.
- The forest soil is an important sink for N from ammonium and nitrate deposition. In a young coniferous stand, the sink strength of the soil is higher than in a mature coniferous stand.

- Litter-released N is much less available to trees than throughfall N. However, after a period of a month (throughfall N) or a year (litter-released N), during which mineral N is available, most of the N input is incorporated in soil organic matter and probably cycles like the bulk of the organic matter. Long-term monitoring of such experiments is needed in order to precise and model the specific fate, if any, of throughfall N and litter-derived N.

References

Aber JD, Nadelhoffer KJ, Steudler P, Melillo JM (1989) Nitrogen saturation in northern forest ecosystems. BioScience 39:378–386

Bauer G, Schulze E-D, Mund M (1997) Nutrient contents and concentrations in relation to growth of *Picea abies* and *Fagus sylvatica* along a European transect. Tree Physiol 17:777–786

Berg B (1988) Dynamics of nitrogen (^{15}N) in decomposing Scots pine (*Pinus sylvestris*) needle litter. Long-term decomposition in a Scots pine forest VI. Can J Bot 66:1539–1546

Berthelin J, Leyval C, Toutain F (1994) Biologie des sol. Rôle des organismes dans l'alteration et l'humification. In: Bonneau M, Souchier B (eds) Pédologie 2. Constituants et propriétés du sol. Masson, Paris, pp 143–237

Blair JM (1988) Nitrogen, sulphur and phosphorus dynamics in decomposing deciduous leaf litter in the southern Appalachians. Soil Biol Biochem 20:693–701

Bottner P, Austrui F, Cortez J, Billes G, Couteaux MM (1998) Decomposition of ^{14}C and ^{15}N-labelled plant material, under controlled conditions, in coniferous forest soils from a north-south climatic sequence in western Europe. Soil Biol Biochem 30:597–610

Buchmann N, Schulze E-D, Gebauer G (1995) ^{15}N-ammonium and ^{15}N-nitrate uptake of a 15-year-old *Picea abies* plantation. Oecologia 102:361–370

Buchmann N, Gebauer G, Schulze E-D (1996) Partitioning of ^{15}N-labeled ammonium and nitrate among soil, litter, below- and above-ground biomass of trees and understory in a 15-year-old *Picea abies* plantation. Biogeochemistry 33:1–23

Casabianca H (1995) La spectrométrie de masse isotopique. Les couplages. La reproductibilité de la technique pour le carbone et l'azote. In: Maillard P, Bonhomme R (eds) Utilisation des isotopes stables pour l'étude du fonctionnement des plantes. INRA, Paris

Durka W, Schulze E-D, Gebauer G, Voerkelius S (1994) Effects of forest decline on uptake and leaching of deposited nitrate determined from ^{15}N and ^{18}O measurements. Nature 372:765–767

FAO (1988) Soil map of the world. Revised legend 1989. Reprint of the World Soil Resources Report 60. FAO, Rome

Gebauer G, Dietrich P (1993) Nitrogen isotope ratios in different compartments of a mixed stand of spruce, larch and beech trees and of understorey vegetation including fungi. Isotopenpraxis 29:35–44

Gebauer G, Schulze E-D (1991) Carbon and nitrogen isotope ratios in different compartments of a healthy and a declining *Picea abies* forest in the Fichtelgebirge, NE Bavaria. Oecologia 87:198–207

Gebauer G, Schulze E-D (1997) Nitrate nutrition of Central European forest trees. In: Rennenberg H, Eschrich W, Ziegler H (eds) Trees – contributions to modern tree physiology. Backhuys, Leiden, pp 273–291

Gebauer G, Hahn G, Rodenkirchen H, Zuleger M (1998) Effects of acid irrigation and liming on nitrate reduction and nitrate content of *Picea abies* (L.) Karst. and *Oxalis acetosella* L. Plant Soil 199:59–70

Joergensen RG, Meyer B (1989) Nutrient changes in decomposing beech leaf litter assessed using a solution flux approach. J Soil Sci 41:279–293

Jussy JH (1998) Minéralisation de l'azote, nitrification et prélèvement radiculaire dans différents écosystèmes forestiers sur sol acide. Effets de l'essence, du stade de développement du peuplement et de l'usage ancien de sols. Thesis, Université Henri Poincaré, Nancy, 161 pp

Koopmans CJ, Tietema A, Boxman AW (1996) The fate of ^{15}N-enriched throughfall in two coniferous forest stands at different nitrogen deposition levels. Biogeochemistry 34:19–44

Kronzucker HJ, Siddiqi MY, Glass ADM (1997) Conifer root discrimination against soil nitrate and the ecology of forest succession. Nature 385:59–61

Marschner H, Häussling M, George E (1991) Ammonium and nitrate uptake rates and rhizosphere pH in non-mycorrhizal roots of Norway spruce [*Picea abies* (L.) Karst.]. Trees 5:14–21

Martin F, Lorillou S (1997) Nitrogen acquisition and assimilation in ectomycorrhizal systems. In: Rennenberg H, Eschrich W, Ziegler H (eds) Trees – contributions to modern tree physiology. Backhuys, Leiden, pp 423–439

Matschonat G, Matzner E (1996) Soil chemical properties affecting NH_4^+ sorption in forest soils. Z Pflanzenernähr Bodenkd 159:505–511

May C (1999) Nutzung von Ammonium und Nitrat durch Rotbuche (*Fagus sylvatica* L.) und Traubeneiche (*Quercus petraea* [L.] Karst.). Thesis, Universität Bayreuth, Bayreuth

May C, Schmidt G, Gebauer G, Schulze E-D (1996) The fate of [^{15}N]ammonium and [^{15}N]nitrate in the soil of a 140-year-old spruce stand (*Picea abies*) in the Fichtelgebirge (NE-Bavaria). Isotopes Environ Health Stud 32:149–158

Nadelhoffer KJ, Fry B (1994) Nitrogen isotope studies in forest ecosystems. In: Lajtha K, Michener R (eds) Isotopes in ecology and environmental science. Blackwell, Boston, pp 23–44

Nadelhoffer KJ, Downs MR, Fry B, Aber JD, Magill AH, Melillo JM (1995) The fate of ^{15}N-labelled nitrate additions to a northern hardwood forest in eastern Maine, USA. Oecologia 103:292–301

Nadelhoffer KJ, Downs MR, Fry B (1999a) Sinks for ^{15}N-enriched additions to an oak forest and a red pine plantation. Ecol Appl 9:72–86

Nadelhoffer KJ, Emmett BA, Gundersen P, Kjonaas OJ, Koopmans CJ, Schleppi P, Tietema A, Wright RF (1999b) Nitrogen deposition makes a minor contribution to carbon sequestration in temperate forests. Nature 398:145–148

Nambiar EKS, Bowen GD (1986) Uptake, distribution and retranslocation of nitrogen by *Pinus radiata* from ^{15}N-labelled fertiliser applied to podzolised sandy soil. For Ecol Manage 15:269–284

Näsholm T, Ekblad A, Nordin A, Giesler R, Högberg M, Högberg P (1998) Boreal forest plants take up organic nitrogen. Nature 392:914–916

Nômmik H, Larsson K (1992) Effects of nitrogen source and placement on fertiliser ^{15}N enrichment in *Pinus sylvestris* foliage. Scand J For Res 7:155–163

Northup RR, Yu Z, Dahlgren RA, Vogt KA (1995) Polyphenol control of nitrogen release from pine litter. Nature 377:227–229

Preston CM, Mead DJ (1994a) A bioassay of the availability of residual ^{15}N fertiliser eight years after application to a forest soil in interior British Columbia. Plant Soil 160:281–285

Preston CM, Mead DJ (1994b) Growth response and recovery of ^{15}N-fertiliser one and eight growing seasons after application to lodgepole pine in British Columbia. For Ecol Manage 65:219–229

Read DJ (1991) Mycorrhizas in ecosystems. Experientia 47:376–396

Schmidt G, May C, Gebauer G, Schulze E-D (1996) Uptake of [^{15}N]ammonium and [^{15}N]nitrate in a 140-year-old spruce stand (*Picea abies*) in the Fichtelgebirge (NE Bavaria). Isotopes Environ Health Stud 32:141–148

Schneider BU, Meyer J, Schulze E-D, Zech W (1989) Root and mycorrhizal development in healthy and declining Norway spruce stands. In: Schulze E-D, Lange OL, Oren R (eds) Forest decline and air pollution. A study of spruce (*Picea abies*) on acid soils. Ecological studies 77. Springer, Berlin Heidelberg New York, pp 370–391

Schulze E-D (1989) Air pollution and forest decline in a spruce (*Picea abies*) forest. Science 244:776–783

Setälä H, Marshall VG, Trofymow JA (1996) Influence of body size of soil fauna on litter decomposition and ^{15}N uptake by poplar in a pot trial. Soil Biol Biochem 28:1661–1675

Smith SE, Read DJ (1997) Mycorrhizal symbiosis, 2nd edn. Academic Press, London

Staaf H (1980) Release of plant nutrients from decomposing leaf litter in a South Swedish beech forest. Holarct Ecol 3:129–136

Tamm CO (1991) Nitrogen in terrestrial ecosystems. Ecological studies 81. Springer, Berlin Heidelberg New York

Tietema A, Wessel WW (1994) Microbial activity and leaching during initial oak leaf litter decomposition. Biol Fertil Soils 18:49–54

Tietema A, Emmett BA, Gundersen P, Kjonaas OJ, Koopmans CJ (1998) The fate of ^{15}N-labelled nitrogen deposition in coniferous forest ecosystems. For Ecol Manage 101:19–27

Vance ED, Brookes PC, Jenkinson DS (1987) Microbial biomass measurements in forest soils: the use of the chloroform-fumigation-incubation method in strongly acid soils. Soil Biol Biochem 19:697–702

Zeller B (1998) Contribution à l'étude de la décomposition d'une litière de hêtre, la libération de l'azote, sa minéralisation et son prélèvement par le hêtre (*Fagus sylvatica* L.) dans une hêtraie de montagne du bassin versant du Strengbach (Haut-Rhin). Thesis, Université Henri Poincaré, Nancy (France), 138 pp

Zeller B, Colin-Belgrand M, Dambrine E, Martin F (1998) ^{15}N partitioning and production of ^{15}N-labelled litter in beech trees following [^{15}N]urea spray. Ann Sci For 55:375–383

8 Canopy Uptake and Utilization of Atmospheric Pollutant Nitrogen

A.F. Harrison, E.-D. Schulze, G. Gebauer, and G. Bruckner

8.1 Introduction

Research on effects of air pollutants on forests concentrated initially on quantifying wet deposition into ecosystems, because of its significance in acidifying soils (Ulrich 1987; Last and Watling 1991) and because it can be easily monitored. By contrast, the deposition of pollutant gases has not received an equivalent amount of attention, even though the ecological importance of this process has long been recognised (Nilgard 1985; Roelofs et al. 1985). Understanding of gas interactions with canopies has led to a general explanation of the processes leading to forest decline (Schulze 1989). The processes involved in deposition and canopy uptake of pollutant N have remained difficult to quantify due to their complexity (e.g. Duyzer et al. 1992; Hanson and Lindberg 1991; Joslin et al. 1990) and a lack of adequate techniques to measure uptake fluxes directly under field conditions. Thus, estimates of the amounts of nitrogen entering into the ecosystem directly via the canopy, bypassing soils and roots, and the induced physiological responses in the trees and ground flora, have been assessed only by indirect methods (Pearson and Stewart 1993; Sutton et al. 1993).

It has not been possible in this study to measure canopy uptake of N at the CANIF forest sites, because of technical difficulties under field conditions and the fact that several uptake pathways are involved, e.g. uptake from wet deposition via foliage and bark, and the absorption of gases by foliage via the stomata. This chapter will therefore review the processes and the factors that are involved in governing the rates of canopy uptake of deposited pollutant N. It will also try to assess quantities of N taken up and their potential importance in forest nutrition.

A key NIPHYS / CANIF project objective was to compare effects of N pollutants on N dynamics in forests of central Europe, characterised by relatively high N deposition rates, with forests in northern and southern Europe where deposition rates are low (Stanners and Bourdeau 1995 based on EMEP data from 1984; see the N deposition map in Persson et al., Chap. 2, this Vol.). More recent assessments of the regional distribution and deposition of nitrogen pollution, however, indicate a much more uniform pattern of emissions (European Environmental Agency 1998), with Italy having similar N emission rates to Germany or Denmark. Thus, only north Scandinavia remains widely unaffected by enhanced N deposition at the present time. Later in this chapter, information will be presented on pollutant N concentrations, deposition inputs and estimates of canopy uptake derived from studies at or near to the German spruce site Waldstein. Given the relative uniformity of N emissions, the estimates for canopy uptake at this site are considered to be relevant and applicable to many other forest areas across central and southern Europe.

8.2 Atmospheric Nitrogen Pollutants

Oxidised and reduced forms of nitrogen are being emitted from both anthropogenic and natural sources in particulate and gaseous forms. These are transformed during transport through the atmosphere (Fig. 8.1) and are deposited to the land surface and to vegetation. Forests are particularly effective in trapping atmospheric pollutants, both as wet and dry deposition. The half-lives of NH_3 and NO_x in the atmosphere are generally <3 days (Stanners and Bourdeau 1990; INDITE 1994). NH_3 usually travels only relatively short distances from the emission sources and can be rapidly codeposited with SO_2 emissions on moist surfaces as ammonium sulphate (Adema et al. 1985; Ineson et al. 1993; Cape et al. 1998). Although NO_x may travel considerably farther than NH_3 (Stanners and Bourdeau 1990), it can be assumed that most of the total N emission is deposited to land during the pass of air masses across Europe (Galloway et al. 1994).

Sixty percent of the total European NO_x emissions of 15.5 million tons originates from transport (traffic), 14% from industry and 21% from energy production. Including households (4%), combustion processes account for 99% of the NO_x production, making biological NO_x sources insignificant (European Environmental Agency 1998). The sources are different for ammonia, where 93% originates from agriculture, and 7% from industry and transport, making natural sources again insignificant. The emissions (NO_x plus NH_3) range between 12 and $120\,kg\,ha^{-1}\,yr^{-1}$ for most of Europe, There are a few hot spots with emissions $>200\,kg\,ha^{-1}\,yr^{-1}$ in some agricultural and

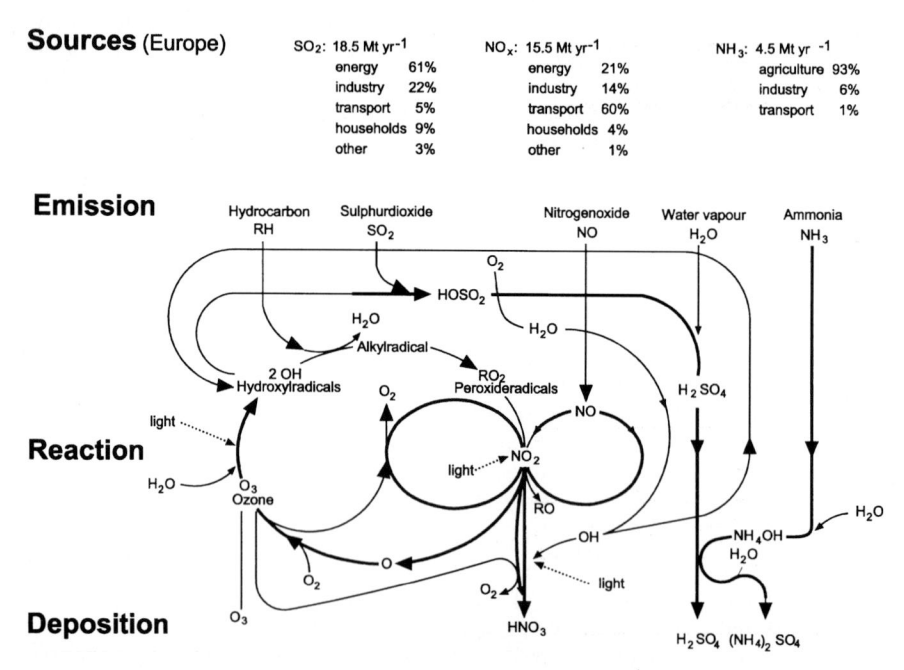

Fig. 8.1. Sources, emissions, chemical reactions in the atmosphere and deposition of air pollutants

industrial centres. Only in some regions of Spain, Greece or North Scandinavia does N deposition remain at <4 kg ha^{-1} yr^{-1}. On a molar basis, the total amount of N emissions (6.01×10^8 kmol N yr^{-1}) exceeds that of sulphur (2.90×10^8 kmol S yr^{-1}) by a factor of 2. It is not clear from the emission data of the European Environmental Agency what relative proportions of these emissions are being deposited in wet and gaseous forms.

N pollutant emissions vary quite significantly between years and with season. The seasonal trends in concentrations of NH$_3$, NO, and NO$_2$ for the Waldstein forest site are shown in Fig. 8.2. In 1997, ammonia showed a three-peaked seasonal trend, with a peak in winter probably caused by emissions from a local power station, and spring

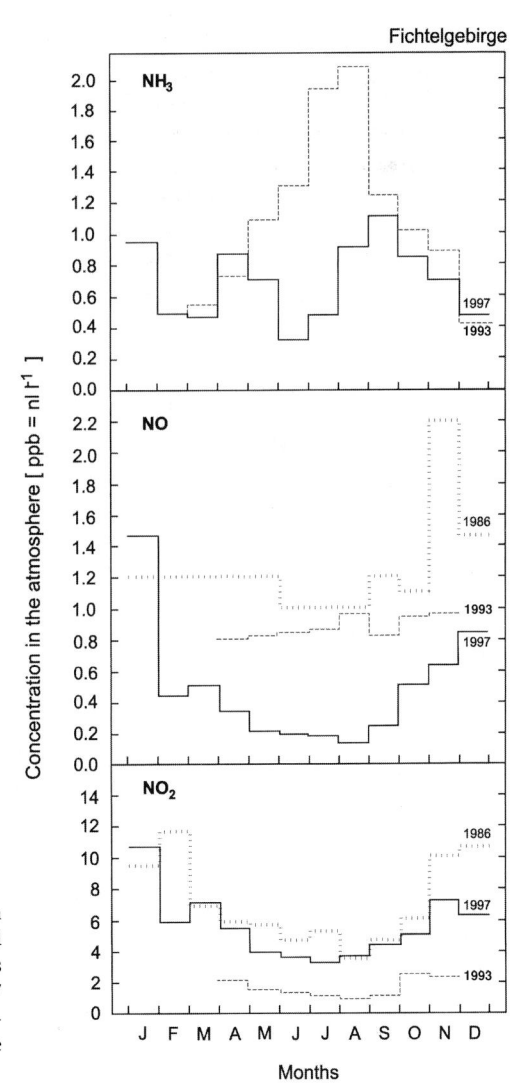

Fig. 8.2. Seasonal variation in ammonia (NH$_3$), nitric oxide (NO) and nitrogen dioxide (NO$_2$) in the years 1986 (Eiden 1989), 1992 and 1997 (BITÖK Data bank) in the Fichtelgebirge, NE Bavaria, the area of the Waldstein and the Schacht site

and autumn peaks probably resulting from spraying of liquid manure on agricultural land. Seasonal trends in atmospheric concentrations of gases may vary depending on changing weather conditions. In 1992, the summer peak was twice as high as that in 1997. The NO concentrations were generally higher in 1986 and 1992 than in 1997, while that of NO_2 was lower in 1992. NO and NO_2 generally show highest concentrations in winter and lower concentrations in summer, but with NO_2 being at ca. 10 times higher concentrations than NO. It is clear that the pollutant concentrations during the growing season will govern canopy uptake potential and that the actual canopy uptake and utilisation of N will vary from year to year as pollutant concentrations vary between years.

8.3 Pathways for Canopy Uptake of Nitrogen

Uptake of atmospheric inputs of N is potentially possible through foliage, twigs, branches and stems (Fig. 8.3). The ability of plants to take up nutrients from solutions via foliar tissues is well known (Tuckey et al. 1962). Uptake into foliage depends largely on the degree of wettability of the foliar surface and may be facilitated by cuticular erosion brought about by acidic deposition (Crossley and Fowler 1986). Ionic transport across foliar cuticles may involve specific membrane carriers (Haynes

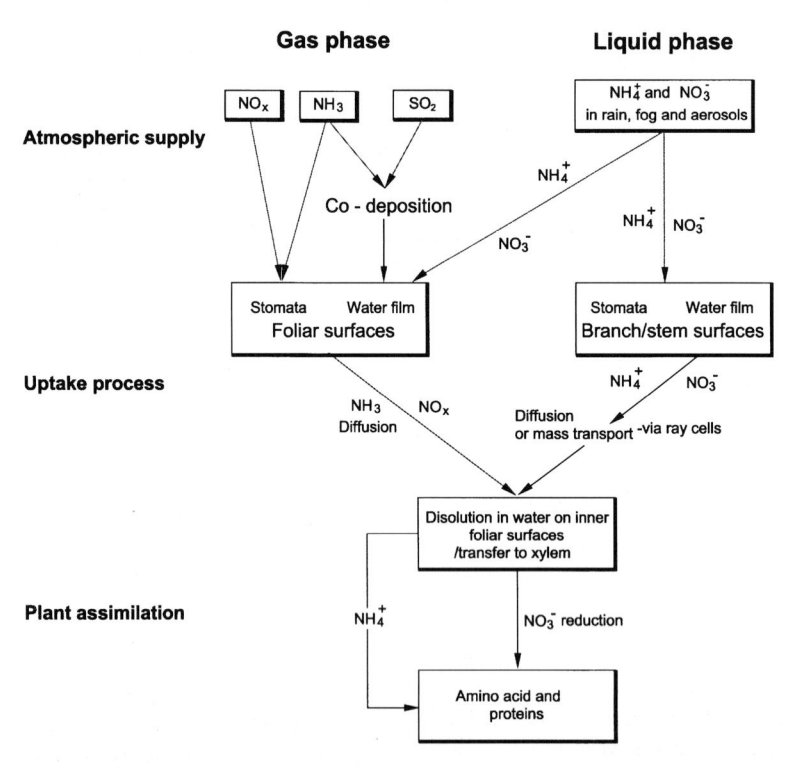

Fig. 8.3. Major pathways for the uptake of gaseous and liquid nitrogenous compounds into the canopy from the atmosphere

1986). There is also evidence of uptake of ions in solution via the stomata (Eichert et al. 1998). Uptake into twigs and stems may be by simple diffusion through bark, which is more permeable than waxy cuticular surfaces (Schoenherr 1982) and has lower resistance to liquid phase diffusion (Schaeffer and Reiners 1990). Movement occurs through the bark and along radially orientated ray tissues into the xylem (Katz et al. 1989; Klemm 1989). The main pathway for uptake of NH_3 and NO_x is by diffusion through stomata in leaves, being largely governed by stomatal conductance, though some gases can even be absorbed by foliar cuticles and bark; for example, NO_2 undergoes a chemical reaction (nitration of phenolics) with compounds of the cuticle and bark (Kisser-Priesack et al. 1990). The incorporation rate of NO_2 into the cuticle and the bark is linearly related to both NO_2 concentration in the atmosphere and the exposure time (Kisser-Priesack and Gebauer 1991).

8.4 Approaches to the Determination of Canopy Uptake of Nitrogen

Several approaches can each provide evidence of uptake of pollutant N by the forest canopy. These are the (1) examination of changes in N composition of rainfall and throughfall, (2) comparisons of N fluxes beneath live and plastic trees of similar "morphology", (3) application of ^{15}N-labelled N solutions or gases, (4) use of the potential in $\delta^{15}N$ techniques, (5) use of canopy conductances as measured by eddy covariance, or (6) examination of changes in concentrations of N in xylem flow related to atmospheric gas concentrations. Each approach has been used to generate estimates of N fluxes into tree or seedling canopies.

Concentrations of N in rainfall are often reduced on passage through the canopy, suggesting that uptake does occur (Parker 1983). However, it is not clear if this results from actual assimilation, the retention on exchange surfaces of foliage and bark, or uptake by epiphytes and microbial populations on canopy surfaces. The interpretation of throughfall data is complicated by the often large amounts of dry-deposited N, which interferes with detection of the uptake processes. However, comparing throughfall of N with sulphate or chloride results in estimates of gross canopy uptake (Horn et al. 1989).

Artificial trees have been used to investigate atmospheric pollutant deposition on forest canopy surfaces (Joslin et al. 1990; Dambrine et al. 1998). Comparison of N deposition behaviour on similar-sized plastic and living trees offers an approximation for estimating potential canopy uptake if these model trees remain without epiphytes and assuming that their structure and roughness resemble those of real trees. A plastic to live tree comparison was used in the NiPHYS project to derive estimates of canopy uptake by Norway spruce at the Aubure site in France (Ignatova and Dambrine 2000). In this approach, the effects of differences in morphology of the living and plastic canopies were adjusted for by reference to the behaviour of the poorly absorbed Na^+ and SO_4^{2-} ions.

The most convincing and sensitive method to determine canopy uptake of N is to apply ^{15}N-labelled sources in solution or as gaseous forms. This enables uptake to be confirmed by "tracing" the ^{15}N in non-exposed plant parts, such as roots, and into cellular metabolites. Care must be exercised if realistic uptake rates are to be derived from ^{15}N use. For example, realistic concentrations of solution or gaseous ^{15}N-labelled sources should be used, but this may generate technical problems. Uptake should be calculated against the $\delta^{15}N$ in control unexposed plants rather than the average 0.3663%, particularly if N fertiliser, often depleted with respect to ^{15}N, is used as part of any treatment effect. In practice, this approach can only be used under controlled conditions. In addition, a number of factors such as (1) short-term exposure, (2) application of ^{15}N to saplings or on a limited scale to specific plant parts, and (3) the often rapid translocation of ^{15}N away from the uptake zone, make difficulties in realistic extrapolation of results to forests exposed to pollutant N for extended periods.

At first sight, there may be potential in using the natural abundance of $\delta^{15}N$ or the enrichment techniques to derive estimates of forest canopy uptake under field conditions. There are, however, many constraints limiting the use of this approach, for example (1) pollutant N deposited can be taken up by the canopy and via throughfall by the root system, (2) a positive $\delta^{15}N$ enrichment effect in forest foliage can result from nitrification of throughfall deposited N in soils (Emmett et al. 1998), and (3) there are significant seasonal and spatial variations both of $\delta^{15}N$ in atmospheric pollutant N inputs to forests (see Bauer et al., Chap. 9, this Vol., on isotopic data) and in tree tissues. A way around these technical difficulties may be:

1. to carry out a $\delta^{15}N$ and N concentration budget study of a suitable forest including its throughfall, then
2. construct a "roof" (cf. those used in the NITREX project: Lamersdorf et al. 1998) to prevent throughfall and stem flow reaching the soil (but the soil would require irrigating with clean water to maintain its moisture content), and
3. releases from gas cylinders of NH_3 or NO_2 (with measured $\delta^{15}N$ signatures) upwind of the forest by an automatically operated valve assembly (Ineson et al. 1996),
4. after release of appropriate pulses of gas and collection any of throughfall and stemflow during and after gas release, the $\delta^{15}N$ and N budget study is repeated.

Canopy uptake is then estimated from changes in the $\delta^{15}N$ of the tree components and the throughfall and stem flow. Controlled releases of NH_3, with an initial $\delta^{15}N$ signature of −3.6‰ and final signature of −0.3‰, onto a spruce forest has resulted in throughfall with a $\delta^{15}N$ signature of +6.2‰. These findings have been interpreted as indicating canopy uptake of the gas (Heaton et al. 1997). Ammann et al. (1999) have also used this approach to estimate NO_x uptake near a motorway. New analytical techniques for future routine on-line isotope analysis of reactive trace gases, like NH_3, NO and NO_2, to study canopy N uptake are under development (Lauf and Gebauer 1998).

The methods described so far have taken an upscaling approach, by extrapolating upwards from detailed measurements to the whole canopy. By contrast, the canopy conductance method is a downscaling approach, i.e. the whole canopy flux is measured and interpretations are made at the detailed process level. Horn et al. (1989) were the first to use canopy conductances as measured by xylem flow to calculate N canopy uptake fluxes from known atmospheric pollutant N concentrations. In this case, canopy conductance integrated not only the stomatal diffusion process, but also the canopy aerodynamic component. The estimate generated represented a total flux for the whole canopy. Generally, this flux is measured as water vapour in xylem flow (Köstner et al. 1992) or by eddy covariance (Hollinger et al. 1994). If differences in molecular weight are allowed for, the conductance can be derived and used as basis for other trace gas flux calculations, provided that the atmospheric concentration and the conductance of the respective gas in the mesophyll (as expressed by the compensation point) are taken into account. In this case, the flux = conductance* (meaning times) its concentration gradient (Pearcy et al. 1989). It is only possible from eddy correlation measurements to derive the aerodynamic component, and therefore the flux of gases to plant external surfaces and soils in the ecosystem. Thus, whilst the eddy correlation approach appears to be the most direct to quantify gaseous deposition, it cannot be used to determine the wet-phase uptake.

8.5 Review of Research

8.5.1 Factors Affecting Uptake from Wet Deposition

Evidence from throughfall studies indicates that canopy uptake of N from wet deposition occurs mainly during the growing season (Parker 1983; Helmisaari and Malkonen 1989; Bauer 1997; Ignatova and Dambrine 2000), i.e. when the canopy is physiologically active. Generally, NH_4^+ is taken up at a significantly greater rate than NO_3^-, as shown by the use of ^{15}N (Bowden et al. 1989; Schulze and Gebauer 1989; Garten and Hanson 1990). However, if the negative charged cell walls are equilibrated by cations, NO_3^- can be taken up as actively as ammonium (Harrison et al. 1991 and unpubl.). Thus the pH of the surface solution can have a significant effect on N uptake, particularly in NH_4^+ form (Bruckner 1995). The potential for uptake is highest in the youngest needles of spruce and declines with age (Eilers et al. 1992; Macklon et al. 1996), though there may be no significant difference between current and 1-year-old needles (Wilson and Tiley 1998). The main route for uptake is considered to be via the foliage, but increasing evidence suggests that uptake via twig, branch and stem surfaces may be at least as important (Bowden et al. 1989; Katz 1991; Bruckner 1995; Boyce et al. 1996; Macklon et al. 1996; Wilson and Tiley 1998), particularly if stem flow N concentrations are relatively high. Rates of uptake of N increase linearly with concentration in the contact solution (Eilers et al. 1992; Bruckner 1995). The efficiency of N uptake by the canopy is significantly increased if the surface remains moist (Harrison et al. 1991; Bruckner 1995). Applications of N fertiliser to roots can have a suppressive effect on uptake via the canopy (A.F. Harrison et al., unpubl.), whereas P and K have no effect (Wilson and Tiley 1998). The rates of canopy uptake of N appear to be faster for beech and birch than for spruce (Brumme et al. 1992; Harrison et al. 1991). This difference may be attributable to a combination of thinner cuticles and higher wettability of foliage and bark of the beech and birch. This advantage may be counterbalanced, in terms of the total potential for canopy uptake, by the fact that both beech and birch have foliage for only part of the year.

8.5.2 Factors Affecting Uptake on N in Gaseous Form

Plants may under some conditions release small quantities of gases (e.g. NH_3) to the atmosphere through stomata in their foliage (Sutton et al. 1993). However, the external concentration of pollutant N gases in the atmosphere at which influx is greater than efflux, the compensation point (Farquhar et al. 1980), is very frequently exceeded, resulting in a potential for net uptake and utilisation of gas by a plant. The compensation point may vary with the solubility of the respective gas in the cell wall water and the rate of assimilation in the mesophyll. Thus, rapid absorption of NH_4^+ by roots or low activity of glutamate synthetase in foliage increases the compensation point (Sutton et al. 1993; Schjoerring et al. 1998). Trees are able to take up and use gaseous pollutant forms of nitrogen, principally as NH_3 and NO_2. NO appears to be taken up much less actively than NO_2 due to the lower water solubility (Skarby et al. 1981). Also, the concentration of NO in the atmosphere is generally low outside urban environments (Warneck 1988). Uptake of NH_3 into needles of spruce is ten times faster than NO_2 due to the higher solubility in cell wall water. It is doubled in the presence of SO_2 as a result of co-deposition under high relative humidity conditions on the foliage surface (Bruckner et al. 1993b; Bruckner 1995). Thus foliar uptake of NH_3 can result in a marked increase in N concentration in the needles of Scots pine (Perez-Soba and van der Eerden 1993). Uptake of gaseous N forms, e.g. HNO_3, NO_x or NH_3, are positively correlated with concentration in the surrounding atmosphere (Skarby et al. 1981; Vose and Swank 1990; Bruckner et al. 1993a) but decrease with needle age. Thus, younger foliage takes up gaseous N (NH_3) from a given concentration much faster than older foliage (Bruckner et al. 1993b). Uptake appears also to be higher in needles of nitrogen-poor spruce than in trees with good N nutrition (Bruckner et al. 1993a). NO_2 and NH_3 deposition to foliage is generally lower for conifers than for broadleaf species, due to differences in stomatal conductance (Schulze and Hall 1992); stomatal conductance is governed by stomatal density of the foliage, stomatal physiology interacting with various environmental factors, such as light intensity and vapour pressure deficit or transpiration rate (Hanson et al. 1992; Thoene et al. 1991).

8.5.3 Evidence for Assimilation in the Canopy

Evidence for assimilation and utilisation within trees is given by the facts that (1) ^{15}N-labelled sources applied to tree foliage and shoots are translocated to "protected" roots (Harrison et al. 1991; Eilers et al. 1992), (2) there is an enhancement of enzyme activity within foliage as a result of the uptake of nitrogen sources by the foliage and (3) it is possible to demonstrate, principally by using the ^{15}N tracer, the conversion to amino acids of the inorganic nitrogen sources taken up (Wingsle et al. 1987; Nussbaum et al. 1993).

8.5.4 Enzyme Activity in Foliage

As a key process in the assimilation of pollutant N via the canopy, the enzymes nitrate reductase (NR) and glutamate synthetase (GS) are induced in the tissues involved in the N uptake. NR has been shown to increase significantly in response to foliar exposure to NO_2 and NO_3^-, whilst GS increases in response to NH_3 or NH_4^+ (Wingsle et al. 1987; Thoene et al. 1991; Perez-Soba et al. 1994). Application of NO_3^- or NH_4^+ to root systems does not induce increases in enzyme activity in foliage of conifers (Wingsle et al. 1987; Nasholm 1991; Pearson and Soares 1998), so that the induction of their activity in foliage is an indicator of foliar response to pollutant N (Tischner et al. 1988). Their induction by N sources is a rapid response and the activity declines equally rapidly following removal of the N source (Egger et al. 1989; Nasholm 1991; Weber et al. 1998). The enzymes can vary significantly through the season (Nasholm 1991; Nussbaum et al. 1993). These enzymes mediate the assimilation of the pollutant N sources into amino acids (Nasholm 1991; Nussbaum et al. 1993), which then are incorporated into the plant's N metabolism. Induced enzyme activity in the foliage of conifers may be useful as an index of pollution levels (Pearson and Soares 1998).

Nitrate reductase activity induced in the foliage of broadleaf trees cannot be used for the same purpose as the enzyme can be induced by nitrate application to their roots (see Gebauer and Schulze 1997 and references therein).

8.5.5 Quantification of Canopy Uptake

Estimates of the potential contributions to forest nutritional demand by canopy uptake of wet deposited and gaseous forms of N are still very approximate. Various studies by throughfall analysis or ^{15}N methods indicate that uptake by trees from wet-deposited N as NH_4^+ and NO_3^- may be in the region of 1–10 kg N ha^{-1} a^{-1} (Matzner 1989; Brumme et al. 1992; Eilers et al. 1992; Lovett 1992; Ingatova and Dambrine 2000). The estimated amounts for canopy uptake range between <5% (Boyce et al. 1996; Wilson and Tiley 1998) and >40% of the N required for annual wood production of a deciduous forest (Lindberg et al. 1986). Uptake from gaseous N sources is potentially more significant. NO_x at 20–30 ppb has been shown to contribute 7% of N in shoots of Scots pine seedlings during the 39-day period of exposure (Nasholm et al. 1991). Estimated uptake of NO_2 by Scots pine from an atmospheric concentration of 1–3 ppb could be between 2.4 and 7.2 kg N ha^{-1} yr^{-1} (Grennfelt et al. 1983). Concentrations of N in Scots pine needles exposed to NH_3 increased by 49% over a 4-month period (Perez-Soba and van der Eerden 1993). The uptake of NO_x at 12–20 ppb and NH_3 at 2–4 ppb could even contribute 8–14% and 11–22%, respectively, of the annual N balance of a 30-year-old spruce forest in the Fichtelgebirge (Bruckner et al. 1993b). The very high estimates in the latter study were, in part, due to the high trace gas concentrations in the atmosphere in the year of measurement. Somewhere between 20 and 40% of the annual N demand by 30-year-old spruce may be satisfied by canopy uptake of both gaseous and wet-deposited N (Katz 1991; Bruckner et al. 1993a). Thus, both atmospheric NO_2 and NH_3 can act as a nitrogen fertiliser (Nussbaum et al. 1993; Perez-Soba and van der Eerden 1993).

To illustrate the potential importance of canopy uptake of deposited pollutant N, a very approximate estimate of canopy uptake for the Waldstein forest has been attempted (Table 8.1). For the Fichtelgebirge, Horn et al. (1989) used the canopy conductance approach, together with the throughfall measurements. They estimated that 2.7 kg N ha^{-1} yr^{-1}, representing only 2% of the total nitrogen requirement of 134 kg N ha^{-1} yr^{-1}, were derived from gaseous uptake by stomata, but that 32.9 kg N ha^{-1} yr^{-1} were derived from uptake from the liquid phase (24%). Thus, the canopy conductance approach appears to result in lower fluxes than upscaling from growth chamber experiments (Bruckner 1995).

In the following we compare both upscaling from chamber experiments and downscaling from canopy flux measurements. For the upscaling exercise, mean NH_3 and NO_x concentrations of 0.7 and 5.9 ppb, respectively, in the air (BITÖK, unpubl.; see Fig. 8.2) and mean NH_4^+ and NO_3^- concentrations in throughfall (1990–1992) of 1.4 and 1.7 mg N l^{-1}, respectively (Manderscheid and Goettlein 1995) were used. Uptake rate of gaseous NH_3 and NO_x through the stomata of 1-year-old needles was based on ^{15}N-labelling experiments where plants had been exposed in special chambers (Katz 1991; Bruckner 1995). Uptake rates of wet-deposited NH_4^+ and NO_3^- through the cuticle of 1-year-old needles and the bark of twigs were based on laboratory experiments with applied ^{15}N-labelled N sources. (Katz 1991). Ranges of uptake rate have been generated for trees of different nutritional status (Schulze and Gebauer

Table 8.1. Estimates for canopy uptake of N from gaseous and wet deposition for the spruce forest at Waldstein, Germany

Uptake pathway	Case 1 Upscaling Chamber experiments at low and high N nutrition ($kg\,N\,ha^{-1}\,yr^{-1}$)		Case 2 Upscaling Chamber experiments + canopy conductance experiments ($kg\,N\,ha^{-1}\,yr^{-1}$)	Case 3 Downscaling Canopy conductance + lab experiments on bark and needle uptake ($kg\,N\,ha^{-1}\,yr^{-1}$)	Case 4 Downscaling Canopy conductance + canopy throughfall ($kg\,N\,ha^{-1}\,yr^{-1}$)
Stomatal uptake					
NH_3	3.02	6.29	21.7	0.53	
NO_x	1.04	3.86	4.2	3.01	
Subtotal	4.06	10.19	25.9	3.54	2.66
Needle surface					
NH_4^+	4.45			4.45	
NO_3^-	0.15			0.15	
Bark					
NH_4^+	5.29			5.29	
NO_3^-	0.15			0.15	
Subtotal	10.05		31.5	10.05	32.9
Total uptake	14.11	20.2	57.4	13.59	35.56
Annual N uptake by forest	88.62		135.8	88.62	134.4
% annual uptake	15.9	22.8	42.2	15.3	26.5

The estimate by upscaling from growth room experiments for case 1 is based on data by Gebauer et al. (1991), Bruckner et al. (1993a) and for case 2 Bruckner (1995). The canopy conductance data for case 3 were supplied by Rebmann (pers. comm.). The canopy throughfall estimate for case 4 is based on data from Horn et al. (1989).

1989). Many assumptions have been made for the calculations derived by upscaling from growth chamber experiments to field conditions: (1) a 12-h day/12-h night cycle, (2) stomatal conductance of needles in the field was similar to that in the exposure chamber, (3) trace gas concentrations do not vary spatially within the forest canopy, (4) ion concentrations on wet canopy surfaces are uniform over time, (5) there are linear relationships between uptake rates and ion concentrations and (6) uptake rates by 1-year-old needles and twig bark are representative of all needles and bark of the canopy. These assumptions are clearly over-simplifications. There is also no way to verify the calculations derived from such an upscaling process. Therefore, it is very important to cross-check this approach with independent measurements, such as canopy conductance. Also the upscaling of wet-phase uptake through cuticles and bark from laboratory experiments needs verification at the canopy scale under field conditions, e.g. by using the canopy throughfall, as suggested by Horn et al. (1989).

For the calculation of the uptake fluxes from the wet-phase, total needle biomass estimate for Waldstein has been taken as 14.96 t ha^{-1} (Mund et al. 2000). The uptake through bark has been assumed to take place only on branches and twigs, because these are more likely to be wettened; it remains unclear if the main stem can absorb surface water. The flux through the bark seems to be related not only to surface area, but also to bark thickness. Since the laboratory experiments related the uptake to bark weight of twigs (Katz 1991), the upscaling was made on a weight basis, which may overestimate the uptake by old branches. The bark weight of branches and twigs at Waldstein was taken as 3.6 t ha^{-1} (Mund et al. 2000). For the uptake of gases through the stomata, the extrapolation was based on response curves relating uptake to atmospheric concentration (Schulze and Gebauer 1989; Bruckner 1995). These response curves were also the basis for estimating the compensation point, which was 0 ppb for NH_3 and 3 ppb for NO_x. The compensation point needed to be taken into account when making uptake estimates based on canopy conductance and atmospheric gaseous concentrations. In case 3, the only assumption made for the wet-phase uptake was that SO_4^{2-} and Cl^- pass through the canopy without interaction with needles and bark, while NH_4^+ and NO_3^- interact with the canopy. However, calculations make no distinction between uptake by trees and epiphytes.

The comparison of different approaches to the estimation of canopy uptake shows that it may range from 15 to 42% of the total nitrogen being taken up by the trees. The high estimate by Bruckner (1995) is in part explained by the high NH_3 concentration in the summer of 1992 (Fig. 8.2). Estimates of the proportion of N uptake via the canopy are sensitive to the estimates of total N taken up and used by the trees. Total N uptake itself may be somewhat variable and be dependent on tree growth responses to the supply via soil availability and to the atmospheric deposition levels.

The comparison detailed above indicates that large differences in estimates of the component fluxes, depending on the method used. The upscaling from chamber experiments resulted in higher estimates than the downscaling estimates from canopy conductance. Although the canopy conductance method integrates over all transport processes, the great unknown remains the assumed compensation point, especially that for NO_x. The lower rate determined by Horn et al. (1989) than by C. Rebmann (unpubl.) was probably due to the fact that xylem flux data for only part of a season was available to Horn et al. (1989). With respect to the wet uptake, the upscaling resulted in lower estimates than those generated by the canopy throughfall method. However, the canopy throughfall method includes co-deposition of ammonium and sulphate on wet surfaces; this aspect is not included in estimates derived from upscaling of growth chamber experiments. Thus, it seems that a combination of canopy conductance measurements by eddy covariance and an analysis of canopy throughfall might yield the most realistic estimates. This might indicate that the uptake of N via the wet phase could be higher by a factor of almost 10 than the gaseous uptake. The reason is that NO_x has a low solubility in water, i.e. the mesophyll internal concentration remains high, especially with conifers which cannot use NO_x because of a lower nitrate reductase activity in foliage (Bauer 1997). The situation may, however, be different for broadleaf trees, which have a higher nitrate reductase activity than conifers (Gebauer and Schulze 1997; Ammann et al. 1999).

We think that the importance of canopy uptake of N as shown in Table. 8.1 holds for conifers over most of Europe, given the even distribution of atmospheric concentrations (European Environment Agency 1998). The actual quantity of N taken up and utilised will be largely determined by the frequency of wet deposition and by leaf and bark biomass, which differ with site quality and by the capacity of tree species for nitrate reduction. The potential contributions for growth and development of N taken up via the canopy appear to be greater for N-deficient and slow-growing tree species (Bruckner et al. 1993a; Muller et al. 1996).

8.6 Role in the Critical Load

At this point, it is unclear to what extent canopy uptake of N relates to the critical load of nitrogen for a forest. Clearly, nitrogen captured by the canopy enters into the nitrogen cycle of an ecosystem, i.e. NH_4^+ and NO_3^- taken up by the canopy are assimilated. N uptake by the canopy, whether from gaseous or wet deposition, is not included in the current calculations of the critical N load for forest ecosystems, as this N does not enter directly into the soil. As it clearly enters the ecosystem and may have a quite significant influence on the growth and nutrition of the trees, it should be taken into consideration. Similarly, the N input to the canopy is also not taken into consideration in terms of the critical load for acidity for forest ecosystems; uptake of reduced N by the canopy is accompanied by a counterbalanced loss of base cations.

The nitrogen taken up by the canopy is eventually returned to the soil in litter bound in an organic rather than inorganic form. The potential influence of this returned organic N will be dependent on the mineralisation processes by which the N is recycled and if the resultant inorganic N acts as an ecosystem stressor. If mineralisation is slow, canopy uptake may, in fact, result in increased accumulations of litter and humus under conditions of higher N deposition (see Harrison et al., Chap. 11, this Vol.). If mineralisation results in formation of NH_4^+, then the recycled N would add to the critical load, because uptake of ammonium would acidify soils (Marschner 1995) or ammonium might displace aluminium from soil exchange sites. If NO_3^- results from the mineralisation process, and the resulting nitrate is not readily taken up by plant cover, this would also clearly add to the critical N deposition (Durka et al. 1994).

8.7 Ecophysiological Consequences of Canopy N Uptake

The ecophysiological consequences of canopy uptake are (1) an induction of canopy growth (Bauer 1997; Mund et al. 2000), which could result (2) in cation deficiencies, particularly if cation uptake is limited by low soil supply or root interactions with ammonium or aluminium (Schulze 1989). The chain of interactions, however, is more complicated. N deposition to the canopy of trees, and presumably canopy uptake of N, has been shown to suppress the rate of uptake of N by roots (Perez-Soba and van der Eerden 1993; Muller et al. 1996; Rennenberg et al. 1998). The mechanism appears to be that amino acids generated from N taken up via foliage are translocated via the phloem to roots, which then interact with inorganic N uptake in the roots (Rennenberg et al. 1998). Thus, canopy uptake of N may result in low Ca:N, K:N, P:N and Mg:N ratios in foliage (Schulze 1989; Eilers et al. 1992; Perez-Soba and van der Eerden 1993) and increases of Ca and Mg in throughfall due to leaching from foliage (Roelofs et al. 1985; Helmisaari and Malkonen 1989; Potter et al. 1991). Canopy uptake of N may be a process inducing nutrient imbalances in forests and therefore a factor in forest decline (Schulze 1989).

At the ecosystem level, the ground vegetation will interact with N deposition via throughfall by more effective N assimilation (see Gebauer et al., Chap. 7, this Vol.). However, as the herb layer does not sequester N in the ways trees do, the N captured by ground cover will mainly lead to a more rapid soil N turnover.

8.8 Conclusions

- Pollutant N can be taken up from wet, dry and gaseous deposition via foliage, twig and stem surfaces, assimilated and utilised by forest trees.
- Uptake of N by trees is much faster in the NH_3 or NH_4^+ form than in the NO_x or NO_3^- form.
- Uptake of N from wet deposition appears to be more important than the uptake from the gas phase. However, the relative contribution of gaseous and wet deposition to canopy uptake may well be related to the prevailing concentrations of gaseous N and N in wet deposition.
- The variability in published estimates of canopy uptake are in part due to the large between-year variability of atmospheric gas concentrations and concentrations of ammonium and nitrate in wet deposition.
- Estimates of canopy uptake of both wet and gaseous N deposition for the spruce forest at Waldstein in Germany suggest that between 16 and 42% of the annual tree demand for N could be provided by canopy uptake of atmospheric pollutant N.
- Uptake of N via the forest canopy may suppress N uptake via the tree root system.
- These findings may be generally applicable to the other NiPHYS / CANIF forests, except for Aheden in northern Sweden, where N concentrations in the atmosphere and in the wet deposition are still relatively low.
- Canopy uptake of atmospheric deposited N may contribute a significant proportion of the N requirements for annual growth by forests in many regions of Europe.

8.9 Way Forward

The main gap in knowledge concerning canopy uptake is related to a lack of direct measurements.

The development of micrometeorological methods to determine canopy conductance has been a great step forward in quantifying gaseous uptake by the forest canopy. However, a key uncertainty remains with regard to the compensation point for NO_x. Thus, only the use of fast sensors for NH_3 and NO_x (such as the tunable diode laser) can verify the estimates. However, direct flux measurements of NH_3 and NO_x above forest canopies cannot quantify canopy uptake, since such measurements will not distinguish between co-deposition of ammonium and sulphate, which may subsequently be washed off canopy surfaces by rain, and the wet-phase uptake.

However, even if we were able to measure the trace gas fluxes directly, other fluxes that contribute to canopy uptake would not be taken into account. The deposition of aerosols follows quite different kinetics from those governing the deposition of gases (Peters and Eiden 1992). Measurements of the wet-phase uptake also remain unsatisfactory.

From the above discussion, it appears that only a combination of methods will be adequate for assessing canopy uptake, and the methodology should involve direct gas flux measurements in combination with throughfall measurements. At present, there is little confidence in computer modelling of canopy uptake, as this requires not only detailed knowledge of the processes involved, but also information on periods when canopy surfaces are wet, and the concentration changes of different N forms in these water films through wetting and evaporation cycles (Klemm 1989).

A further important point remaining is the extent to which canopy uptake drives the N cycle. The assessments range from a dominant role (Bruckner 1995) supported by measurements of nitrate reductase in leaves (Bauer 1997) to one where the canopy uptake is of little significance. Models of biogeochemical cycles, however, operate without taking the potential for canopy uptake into consideration, but cover only canopy throughfall and root uptake. However, there are forests along the CANIF transect (see Van Oene et al., Chap. 20, this Vol.) where the estimated N demands of the tree cover exceeds the modelled soil N supply. This finding suggests that another "available" N source makes a significant contribution to forest growth, adding support to the conclusion that canopy uptake is important in the forest N cycle.

8.10 Policy Implications

Despite all uncertainties concerning the exact amounts of N involved in canopy uptake, there is little doubt that processes at the canopy level affect forest health (Last and Watling 1991), ranging from affects on pests and diseases (mildew, aphid attacks to bark beetle infestations) to ecosystem level consequences. These impacts depend strongly on the conditions under which canopy N uptake occurs. The effect may range from accelerated release of aluminium from soils after release of ammonium to flush events of nitrate leaching after logging or liming (Durka et al. 1994). However, it will remain difficult to separate specific effects originating from canopy N uptake from other effects operating via the overall ecosystem nitrogen cycle. Although in some circumstances canopy uptake of N may result in positive effects on forest productivity, there is the overall potential for forest ecosystem destabilisation by N deposition. In view of this, a precautionary measure would be to reduce N deposition, as much as possible.

References

Adema EH, Heeres P, Hulskotte J (1985) On the dry deposition of NH_3, SO_2 and NO_2 in wet surfaces in a small scale windtunnel. 7th World Clear Air Congress, Sydney 1986

Ammann M, Siegwolf R, Pichlmayer F, Suter M, Saurer M, Brunold C (1999) Estimating the uptake of traffic-derived NO_2 from ^{15}N abundance in Norway spruce. Oecologia 118:124–131

Bauer G (1997) Stickstoffhaushalt und Wachstum von Fichten- und Buchenwäldern entlang eines europäischen Nord-Süd-Transektes. Bayreuther Forum Ökologie Bd 53, BITÖK, Bayreuth

Bowden RD, Geballe GT, Bowden WB (1989) Foliar uptake of ^{15}N from simulated cloud water by red spruce (*Picea rubens*) seedlings. Can J For Res 19:382–386

Boyce R, Friedland AJ, Chamberlain CP, Poulson SR (1996) Direct canopy uptake from ^{15}N-labeled wet deposition by mature red spruce. Can J For Res 26:1539–1547

Bruckner G (1995) Deposition und oberirdische Aufnahme von gas- und partikelförmigem Stickstoff aus verschiedenen Emissionquellen in ein Fichtenökosystem. PhD Thesis, University of Bayreuth, Bayreuth

Bruckner G, Gebauer G, Schulze E-D (1993a) Uptake of $^{15}NH_3$ by *Picea abies* in closed chamber experiments. Isotopenpraxis Environ Health Stud 29:71–76

Bruckner G, Schulze E-D, Gebauer G (1993b) ^{15}N labelled NH$_3$ uptake experiments and their relation to natural conditions. Air Pollution Research Report 47 to the CEC. E Guyot, Brussels, pp 305–311

Brumme R, Leimcke U, Matzner E (1992) Interception and uptake of NH$_4$ and NO$_3$ from wet deposition by above-ground parts of young beech (*Fagus sylvatica* L.) trees. Plant Soil 142:273–279

Cape JN, Sheppard LJ, Binnie J, Dickinson AL (1998) Enhancement of the dry deposition of sulphur dioxide to a forest in the presence of ammonia. Atmos Environ 32:519–524

Crossley A, Fowler D (1986) The weathering of Scots pine epicuticular wax in polluted and clean air. New Phytol 103:207–218

Dambrine E, Pollier B, Bonneau M, Ignatova N (1998) Use of artificial trees to assess dry deposition in spruce stands. Atmos Environ 32:1817–1824

Durka W, Schulze E-D, Gebauer G, Voerkelius S (1994) Effects of forest decline and leaching of deposited nitrate determined from ^{15}N and ^{18}O measurements. Nature 372:765–767

Duyzer JH, Verhagen HLM, Westrate JH (1992) Measurement of the dry deposition flux of NH$_3$ on to a coniferous forest. Environ Pollut 75:3–13

Eichert T, Goldbach HE, Buckhardt J (1998) Evidence of the uptake of large anions through stomatal pores. Bot Acta 111:461–466

Eiden R (1989) Air pollution and deposition. In: Schulze E-D, Lange OL, Oren R (eds) Forest decline and air pollution. Ecological Studies 77. Springer, Berlin Heidelberg New York, pp 57–103

Eilers G, Brumme R, Matzner E (1992) Above-ground N-uptake from wet deposition by Norway spruce (*Picea abies* Karst.) For Ecol Manage 51:239–249

Egger A, Landolt W, Brunold Ch (1989) Effects of NO$_2$ on assimilatory nitrate and sulfate reduction in needles from spruce trees (*Picea abies* L.) In: Bucher JB, Bucher-Wallin I (eds) Air pollution and forest decline. Proc 14th Meeting Interlaken Switzerland, IUFRO, Birmensdorf, Switzerland, pp 401–403

Emmett BA, Kjonas OJ, Gundersen P, Koopmans C, Tietema A, Sleep D (1998) Natural abundance of ^{15}N in forests across a nitrogen deposition gradient. For Ecol Manage 101:9–18

European Environment Agency (1998) Europe's environment: the second assessment. European Environment Agency, Copenhagen, 293 pp

Farquhar GD, Firth PM, Wetselaar R, Wier B (1980) On the gaseous exchange of ammonia between leaves and the environment: determination of the compensation point. Plant Physiol 66:710–714

Galloway JN, Levy H, Kasibhatia PS (1994) Year 2020: Consequences of population growth and development on deposition of oxidized nitrogen. Ambio 23:120–123

Garten CT, Hanson PJ (1990) Foliar retention of ^{15}N-nitrate and ^{15}N-ammonium by red maple (*Acer rubrum*) and white oak (*Quercus alba*) leaves from simulated rain. Environ Exp Bot 30:333–342

Gebauer G, Schulze E-D (1997) Nitrate nutrition of central European forest trees. In: Rennenberg H, Eschrich W, Ziegler H (eds) Trees – contributions to modern tree physiology. Backhuys, Leiden, The Netherlands, pp 273–291

Gebauer G, Katz C, Schulze E-D (1991) Uptake of gaseous and liquid nitrogen depositions and influence on the nutritional status of Norway spruce. In: Hantschel R, Beese F (eds) Effects of forest management on the nitrogen cycle with respect to changing environmental conditions. GSF-Bericht 43/91 GSF, Muenchen, pp 83–92

Grennfelt P, Bengtson C, Skarby L (1983) Dry deposition of nitrogen dioxide to Scots pine needles. In: Pruppacher HR, Semonin RG, Slinn WGN (eds) Precipitation, scavenging, dry deposition and resuspension, vol 2. Elsevier, New York, pp 753–761

Hanson PJ, Lindberg SE (1991) Dry deposition of reactive nitrogen compounds: a review of leaf canopy and non-foliar measurements. Atmos Environ 25A:1615–1634

Hanson PJ, Rott K, Taylor GE et al. (1990) NO$_2$ deposition to elements representative of a forest landscape. Atmos Environ 23:1783–1794

Hanson PJ, Taylor GE, Vose J (1992) Experimental laboratory measurements of reactive N gas deposition to forest landscape surfaces: biological and environmental controls. In: Johnson DW, Lindberg SE (eds) Atmospheric deposition and forest nutrient cycling. A synthesis is the integrated forest study. Ecological Studies 91. Springer, Berlin Heidelberg New York, pp 166–213

Harrison AF, Taylor K, Chadwick D (1991) Foliar uptake of nitrogen and its translocation in trees. In: Effects of atmospheric pollutants on forests and crops. Natural Environment Research Council, Swindon, 28 pp

Haynes RJ (1986) Mineral nutrition in the plant-soil system. Academic Press, New York

Heaton THE, Spiro B, Robertson SMC (1997) Potential canopy influences on the isotopic composition of nitrogen and sulphur in atmospheric deposition. Oecologia 109:600–607

Helmisaari H-S, Malkonen E (1989) Acidity and nutrient content of throughfall and soil leachate in three *Pinus sylvestris* stands. Scand J For Res 4:13–28

Hollinger DY, Kelliher FM, Schulze E-D, Köstner BMM (1994) Coupling tree transpiration to atmospheric turbulence. Nature 371:60–62

Horn R, Schulze E-D, Hantschel R (1989) Nutrient balance and element cycling in healthy and declining Norway spruce stands. In: Schulze E-D, Lange OL, Oren R (eds) Forest decline and air pollution. Ecological Studies 77. Springer, Berlin Heidelberg New York, pp 444–455

Ignatova N, Dambrine E (2000) Forest canopy uptake of N deposition. For Ann For Sci 57:113–120

INDITE (1994) Impacts of nitrogen deposition in terrestrial ecosystems. Report for the UK Department of the Environment, London

Ineson P, Robertson SMC, Thomson P (1993) Aerial transport of ammonia from agriculture to forest. Report of the Institute of Terrestrial Ecology for 1992–93, Abbots Ripton, Huntingdon, UK, pp 57–59

Ineson P, Robertson SMC, Hornung M, Jones HE, Benham DG, Heaton THE (1996) Nitrogen critical loads: N deposition around a point source. Final Report by ITE Merlewood, Erangeover-Sands to Department of the Environment, London

Joslin JD, Muller SF, Wolfe MH (1990) Test of models of cloud water deposition to forest canopies using artificial and living collectors. Atmos Environ 24A:3007–3019

Katz C (1991) Die Aufnahme gasförmiger und gelöster anorganischer Stickstoff-Verbindungen über Nadeln und Zweige der Fichte (*Picea abies* L. (Karst). Dissertation, Uni Bayreuth, Bayreuth, 113 pp

Katz C, Oren R, Schulze E-D, Milburn JA (1989) Uptake of water and solutes through twigs of *Picea abies* (L.) Karst Trees 3:33–37

Kisser-Priesack G, Bieniek D, Ziegler H (1990) NO_2 binding to defined phenolics in the plant cuticle. Naturwissenschaften 77:492–493

Kisser-Priesack G, Gebauer G (1991) Kinetics of $^{15}NO_x$ uptake by plant cuticles. In: IAEA (ed) Stable isotopes in plant nutrition, soil fertility and environmental studies. Vienna, IAEA-SM-313/18, pp 619–625

Klemm O (1989) Leaching and uptake of ions through above-ground Norway spruce tree parts. In: Schulze E-D, Lange O, Oren R (eds) Forest decline and air pollution. A study of spruce on acid soils. Ecological Studies 77. Springer, Berlin Heidelberg New York, pp 210–233

Köstner BMM, Schulze E-D, Kelliher FM, Hollinger DY, Byers JN, Hunt JE, McSeveny TM, Meserth R, Weir PL (1992) Transpiration and canopy conductance in a pristine broad-leaved forest of *Nothofagus*: an analysis of xylem sap flow and eddy correlation measurements. Oecologia 91:350–359

Lamersdorf NP, Beier C, Blanck K, Bredemeier M et al. (1998) Effect of drought experiments using roof installations on acidification/nitrification of soils. For Ecol Manage 101:95–109

Last FT, Watling R (eds) (1991) Acid deposition: its nature and impacts. Proc R Soc Edinb Sect B 97:1–326

Lauf J, Gebauer G (1998) On-line analysis of stable isotopes of nitrogen in NH_3, NO and NO_2 at natural abundance levels. Anal Chem 70:2750–2756

Lindberg SE, Lovett GM, Richter DD, Johnson DW (1986) Atmospheric deposition and canopy interactions of major ions in a forest. Science 231:141–144

Lovett GM (1992) Atmospheric deposition and canopy interactions of nitrogen. In: Johnson DW, Lindberg SE (eds) Atmospheric deposition and forest nutrient cycling. A synthesis is the integrated forest study. Ecological Studies 91. Springer, Berlin Heidelberg New York, pp 152–166

Macklon AES, Sheppard LJ, Sim S, Leith ID (1996) Uptake of ammonium and nitrate ions from acid mist applied to Sitka spruce [*Picea sitchensis* (Bong.) Carr.] grafts over the course of one growing season. Trees 10:261–267

Manderscheid B, Goettlein A (1995) Wassereinzugsgebiet "Lehsteinbach" – das BITÖK – Untersuchungsgebiet am Waldstein (Fichtelgebirge NO-Bayern). Bayreuther Forum Okologie 18, BITÖK, Bayreuth

Marschner H (1995) Mineral nutrition of higher plants. Academic Press, London, 889 pp

Matzner E (1989) Acidic precipitation case study Solling. In: Adriano DC, Havas M (eds) Acidic precipitation. Advances Environmental Sciences, vol 1. Springer, Berlin Heidelberg New York, pp 39–84

Muller B, Touraine B, Rennenberg H (1996) Interaction between atmospheric and pedospheric nitrogen nutrition in spruce (*Picea abies* L. Karst) seedlings. Plant Cell Environ 19:345–355

Mund M, Kummetz E, Hein M, Bauer GA, Schulze E-D (2000) Growth and carbon stocks of a spruce forest chronosequence in central Europe. For Ecol Manage (in press)

Nasholm T (1991) Aspects of nitrogen metabolism in Scots pine, Norway spruce and birch as influenced by the availability of nitrogen in pedosphere and atmosphere. Swedish University, Agriculteral Sciences, Umea, 514 pp

Nasholm T, Högberg P, Edfast A-B (1991) Uptake of NO_x by mycorrhizal and non- mycorrhizal Scots pine seedlings: quantities and effects on amino-acid and protein concentrations. New Phytol 119:83–92

Nilgard B (1985) The ammonium hypothesis – an additional explanation to the forest die-back in Europe. Ambio 14:1–8

Nussbaum S, von Ballmoos P, Gfeller H et al. (1993) Incorporation of atmospheric $^{15}NO_2$-nitrogen into free amino acids by Norway spruce [*Picea abies* (L.) Karst]. Oecologia 94:408–414

Parker GG (1983) Throughfall and stemflow in the forest nutrient cycle. Adv Ecol Res 13:58–135

Pearson J, Soares A (1998) Physiological responses of plant leaves to atmospheric ammonia and ammonium. Atmos Environ 32:533–538

Pearson J, Stewart GR (1993) The deposition of atmospheric ammonia and its effects on plants. New Phytol 125:283–305

Pearcy RW, Ehleringer JR, Mooney HA, Rundel PW (1989) Plant physiological ecology, field methods and instrumentation. Chapman and Hall, London, 550 pp

Perez-Soba M, van der Eerden LJM (1993) Nitrogen uptake in needles of Scots pine (*Pinus sylvestris* L.) when exposed to gaseous ammonia and ammonium fertilizer in the soil. Plant Soil 153:231–242

Perez-Soba M, Stulen I, van der Eerden LJM (1994) Effect of atmospheric ammonia on the nitrogen metabolism of Scots pine (*Pinus sylvestris*) needles. Physiol Plant 90:629–636

Peters K, Eiden R (1992) Modelling the dry deposition velocity of aerosol particles to a spruce forest. Atmos Environ 26a:2555–2564

Potter CS, Ragsdale HL, Swank WT (1991) Atmospheric deposition and foliar leaching in a regenerating southern Appalachian forest canopy. J Ecol 79:97–115

Rennenberg H, Kreutzer K, Papen H, Weber P (1998) Consequences of high loads of nitrogen for spruce (*Picea abies*) and beech (*Fagus sylvatica*) forests. New Phytol 139:71–86

Roelofs JGM, Kempers AJ, Houdijk LFM, Jansen J (1985) The effect of airborne ammonium sulphate on *Pinus nigra* var. maritima in the Netherlands. Plant Soil 84:45–56

Schaeffer DA, Reiners WA (1990) Throughfall chemistry and canopy processing mechanisms. In: Lindberg SE, Page AL, Norton SA (eds) Acidic precipitation. Sources, deposition and canopy interactions, vol 3. Springer, Berlin Heidelberg New York, pp 241–273

Schjoerring JK, Husted S, Mattsson M (1998) Physiological parameters controlling plant-atmosphere ammonia exchange. Atmos Environ 32:491–498

Schoenherr J (1982) Resistance of plant surfaces to water loss: transport properties of cutin, suberin and associated lipids. In: Lange OL, Nobel PS, Osmond CB, Ziegler H (eds) Physiological plant ecology II. Encyclopedia of plant physiology new series 12B. Springer, Berlin Heidelberg New York, pp 154–179

Schulze E-D (1989) Air pollution and forest decline in a spruce (*Picea abies*) forest. Science 244:776–783

Schulze E-D, Gebauer G (1989) Aufnahme, Abgabe und Umsatz von Stickoxiden, NH_4^+ and Nitrat bei Waldbäumen, insb. der Fichte. Proc: 1. Statusseminar der PBWU, GSF-Bericht 6:119–133, GSF, München

Schulze E-D, Hall AE (1992) Stomatal responses and water loss and CO_2 assissimilation rates of plants in contrasting environments. In: Lange OL, Nobel PS, Osmond CB, Ziegler H (eds) Physiological plant ecology. II. Water relations and carbon assimilation. Encyclopedia of plant physiology, vol 12B. Springer, Berlin Heidelberg New York, pp 181–230

Skarby L, Bengtson C, Bostrom C-A, Grennfelt P, Troeng E (1981) Uptake of NO_x in Scots pine. Silva Fenn 15:396–398

Stanners D, Bourdeau P (1990) Europe's environment. European Environment Agency, Copenhagen, 676 pp

Stanners D, Bourdeau P (eds) (1995) Europe's environment; the Dobris assessment. European Environment Agency, Copenhagen, 676 pp

Sutton MA, Pitcairn CER, Fowler D (1993) The exchange of ammonia between the atmosphere and plant communities. Adv Ecol Res 24:301–393

Thoene B, Schroder P, Papen H, Egger A, Rennenberg H (1991) Absorption of atmospheric NO_2 by spruce (*Picea abies* Karst.) trees. I NO_2 influx and its correlation with nitrate reduction. New Phytol 117:575–585

Tischner R, Peuke A, Godbold DL, Feig R, Merg G, Hüttermann A (1988) The effect of NO_2 fumigation on asceptically grown spruce seedlings. J Plant Physiol 133:243–246

Tuckey HB, Wittwer SH, Bukovac MJ (1962) The uptake and loss of materials by leaves and other above ground parts with special reference to plant nutrition. Agrochimica 7:1–28

Ulrich B (1987) Stability, elasticity and resilience of terrestrial ecosystems with respect to matter balance. In: Schulze E-D, Zwölfer H (eds) Potentials and limitations of ecosystem analysis. Ecological studies 61. Springer, Berlin Heidelberg New York, pp 11–49

Vose JM, Swank WT (1990) Preliminary estimates of foliar absorption of ^{15}N-labeled nitric acid vapor (HNO_3) by mature eastern white pine (*Pinus strobus*) Can J For Res 20:857–860

Warneck P (1988) Chemistry of the natural atmosphere. Academic Press, New York

Weber P, Thoene B, Rennenberg H (1998) Absorption of atmospheric NO_2 by spruce (*Picea abies*) trees. III Interaction with nitrate reductase activity in the needles and phloem transport. Bot Acta 111:377–382

Wilson EJ, Tiley C (1998) Foliar uptake of wet-deposited nitrogen by Norway spruce: an experiment using ^{15}N. Atmos Environ 32:513–518

Wingsle G, Nasholm T, Lindmark T, Ericsson A (1987) Induction of nitrate reductase in needles of Scots pine seedlings by NO_x and NO_3. Physiol Plant 70:399–403

9 Biotic and Abiotic Controls Over Ecosystem Cycling of Stable Natural Nitrogen, Carbon and Sulphur Isotopes

G.A. BAUER, G. GEBAUER, A.F. HARRISON, P. HÖGBERG, L. HÖGBOM, H. SCHINKEL, A.F.S. TAYLOR, M. NOVAK, F. BUZEK, D. HARKNESS, T. PERSSON, and E.-D. SCHULZE

9.1 Introduction

In forest ecosystems, physical, chemical and biological processes which regulate the uptake and flow of matter cause differences in the isotopic signatures of N, C and S compounds due to fractionation. Thus, tracing processes of isotopic fractionations in forest ecosystems can help identify transfer pathways and capacities. Atmospheric nitrogen pollution has been recognised as the cause for formerly N-limited forests to approach N saturation (Aber et al. 1989, 1998). The increase in N availability in forest ecosystems may have consequences for net ecosystem productivity on a local scale, but also for the carbon budget of terrestrial ecosystems on a global scale (Schulze 1994; Schimel 1995). Clearly, knowledge about the amount of N being deposited and cycling through forest ecosystems is of paramount importance for the carbon assimilation of the terrestrial biosphere (Lloyd and Farquhar 1996). In a similar manner, information on stable sulphur isotopes is of importance regarding pollution inputs and their effect on ecosystem health (Krouse 1989; Gebauer et al. 1994).

We have investigated different aspects of stable isotope signatures for N, C and S in forests along a European transect to identify key processes in the turnover of these major elements. Initially, we will compare nitrogen isotope signatures in wet deposition, followed by an analysis of N, C and S isotope signatures in the vegetation and soil, and finally, we will compare whole ecosystem profiles of $\delta^{15}N$ values. Specifically, we address the following questions: Do the same plant species show different isotopic signatures across a range of environmental conditions? Are there differences in the predominant N form which is taken up by plants in forest ecosystems across Europe with a similar species composition? Is there a seasonal variability in the isotopic signature of wet deposition? Can we separate agricultural and industrial N pollution sources via their $\delta^{15}N$ signature?

9.2 Approaches to the Study of Stable Isotopes in the Field

9.2.1 Field Sampling

Plant and xylem sap samples across the European transect were collected during several site visits from 1993 to 1995; data for the two Czech sites are according Novak et al. (1999). Apart from the two Italian forest stands, which where sampled in 1994, all plant $\delta^{15}N$ values originated from samples taken in 1993. Xylem sap was collected with the displacement method (Schurr 1998) from sun crown branches of the main canopy trees in 1994. At each site, above-ground plant material was collected from $n = 4$ individuals per species, whereas for roots and soil generally more than four samples were collected and analysed. Soil sampling for the carbon isotope analysis was carried out in 1996 and 1997. Soil samples from the organic layers were initially collected in 1993 and 1994 (Högberg et al. 1996). An additional set of soil samples were collected for $\delta^{15}N$ and radiocarbon analysis (see Harrison et al., Chap. 11, this Vol.) covering organic layers and mineral soil to a depth of 20 cm. Both data sets on $\delta^{15}N$ values are combined in this chapter.

9.2.2 Local and Regional Assessment of Wet Deposition in the UK

For the study of seasonal variation of $\delta^{15}N$ in NH_4-N and NO_3-N in rainfall and throughfall, rain collectors were established in the surroundings of the Institute of Terrestrial Ecology (Merlewood, UK). This area is surrounded to the west by deciduous woodland, but faces an open grassland on its eastern boundary along the main drive to the institute, which at intervals receives animal slurries and represents a local emission source. In total, six collectors were established, two in the open in the garden (further referred to as rain), two in the woodland to the west on the side of a hill (further referred to as wood) and two along the drive in the grassland (further referred to as drive). One of the two collectors in the woodland and along the drive was placed under a yew tree (*Taxus cf. baccata*) and the other under a beech (*Fagus sylvatica*). The experimental layout was designed to determine if tree species or sampling position was more important in governing the $\delta^{15}N$ signature of the NH_4-N or NO_3-N. The water samples were collected at roughly monthly intervals from 1995 to 1996, except for a period between July and early October 1995, when little rain occurred. The frequent collection pattern has also allowed to investigate the seasonal pattern of the $\delta^{15}N$ signature.

In order to investigate the $\delta^{15}N$ signature in N pollution from different sources, a regional transect was established. The four sites were located on a south-north line from Liverpool to Edinburgh in the UK. The prevailing wind direction along this transect is from the southwest, so the sites were located to the northeast of the perceived N source. Their locations are (I) Knowsley Park, NE of the Liverpool conurbation, (II) Shap, east of junction 39 along the M6 motorway, (III) Glassonby, surrounded by dairy farms and (IV) West Linton, beside a large chicken farm battery. Sites I and II were thought to receive mainly industrial N inputs and sites III and IV mainly agricultural N inputs, though few sites in the UK will have a source-specific N input, because of the patchy land-use matrix. Rainfall and throughfall collections, usually over a 3- to 4-week period, were made during the winter period (Jan.–Feb., 1994) and in the summer (June–July, 1994), to test between season for the $\delta^{15}N$ signature in NH_4-N. Another series of samplings were carried out during February 1995 for both NH_4-N and NO_3-N.

9.2.3 Collection of Wet Deposition Along the European Transect

From 1994 to 1995 rain collectors were set up at the spruce sites Åheden, Klosterhede, Waldstein, Jezeri and Aubure, and at the mediterranean site in Thezan. At each site three collectors were set up in the open (rainfall) and three were placed underneath the canopy (throughfall). Water was sampled in the autumn periods over approximately 3 months in order to collect at least 3 to 4 l of rain, which generally contained less N at that time of the year.

9.2.4 Laboratory Analysis

The $\delta^{15}N$ values for plant and soil samples were measured on different mass spectrometers among the involved laboratories, which all routinely measure stable nitrogen isotopes. For further details on the specifications of the equipment refer to Gebauer and Schulze (1991), Högberg et al. (1996), and Novak et al. (1999). Carbon, nitrogen and sulphur concentrations were determined on elemental analysers. The reproducibility was better than 5% for all three elements. Sulphur was extracted from peat as $BaSO_4$ (Novák et al. 1994) and converted to SO_2 in a vacuum line (Yanagisawa and Sakai 1983). Inorganic free plus adsorbed sulphate S was extracted from soil using 16.1 mM solution of NaH_2PO_4 in H_2O. Stable isotope compositions were expressed in the delta notation (i.e. $\delta^{13}C$, $\delta^{15}N$ and $\delta^{34}S$) as a ‰ deviation of the heavy-to-light isotope abundance ratio ($^{13}C/^{12}C$, $^{15}N/^{14}N$ and $^{34}S/^{32}S$) in the sample from a standard. The standards used were PeeDee Belemnite, atmospheric N_2 and Cañon Diabolo Troilite for C, N, and S, respectively. In each case, the reproducibility was better than 0.3‰.

Prior to the determination of $\delta^{15}N$ in rainfall and throughfall, all samples were tested for contamination by phosphate due to bird droppings; contaminated samples were rejected. Bulk samples were filtered through GF/D and GF/F glass papers to remove small particulate materials and a subsample was analysed for NH_4-N and NO_3-N concentration to determine the volume of sample needed. Water samples were then passed slowly (30 h for 5 l) through 3–4-ml wet volume of Amerlite IR-120 cation exchanger followed by Amberlite IRA-420 or 458 for the anion exchange. After rinsing the resins with distilled water, the NH_4-N and NO_3-N were eluted using 2 M KCl. A small portion of the NH_4-containing eluate was steam-distilled and the distillate titrated to determine the N concentration; this distillation also served to eliminate the isotopic memory of the previous sample on the glassware. Knowing the nitrogen concentration, a second aliquot was distilled directly into 0.025 M H_2SO_4 allowing 1.5 ml for each mg N, enough to keep the acid in very slight excess. The second distillate was evaporated down to about 3–4 ml, transferred to a small vial and then dried at 75 °C over 48 h. The dry $(NH_4)_2SO_4$ residue was then weighed into tin capsules for isotopic anslysis or, if insufficient material was available, the residue was taken up in 100 µl water and a suitable aliquot of this solution added to about 5 mg Ultragel in the tin cup. Basically, the same procedure was adopted for the NO_3-N but with the addition of Devardas alloy at the distillation stage to reduce NO_3 to NH_3.

9.3 $\delta^{15}N$ of Ammonium and Nitrate in Wet Deposition

9.3.1 Seasonal Variation in the $\delta^{15}N$ of Nitrate and Ammonium in Wet Deposition

The deposition data collected on the Merlewood grounds showed two general trends. First, the concentration of NH_4-N and NO_3-N in throughfall were generally higher than the concentrations in the rainfall (Figs. 9.1 and 9.2). Second, the variation of both concentration and $\delta^{15}N$ signatures in throughfall were much larger than in the rainfall.

The $\delta^{15}N$ signature of ammonium was generally negative, but there was a change to positive values in the rainfall at both sampling stations along the drive in

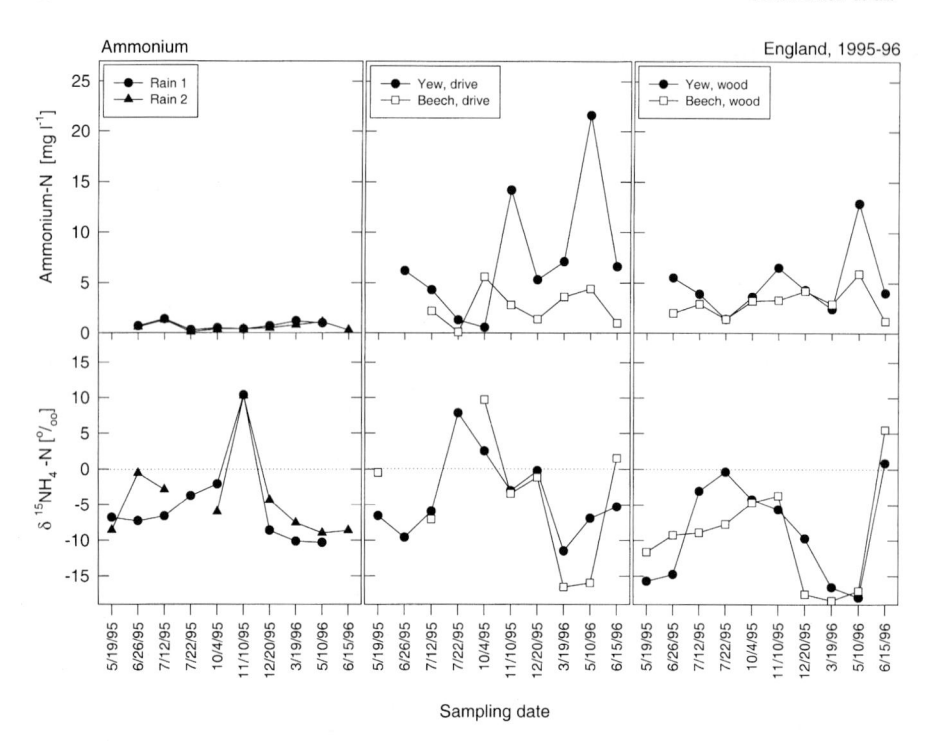

Fig. 9.1. Seasonality of NH_4-N concentration ($mg\,l^{-1}$) and $\delta^{15}N$ values of NH_4-N (‰) in rainfall and throughfall at two sites adjacent to the grounds of the ITE in Merlewood, UK. Throughfall was collected underneath a yew (*Taxus* cf. *baccata*) and a beech (*Fagus sylvatica*) tree. *Each data point* in the figure represents one single observation for the N concentration and the $\delta^{15}N$ value. (A.F. Harrison, unpubl. data)

November (Fig. 9.1). This change in isotopic signature coincided with a period of active slurry spraying on the adjacent grass fields. From the seasonal change in the $\delta^{15}N$ signature, it can be seen that there was a good agreement between the same sampling positions regardless of tree species, despite the fact that the tree species have very different canopy structures. This suggests that the distance and relative position to the local source is more important for the isotopic signature in the throughfall than the intercepting species.

The results from both sampling points also showed a clear seasonal pattern. It appeared that the $\delta^{15}N$ signature in the NH_4-N became more negative during autumn and winter months compared with summer and spring. The more negative $\delta^{15}N$ values may well have been associated with leaching of NH_4-N from the canopy. The less negative values of throughfall during the summer could have been due to foliar uptake of NH_4-N, assuming a preferential uptake of the lighter isotope. This is supported by several other findings for the *Picea* sites (see Harrison et al., Chap. 8, this Vol.) showing a strong preference in N uptake of ammonium over nitrate. It also matches results for conifers which emphasised a strong potential for foliar uptake of NH_4-N and subsequent metabolisation through glutamine-synthetase (Pearson and Stewart 1993;

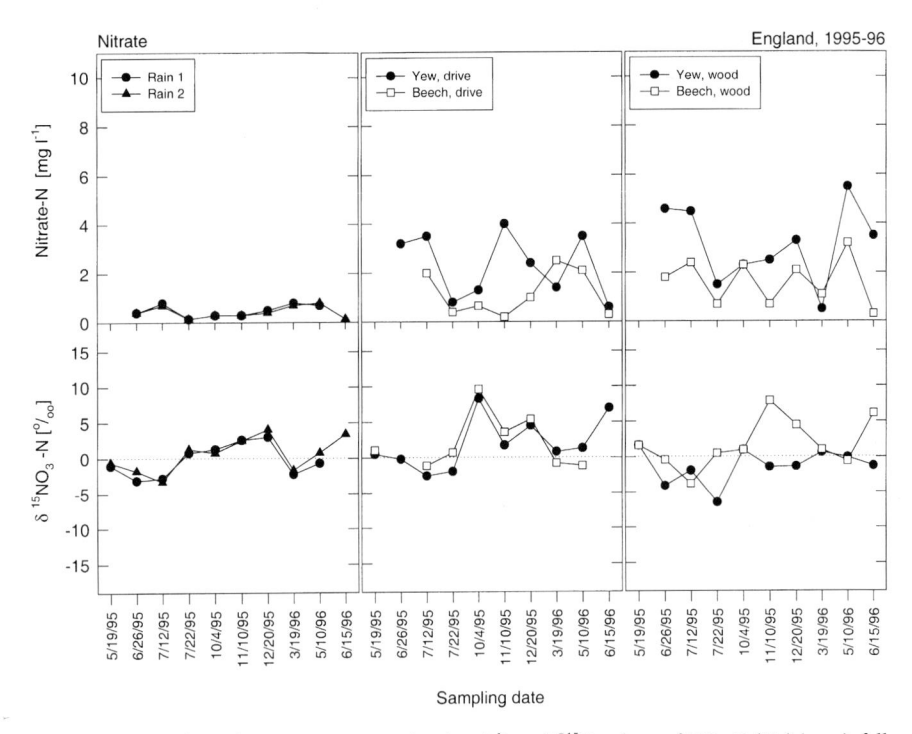

Fig. 9.2. Seasonality of NO_3-N concentration (mg l^{-1}) and $\delta^{15}N$ values of NO_3-N (‰) in rainfall and throughfall at two sites adjacent to the grounds of the ITE in Merlewood, UK. Throughfall was collected underneath a yew (*Taxus* cf. *baccata*) and a beech (*Fagus sylvatica*) tree. *Each data point* in the figure represents one single observation for the N concentration and the $\delta^{15}N$ value. (A.F. Harrison, unpubl. data)

Bruckner et al. 1993; Perez-Soba et al. 1994). Results from a grassland study showed that $\delta^{15}N$ values of NH_4-N change from negative to positive with the duration of the NH_3-N volatilisation after slurry application (Bruckner 1996). Thus, isotopic fractionation during atmospheric transport could have taken place prior to canopy uptake.

The $\delta^{15}N$ signatures in NO_3-N showed a strong effect of distance to the local emittent (Fig. 9.2), i.e. the nearby grassland. The replicate collections of rainfall and throughfall were in good agreement. However, the results for the two woodland positions showed marked variations. The seasonal pattern of NO_3-N was different from that of NH_4-N. The $\delta^{15}N$ signature in the rainfall was negative in summer and changed to positive values in the winter. The same pattern was seen for the throughfall along the driveway, but with a stronger shift toward more positive $\delta^{15}N$ values. The throughfall data beneath the yew tree in the woodland tended to be somewhat depleted relative to the $\delta^{15}N$ values of the other two sampling locations, whereas no obvious pattern for the beech collectors was seen. The pattern in these results seemed to reflect changes in the N pollution sources and deposition processes within the canopy rather than uptake by the canopy itself.

9.3.2 Influence of Different N Pollution Sources on the Isotopic Signature in Rainfall and Throughfall

The sites selected for this study were situated in the vicinity of different industrial or agricultural pollution sources. One industrial site was located at Knowsley Park near Liverpool and the other beside junction 39 (Shap) of the M6 motorway in Cumbria; the two agricultural sites were located at a dairy farm (Glassonby) and near to a large chicken farm (West Linton, Edinburgh).

The variations in the $\delta^{15}N$ signature in NH_4-N for the UK transect (Fig. 9.3, Table 9.1) showed a range similar to that found for the local scale study around the Merlewood grounds. The industrial site (Knowsley Park) always had more negative inputs for NH_4-N than the agricultural site at West Linton (chicken farm) and the location along the motorway, thus corresponding to findings for isotopic signatures in NO_2 from highway measurements in Switzerland (Ammann et al. 1999). Throughfall col-

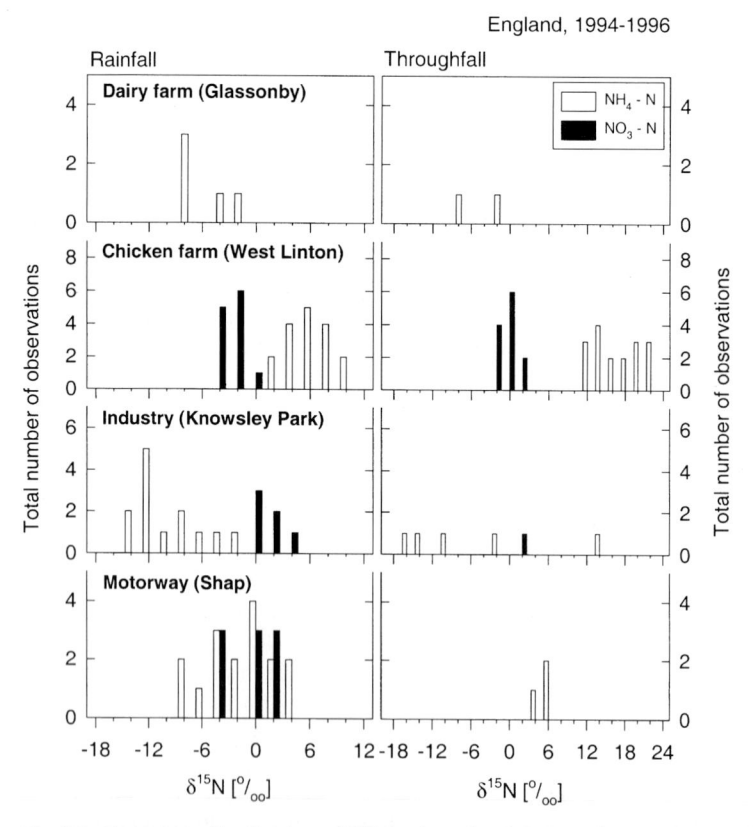

Fig. 9.3. Frequency distribution of $\delta^{15}N$ values for NH_4-N and NO_3-N in rainfall (*left-hand panel*) and throughfall (*right-hand panel*) at four sites which are under the influence of different local pollution sources in the UK. The sites were located between Liverpool and Edinburgh, UK. (A.F. Harrison, unpubl. data)

Biotic and Abiotic Controls over Ecosystem Cycling

Table 9.1. Minimum and maximum values of $\delta^{15}N$ (‰) in wet-deposited ammonium and nitrate and average nitrogen concentrations (ppm) for NH_4-N and NO_3-N in rainfall and throughfall on four sites adjacent to different pollution sources along a north-south line between Liverpool and Edinburgh, UK. (A.F. Harrison unpubl. data)

Site		NO_3-N $(mg\,N\,l^{-1})$	$\delta^{15}N$ of NO_3-N Min.	Max.	NH_4-N $(mg\,N\,l^{-1})$	$\delta^{15}N$ of NH_4-N Min.	Max.
Glassonby	RF				0.94 ± 0.44	−7.92	−2.25
	ThF				7.05 ± 1.06	−8.10	−2.68
Knowsley Park	RF	0.31 ± 0.24	0.63	4.49	0.81 ± 0.63	−15.40	−1.69
	ThF	0.40 ± 0.10	3.79	3.82	1.85 ± 0.82	−17.43	14.63
Shap	RF	0.34 ± 0.22	−2.97	2.20	0.51 ± 0.26	−6.57	5.14
	ThF				1.37 ± 1.35	5.61	7.18
West Linton	RF	0.53 ± 0.20	−3.40	0.15	1.31 ± 0.43	2.55	10.97
	ThF	3.39 ± 1.11	−0.88	4.22	11.58 ± 4.25	12.96	22.74

lected at Knowsley Park was more negative than the rainfall, which is in contrast to throughfall isotopic signature at the chicken farm and along the motorway, which were much more enriched in $\delta^{15}N$. Especially the forest throughfall adjacent to the chicken farm showed a strong enrichment in the $\delta^{15}N$ of the NH_4-N fraction.

The concentration of N in all throughfall samples was significantly higher than in the corresponding rainfall samples (Table 9.1). The positive shift in $\delta^{15}N$ for the West Linton site was probably a result of a large change in concentration, indicating that the positive shift was related to high levels of dry deposition from the adjacent chicken farm. This finding conformed with the fact that dry-deposited N is enriched in $\delta^{15}N$ relative to its source (Bruckner 1996).

The few data obtained for NO_3-N suggest that there may also have been a difference in the $\delta^{15}N$ signature between the industrial site (Knowsley Park) and the chicken farm (West Linton). Generally, the range of observations for the $\delta^{15}N$ in nitrate was smaller compared to the signature in NH_4-N (Table 9.1). However, at both sampling locations, the concentrations of NO_3-N in the throughfall samples also were higher compared to the corresponding rainfall data.

9.3.3 Isotopic Signature of Ammonium- and Nitrate-N in Wet Deposition Along the European Transect

The $\delta^{15}N$ data for the conifer sites along the European transect indicated consistent patterns for the individual forest sites, though for nitrate-N more samples would have been needed to confirm this result (Fig. 9.4). The ammonium data indicate that there were different $\delta^{15}N$ signatures for each site with respect to the NH_4-N inputs (Fig. 9.4; Bruckner 1996). Also, it appears that the range of isotopic signatures across the European transect was not larger than the variation at the local and regional scales in the UK. This emphasises that the isotopic fractionation of pollutant inputs may have been governed more by local conditions like stand structure, canopy roughness and type

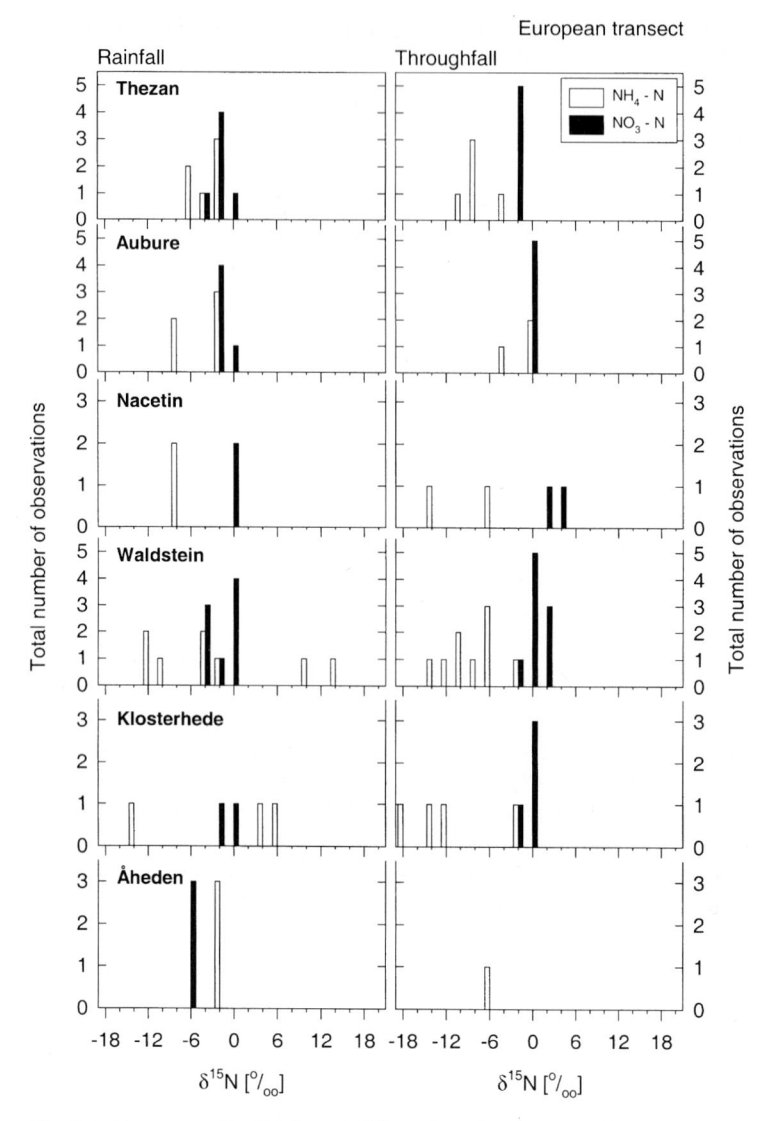

Fig. 9.4. Frequency distribution of $\delta^{15}N$ values for NH_4-N and NO_3-N in rainfall (*left-hand panel*) and throughfall (*right-hand panel*) at six coniferous forests across the European transect. (A.F. Harrison, unpubl. data)

of source than by broader geographical conditions such as long-range pollution transport (Heaton et al. 1997; Ammann et al. 1999).

From the relationship of $\delta^{15}N$ signature in ammonium versus NH_4-N concentration (Fig. 9.5A), it becomes clear that almost all throughfall samples have significantly higher concentrations than rainfall samples. Also the $\delta^{15}N$ signatures of throughfall ammonium had shifted significantly towards more negative values compared to rainfall. Theoretically, the physical processes of deposition to and subsequent leaching

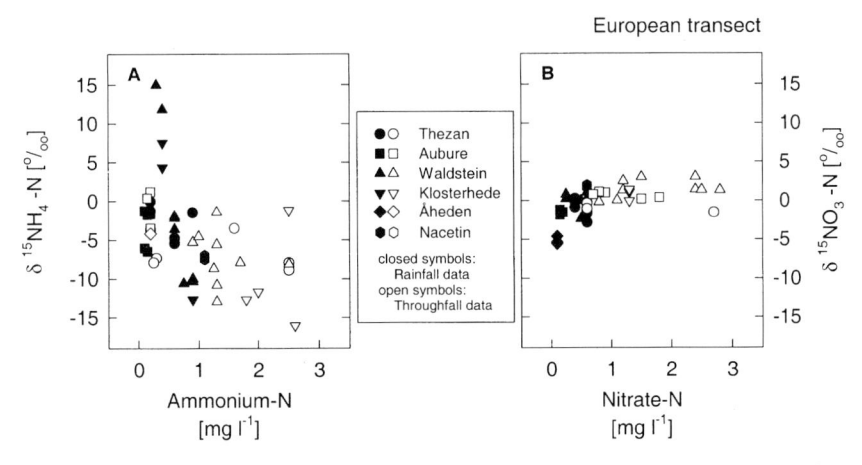

Fig. 9.5A,B. $\delta^{15}N$ values (‰) in NH_4-N (A) and NO_3-N (B) versus nitrogen concentration (mg l^{-1}) in nitrate and ammonium of rainfall and throughfall from six coniferous forests across the European transect. Rainfall data in *black symbols*, throughfall data in *open symbols*. (A.F. Harrison, unpubl. data)

from needle surfaces should have resulted in more positive $\delta^{15}N$ throughfall than rainfall (Heaton 1987). If, however, canopy uptake was also taking place, there should have been a further shift in $\delta^{15}N$ to the more positive range, as the lighter isotope would have been preferentially taken up by the trees. Neither of these shifts toward more positive isotope ratios was observed. An important point is that the rainfall and throughfall samples were collected during the autumn/winter period, when needle metabolism and N demand were low. Nevertheless, there are suggestions from the concentration data alone that there might have been uptake by the canopies at Åheden and Aubure (Fig. 9.5A) as throughfall concentrations were lower than those of the rainfall.

The $\delta^{15}N$ values for NO_3-N, though varying overall less than for NH_4-N, also appeared to differ between the forest sites of the transect (Fig. 9.4). There were also differences in the isotopic signature between rainfall and throughfall. The throughfall samples generally had higher NO_3-N concentrations than the rainfall samples, but in contrast to ammonium, the $\delta^{15}N$ signatures in the NO_3-N of the throughfall showed a clear shift toward more positive values (Fig. 9.5B). This change in isotopic signature from rainfall to throughfall would confirm isotopic fractionation during dry deposition or canopy uptake of oxidised nitrogen forms (Ammann et al. 1999).

9.4 Stable Isotope Signatures in Different Ecosystem Compartments

9.4.1 Soil Depth Profiles for $\delta^{15}N$, $\delta^{13}C$ and $\delta^{34}S$ Along the European Transect

Soil profiles of the spruce (*Picea abies*) stands along the European transect all displayed an increase in total nitrogen $\delta^{15}N$ with increasing soil depth (Fig. 9.6). The

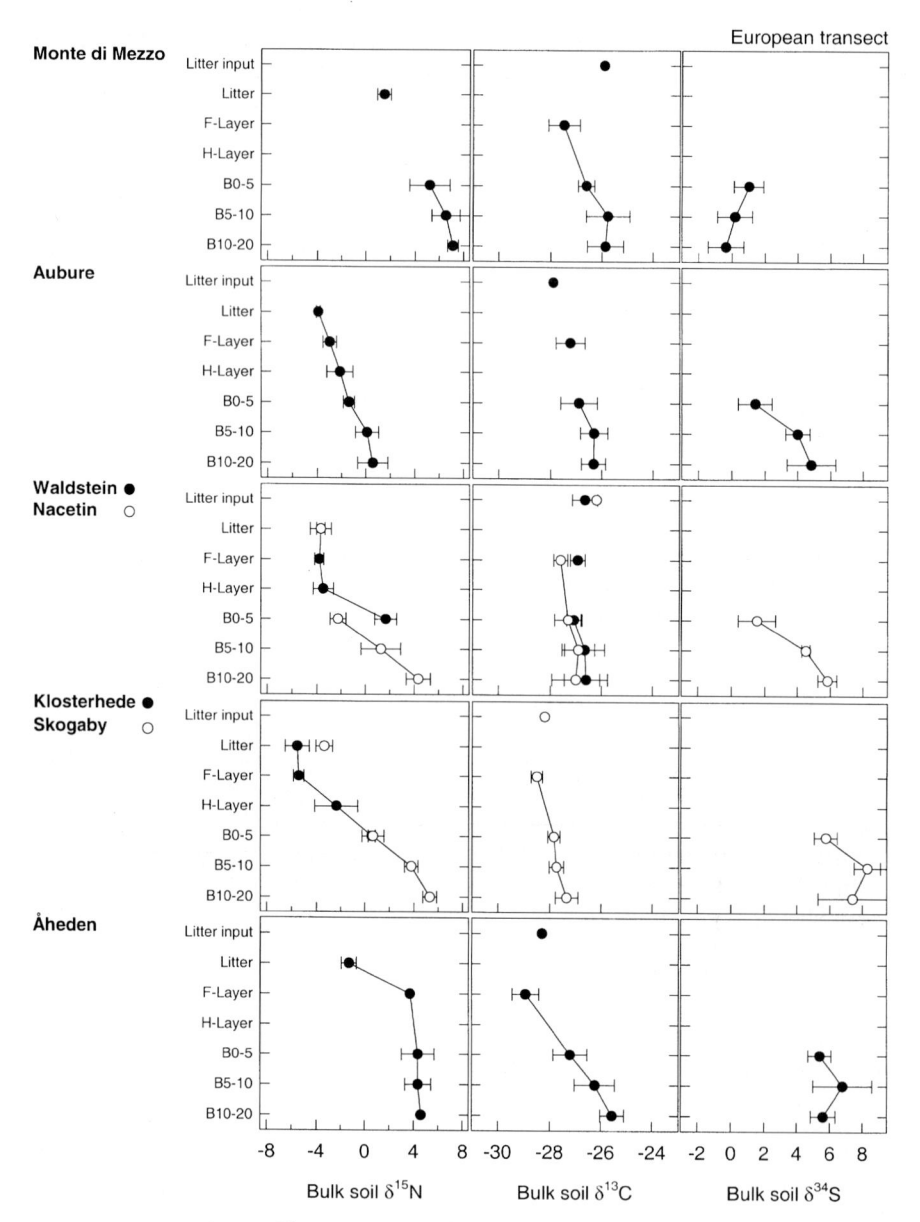

Fig. 9.6. Depth profiles of $\delta^{15}N$ (*left panel*), $\delta^{13}C$ (*middle panel*) and $\delta^{34}S$ (*right panel*; all data in ‰) for bulk soil of seven coniferous forests across the European transect. *Data points* represent means ± standard deviation of $n > 4$ repetitions. $\delta^{15}N$ values for litter, F layer and H layer: Högberg et al. (1996); nitrogen and sulphur isotope data for mineral soil: Novak et al. (1999); carbon isotope data: D. Harkness and A.F. Harrison, unpubl.

range in soil $\delta^{15}N$ values from about −6‰ in the organic layer to about +8‰ at 20 cm depth covered the whole range of observations made in other forest studies (Gebauer and Schulze 1991; Garten 1993; Garten and van Miegroet 1994; Handley and Scrimgeour 1997). The stand at Monte di Mezzo, although showing the same increase in $\delta^{15}N$ with soil depth as the other sites, had higher $\delta^{15}N$ values, which probably reflect the loss of volatile N from the former pasture on which this stand was planted.

At Åheden the increase in soil $\delta^{15}N$ from the litter layer to the H layer was greatest, with a difference of about 4.5‰, whereas at Nacetin, Waldstein and Aubure the organic layers did not vary significantly in their $\delta^{15}N$. Except for the stand in Aubure, all the spruce stands had positive $\delta^{15}N$ values in the mineral soil layers. The beech (*Fagus sylvatica*) stands showed stronger increases towards positive $\delta^{15}N$ values in the two uppermost layers (Fig. 9.7), but the $\delta^{15}N$ values of the top mineral soil layers were not as negative as those at the spruce sites. This observation may be an indication for denitrification in the densely packed beech litter.

The $\delta^{13}C$ signature of the spruce soils showed relatively little variation with depth. The strongest increase in $\delta^{13}C$ was found at Åheden from about −28‰ in the F layer to about −26.5‰ in the mineral soil at a depth of 10–20 cm. At the other sites the depth increase in $\delta^{13}C$ was much more shallow compared to Åheden. Except for Aubure, fresh litter inputs were about 1‰ more positive than the $\delta^{13}C$ value of the F layer. Among the beech stands, only the site at Collelongo had isotopically enriched litter input compared to the F layer, and also had the greatest increase in $\delta^{13}C$ with soil depth (Fig. 9.7). At Aubure, Schacht, Jezeri and Gribskov fresh leaf litter was slightly more negative than the $\delta^{13}C$ of the F layer. Also the $\delta^{13}C$ values at these sites showed only minor changes with increasing depth. Similarly to the spruce stands, the lack of a depth gradient in soil $\delta^{13}C$ could indicate that the excess amount of N in these soils triggers higher mineralisation rates by soil microorganisms which, in turn, consume most of the available soil C, resulting in only a small isotopic fractionation. This is also supported by the fact that the mean residence time of C in the L + F layer represents very young carbon (see Harrison et al., Chap. 11, this Vol.) and that C export from this horizon into deeper soil layers is very limited.

The rates of mineralisation of C, N, and S in forest soils may vary. However, all the forest soils from the present study displayed a striking common feature in their stable isotope patterns. All sites exhibited an increase in $\delta^{13}C$ and $\delta^{15}N$ with soil depth with a similar pattern for the $\delta^{34}S$ signature (Fig. 9.6). A positive $\delta^{34}S$ shift down the soil profile had been previously reported for 18 different spruce forests in Central Europe (Novak et al. 1996). Positive $\delta^{13}C$ and $\delta^{15}N$ shifts have long been known (Melillo et al. 1989; Högberg et al. 1996). Progressively higher $\delta^{13}C$ and $\delta^{15}N$ ratios at greater depth have been correlated with increasing age and degree of decay of organic matter. Mineralisation-related C and N isotope fractionation may have caused the observed patterns, but a number of complementary mechanisms have been considered (e.g. change in atmospheric CO_2 signature due to fossil fuel burning, plant uptake of low amounts of ^{15}N in combination with redeposition of the litterfall on the soil surface). Mineralisation of soil organic matter could cause isotopic discrimination for C, N and S. Isotopically light products of decomposition are preferentially taken up by plant roots, leached out of the soil or consumed by soil microorganisms, while the enriched residuals tend to remain in the soil.

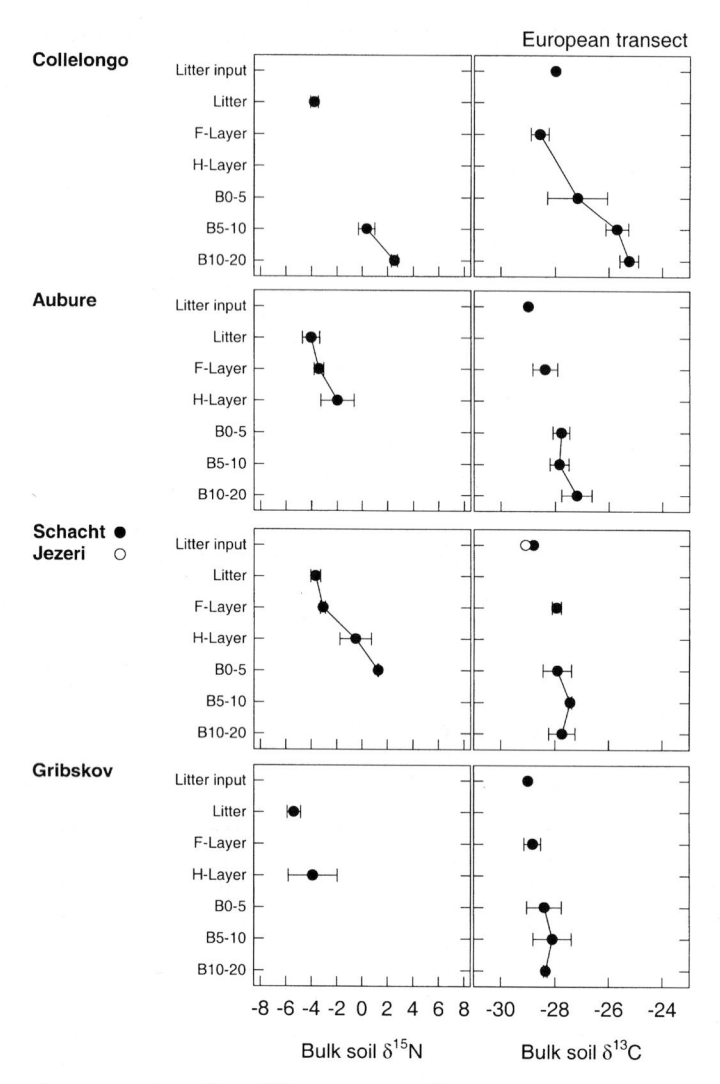

Fig. 9.7. Depth profiles of $\delta^{15}N$ (*left panel*), $\delta^{13}C$ (*right panel*; all data in ‰) for bulk soil of five *Fagus sylvatica* forests across the European transect. *Data points* represent means ± standard deviation of $n > 4$ repetitions. $\delta^{15}N$ values for litter, F layer and H layer: Högberg et al. (1996); carbon isotope data: D. Harkness and A.F. Harrison, unpubl.

9.4.2 Sulphur Isotope Signatures in Soil Solution

To better understand the behaviour of S in the soil, $\delta^{34}S$-SO_4 signatures of soil input (spruce canopy throughfall) and deep soil water (suction lysimeters installed deeper than 50 cm below soil surface; Fig. 9.8) were monitored in 1997 (Åheden, Skogaby and Aubure) and 1998 (Nacetin and Monte di Mezzo). The $\delta^{34}S$ ratio of soil water was generally lower compared to throughfall input, on average by 2.8‰. Sulphur isotopes

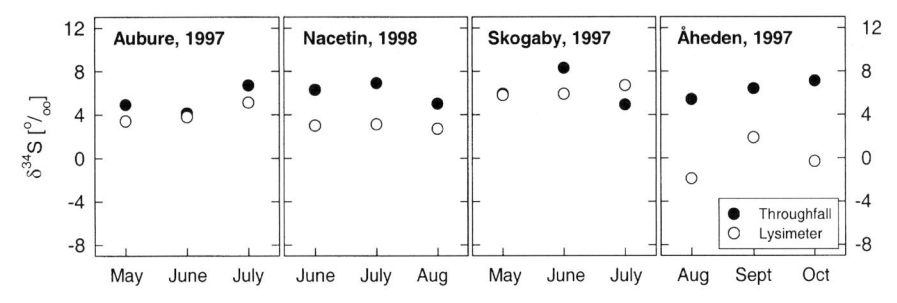

Fig. 9.8. $\delta^{34}S$ values (‰) in canopy throughfall and soil lysimeter water (depth 90 cm) for four coniferous forests across the European transect. (Novak et al. 1999)

Table 9.2. Total sulphur concentration ($\mu g\, g^{-1}$), isotopic composition (‰), sulphur content (%) of organic and inorganic sulphur in different layers of *Picea abies* soils at five stands along the European transect. (Novak et al. 1999)

Site	Depth (cm)	Organic S (%)	$\delta^{34}S_{organic}$ (‰)	Inorganic S (%)	$\delta^{34}S_{sulphate}$ (‰)	Content of total S ($\mu g\, g^{-1}$)	$\delta^{34}S_{total}$ (‰)
Åheden	0–5	86	5.4	14	5.4	100	5.4
	5–10	68	8.2	32	3.8	100	6.8
	10–20	59	6.6	41	4.1	100	5.6
Skogaby	0–5	76	6.4	24	4.0	230	5.8
	5–10	67	10.2	33	4.5	100	8.3
	10–20	47	8.8	53	6.2	100	7.4
Nacetín	0–5	95	2.2	5	1.6	1500	2.2
	5–10	88	4.7	12	3.2	500	4.5
	10–20	53	7.8	47	3.6	400	5.5
Aubure	0–5	84	1.6	16	0.1	330	1.4
	5–10	56	5.3	44	2.3	100	4.0
	10–20	11	12.1	89	3.9	100	4.8
M. di Mezzo	0–5	94	n.d.	6	n.d.	430	1.0
	5–10	83	−0.3	17	2.7	330	0.2
	10–20	81	−0.7	19	0.9	330	−0.4

clearly showed that sulphate entering the soil is different from sulphate leaving the system in the deep soil compartments. Bedrock sulphides are generally not preserved in the upper 50 cm of common forest soils (Likens et al. 1977). Consequently, low $\delta^{34}S$ ratios of deep soil water are not a result of mixing of atmospheric sulphate with sulphate supplied by weathering of bedrock sulphides. The isotope shift toward lower $\delta^{34}S$ ratios between input and output was most likely caused by a real isotope fractionation. Van Stempvoort et al. (1990) have shown that sorption of inorganic sulphate onto soil particles has negligible isotope effect on sulphate-S. Therefore, the observed isotope fractionation could have been associated with biologically mediated processes.

Table 9.2 gives the ratio of organic (C-bonded + ester-bonded) to inorganic (free and adsorbed sulphate) sulphur in all five soil profiles, along with $\delta^{34}S$ speciation. Organic S constituted more than 50% of total S throughout all profiles, except for the deepest layer at Skogaby (47%) and Aubure (11%). With increasing depth, the percentage of inorganic sulphate S increased at the expense of organic S. At sites where total soil S became isotopically heavier (e.g. Aubure, Nacetin) with increasing depth (i.e. has higher $\delta^{34}S$), the $\delta^{34}S_{organic}$ was greater than the $\delta^{34}S_{inorganic}$ (Gebauer et al. 1994). Moreover, $\delta^{34}S$ of atmospheric input at all sites was higher than $\delta^{34}S_{total}$ of the topmost analysed soil layer (cf. Fig. 9.7 and Table 9.2). These findings were consistent with the fact that deeper soil layers contain older and more decomposed organic matter (see Harrison et al., Chap. 11, this Vol.). As mineralisation has proceeded, newly formed "secondary" sulphate was enriched in low-$\delta^{34}S$ sulphur. Sulphur of organic compounds remaining in situ becomes isotopically heavier. By such a process, soil water sulphate collected by lysimeters contains isotopically lighter S than atmospheric deposition, while total soil sulphate is generally isotopically lighter than organic S from the same depth. The role of sulphur assimilation, seepage of isotopically distinct S compounds, as well as the uniqueness of $\delta^{34}S$ patterns at Monte di Mezzo have been discussed in greater detail by Novák et al. (1999). The apparent isotope fractionation effect of S mineralisation has been robust enough to occur at a variety of mean annual temperatures (from +0.8 to +5.9 °C) and atmospheric pollution levels (from 6 to 64 kg S ha^{-1} a^{-1}) along the European transect.

9.4.3 Carbon Isotope Ratios in Foliage

Across the European transect the highest $\delta^{13}C$ value for *Picea* needles were found at Waldstein (−24.8‰; Table 9.3), whereas needles at Åheden were lowest in $\delta^{13}C$ (−28.0‰). For *Fagus* the differences between the Danish site at Gribskov and Aubure in Central Europe was in the same range. The large geographical variation in $\delta^{13}C$ indicates a lower C_i at Aubure and Waldstein compared to northern Sweden. However, the fact that boreal forests suffer from N limitation might have had a strong influence on canopy photosynthesis in this case. The negative $\delta^{13}C$ values of the trees in Thezan, a mediterranean forest with summer drought, seem to be the result of a very wet spring at the time of sampling.

Table 9.3. $\delta^{13}C$ values (‰) of needles and leaves from main canopy trees across the European transect. Leaf material was collected during spring 1993. Values represent means ± standard deviation of $n > 2$ repetitions. (G.A. Bauer, unpubl. data)

Thezan	Aubure	Schacht Waldstein	Gribskov Klosterhede	Åheden
C. sativa −25.6 ± 0.87	F. sylvatica −25.1 ± 0.12	F. sylvatica −27.8 ± 0.66	F. sylvatica −28.9 ± 1.44	B. pendula −28.9 ± 0.35
Q. pubescens −26.9 ± 0.81				
P. pinaster −25.9 ± 1.10	P. abies −25.4 ± 0.15	P. abies −24.8 ± 0.68	P. abies −27.3 ± 0.95	P. abies −28.0 ± 0.85
				P. sylvestris −26.8 ± 0.35

Biotic and Abiotic Controls over Ecosystem Cycling 203

9.4.4 $\delta^{15}N$ Patterns in the Vegetation

9.4.4.1 Seasonal and Age-Dependent Changes in $\delta^{15}N$ Values of Foliage

In spruce, the $\delta^{15}N$ values of needles and twigs showed considerable variation early in the growing season (Fig. 9.9; $p < 0.001$), but these shifts were larger in 1-year-old needles and twigs and significantly different ($p < 0.001$) from the developing 0-year-old needles. Within 1 month the $\delta^{15}N$ value of 1-year-old needles increased from about −2‰ to about −0.5‰, whereas in the twigs the $\delta^{15}N$ value dropped from an initial value of about −2.0‰ to a minimum of about −3.5‰, but by July reached a value similar to that of the needles. These differences in the $\delta^{15}N$ values within a few centimetres of a single shoot probably reflect translocation of N as well as different origins of the N, which could have been N from storage pools in case of the 0-year-old needles and atmospheric N from agricultural sources surrounding the spruce stand at Waldstein in case of the 1-year-old needles (see Harrison et al., Chap. 8, this Vol.; Bruckner 1996). Changes in needle $\delta^{15}N$ occurred at a time when soils were still cold and therefore root uptake might have been low. Also, the shift in the $\delta^{15}N$ values in 1-year-old needles occurred at a time when these needles exported N, which presumably served growth of the current year shoot (Bauer et al. 1997).

In contrast to spruce, the seasonal variation in the $\delta^{15}N$ values in leaves and twigs of beech were much smaller (± 0.5‰; $p = 0.206$) and leaf and twig $\delta^{15}N$ changed in

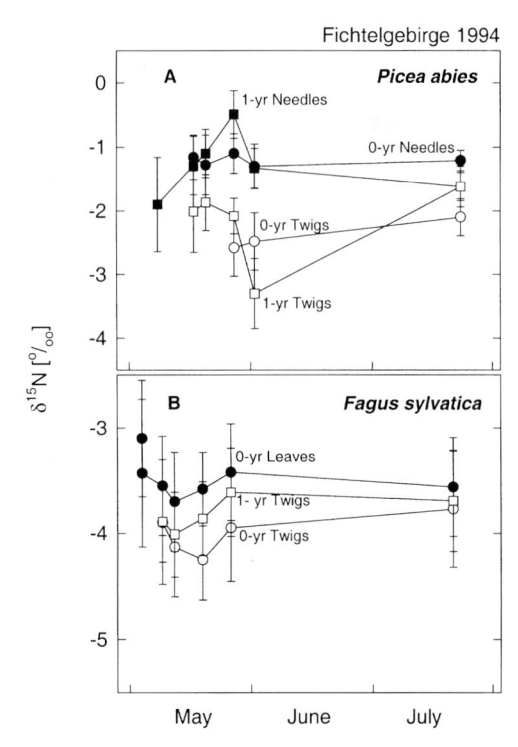

Fig. 9.9A,B. Seasonal change in $\delta^{15}N$ values (‰) for needles, leaves and twigs of *Picea abies* (Waldstein site, **A**) and *Fagus sylvatica* (Schacht site, **B**) during the growing season in 1994 in the Fichtelgebirge (NE Bavaria). (G.A. Bauer, unpubl. data)

parallel with only a minor decrease in mid-May. The pattern in beech coincides with the fact that early leaf development in deciduous trees depends on stored N rather than on actual uptake (Millard 1996).

The $\delta^{15}N$ values in different age classes of coniferous needles during the period of needle growth in spring 1993 showed a larger variation along the transect ($p < 0.001$) than between needle ages ($p = 0.064$; Fig. 10A/B). Current-year to 3-year-old needles at Waldstein, Schacht (spruce trees adjacent to the beech stand) and at Aubure had $\delta^{15}N$ values of about −2.5‰, thus confirming the increase in $\delta^{15}N$ during early needle growth at Waldstein (Fig. 9.9). Needles at the boreal site in Åheden had $\delta^{15}N$ values of about −6.0‰, whereas *Pinus* at Thezan showed the most negative $\delta^{15}N$ values at an average of about −7.5‰.

The large differences in needle $\delta^{15}N$ between the different sites also holds true for the autumn period after termination of growth (Fig. 10B; $p < 0.001$), with a range in $\delta^{15}N$ between −2.0‰ at the Schacht and about −6.0‰ for needles at Åheden. Irrespective of the site, needle $\delta^{15}N$ values remained constant across all needle age classes ($p = 0.677$), except for the stand at Aubure where needle $\delta^{15}N$ values decreased with increasing needle age. Generally, the differences in needle $\delta^{15}N$ signature either in

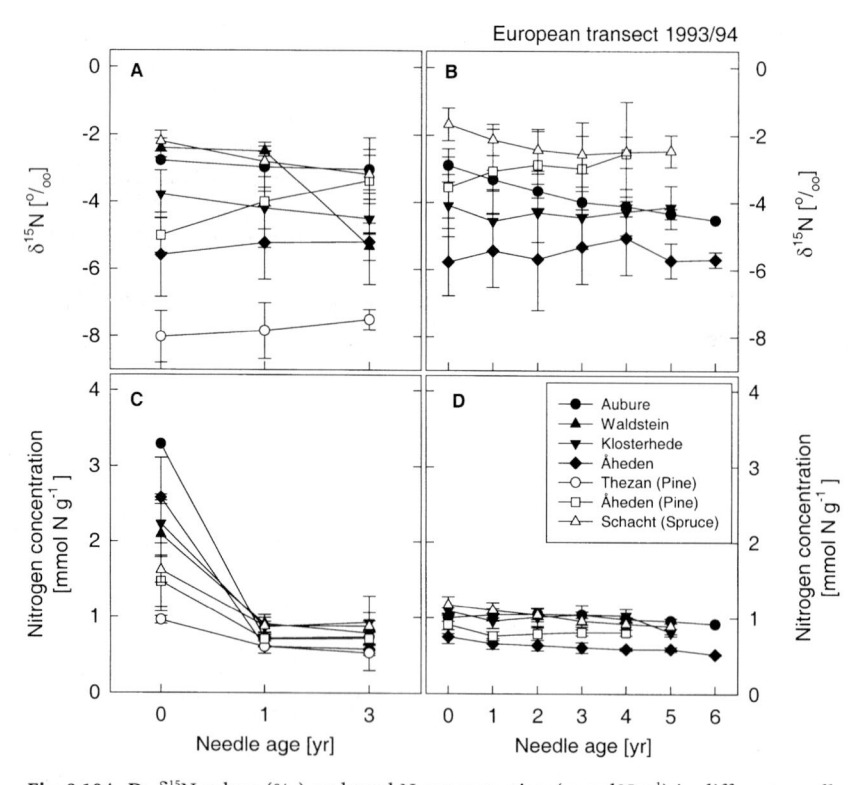

Fig. 9.10A–D. $\delta^{15}N$ values (‰) and total N concentration (mmol N g^{-1}) in different needle age classes of *Picea abies*, *Pinus sylvestris* and *Pinus pinaster* during bud break in 1993 (**A, C**) and during autumn in 1994 (**B, D**) for the coniferous forests across the European transect. Data represent means ± standard deviation of $n = 4$ repetitions. (Bauer 1997; Bauer et al. 1997)

spring or in autumn are not related to changes in total nitrogen concentration (Fig. 9.10C,D). Except for a large variation at the beginning of needle growth due to rapid increase in dry weight (Bauer et al. 1997), N concentration was essentially constant over needle age ($p = 0.028$). This clearly indicates large differences in the quality of N being used for needle growth which contrasts with the more or less constant N to dry weight ratio across the entire European transect (see Bauer et al., Chap. 4, this Vol.).

9.4.4.2 Comparison of $\delta^{15}N$ Signatures Across Different Life Forms

Across the European transect, there was an overlap in $\delta^{15}N$ signatures in current-year foliage for different groups of higher plants (Fig. 9.11). Among the canopy trees, conifers showed the largest range in $\delta^{15}N$ values between −9 and −2‰. Deciduous trees varied over a much smaller range of $\delta^{15}N$ values (−7 to −4‰). Canopy trees and plants from the understorey vegetation overlapped between −7.0 and −2.0‰. However, at Klosterhede and Åheden *Deschampsia* and ericaceous shrubs had $\delta^{15}N$ values close to or even above zero. Apart from these two sites, the majority of the $\delta^{15}N$ values in the understorey layer showed a smaller range than the canopy trees. This is especially true for the deciduous trees and *Deschampsia*. However, *Deschampsia* had the largest variability in the understorey, ranging from about −7 to +1‰, which might mirror its wide range of habitats (Grime et al. 1988) and the broad spectrum of N sources which can be used for growth.

Despite overlaps in the $\delta^{15}N$ values among species there was a clear separation at some sites among cooccurring plant lifeforms. At Thezan, conifers had $\delta^{15}N$ values between −9 and −7‰, whereas the values for deciduous trees and ericaceaous plants were between −7 and −4‰. At Åheden, the trees and their seedlings in the understorey seem to tap from the same N source, whereas *Deschampsia* and ericaceous plants had more positive $\delta^{15}N$ values.

The wide range of $\delta^{15}N$ values in plant foliage is likely to be an indicator that there are differences in the quality of nitrogen taken up and transported to the leaves, since, for example, *Fagus* in Gribskov had more negative $\delta^{15}N$ values than *Fagus* foliage in Collelongo. The deposition data for Central Europe clearly indicated that NH_4-N and NO_3-N in the throughfall were more positive in $\delta^{15}N$ than in the rainfall (Fig. 9.4). This makes it likely that the enriched N in the throughfall might be the preferred form of N taken up by understorey plants at those sites (see Gebauer et al., Chap. 7 and Harrison et al., Chap. 8, this Vol.). On the other hand, for example, *Deschampsia flexuosa* at Åheden seems to tap nitrogen from a strongly enriched source other than N from rainfall, which at Åheden was much more negative and probably far too low in concentration to meet plant N demand at this site. Additionally, trees and understorey plants differ in their rooting depth, which, based on the $\delta^{15}N$ depth profile in the soils would cause another change in the isotopic signal.

Fruit bodies of ectomycorrhizal (ECM) fungi collected at Wülfersreuth (Fichtelgebirge, close to the Waldstein site) and Åheden showed a considerable difference in $\delta^{15}N$ compared to soil and tree foliage (Table 9.4), which is in line with observations made for several fungal taxa at Åheden (Taylor et al. 1997). At Wülfersreuth ECM fruitbodies had $\delta^{15}N$ values in between soil and spruce needles, whereas at Åheden, fruitbodies and soil were enriched by about 8‰ compared to needles. $\delta^{15}N$ values of fruit bodies of fungi collected at Wülfersreuth and in Åheden ranged from −3.8 to

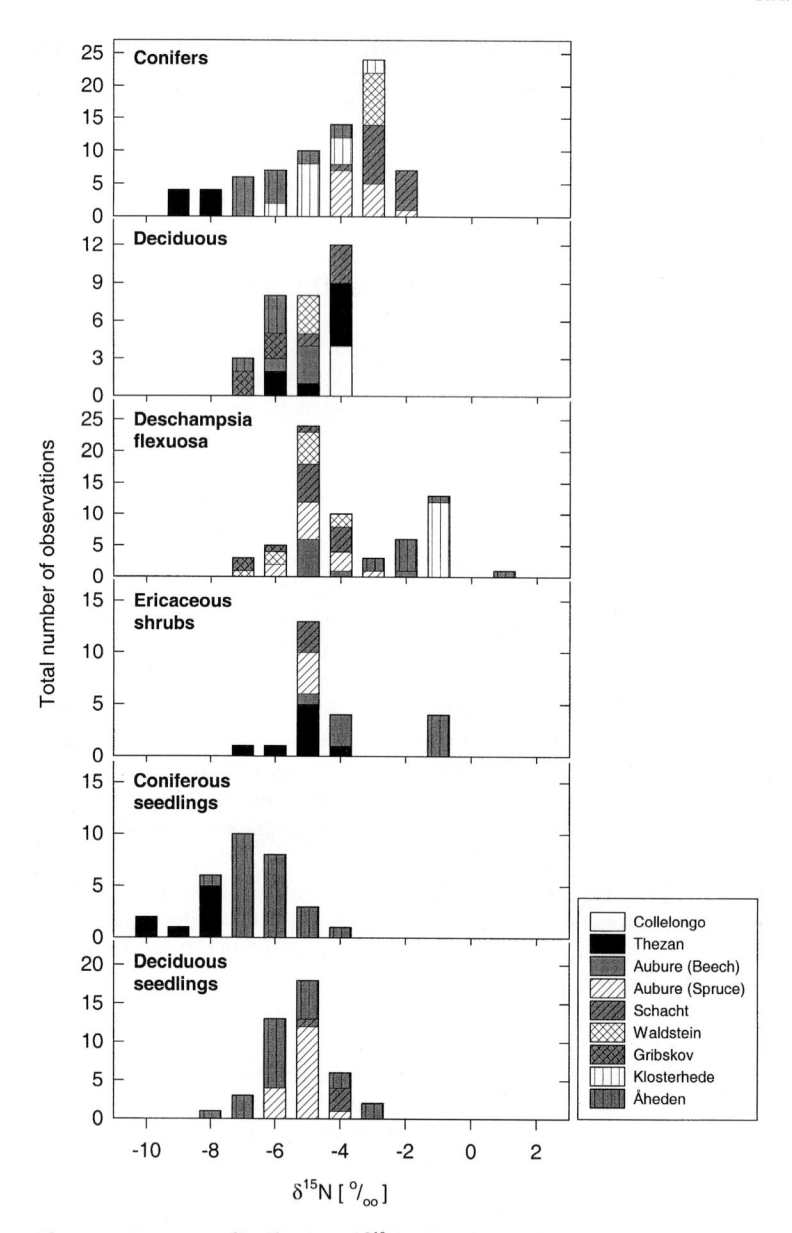

Fig. 9.11. Frequency distribution of $\delta^{15}N$ values (‰) in foliage of different plant life forms from nine forests across the European transect. (Bauer 1997)

4.2‰ or from −0.8 to 15.4‰, respectively (Gebauer and Dietrich 1993; Taylor et al. 1997; Gebauer and Taylor 1999). Some of the variability in the relative ^{15}N abundance amongst the fungi is related to the substrate of the fungal mycelium growth (Table 9.4). N in the humus layer of the Åheden site is more enriched in ^{15}N by about 4.4‰ than N in the humus layer of the Fichtelgebirge site and thus ECM fungi collected in

Table 9.4. Average $\delta^{15}N$ values ± SE (‰) (ranges in brackets) and enrichment factors $\varepsilon_{f\text{-}s}$ ± SE (‰) of fruit bodies of four functional groups of fungi collected in the Fichtelgebirge (Wülfersreuth site, NE Bavaria, Germany) and at Åheden (N Sweden). Data for the fruitbodies are according Taylor et al. (1997) and Gebauer and Taylor (1999)

Functional group	Fichtelgebirge (Wuelfersreuth)			Åheden		
	n	$\delta^{15}N$	$\varepsilon_{f\text{-}s}$	n	$\delta^{15}N$	$\varepsilon_{f\text{-}s}$
ECM fungi capable of utilizing organic N from humus	9	1.5 ± 0.6 (−1.4 to 4.2)	2.2 ± 0.5	54	6.2 ± 0.5 (−0.8 to 15.4)	2.5
ECM fungi utilizing inorg. N in the soil (*Laccaria* ssp.)	3	−2.0 ± 0.3 (−2.3 to −1.5)	n.d.	2	2.2 (1.8 to 2.6)	n.d.
Saprophytic fungi capable of utilizing organic N from humus	9	1.9 ± 0.3 (0.2 to 3.0)	2.5 ± 0.4			
Saprophytic fungi utilizing N from litter/wood	13	−2.0 ± 0.3 (−3.8 to 0.5)	0.9 ± 0.4			

n.d. Not determined.

Åheden were as a mean more enriched in ^{15}N by a similar factor than fungi from the Fichtelgebirge site. ECM and saprophytic fungi living on humus and presumably capable of utilizing organic N were more enriched in ^{15}N than ECM fungi of the genus *Laccaria*, which are known to depend on soil inorganic N (Abuzinadah and Read 1986), or saprophytic fungi living on litter or wood. Positive enrichment factors $\varepsilon_{f\text{-}s}$ (differences in $\delta^{15}N$ between fungi and their substrates) indicate a general ^{15}N enrichment of fungi by 0.9 to 2.5‰ when compared to their respective substrates. This finding distinguishes fungi from trees or understorey plants, which were mostly more depleted in ^{15}N than their substrates. Recently, it was presumed that the transfer of N from ECM fungi to plant hosts might be responsible for the relative ^{15}N enrichments of ECM fungi (Hobbie et al. 1999 and references therein). The ^{15}N enrichment of fungi relative to their substrate, however, seems to occur irrespective of the fungal life form. On the Fichtelgebirge site, ECM fungi and saprophytes living on humus and presumably capable of utilizing organic N could not be distinguished by their $\delta^{15}N$ values or by their enrichment factors (Gebauer and Taylor 1999). Thus, relative ^{15}N enrichment in fungal fruit bodies seems to be a more general feature, occurring independently of fungal life forms. A strong taxonomic influence on the ^{15}N abundance was found for the ECM fungi collected in Åheden (Taylor et al. 1997). In particular, members of the genus *Cortinarius* and many boletoid species tend to have highly ^{15}N-enriched fruit bodies. Highly ^{15}N-enriched fungi might be specialised to mobilise N from recalcitrant organic sources. Recalcitrant N compounds in the soil are expected to be enriched in ^{15}N.

9.5 $\delta^{15}N$ Signatures as Indicators of N Saturation in Forest Ecosystems

The overlap in $\delta^{15}N$ signatures across the various vegetation layers (Fig. 9.11) indicated that cooccurring life forms may compete for similar resources of N. However, as at the boreal site Åheden, trees seem to tap from different N sources than ericaceous plants and tree seedlings. The $\delta^{15}N$ signatures in NH_4-N and NO_3-N in the wet deposition partly displayed a site-specific pattern and also soil N mineralisation differed strongly across the transect (see Persson et al., Chap. 14, this Vol.), which would cause different isotopic signatures depending on rooting depth. Therefore positive shifts in the $\delta^{15}N$ signatures throughout the soil-root-canopy continuum due to isotopic fractionation might indicate changes in the preferred N source which is taken up, assuming no fractionation during long-range transport in the xylem.

Soil and root $\delta^{15}N$ values of all the conifer sites were consistently mirrored in the isotopic signature of the xylem sap (Fig. 9.12). Also the $\delta^{15}N$ values of understorey plants matched the isotopic signature of the soil. The largest increase in $\delta^{15}N$ within the humus layer was found at Åheden and Monte di Mezzo, while at Aubure and Waldstein no differences between F and H layer were found. Similarly to the xylem sap, the root $\delta^{15}N$ values closely mirrored the isotopic signature of the soil without significant differences for the two root types. Only at Åheden did a clear difference in $\delta^{15}N$ between ectomycorrhizal, nonmycorrhizal roots and the soil emerge. At Aubure and Waldstein, needle $\delta^{15}N$ values were close to or even somewhat more positive than soil $\delta^{15}N$ values. As indicated above, this could have been due to a mixing of the different N sources from which these trees feed. Canopy uptake seems to be a likely source (Ammann et al. 1999) and the high degree of mycorrhizal association also favours the uptake of organic N via the fungal mycelium. However, mycorrhizal uptake should leave the mycelium enriched and the host depleted.

The $\delta^{15}N$ profiles among the deciduous stands was very different from the geographical corresponding conifer stands (Fig. 9.13). Except for the beech stand in Collelongo there was almost no difference in $\delta^{15}N$ between roots and the soil with increasing depth. Also in beech, xylem sap $\delta^{15}N$ closely matched soil and root isotopic signatures, but in contrast to spruce was generally enriched in $\delta^{15}N$ compared to the leaves. Although roots and soil exhibited no differences between F and H layer, there was no indication that leaves at Aubure and Schacht were enriched in $\delta^{15}N$.

Given the spatial distance between N uptake from the soil and the sites of unloading in the foliage it is interesting that during xylem transport no major shifts in the $\delta^{15}N$ of the xylem sap occur. A similar result was found in profiles of xylem sap isotopic signatures for several spruce stands surrounding the Waldstein site (Hein 1996), where the $\delta^{15}N$ in the xylem sap remained constant during the passage from the roots to the canopy. This indicates that the uptake along the main axis between conducting xylem and surrounding parenchyma is low. Thus, the $\delta^{15}N$ value of the xylem water might be indicative of either soil or mycorrhizal nitrogen.

Even though the $\delta^{15}N$ values of the wet deposition are partly mirrored in the isotopic signatures of the vegetation, we cannot detect an atmospheric N input from a single, specific source as being the predominant cause for the observed pattern or as the preferred source of above-ground N uptake, in part because the $\delta^{15}N$ ranges of deposition and soil supply overlap (Durka et al. 1994). Root uptake of nitrogen dominates over canopy uptake, but foliar uptake can substantially contribute to whole-stand N demand (Pearson and Stewart 1993). As a result, negative feedback regulation between canopy and root N uptake could take place (Rennenberg et al. 1998). Also, soil N mineralisation across the European transect was highest on those sites with high N input from the atmosphere (see Persson et al., Chap. 14, this Vol.). Therefore

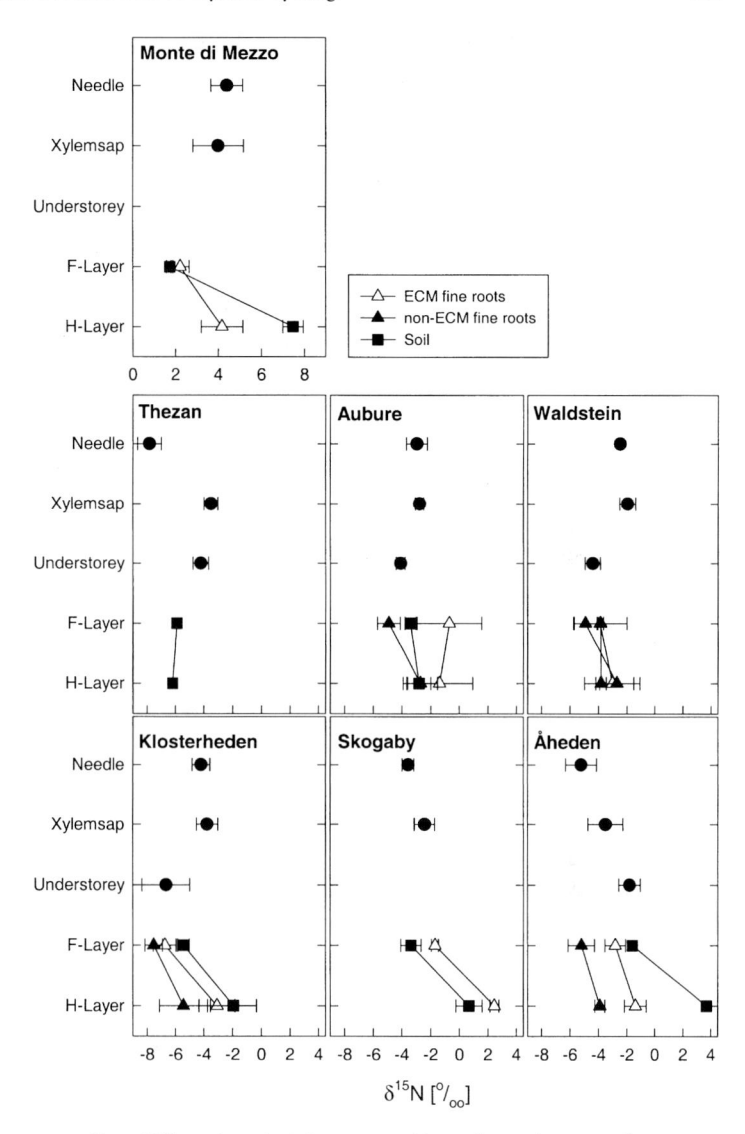

Fig. 9.12. Ecosystem profiles of $\delta^{15}N$ values (‰) for 1-year-old needles, xylem sap of suncrown branches, understorey plants, mycorrhizal and non-mycorrhizal fine roots and soils for seven coniferous forests across the European transect. Understorey values for Aubure, Waldstein, Klosterdede and Åheden represent $\delta^{15}N$ values for *Deschampsia flexuosa* and for Thezan the $\delta^{15}N$ value for *Quercus coccifera*. Values represent the means ± standard deviation of n = 4 repetitions. (Bauer 1997; Högberg et al. 1996)

the $\delta^{15}N$ values within the vegetation are likely to be a mixture of $\delta^{15}N$ signatures from a wide variety of N sources rather than the result of one prevailing N form.

High N availability in forest ecosystems has been found to correlate with a positive shift in the isotopic signature of foliage relative to the soil, i.e. the enrichment

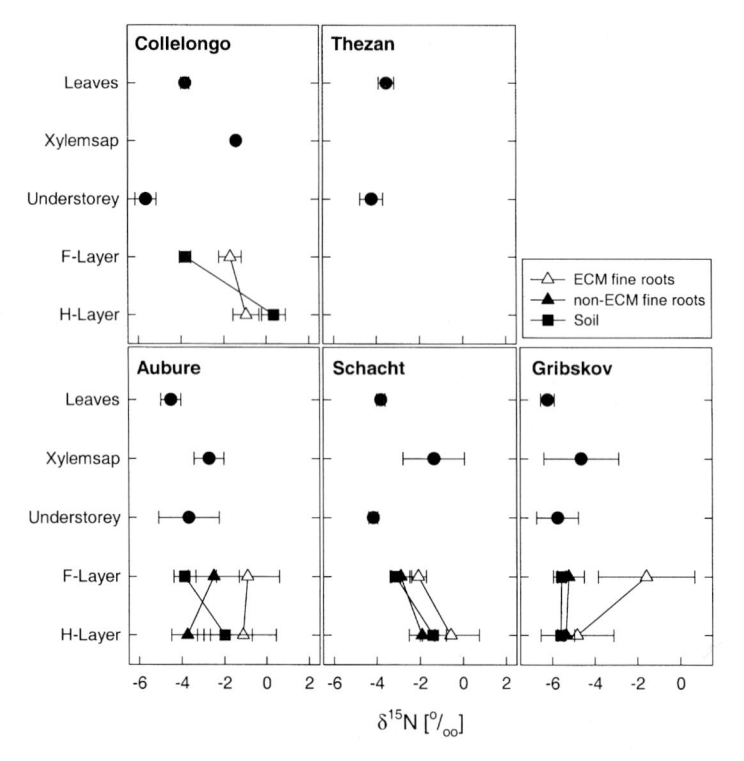

Fig. 9.13. Ecosystem profiles of $\delta^{15}N$ values (‰) for leaves of trees, xylem sap of suncrown branches, understorey plants, mycorrhizal and non-mycorrhizal fine roots and soils for five broadleaf forests across the European transect. Understorey values for Aubure, Schacht and Gribskov represent $\delta^{15}N$ values of *Deschampsia flexuosa*, for Thezan the $\delta^{15}N$ value of *Quercus ilex* and for Collelongo the $\delta^{15}N$ value of *Galium* cf. *sylvaticum*. Values represent the means ± standard deviation of $n = 4$ repetitions. (Bauer 1997; Högberg et al. 1996)

factor ε, defined as $\delta^{15}N_{leaf} - \delta^{15}N_{soil}$, increases (Garten 1993; Garten and van Miegroet 1994; Naesholm et al. 1997; Emmett et al. 1998). This shift in isotopic signature toward more positive values is assumed to be due to a surplus of N within a forest ecosystem causing a decrease in N uptake. As a result, ^{15}N-depleted nitrate is lost through leaching and denitrification, while $\delta^{15}N$-enriched ammonium is preferentially taken up by the plants. As a result, mycorrhiza and leaves remain enriched in $\delta^{15}N$. The isotopically heavier N in the leaf litter in turn results in an enrichment of the upper soil layers, since this heavier N remains during decomposition (Nadelhoffer and Fry 1994; Högberg 1997). For the spruce stands across the European transect, we found a similar relationship for ε to increase at high mineralisation rates (Fig. 9.14B). Foliar $\delta^{15}N$ values of spruce, however, did not parallel the increase in net N mineralisation (Garten and van Miegroet 1994), but rather levelled off at a $\delta^{15}N$ value of about –2.0‰ (Fig. 9.14A). Leaf $\delta^{15}N$ values of beech showed no correlation with the mineralisation rate because of their much smaller range (Fig. 9.11).

For both groups of trees no relationship was found for leaf $\delta^{15}N$ values or enrichment factor with net N nitrification (Fig. 9.15), which for spruce as a predominant

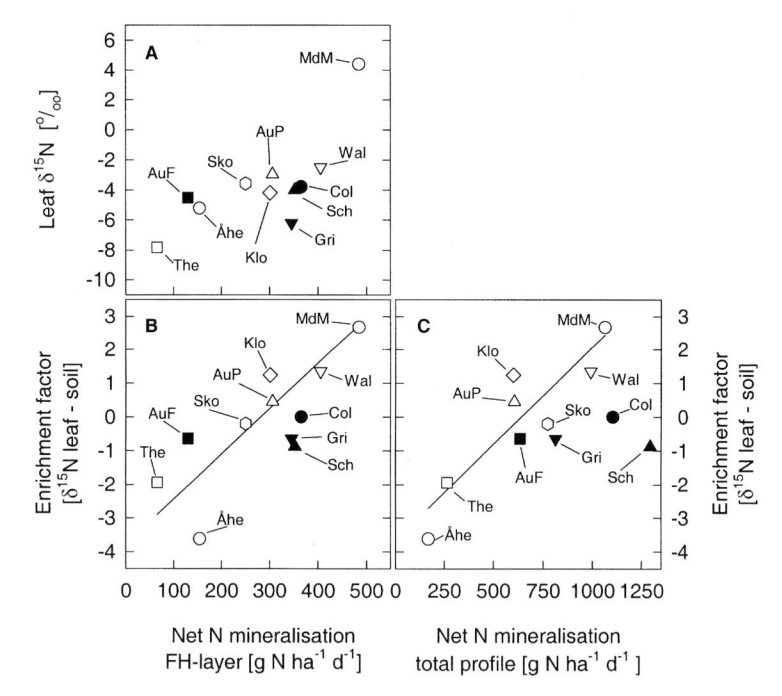

Biotic and Abiotic Controls over Ecosystem Cycling 211

Fig. 9.14A–C. Leaf $\delta^{15}N$ values (‰) and enrichment factor for $\delta^{15}N$ versus net N mineralisation rates (g N ha^{-1} day^{-1}; data according Persson et al., Chap. 14, this Vol.) in the FH layer (A, B) and versus total net N mineralisation rates over the whole soil profile (C) for spruce and beech stands across the European transect. The enrichment factor is calculated as the difference between the foliage $\delta^{15}N$ and the $\delta^{15}N$ values of the F layer. Linear regression equation for A: $y = -3.79 + 0.013*x$ ($r^2 = 0.795$, $p < 0.01$) and for C: $y = -3.675 + 0.005*x$ ($r^2 = 0.824$, $p < 0.01$)

ammonium user could be expected. For beech, also no apparent correlation of leaf $\delta^{15}N$ and ε was found with nitrification rates, although the four beech stands which were included in this analysis represent a broad range of conditions regarding nutrient supply, productivity and climate. However, this does not imply that the beech stands have a greater tolerance with regard to high N availability. Many steps along the soil-root interface and within the trees could change the isotopic signature of nitrogen compounds. Additionally, seasonal variation in N uptake and allocation could mask a species-specific relationship between the enrichment factor and soil processes.

9.6 Conclusions

- Natural abundance of stable isotopes for N, C and S in different ecosystem components across the European transect strongly supported our assumption that atmospheric N and S deposition has an impact on the biogeochemical cycling in the forest ecosystems studied. In general, we found that throughout this transect natural abundance measurements indicated that pollution stress affects forest ecosystems, which probably could only be detected in a large transcontinental comparison.

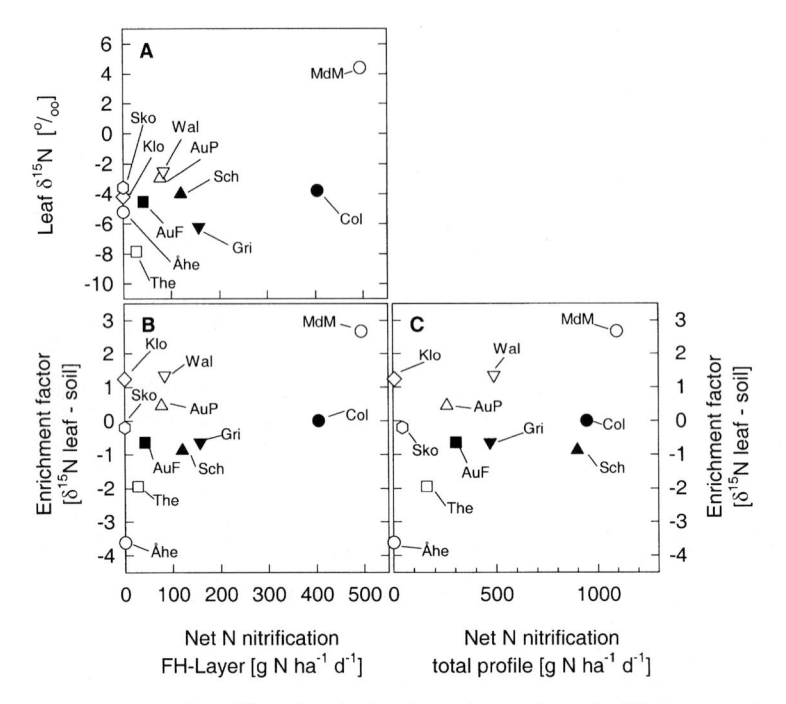

Fig. 9.15A–C. Foliage $\delta^{15}N$ values (‰) and enrichment factor for $\delta^{15}N$ between plant-soil versus net N nitrification rates (g N ha^{-1} day^{-1}; data according Persson et al., Chap. 14, this Vol.) in the FH layer (**A, B**) and versus total net N nitrification rates over the whole soil profile (**C**) for spruce and beech stands across the European transect. The enrichment factor is calculated as the difference between the foliage $\delta^{15}N$ and the $\delta^{15}N$ values of the F layer

- Nitrogen isotope analysis in wet deposition indicated that for some of the forests the $\delta^{15}N$ signals can be traced back to the pollution sources. However, the influence of local-scale parameters like canopy structure in combination with micrometeorological conditions was much stronger than expected. Additionally, interannual variation in the isotopic signature of nitrogen in the wet deposition is large and potentially could out-date current regional and/or continental N budgets based on modelled deposition estimates.
- Forest ecosystems which are exposed to N pollution show a significant enrichment in $\delta^{15}N$ of foliage and the upper soil horizons. In this regard, our results support earlier observations, that N pollution leads to a retention of isotopically enriched nitrogen. Overall, site-specific patterns for natural nitrogen abundance among deposition, plants and soils emerged.
- Sulphur isotope signatures also indicated that S which is deposited on forest ecosystems is different from the S which leaks out of the forest soils. This indicates a break in the terrestrial S cycle, where storage of organic S in the soil is of importance.
- Soil carbon and nitrogen isotope signatures, together with radiocarbon data (see Harrison et al., Chap. 11, this Vol.) indicated that at high atmospheric N deposition the increased rates of N transformations in the soil might effect C storage and turnover in the organic soil horizons.

References

Aber JD, Nadelhoffer KJ, Steudler P, Melillo JM (1989) Nitrogen saturation in northern forest ecosystems. BioScience 39:378-386

Aber JD, McDowell W, Nadelhoffer KJ, Magill A, Berntson GM, Kamakea M, McNulty S, Currie W, Rustad L, Fernandez I (1998) Nitrogen saturation in temperate forest ecosystems – hypothesis revisited. BioScience 48:921-934

Abuzinadah RA, Read DJ (1986) The role of proteins in the nitrogen nutrition of ectomycorrhizal plants. I. Utilization of peptides and proteins by ecotmycorrhizal fungi. New Phytol 103:481-493

Ammann M, Siegwolf R, Pichlmayer F, Suter M, Saurer M, Brunold C (1999) Estimating the uptake of traffic-derived NO^2 from ^{15}N abundance in Norway spruce needles. Oecologia 118:124-131

Bauer GA (1997) Stickstoffhaushalt und Wachstum von Fichten- und Buchenwäldern entlang eines Europäischen Nord-Süd-Transektes. PhD Thesis, University of Bayreuth, Bayreuther Forum Ökologie Vol 53, Bayreuth 176 pp

Bauer G, Mund M, Schulze E-D (1997) Nutrient contents and concentrations in relation to growth of *Picea abies* and *Fagus sylvatica* along a European transect. Tree Physiol 17:777-786

Bruckner G (1996) Deposition und oberirdische Aufnahme von gas- und partikelförmigem Stickstoff aus verschiedenen Emissionsquellen in ein Fichtenökosystem. PhD Thesis, University of Bayreuth, Bayreuther Forum Ökologie Vol 29, Bayreuth 230 pp

Bruckner G, Gebauer G, Schulze E-D (1993) Uptake of $^{15}NH_3$ by *Picea abies* in closed chamber experiments. Isotopenpraxis Environ Health Stud 29:76-71

Durka W, Schulze E-D, Gebauer G, Voerkelius S (1994) Effects of forest fecline on uptake and leaching of deposited nitrate determined from ^{15}N and ^{18}O measurements. Nature 372:765-767

Emmett BA, Kjonaas OJ, Gundersen P, Koopmans C, Tietema A, Sleep D (1998) Natural abundance of ^{15}N in forest across a nitrogen deposition gradient. For Ecol Manage 101:9-18

Garten CT (1993) Variation in foliar ^{15}N abundance and the availability of soil nitrogen on Walker Branch watershed. Ecology 74:2098-2113

Garten CT, van Miegroet H (1994) Relationships between soil nitrogen dynamics and natural ^{15}N abundance in plant foliage from Great Smoky Mountains National Park. Can J For Res 24:1636-1645

Gebauer G, Dietrich P (1993) Nitrogen isotope ratios in different compartments of a mixed stand of spruce, larch and beech trees and of understorey vegetation including fungi. Isotopenpraxis 29:35-44

Gebauer G, Schulze E-D (1991) Carbon and nitrogen isotope ratios in different compartments of a healthy and a declining *Picea abies* forest in the Fichtelgebirge, NE Bavaria. Oecologia 87:198-207

Gebauer G, Taylor AFS (1999) ^{15}N natural abundance in fruitbodies of different functional groups of fungi in relation to substrate utilisation. New Phytol 142:93-101

Gebauer G, Giesemann A, Schulze E-D, Jaeger HJ (1994) Isotope ratios and concentrations of sulfur and nitrogen in needles and soils of *Picea abies* stands as influenced by atmospheric deposition of sulfur and nitrogen compounds. Plant Soil 164:267

Grime JP, Hodgson JG, Hunt R (1988) Comparative plant ecology: a functional approach to common British species. Unwin Hyman, London

Handley LL, Scrimgeour CM (1997) Terrestrial plant ecology and ^{15}N natural abundance: The present limits to interpretation for uncultivated systems with original data from a Scottish old field. In: Begon M, Fitter AH (eds) Advances in ecological research. Academic Press, London, pp 133-212

Heaton THE (1987) $^{15}N/^{14}N$ ratios of nitrate and ammonium in rain at Pretoria, South Africa. Atmos Environ 21:843-852

Heaton THE, Spiro B, Madeline S, Robertson C (1997) Potential canopy influences on the isotopic composition of nitrogen and sulphur in atmospheric deposition. Oecologia 109:600-607

Hein M (1996) Aminosäuren, Kationen und Stickstoffisotope im Xylemsaft der Fichte (*Picea abies*) auf saurem Substrat in Abhängigkeit vom Bestandesalter. Diplom Thesis University of Bayreuth, Bayreuth

Hobbie EA, Macko SA, Shugart HH (1999) Insights into nitrogen and carbon dynamics of ecotmycorrhizal and saprotropic fungi from isotopic evidence. Oecologia 118:353–360

Högberg P (1997) ^{15}N natural abundance in soil-plant systems. New Phytol 137:179–203

Högberg P, Högbom L, Schinkel H, Högberg M, Johannisson C, Wallmark H (1996) ^{15}N abundance of surface soils, roots and mycorrhizas in profiles of European forest soils. Oecologia 108:207

Krouse HR (1989) Sulfur isotope studies of the pedosphere and biosphere. In: Rundel PW, Ehleringer JR, Nagy KA (eds) Stable isotopes in ecological research. Springer, Berlin Heidelberg New York, pp 424–444

Likens GE, Bormann FH, Pierce RS, Eaton JS, Johnson NM (1977) Biogeochemistry of forested ecosystems. Springer, Berlin Heidelberg New York

Lloyd J, Farquhar GD (1996) The CO_2 dependence of photosynthesis, plant growth responses to elevated atmospheric CO_2 concentrations and their interaction with soil nutrient status. I. General principles and forest ecosystems. Funct Ecol 10:4–32

Melillo JM, Aber JD, Linkins AE, Ricca A, Fry B, Nadelhoffer KJ (1989) Carbon and nitrogen dynamics along the decay continuum: plant litter to soil organic matter. Plant Soil 115:189–198

Millard P (1996) Ecophysiology of the internal cycling of nitrogen for tree growth. Z Pflanzenernaehr Bodenkd 159:1–10

Nadelhoffer KJ, Fry B (1994) Nitrogen isotope studies in forest ecosystems. In: Lajtha K, Michener R (eds) Methods in ecology – stable isotopes in ecology and environmental science. Blackwell Oxford, pp 22–44

Naesholm T, Nordin A, Edfast A-B, Högberg P (1997) Identification of coniferous forests with incipient nitrogen saturation through analysis of arginine and nitrogen-15 abundance of trees. J Environ Qual 26:302–309

Novák M, Wieder RK, Schell WR (1994) Sulfur during early diagenesis in *Sphagnum* peat: Insights from $\delta^{34}S$ ratio profiles in ^{210}Pb-dated peat cores. Limnol Oceanogr 39:1172–1185

Novak M, Bottrell SH, Fottova D, Buzek F, Groscheova H, Zak K (1996) Sulfur isotope signals in forest soils of Central Europe along an air pollution gradient. Environ Sci Technol 30:3473–3476

Novák M, Buzek F, Harrison AF, Plechová E, Jaãková I (1999) $\delta^{34}S$ speciation in five European spruce forest soils in relation to vertical $\delta^{13}C$ and $\delta^{15}N$ ratio profiles. Oecologia (submitted)

Pearson J, Stewart GR (1993) The deposition of atmospheric ammonia and its effects on plants. New Phytol 125:283–305

Perez-Soba M, Stulen I, van der Eerden LJM (1994) Effect of atmospheric ammonia on the nitrogen metabolism of Scots Pine (*Pinus sylvestris*) needles. Physiol Plant 90:629–636

Rennenberg H, Kreutzer K, Papen H, Weber P (1998) Consequences of high loads of nitrogen for spruce (*Picea abies*) and beech (*Fagus sylvatica*) forests. New Phytol 139:71–86

Schimel DS (1995) Terrestrial ecosystems and the carbon cycle. Global Change Biol 1:77–91

Schulze E-D (1994) The impact of increased nitrogen deposition on forests and aquatic ecosystems. Round Table Talk. Nova Acta Leopoldina 288:415–436

Schurr U (1998) Xylem sap sampling – new approaches to an old topic. Trends Plant Sci 3:293–298

Taylor AFS, Högbom L, Högberg M, Lyon AJE, Näsholm T, Högberg P (1997) Natural ^{15}N abundance in fruitbodies of ectomycorrhizal fungi from boreal forests. New Phytol 136:713–720

Van Stempvoort DR, Reardon EJ, Fritz P (1990) Fractionation of sulfur and oxygen isotopes in sulphate by soil sorption. Geochim Cosmochim Acta 54:2817–2826

Yanagisawa F, Sakai H (1983) Precipitation of SO_2 for sulphur isotope ratio measurements by the thermal decomposition of $BaSO_4$-V_2O_5-SiO_2 mixtures. Anal Chem 55:985–987

Part C
Heterotrophic Processes

10 Soil Respiration in Beech and Spruce Forests in Europe: Trends, Controlling Factors, Annual Budgets and Implications for the Ecosystem Carbon Balance

G. Matteucci, S. Dore, S. Stivanello, C. Rebmann, and N. Buchmann

10.1 Introduction

The two main processes involved in forest growth are photosynthesis and respiration. During photosynthesis, forest ecosystems absorb carbon dioxide from the atmosphere, while respiration releases CO_2 into the atmosphere. Ecosystem respiration includes autotrophic and heterotrophic processes: the first is caused by the growth and maintenance of plant tissues, while the second is mainly the result of the decomposition of litter and soil organic matter brought about by microbial biomass.

At the level of soils, CO_2 emission is caused by both plant and microbial processes, namely root respiration and soil/litter organic matter decomposition, and has been reported to make up 60–80% of total ecosystem respiration (Meir et al. 1996; Janssens et al. 2000a; Law et al. 1999). Generally, the majority of heterotrophic respiration in forest ecosystem is caused by soil and litter microorganisms (Waring and Schlesinger 1985). At the global level, total soil carbon efflux has been estimated to range from 50 and 75 $Gt\,C\,a^{-1}$ (Raich and Schlesinger 1992), an amount comparable to that of global gross primary production. Furthermore, soils have been reported to be a key factor for the overall response of the terrestrial biosphere to global change (Anderson 1992), particularly in the northern region, where they contain a large fraction of soil organic matter (SOM) as labile fraction, more prone to decomposition (Anderson 1992; Vogt et al. 1995; Schlesinger 1997).

Recent findings on annual ecosystem fluxes, measured above 17 European forests within the EUROFLUX network, have shown that while gross primary production tended to be constant across sites, annual ecosystem respiration increased with latitude despite the general decrease of mean annual air temperature (Valentini et al. 2000) and resulted in being driven by multiple factors. Nevertheless, the effective temperature sensitivity (Q_{10}) of soil organic matter decomposition is much higher in colder than in warmer climates and temperature increases in cold regions are likely to affect decomposition rates more than net primary productivity (Kirschbaum 1995). In these regions, forest ecosystems seem to be very vulnerable to changing climate: for example, warm winters tend to turn old boreal stands from a carbon sink to a carbon source due to increased annual respiration (Lindroth et al. 1998), while year-to-year variation in the timing of thawing of the soil in spring have been reported to play a major role for ecosystem carbon balance (Goulden et al. 1998).

In individual forest ecosystems, the range of annual soil respiration has been reported to vary between 180 and 1510 $g\,C\,m^{-2}\,a^{-1}$ (Raich and Nadelhoffer 1989). At the annual or seasonal time scales, soil respiration is driven mainly by climatic variables, such as soil temperature and soil moisture. Nevertheless, soil CO_2 efflux shows a large variability, linked to the seasonality of root and microbial processes and of litter input, that can confound the temperature/moisture dependence of soil respiration. Furthermore, soil respiration frequently shows a very pronounced microsite variability, probably linked to the heterogeneity of soil conditions within a system (Dugas 1993; Jensen et al. 1996; Janssens et al. 1998).

Hence, it is important to investigate soil CO_2 efflux and to understand better the factors that drive the most relevant respiration process in forest ecosystems. Furthermore, its determination is crucial to address the issues of ecosystem carbon exchange, particularly if "comprehensive carbon budgets" have to be estimated (IGBP Terrestrial carbon working group 1998).

Three objectives were addressed in the present study: (1) to investigate the processes underlying soil respiration; (2) to identify climatic, physical and biological variables that determine carbon dioxide fluxes from soil and litter in beech and spruce stands across Europe; (3) to calculate annual budgets of soil respiration and to compare them with the available total net ecosystem exchange fluxes and with values of soil carbon mineralisation determined with different techniques (see Harrison et al., Chap. 11 and Persson et al., Chap. 12, this Vol.).

10.2 Approaches to Measuring Soil Respiration

Soil respiration fluxes were measured at three sites, two spruce stands (*Picea abies*; WAL, Waldstein, Germany; MDM, Monte di Mezzo, Italy) and one beech forest (*Fagus sylvatica* L.; COL, Collelongo, Italy) of the CANIF transect. In the present study, another site was used: Renon, a mixed coniferous forest dominated by spruce in northern Italy, where concurrent measurements of net ecosystem fluxes and soil respiration were available.

All sites are described in detail in Persson et al. (Chap. 2, this Vol.), but here it should be mentioned that soil respiration rates at Waldstein were measured mostly in a 47-year-old stand, near the flux tower of the EUROFLUX project. The other measurements performed within the CANIF project at WAL and presented in other chapters of this book were carried out in a 142-year-old nearby spruce stand (200 m distance).

At all sites, soil respiration was measured with portable infrared gas analysers and chambers based on the principle of closed dynamic systems. At the three Italian sites, the EGM-1 analyser and a SRC-1 chamber were used (PP-Systems, Hitchin, UK), while at the German site, measurements were performed with a Li-6400 analyser and the corresponding soil chamber (LI-6400-09, LiCor Inc., Lincoln, Nebraska, USA).

The EGM-1 is a portable infrared gas analyser connected to the SRC-1 soil respiration chamber and to the STP-1 sensor for soil temperature. The chamber is cylindrical, with a volume of 1170 cm^3, a base area of 78.5 cm^2 and a diameter of 10 cm. A small fan in the chamber ensures air mixing. Though a stainless steel perimeter ring at the bottom of the chamber should ensure a good sealing between chamber and soil during measurements, in the present study, rigid PVC collars (diameter = 10 cm; height = 8 cm) were inserted in the soil until the upper end reached the humus layer (just below the litter layer). At two Italian sites, COL and MDM, the collars were inserted after the first year of measurements, while at the third site (Renon), the collars were used from the beginning of the experiment. At all three sites, after the insertion, the collars remained in place for the duration of the experiment.

At the German site (WAL), soil respiration measurements were performed with the LI-6400, a portable infrared gas analyser that can be connected to the LI-6400-09 soil CO_2 flux chamber and a temperature probe. The chamber resembles the one used in Italy, with a volume of 991 cm^3, a base area of 71.6 cm^2 and a diameter of 9.55 cm. It is a closed dynamic system, though the chamber has a pressure relief valve to provide pressure equilibration with the atmosphere when the chamber is inserted into the soil or installed on a soil collar. The infrared gas analyser is on the top of the chamber in a sensor head. PVC collars (diameter and height 10 cm) were inserted in the soil 24 h before each measuring date.

10.2.1 Auxiliary Measurements

Concurrently with soil respiration, soil temperature and soil water content were measured at all sites. Soil temperature (Ts) was determined with the built-in temperature probes of the analysers, usually at a depth of 5 cm. At Waldstein, soil temperatures at 10 and 15 cm depth were also measured.

Additionally, soil temperature was measured continuously by sensors and data loggers at a 30-min time step at COL and WAL and daily averages were calculated. At MDM, only air temperature was measured continuously. Hence, a continuous record of soil temperature was obtained applying a linear regression between the air temperature and the available measurements of Ts (Ts = $0.6462 \cdot$ Tair $- 0.1725$, $r^2 = 0.90$, $p < 0.05$).

At COL, MDM and Renon, during respiration measurements, soil water content (vol%, 0–20 cm soil depth) was measured using a portable instrument based on the TDR technique (Trime, IMKO GmbH, one measurement for each sampling point). At WAL, gravimetric water content was determined by weighing and oven-drying soil samples to weight constancy (three replicates). For this site, data are presented as percentage of water weight over wet soil weight, that can be converted to volume content using soil bulk density. In addition, at COL, soil water content measurements (TDR) were available independently on a fortnightly basis.

10.2.2 Sampling Schemes

At COL, soil respiration was monitored between 1996 and 1998. Measurements were performed at 10 dates in 1996 (from May to November, minimum and maximum intervals were 14 and 30 days, respectively); 15 dates in 1997 (Feb.–Dec., min-max intervals 8 and 45 days) and 10 dates in 1998 (April–Nov., min-max. intervals 8 and 40 days). To account for the spatial heterogeneity of the soil, the number of sampling points varied between 30 (1996) and 18–20 (1997–98), although replicates were less during snow cover. At MDM, a spruce plantation on former agricultural land with sufficiently homogenous soil conditions, 10 sampling points were selected, and a total of 13 campaigns were performed between June 1996 and October 1998. Ten sampling points were selected at Renon, where soil respiration was monitored monthly during 1998. At these three Italian sites, single measurements were made at each of the sampling points usually during mid-morning. Daily courses of soil respiration were measured at all dates at Renon and occasionally at COL and MDM.

At WAL, four sampling points were selected the day before measurements, and 24 h after collar installation a daily course was monitored on each point. Each flux rate was calculated as the mean of five observations per measurement. At this site, soil respiration rates were determined monthly between April and October 1998.

10.2.3 Flux Measurement Using the Eddy Covariance Technique

WAL (spruce), COL (beech) and Renon (mixed conifers) are also flux stations of the EUROFLUX network. At these sites, ecosystem fluxes of CO_2 and H_2O are measured using the eddy covariance technique. Details of the technique can be found elsewhere (Baldocchi et al. 1988); theory and experimental setup have been described in detail by Aubinet and colleagues (2000). The flux stations measure the net flux of carbon entering or leaving the ecosystem, providing a measure of net ecosystem exchange (NEE), and, if summed annually, providing a direct estimate of the annual ecosystem carbon balance. As the eddy covariance system measures the net flux across a certain measurement plane, it is difficult to separate the contribution of ecosystem components to these fluxes. However, at night, only respiration fluxes are measured, which can be correlated with climatic variables to derive regressions and calculate total ecosystem respiration on a daily and/or seasonal/annual basis.

In this respect, it must be indicated that one of the major limitations of the eddy covariance technique is the measurement of the carbon exchange during periods of atmospheric stability, such as those that can be present during calm nights and during the transitions to dawn and sunset. Under these conditions, the flux measured by the eddy covariance (EC) system may be different from the real net ecosystem exchange (Baldocchi et al. 1996; Goulden et al. 1996; Grace et al. 1996). In practice, it can be possible that the CO_2 respired by the system (particularly that of the soil) accumulates below the EC system and is only partly measured by it. This can be due to instrumental reasons (insufficient frequency response of the analysers during the transition phases from atmospheric stability to instability) or to site topographic conditions, when they are suitable to phenomena such as katabatic flows that are not detected by the instrument (Moncrieff et al. 1996). The above problems can be partly overcome by adding the CO_2 storage flux, that can be measured thorough concentration vertical profiles, to the flux measured by the EC system (Baldocchi et al. 1996; Moncrieff et al. 1996). This flux is usually negligible during daytime and windy nights, but can be relevant under atmospheric stability and at dawn and sunset. The atmospheric turbulence can be expressed by the wind friction velocity (u^*, $m\,s^{-1}$), that indicates the turbulence exchange between the air and the canopy. Usually, for $u^* > 0.2–0.4\,m\,s^{-1}$, the conditions are favourable for EC measurements, while below that threshold, measurements often need to be corrected. In all the sites of the EUROFLUX network, corrections have been applied during non-optimal turbulence conditions, in order to arrive at reliable respiration and net ecosystem budgets (Aubinet et al. 2000; Valentini et al. 2000).

The eddy covariance technique allows measurements of surface fluxes from a relatively large area, dependent on the height at which the sonic anemometer is placed. If an eddy covariance system is placed near the ground (at 1.5–2.5 m height), CO_2 fluxes from that surface can be estimated. In forests with absent or negligible understorey vegetation and herb layer, below tree canopy measurements can give an area-integrated value of soil respiration that can be compared to the values obtained by aggregation of soil chamber measurements (Goulden et al. 1996; Law et al. 1999). Such an experiment was performed in May 1997 at Collelongo, where a sonic anemometer was placed at 2 m height for 3 days (below-canopy system), while another EC system measured fluxes above the canopy. Concurrently, soil respiration rates were measured using the EGM system, approximately every hour at 13 locations randomly distributed in the footprint area of the below-canopy system.

10.3 Daily and Seasonal Trends in Soil Respiration and Climatic Variables

Typical daily trends in soil respiration rates (Rs) measured at single sampling points (*microsite*) are presented in Fig. 10.1. At all sites (Renon, WAL and COL), the daily variation in Rs at each sampling point was much smaller than the variation among sampling points at the same sites, and the variation among sampling points at the same site (local variability) was often as high as the variability among sites. This pattern was confirmed at all dates and sampling hours. At WAL and Renon, for the day presented in Fig. 10.1, the average coefficient of variation of single sampling points was respectively, 2.6% (range 1.5–3.5%, four points) and 6.1% (range 5–10.2%, ten points), while the coefficient of variation of all the measurements and all points was 21.1% (WAL, $n = 20$) and 49.2% (Renon, $n = 40$). At the beech forest (COL), soil respiration rates were measured from dawn to complete darkness in spring, and the coefficient of variation of single sampling points was larger than in the spruce sites (20.7%, range 14.5–27%, eight microsites), but the coefficient of variation across all measurements was still larger than that of single sampling points (31%, $n = 64$).

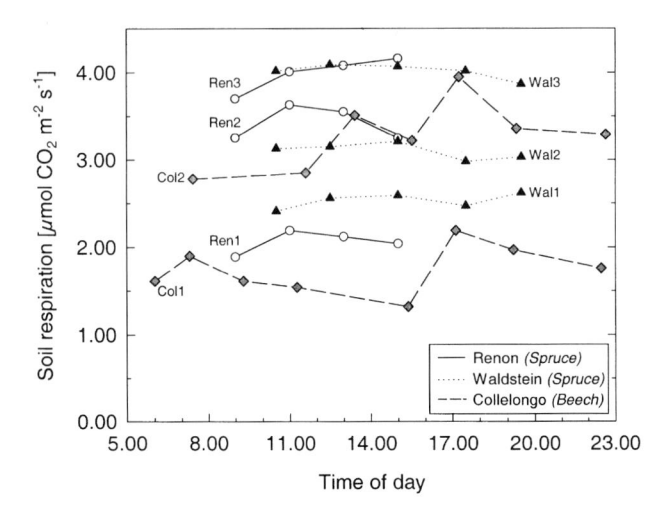

Fig. 10.1. Typical daily trends of soil respiration (Rs) for single sampling points at Renon (spruce, *open symbols, continuous line*), Waldstein (WAL, spruce, *closed symbols dotted lines*) and Collelongo (COL, beech, *closed symbols, dashed line*). Measurements performed on 17/10/98 (Renon), 14/07/98 (WAL) and 22/05/97 (COL). For all the microsites, soil respiration ranges were as follows: 1.9–10.1 µmol CO_2 m^{-2}s^{-1} for Renon (ten sampling points, $n = 40$); 2.4–4.1 µmol CO_2 m^{-2}s^{-1} for WAL (four sampling points, $n = 20$) and 1.3–6.4 µmol CO_2 m^{-2}s^{-1} for COL (eight sampling points, $n = 64$)

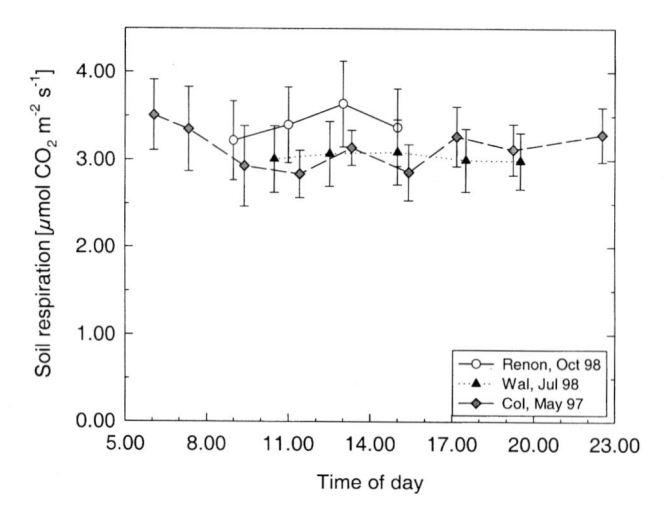

Fig. 10.2. Average daily trends of soil respiration rates (Rs) for Renon (*open circles, continuous line*), Waldstein (WAL, *closed triangles, dotted lines*) and Collelongo (COL, *closed diamonds, dashed line*). Each data point is the average of all sampling points measured at the sampling hour. *Error bars* represent one standard error. Measurements were performed on 17/10/98 (Renon), 14/07/98 (WAL) and 22/05/97 (COL)

For all sites, the local variability of Rs could not be explained by the corresponding differences in soil temperatures, which were limited to a maximum of 1.5 °C. A temperature variation of this magnitude was found at single points along the daily course, explaining the limited temporal variability at this scale. Local variability tended to be larger in spring and summer than in winter and, at all sampling dates, it was larger than the daily variability at single points (data not shown).

Mean daily courses of soil respiration for all sites reflected the limited temporal and the large spatial variability of the measurements performed at single points (Fig. 10.2). Over the day, the daily coefficient of variation was 1.5% (WAL), 5.1% (Renon) and 7.4% (COL), while for individual hours it was, on average, 23.6, 35.1 and 30.1% for the spruce sites of Renon and WAL and for the beech forest of COL, respectively. These results are in agreement with other reports (Dugas 1993; Jensen et al. 1996; Janssens et al. 1998; Buchmann 2000), confirming the need for a large number of replicates to characterise the site-specific soil respiration rates and to scale it up to ecosystem level (Jensen et al. 1996). The spatial variability of soil respiration rates within a site is not fully explained by temperature, hence other factors must be considered, such as soil moisture, changes in root density (Janssens et al. 1998), soil carbon content, litter layer thickness and litter C/N ratios (Janssens et al. 2000a).

The number and location of sampling points at the four sites were determined after a detailed survey of site conditions and their variability. Hence, it was possible to calculate, for single dates, the average in soil respiration of the sampling points that should reasonably represent the soil respiration of a site. It is then possible to present the seasonal trends of soil respiration, where each point represents the sampling date

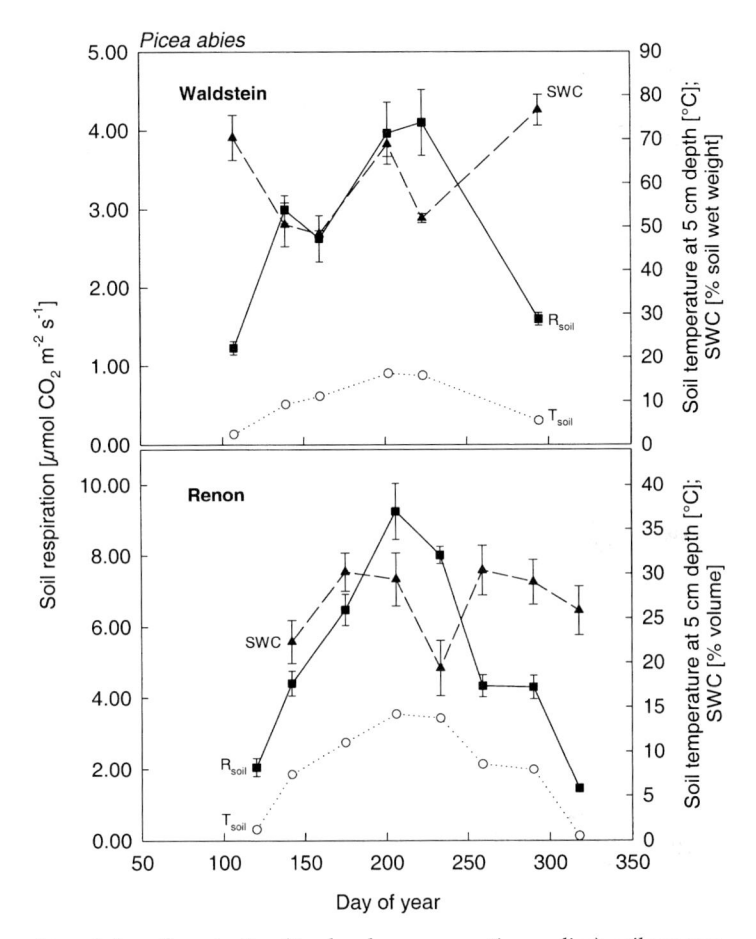

Fig. 10.3. Seasonal trends in soil respiration (Rs *closed squares, continuous line*), soil temperature at 5 cm depth (Ts *open circles, dotted line*) and soil water content (SWC *closed triangles, dashed line*) for the spruce stands of Waldstein (WAL, *top panel*) and Renon (*bottom panel*). Each point is the average of bi-hourly measurements performed at four to five different times of day on four (WAL) and ten (Renon) microsites. Soil water content was measured by gravimetric method on the Of and Oh soil layers at WAL and by the TDR technique in the first 20 cm. *Error bars* represent 1 standard error

average. Furthermore, providing that the number of sampling points was representative of the site characteristics, the limited diurnal variability of Rs (see Figs. 10.1 and 10.2) permits the calculation of a daily average even though the measurements were not always performed as part of daily courses.

The seasonal trends in soil respiration (Rs), soil temperature (Ts) and soil water content (SWC) resulted very similar at WAL and Renon (Fig. 10.3) with soil respiration peaking in summer between mid July and mid August (day 200–250), and showing autumn-winter minima. Nevertheless, soil CO_2 fluxes values are very

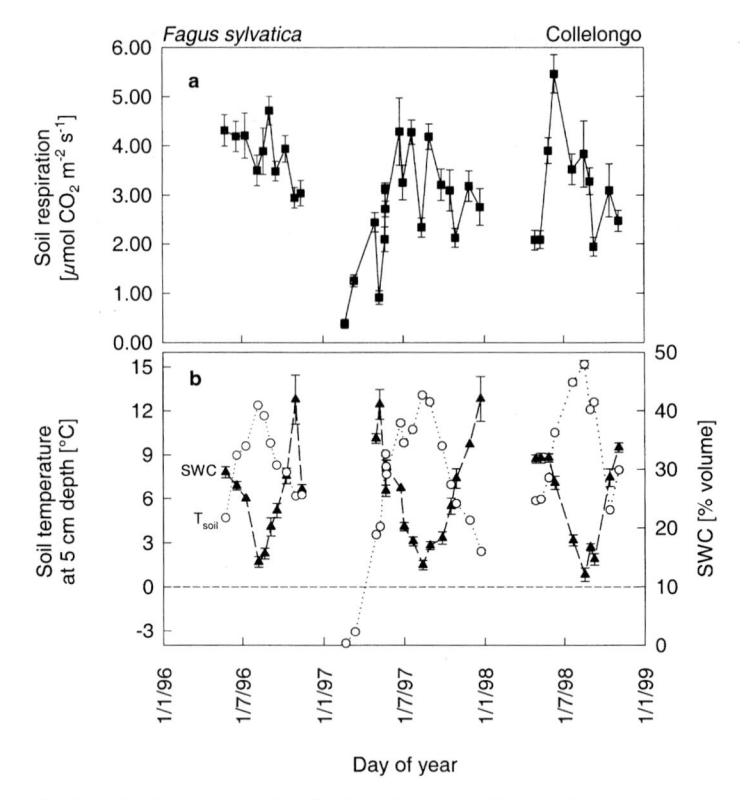

Fig. 10.4a,b. Seasonal trends of soil respiration (a: Rs *closed squares, continuous line*), soil temperature at 5 cm depth (panel **b**, Ts, *open circles, dotted line*) and soil water content (panel **b**, SWC, *closed triangles, dashed line*) for the beech forest of Collelongo (COL). Generally, each point is the average of 15–30 microsites measured at mid-morning. Soil water content was measured by the TDR technique in the first 20 cm of soil. *Error bars* represent 1 standard error

different ranging from 1.23 ± 0.08 to $4.1 \pm 0.41\,\mu\mathrm{mol}\,CO_2\,m^{-2}\,s^{-1}$ at WAL and from 1.45 ± 0.11 to $9.24 \pm 0.79\,\mu\mathrm{mol}\,CO_2\,m^{-2}\,s^{-1}$ at Renon. At each sampling date, Renon was characterised by nearly doubled Rs compared to WAL, probably due to the important contribution of sampling points with a thick litter layer, which, on average, increased the daily values at Renon by 30%. At both sites, Rs appears to be controlled mainly by Ts, as shown by the parallel evolution of the respective trends. Soil water content was always non-limiting, apart from minor effects at minima in June, at WAL (day 160) and in August (day 233) at Renon. At those dates, Rs showed a slight decrease despite an increase in temperature (WAL) or a constant temperature (Renon) compared to previous dates.

The trend of Rs for the beech site (COL, Fig. 10.4a) shows that, when this parameter is measured frequently, the seasonal trend is confounded by a pronounced variability at shorter time scales. During the 3 years of measurements, soil respiration at COL ranged from $0.38 \pm 0.09\,\mu\mathrm{mol}\,CO_2\,m^{-2}\,s^{-1}$ when measured over a complete snow

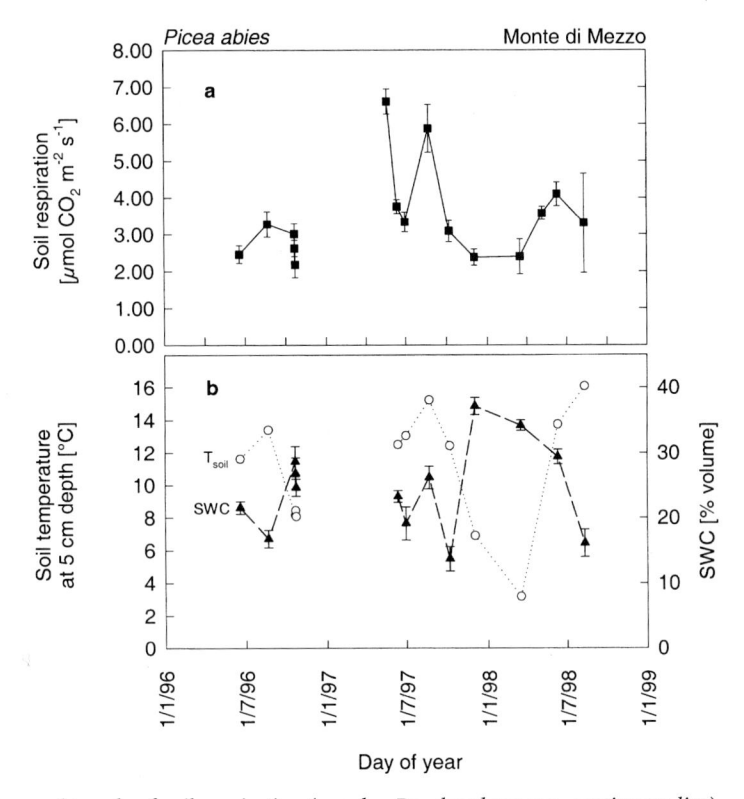

Soil Respiration in Beech and Spruce Forests in Europe 225

Fig. 10.5a,b. Seasonal trends of soil respiration (panel **a**, Rs, *closed squares, continuous line*), soil temperature at 5 cm depth (panel **b**, Ts, *open circles, dotted line*) and soil water content (panel **b**, SWC, *closed triangles, dashed line*) for the spruce plantation of Monte di Mezzo (MDM). Generally, each point is the average of ten microsites measured at mid-morning. Soil water content was measured by the TDR technique in the first 20 cm of soil. *Error bars* represent 1 standard error

cover (February 1997), to 5.47 ± 0.39 µmol CO_2 $m^{-2} s^{-1}$ in early June 1998 when the canopy was in full development. Throughout the 3 years, Rs showed spring and summer peaks, although the latter occurred usually after rain events. Minimum values of soil respiration were measured in winter above the snow cover, as well as in early spring and autumn. The trend of Ts (Fig. 10.4b) was less variable than that of Rs, indicating that not only Ts was influencing Rs at this site. Nevertheless, COL differed from the northern spruce sites (WAL and Renon), in that soil water content exerted a strong influence on Rs. Low respiration values (around 2 µmol CO_2 $m^{-2} s^{-1}$) were measured in summer 1997 and 1998, when SWC was below 20% and Ts was at its maximum (Fig. 10.4b). Compared to the following years, soil respiration during summer 1996 was less variable, probably due to more favourable SWC after the frequent afternoon rains.

At the spruce site of Monte di Mezzo (MDM), soil respiration (Fig. 10.5a) peaked in spring (6.61 ± 0.34 µmol CO_2 $m^{-2} s^{-1}$ in May 1997; 4.10 ± 0.32 µmol CO_2 $m^{-2} s^{-1}$ in

June 1998) and in summer after rain events (August 1997), whereas minimum values were observed in autumn and winter (around 2.3 µmol CO_2 m^{-2} s^{-1}). At the MDM site, Rs seemed to be driven primarily by soil temperature, while the role of SWC appeared less important compared to the beech site of COL (Fig. 10.5b). Moreover, it must be stressed that the summer sampling at MDM was less frequent than at COL, thus final conclusions about the importance of SWC in influencing Rs at this site cannot be drawn.

The soil CO_2 efflux data reported here are in line with previous results, e.g. from a mixed forest in Tennessee (0.8–5.7 µmol CO_2 m^{-2} s^{-1}, Hanson et al. 1993). Generally, for temperate forests, soil CO_2 emissions can vary between 0.6 and 14.8 µmol CO_2 m^{-2} s^{-1}, with the majority of sites ranging between 0.6 and 3.2 µmol CO_2 m^{-2} s^{-1} (Singh and Gupta 1977). Obviously, the seasonality of root and microbial activity plays a major role in the seasonal patterns of soil respiration, as does the input of litter. Both roots and microorganisms respond to climatic fluctuations, but temporal changes in their physiological activity may confound the relationships between Rs, temperature and soil water content.

10.4 Factors Controlling Soil Respiration

Even if soil respiration is dependent on a number of factors, soil temperature and water content are the two parameters most closely related to it. Therefore, soil respiration has been frequently modelled using these two climatic variables, with a major emphasis on temperature (Singh and Gupta 1977; Schlenter and Van Cleve 1985; Raich and Schlesinger 1992; Hanson et al. 1993; Lloyd and Taylor 1994; Davidson et al. 1998; Epron et al. 1999).

In forests where the influence of water stress is negligible, soil temperature was reported to explain up to 90% of soil respiration variability and exponential equations are generally used to describe the Rs vs. Ts relationship, particularly in the form of the Q_{10} function (Boone et al. 1998; Davidson et al. 1998; Epron et al. 1999; Janssens et al. 2000a; but see also discussion in Lloyd and Taylor 1994).

The regressions between soil respiration rates and soil temperature (at 5-cm depth) differ among sites (Fig. 10.6). To exclude confounding effects and to allow for a better comparison among data sets, data under limiting soil water content were not included in this analysis (for COL). In the absence of water stress, Q_{10} exponential regressions in the form:

$$Rs = R_{10} \cdot Q_{10}^{\frac{(Ts-10)}{10}} \tag{1}$$

proved to explain a large part of the variance in soil respiration rates at all sites, with r^2 ranging from 0.69 (MDM) to 0.89 (COL). Only for the WAL site, a simple exponential regression [$Rs = a \cdot e^{(b \cdot Ts)}$] did fit slightly the data better than a Q_{10} relationship ($r^2 = 0.80$ vs. 0.72). The parameters of the Q_{10} regression equations are presented in Table 10.1. The Q_{10} values, ranging between 2.09 (MDM) and 2.21 (COL), fall well within the reported range of soil respiration Q_{10}, that, for a range of ecosystems, most of which were forests, resulted in a median of 2.4 (range 1.3–3.3, Raich and Schlesinger 1992). Recently, higher Q_{10} values were reported for six different sites in a temperate forest in the USA (3.4–5.6, Davidson et al. 1998), while Q_{10} values between 2 and 6

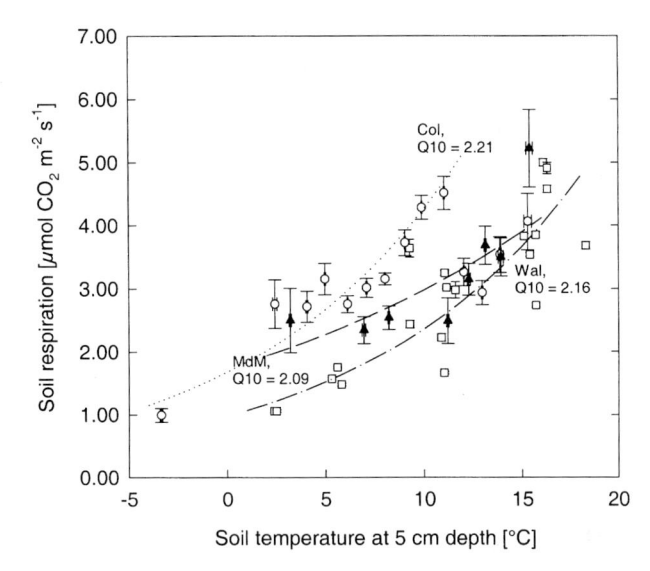

Fig. 10.6. Relationships between soil respiration and soil temperature (−5 cm depth) for the spruce sites of Waldstein (WAL, *open squares, dotted dashed line*, $r^2 = 0.80$) and Monte di Mezzo (MDM, *closed triangles, dashed line*, $r^2 = 0.64$) and for the beech site of Collelongo (COL, *open circles, dotted line*, $r^2 = 0.88$). Data collected in 1998 for WAL and in 1996–1998 for MDM and COL. The *single points* represent average of hourly measurements (WAL) or data averaged according to soil temperature classes of 1–2 °C (Renon, MDM and COL). *Error bars* represent 1 standard error. Lines are exponential regressions in the form $Rs = a \cdot e^{(b \cdot Ts)}$

Table 10.1. Parameters of the Q_{10} regressions fitted on the soil respiration data of three sites. Data are presented in Fig. 10.6. The equation used has the form $Rs = R10 \ast Q10^{[(Ts-10)/10]}$ and was calculated using the SYSTAT statistical package

Site	R_{10} μmol CO$_2$ m^{-2}s^{-1}	Q_{10}	r^2
Waldstein, spruce	2.50 ± 0.18	2.16 ± 0.29	0.72
Collelongo, beech	4.11 ± 0.13	2.21 ± 0.25	0.89
Monte di Mezzo, spruce	2.93 ± 0.36	2.09 ± 0.48	0.69

have been reported for a number of European forest ecosystems studied within the EUROFLUX project (Janssens et al. 2000a).

However, in summer, soil water content was a significant determinant of soil respiration rates for the Italian beech (COL) and, partially, spruce (MDM) sites. At COL, soil respiration increased with soil temperature until it reached 11 °C (Fig. 10.6), while measurements made in the summer at higher soil temperature (12–15 °C), when soil water content was below 20%, were from 10 to 35% lower than those measured at 11 °C. Considering the entire soil temperature and soil moisture ranges, the relationship of Rs and Ts was better explained by a sigmoid function ($r^2 = 0.72$) than using

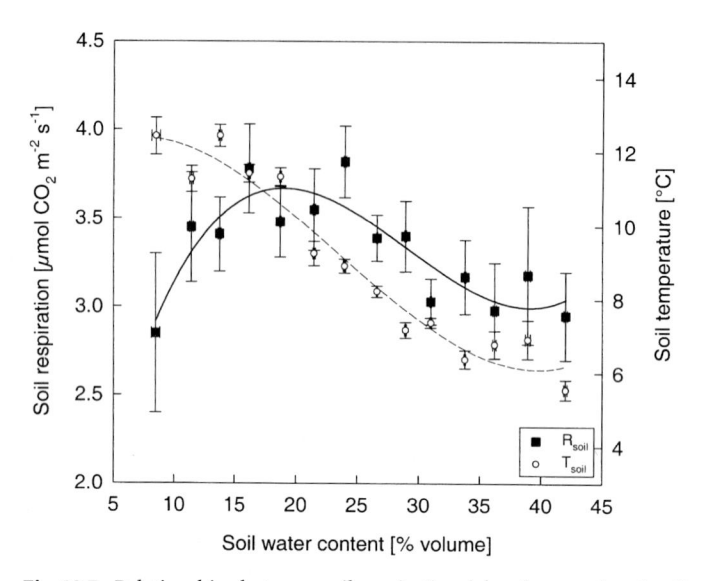

Fig. 10.7. Relationships between soil respiration (*closed squares*) and soil water content for the beech site at COL. Data were collected during 3 years (1996–1998). The *single points* represent data averaged according to soil water content classes of 2.5%. Soil temperatures corresponding to each SWC class are also shown (*open circles*). *Error bars* represent 1 standard error. Lines are freely drawn to enhance the trend

a Q_{10} one ($r^2 = 0.55$). At the COL site, Rs showed a positive relationship with increasing SWC until it reached 22.5% (Fig. 10.7). This can be considered a direct effect of SWC on Rs, as Ts was almost constant or decreasing in the range of 0–22.5% SWC range. At higher SWC values, Rs tended to decrease to an almost constant level for SWC greater than 30%. However, this decrease could not be fully ascribed to an effect of SWC, as, in that range of SWC, Ts was concurrently decreasing with increasing SWC (Fig. 10.7), in a way similar to that described by Davidson and colleagues (1998). For the COL beech site, during periods of water stress, the use of models that include both soil temperature and soil water content increased the explained variance of Rs. Anyhow, for the entire data range, at COL, the high variability in Rs, Ts and SWC at the short time scale (see Fig. 10.4a) resulted in a general modest fit of models such as those proposed by Hanson ($r^2 = 0.52$, Hanson et al. 1993) and Epron ($r^2 = 0.49$, Epron et al. 1999). At this site, the better fitting of Rs data was obtained by the concurrent use of a sigmoid function for the period with non-limiting SWC and the Epron model for the periods with water stress. At the southern spruce site (MDM), the role of SWC in governing soil respiration was not as evident as in the beech forest, maybe for the particular soil conditions and the reduced sampling intensity. In any case, at MDM, Rs data measured in the growing season and averaged for SWC increments of 2.5% showed an increase of Rs with SWC until it reached 40% volume. Nevertheless, the changes in SWC almost paralleled that of Ts (Fig. 10.5b), so that direct effects of Ts on Rs could not be ruled out. At this site, the fitting of the data of MDM with a model that included also soil water content (Hanson et al.

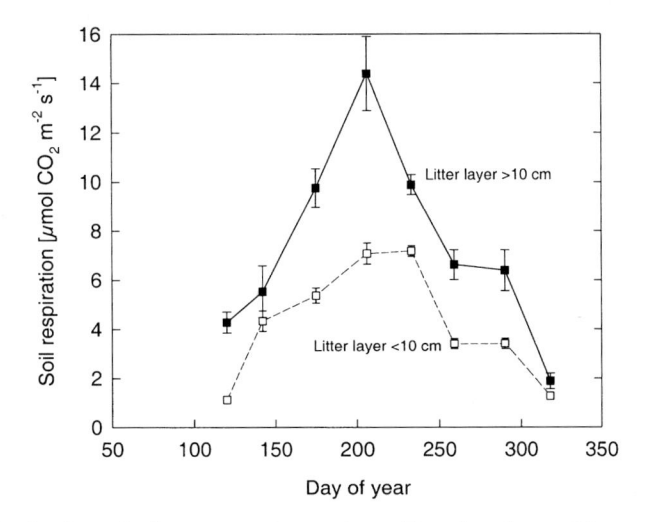

Fig. 10.8. Seasonal trends of soil respiration measured on points with a litter layer >10 cm (*closed squares*) and with an average litter layer (<10 cm, *open squares*) at the Italian spruce site at Renon. *Each point* is the average of bi-hourly measurements performed at four to five different times of day on three points. *Error bars* represent 1 standard error

1993) explained soil respiration data better than the simple Q_{10} function ($r^2 = 0.75$ vs. 0.69).

The interaction of soil temperature and soil water content on soil respiration has been pointed out by several authors and, often, the seasonal course of soil respiration was better explained with models that were dependent on both parameters (Schlenter and Van Cleve 1985; Hanson et al. 1993; Davidson et al. 1998; Epron et al. 1999). The inclusion of soil moisture in models explaining soil respiration is particularly important in summer and for mediterranean sites. The results reported here for the beech forest, with low Rs values in summer associated to dry soil conditions, are similar to those frequently observed in other temperate (Froment 1972; Burton et al. 1998; Davidson et al. 1998; Epron et al. 1999) and mediterranean forests (Virzo de Santo et al. 1976; Dore et al. 1998).

There are a number of additional factors which can influence soil CO_2 efflux, one of which being the amount organic carbon which is available for decomposition. An important part of this carbon is the labile fraction that is prone to rapid turnover and to which the litter layer contributes mostly (Vogt et al. 1995).

At Renon, where three of the ten sampling points were located in areas characterised by a thicker than average litter layer (>10 cm), an analysis of the effect of the litter layer thickness on soil respiration is made possible (Fig. 10.8). Soil respiration at sampling points with a thicker litter layer were systematically higher than the Rs of the others. The differences were statistically significant at $p < 0.05$ for all but the last sampling date. Similar results were found in a temperate forest, where plots with a double litter input showed systematically higher soil respiration rates than control plots (Boone et al. 1998).

10.5 Comparison of Chamber Measurements with the Eddy Covariance Measurements Below the Canopy

During May 1997, an experiment was performed at COL comparing below-canopy eddy flux measurements with soil chambers. At this site, the understorey vegetation consists of a sparse herb layer, which most probably gives a negligible contribution to the overall ecosystem flux.

Measurements were taken for 3 days (Fig. 10.9) and the range of variation of respiration rates was larger for the eddy covariance system (EC, 0.25–4.47 µmol CO_2 $m^{-2} s^{-1}$) than for the average of soil chambers (SC, 1.45–3.51 µmol CO_2 $m^{-2} s^{-1}$). This can be caused by the different area of integration of the two systems. Furthermore, for the EC, the source area of the fluxes changes according to wind direction, while the soil chambers are always measuring, time by time, at the same 13 sampling points. For the entire period, the average respiration rate was 1.70 ± 0.09 µmol CO_2 $m^{-2} s^{-1}$ for the EC ($n = 60$) and 2.82 ± 0.10 µmol CO_2 $m^{-2} s^{-1}$ for the SC (SC, $n = 20$). Hence, EC measured a 40% lower respiration rate than SC. These results agree perfectly with another study performed in a Belgian forest using a similar setup for both EC and SC measurements (Janssens et al. 2000b). However, a more thorough data analysis showed that the two systems measured similar Rs rates when the turbulence, estimated by friction velocity, mixed the air layers sufficiently above the soil (Fig. 10.9).

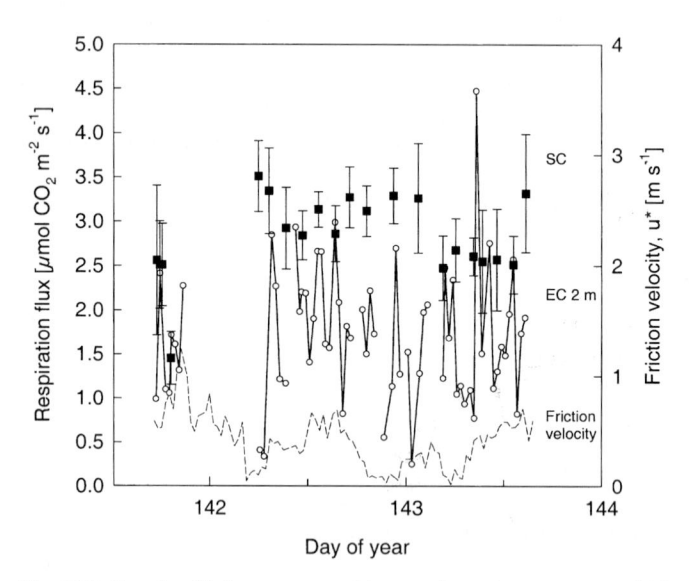

Fig. 10.9. Trends of below-canopy eddy covariance measurements (EC, *open circles*), soil respiration measured with chambers (SC, *closed squares*) and friction velocity (*dotted line*) for the period 21–23/05/1997 at the Collelongo beech forest. Below-canopy fluxes were measured with a 30-min time step, and each point is an average for this period. SC measurements were performed at 13 locations within the footprint area of the EC system. The measurements took approximately 30 min and each point is the average over all 13 locations. Friction velocity gives an estimate of the turbulence. For SC measurements, *error bars* represent 1 standard error

This is particularly true at the beginning and end of the experiment and confirms results from other groups (Black et al. 1996; Goulden et al. 1996; Lindroth et al. 1998). Differences between estimates of soil respiration using scaled chamber measurements and eddy covariance have been found in other studies. Generally, these differences are low under turbulent conditions and more important under conditions of CO_2 accumulation within the canopy. Under these conditions, on one hand the eddy covariance systems may not measure the real flux, while, on the other hand, the ventilated soil chambers can overestimate the fluxes because they may enhance the usual low diffusion rate of CO_2 from the litter. It has been argued that, at the same site, also with differences present, both measurements could be correct if the EC system integration is not matched by the aggregation scheme of SC measurements (Goulden et al. 1996).

10.6 Annual Budgets of Soil Respiration

From seasonal measurements of soil respiration, it is possible to calculate an annual budget of the amount of carbon lost by the soil respiration processes by interpolation of the measuring points (Hanson et al. 1993; Davidson et al. 1998) or by using regressions with temperature alone or in combination with soil moisture (Davidson et al. 1998; Epron et al. 1999). For a proper interpolation, measurements of soil respiration by SC need to be spaced representatively in time, to reflect the real seasonal trend of soil respiration rates. At sites where soil respiration is largely driven by soil temperature, the difference between annual totals calculated by interpolation or by regression is usually minor (Davidson et al. 1998). However, at sites where soil water content plays an important role, differences between the two methods can be important, particularly in periods of water limitation (Schlenter and Van Cleve 1985; Hanson et al. 1993; Epron et al. 1999).

Annual budgets of soil respiration measured by SC, have been calculated for the three sites for which soil temperature was available all year round (Table 10.2). Depending on the availability of climatic data, the amount of CO_2 emitted by the soils has been calculated by regressions based on soil temperature alone (WAL) or in combination with water content (COL and MDM).

The values for the spruce sites WAL and MDM as well as for the beech site COL fall within the reported ranges for temperate coniferous (250–1300 $g C m^{-2} a^{-1}$) and deciduous forests (300–1414 $g C m^{-2} a^{-1}$, Raich and Schlesinger 1992). The value of WAL (711 $g C m^{-2} a^{-1}$) is very close to the average of coniferous forests (681 ± 95 $g C m^{-2} a^{-1}$), while the central Italy spruce (MDM, 803 $g C m^{-2} a^{-1}$) and beech sites (COL, 879 $g C m^{-2} a^{-1}$) show values larger than the average (645 ± 51 $g C m^{-2} a^{-1}$). For these three sites, the comparison between interpolated and modelled values of total soil respiration resulted always in a 20–40% overestimation of the yearly budget by the simple interpolation method. It should be also mentioned that, in those sites where the daily course is significant, measuring Rs at mid-morning and extrapolating the obtained value to the entire day could lead to an overestimation of soil respiration budgets (Davidson et al. 1998).

Differences between the soil respiration budgets could be site-specific or methodological. At WAL, soil respiration was measured with a LiCor system, while at the other two sites, an EGM-1 from PP-Systems was used. The two systems have been reported to differ systematically when measuring at the same points sequentially. In fact, in the

Table 10.2. Annual totals of soil respiration ($gCm^{-2}a^{-1}$) for three sites. The budgets were calculated by exponential regression with soil temperature (WAL) and regressions with both soil temperature and soil water content (COL and MDM). At COL, two regressions were applied, a sigmoid function with temperature during the dormant season and the model from Epron (Ts and SWC) for the growing season. Total ecosystem respiration values (TER) measured by eddy covariance (EC) above the canopy and of soil carbon mineralisation (average by the ^{14}C technique, Harrison et al., Chap. 11, this Vol., and the incubation method, Persson et al., Chap. 12, this Vol.) are also presented. The values of TER are for the same period of total Rs for COL, for 1997 for WAL

Site	Total Rs $gCm^{-2}a^{-1}$	Means Ts (°C)	Method to calculate the annual budget	TER, EC $gCm^{-2}a^{-1}$	Year	C mineralisation $gCm^{-2}a^{-1}$
Waldstein, spruce	711.1	6.5	T function (exponen.)	1300	1998	262
Collelongo, beech	879.2	2.5	Mixed (T, T + SWC)	640	96–97	205
Monte di Mezzo, spruce	803.2	5.6	Hanson model	n.a.	1997	406

comparison made in summer in a Belgian pine-oak forest, EGM-1 always measured a respiration rate nearly double that of the LiCor analyser (Janssens et al. 2000b). It is not certain if such a difference can be extended to other sites and periods of the year; but if this is the case, the yearly totals of the Italian sites would have been significantly diminished if determined by the LiCor system.

Yet the annual soil respiration budgets can also be compared with the total ecosystem respiration (TER) measured by the eddy covariance system above the canopy (Table 10.2). Two of the four sites (WAL and COL) were part of the EUROflUX network, and TER was calculated by extrapolation of night-time measurements to the entire day using soil temperature, followed by summation of all daily values (Valentini et al. 2000). For this purpose, all data had been corrected when measured under stable conditions (see Aubinet et al. 2000 for details).

For WAL, the total soil respiration estimate based on aggregated soil chamber measurements is lower than TER, contributing to it by 55%. For the beech sites of COL, total Rs is 20% larger than TER. In this site, the reported difference between respiration totals estimated by aggregated chamber measurements and eddy covariance has been confirmed (Davidson et al. 1998; Law et al. 1999). However, since the same procedure was used to calculate TER both at WAL and COL, it is likely that total Rs by soil chambers is somewhat overestimated at the beech site. As an example, winter soil respiration added up to $133 gCm^{-2}$, while, in the same period, TER, which includes also a contribution from plant biomass, was $110 gCm^{-2}$ (Matteucci 1998). Another clue about the possible overestimation of Rs budget by soil chambers is that COL, in the year of interest, resulted in active uptake of carbon, at a rate of $660 gCm^{-2}a^{-1}$ (Matteucci 1998; Valentini et al. 2000). This would mean that its total photosynthesis (gross primary production) should have been too high with respect to reported values for the site type to account for the estimated soil respiration, the measured carbon uptake and the above-ground biomass respiration (usually 20–30% of TER).

The values of total soil respiration can be also compared to those of soil carbon mineralisation (root activity excluded), that were measured by the ^{14}C bomb carbon techniques (see Harrison et al., Chap. 11, this Vol.) and by the incubation method (see Persson et al., Chap. 12, this Vol.) and then scaled-up to ecosystem scale. For the three sites, the average carbon mineralisation by the two methods is 262, 205 and 406 g C $m^{-2} a^{-1}$ at WAL, COL and MDM, respectively (Table 10.2). These values do not include any root activity and respiration, which have been reported to contribute between 30 and 60% to soil respiration in closed forest ecosystems (Bowden et al. 1993; Kelting et al. 1998).

The values of annual soil respiration and soil carbon mineralisation calculated here indicate a contribution of root respiration of 50% for MDM and 63% for WAL, in line with the reported range. It must be remembered that, at WAL, the site of soil respiration measurements is different from that of carbon mineralisation estimates, thus preventing a more detailed comparison of the two techniques for this site. Nevertheless, in July 1998, soil respiration rates measured at the two WAL stands (47 and 142 years old) were not statistically different (two-tail T test, $p > 0.3$), although the older stands showed a higher mean respiration.

For COL, the contribution of roots to annual soil respiration is in the order of 77%. The value of COL appears larger than expected, possibly due to an overestimation of annual soil respiration by soil chambers.

These findings point to the need of being cautious when scaling up and aggregating measurements to yield ecosystem-valid soil respiration budgets, by whatever method they are made (chambers, eddy covariance or carbon mineralisation estimates). However, the use of different techniques can help in understanding the process and in limiting the uncertainties linked to its measurement (Janssens et al. 1999b).

10.7 Conclusions

The work performed within the CANIF project on soil respiration has brought a better understanding of the process, its determinants and of the uncertainties linked to the technique used.

At all sites, soil respiration showed a larger spatial (among sampling points) than temporal (over the day) variability. Sometimes, spatial variability exceeded also differences between sampling dates. When seasonal changes in soil CO_2 fluxes were dependent mainly on temperature a single summer maximum was observed. At sites where also soil water content exerted a role, the seasonal trend was much more variable, with a succession of higher and lower values in spring and summer. At all sites, soil respiration was minimal in winter.

When the measurement campaigns were spaced at shorter time intervals, a larger variability was detected, pointing to the need of a careful planning of sampling campaigns in order to detect that variability.

Soil temperature and soil water content were found to be the major determinants of soil respiration. At the northern sites, 80–90% variations in soil respiration rates were explained by soil temperature, while at the two sites in central Italy, soil water content was also important, particularly in summer. At one of the sites, the amount of litter strongly affected soil respiration, resulting in higher rates where the litter layer was thicker.

The comparison of chamber methods and below-canopy eddy covariance resulted in an overall difference of fluxes of the order of 40%, with the chamber method giving higher soil respiration rates. When the turbulence conditions were more suitable for eddy covariance measurements, the two methods agreed reasonably well.

Annual totals were calculated using regressions. A comparison of the soil respiration data with those of carbon mineralisation seems to indicate a possible overestimation of soil respiration budgets when aggregating the results from soil chambers, at least at one of the sites (COL). Several reasons may lead to this overestimation, and the general conclusion is that selection of sampling points, frequency of campaigns, auxiliary parameters to be measured and aggregation scheme used to scale to the ecosystem level must be designed with attention and profound knowledge of the sites, if an accurate estimation is to be obtained.

Acknowledgments. The help and assistance of the National Forest Service department of Avezzano (AQ) and of people on duty at the Collelongo Forest Service station and at the Collelongo township administration are acknowledged. Research at the MdM site was possible with the assistance of the ex-ASFD administration of Castel di Sangro (AQ). The technical help of Mr. Renato Zompanti, Mr. Tullio Oro (DISAFRI – University of Tuscia, Viterbo, Italy) and Mr. Roberto Bimbi (CNR-IFA, Rome, Italy) is also gratefully acknowledged.

References

Anderson JM (1992) Responses of soils to climate change. Adv Ecol Res 22:163–210

Aubinet M, Grelle A, Ibrom A et al. (1999) Estimates of the annual net carbon and water exchange of forests: the EUROFLUX methodology. Adv Ecol Res 30:113–175

Baldocchi DD, Hicks BB, Meyers TP (1988) Measuring biosphere-atmosphere exchanges of biologically related gases with micrometeorological methods. Ecology 69(5):1331–1340

Baldocchi DD, Valentini R, Running S, Oechel W, Dahlman R (1996) Strategies for measuring and modelling carbon dioxide and water vapour fluxes over terrestrial ecosystems. Global Change Biol 2:159–167

Black TA, den Hartog G, Neumann HH, Blanken PD, Yang PC, Russell C, Nesic Z, Lee X, Chen SG, Staebler R, Novak MD (1996) Annual cycles of water vapour and carbon dioxide fluxes in and above a boreal aspen forest. Global Change Biol 2:219–229

Boone RD, Nadelhoffer KJ, Canary JD, Kaye JP (1998) Root exert a strong influence on the temperature sensitivity of soil respiration. Nature 396:570–572

Bowden RD, Nadelhoffer KJ, Boone RD, Melillo JM, Garrison JB (1993) Contributions of aboveground litter, belowground litter, and root respiration to total soil respiration in a temperate mixed hardwood forest. Can J For Res 23:1402–1407

Buchmann N (2000) Biotic and abiotic factors modulating soil respiration rates in *Picea abies* stands. Soil Biol Biochem (in press)

Burton AJ, Pregitzer KS, Zogg GP, Zak DR (1998) Drought reduces root respiration in sugar maple forests. Ecol Appl 8:771–778

Davidson EA, Belk E, Boone RD (1998) Soil water content and temperature as independent or confounded factors controlling soil respiration in a temperate mixed hardwood forest. Global Change Biol 4:217–227

Dore S, Muratore G, Tirone G (1998) Emissioni di anidride carbonica dal suolo: confronto tra una faggeta e una lecceta. In: Borghetti M (ed) La ricerca italiana in selvicoltura ed ecologia forestale. Proc Ist Congr of SISEF, Padova, 4–6 Giugno 1997, pp 57–62

Dugas WA (1993) Micrometeorological and chamber measurements of CO_2 flux from bare soil. Agric For Meteorol 67:115–128

Epron D, Farque L, Lucot E, Badot P-M (1999) Soil CO_2 efflux in a beech forest: dependence on soil temperature and soil water content. Ann For Sci 56:221–226

Froment A (1972) Soil respiration in a mixed oak forest. Oikos 23:273–277
Goulden ML, Munger JW, Fan S-M, Daube BC, Wosfy WC (1996) Measurements of carbon sequestration by long-term eddy covariance: methods and critical evaluation of accuracy. Global Change Biol 2:169–181
Goulden ML, Wosfy WC, Harden JW, Trumbore SE, Crill PM, Gower ST, Fries T, Daube BC, Fan S-M, Sutton DJ, Bazzaz A, Munger JW (1998) Sensitivity of boreal forest carbon balance to soil thaw. Science 279:214–217
Grace J, Malhi Y, Lloyd J, McIntyre J, Miranda AC, Meir P, Miranda HS (1996) The use of eddy covariance to infer the net carbon dioxide uptake of Brazilian rain forest. Global Change Biol 2:209–217
Hanson PJ, Wullschleger SD, Bohlmann SA, Todd DE (1993) Seasonal and topographic patterns of forest floor CO_2 efflux from an upland oak forest. Tree Physiol 13:1–15
IGBP Terrestrial carbon working group (1998) The terrestrial carbon cycle: implications for the Kyoto protocol. Science 280:1393–1394
Janssens IA, Barigah T, Ceulemans R (1998) Soil CO_2 efflux rates in different tropical vegetation types in French Guyana. Ann Sci For 55:671–680
Janssens IA, Dore S, Epron D, Lankreijer H, Buchmann N, Longdoz B, Montagnani L (2000a) Soil respiration: a summary of results from the EUROflUX sites. In: Valentini R (ed) Biospheric exchanges of carbon, water and energy from European forests. Final Report of the EUROFLUX project, EC, Brussels
Janssens IA, Kowalski AS, Longdoz B, Ceulemans R (2000b) Assessing forest soil CO_2 efflux: an in situ comparison of four techniques. Tree Physiol 20:23–32
Jensen LS, Mueller T, Tate KR, Ross DJ, Magid J, Nielsen NE (1996) Soil surface CO_2 flux as an index of soil respiration in situ: a comparison of two chamber methods. Soil Biol Biochem 28:1297–1306
Kelting DL, Burger JA, Edwards GS (1998) Estimating root respiration, microbial respiration in the rhizosphere, and root-free soil respiration in forest soils. Soil Biol Biochem 30:961–968
Kirschbaum MU (1995) The temperature dependence of soil organic matter decomposition, and the effect of global warming on soil organic C storage. Soil Biol Biochem 6:753–760
Law BE, Ryan MG, Anthoni PM (1999) Seasonal and annual respiration of a ponderosa pine ecosystem. Global Change Biol 5:169–182
Lindroth A, Grelle A, Morén A-S (1998) Long-term measurements of boreal forest carbon balance reveal large temperature sensitivity. Global Change Biol 4:443–450
Lloyd J, Taylor JA (1994) On the temperature dependence of soil respiration. Funct Ecol 8:315–323
Matteucci G (1998) Bilancio del Carbonio in una Faggeta dell'Italia Centro-Meridionale: Determinanti Ecofisiologici, Integrazione a Livello di Copertura e Simulazione dell'Impatto dei Cambiamenti Ambientali. PhD Thesis, Università degli Studi di Padova. Padova
Meir P, Grace J, Miranda A, Lloyd J (1996) Soil respiration in a rainforest in Amazonia and in cerrado in central Brazil. In: Gash JHC, Nobre CA, Roberts JM, Victoria RL (eds) Amazonian deforestation and climate. Wiley, New York, pp 319–329
Moncrieff JB, Malhi Y, Leuning R (1996) The propagation of errors in long-term measurements of land atmosphere fluxes of carbon and water. Global Change Biol 2:231–240
Raich JW, Nadelhoffer KJ (1989) Belowground carbon allocation in forest ecosystems: global trends. Ecology 70(5):1346–1354
Raich JW, Schlesinger WH (1992) The global carbon dioxide flux in soil respiration and its relationship to vegetation and climate. Tellus 44B:81–99
Schlenter RE, Van Cleve K (1985) Relationship between CO_2 evolution from soil, substrate temperature, and substrate moisture in four mature forest types in interior Alaska. Can J For Res 15:97–106
Schlesinger WH (1997) Biogeochemistry: an analysis of global change. Academic Press, San Diego, California, pp 161–165
Singh JS, Gupta SR (1977) Plant decomposition and soil respiration in terrestrial ecosystems. Bot Rev 53(4):449–528

Valentini R, Matteucci G, Dolman AJ et al. (2000) Respiration as the main determinant of European forests carbon balance. Nature (in press)

Virzo de Santo A, Alfani A, Sapio S (1976) Soil metabolism in beech forests of Monte Taburno (Campania Apennines). Oikos 27:144-152

Vogt KA, Vogt DJ, Brown S, Tilley JP, Edmonds RL, Silver WL, Siccama TG (1995) Dynamics of forest floor and soil organic matter accumulation in boreal, temperate, and tropical forests. In: Lal R, Kimble J, Levine E, Stewart BA (eds) Soil management and greenhouse effect. CRC Press, Boca Raton, pp 159-178

Waring RH, Schlesinger WH (1985) Forest productivity and succession. In: Waring RH, Schlesinger WH (eds) Forest ecosystems. Concepts and management. Academic Press, San Diego, pp 38-69

11 Annual Carbon and Nitrogen Fluxes in Soils Along the European Forest Transect, Determined Using ^{14}C-Bomb

A.F. Harrison, D.D. Harkness, A.P. Rowland, J.S. Garnett, and P.J. Bacon

11.1 Introduction

The production of forest ecosystems and their responses to environmental factors are highly dependent on the sustainable functioning of the soils on which they grow, in particular the processes of organic matter and nutrient recycling through the soil component of the ecosystem. Soils in forest ecosystems are important reservoirs of C and N. In the contexts of two major environmental issues of (1) climate change feedback effects on carbon sequestration and (2) pollutant deposition effects on forest ecosystem function, it is essential therefore to be able to quantify the fluxes of C and N through the biologically active layers of the whole soil profile (not just the litter layer) in forests and relate the fluxes to climate and pollutant deposition. The key questions are: how can the C and N fluxes of the biologically active layers of the whole soil profile be determined and the methodology be carried out without inducing experimental disturbance effects?

During the period from ca. 1954 to 1962, there was a massive input of ^{14}C into the earth's atmosphere, resulting from nuclear weapons testing. This input nearly doubled the concentration of ^{14}C in the atmosphere, thus creating a global-scale tracer for monitoring carbon dynamics in the terrestrial and oceanic environment. Through this, it is now possible to compute mean residence times of young contemporary soil organic matter, because of the steady decrease in the ^{14}C-bomb concentration in the atmosphere, largely governed by its uptake into the oceans (Broecker and Olson 1960; Nydal et al. 1980). Changes in ^{14}C-bomb concentrations over time have been well documented (Walton et al. 1970; Baxter and Walton 1970; Nydal et al. 1980; Ingeborg and Kromer 1997) and this variation is accurately and rapidly reflected in the ^{14}C-specific activity of the contemporary plant tissues (Baxter and Walton 1971).

As was first recognised by Jenkinson (1963), this ^{14}C-bomb has offered a unique opportunity to be used as a tracer in studies of soil organic matter dynamics (e.g. Rafter and Stout 1970; Martel and Paul 1974; Jenkinson and Rayner 1977; O'Brien and Stout 1978; Ladyman and Harkness 1980; Stout and Goh 1980; O'Brien 1984).

In the early 1970s, we began monitoring the incorporation of ^{14}C-bomb into the litter and organic matter of a brown earth soil profile in a mixed deciduous woodland, Meathop wood; this wood was studied intensively as part of the International Biological Programme (e.g. Satchell 1971). The objective of the research was to explore the potential of ^{14}C-bomb incorporation to reveal mean residence times of soil organic matter (Harkness et al. 1986). By simply modelling the rates of changes of ^{14}C-bomb in the organic matter of different soil layers over time (Fig. 11.1), it was

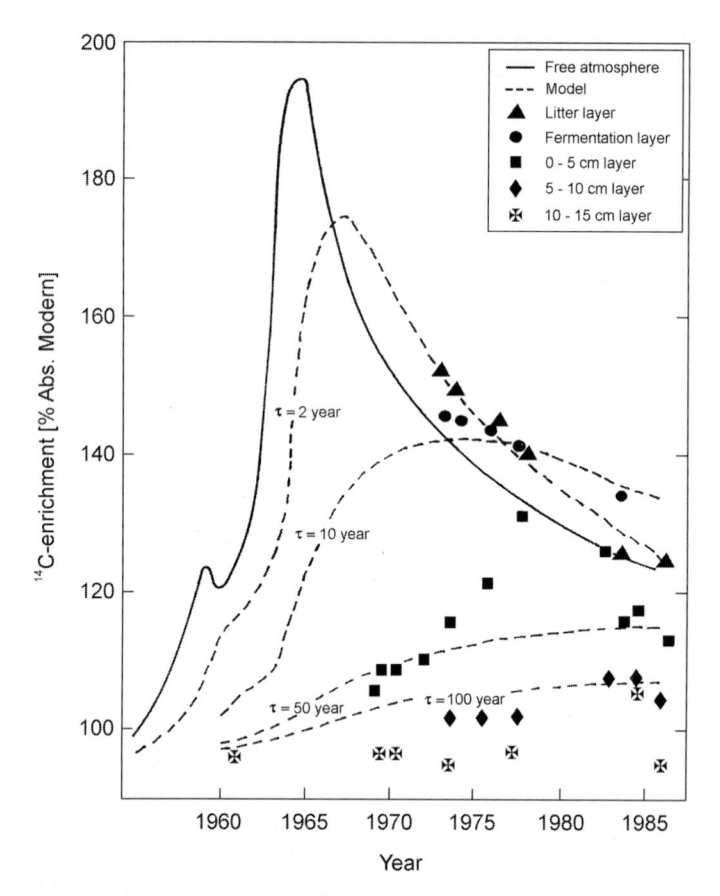

Fig. 11.1. Temporal changes in ^{14}C-bomb in the soil profile of Meathop Wood, UK (Harrison and Harkness 1993); *solid line* free atmosphere; *dotted line* model evaluation for specific mean residence times (MRT; τ); *triangles* litter layer; *circles* fermentation layer; *squares* 0–5-cm layer; *diamonds* 5–10 cm layer; *crosses* 10–15 cm layer

possible to generate mean residence times (MRTs) for the forest litter and surface mineral soil layers with a precision of a few years. Using these MRTs, estimates of the net annual fluxes of C, N and P through the soil profile were calculated, which very closely agreed with those derived from studies of ecosystem budgets (Harrison et al. 1990; Harrison and Harkness 1993). We have extended this approach to quantify the annual fluxes of C and N through the active layers of the soil profile in the CANIF (Project Acronyon) forests.

The key objectives of the research have been to:

1. estimate the total soil C and N fluxes under spruce and beech sites,
2. relate C and N fluxes to trends in climate, pollutant inputs and site characteristics across the transect, and
3. provide data for application in forest modelling, either for soil subroutine development or for model validation (see Harrison and Harkness 1993).

11.2 Forests, Sampling Procedure and Analysis

11.2.1 Forests

All forests in the project transect were sampled, namely Aheden (AhP), Skogaby (SkP), Gribskov (GrF), Waldstein (WlP), Schacht (ScF), Nacetin (NaP), Aubure spruce (AuP-90 years old), Aubure beech (AuF), Monte di Mezzo (MdMP) and Collelongo (CoF); P = *Picea abies*, F = *Fagus sylvatica*. Some of the key data and characteristics of these sites are given in Table 11.1. See Chapter 2 (this volume) for detailed descriptions of the sites.

11.2.2 Sampling Procedure

Nine 10.3-cm diameter cores to 20-cm depth within the mineral profile were sampled at random in the experimental area of each site, except for Skogaby, where three cores were sampled from each of the control plots in blocks 2, 3 and 4. The CV% for total carbon content for eight out of the ten sites was between 11.8 and 19.1, with extremes of 5.5 (Aheden) and 25.4 (Skogaby); the relative uniformity of the sites justified, retrospectively, the standardised sampling intensity. To increase the accuracy of the estimates of the L+F layer mass, where the majority of soil carbon and nitrogen turnover occurs, all L+F material was also removed from nine random 0.25-m² (0.5 × 0.5 m) plots on each site, hand-sorted, dried and weighed. Fresh litter samples were selected from material caught in nets or litter traps and provided by local researchers. In some instances some samples were collected in a year prior to the soil sampling. Allowance was made for the difference in the ^{14}C signature for the year of sampling prior to use in the model.

Table 11.1. Forest site attributes used for interpreting the patterns in the soil C and N pools and fluxes

Forest	Latitude (°N) LAT	Altitude (m asl*) ALT	Mean annual Temperature (°C) MAT	Rainfall (mm) RI	Total N deposition (kg ha⁻¹ a⁻¹) M_{dep}	Total S deposition (kg ha⁻¹ a⁻¹) S_{dep}	Soil pH (0–10 cm depth)
Spruce							
AhP	64°13′	175	1.0	588	1.7	5.8	4.45
SkP	56°33′	115	7.6	1240	18.9	13.8	4.01
WlP	50°12′	700	5.5	890	20.1	17.0	3.52
NaP	50°36′	775	5.9	935	18.2	42.4	3.47
AuP	48°12′	1050	5.4	1190	14.7	11.8	3.71
MdMP	41°45′	905	8.5	1030	10.8	10.7	6.92
Beech							
GrF	55°58′	45	7.5	630	11.6	10.8	4.07
ScF	50°12′	850	5.5	890	20.1	17.0	3.99
AuF	48°12′	1000	5.4	1190	14.7	11.8	3.57
CoF	41°52′	1560	6.4	1010	10.8	10.7	5.69

* asl = above sea level.

11.2.3 Preparation of Soil and Calculation of Mass

Each core was separated into the L+F, 0–5, 5–10 and 10–20 cm layers. Stones (>2 mm) were removed by sieving to leave the <2-mm fraction, which was air-dried at room temperature before analysis. Prior to drying, living plant material was removed from the L+F layer samples and all visible root material was removed from mineral soil samples; this procedure was carried out very thoroughly by hand-picking on soil sub-samples (10–20 g) analysed for ^{13}C and ^{14}C enrichments. Weight and volume of stones and dry weight of root mass were recorded. Moisture content of the air-dried soil was determined on subsamples oven-dried at 105 °C. Bulk density was calculated as the mass of oven-dry <2-mm fraction divided by the volume of the soil layer minus the volume of the stones contained in that layer. Estimates of soil mass in 10^3 kg ha^{-1} were calculated for each soil layer. As estimates of the masses of the L+F layer derived from the two sets of samples differed, weighted mean values were calculated on the basis of the relative area sampled and were used in all computations.

11.2.4 Chemical Analysis

The milled L+F material and <2-mm fraction for each layer of all nine cores were analysed for organic carbon content using the modified Tinsley method (Kalembasa and Jenkinson 1973) and for nitrogen using the H_2O_2/conc. H_2SO_4 digestion procedure (Rowland and Grimshaw, 1985) followed by continuous-flow colorimetric analysis using the indophenol blue method (Allen 1989).

11.2.5 Isotope Analyses

Determinations of ^{14}C and ^{13}C concentrations were carried out on four random replicate cores for the L+F and 0–5-cm layers, where the relative variability among soil replicates was expected to be greatest, and two replicate cores from the 5–10 and 10–20 cm layers where variation was expected to be lower, giving a total of 12 analyses per site. All samples were analysed for ^{14}C concentration by accelerator mass spectrometry (AMS) (Pilcher 1991). Subsamples of dried sample were oxidised to CO_2 in evacuated flame-cleaned quartz combustion tubes with CuO and Ag foil in a muffle furnace over 12 h, with a temperature peak of 950 °C. The CO_2 was recovered and purified by selective cryogenic trapping (Boutton et al. 1983). Aliquots of the CO_2 were converted to an iron/graphite mixture using a Fe/Zn reduction procedure (Slota et al. 1987) and prepared as accelerator targets for ^{14}C measurement at the University of Arizona NSF Accelerator Facility. Subsamples of the CO_2 were analysed for $\delta^{13}C_{PDB}$ using a VG Optima mass spectrometer. The ^{13}C values were used for normalisation of the measured ^{14}C compensating for isotopic fractionation effects prior to the calculation of the ^{14}C concentrations (Stuiver and Polach 1977; Donahue et al. 1990).

In addition, for all sites ^{14}C concentrations were measured for each freshly fallen tree litter sample collected. The ^{14}C concentrations of these fresh fall samples were determined by radiometric counting of benzene synthesised from the component carbon (Harkness and Wilson 1972), to allow analysis of relatively large samples to ensure representative compositions and high analytical precision in the estimations of carbon age in the litter input to the forest floor, the second input function of the MRT model.

11.3 Model Description

The incorporation of bomb ^{14}C within the soil matrix is a dynamic process. At a particular time the ^{14}C enrichment is determined by (1) the previous temporal variations in the atmospheric ^{14}C which condition the ^{14}C activity in the annual litter inputs, and (2) the progressive delays incurred during the passage of the ^{14}C sequentially downwards through the successive soil compartments or boxes. A four-box model (Fig. 11.2) was developed to represent the flow of carbon through each CANIF forest soil from the tree canopy input down to the 10–20 cm soil layer, initially presumed to be the lowest depth where significant C flux activity would occur. A series of mean residence time curves reflecting the time courses of change in ^{14}C concentration within each component carbon box over previous years including the years of sampling were generated, as in the Meathop Wood study model (Harkness et al. 1986), using the equation:

$$A_t = A_{(t-1)} e^{-k} + (1-e^{-k}) A_i, \qquad \text{(Eq. 1)}$$

where A_t is the ^{14}C enrichment in a given year, $A_{(t-1)}$ that established for the previous year; A_i is the contemporaneous activity established for the preceding carbon box and k the corresponding exchange rate constant for carbon leaving the box, i.e. $1/k = T$ (years). The appropriate MRT curve for the first box (fresh tree litter) was determined by matching the ^{14}C activity of the litter input with a series of options generated using the atmospheric ^{14}C curve as the input function with different values of k. The MRT curve for the succeeding box (L+F layer) was similarly determined by matching the measured ^{14}C activity with the appropriate MRT from the series of curves generated, using the MRT trend established for the preceding (tree litter) box as the input func-

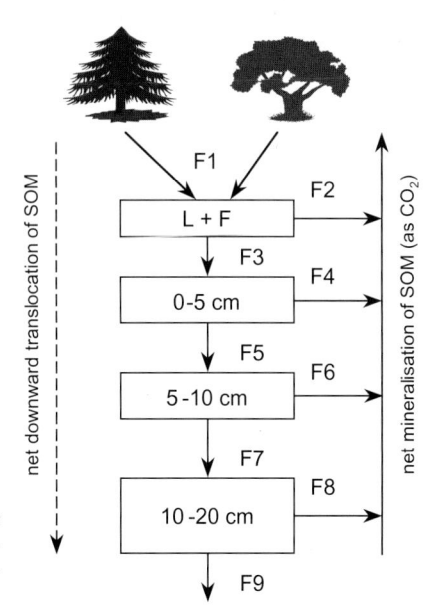

Fig. 11.2. Diagrammatic representation of the model used to compute soil carbon mean residence times (MRT) from ^{14}C-bomb concentrations and carbon fluxes in and between layers

tion, i.e. to define A_i in Eq. (1). The temporal trend in ^{14}C concentrations thus established for the L+F box was then taken as the input function in determining the MRT for the 0–5 cm layer of the mineral soil profile. The process was repeated sequentially down through the soil profile layers. For each of the forest sites, there was only one combination of MRTs derived from the model which could account for its ^{14}C concentrations measured in the respective soil layers, for the year in which sampling took place, i.e. the model outputs were unambiguous.

For three of the forests, Gribskov, Aubure spruce and Monte di Mezzo, the pattern of ^{14}C bomb concentration with depth indicated some penetration below 20 cm (see MRT values versus depth in Table 11.2). For these sites only, MRTs for the 20–50 cm soil layer were derived using estimates of the ^{14}C concentration in the layer produced by extrapolation of relationships of ^{14}C concentration to soil depth.

11.4 Estimations of C and N Pools and Fluxes

Estimates of the carbon and nitrogen pool in each soil layer, expressed as kg ha^{-1}, were generated by multiplying its %C and %N concentrations by the estimated soil (L+F or <2-mm fraction) mass, respectively. Estimates of net annual C and N flux in each soil layer for each of the forest sites were derived according to Eq. 2 and Eq. 3 respectively:

$$\text{C flux (kg C ha}^{-1} \cdot \text{a}^{-1}) = \text{C pool (kg C ha}^{-1}) / \text{C mean residence time (years)} \quad \text{Eq. 2}$$

$$\text{N flux (kg N ha}^{-1} \cdot \text{a}^{-1}) = \text{N pool (kg N ha}^{-1}) / \text{C mean residence time (years)} \quad \text{Eq. 3}$$

Throughout the text, the term flux is used to denote turnover in, or flow through, the organic matter in a soil layer. In deriving estimates of fluxes, it has been assumed that the soil is in steady state and therefore inputs to soil layers equal the export from those layers, such that:

Box	Input flux	Export fluxes	Defined turnover rate
L + F	F1	F2 + F3	$M_{(L+F)}/MRT_{(L+F)}$
0–5 cm	F3	F4 + F5	$M_{(0-5\,cm)}/MRT_{(0-5\,cm)}$
5–10 cm	F5	F6 + F7	$M_{(5-10\,cm)}/MRT_{(5-10\,cm)}$
10–20 cm	F7	F8 + F9	$M_{(10-20\,cm)}/MRT_{(10-20\,cm)}$

where F1 = flux of C falling as fresh litter onto the forest floor; F2 = flux of C respired from the L+F layer to the atmosphere; F3 flux of C translocated downwards from the forest floor into the 0–5 cm soil layer; the logic is repeated for each soil layer. M = mass of carbon and MRT = carbon mean resedence time. To obtain the estimate of the total flux of carbon or nitrogen pool in the whole soil profile, the fluxes of all the soil layers are summed.

11.5 Pools and Distribution of Carbon and Nitrogen in Soil Profiles

The total carbon pool in soil profiles to 20-cm depth ranges from 31.6×10^3 (Aheden) to ca. 104×10^3 (Nacetin) $kg\,C\,ha^{-1}$ (Fig. 11.3). There are no statistically significant trends with latitude, altitude or mean annual temperature across all the forest sites, but there is with annual rainfall (Fig. 11.4a). The total nitrogen content in the soil profiles to the same depth, ranges from 1.34×10^3 (Aubure) to 7.6×10^3 (Monte-di-Mezzo) $10^3\,kg\,N\,ha^{-1}$ (Fig. 11.3). In contrast to carbon, there is a significant ($p < 0.01$) increase in the soil total N content in a southerly direction (Fig. 11.4b) and with altitude ($TN_{20\,cm} = 0.0027\,ALT + 2.0$; $r^2 = 0.43, p < 0.05$) along the transect, but not with either mean annual temperature or rainfall. The relationship between the total amounts of carbon and total nitrogen in the soil profiles to 20-cm depth is not significant, because the data sets for Monte di Mezzo and Collelongo deviate from the general trend. If the data for these two forests are excluded, the relationship between total carbon and nitrogen to 20 cm depth becomes highly significant ($C_{20\,cm} = 23.2\,TN_{20\,cm} - 1.77$, $r^2 = 0.85$, $p < 0.01$). The deviation of Monte di Mezzo and Collelongo from the general trend is probable determined by their higher soil pH (Table 11.1). The C:N ratio for the whole soil profile to 20-cm depth, however, significantly ($p < 0.01$) decreases with latitude in a southerly direction (Fig. 11.4c) and with altitude ($C:N_{20\,cm} = 0.0081\,ALT + 26.26$; $r^2 = 0.49$, $p < 0.05$) along the transect. The N content of the soil to 20-cm depth both increases significantly with increasing soil pH ($TN_{20\,cm} = 1.33\,pH - 1.78$; $r^2 = 0.55, p < 0.05$), whilst the C:N ratio of the soil to 20 cm decreases with pH ($C:N_{20\,cm} = 37.0 - 3.84\,pH$, $r^2 = 0.59, p < 0.01$); these relationships are mainly determined by the higher pHs of soils at the Collelongo and Monte di Mezzo sites. The total carbon in the L+F plus soil to 5-cm depth was related ($p < 0.05$) to the total N deposition, but the trend was not significant when the data point for Aheden was omitted.

The distribution of carbon and nitrogen pools is somewhat different between the forests (Fig. 11.3). The most obvious difference is that there is a greater proportion of the C and N deeper in the soil profile at Collelongo and Monte di Mezzo sites.

The key finding is that the N contents and C:N ratios of these mainly acidic generally freely drained forest soils are associated with a latitude/altitude factor and this concurs with results found elsewhere for latitudinal forest transects, where these factors are related to the quantity and nature of the organic matter returns to soil (Van Cleve and Alexander 1981; Van Cleve et al. 1983; Bird et al. 1996). The latitudinal trend in the amount of organic matter accumulated in soils is generally linked to site mean annual temperature (Van Cleve and Powers 1995). However, the differences in mean annual temperature between the CANIF forest sites, with the exception of the Aheden site, are quite small. The reason for the narrow temperature range may be that there is a strong negative correlation between site latitude and altitude along the transect (Fig. 11.4d). Much of the latitude effect on soil organic matter accumulation, as indicated above, may well be due to altitude, which is known to have a major impact on soil pedogenetic processes and the C:N ratio of the soil (Jenny 1980); in this study separating latitudinal from altitudinal effects is difficult, because of the strong relationship between these two factors across the forest sites (Fig. 11.4d).

The effects of tree species on the amounts of C and N in the soil profile has been examined, using the nine cores as replicates in an analysis of variance, to compare

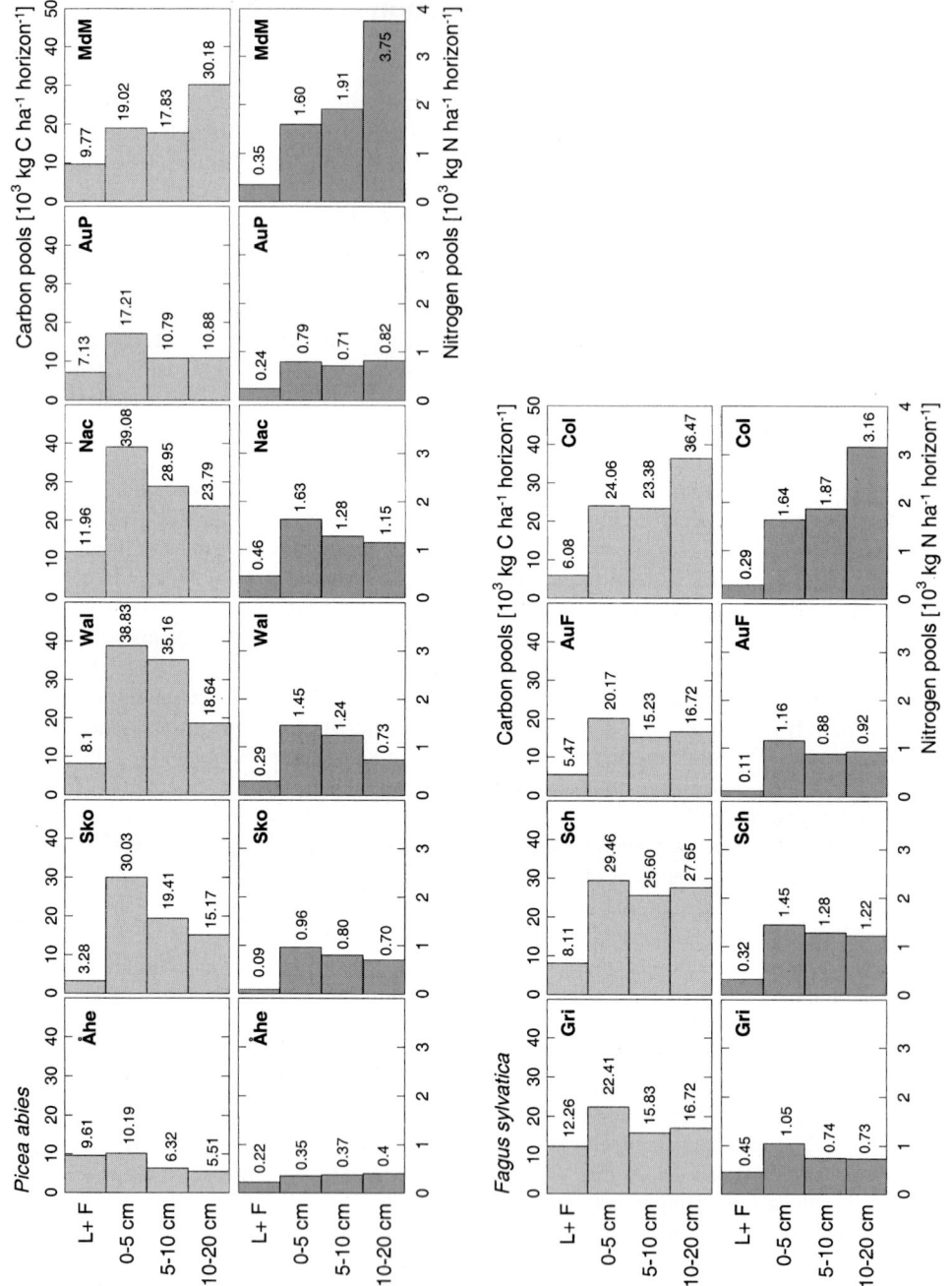

Fig. 11.3 Variations of carbon and nitrogen pools (10^3 kg ha^{-1}) in soil layers of the spruce and beech forests

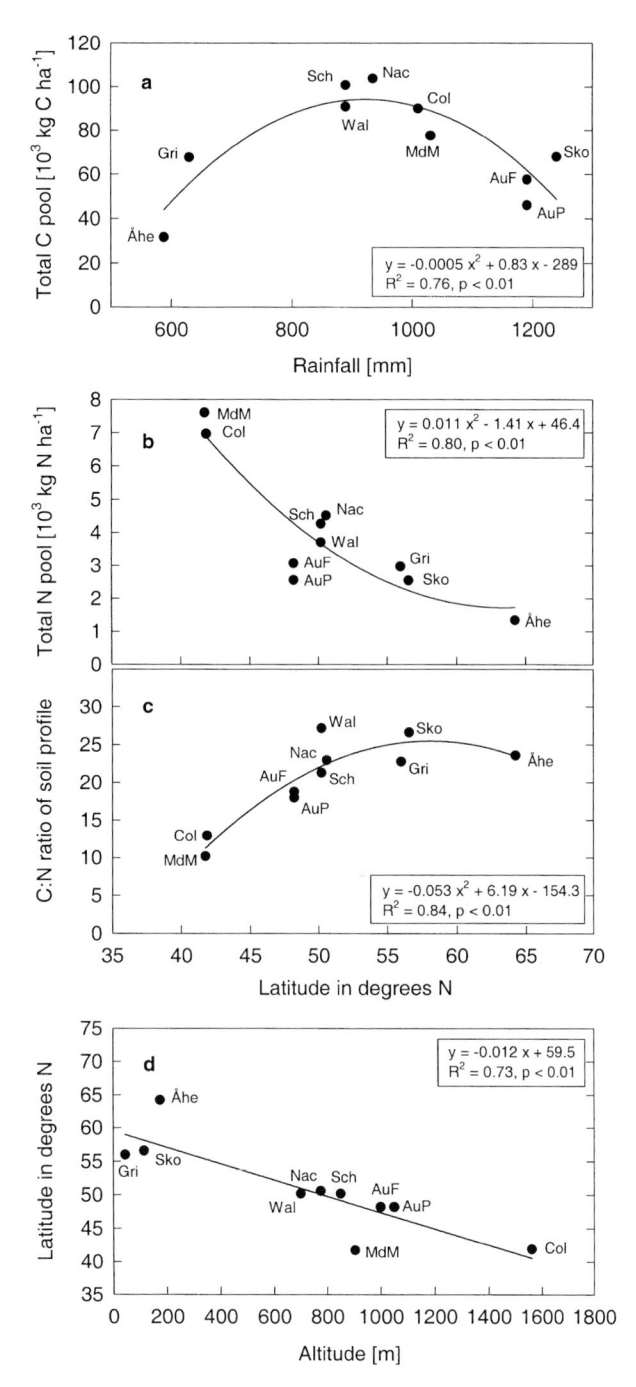

Fig. 11.4a–d. Relationships between **a** the total carbon pool (10^3 kg C ha^{-1}) to 20-cm depth in the soil profile and annual rainfall (mm) to forests; **b** total N pool (10^3 kg N ha^{-1}) in the soil profile to 20-cm depth and forest latitude; **c** the C:N ratio of soil profile to 20-cm depth and forest latitude (°N); **d** forest latitude (°N) and altitude (m)

the four pairs of forests at similar latitudes, namely Skogaby / Gribskov, Waldstein / Schacht, Aubure / Aubure and Monti-di-Mezzo / Collelongo. No consistent differences have been found for the amounts of either C or N accumulated in soils across all pairs of sites, i.e. there was no obvious tendency for organic matter to accumulate more under one species than the other.

11.6 Variations in the Carbon Age and Mean Residence Times (MRTs)

The age of the carbon in the fresh litter input and MRTs of the carbon in the various soil layers at each forest site are presented in Table 11.2. The consistency of the age of carbon in the freshly fallen tree litters (AGE_{TLI}) across all spruce and beech sites, respectively, is remarkable. The carbon in freshly fallen spruce tree litter is 5–6 years old. This age is probably a function of both (1) the different components making up the litter inputs, e.g. spruce retains its needles over many years, and (2) recycling of several years' photosynthate in the growth of new canopy tissues. The relatively young carbon in beech tree litter inputs is not surprising, as the species is deciduous.

The consistency of the MRTs of the L+F layer (MRT_{L+F}) even across tree species (Table 11.2), despite differences in the quantity and nature of the biomass in the layers across forest sites, is equally remarkable, given the variation in the latitude of the sites. The similarity of MRT_{L+F} values may be due to the narrow mean annual temperature range across the forest sites (Table 11.1).

Unlike the tree litter inputs and the L+F layers, the MRTs of the carbon in the mineral soil layers vary very considerably between layer and forest site (Table 11.2). There is a common tendency for the MRT of the carbon to increase with soil depth. However, in some of the forests (Skogaby, Waldstein, Aubure P, Monte di Mezzo, Gribskov, Aubure F and Collelongo) there is clear evidence of a deposition of carbon with a relatively low MRT (more rapidly turning over) in the 5–10 and 10–20 cm layers. This incorporation of young carbon into the soil humus cannot be from aboveground litter (see Table 11.3), but very likely originates from direct root inputs, via root secretions, sloughed off surface tissue, root hairs, mycorrhizal biomass or decaying root material. This was found despite the fact that an important part of the preparation of soil samples for isotope analysis was the careful removal of all recognisable root material. Estimates of the relative proportion of the total carbon fluxes that can be ascribed to root inputs are discussed later in this chapter.

The MRT 0–5-cm soil carbon is positively correlated with latitude ($p < 0.05$), as has been found elsewhere (Bird et al. 1996), and is also negatively correlated with both mean annual temperature ($p < 0.05$) and rainfall ($p < 0.05$), though the latter relationships are heavily dependent on the data points for the Aheden site (Fig. 11.5). The first relationship shows that 0–5 cm soil carbon at Aheden turns over much more slowly than that in Collelongo and Monte di Mezzo (Fig. 11.5) and the second relationship suggests that these differences in turnover rate are related to site mean annual temperature. A positive correlation between MRT and quantity of organic carbon in the 0–5-cm soil (Fig. 11.5) indicates a not surprising conclusion that organic matter accumulates where the MRT is greater. This finding concurs with the known soil processes, in which decay-resistant organic matter components tend to accumulate, organic matter becomes tightly bound onto silt or clay particles or transformed into stable Fe and Al complexes (Oades 1988; Sollins et al. 1996), and during soil devel-

Table 11.2. Mean residence times (years) of carbon in organic matter of fresh tree litter input and in the individual soil layers at each of the forest sites derived, using the model described above

Forest	Tree litter input AGE_{TLI}[a]	L+F layer MRT_{L+F}	0–5-cm layer $MRT_{0-5\,cm}$	5–10-cm layer $MRT_{5-10\,cm}$	10–20-cm layer $MRT_{10-20\,cm}$	20–50-cm layer MRT[b]
Spruce						
AhP	5	5	340	690	1460	–
SkP	6	5	130	45	145	–
WlP	6	5	260	20	230	–
NaP	6	5	150	160	450	–
AuP	5	6	35	35	45	140
MdMP	5	5	55	10	35	460
Beech						
GrF	2	4	70	30	50	255
ScF	2	4	140	150	370	–
AuF	2	5	80	50	410	–
CoF	2	5	65	55	440	–

[a] Mean age of carbon since photosynthesis.
[b] Derived from extrapolation of the correlation between ^{14}C enrichment and soil depth (see text).

Table 11.3. Estimates of carbon fluxes (10^3 kg C ha^{-1} a^{-1}) of C respired to atmosphere in the L+F layer, transfer from L+F to mineral soil, total flux of C through the mineral soil, the input of root carbon to soil organic matter and the percentage of C flux in mineral soil attributable to root inputs

Forest	C flux through the L+F layer	C respired *in situ* in L+F layer (F2)	C transferred from L+F layer to soil mineral layers (F3)	Total C flux through the mineral soil (see Fig. 11.6a)	C flux in mineral soil derived from root carbon[a]	Percentage of C flux in mineral soil due to direct root carbon input
Spruce						
AhP	1.92	1.90	0.02	0.04	0.02	50
SkP	0.66	0.38	0.28	0.77	0.54	70
WlP	1.62	1.47	0.15	1.99	1.84	92
NaP	2.39	1.73	0.26	0.49	0.23	47
AuP	1.18	0.93	0.49	1.15	0.66	57
MdMP	1.95	1.61	0.35	3.18	2.83	89
Beech						
GrF	3.07	2.75	0.32	1.34	1.02	76
ScF	2.03	1.82	0.21	0.46	0.25	54
AuF	1.09	0.84	0.25	0.59	0.34	58
CoF	1.22	0.88	0.34	0.88	0.44	50

[a] Estimates of root-derived carbon input to soil have been calculated as the difference between (1) the total C flux through the mineral soil [column 4], minus (2) the input to the mineral soil from the aboveground litter [column 3]. All estimates are representative mean values and will not reflect annual variation that may be inherent in the litter production and decomposition.

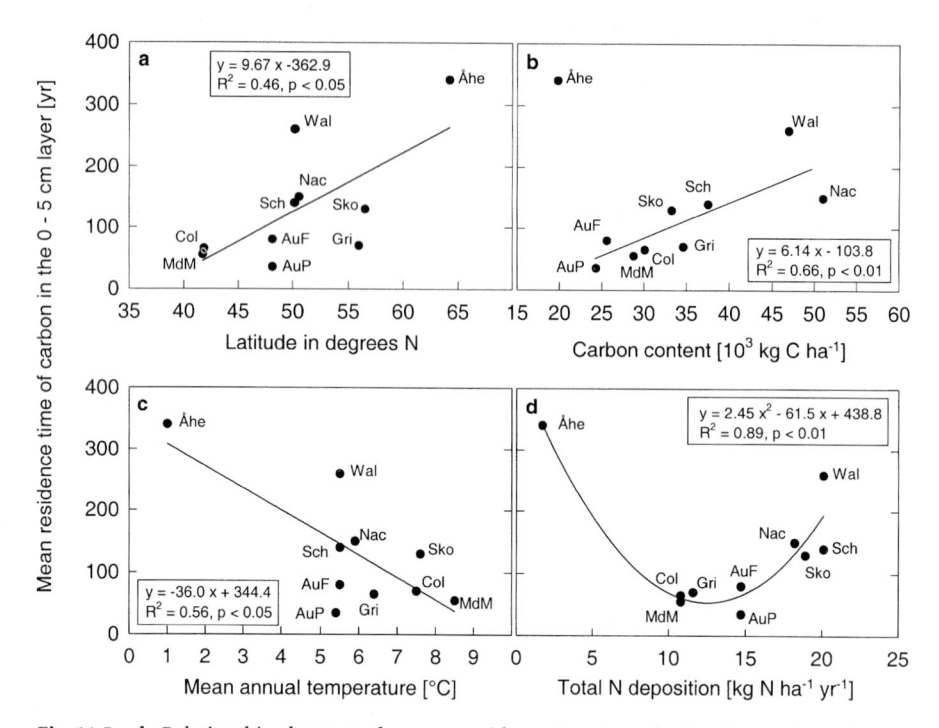

Fig. 11.5a–d. Relationships between the mean residence time (years) of carbon in the 0–5-cm layer and **a** forest latitude (°N); **b** carbon content of soil to 5-cm depth (10^3 kg C ha^{-1}), excluding Aheden; **c** mean annual temperature (°C) of forests; **d** total N deposition (kg N ha^{-1} yr^{-1}) to the forests

opment, via its effects on productivity, an increasing proportion of the organic matter becomes fixed in these stable forms (Van Cleve and Powers 1995). The Aheden site does not fit this trend, probably because its low mean annual temperature, low rainfall and low nitrogen availability have severely limited organic matter accumulation in the soil at this site, via their adverse effects on tree productivity. The significant increases in carbon MRTs with depth in the soil profile, which were found at all sites, also indicate that a significant proportion of the carbon below 10 cm is locked up in stable, non-recycling forms.

11.7 Annual Carbon and Nitrogen Fluxes

The calculated annual carbon and nitrogen fluxes, expressed as 10^3 kg C ha^{-1} a^{-1} or kg N ha^{-1} a^{-1} for all forest sites are presented in Fig. 11.6. With three out of four of the spruce/beech pairs of sites at similar latitudes, the spruce forest tends to turn over annually more carbon and nitrogen in the soil than its beech counter part. Only with the Skogaby / Gribskov pair did the beech site turn over more carbon and nitrogen; this is probably related to the fact that the Skogaby site comprised control plots of a non-fertilised N-deficient forest experiment.

The results show that at most sites the majority of the soil C and N flux takes place in the L+F layer. However, substantial additional fluxes take place in the 5–10 cm layer

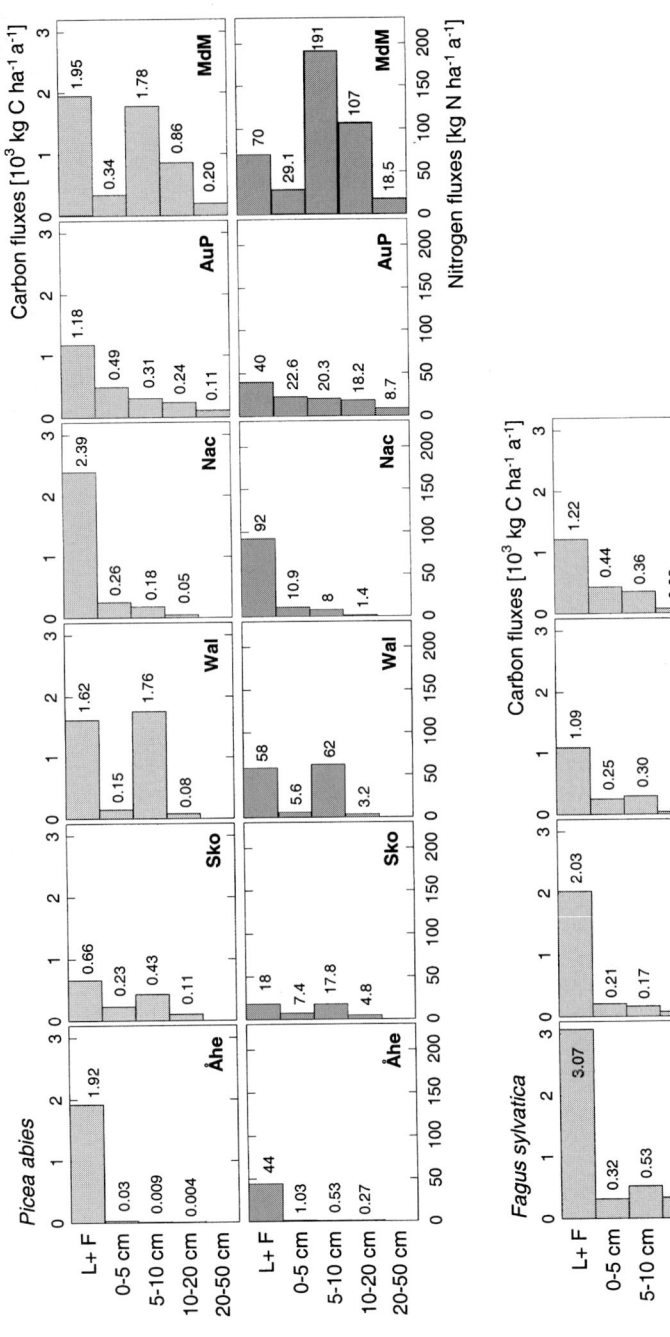

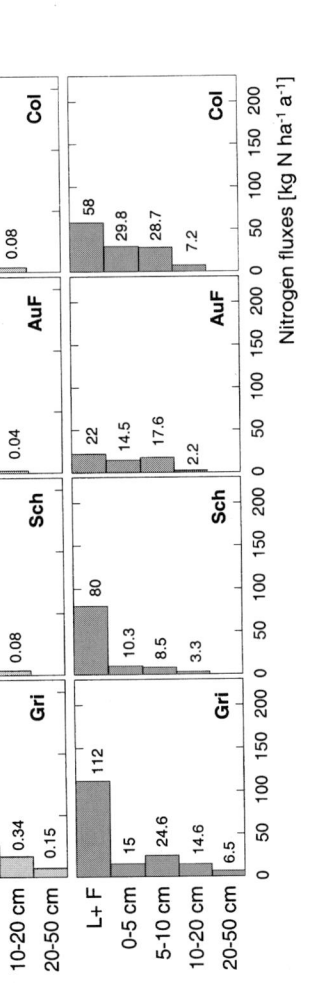

Fig. 11.6. Variation of carbon (10^3 kg C ha^{-1} a^{-1}) and nitrogen fluxes (kg N ha^{-1} a^{-1}) from soil layers of the spruce and beech forests

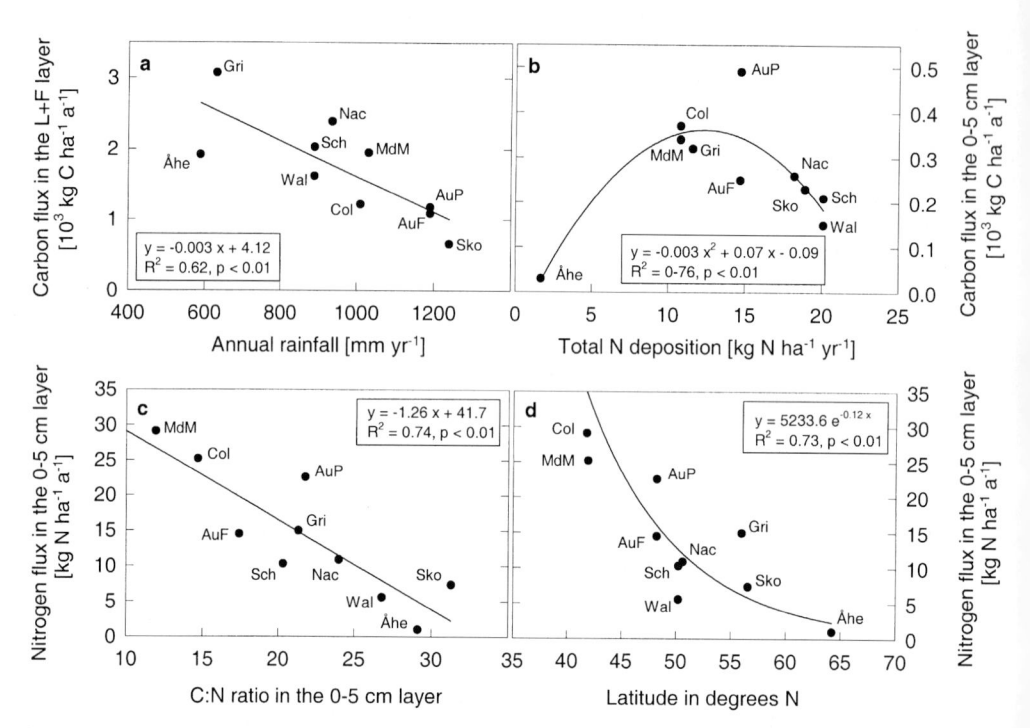

Fig. 11.7a–d. Relationships between **a** carbon flux in the L+F layer (10^3 kg C ha^{-1} a^{-1}) and annual rainfall (mm) to forests; **b** carbon flux in the 0–5-cm layer (10^3 kg C ha^{-1} a^{-1}) and total N deposition (kg N ha^{-1} a^{-1}) to the forests; **c** nitrogen flux (kg N ha^{-1} a^{-1}) and the C:N ratio in the 0–5 cm layer; **d** nitrogen flux in the 0–5 cm layer (kg N ha^{-1} a^{-1}) and forest latitude (°N)

in the soils at Waldstein, Monte di Mezzo and possibly Skogaby and Gribskov sites. A key finding is that the estimate of total C flux (turnover) in the soil of 2.1×10^3 kg C ha^{-1} a^{-1} of the Collelongo forest (also a Euroflux project site) closely equates with the estimate 2.03×10^3 kg C ha^{-1} a^{-1} derived from the difference between estimates of total productivity and net primary productivity of 8.22×10^3 and 6.19×10^3 kg C ha^{-1} a^{-1}, respectively, for the site (Matteucci 1998); this result concurs with the findings from the Meathop Wood study that this ^{14}C-bomb approach can generate realistic estimates of annual carbon and nitrogen fluxes (Harrison et al. 1990; Harrison and Harkness 1993).

Total C and N fluxes for the whole active soil profile showed no significant relationships with latitude, altitude, rainfall, mean annual temperature or total N deposition across all spruce and beech sites. The trends improve, but not to statistical significance, if the data for the spruce and beech forest sites are analysed separately. However, relationships of carbon and nitrogen fluxes to site factors do become significant if the fluxes are considered at the soil layer level rather than at whole-profile level, which indicates that different factors may govern carbon and nitrogen fluxes in the various soil layers. Carbon flux in the L+F layer is negatively correlated with rainfall input to the forests (Fig. 11.7a). Nitrogen flux in the L+F layer is correlated to rainfall in a manner similar to that of carbon flux (N flux$_{L+F}$ = −0.0003 RI^2 + 0.49 RI − 108, r^2 = 0.62, p < 0.05), but also to the C:N ratio of the layer (N flux$_{L+F}$ = −2.17 C:N +127.8, r^2 = 0.42, p < 0.05). Carbon flux in the 0–5-cm layer is correlated both to the forest latitude (C flux$_{0-5cm}$ = 20.3 e$^{-0.089 lat}$, r^2 = 0.58, p < 0.05) and the total N deposition to the forest (Fig. 11.7). The nitrogen flux for the 0–5-cm layer is also correlated with forest latitude (Fig. 11.7), total N deposition to the forest (N flux$_{L+F}$ = −0.23 N$_{dep}$^{2} + 5.14 N$_{dep}$ − 5.76, r^2 = 0.74, p < 0.01) and the C:N ratio of the soil (Fig. 11.7). Though the MRT of soil carbon in the 0–5 cm layer weakly correlated with site mean annual temperature, there were no significant relationships between the carbon and nitrogen fluxes in these surface soil layers with mean annual temperature. The lack of significant relationships may be attributed to the strong association, identified earlier (Fig. 11.4d), between latitude and altitude of the forests, which has attenuated differences in mean annual temperature between sites.

These results indicate that the trends in carbon and nitrogen fluxes in the surface soil layers decrease with soil C:N ratio, rainfall input to the forest and latitude. These latter relationships concur with the general pattern for C:N ratio and C and N fluxes found along latitudinal forest gradients (Van Cleve et al. 1983). The relationships between carbon and nitrogen fluxes in the 0–5-cm layer and the total N deposition to forests are curvilinear and quite complex, and are similar to those found for the soil carbon mean residence times discussed earlier. They suggest that at the Aheden site the addition of nitrogen might stimulate soil carbon fluxes, whilst at Skogaby, Nacetin, Waldstein and Schacht N deposition is actually reducing both carbon and nitrogen turnover in the 0–5-cm soil organic matter. The MRTs of 0–5 cm carbon of the sites Gribskov, Aubure P, Aubure F, Monte di Mezzo and Collelongo range from 35 to 80 years, whereas those for Skogaby, Nacetin, Schacht and Waldstein vary from 130 to 260 years. The average MRT for the latter group of sites is more than twice that of the former group, suggesting that even small increases in N deposition may have large effects on carbon turnover in soils. Whilst N deposition may stimulate the initial decomposition of litter, it appears to suppress decomposition of humus; significant correlations have been found between N concentrations in humus and its respiration (Berg and Matzner 1997). The N deposition effect may be via induced soil acidification, as there was also a highly significant relationship between the MRT of carbon in the 0–5 cm and S deposition (MRT$_{0-5 cm}$ = 6.54 S$_{dep}$^{2} − 160 S$_{dep}$ + 1039, r^2 = 0.86, p < 0.01), if the data for the Nacetin site (where there is a much higher total S deposition) is excluded from the regression. These latter findings suggest that pollutant inputs may have damaging influences on organic matter decomposition and that a critical load for N with respect to soil carbon and nitrogen flux has been exceeded in some of the forests; the critical load appears to be between 10 and 15 kg N ha^{-1} a^{-1} (Figs. 11.5d and 11.7b). Though this possible impact of pollutant N inputs on forests has been raised previously (Berg and Matzner 1997), it has not been demonstrated quantitatively in this novel manner.

From the ^{14}C modelling of the tree litter input and L+F layer data, it was possible to generate estimates of the mass of the litter carbon that is decomposed annually *in situ* on the soil surface and how much of this carbon is annually translocated down into the mineral soil. Some 58–99% (average 81%) of the litter input appears to be decomposed *in situ*, and approximately 1 to 42% (average 19%) enters the mineral soil profile (Table 11.3). Since for all sites, the modelled estimate of litter-derived soil carbon entering the mineral soil is significantly less than the total carbon flux within the mineral soil profile (Fig. 11.3), there must be an additional input of carbon directly to the mineral soil profile. This is most likely of root origin and has been quantified as such in Table 11.3. The calculated contributions from root-derived carbon within the soil is positively correlated with both the mean annual temperature (C flux$_{root}$ = $0.016\,e^{0.57MAT}$, $r^2 = 0.72$, $p < 0.01$) and the total nitrogen flux in the mineral soil profile (C flux$_{root}$ = 0.008N flux$_{min\ soil}$ + 0.25, $r^2 = 0.79$, $p < 0.01$) of the forests. Root growth is generally related to soil temperature and nitrogen availability, so the pattern found here conforms with expectation. This direct input of root carbon dominates the carbon and nitrogen fluxes in the 5–10 cm layer and has a major influence in the fluxes of the 10–20 cm layer at Gribskov, Aubure P, and Monte di Mezzo sites (Table 11.2).

11.8 General Discussion

The isotope geochemical modelling of carbon dynamics, based on incorporation of ^{14}C-bomb into soil organic matter, provides a novel and integrated approach to estimating the annual C and N fluxes of all soil layers in which there is some active turnover of organic matter. The ^{14}C-bomb tracer is introduced naturally into every organic component of each ecosystem via photosynthesis, and so becomes incorporated into all living plant tissues initially at the level of enrichment present in the atmosphere at the time of photosynthesis. Dead plant material and litters are, in turn, incorporated into soil organic matter and so the amplitude of the ^{14}C-bomb signal that passes through the soil carbon pools provides a proximate measure of turnover activity that is occurring *in situ* and under the actual site environmental conditions. The procedure is non-intrusive, so therefore does not induce artefacts, which may occur as a result of laboratory incubations or field manipulation of soils or ecosystems.

At the current stage of its development, our model assumes that the soil organic matter pools are in a quasi-steady-state condition and that the carbon and nitrogen present within the bulk organic matter reservoirs cycle in tandem. Both seem reasonable assumptions. Although the mathematical model construction used to interrogate the available CANIF carbon isotope data set is little more than a first-order approximation of the natural system, it nevertheless seems to offer an effective approach towards the generation of realistic estimates of mean annual fluxes of C and N through forest soils. Several of these quantitative assessments are difficult or indeed impossible to obtain otherwise.

The analogue flow pathways as depicted in Fig. 11.2 provide an accurate representation of the quantitative transfer of organic carbon that overlies the mineral soil, i.e. the MRTs and their derived flux values F1, F2 and F3 are defined unambiguously. However, the calculated flux values F5 through F9 cannot be ascribed entirely in terms of net downward translocation of organic matter originating in the L+F layer. An additional carbon input has to be invoked, accounting for somewhere between 45 and 92% of the humus carbon that is annually turned over in the mineral soil at the various sites. Direct root inputs of carbon into the soil organic matter, as discussed earlier, seem to be the obvious source. The previous experience derived from the earlier Meathop Wood studies had shown the necessity to simplify the soil system by deliberately removing the recognisable root material before soil isotopic analysis. The finer root fragments, including mycorrhizal hyphae and sloughed off root cells not removed during sample processing, are in any case likely to be components of this rapidly turning over humus carbon in the natural soil system. A further development of the model to allow for direct inputs of root carbon is necessary. This would have required an additional data set of the ^{14}C enrichments of root material for the forests, as previously carried out in the Meathop Wood study (Harrison et al. 1990), but costs of AMS analyses precluded further analyses as part of the CANIF project.

There can be no question that the research progress attained during the CANIF project has further demonstrated the considerable potential of the ^{14}C-bomb approach for determining soil carbon MRTs and C and N fluxes in forest ecosystems. We see considerable potential for extending the technology to other ecosystem types. e.g. grasslands (Harrison and Harkness 1994), and to address environmental questions, such as quantifying the impacts of climate change and pollutant inputs. There are opportunities for a new basis for structuring model subroutines to simulate soil C and N cycling processes, taking account of differences in the biological activity at different depths in a soil profile, and to run the models within the constraints of the determined organic matter mean residence times. It is also possible to model carbon fluxes through individual organic matter components, as ^{14}C-bomb concentrations can now be determined in twig, leaf, root and H-layer materials (Harrison et al. 1990), physically separated organic matter fractions (Oades et al. 1987), soil mesofauna, microbial biomass and even DOC in soil leachates (Tipping et al. 1999).

11.9 Conclusions

- Total organic carbon pools to 20-cm depth in the mineral soil varied from 31600 to 103800 kg C ha^{-1}, whilst the nitrogen pools varied from 1300 to 7600 kg N ha^{-1}.
- Total organic carbon pools were strongly curvilinearly related to annual rainfall to the forest ($p < 0.01$), with the peak at ca. 950 mm.
- Total nitrogen pools were significantly negatively related to forest latitude ($p < 0.01$), but the C:N ratio of the whole soil profile to 20-cm depth varying from 10.2 to 27.2 was positively related to latitude ($p < 0.01$).
- Mean residence times (MRTs) of carbon in the L+F, 0–5, 5–10 and 10–20 cm soil layers in all CANIF forests can be modelled from ^{14}C-bomb concentrations in the respective layers. For each forest site, one model solution only fitted the set of the measured ^{14}C concentrations present in the soil layers in the year of sampling, i.e. the model outputs were unambiguous.
- MRTs for the L+F layer carbon are fairly uniform at between 4 to 6 years across all sites.
- MRTs for carbon in the 0–5 cm layer range from 35 to 340 years, and show a positive correlation with forest latitude ($p < 0.05$) and a negative correlation with both rainfall input ($p < 0.05$) and mean annual temperature ($p < 0.05$) of the forests.
- MRTs for carbon in the 0–5 cm layer is strongly correlated with total N deposition ($p < 0.01$) across all sites. The MRTs of 0–5-cm carbon of the sites Gribskov, Aubure P, Aubure F, Monte di Mezzo and Collelongo with N deposition of 10–15 kg N ha^{-1} a^{-1} range from 35 to 80 years, whereas those for Skogaby, Nacetin, Schacht and Waldstein receiving N inputs of 15–20 kg N ha^{-1} a^{-1} vary from 130–260 years. The average MRT for the latter group of sites is more than twice that of the

- former group, suggesting even small increases in N deposition may have large effects on carbon turnover in soils.
- MRTs for carbon in 5–10 cm layer range from 10–690 years and indicate, in some forests, substantial inputs of rapidly turning over carbon from roots to soil organic matter in this layer.
- These MRTs have been used to calculate annual C and N fluxes for each of the soil layers, by dividing the pool of C or N by the mean residence time of the soil carbon.
- Total annual C and N fluxes for the biologically active part of the soil profiles range from ca. 1400 to 5100 kg C ha^{-1} a^{-1} and 46 to 416 kg N ha^{-1} a^{-1}, respectively. The C flux of 2100 kg C ha^{-1} a^{-1} for one site, Collelongo, closely agrees with the estimate of 2030 kg C ha^{-1} a^{-1} derived from other studies at that site, giving validity to the flux estimates derived from the ^{14}C-bomb method.
- With three out of four of the spruce/beech pairs of sites at similar latitudes, the spruce forest turns over annually more carbon and nitrogen in the soil than its beech counterpart.
- Total C and N fluxes for the whole active soil profile at all sites were only poorly correlated with site variables, such as latitude, altitude, mean annual temperature and rainfall. Significant relationships with site variables were, however, found, when C and N fluxes were examined at the individual soil layer level, indicating fluxes in each of the soil layers are governed by different factors.
- C and N fluxes in the L+F layer along the transect range from 660 to 3070 kg C ha^{-1} a^{-1} and 18 to 112 kg N ha^{-1} a^{-1}, respectively. Both the C and N fluxes in the L+F layer are related negatively to annual rainfall input ($p < 0.01$) to the forests and N fluxes are also related negatively to the C:N ratio ($p < 0.05$) of the L+F layer material.
- An average of 81% of C in the annual above-ground litter input is decomposed in situ in the L+F layer and an average of 19% enters the soil mineral layers.
- C and N fluxes in the 0–5 cm layer ranged from 30 to 490 kg ha^{-1} a^{-1} and 1.03 to 29.8 kg ha^{-1} a^{-1}, respectively. These C and N fluxes are related negatively and exponentially to forest latitude ($p < 0.01$) and related curvilinearly to total N deposition ($p < 0.01$), but N fluxes are also related negatively to the C:N ratio ($p < 0.01$) of the 0–5 cm soil.
- The reduction in C and N fluxes with increasing latitude and with soil C:N ratio concurs with results from other studies of latitudinal transects of forests.
- The relationship of C and N fluxes in the 0–5 cm layer with total N deposition suggests that N deposition is retarding C and N cycling rate in the four sites with the highest N inputs. The critical load for N deposition with respect to soil organic matter turnover appears to be between 10–15 kg N ha^{-1} a^{-1}.
- Estimates of root-derived C inputs, mainly to the 5–10 cm layer of the mineral soil, range from 20 to 2830 kg C ha^{-1} a^{-1}. These C fluxes derived from root carbon inputs to soil are positively related to forest mean annual temperature ($p < 0.01$) and the total N fluxes ($p < 0.01$) in the mineral part of the soil profile.
- The ^{14}C-bomb methodology has considerable potential for investigations of C and N fluxes, as demonstrated by the CANIF project application, in forest and other ecosystems, and for enabling the development and application of much-improved dynamic process-based soil computer models or soil subroutines in ecosystem models.

References

Allen SE (ed) (1989) Chemical analysis of ecological materials 2nd edn. Blackwell, Oxford, 368 pp

Baxter MS, Walton A (1970) A theoretical approach to the Suess effect. Proc R Soc Lond Ser A 318:213-230

Baxter MS, Walton A (1971) Fluctuations of atmospheric C-14 concentrations during the past century. Proc R Soc Ser A 321:105-127

Berg B, Matzner E (1997) Effect of N deposition on decomposition of plant litter and soil organic matter in forest systems. Environ Rev 5:1-25

Bird MI, Chivas AR, Head J (1996) A latitudinal gradient in carbon turnover times in forest soils. Nature 381:143-146

Boutton TW, Wong WW, Hachley DL, Lee LS et al. (1983) Comparison of quartz and pyrex tubes for combustion of organic samples for stable isotope analysis. Anal Chem 55:1832-1833

Broecker WS, Olson EA (1960) Radiocarbon from nuclear tests. II. Future concentrations predicted for this isotope in the global carbon cycles suggest its use in tracer studies. Science 132:712-721

Donahue DJ, Linick TW, Jull AJT (1990) Isotope ratio and background corrections for accelerator mass spectrometry radiocarbon measurements. Radiocarbon 32:135-142

Harkness DD, Wilson HW (1972) Some applications in radiocarbon measurement at the Scottish Research Reactor Centre. In: Proc 8th Int Conf Radiocarbon Dating. B101-B115. Royal Society, New Zealand

Harkness DD, Harrison AF, Bacon PJ (1986) Temporal distribution of "bomb" ^{14}C in a forest soil. Radiocarbon 28:328-337

Harrison AF, Harkness DD (1993) The potential for estimating carbon fluxes in forest soils using ^{14}C techniques. NZ J For Sci 23:367-379

Harrison AF, Harkness DD (1994) Potential for measurement of climatic effects on soil carbon turnover. Carbon sequestration by soils in the UK. Final report to UK Department of the Environment, London, pp 109-134

Harrison AF, Harkness DD, Bacon PJ (1990) The use of bomb-^{14}C for studying organic matter and N and P dynamics in a woodland soil. In: Harrison AF, Ineson P, Heal OW (eds) Nutrient cycling in terrestrial ecosystems: field methods, application and interpretation. Elsevier, London, pp 246-258

Ingeborg L, Kromer B (1997) Twenty years of atmospheric $^{14}CO_2$ observations at Schaumsland station, Germany. Radiocarbon 39:205-218

Jenkinson DS (1963) The turnover of organic matter in soil. In: The use of isotopes in soil organic matter studies. FAO/IAEA Rep Tech Mtg (Volkenrode) Pergamon Press, Oxford, pp 187-198

Jenkinson DS, Rayner JH (1977) The turnover of soil organic matter in some of the Rothamsted classical experiments. Soil Sci 123:298-305

Jenny H (1980) The soil resource: origin and behavior. Ecological Studies 37. Springer, Berlin Heidelberg New York, 377 pp

Kalembasa SJ, Jenkinson DS (1973) A comparative study of titrimetric and gravimetric methods for determining organic carbon in soils. J Sci Food Agric 24:1085-1090

Ladyman SJ, Harkness DD (1980) Carbon isotope measurement as an index of soil development. In: Stuiver M, Kra RS (eds) Int Conf 10th Proc. Radiocarbon 22(5):885-891

Martel YA, Paul EA (1974) Use of radiocarbon dating of organic matter in a study of soil genesis. Can J Soil Sci 38:501-506

Matteucci G (1998) Bilancio del carbonio in una faggeta dell'Italia Centro-Meridionale: determinanti ecofisiologici, integrazione a livello di copertura e simulazione dell'impatto dei cambiamenti ambientali. PhD Thesis, University of Padova, Italy, 220 pp

Nydal R, Louseth K, Skogseth FH (1980) Transfer of bomb ^{14}C to the ocean surface. In: Stuiver M, Kra RS (eds) Int Conf 10th Proc. Radiocarbon 22(5):626-635

Oades JM (1988) The retention of organic matter in soils. Biogeochemistry 5:35-70

Oades JM, Vassalo AM, Waters AG, Wilson MA (1987) Characterisation of organic matter in particle size and density fractions from a red-brown earth by solid state ^{13}C NMR. Aust J Soil Res 25:71–82

O'Brien BJ (1984) Soil organic carbon fluxes and turnover rates estimated from radiocarbon enrichments. Soil Biol Biochem 16:115–120

O'Brien BJ, Stout JD (1978) Movement and turnover of soil organic matter as indicated by carbon isotope measurements. Soil Biol Biochem 10:309–317

Pilcher J (1991) Radiocarbon dating. In: Smart PL, Francis PD (eds) Quaternary dating methods – a user's guide. Tech Guide No 4. Quat Res Assn, London, pp 16–36

Rafter TA, Stout JD (1970) Radiocarbon measurements as an index of the rate of turnover of organic matter in forest and grassland ecosystems in New Zealand. In: Olsson IU (ed) Proc 12th Nobel Symp. Wiley, New York, pp 401–417

Rowland AP, Grimshaw HM (1985) A wet oxidation procedure suitable for total nitrogen and phosphorus in soil. Comm Soil Sci & Plant Anal 16:551

Satchell JE (1971) Feasibility study of an energy budget for Meathop Wood. In: Duvigneaud P (ed) Productivity of forest ecosystems. UNESCO, Paris, pp 619–630

Slota PJ, Jull AJT, Linick TW, Toolin LJ (1987) Preparation of small samples for ^{14}C accelerator targets by catalytic reduction of CO. Radiocarbon 29:303–306

Sollins P, Homann P, Caldwell BA (1996) Stabilisation and destabilisation of soil organic matter: mechanisms and controls. Geoderma 74:65–105

Stout JD, Goh KM (1980) The use of radiocarbon to measure the effects of earthworms on soil development. In: Stuiver M, Kra RS (eds) Int Conf 10th Proc. Radiocarbon 22(5):892–896

Stuiver M, Polach HA (1977) Discussion: reporting of ^{14}C data. Radiocarbon 19(3):355–363

Tipping E, Woof C, Harrison AF, Ineson P et al. (1999) Climatic influences on the leaching of dissolved organic matter from upland UK moorland soils, investigated by a field manipulation experiment. Environ Int 25:83–95

Van Cleve K, Alexander V (1981) Nitrogen cycling in tundra and boreal ecosystems. Ecol Bull 33:375–404

Van Cleve K, Powers R (1995) Soil carbon, soil formation and ecosystem development. In: McFee W, Kelly JM (eds) Carbon forms and functions in forest soils. SSSA, Madison, Wisconsin, pp 155–200

Van Cleve K, Oliver L, Schlentner R, Viereck LA, Dyrness CT (1983) Productivity and nutrient cycling in taiga forest ecosystems. Can J For Res 13:747–766

Walton A, Ergin M, Harkness DD (1970) Carbon-14 concentrations in the atmosphere and carbon exchange rates. J Geophys Res 75:3089–3098

12 Carbon Mineralisation in European Forest Soils

T. Persson, P.S. Karlsson, U. Seyferth, R.M. Sjöberg, and A. Rudebeck

12.1 Introduction

Most decomposition studies have focussed on litter decomposition (Aerts 1997), and few studies concern decomposition of organic matter in humus layer and mineral soil. This is especially true for deeper soil horizons. In materials with a low decomposition rate, determination of CO_2 evolution rate is a more sensitive way of estimating decomposition during a limited period of time than determining weight loss in litterbags. The efflux of CO_2 from soil (soil respiration) originates not only from decomposer organisms assimilating organic matter but also from root and mycorrhizal respiration. To estimate the heterotrophic soil respiration, roots and mycorrhizal respiration must be excluded, for example, by removing roots and mycorrhiza through sieving. Although sieving disturbs the soil structure, we considered sieving as the best practical method for comparison of heterotrophic soil respiration [here called carbon (C) mineralisation] in different soil layers and at different sites. The sites are presented in Persson et al. (Chap. 2, this Vol.), and detailed site characteristics are given in Tables 2.1 and 2.3.

The main aim in the present study was to determine C mineralisation rates in different soil horizons (to a depth of 50 cm) and at different sites to assess the effects of single factors like temperature, moisture, soil N status, tree species and acidity on the regulation of this process. The sites consisted of coniferous (mainly Norway spruce, *Picea abies*) and deciduous (mainly European beech, *Fagus sylvatica*) stands along a geographical, climatical and depositional transect.

More specific aims were (1) to determine C mineralisation rates in relation to soil temperature and moisture in the laboratory to obtain response functions that could be used to extrapolate estimates obtained in the laboratory to the field, (2) to compare C mineralisation rates in conifer and broadleaf forest stands, (3) to determine C mineralisation rates in long-term field N-fertilisation experiments to assess the impact of increased N addition, and (4) to calculate turnover time of carbon for different soil layers and sites.

12.2 Experimental Background

12.2.1 Field Sampling

Soil samples, including the litter and humus layers, were taken from all sites in the NIPHYS and CANIF transect (Åheden, Skogaby, Klosterhede, Gribskov, Sorø, Nacetin, Jezeri, Waldstein, Schacht, Aubure, Thezan, Collelongo and Monte di Mezzo) for the overall C mineralisation studies. In addition, samples were also taken from control and fertilised plots in the long-term N-fertilisation experiments Norrliden (northern Sweden), Stråsan (central Sweden) and Skogaby (southern Sweden) to assess the effects of N enrichment on C mineralisation. At Norrliden, the fertilised plots ($n = 2$) were treated with ammonium nitrate at a mean dose of 73 kg N ha^{-1} a^{-1} for 23 years, 1971–1994, no fertilisation in 1990 (Tamm et al. 1995), and the sampling was performed in 1994. At Stråsan, the fertilised plots ($n = 2$) were treated with ammonium nitrate at a mean dose of 74 kg N ha^{-1} a^{-1} for 23 years (1967–1989, Berdén 1994). The sampling was made in 1994, i.e. 5 years after last fertilisation. At Skogaby, the fertilised plots ($n = 4$) were treated with ammonium sulphate at a dose of 100 kg N ha^{-1} a^{-1} for 9 years (1988–1996) and the sampling was made before fertilisation in early 1997. The characteristics of each site are given in Persson et al. (Chap. 2, this Vol.).

Litter (L) and humus (FH) samples were quantitatively sampled with a cylindrical 250-cm^2 frame, while the mineral soil was sampled to a depth of 50 cm (most sites) or 30 cm (Schacht and Stråsan) with corers of different sizes (see Persson et al., Chap. 2, this Vol.). The mineral soil was divided in the field into 10-cm-thick sublayers with the exception of the 30–50-cm depth, that was not subdivided. At sites where a humus layer was lacking, the 0–10-cm mineral soil layer was subdivided into 0–5- and 5–10-cm depth, of which the humus-rich 0–5-cm layer was often compared with the FH layers at other sites. The samples were always taken from four plots per stand/site, and within each plot five samples were taken systematically. Samples from the same soil layer were pooled for each plot.

12.2.2 Laboratory Treatment

The samples were immediately transported to the laboratory, where they were treated without being dried or frozen. Litter samples were separated from living plants and twig fragments, so the remaining material mainly consisted of leaf or needle litter. Humus samples were passed through a 5-mm sieve and the mineral soil through a 2-mm sieve. We determined the fresh weight/dry weight ratios at 105 °C and the ash content/loss-on-ignition (LOI) at 550 °C for 3 h. Based on the former determination, portions corresponding to 6 g dw (dry weight) of litter, 16 g dw of humus material and 100 g dw of sieved mineral soil were placed in plastic containers (50 cm^2 surface area, 466 cm^3 volume). The containers had a lid with a 5-mm-diameter aperture for gas exchange. These soil microcosms were incubated at a constant temperature of 15 °C. Distilled water was added once a month to keep the water content in the samples at 50–60% of water-holding capacity, WHC (100% WHC was defined as the water content of saturated soil allowed to drain for 12 h in a 3-cm-high cylinder). A whole incubation period lasted from 130 to 210 days.

To determine C mineralisation in the litter and soil materials, the containers were periodically closed with airtight lids with a rubber septum. Background gas samples were taken after 15 min from the headspace with a syringe and were injected into a gas chromatograph (Hewlett Packard 5890, H.P. Company, Avondale, PA, USA). The measurement was repeated when an appropriate amount of CO_2 had accumulated in the containers, from 120 min to 24 h, depending on the respiration rate. The mass of C evolved per container and hour was calculated according to Persson et al. (1989) and Persson and Wirén (1993), taking the pH-dependent solubility of CO_2 in the soil water into account. CO_2 measurements were performed once a week during the first month and every fourth week thereafter. To make the estimates as comparable as possible, an incubation period of 150 days was used to calculate mean C mineralisation rates in this study. However, to extrapolate the laboratory data to the field, it was also necessary to estimate the length of time a substrate should be incubated in the laboratory to decompose equally as much as during 1 year in the field. Based on temperature and soil moisture in the field and the correction factors for temperature and moisture (see below), 1 year in the field corresponded to between 73 days (Åheden) and 209 days (Thezan) at 15 °C and optimal soil moisture in the laboratory. For all other sites, the corresponding number of days in the laboratory varied between 104 (Collelongo) and 143 days (Monte di Mezzo). The measurements were made on the same samples as were destructively sampled on the last sampling occasion in the N mineralisation study (see Persson et al., Chap. 14, this Vol.). The samples from the four plots at each site acted as replicates. C mineralisation was usually expressed per g of organic C (analysed in a Carlo-Erba NA 1500 Analyzer), or per ha.

12.2.3 Extrapolation to the Field

To obtain a rough estimate of the annual C mineralisation in the field, the rates obtained (per g C per soil layer and day) in the laboratory at 15 °C and optimal moisture (50–60% WHC) were multiplied by (1) the amount of soil per soil layer (g C ha^{-1}) estimated from the quantitative samplings described above and corrected for stoniness, and (2) a correction factor obtained from response functions for temperature and moisture determined in the laboratory (see below) and data on soil temperature and moisture in the field. The response functions were based on laboratory studies of temperature and moisture relations in a humus layer from a site close to Skogaby. The latter study was done as part of the NIPHYS/CANIF project, but the results are presented in detail elsewhere (Seyferth 1998; U. Seyferth and T. Persson, submitted). Seyferth (1998) found that the temperature dependence of respiration rates in forest soils could be well described by a quadratic function (Ratkowsky et al. 1982):

$$A = b^2(T - T_{min})^2, \qquad (1)$$

where A denotes the activity (C mineralisation rate) at temperature T (in our case 15 °C), T_{min} the minimum temperature at which the activity starts, and b the slope. Seyferth (1998) estimated T_{min} to be −6.2 °C for all moisture levels, and b was moisture-dependent according to the log-linear function:

$$b = -0.11 \log(\psi) + 0.48, \qquad (2)$$

where the water potential ψ was expressed in -MPa. The combined effect of temperature and moisture in relation to a reference temperature and a reference moisture level is given by the correction factor c_C

$$c_C = b^2(T - T_{min})^2 / b_{ref}^2 (T_{ref} - T_{min})^2. \qquad (3)$$

For the practical calculations, we used the following equation:

$$c_C = (T - T_{min})^2 / (T_{ref} - T_{min})^2 \cdot [-0.11 \log(\psi) + 0.48]^2 / [-0.11 \log(\psi_{ref}) + 0.48]^2, \qquad (4)$$

where T is the soil temperature (°C) in the field, T_{min} -6.2 °C, T_{ref} 15 °C, ψ the soil water potential (MPa) in the field and ψ_{ref} the soil water potential (MPa) at 60% WHC. Where the latter value was not available for specific sites (most sites), the ψ_{ref} for Skogaby (0.0035 MPa) (Seyferth 1998) was used.

Daily soil temperatures and water potentials at 5-10-cm depth were recorded and/or simulated at Skogaby (data for 1993-1997 from G. Alavi), Klosterhede (data for 1989-1994 from C. Beier) and Thezan (data for 1993-1995 from M.-M. Coûteaux and P. Bottner) during a minimum of 3 years. For the other sites, monthly mean air temperatures were recorded (see Persson et al., Chap. 2, this Vol.), and these values were used for extrapolation to the field. At many sites, the water content (kg kg^{-1} fresh weight or m^3 m^{-3} soil layer) and not the water potential had been determined (see CD-ROM). The water potential was roughly calculated by using the relationship between actual water content and water content at saturation and converting this relationship into water potentials using the pF curve from Skogaby. The use of air temperatures instead of soil temperatures meant higher amplitudes over the year, but the mean correction factor (c_C) for the whole year was only marginally affected where a comparison could be made.

C mineralisation rates in the litter layer, but also in some other soil layers, decreased with time in the laboratory, indicating changes in substrate quality. Consequently, the mean C mineralisation rate per day calculated in the laboratory was dependent on the length of the incubation period. Therefore, the mean annual correction factor (c_C) for a site multiplied by 365 (days) gave the number of days needed in the laboratory to evolve the same amount of CO_2 as for an entire year in the field. The mean C mineralisation rates in the laboratory were thus calculated based on this incubation time.

12.3 C Mineralisation in the North-South Transect

12.3.1 C Mineralisation Rates in the Laboratory

At all sites the carbon mineralisation g^{-1}C decreased with increasing depth (Table 12.1). As an example, results on cumulative C mineralisation in different soil layers are given for Skogaby, where fresh needle litter was also sampled (Fig. 12.1). Fresh needle litter had higher C mineralisation than the litter (L) layer consisting of a mixture of ages. However, when the fresh litter had been incubated for 30-40 days, the C mineralisation rate was equal to the initial C mineralisation rate estimated for the L layer. The C mineralisation rate in the L layer was often ten fold higher than at 0-10-cm depth in the mineral soil and 40-50-fold higher than in the 30-50-cm soil layer (Table 12.1).

C mineralisation rate in the L layer, both in conifer and beech stands, declined with time indicating that the substrate quality decreased as the decomposition proceeded (Fig. 12.2a and b). In the humus layer from conifer stands, C mineralisation rates mostly did not change with time for a period of 150 days (Fig. 12.3a). In conifer stands lacking a humus layer (Thezan and Monte di Mezzo), C mineralisation rates in the 0-5-cm mineral soil had a tendency to decrease during the incubation period. This was also the case for humus layers and 0-5-cm mineral soil in beech stands (Fig. 12.3b). In deeper mineral soil layers, C mineralisation rates decreased by 10-50% during the first 2-3 weeks of incubation, whereafter the rates became stable at a lower level (similar curve as for the 0-5-cm layer at Thezan in Fig. 12.3a).

Table 12.1. Mean C mineralisation rate ($\mu g\,C\,g^{-1}\,C\,day^{-1}$) during a period of 150 days for each soil layer and site as calculated from laboratory incubations at 15 °C and 50–60% WHC

Soil layer	Åhe	Sko	Klo	Gri	Sor	Nac	Jez	Wal	Sch	AuP90	AuP45	AuF	The	Col	MdM
L	1490	1319	750	759	1087	1246	1127	747	1022	Incl FH	953	768	1495	892	627
FH	759	295	132	165a	287a	193	186	101	154	341	342	224	552a	178a	207a
0–10 cm	142	191	119	64b	170b	65	70	55	52	86	157	89	360b	83b	105b
10–20 cm	92	99	53	75	128	48	51	52	55	49	51	42	343	36	69
20–30 cm	52	41	26	49	71	30	41	21	34	32	33	30	215	23	48
30–50 cm	n.e.	29	n.e.	62	95	21	42	13	n.e.	40	34	19	130	8	29

Where no FH layer was found, the 0–10 cm layer was subdivided into 0–5 (a) and 5–10 cm (b). n.e. Not estimated. See Tables 2.1 and 2.3 in Persson et al., Chapter 2, this Vol.) for further explanations.

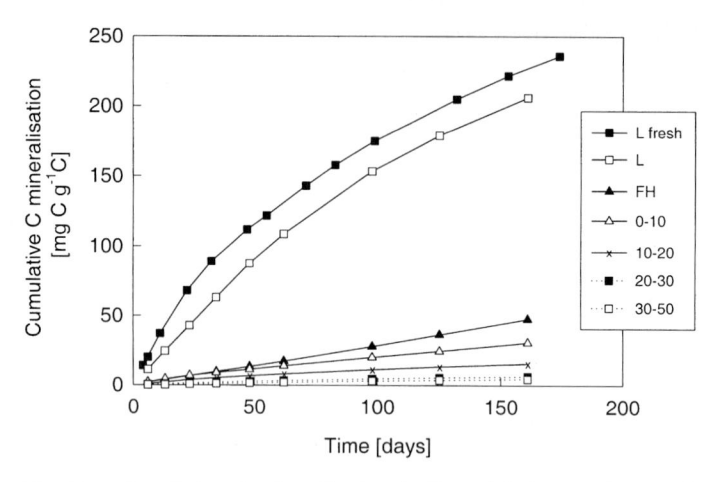

Fig. 12.1. Cumulative C mineralisation in litter (*L*), humus (FH) and mineral soil layers (0–50 cm depth) from the 33-year-old spruce stand at Skogaby sampled in May 1997, and incubated in the laboratory at 15 °C and 50–60% WHC. *L* fresh denotes fresh needle litter sampled in litter traps followed by incubation in the laboratory

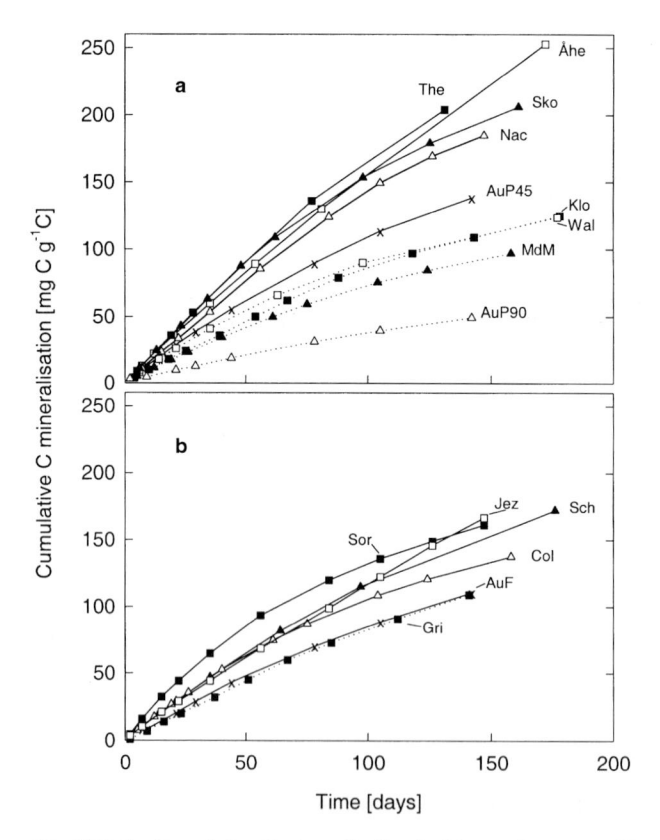

Fig. 12.2a,b. Cumulative C mineralisation in L layers from **a** conifer stands and **b** beech stands incubated in the laboratory at 15 °C and 50–60% WHC. For AuP90, a mixture of L and FH materials was incubated, which explains the low mineralisation. For abbreviations, see Table 2.1

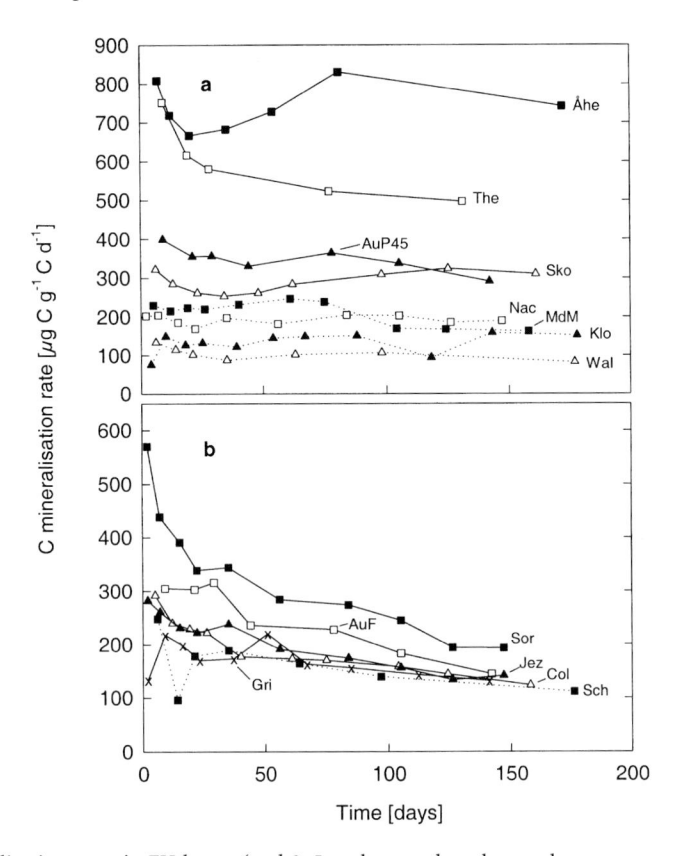

Carbon Mineralisation in European Forest Soils   263

Fig. 12.3a,b. C mineralisation rates in FH layers (and 0–5 cm layers where humus layers were lacking) from **a** conifer stands and **b** beech stands incubated in the laboratory at 15 °C and 50–60% WHC

C mineralisation rates measured in the laboratory at 15 °C and 50–60% WHC differed significantly between sites (Table 12.1). The mean rates in the litter layer varied between about 600 (Monte di Mezzo) and 1500 (Thezan and Åheden) $\mu g\,C\,g^{-1}\,C\,day^{-1}$. In the humus layer the Åheden site had considerably higher rates ($760\,\mu g\,C\,g^{-1}\,C\,day^{-1}$) than any of the other coniferous sites (100–$550\,\mu g\,C\,g^{-1}\,C\,day^{-1}$); (Fig. 12.3a). The corresponding layers at the beech sites had rates with a range of 150–$290\,\mu g\,C\,g^{-1}\,C\,day^{-1}$ (Fig. 12.3b, Table 12.1). In the mineral soil, the differences were very pronounced between sites, of which the Thezan soil had almost ten times higher C mineralisation rates than the Collelongo and Waldstein soils at 20–50-cm depth (Table 12.1). There were no clear differences in C mineralisation rate between conifer and beech stands.

12.3.2 C Mineralisation Extrapolated to the Field

The laboratory estimates were extrapolated to the field by using the monthly correction factors calculated for each site (Fig. 12.4) and by taking the soil layer pools (see

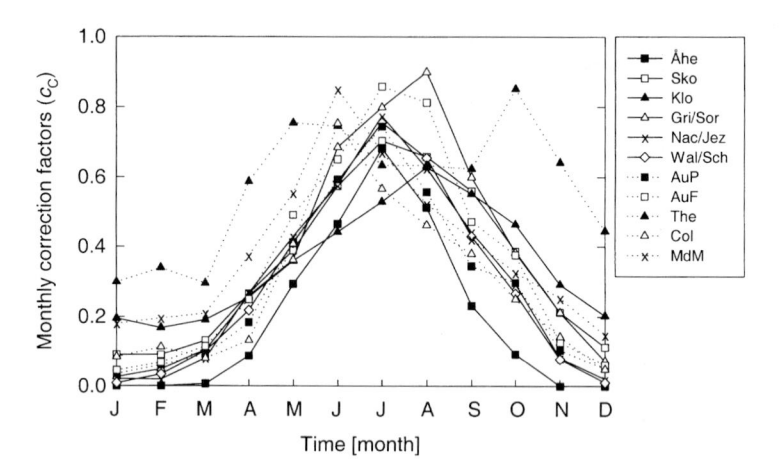

Fig. 12.4. Monthly correction factors (c_C) used to convert C mineralisation rates obtained in the laboratory at 15 °C and 50–60% WHC to C mineralisation rates in the field

Tables 2.5 and 2.8 in Persson et al., Chap. 2, this Vol.) into consideration. Åheden had virtually no C mineralisation during 5 months (November to March) because of low winter temperatures (see Persson et al., Table 2.2 in Chap. 2, this Vol.), whereas the estimate of C mineralisation in July was relatively high because of moderately high temperatures and high moisture levels (Fig. 12.4). Thezan had comparatively high C mineralisation g^{-1} C during the mild winter, while the C mineralisation was clearly limited by drought during July and August. The annual mean correction factor in relation to the laboratory situation ranged from 0.20 for Åheden to 0.57 for Thezan. At all other sites, the correction factors varied between 0.29 (Aubure) to 0.39 (Monte di Mezzo).

The estimated field C mineralisation varied between 900 (AuP90) and 3200 kg C ha^{-1} a^{-1} (Thezan) in the north-south transect (Table 12.2, Fig. 12.5). A considerable fraction of mineralised C originated from the organic soil layers (L and FH) where a humus layer had developed. As much as 71–78% of total C mineralisation (to a depth of 50 cm) evolved from the LFH layers at AuP90, Åheden, Nacetin and Jezeri (Table 12.2). At the other sites with a humus layer, the organic soil layers contributed 53–65% to total C mineralisation. At sites where a humus layer was not present (Thezan, Collelongo, Monte di Mezzo, Gribskov and Sorø), the litter layer accounted for 25–41% of total C mineralisation estimated. Consequently, the organic soil layers had a proportionally high contribution to C turnover, whereas the large C pools in the mineral soil and especially below a depth of 10 cm had a low contribution.

There was no clear relationship between latitude and C mineralisation in the field (Fig. 12.5). This was not surprising, because most of the sites at low latitudes were located at high altitudes. There was, as expected, a tendency to a positive correlation between mean annual air temperature and C mineralisation in the field (Fig. 12.6). However, in the temperature interval between 5.5 and 8.5 °C, there was no clear correlation, indicating that factors other than mean temperature were also important.

Table 12.2. Annual C mineralisation (kg C ha^{-1} a^{-1}) for each soil layer and site calculated from laboratory incubations, soil pools in the field (see Persson et al., Chap. 2, this Vol.) and corrected for soil temperature and moisture (Fig. 12.4). Mean annual correction factors (c_C) used to extrapolate from laboratory to field temperature and moisture included

Soil layer c_C	Åhe 0.20	Sko 0.35	Klo 0.36	Gri 0.37	Sor 0.37	Nac 0.30	Jez 0.30	Wal 0.30	Sch 0.30	AuP90 0.29	AuP45 0.29	AuF 0.36	The 0.57	Col 0.29	MdM 0.39
L	399	599	477	548	550	979	620	405	857	Incl FH	431	355	1249	818	1150
FH	577	1017	857	443a	601a	1059	833	607	514	639	454	232	794a	539a	641a
0–10 cm	250	937	430	115b	329b	349	208	294	384	174	353	324	315b	266b	317b
10–20 cm	56	285	211	168	296	147	63	201	212	33	66	83	368	219	404
20–30 cm	25	107	62	108	127	68	38	75	128	18	44	51	255	86	256
30–50 cm	22	86	62	162	261	55	62	49	n.e.	38	62	65	251	63	222
Total	1330	3031	2099	1543	2165	2656	1854	1631	2095	902	1411	1109	3233	1992	2991
% min in LFH	73	53	64	36	25	77	78	62	65	71	63	53	39	41	38

SE is approximately 10–20% of the mean values.
Where no FH layer was found, the 0–10 cm layer was subdivided into 0–5 (a) and 5–10 cm (b).

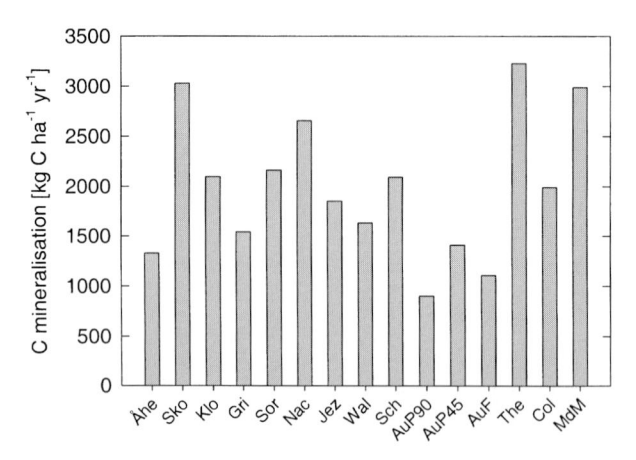

Fig. 12.5. Annual C mineralisation in the field estimated for the soil profile to a depth of 50 cm (30-cm depth at Schacht) at different sites from northern (Åhe) to southern (MdM) latitudes. For details, see Table 12.2

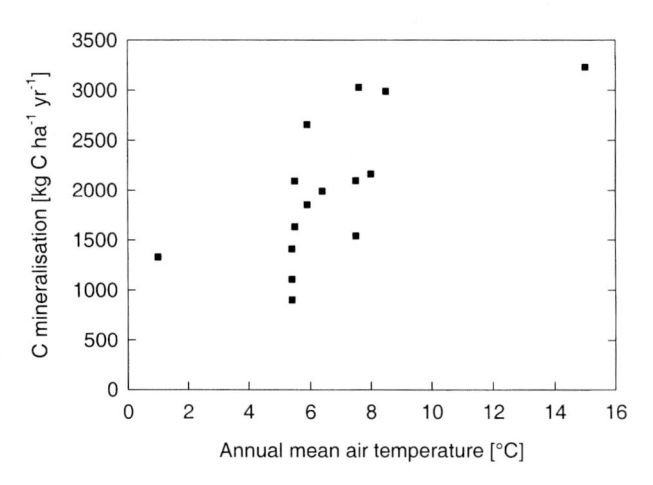

Fig. 12.6. Relationships between mean air temperature and estimated annual field C mineralisation in the European transect

No clear correlation was found between soil C pool and annual soil C mineralisation (Fig. 2.7). For most of the sites, C mineralisation seemed to increase with increasing soil C pool, but this pattern was obscured by the results from Thezan (low C pool and high C mineralisation) and Waldstein and Collelongo (high C pool and low C mineralisation). The high C mineralisation at Thezan can partly be explained by high soil temperature, but C mineralisation rates were also high at comparable temperatures (Table 12.1). It is possible that herbs and shrubs, that were abundant at that site, produced root litter of better quality than at the other sites with a poor influence of understorey. Low C mineralisation at Waldstein and Collelongo seemed to be dependent on soil properties and not on low soil temperatures (see also below).

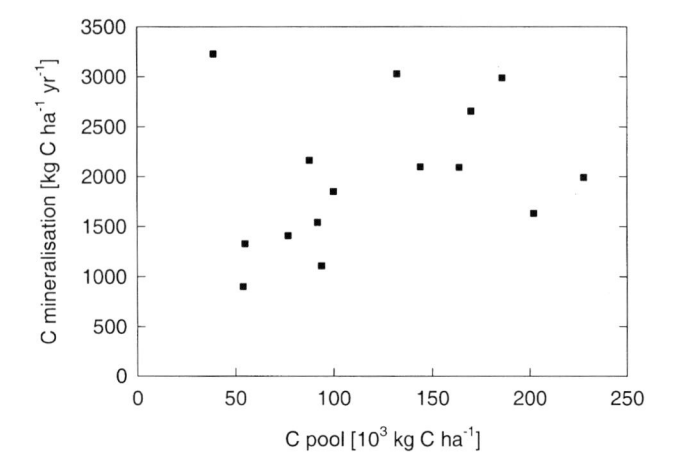

Carbon Mineralisation in European Forest Soils

Fig. 12.7. Relationships between mean soil C pool and estimated annual field C mineralisation in the European transect

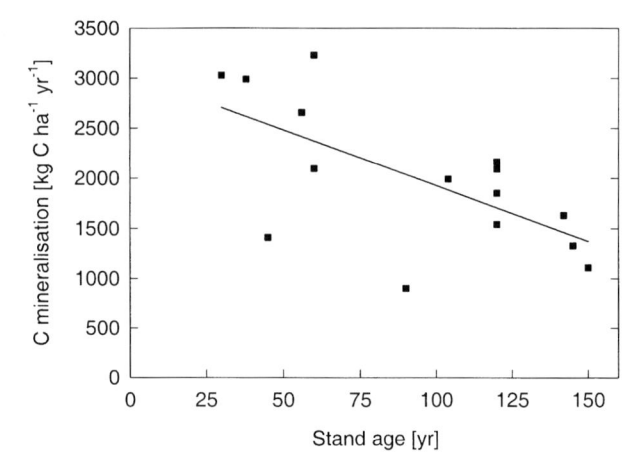

Fig. 12.8. Relationship ($R^2 = 0.41, p = 0.01$) between stand age and estimated annual field C mineralisation in the European transect

There seemed to be a negative correlation between stand age and annual C mineralisation (Fig. 12.8). A possible explanation could be that 30–50-year-old stands are often denser than old stands, and dense stands often have a higher litter production than sparse stands. Data from Table 12.2 also indicate that the organic soil layers contribute a lot to total C mineralisation, for example in the 30–40-year-old stands at Skogaby and Monte di Mezzo (see Persson et al., Table 2.1, Chap. 2, this Vol.). However, there are several exceptions to the main relation between stand age and C mineralisation. To test the hypothesis of such a relation, data sets consisting of chronosequences of nearby stands should be studied.

12.3.3 Relation Between Total Soil Respiration and Heterotrophic Respiration

The estimated field C mineralisation ranging from 900 (AuP90) to 3200 kg C ha^{-1} a^{-1} (Thezan) in the north-south transect (Table 12.2, Fig. 12.5) was considerably lower than the estimates of total C efflux from the soil. Matteucci et al., Chap. 10, this Vol. estimated the C efflux to be 7100, 8000 and 8800 kg C ha^{-1} a^{-1} for Waldstein, Monte di Mezzo and Collelongo, respectively, whereas the corresponding estimates of C mineralisation was 1600, 3000 and 2000 kg C ha^{-1} a^{-1}, respectively. The difference between the estimates indicates a very high efflux of C from roots and mycorrhiza. However, Matteucci et al. (Chap. 10, this Vol.) considered their results to be overestimates, which means that the heterotrophic contribution to total soil CO$_2$-C efflux was rather in the order of 30–50%. The data from the north-south transect for beech stands ranged from 1100 to 2100 kg C ha^{-1} a^{-1} (Table 12.2). If these estimates make up 30% of the total soil respiration, the total efflux should be 3700 to 7000 kg C ha^{-1} a^{-1}, which is close to the figure (5750 to 6630 kg C ha^{-1} a^{-1}) for a French beech forest estimated by Epron et al. (1999). Other studies give an even wider range of estimates. For example, Hanson et al. (1993) report figures of total soil efflux to be 1600 to 10600 kg C ha^{-1} a^{-1} in various oak forests in the United States and Europe.

12.3.4 Influence of Soil N Status on C Mineralisation

An examination of individual soil layers indicates that there was a reasonably good correlation between C:N ratio and C mineralisation rate (g^{-1} C at constant temperature and moisture) in L and FH layers (Fig. 12.9). The regressions indicated that C mineralisation is reduced by a factor of 2 when C:N ratio is decreased from 45 to 20

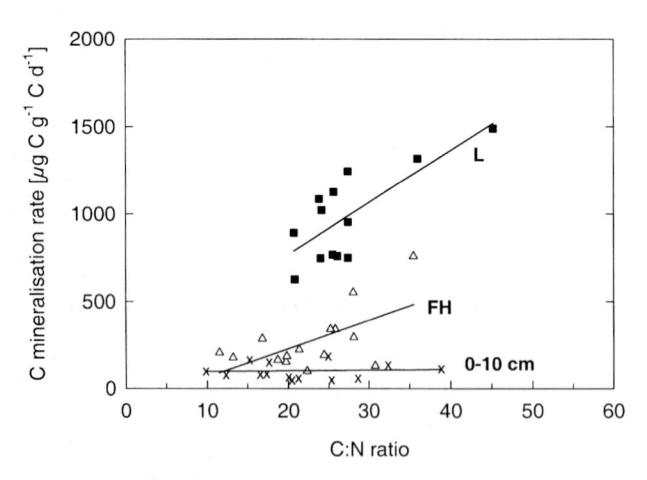

Fig. 12.9. Relationships between C:N ratio and C mineralisation rate at 15 °C and 50–60% WHC for litter (L, $R^2 = 0.57$, $p = 0.003$), humus (FH, $R^2 = 0.37$, $p = 0.02$) and 0–10-cm mineral soil ($R^2 = 0.00$) layers in spruce and beech stands in the European transect. At sites without a humus layer, the humus-rich 0–5-cm layer has been included in the FH estimates. Data for the L layer in the *P. pinaster* stand at Thezan (C:N = 94, C$_{min}$ = 1495) not included

in the L layer, and by a factor of 3 when C:N ratio is decreased from 35 to 15 in the FH layer. However, at 0–10-cm depth in the mineral soil, there was no such correlation. The relationships are in agreement with data from unfertilised coniferous stands in Sweden (Persson and Wirén 1989), which also showed a negative relationship between N concentration and C mineralisation rate in the organic layers. Two sites had exceptionally low C mineralisation rates in relation to the regression line for FH layers, namely Waldstein ($100\,\mu g\,C\,g^{-1}\,C\,day^{-1}$) and Klosterhede ($130\,\mu g\,C\,g^{-1}\,C\,day^{-1}$). Also Nacetin had comparatively low rates ($190\,\mu g\,C\,g^{-1}\,C\,day^{-1}$). These three sites had also the largest FH-layer pools of C (50–$55\,kg\,C\,ha^{-1}$) in the whole transect. Consequently, some factor other than the C:N ratio (and factors associated with this ratio) seems to be responsible. Together with AuP45, these sites had a markedly low pH in the FH layer, which might be an additional explanation of the low C mineralisation rates.

The relationships in Fig. 12.9 are just indications that C mineralisation rate is affected by the soil's N status. Another interpretation is, for example, that low temperature in the field can result in low decomposition rate and high C:N ratio just because the substrate is less decomposed. When incubated at the same temperature as organic matter from warm sites, the less decomposed substrates from cold sites can mineralise more C than more decomposed substrates from warm sites.

12.4 Long-Term Fertilisation Experiments

To test whether C mineralisation in litter and humus materials was affected by the N factor itself and not by other factors, C mineralisation rates in materials from control and long-term N-fertilised plots were compared in an incubation experiment. Figure 12.10 shows that fresh needle litter from N-fertilised plots at Skogaby had initially much lower C:N ratio (C:N 22) and higher C mineralisation rate than the corresponding litter from control plots (C:N 54). This is in agreement with a number of decomposition studies, which show that litter with high N concentrations may have rapid initial decomposition rates (Cotrufo et al., Chap. 13, this Vol.; Berg et al. 1982).

In the litter and humus layers, however, repeated N fertilisation (in the field) with mineral fertilisers resulted in decreased C:N ratios and lower C mineralisation rates than in the corresponding control materials. Only for the Skogaby FH layer, the N addition had not significantly reduced the C:N ratio, probably because a period of 9 years is too short. To affect the FH layer, the N addition must probably result in an uptake of N, an allocation of N to old needles forming litterfall, and a transport of fragmented litter to the FH layer. Despite no clear change in C:N ratio in the FH layer in N-fertilised plots, the C mineralisation rate was considerably lower than in the control plots.

The results fit with the regressions for the European transect (Fig. 12.10). The data for the L layer at Skogaby and Stråsan and those for the FH layer at Stråsan and Norrliden are close to the mean regression lines. N fertilisation at Skogaby and Stråsan reduced the C:N ratio in the L layer to be similar to that at AuP45 and Waldstein, respectively. C mineralisation rates were also similar to those at AuP45 and Waldstein. The same tendency was found for the FH layers at Stråsan and Norrliden.

The close relation between C:N ratio and C mineralisation rate at most sites can give rise to the hypothesis that addition of N in the form of N fertiliser (and probably also N deposition) might result in organic matter which is more difficult to decom-

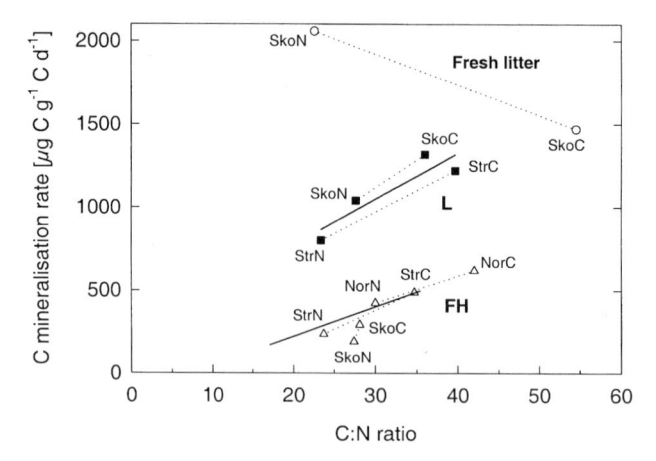

Fig. 12.10. Relationships between C:N ratio and C mineralisation rate at 15 °C and 50–60% WHC for litter (*L*) and humus (*FH*) layer materials from control plots (*C*) and N-fertilised plots (*N*) at Skogaby (*Sko*), Stråsan (*Str*) and Norrliden (*Nor*). The litter layer at Nor was quantitatively small and included in the FH layer. *Solid lines* denote the regressions from the European transect. *Broken lines* link values for control and fertilised plots at the same sites

pose, either because of a direct effect in the soil or through the formation of recalcitrant substances that will appear in the litterfall. However, the observations also fit the explanation that high N concentrations increase C mineralisation rate in fresh litter, and thereby allow N-rich litter to be more decomposed and have a lower C mineralisation rate than N-poor litter after a given time. This is also supported by data from Berg and Ekbohm (1991), who found that N-rich litter of birch and pine initially had high decomposition rate, but at a later stage of decomposition, the rates declined. This decline was, however, so pronounced that the cumulated mass loss became lower after 3–4 years in the field than for the N-poor litter. Thus, high initial decomposition in N-rich litter can probably explain the low decomposition in moderately old L layers, but not in old litter and fragmented litter belonging to the FH layer.

C mineralisation rate in the FH layer in N-fertilised plots at Skogaby was reduced to 60–70% of that in the control plots, despite the fact that there was no significant reduction in C:N ratio. This indicates that C mineralisation rate is restricted not only by high concentrations of organic N in the substrate but also by high concentrations of inorganic N in the soil. The latter argument is supported by results of Keyser et al. (1978) and Kirk et al. (1978), who found that the ammonium ion can suppress the lignolytic activity. Because lignin may stochiometrically protect holocelluloses, a suppression of lignin decomposition may have marked effects on the overall decomposition rate (Berg and Ekbohm 1991). In the fertilisation experiment at Skogaby, the ammonium level was very high (600 µg N g^{-1} organic matter) in the FH layer, especially in comparison with that in the control plots (16 µg N g^{-1} organic matter, Persson et al., Table 2.9, Chap. 2, this Vol.).

In conclusion, repeated N fertilisation as a simulation of high N deposition will, after some delay, result in increased N concentrations and reduced C:N ratios in needles/leaves, litterfall, litter layer and underlying soil layers. High N concentrations are correlated with high C mineralisation and decomposition rates in the fresh litter, but as soon as the initial decomposition phase has passed, a low C mineralisation is associated with high N concentrations. This was evident for the L and FH layers, in both the European transect and the fertilisation experiments. For the L layers, it is still an open question whether the inverse correlation between N concentration and C mineralisation rate can be explained by a suppressing effect exerted by N or simply by the high decomposition rate in N-rich litter during the initial stages of decomposition. However, the inverse correlation between N concentration and C mineralisation in FH layers can probably not be explained by high initial decomposition. Likely explanations, suggested by Fog (1988), are (1) that N additions can result in more stable N-containing substances, (2) that presence of inorganic N can suppress lignin decomposition and (3) that the microbial community can change.

12.5 Mean Residence Time

Mean residence time (MRT) for a carbon atom in a soil layer can be calculated if the in- or outflows are known. CO_2 efflux following C mineralisation is probably the dominant outflow of C from each soil layer, although DOC leaching and particulate transports should not be neglected. However, in Table 12.3 MRT for each soil layer has been calculated only as mean soil-layer C pools (Table 2.8, Persson et al., Chap. 2 this Vol.) divided by annual C mineralisation (Table 12.2). The data in Table 12.3 show that MRTs for L, FH, 0–10 cm, 10–20 cm and 20–30 cm soil layers, calculated in this manner, were estimated to be 6–11, 26–91, 41–172, 64–246 and 112–411 years, respectively, when Thezan with extremely short MRT was excluded (Table 12.3). In the following comparison, the results from Thezan are not included.

The MRTs for C in the L layers were longer than compared with the data from Harrison et al. (Chap. 11, this Vol.). This can be explained by the fact that the L layers were kept inside a container, and old L-layer material that normally is incorporated into the F layer in the field was kept in the container. The transfer of L materials to F materials is quantitatively important, implying that the calculation of C mineralisation rates underestimated total L-layer outputs and, therefore, overestimated MRTs. For other soil layers, MRT was probably more accurately estimated.

MRTs for C in FH layers were short (<30 years) for Åheden, Skogaby and the spruce sites at Aubure and long (>50 years) for Waldstein, Klosterhede, Schacht and Nacetin (Table 12.3). MRTs for C increased pronouncedly with increasing soil depth from the litter layer to the 30–50-cm depth in the mineral soil. Although the MRTs below 10-cm depth were normally above 100 years and often 200–400 years, the mean residence time of C for the whole soil profile was sometimes lower than 50 years (Sorø, Åheden and Skogaby) and only seldom more than 100 years (Waldstein and Collelongo). This can be explained by the fact that most of the C turnover occurs in the L and FH layers (see above). The MRT estimates for Thezan was extremely short, indicating that the organic matter, also at great depths, was easy to decompose.

12.6 Comparison of Intact and Sieved Soil Cores

The studies reported above were based on sorted L layers and sieved FH and mineral soil layers. A drawback with sieving, although necessary for removing roots and myc-

Table 12.3. Mean residence time (years) of C as calculated from C pools (Table 2.8, in Persson et al., Chap. 2, this Vol.) and estimated annual C mineralisation in the field (Table 12.2) for each soil layer and site

Soil layer	Åhe	Sko	Klo	Gri	Sor	Nac	Jez	Wal	Sch	AuP90	AuP45	AuF	The	Col	MdM
L	9	6	10	10	8	7	10	11	8	Incl FH	9	10	3	10	11
FH	19	27	59	44a	27a	54	46	91	57	26	27	36	9a	50a	37a
0–10 cm	91	41	65	115b	46b	151	123	160	172	121	66	86	14b	112b	69b
10–20 cm	218	78	142	141	64	177	167	167	157	232	169	178	15	246	106
20–30 cm	256	200	294	168	112	326	203	400	257	320	245	270	26	411	163
30–50 cm	260	315	300	129	115	413	199	795	n.e.	266	329	458	53	1352	274
Total	43	43	69	60	42	68	56	125	79	60	56	85	12	115	66

Where no FH layer was found, the 0–10 cm layer was subdivided into 0–5 (a) and 5–10 cm (b).

orrhiza, is that this procedure can induce an anormal stimulation of microbial activity through aeration and exposition of new surfaces. To test whether sieving affects C mineralisation rates, 8-cm-diameter cores were cut out of the FH layer in control and N-fertilised plots at Skogaby. Four of the eight cores per plot (four replicate plots) were sieved and four were kept intact and placed in the same type of containers as previously described.

Periodical measurements of C mineralisation rates showed that sieved cores from the control plots had lower C mineralisation than intact cores during day 20 to 50 (Fig. 12.11). After this period, there was no difference between sieved and intact cores. A similar tendency was found for the cores from the fertilised plots. The results show that sieving had a tendency to reduce rather than stimulate C mineralisation rate. Our interpretation of these results is that presence of fine roots and mycorrhizal fungi in the intact cores increase C mineralisation through their respiration or by providing decomposer microorganisms with easily available carbon compounds. Fine-root biomass has been estimated to be smaller in fertilised than in control plots at Skogaby (Majdi and Kangas 1997). The smaller difference in C mineralisation between intact and sieved soil cores in the N-fertilised plots than in control plots can thus reflect the lesser amount of fine-root biomass remaining in the intact soil from the fertilised plots.

In conclusion, the results indicate that sieving of humus layers does not appear to cause unnatural disturbances of C mineralisation rates. The results from the FH layers can probably not be uncritically extrapolated to mineral soil layers, where sieving might have had greater influence. However, comparison of intact and sieved mineral soils (sand, loam and clay) indicate that sieving stimulates C mineralisation only during the first week after disturbance, whereafter no clear effects can be seen (Hassink 1992).

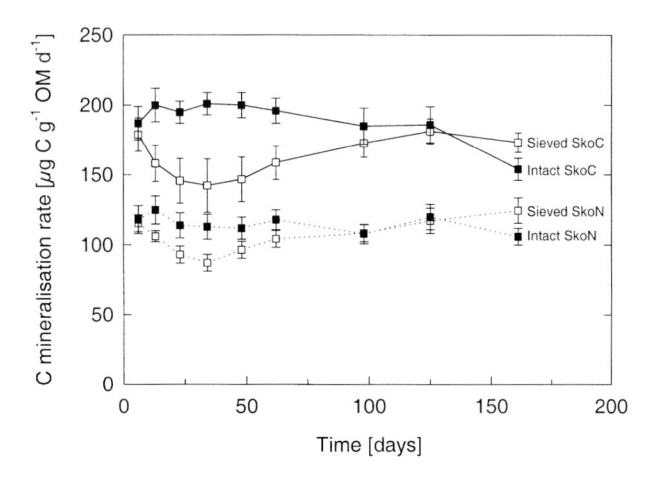

Fig. 12.11. Comparison of C mineralisation rates (mean ± SE) during an incubation at 15°C and 50–60% WHC in sieved and 'intact' FH layers from control (*C*) and N-fertilised (*N*) plots at Skogaby. *OM* organic matter

12.7 Conclusions

Soil sampling of six soil layers, generally to a depth of 50 cm, was performed at 15 different sites/stands in the NIPHYS/CANIF transect with coniferous and broadleaf stands. CO_2 evolution rate (C mineralisation rate) was determined in the laboratory for each of these soil layers at constant temperature (15 °C) and near-optimal moisture. At the same temperature and moisture, C mineralisation rate was always highest in the litter layer and decreased with increasing soil depth, in both coniferous and broadleaf stands. There was no clear effect of spruce or beech at comparable sites on C mineralisation rates in, for example, the FH layer or in the 0–5-cm layer where a humus layer was lacking. C mineralisation rate was positively correlated with the C:N ratio in the litter and humus layers but not in the mineral soil layers, indicating that high N concentrations might suppress decomposition during the phases when lignin decomposition is determining total decomposition rate. The correlation between C:N ratio and C mineralisation rate was supported by data from long-term N-fertilisation experiments, which showed that fertilised plots in Scandinavia with a low N deposition had properties similar to those in central Europe with high N deposition. C:N ratio (or factors associated with this relation) could, however, only explain some variation. At some sites, pH also seemed to be an important factor.

Results obtained in the laboratory were extrapolated to the field by taking soil pools as well as temperature and moisture into consideration. Temperature and moisture response functions were constructed based on a laboratory study, and these functions, in combination with temperature and moisture data from the sites, were used to extrapolate the data to the field. C mineralisation data from the laboratory had to be reduced by 80% to fit the temperature and moisture conditions at the northernmost site (Åheden), whereas the corresponding reduction for the Mediterranean site, Thezan, was only 43%. Annual field C mineralisation estimated for the whole soil profile varied between 900 and 3200 kg C ha^{-1} a^{-1} and showed no clear latitudinal trend, because many southern sites were situated at high altitudes. There was a tendency of increasing C mineralisation with increasing temperature, but this trend was obscured by, for example, the influence of N status. There was no clear relation between soil C pool and annual C mineralisation, mainly because the Mediterranean site Thezan had high turnover of C and two other sites had exceptionally low turnover.

In conclusion, soil temperature and moisture had a combined effect on C mineralisation, resulting in a tendency to higher C mineralisation at higher temperatures. There was no clear effect of tree species on C mineralisation, but a higher proportion of annual field C mineralisation originated from the organic soil layers which were often more developed in spruce than in beech stands. Mean residence time (MRT) of carbon was always shorter than 100 and often shorter than 50 years in the organic layers. This indicates that the present tree stand has a marked influence on C turnover in these soil layers. However, below a depth of 10 cm in the mineral soil, MRT was normally 100–400 years, sometimes more, indicating that this carbon must have originated from vegetation periods often long before the present tree stand.

Acknowledgments. The study was supported by grants from EEC (ENV5V-CT92-1043 and ENVH-CT95-0053) and the Swedish Environmental Protection Agency.

References

Aerts R (1997) Climate, leaf litter chemistry and leaf litter decomposition in terrestrial ecosystems: a triangular relationship. Oikos 79:439-449

Berdén M (1994) Ion leaching and soil acidification in a forest Haplic podzol: effects of nitrogen application and clear-cutting. PhD Thesis, Swedish University of Agricultural Sciences, Department of Ecology and Environmental Research, Uppsala, Report 73:1-22

Berg B, Ekbohm G (1991) Litter mass-loss rates and decomposition patterns in some needle and leaf litter types. Long-term decomposition in a Scots pine forest. VII. Can J Bot 69:1449-1456

Berg B, Wessén B, Ekbohm G (1982) Nitrogen level and decomposition of Scots pine needle litter. Oikos 38:291-296

Epron D, Farque L, Lucot E, Badot P-M (1999) Soil CO_2 efflux in a beech forest: dependence on soil temperature and soil water content. Ann For Sci 56:221-226

Fog K (1988) The effect of added nitrogen on the rate of decomposition of organic matter. Biol Rev 63:433-462

Hanson PJ, Wullschleger SD, Bohlman SA, Todd DE (1993) Seasonal and topographic patterns of forest floor CO_2 efflux from an upland oak forest. Tree Physiol 13:1-15

Hassink J (1992) Effects of soil texture and structure on carbon and nitrogen mineralization in grassland soils. Biol Fertil Soils 14:126-134

Keyser P, Kirk TK, Zeikus IG (1978) Ligninolytic enzyme of *Phanerochaete chrysosporium*: synthesized in absence of lignin in response to nitrogen starvation. J Bacteriol 135:790-797

Kirk TU, Schulz E, Connors WI, Lorbaz LF, Brauns IG (1978) Influence of culture parameters on lignin metabolism by *Phanerochaete chrysosporium*. Arch Microbiol 117:227-285

Majdi H, Kangas K (1997) Demography of fine roots in response to nutrient applications in a Norway spruce stand in southwestern Sweden. Ecoscience 4:199-205

Persson T, Wirén A (1989) Microbial activity in forest soils in relation to acid/base and carbon/nitrogen status. Medd Nor Inst Skogforsk 42:83-94

Persson T, Wirén A (1993) Effects of experimental acidification on C and N mineralization in forest soils. Agric Ecosyst Environ 47:159-174

Persson T, Lundkvist H, Wirén A, Hyvönen R, Wessén B (1989) Effects of acidification and liming on carbon and nitrogen mineralization and soil organisms in mor humus. Water Air Soil Pollut 45:77-96

Ratkowsky DA, Olley J, McMeekin TA, Ball A (1982) Relationship between temperature and growth rate of bacterial cultures. J Bacteriol 149:1-5

Seyferth U (1998) Effects of soil temperature and moisture on carbon and nitrogen mineralisation in coniferous forests. Dept of Ecology and Environmental Research, Swedish University of Agricultural Sciences, Uppsala. Licentiate Thesis, No 1

Seyferth U, Persson T. Effects of soil temperature and moisture on carbon and nitrogen mineralisation in a coniferous forest soil. Soil Biol Biochem (submitted)

Tamm CO, Aronsson A, Popovic B (1995) Nitrogen saturation in a long-term forest experiment with annual additions of nitrogen. Water Air Soil Pollut 85:1683-1688

13 Litter Decomposition

M.F. Cotrufo, M. Miller, and B. Zeller

13.1 Introduction

Through the process of decomposition, dead organic material is broken down into particles of progressively smaller size, until the structure can no longer be recognised, and organic molecules are mineralised to their prime constituents: H_2O, CO_2 and recalcitrant organic as well as mineral components. During litter decomposition, C may additionally be leached as dissolved organic C to the mineral soil. Other mineral components, like K and Mn, are released from the original tissue.

The decomposition of plant litter is a key process in the nutrient dynamics of forest ecosystems, and it is through this process that nutrients immobilised in the detritus are mineralised and released into the soil in a form suitable for plant uptake. In nutrient-poor ecosystems, litter decomposition becomes the controlling step of nutrient cycles and forest productivity (Flanagan and Van Cleve 1983). The second major output of the decomposition process is the formation of soil organic matter, including both cellular and humic components. The rate of accumulation of organic C in soils is a function of primary production and decomposition. An accumulation of organic C at the soil surface is generally observed in the early stage of forest development, whilst at maturity, the amount of organic matter in soil tends to be more constant and it is distributed to a greater depth in the soil profile (Dickson and Crocker 1953).

In forest ecosystems the main source of organic matter entering the decomposition subsystem is represented by plant litter, with leaf litter accounting for from 70 to 90% of total litter annual production (see Stober et al., Chap. 5, this Vol.). The fine root system is also a major component of forest production and turnover, estimated to contribute between 20 and 26% of primary production in European beech forests (see Stober et al., Chap. 5, this Vol.).

13.2 Factors Affecting the Decomposition Process

Decomposition is both an abiotic and a biotic process and it is the result of three combined phenomena: leaching, fragmentation and catabolism. Leaching is the removal of soluble compounds from the detritus by water and it is particularly significant in the early stages of decomposition when nutrients and soluble carbohydrates are still abundant in the litter. Although not a decomposition process per se, the fragmentation of litter into smaller pieces caused by both physical action and the action of soil fauna accelerates both leaching and catabolism, by increasing the surface area for catabolic processes operated mainly by microorganisms (fungi and bacteria). Organic compounds are transformed into simpler molecules, with an energy gain for the consumer.

The rate of decomposition is regulated by three main driving variables and their interactions: the physicochemical environment, the resource quality and the decomposer organisms (Swift et al. 1979). To evaluate the effects of the physicochemical environment on decomposition rates it is necessary to distinguish between macroclimatic variables (such as temperature and precipitation), and the characteristics of the microenvironment in which decomposition takes place. In a recent review, Aerts (1997) showed that on a global scale, climate [expressed as annual actual evapotranspiration (AET)] is the factor that can best predict first-year leaf litter decay rates. AET was also shown to be a good predictor of first-year decay of Scots pine needle litter, over a transect that ranged from the subarctic to the subtropic climate, although other climatic variables, linked to seasonal patterns of precipitation and temperature, could explain part of the variability in the decay rates (Berg et al. 1993).

Any environmental parameter that affects the distribution and activity of decomposer organisms will affect decomposition rates. Soil pH is important because it is correlated with other soil properties such as cation exchange capacity, base saturation and the availability of metal ions (Al^{3+} and heavy metals such as Pb). However, many soil microorganisms seem to tolerate a broad range of soil pH; only pH values <2 have been reported to strongly inhibit microbial activity (Myrold 1990).

Under similar environmental conditions, litter decomposition rates are related to resource quality. However, the concept of resource quality is often difficult to define, as it contains chemical and physical properties of litter material, and their interactions. Thus, resource quality is generally described as the relative decomposability of litters, depending on the relation between labile and recalcitrant compounds, which defines the nature of energy sources, concentrations of nutrients and modifier compounds and the physical structure of the decomposing substrate (Swift et al. 1979). The variable importance of these parameters cannot, at present, be fully quantified because they vary mainly with the type of vegetation, and the part of the plant from which the litter was derived (branches, leaves, needles, roots etc.).

Decomposition of plant material is largely mediated by fungi and bacteria, which have lower C/N ratios than the vegetation on which they feed. They therefore have a high demand for N and it may be predicted that detritus with high concentrations of N will decompose faster because of the associated faster growth of microbial populations. Therefore, the C/N ratio of plant litter has often been negatively correlated with litter decomposition rates (Edmonds 1980; Taylor et al. 1989), and the general concept that the addition of N to decomposing material enhances decomposition rates is widely accepted. However, Fog (1988) demonstrated that such correlation cannot be generalised and that a strong distinction between resource types needs to be made, with a positive effect of N addition commonly being observed for materials with low C/N ratio, whilst the reverse is often the case for resources with high C/N ratios.

Berg et al. (1982) observed that initial N litter content is an important rate-regulating factor of decomposition in the first stages of the process, while at later stages lignin concentration becomes a better predictor. Lignin is a recalcitrant compound for many organisms, and thus influences decomposition by resisting enzymatic degradation, reacting with other litter components and building up more recalcitrant complexes. Therefore, lignin and lignin-derived indices have often been used to quantify substrate quality, and correlated with litter decomposition rates (Melillo et al. 1982). However, the relation between lignin, N and polymeric carbohydrates in controlling decomposition rates is extremely complex because of the different effect that these compounds exert on mass loss rates individually or in combination. Litter with high N concentrations may have rapid initial decomposition rates, but in later stages, high N concentration may suppress the formation of ligninolytic enzymes, enhancing the formation of recalcitrant N-phenolic complexes, and inducing a slow down in lignin degradation and litter decomposition rates (Berg et al. 1982). In the later stages of decomposition, McClaugherty and Berg (1987) found a positive correlation between mass loss of Scots pine needle litter and holocellulose to lignocellulose ratios. In this case, litter mass loss was inversely correlated with N concentrations.

It has been suggested that decomposition models must recognise qualitative differences between resource types, and the need to consider the interactions between substrate quality and environmental conditions (Hunt et al. 1988). A higher percentage of variance in decomposition rates was explained by the combined effects of quality and climatic factors, rather than the effect of these factors in isolation (Meentemeyer 1978; Meentemeyer and Berg 1986; Upadhyay and Singh 1989; Berg et al. 1993).

Besides physicochemical parameters, it has often been neglected that the decomposer community is very diverse, including a wide range of bacteria, fungi, protozoa and invertebrates. It is complex because of the multitude of interactions between trophic levels and organisms involved (Swift et al. 1979). The role of soil fauna in C and N cycling of forest ecosystems is discussed in details in Chapters 16, 17 and 18 of this Volume. Estimates of the relative contribution of fungi and bacteria to microbial biomass, using selective inhibitors in combination with substrate-induced respiration measurements, have suggested a fungal dominance in both agricultural and forest soils (Andersson and Domsch 1975) and decomposing litter residues (Beare et al. 1990). Where resource quality is low and climatic constraints are present, fungi tend to play a more important role in the decomposition process than bacteria (Dighton 1995).

13.3 Enzymatic Activity

Litter decomposition is ultimately a result of the catabolic activities of saprotrophic communities associated with litter resources. At the molecular level, these catabolic activities are directly mediated by extracellular enzymes. The first action in enzyme-mediated decomposition of complex organic substrates is their depolymerisation by cleavage of intermonomeric bonds, leading to the production of low molecular weight molecules (amino acids, disaccharides, dipeptides etc.) which can subsequently be assimilated and mineralised by the saprotrophic communities (Mayer 1993).

Because of the complex chemical nature of plant tissue, total degradation requires the interaction between many classes of enzymes (e.g. Ljungdahl and Eriksson 1985; Kirk and Farell 1987; Klyosov 1990). As no single microorganism is able to produce all the enzymes necessary for the complete degradation of plant residues, the decomposition process requires the succession and interaction of many different species of microorganisms. The role of functional diversity in decomposition processes is discussed in more detail in Kjøller et al. (Chap. 18, this Vol.).

The production of enzymes by the soil microbial community is dependent on site-specific factors (i.e. temperature, moisture and nutrient availability), which control microbial growth. Once enzymes are released into the environment, litter chemistry becomes the primary determinant of enzyme activity, through its effects on adsorption, inhibition and stabilisation processes (Sinsabough et al. 1992). Thus, physico-chemical conditions and resource quality affect enzyme-mediated catabolic activities of decomposer organisms at scales much smaller than the regional scales at which climatic parameters are usually monitored.

Litter decomposition is generally measured by means of mass loss, and substrate weight loss is currently being correlated to climate and/or resource quality for modelling purposes. Parnas (1975) suggested that plant litter decomposition could be directly related to the catabolic activities of saprotrophic communities. Bunnel et al. (1977) successfully used soil temperature and moisture in models to explain the variability of microbial respiration. More recently, Sinsabough and Moorhead (1997) proposed the use of an enzymic approach to decomposition modelling, based on determination of key enzyme activities involved in the nutrient acquisition of saprotrophic communities.

13.4 Nitrogen Dynamics in Decomposing Litter

Nutrient release from litter is an important process for the recycling of nutrients in forests. During litter decomposition, plant nutrients are released either by leaching or through decomposition of complex organic structures by the soil biota. In plant litter most N forms complexes with organic compounds such as polyphenols; only a small proportion of N occurs in soluble compounds. Berg and Staaf (1981), distinguished three phases which describe the evolution of the absolute amounts of N in decomposing litter. The first phase corresponds to the leaching of soluble N compounds from the litter. This is followed by a second phase in which incorporation of new N by decomposing organisms can be observed, and finally a release of N. However, depending on initial litter chemistry, phase one and two could be very short or even non-existent. Incorporation of external N in decomposing litter was observed in studies on beech litter decomposition (Staaf 1980; Joergensen and Meyer 1989; Colpaert and van Tichelen 1996) and related to the initial N concentration (Mellilo et al. 1982). As a consequence of this incorporation of external N into decomposing litter, the C/N ratio decreased initially and, after reaching a critical value, a net release of N occurred.

Considerable progress in understanding the N dynamic in decomposing litter was obtained by using ^{15}N-labelled litter. Labelling of litter N pools allows a quantification of a simultaneous input and output of N from decaying litter across a variety of litter types (Berg 1988; Blair et al. 1992; Hart et al. 1993; Setälä et al. 1996). However, even in forests, possible mechanisms by which these N transfers occur and the form of N during these transfers are not very well understood. External N incorporated in the litter may be provided by throughfall (Downs et al. 1996; Koopmans et al. 1998), by passive diffusion of N compounds from the organic soil layer (Berg 1988) and by import of N through soil fauna, fungi and bacteria (Setälä et al. 1996). However, the degree of incorporation of each N source may vary between litter types and forest ecosystems, because of differences in decomposer communities and N deposition.

Despite a few studies which investigated the role of throughfall N on litter decomposition (Downs et al. 1996; Koopmans et al. 1998), there is a lack of information about the participation of the different sources of N incorporated into the litter.

Berg (1988) and Zeller (1998) showed that ^{15}N was continuously released from decomposing needle and leaf litter, but this release was balanced for 2 years by the incorporation of external N into the litter. Possible mechanisms for this release of ^{15}N from the litter were: leaching of soluble litter N (Joergensen and Meyer 1989; Tietema and Wessel 1994), biochemical degradation by bacteria, saprophytic and symbiotic fungi and fragmentation and consumption by soil fauna (Setälä et al. 1996). During initial litter decomposition, mechanical leaching of soluble or easily degradable N compounds occurred, followed by a release of N due to the activity of soil fauna, fungi and bacteria, which is site-specific. According to the release mechanisms, the size of the labile and recalcitrant N pools in the litter will affect the rate of N release.

13.5 Decomposition Studies in Europe: from DECO, VAMOS, MICS to CANIF

Decomposition processes in terrestrial ecosystems have been investigated in the past under several research projects within the European Community Framework. The MICS (Decomposition of Organic Matter in Terrestrial Ecosystems: Microbial Communities in Litter and Soil) project studied how to integrate results derived from different methods of analysing organic matter decomposition, with the overall objective of reaching an understanding of the activity of soil microflora during decomposition, under varying soil management and impacts of land use on decomposer communities. Other projects focused on decomposition processes in forest ecosystems. The DECO (organic matter turnover in a west European climatic transect of coniferous forests) project aimed to estimate the decomposition rate of coniferous litter as dependent on climate along a European transect that ranged from Scandinavia to the Mediterranean region, and to make a budget of organic matter in each site studied. More recently, the VAMOS (Variation of Soil Organic Matter Reservoir) project was performed to investigate decomposition under the perspective of global change, aiming to understand the extent to which climate changes will affect the soil organic matter reservoir, and possible changes in the C balance will modify N availability to plants. Results from these projects highlighted the importance of climate on decomposition processes, via direct effects (i.e. higher AET resulted in higher decay rates of needle litter and soil organic matter) and indirect effects via changes in litter quality. An important conclusion of the DECO project was the observation of increased nutrient concentrations in coniferous litters produced at lower latitudes with higher AET (Berg et al. 1995). They also helped to highlight gaps in the knowledge of decomposition processes in forest ecosystems. In particular, the importance of clarifying the interactions between C and N that control decomposition was recognised, focusing on C availability and on litter decomposition as a source of inorganic N, or as a temporary sink of soil N (Coûteaux et al. 1995). Both these studies were performed across a broad European latitudinal transect, but focused only on coniferous forests. It was due to these recognised knowledge gaps that the following decomposition studies were designed and conducted on European broadleaf forests (i.e. beech forest).

13.6 Decomposition Studies Within a Latitudinal Transect of European Beech Forests

13.6.1 Hypotheses to Test

A range of experiments on decomposition processes were performed in the present study to determine: (1) turnover rates of soil organic matter (see Harrison et al., Chap. 11, this Vol.); (2) dynamics of N mineralisation (see Persson et al., Chap. 14, this Vol.); (3) activity of both microorganisms and soil fauna (see Wolters et al., Chap. 17 and Kjøller et al., Chap. 18, this Vol.) in European forest ecosystems. Here, we refer to leaf litter decomposition studies which were performed only at the beech forest sites, in order to test the following hypotheses:

1. Mass loss rate of a standard leaf litter differs when the litter is incubated at different forest sites along a latitudinal and altitudinal European transect (55°58'N–41°52'N; 45–1560 m asl), with decay rates being controlled by climate. Varying N deposition along the transect may explain the residual.
2. Elemental composition, in particular N concentration, of leaf litter differs along a latitudinal and altitudinal European transect (55°58'N–41°52'N; 45–1560 m asl), and affects subsequent litter decomposition rate.
3. Leaf litter mass loss rates relate to fungal mediated cellulolytic activity and the climatic and quality factors that control mass loss rate also control cellulolytic activity.
4. Nitrogen dynamics in decomposing beech litter is characterised by simultaneous release of litter N and incorporation of external N.

13.6.2 Approaches to Test the Hypotheses and Results

13.6.2.1 Experiment 1: The Transect Study

With the aim of testing the first hypothesis, the following experiment (hereafter referred to as the transect study) was established in order to study the effects of climatic variables on litter decomposition under natural environmental conditions. A standard beech leaf litter was incubated at four different forest sites of the CANIF transect.

Beech leaf litter collected from traps at Collelongo (Italy) in autumn 1996 was used as standard litter for the transect study. The material was incubated for 2 years by using litter bags (Cotrufo and Ineson 1996), at four beech forests of the CANIF transect: Collelongo (Italy), Aubure (France), Schacht (Germany) and Sorø (Denmark). For a full description of sites refer to Persson et al. (Chap. 2, this Vol.).

By the end of the experiment, rates of litter mass loss were significantly affected by the site of incubation (ANOVA; $p < 0.0001$), and the standard beech leaf litter incubated at Sorø, had lost almost three times as much mass as the same litter incubated at Collelongo (Fig. 13.1). During the incubation period, the dynamics of mass loss also differed between sites and a statistically significant interaction (two-way ANOVA; $p < 0.0001$) was observed between site of incubation and time. For instance, at Aubure, litter decomposed faster during the first year, but decay rates decreased sharply in the second year. By contrast, litter incubated at Sorø decomposed faster during the second

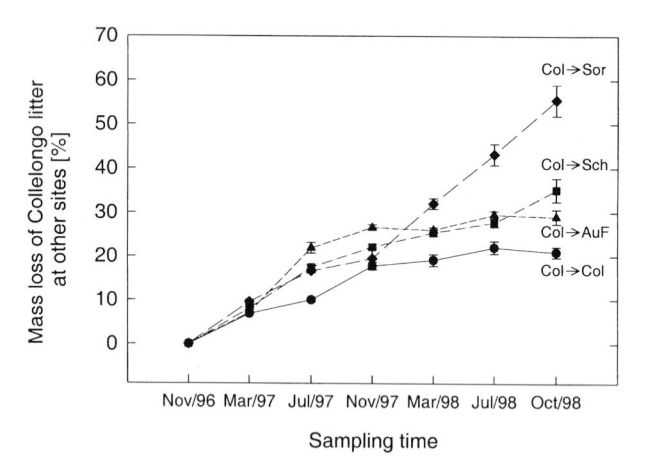

Fig. 13.1. Mass loss of a standard beech leaf litter incubated at four sites along a European transect of beech forests. A two-name nomenclature is used to describe litter types, with the names Col, AuF, Sch and Sor representing Collelongo (Italy), Aubure (France), Schacht (Germany) and Sorø (Denmark), respectively. The first name of the nomenclature gives the source of the litter, whilst the second name identifies the site where the litter was decomposed; hence, for example, Col-AuF denotes litter grown at Collelongo and decomposed at Aubure. Values are means ($n = 7$), with standard errors as vertical bars

than during the first year (Fig. 13.1; Table 13.2). In this case, the very high second year litter decay rate at Sorø was probably a consequence of the warm winter months in 1998. At Schacht, litter decomposed at constant rates throughout the incubation period, whereas at Collelongo decomposition appeared to suffer periods of very low decomposition rates, corresponding to periods of unfavourable weather conditions both in the winter, due to prolonged periods of low soil temperatures, and in the summer, due to low levels of precipitation (Fig. 13.1, Table 13.2). By the end of the experiment the litter incubated at Sorø had reached the highest value of mass loss (55%) whilst the litter at Collelongo showed the lowest mass loss value (21%). At both the central European locations (Aubure and Schacht) litter had lost 36% of its original weight by the end of the experiment (Fig. 13.1).

Final mass loss values were regressed against geographical parameters (i.e. latitude and altitude), selected climatic variables (i.e. AET and PET – computed from Thornthwaite and Mather 1955 – mean annual temperature and precipitation) and N deposition. Decomposition rates appeared to follow a clear geographical pattern, with lower decay rates being observed at high (Collelongo) compared to low (Sorø) elevation sites, and the same pattern from south to north. Altitude had the closest correlation to final litter mass loss (Fig. 13.2; $p < 0.01$; $R^2 = 0.99$), but latitude was also significantly correlated with litter decomposition ($p < 0.05$; $R^2 = 0.94$). However, none of the climatic variables tested could explain the variation in final litter mass loss between different sites, with the exception of PET which explained 88% of the variation ($p = 0.052$). No significant correlation was obtained with N deposition. These results contrast with studies on coniferous sites which reported increasing decay rates moving from northern to southern locations, with AET being the leading climatic

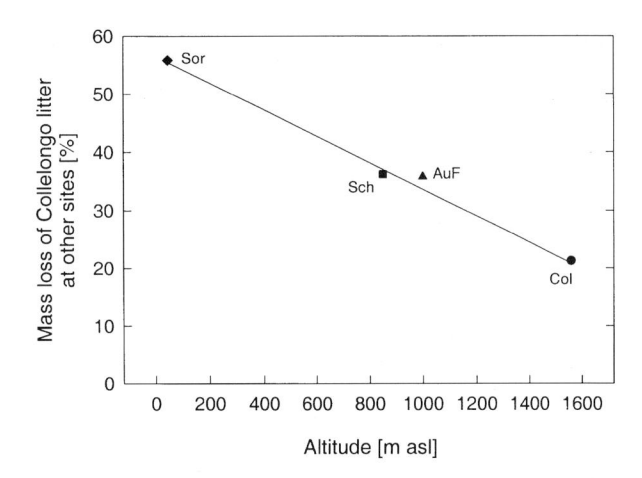

Fig. 13.2. Final year mass loss values of a standard beech leaf litter from litter bags regressed against the elevation of the four incubation sites, at Aubure (France), Collelongo (Italy), Sorø (Denmark) and Schacht (Germany). $p < 0.01$, $R^2 = 0.99$, $y = -0.22x + 57$

constraint on litter decay rates (Meentemeyer 1978; Berg et al. 1993). However, within each regional climate, different degree of relation between AET and decay rates were observed (Berg et al. 1993). In our study, the altitudinal gradient appeared to play a more important role. Since the most northern site (Sorø) is also the one at the lowest altitude, and the most southern (Collelongo) is at the highest altitude, a positive correlation between decay rates and latitude was observed. At Collelongo, the combination of both cold winter and dry summer probably constrained decomposition when compared to the other sites (Fig. 13.1).

13.6.2.2 Experiment 2: The Transplant Study

To determine whether beech leaf litter grown at different forest sites along the CANIF transect differ in their elemental composition and subsequent decomposition rates (i.e. to test hypothesis 2), beech leaf litters, collected from traps in October 1996 at Aubure (France), Collelongo (Italy), Schacht (Germany) and Sorø (Denmark), were analysed for their elemental composition and incubated in litter bags at a standard forest site (Collelongo, Italy).

The chemical composition of the beech leaf litters collected at the different sites along the European transect, is reported in Table 13.1. In contrast to the original hypothesis, no significant differences were observed between the sites of litter collection for most of the chemical parameters studied, and the litter of the four locations had similar N concentrations and C/N ratios. Beech litter derived from Aubure had higher values of P and Mn when compared to the other litter types, whereas litter from Collelongo had an Al concentration twice that of the other litter types (Table 13.1). No consistent patterns of variation were observed among nutrients; neither geographical nor climatic factors appeared to be correlated to litter elemental components in this study.

Table 13.1. Initial elemental composition (mg g^{-1}) and C/N ratio for the beech leaf litters used for the transplant decomposition experiment

Site	N	C	C/N	P	S	K	Ca	Mg	Mn	Na	Al	Fe	Cu	Zn	B
AuF	11.8	503	43	0.857	1.07	2.74	7.82	0.617	0.998	0.140	0.098	0.10	0.042	0.056	0.042
Col	10.8	494	45	0.571	1.12	2.40	17.1	1.53	0.310	0.213	0.206	0.145	0.009	0.040	0.033
Sch	11.3	499	44	0.426	1.03	0.902	10.5	0.927	0.420	0.054	0.093	0.096	0.009	0.053	0.053
Sor	10.8	501	46	0.519	1.21	1.29	16.4	1.58	0.141	0.164	0.061	0.069	0.009	0.049	0.035

Table 13.2. Litter decomposition rate constants (k) derived from fitting the single exponential Olson's (1963) model (mass loss = e^{-kt}; with t = days of incubation) to the data from the transect and the transplant experiment, using SAS (version 6). Additionally, first and second year decomposition rate constants were calculated individually. A two-name nomenclature is used to describe litter types, with the names Col, AuF, Sch and Sor representing Collelongo (Italy), Aubure (France), Schacht (Germany) and Sorø (Denmark), respectively. The first name of the nomenclature gives the source of the litter, whilst the second name identifies the site where the litter was decomposed; hence, for example, Col-AuF denotes litter grown at Collelongo and decomposed at Aubure

Litter type	Daily K values	Yearly K values (1st year)	Yearly K values (2nd year)
AuF – Col	0.00022	0.086	0.053
Sch – Col	0.00028	0.106	0.097
Sor – Col	0.00036	0.132	0.106
Col – Col	0.00043	0.217	0.046
Col – Sch	0.00062	0.247	0.228
Col – AuF	0.00066	0.308	0.148
Col – Sor	0.00090	0.233	0.663

Rates of mass loss for the four beech litters incubated at the standard site (Collelongo) were slow and the native litter decomposed faster than the transplanted ones (Fig. 13.3). In particular, the litter derived from Aubure showed the lowest decay rates (Fig. 13.3; Table 13.2), but after the 2 years of incubation these differences were no longer statistically significant, in part due to an increased variability in mass loss between replicate bags at advanced decomposition stages, which is a common problem in litterbag studies (Latter et al. 1998). A significant negative correlation was observed between litter N concentrations and final mass loss ($p < 0.05$, $R^2 = 0.97$, mass loss = −90N + 118). However, given the narrow range of N concentrations in the four litter types (1.08–1.17%), generalisations cannot be made. None of the other nutrient concentrations was significantly correlated to final mass loss values nor to decay rates.

Along a broad north-south latitudinal transect (69°45′N–49°53′N), Berg et al. (1995) observed for pine species that concentrations of N, P, S and K in needle litters increased with decreasing latitude and increasing AET. These findings were attributed to increased nutrient availability in soils in warmer and wetter climates (lower latitudes of the transect) and as a result of increased microbial activity. Litter concentrations of Mg and Mn appeared to be related more to site-specific soil properties

Litter Decomposition

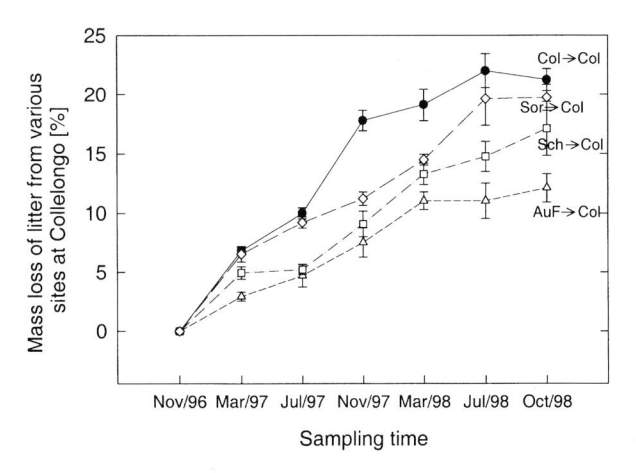

Fig. 13.3. Mass loss of beech leaf litters derived from four sites along a European transect of beech forests and incubated at a standard site. The same nomenclature as for Fig. 13.1 is used. Values are means ($n = 7$), with standard errors as vertical bars

than to climate. The effect of different N deposition on litter N concentrations was considered to be of limited importance (Berg et al. 1995). The transect used in our study had a much smaller latitudinal and climatic range than the one used by Berg et al. (1995), and within this range the elemental composition of beech litters appeared to be mostly similar (Table 13.1) and not related to geographical location, climate and N deposition. These contrasting results could be interpreted as being the consequences of: (1) less extreme climatic conditions at the sites of the present transect; (2) different behaviour among broadleaf and coniferous litter; (3) a smaller data set (i.e. 4 compared to 59 sites) in the current study. In fact, within smaller latitudinal and AET ranges of the pine transect, correlations between AET and latitude with litter nutrient concentrations were less clear (Berg et al. 1995).

Nutrient concentrations did not seem to play an important role in determining litter decomposition rates in the current study, possibly because all litter types had concentrations of essential nutrients above limiting values. No other quality parameters which might have played a more important role were investigated. The fact that the native litter appeared to decompose faster, especially at the beginning of the decomposition process (Fig. 13.2), could be due to an adapted edaphic community of microorganisms which preferentially digested the native litter, or to some quality factors (i.e. lignin, toughness etc.) which were not measured. Work in progress will help to clarify these uncertainties.

13.6.2.3 Experiment 3: Cellulase Activity

Initial cellulose concentrations have been reported to make up half of the initial dry weight of beech leaf litter (Downs et al. 1996). The rate-limiting step in cellulose turnover is the cellulase-mediated hydrolysis of glycosidic bonds leading to the production of low molecular weight subunits, which are subsequently assimilated and mineralised by the saprotrophic community. The intimate association of cellulose

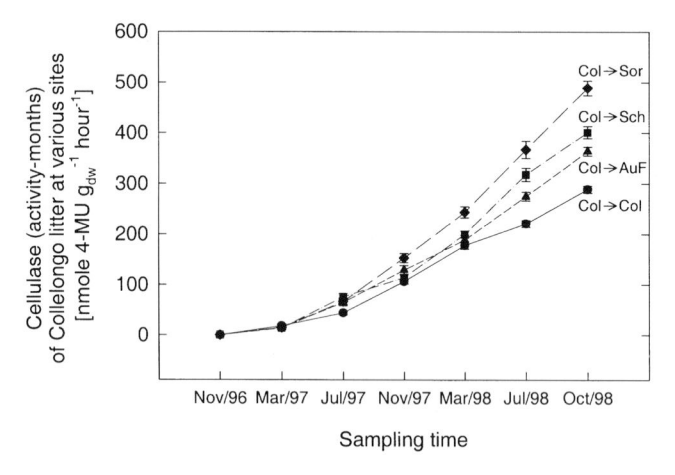

Fig. 13.4. Cumulative cellulase activity of a standard beech leaf litter incubated at four sites along a European transect. The same nomenclature as for Fig. 13.1 is used. Values are means ($n = 9$), with standard errors as *vertical bars*

with lignin and other structural polymers in the plant cell wall makes the cellulase activity a potentially rate limiting process in beech litter decomposition.

Fungi have been reported to be the predominant source of cellulolytic enzymes in soils (Hayano and Katami 1985; Hayano 1986; Rhee et al. 1994; Miller et al. 1998; Møller et al. 1999) and in beech litter (Miller et al. 1998). A novel method based on the use of specific enzyme activities to selectively determine fungal presence and activity was recently proposed for use in soil (Miller et al. 1998) and beech litter (Møller et al. 1999). In the present study we used this technique to target fungal cellulolytic activity in direct enzyme activity determinations on beech litter samples, with the aim to test hypothesis 3.

Cellulase activity was measured in the decomposing beech litters derived from the experiments described above (i.e. the transect and the transplant studies). In Fig. 13.4, the integrated activity showed a clear north-south geographical differentiation between sites. Thus, cumulative cellulase activity increased with increasing northern latitude. At the end of the experiment the activity was approximately 1.5 times higher at Sorø as compared to that of the same litter incubated at Collelongo. Results from the transplant study showed that cumulative cellulase activity for the transplanted litters, incubated at the standard site (Collelongo), showed no statistically significant difference among litter types, although the transplanted litters from Aubure, Schacht and Sorø, were characterised by a consistently lower activity as compared to the native litter (Fig. 13.5).

Linear regression analysis on the pooled data set (transect and transplant experiments), showed that 73% of the total variation in mass loss could be explained by regression on cumulative cellulase activity ($R^2 = 0.73$, $p < 0.001$; Fig. 13.6). The observations of a high correlation between temporally integrated cellulase activity and mass loss seem to reflect a unique relationship. The cellulase activity reflected the same overall dynamics as did mass loss data in the two experiments. These results are consistent with the findings of Sinsabough et al. (1992), reporting that 83% of the

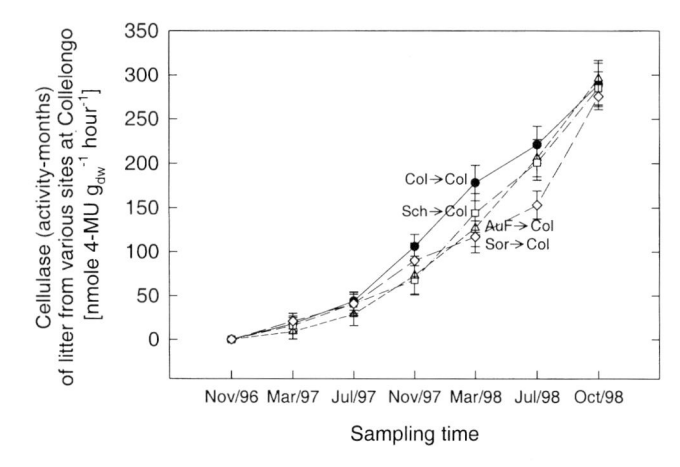

Fig. 13.5. Cumulative cellulase activity of beech leaf litters derived from four sites along a European transect of beech forests and incubated at a standard site. The same nomenclature as for Fig. 13.1. Values are means ($n = 9$), with standard errors as vertical bars

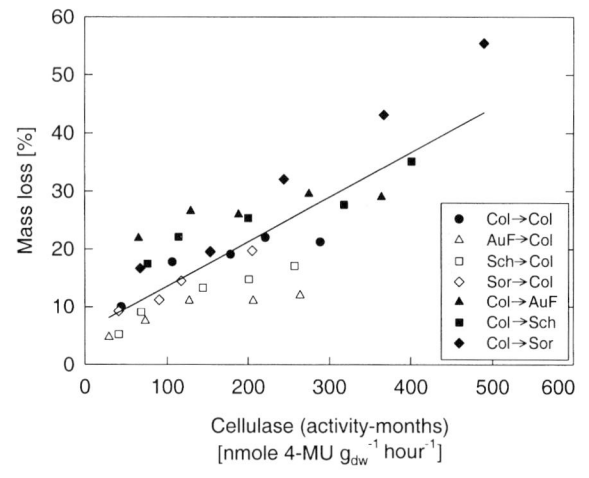

Fig. 13.6. Linear regression model of mass loss as a function of cumulative cellulase activity for pooled data on decomposing beech litter ($n = 35$, $p < 0.001$, $R^2 = 0.73$, $y = 0.077x + 5.87$)

total variation in mass loss could be explained by regression on cumulative lignocellulase activity, in decomposing birch sticks.

The data indicate that initial beech litter mass loss is strongly related to fungal-mediated cellulase activity. The fact that the relationship between fungal cellulase activity and litter mass loss transcended site specific differences on a large geographical scale, is consistent with the emerging similarity in decomposition processes among different landscape patches described by Sinsabough et al. (1992). The data have shown that the three main regulators of decomposition processes: the physico-chemical environment, the resource and the catabolic activities of decomposer organ-

Table 13.3. Characterisation of the forest stands (humus type, C/N ratio, age, height, DBH), surface of each of the experimental plots, amount of ^{15}N-labelled litter deposited, N concentration and ^{15}N excess of the ^{15}N-labelled litter distributed at Aubure (France), Ebrach (Germany) and Collelongo (Italy)

Location	Soil		Beech stand			Plots		^{15}N-labelled litter		
	Humus type	C/N (0–10 cm)	Age (years)	Height (m)	DBH (cm)	Surface (m^2)	Number	Amount (g m^2)	N (%)	^{15}N excess (atom%)
Aubure	Moder	21	50	13.4	7.9	4	5	200	8.7	3.24
Ebrach	Moder	20	15	3.7	2.5	1	9	250	8.5	2.13
Collelongo	Mull	15	30	6.2	4.4	1	9	250	8.5	2.13

isms seem to converge in the activity of a C-acquiring key-enzyme activity. This study thus corroborates the utility of an enzymatic approach for monitoring and modelling decomposition processes as recently proposed by Sinsabough and Moorhead (1997).

13.6.2.4 Experiment 4: ^{15}N-Labelled Litter Decomposition

The N dynamic in decomposing ^{15}N-labelled litter was studied at selected beech forest sites along the CANIF transect: Aubure (France), Ebrach (Germany) and Collelongo (Italy). At each site, between 5 (Aubure) or 9 circular plots around single trees were isolated by digging a trench and inserting a plastic ring down to 30 cm soil depth. Plot size were in relation to tree age and height (Table 13.3). After removal of the F layer, the local annual litterfall was replaced by 200 or 250 g m^{-2} of labelled beech litter (see methodology in Zeller et al. 1998), which was protected by a nylon net. The labelled litter had N and P concentrations of 0.9 and 0.4%, respectively and contained 2.1% (Ebrach and Collelongo) or 3.2% (Aubure) excess ^{15}N (Table 13.4). A parallel litterbag experiment was established at Aubure, with a labelled litter whose ^{15}N excess was similar to that in the other experiment but with a higher N content (1.5%). Labelled litter was deposited in November or December (litterbags) at Aubure, in March at Ebrach and in May at Collelongo. Samples of litter were collected at each site with different frequencies, weighted and analysed for N and ^{15}N concentrations.

Mass Loss. At all sites, the amount of ^{15}N-labelled litter decreased continuously during decomposition (Fig. 13.7) following a negative exponential function. During the first year, litter mass loss values at Aubure and Ebrach were quite similar, and higher than at Collelongo. At this last site, the deposition of litter before summer, which was quite dry, may have limited the activity of decomposers. During the second year, mass loss was slightly higher at Ebrach than at Aubure but the differences between the three sites decreased in the third year, consistently with results observed for the transect experiment described above.

N Dynamic. N concentrations in the ^{15}N-labelled litter increased at all sites, sharply during the first year, slightly during the second year and levelled off in the third year. At Aubure, in relation to the original N content, the increase was proportionally higher in the N-poor litter in comparison to the N-rich litter. The final N concentration was somewhat lower in Collelongo (Fig. 13.8).

Litter Decomposition

Table 13.4. C/N ratio and nutrient content in the annual litterfall collected at the Aubure, Ebrach and Collelongo experimental sites and in the provided ^{15}N-labelled litter at beginning of the experiments

Location	Nutrient concentration in mg g^{-1} litter dry weight						
	C/N	N	S	P	Mg	Ca	K
Aubure (original litter)	35	14.2	1.2	1.2	1.0	7.9	6.2
Ebrach (original litter)	39	12.8	1.3	0.9	1.3	6.3	4.7
Collelongo (original litter)	38	13.0	1.5	0.8	1.6	23.8	2.2
N-rich ^{15}N-labelled litter deposited at Aubure (litterbag experiment)	32	14.8	1.2	0.6	1.3	14.7	2.8
N-poor ^{15}N-labelled litter deposited at Aubure	55	8.7	1.1	0.3	1.2	15.1	3.1
N-poor ^{15}N-labelled litter deposited at Ebrach and Collelongo	58	8.5	1.2	0.4	1.4	14.8	2.9

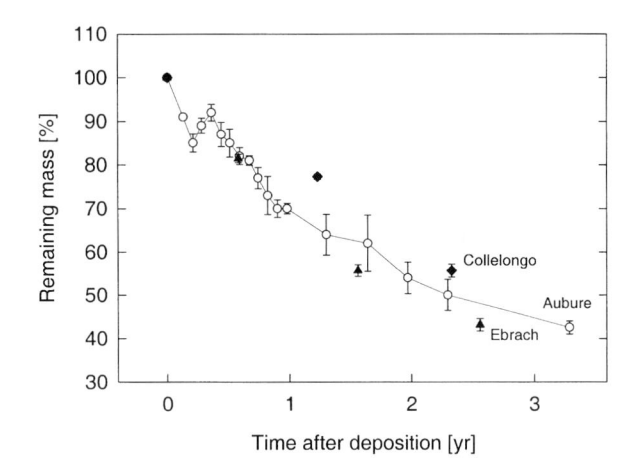

Fig. 13.7. Remaining mass of ^{15}N-labelled beech litter (15 mg N g^{-1}) incubated at Aubure, Ebrach and Collelongo along a European transect

At all sites, the ^{15}N excess of the litter decreased. The slope of the decrease was higher for the N-rich litter. The combination of N concentration increase and ^{15}N excess decrease, suggested that external N was incorporated while litter N was released. Incorporated and released N were calculated according to Berg (1988):

$$\text{released N} = \frac{(\text{mass}^{15}\text{N}_{T0} - \text{mass}^{15}\text{N}_{T1}) - R_{std}(\text{massN}_{T1} - \text{massN}_{T0})}{R_{litter} - R_{std}}$$

$$\text{incorporated N} = \text{massN}_{T1} - (\text{massN}_{T0} + \text{massN}_{released})$$

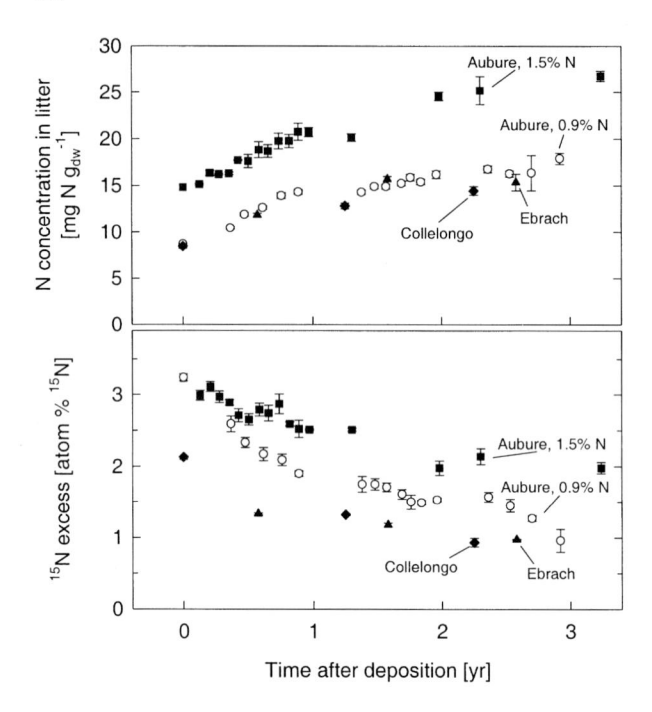

Fig. 13.8. Time variation in N concentration and ^{15}N excess in the ^{15}N-labelled beech litter during litter decomposition. The same nomenclature as for Fig. 13.7 is used, excepting for Aubure, where two litters with high (15 mg N g^{-1}) and low (9 mg N g^{-1}) N concentration were used

where, $R = \dfrac{^{15}N}{^{14}N + ^{15}N}$.

At Aubure, about 16, 25 and 64% of the initial N amount of the litter poor in N had been released after 4 months, 1 year and 3 years, respectively (Fig. 13.9). While N was released from the litter, external N was simultaneously incorporated into the decomposing litter. Almost all external N was incorporated during the first year. Over a 2-year period, the release of litter N balanced the amount of external N incorporated. Similar results were obtained at Ebrach, while a lower N release and N incorporation occurred at Collelongo during the second year.

In comparison, at Aubure the N-rich litter released a much larger amount of N, which is most likely soluble N, during the first months of decomposition. Later, the release rate of N was similar to that of the other litter types. Measurements of ergosterol concentration in the litter at Aubure, as an indicator for fungal biomass (Martin et al. 1990; Djajakirana et al. 1996), showed that fungal N accounted for approximately 30% of the external N incorporated into the litter. Additional N might be provided by throughfall and faecal pellets. At Aubure, deposition of throughfall N varied between 6 to 8 kg N ha^{-1} a^{-1}, with about 40% as NH$_4^+$. As throughfall nitrate and ammonium can be incorporated at rates of approximately 1 kg N ha^{-1} a^{-1} in decomposing litter (Buchmann et al. 1995; Downs et al. 1996; Koopmans et al. 1998), throughfall N could account for approximately 30–50% of the external N incorporation at this site. ^{15}N-labelled nitrate or ammonium applied to beech litter is incorporated mostly via bio-

Litter Decomposition

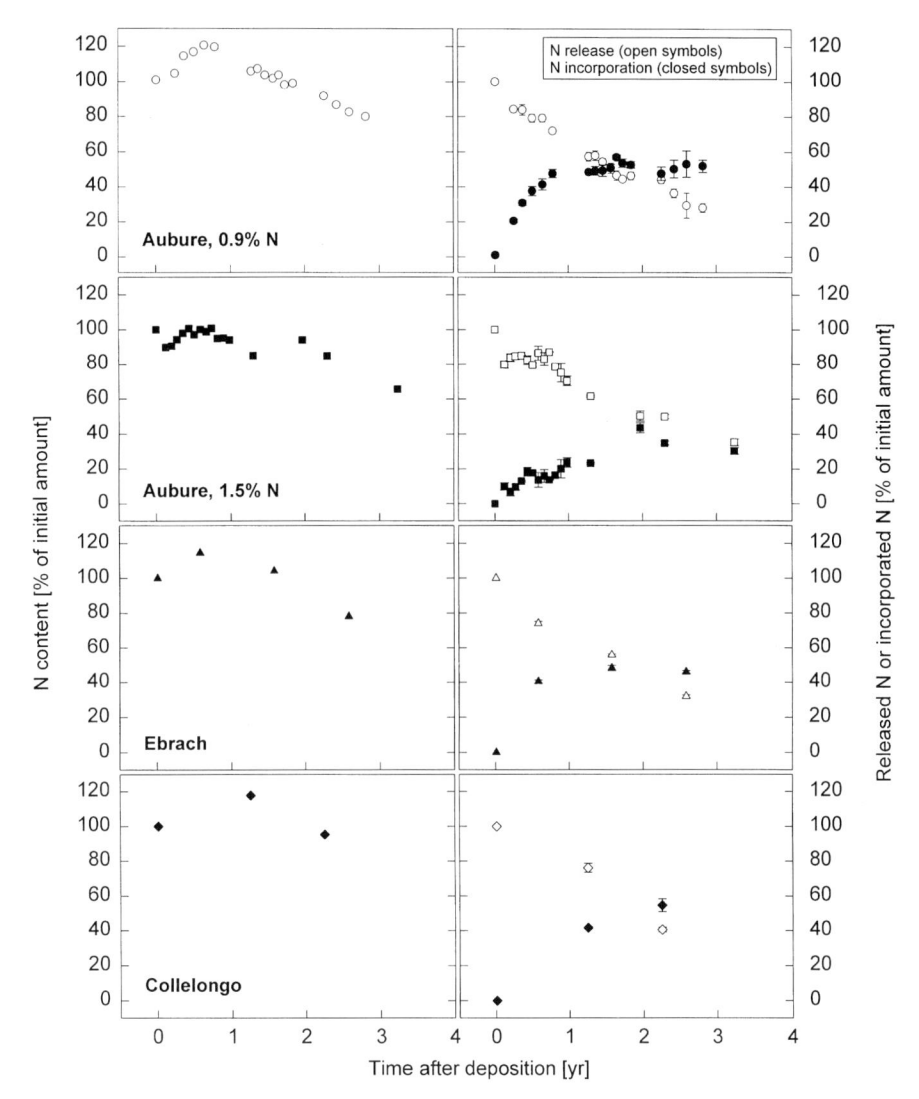

Fig. 13.9. Evolution of the N content in the ^{15}N-labelled litter for each site. Accumulated N incorporation (*closed symbols*) and N release (*open symbols*) in the ^{15}N-labelled litter during litter decomposition at each site

logical processes into peptides and other compounds, as detected by ^{15}N NMR (Clinton et al. 1995).

At all sites, a close linear correlation ($R^2 = 0.98$, for Aubure) between N release and mass loss of the litter poor in N was observed (Fig. 13.10). A similar linear relationship was observed for the N-rich litter, except for the first 6-month period, where the loss in N was not related to an equivalent loss in mass. These strong correlations exist independently of the large climatic and site-specific differences between the three sites. N release during litter decomposition results from leaching of soluble N com-

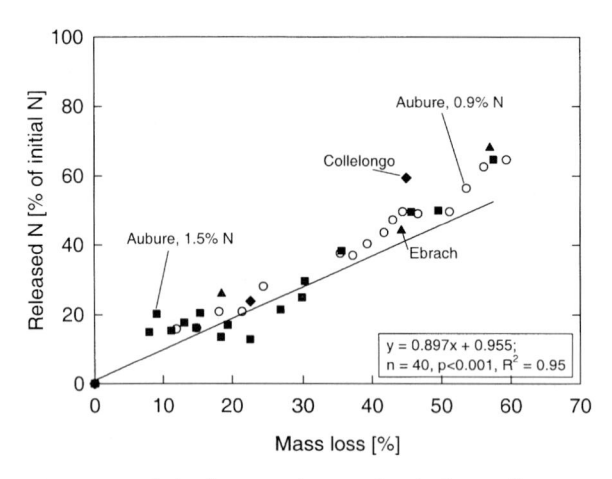

Fig. 13.10. Relation between the cumulated release of litter N and litter mass loss measured at Aubure (litters with high, 15 mg N g^{-1}, and low, 9 mg N g^{-1}, N), Ebrach and Collelongo

pounds, followed by consumption and degradation of polyphenol protein complexes by fungi and soil microorganisms (Joergensen and Meyer 1989; Tietema and Wessel 1994; Setälä et al. 1996) and final consumption by soil animals. The comparison between N-rich and N-poor litter types suggests that when the litter is rich in N, the rate of soluble N leaching is higher than the rate of C release by leaching of soluble compounds (sugars) or mineralisation. Then, C and N were lost at similar rates. As N is mainly incorporated in polyphenols and proteins, and C in cellulose, it should be assumed that their common release reflects the parallel invasion and dissolution by microorganisms and consumption by animals. As none of the sites had a real raw humus, it cannot be stated that this parallelism is true under all conditions, but it suggests that among different humus and site conditions, differences in the fate of litter derived N and C occurs mainly after the physical disappearance of the litter.

13.7 Conclusions

Leaf litter decomposition in European beech forests appears to be dependent on site-specific characteristics, with local climate being the most important rate-constraining factor. In the European transect studied here, beech forests occurred along a north-south (55°58′–41°52′) latitudinal transect and at progressively higher altitude (from 45 to 1560 m asl). The Danish site was not only the most northern, but also the closest to the sea. Therefore, local climates followed site-specific seasonal patterns, which resulted in high decomposition rates at the Danish site, where winters were relatively mild and decomposition could occur all year long at high rates. Despite this, mean annual climatic parameters for the sites, such as temperatures, precipitation and computed PET and AET (Thornthwaite and Mather 1955), did not appear to be good predictors of decomposition in this study. Also levels of N deposition at the different sites seemed not to affect litter decay rates.

Independently of the studied site, N dynamics in decomposing beech litters followed a general pattern, with both N release and incorporation occurring simultaneously. Large fluxes of external N from different pools in decomposing beech litter were observed, with the amount of external N incorporation remaining almost the same, independent of the initial litter N concentration. This finding suggests that litter N dynamics was driven by the same mechanism at all sites. Additionally, rates of N mineralisation were strictly related to rates of mass loss, which makes it possible to predict litter N release from mass loss data. In fact, after a first N-leaching phase whose entity depended on the initial litter N concentration, rates of decomposition appeared to constrain N mineralisation, more than the reverse. Collelongo (the Italian beech forest site) had the lowest rates of decomposition and N transfers. However, N concentrations in leaf litters were similar to those of the other sites studied, and in particular to the N concentration of the beech leaf litter at the Danish site, which was characterised by fast decomposition rates. Beech forests of France and Germany showed very similar patterns of litter decomposition and N dynamics, and produced litters with the same N concentrations. Across the studied transect differences in litter elemental composition appeared to be minor. The elemental composition of litter derived from the different sites did not affect subsequent litter decomposition rates.

Decomposition of this study was measured as weight loss from litter bags. Although this is a widely used method for field decomposition study, the litterbag method has major drawbacks, including a lack of accuracy due to loss of fragments and soil contamination at later stages of sampling. Additionally, the litter bag method is based on the assumption that all the mass that is lost from bags is fully mineralised. Therefore, it does not account for partially decomposed organic compounds, which may be lost from bags, but immobilised in the soil organic matter with longer turnover times. There is an urgency to find more accurate ways to determine field decay rates, and the use of stable isotopes appears to be the most promising approach. We used ^{15}N techniques to follow patterns of N release and incorporation from litters, and give evidence of the usefulness of this now broadly accepted method. Similar studies need to be performed using C isotopes to accurately determine C mineralisation processes.

The present study emphasises the utility of an enzymatic approach in decomposition studies by providing a tool that can be used to link abiotic and biotic activities and increase our understanding of decomposition processes at the molecular level.

Acknowledgements. The authors are grateful to Dr. G. Matteucci, Dr. G. Bauer and M. Hein for their assistance throughout the project and to A. Spatola and E. Pantschitz for their technical help.

References

Aerts R (1997) Climate, leaf litter chemistry and leaf litter decomposition in terrestrial ecosystems: a triangular relationship. Oikos 79:439–449

Andersson JPE, Domsch KH (1975) Measurement of bacterial and fungal contributions to respiration of selected agricultural and forest soils. Can J Microbiol 21:314–322

Beare MH, Neely CL, Coleman DC, Hargrove WL (1990) A substrate-induced respiration (SIR) method for measurement of fungal and bacterial biomass on plant residues. Soil Biol Biochem 22:585–594

Berg B (1988) Dynamics of nitrogen (^{15}N) in decomposing Scots pine (*Pinus sylvestris*) needle litter. Long-term decomposition in a Scots pine forest VI. Can J Bot 66:1539–1546

Berg B, Staaf H (1981) Leaching, accumulation and release of nitrogen in decomposing forest litter. In: Clark FT, Rosswall T (eds) Terrestrial nitrogen cycle. Ecol Bull 33:163–178

Berg B, Wessén B, Ekbohm G (1982) Nitrogen level and decomposition of Scots pine needle litter. Oikos 38:291–296

Berg B, Berg MP, Bottner P, Box E, Breymeyer A, Calvo de Anta RC, Coûteaux MM, Escudero A, Gallardo A, Kratz W, Madeira M, McClaugherty C, Meentemeyer V, Muñoz F, Piussi P, Remacle J, Virzo De Santo A (1993) Litter mass-loss rates in pine forests of Europe and eastern United States: some relationships with climate and litter quality. Biogeochemistry 20:127–159

Berg B, Calvo de Anta R, Escudero A, Gärdenäs A, Johansson MB, Laskowski R, Madeira M, Mälkönen, McClaugherty C, Meentemeyer V, Virzo De Santo A (1995) The chemical composition of newly shed needle litter of Scots pine and some other pine species in a climatic transect. X Long-term decomposition in Scots pine forest. Can J Bot 73:1423–1435

Blair JM, Crossley DA Jr, Callagham LC (1992) Effects of litter quality and micro-arthropods on N dynamics and retention of exogenous ^{15}N in decomposing litter. Biol Fertil Soils 12:241–252

Buchmann N, Schulze E-D, Gebauer G (1995) ^{15}N-ammonium and ^{15}N-nitrate uptake of a 15-year-old *Picea abies* plantation. Oecologia 102:361–370

Bunnel FL, Tait DEN, Flanagan PW, van Cleve K (1977) Microbial respiration and substrate weight loss. I. A general model of the influence of abiotic variables. Soil Biol Biochem 9:33–40

Clinton PW, Newman RH, Allen RB (1995) Immobilisation of ^{15}N in forest litter studied by ^{15}N CPMAS-NMR spectroscopy. Eur J Soil Sci 46:551–556

Colpaert JV, van Tichelen KK (1996) Decomposition, nitrogen and phosphorus mineralisation from beech leaf litter colonised with ectomycorrhizal or litter decomposing basidiomycetes. New Phytol 134:123–132

Cotrufo MF, Ineson P (1996) Elevated CO_2 reduces field decomposition rates of *Betula pendula* Roth. leaf litter. Oecologia 106:525–530

Coûteaux MM, Bottner P, Berg B (1995) Litter decomposition, climate and litter quality. Tree 10:63–66

Dickson BA, Crocker RL (1953) A chronosequence of soils and vegetation near Mt. Shasta, California. II. The development of the forest floors and the carbon and nitrogen profiles of the soils. J Soil Sci 4:142–154

Dighton J (1995) Nutrient cycling in different terrestrial ecosystems in relation to fungi. Can J Bot 73:1349–1360

Djajakirana G, Joergensen RG, Meyer B (1996) Ergosterol and microbial biomass relationship in soil. Biol Fertil Soils 22:299–304

Downs M, Nadelhoffer K, Melillo JM, Aber JD (1996) Immobilization of a ^{15}N-labelled nitrate addition by decomposing forest litter. Oecologia 105:141–150

Edmonds RL (1980) Litter decomposition and nutrient release in Douglas-fir, red alder, western hemlock, and Pacific silver fir ecosystems in western Washington. Can J For Res 10:327–337

Flanagan PW, Van Cleve K (1983) Nutrient cycling in relation to decomposition and organic-matter quality in taiga ecosystems. Can J For Res 13:795–817

Fog K (1988) The effect of added nitrogen on the rate of decomposition of organic matter. Biol Rev 63:433–462

Hart SC, Firestone MK, Paul EA, Smith JL (1993) Flow and fate of soil nitrogen in an annual grassland and a young mixed conifer-forest. Soil Biol Biochem 25:431–442

Hayano K (1986) Cellulase complex in a tomato field soil: induction, localization and some properties. Soil Biol Biochem 18:215–219

Hayano K, Katami A (1985) Origin and properties of β-glucosidase activity of tomato-field soil. Soil Biol Biochem 17:553–557

Howard DM, Howard PJA (1993) Relationships between CO_2 evolution, moisture content and temperature for a range of soil types. Soil Biol Biochem 25:1537–1546

Hunt HW, Ingham ER, Coleman DC, Elliott DC, Reid CPP (1988) Nitrogen limitation of production and decomposition in prairie, mountain meadow and Pine forest. Ecology 69:1009–1016

Joergensen RG, Meyer B (1989) Nutrient changes in decomposing beech leaf litter assessed using a solution flux approach. J Soil Sci 41:279-293

Kirk TK, Farell RL (1987) Enzymatic "combustion": the microbial degradation of lignin. Annu Rev Microbiol 41:465-505

Klyosov AA (1990) Trends in biochemistry and enzymology of cellulose degradation. Biochemistry 47:10577-10585

Koopmans CJ, Tietema A, Verstraten JM (1998) Effects of reduced N deposition on litter decomposition and cycling in two N-saturated forests in the Netherlands. Soil Biol Biochem 30:141-151

Latter PM, Howson G, Howard DM, Scott WA (1998) Long-term study of litter decomposition on a Pennine peat bog: which regression? Oecologia 113:94-103

Ljungdahl LG, Eriksson K-E (1985) Ecology of microbial cellulose degradation. Adv Microbiol Ecol 8:237-299

Martin F, Delaruelle C, Hilbert JL (1990) An improved ergosterol assay to estimate fungal biomass in ectomycorrhizas. Mycol Res 94:1049-1064

Mason CF (1977) Decomposition. The Camelot Press, Southampton

Mayer O (1993) Functional groups of microorganisms. In: Schulze E-D, Mooney HA (eds) Biodiversity and ecosystem function. Ecological Studies 99. Springer, Berlin Heidelberg New York, pp 67-96

McClaugherty CA, Berg B (1987) Cellulose, lignin and nitrogen concentrations as rate-regulating factors in late stages of forest litter decomposition. Pedobiologia 30:101-112

Meentemeyer V (1978) Macroclimate and lignin control of litter decomposition rates. Ecology 59:465-472

Meentemeyer V, Berg B (1986) Regional variation in rate mass loss of *Pinus sylvestris* needle litter in Swedish pine forest as influenced by climate and litter quality. Scand J For Res 1:167-180

Mellilo JM, Aber JD, Muratore JF (1982) Nitrogen and lignin control of hardwood leaf litter decomposition dynamics. Ecology 63:621-626

Miller M, Paloj A, Rangger A, Reeslev M, Kjøller A (1998) The use of fluorogenic substrates to measure fungal presence and activity in soil. Appl Environ Microbiol 64:613-617

Møller J, Miller M, Kjøller A (1999) Fungal-bacterial interaction on beech leaves: influence on decomposition and dissolved organic carbon quality. Soil Biol Biochem 31:367-374

Myrold DD (1990) Effects of acid deposition on soil organisms. In: Lucier AA, Haines SG (eds) Mechanisms of forest response to acidic deposition. Springer Berlin Heidelberg New York, pp 163-187

Olson JS (1963) Energy storage and the balance of producers and decomposers in ecological systems. Ecology 44:322-331

Parnas H (1975) Model for decomposition of organic material by microorganisms. Soil Biol Biochem 7:161-169

Rhee YH, Hah YC, Hong W (1987) Relative contribution of fungi and bacteria to soil carboxymethylcellulase activity. Soil Biol Biochem 19:479-481

Rodin LE, Bazilevich NI (1967) Production and mineral cycling in terrestrial vegetation. Oliver & Boyd, London

Setälä H, Marshall VG, Trofymow JA (1996) Influence of body size of soil fauna on litter decomposition and ^{15}N uptake by poplar in a pot trial. Soil Biol Biochem 28:1661-1675

Sinsabough RL, Moorhead (1997) Synthesis of litter quality and enzymic approaches to decomposition modelling. In: Cadish G, Giller KE (eds) Driven by nature: plant litter quality and decomposition. CAB International, Wallingford, UK, pp 363-375

Sinsabough RL, Antibus RK, Linkins AE (1991) An enzymic approach in to the analysis of microbial activity during plant litter decomposition. Agric Ecosyst Environ 34:43-54

Sinsabough RL, Antibus RK, Linkins AE, McClaugherty CA, Rayburn DR, Weiland T (1992) Wood decomposition over a first-order watershed: mass loss as a function of lignocellulase activity. Soil Biol Biochem 24:743-749

Staaf H (1980) Release of plant nutrients from decomposing leaf litter in a South Swedish beech forest. Holarct Ecol 3:129-136

Swift MJ, Heal OW, Anderson JM (1979) Decomposition in terrestrial ecosystems. Blackwell, Oxford

Taylor BR, Parkinson D, Parsons WFJ (1989) Nitrogen and lignin content as predictor of litter decay rates: a microcosm test. Ecology 70:97–104

Thornthwaite CW, Mather JR (1955) The water balance. Publ Climatol 8:5–104

Tietema A, Wessel WW (1994) Microbial activity and leaching during initial oak leaf litter decomposition. Biol Fertil Soils 18:49–54

Upadhyay VP, Singh JS (1989) Patterns of nutrient immobilization and release in decomposing forest litter in central Himalaya, India. J Ecol 77:127–146

Zeller B (1998) Contribution à l'étude de la décomposition d'une litière de hêtre, la libération de l'azote, sa minéralization et son prélèvement par le hêtre (*Fagus sylvatica* L.) dans une hêtraie de montagne du bassin versant du Strengbach (Haut-Rhin). Thesis, Université Henri Poincaré, Nancy, France, 138 pp

Zeller B, Colin-Belgrand M, Dambrine E, Martin F (1998) ^{15}N partitioning and production of ^{15}N-labelled litter in beech trees following [^{15}N]urea spray. Ann Sci For 55:375–383

14 Soil Nitrogen Turnover – Mineralisation, Nitrification and Denitrification in European Forest Soils

T. Persson, A. Rudebeck, J.H. Jussy, M. Colin-Belgrand, A. Priemé, E. Dambrine, P.S. Karlsson, and R.M. Sjöberg

14.1 Background and Aim of the Study

Mineralisation, nitrification and denitrification are major processes in the soil nitrogen cycle (see Schulze, Fig. 1.2, this Vol. Chap. 1). Nitrogen (N) mineralisation is by definition the biotic conversion of organic N into inorganic N, mainly NH_4^+ and NO_3^-. Many organism groups, bacteria, fungi and certain soil animals, possess this capability, and the product formed is generally ammonia/ammonium, depending on the soil pH. Ammonium (NH_4^+) can be oxidised via nitrite (NO_2^-) to nitrate (NO_3^-). Both chemoautotrophic (using CO_2 as a C source) and heterotrophic (using organic C as a C source) microorganisms can perform this nitrification process (Prosser 1986). Heterotrophic microorganisms can also produce nitrate directly from organic N sources (Killham 1986). Nitrate can be reduced via nitrite to nitric oxide (NO), nitrous oxide (N_2O) and/or dinitrogen (N_2) (denitrification). In addition to these processes, ammonium/ammonia, nitrite and nitrate can be assimilated (immobilised) by microorganisms, mycorrhizal fungi and roots. Further, ammonia and nitrite are known to react with certain organic compounds and, thus, be chemically immobilised (Nömmik and Vahtras 1982; Azhar El Sayed et al. 1986).

Many of these processes occur simultaneously, which means that the determination of N mineralisation, nitrification and denitrification is problematic. A commonly used method is to remove the root uptake (in the field or in the laboratory) by sieving or by cutting the roots, and let ammonium and nitrate accumulate without any leaching or denitrification. In so doing, the result of gross N mineralisation minus microbial and abiotic immobilisation is calculated, i.e. net N mineralisation (Hart et al. 1994). Net N-mineralisation rate determined in this manner is sometimes assumed to be the flux of inorganic N available to plant uptake. This presupposes, however, that the immobilising soil microorganisms are more competitive with regard to inorganic N than the plant/mycorrhizal association (Zak et al. 1990). If this is not the case, the net N mineralisation estimate is an underestimate of the flux available to the plants. On the other hand, living roots can exude C-rich substances, whereby more N can be assimilated and immobilised by the microorganisms. Thus, the estimate of net N mineralisation determined in the absence of roots can also be higher than when roots are present.

Gross N mineralisation, which is the estimate closest to the definition of N mineralisation, can be determined by the isotopic dilution technique (Hart et al. 1994). Also this method has drawbacks, since $^{15}NH_4^+$ added is probably more available to microbial assimilation than indigenous NH_4^+ at the exchange sites in the soil. This seems most serious when the pool of unlabelled inorganic N is small (Jenkinson et al. 1985).

Ecological Studies, Vol. 142
E.-D. Schulze (ed.) Carbon and Nitrogen Cycling in European Forest Ecosystems
© Springer-Verlag Berlin Heidelberg 2000

Nitrification is an ecological key process, since it converts less readily mobile ammonium into mobile nitrate that is easily leached out of the soil profile. Furthermore, the process produces protons that acidify the soil. Because microorganisms can assimilate nitrate, and some microorganisms even seem to prefer nitrate to ammonium, estimates of net and gross nitrification rates can be quite different (Stark and Hart 1997). Although net nitrification rates are lower than gross nitrification rates, long-term incubations without roots can indicate unnaturally high net nitrification rates because of the abundance of ammonium for the nitrifiers. Therefore, nitrification rates obtained from long-term incubations cannot be extrapolated to the field with the intention of simulating the nitrate situation in vegetated plots.

Nitrous oxide (N_2O) can be produced during nitrification, but denitrification and not nitrification is often the major source of nitrous oxide emission in forest soils (e.g. Ambus 1998). During denitrification, N may be lost as nitric oxide (NO), nitrous oxide or pure nitrogen (N_2). In most forest soils, nitrous oxide is the dominant source of N loss through denitrification, because the emission of nitric oxide is usually very low ($<0.2\,kg\,N\,ha^{-1}\,a^{-1}$; Davidson and Kingerlee 1997) and the N_2O/N_2 ratio of denitrification is usually shifted in favour of nitrous oxide (e.g. Ambus 1998).

Conclusions about factors regulating N mineralisation, nitrification and denitrification have often been made at the site, region or country scale. Few studies have, so far, been designed to compare the regulation of processes over a whole continent. Such studies are needed to obtain generalised conclusions, because plant, faunal and microbial communities have adapted to different physical and chemical climate and, thereby, probably have also affected N turnover.

The main aim of this part of the study was to quantify net N mineralisation, net nitrification and denitrification in the NIPHYS/CANIF transect through Europe and try to assess the effects of single factors like soil N status, tree species (mainly *Picea abies* and *Fagus sylvatica*) and acidity on the regulation of these processes. Soil depth/soil horizon was considered to be a very determining factor as regards N turnover. Studies of gross N mineralisation and gross nitrification were performed at selected sites and soil depths but are not entirely finished and will only briefly be reported here.

More specific aims were (1) to determine net N mineralisation (for all sites and soil layers to a depth of 50 cm) in the laboratory and extrapolate these data to the field, and (2) to determine net nitrification in the laboratory for the same sites and soil layers. Other aims were (3) to manipulate pH in the laboratory to test the acid sensitivity of net nitrification, (4) to determine which of the factors low pH, low NH_4^+ availability and/or low nitrifier populations was responsible for the absence of measurable net nitrification at some northern sites, and (5) to study the importance of autotrophic versus heterotrophic nitrification in the European transect. Further aims were (6) to determine net N mineralisation and nitrification rates in long-term field N-fertilisation experiments to assess the impact of increased N addition, (7) to compare N turnover in soils with similar properties but situated at different temperature climate, and (8) to compare ammonium and nitrate formation rates in undisturbed soil monoliths with those in sieved soil horizons to test the methods used. Finally, (9) we compared the estimates obtained from laboratory incubations with in situ measurements of net N mineralisation and net nitrification in three stands at one site (Aubure) and (10) determined in situ nitrous oxide emission from the soil surface and potential denitrification enzyme activity in soil samples from beech sites.

14.2 Methods Used to Study N Turnover

14.2.1 Determination of Net N Mineralisation and Net Nitrification in the Laboratory

Soil samples including the litter and humus layers were taken from all sites in the NIPHYS and CANIF transect (see Persson et al., Chap. 2, this Vol.) for the overall net mineralisation and net nitrification studies. At all sites and stands, four plots were sampled. In addition, soil samples were also taken from control and fertilised plots at Skogaby, Stråsan and Norrliden in Sweden to check whether N processes found in areas with high N deposition resembled those of N-fertilised plots in areas with low N deposition. Furthermore, soil samples were also taken from Andersby in central Sweden as well as La Clape and Thezan in southern France to elucidate whether different climate can affect the N processes despite similar soil characteristics. The characteristics of each site are given in Persson et al. (Chap. 2, this Vol.)

Litter (L) and humus (FH) layers were quantitatively sampled with a ring-formed 250-cm^2 frame, five units per plot. Just below the humus layer, the upper mineral soil was sampled with an 83-cm^2 corer (three to four units per plot), and the soil was divided into 0–10- and 10–20-cm layers independent of generic horizons. Below 20 cm, the 20–30- and 30–50-cm soil layers were collected with a 16-cm^2 corer (three cores per plot). The 30–50-cm layer was not sampled at Schacht, due to high stoniness. Samples from the same soil layer were pooled for each plot. At sites without a visible humus layer, the 0–10-cm mineral soil layer was subdivided into 0–5 cm (sampled by the 250-cm^2 frame) and 5–10 cm (sampled by the 83-cm^2 corer). For the extra study sites used for specific questions (Norrliden, Stråsan, Andersby, La Clape and the oak stand at Thezan), the number of soil layers sampled was generally smaller.

Within 1–4 days, depending on the distance, the samples were transported to the laboratory, where they were treated without being dried or frozen. In the litter samples, living plants and twig fragments were removed, and the remaining litter material used for the study mainly consisted of leaf or needle litter. Humus samples were passed through a 5-mm sieve, and the mineral soil through a 2-mm sieve to remove roots and stones. Fresh weight/dry weight ratios were determined after the samples were dried at 105 °C for 24 h. Loss-on-ignition (LOI) was determined after combustion at 550 °C for 3 h. Based on the former determination, fresh litter/soil portions corresponding to 6 g dry weight (dw) of litter, 16 g dw of humus layer material and 100 g dw of sieved mineral soil were placed in plastic containers (50 cm^2 surface area, 466 cm^3 volume). The containers had a lid with a 5-mm diameter aperture for gas exchange and were placed in a room with constant temperature of 15 °C. The water content of the samples was kept at 50–60% water-holding capacity, WHC (100% WHC was defined as the water content of inundated soil allowed to drain for 12 h in a 3-cm-high cylinder). The whole procedure before the onset of the incubation (sampling, transport and pretreatment in the laboratory) lasted for about 14 days. During this period, the samples were normally kept at +5 °C.

Because no leaching could occur, inorganic nitrogen accumulated in the system and the accumulation rate was considered as the net N mineralisation rate. A whole incubation period lasted 150 to 210 days. Destructive samplings were made at certain intervals (0, 50, 100 and 150–210 days from the start of the incubation) to characterise the temporal pattern of mineralisation. To make the estimates as comparable as possible and to avoid unnatural ammonium and nitrate levels, only the first 50 days of the incubation period were mostly used in this study (for exceptions, see text). On each sampling occasion, samples from the four plots acted as replicates. The net N-mineralisation rate in a given soil layer was calculated as the inorganic (NH_4^+ + NO_3^-)-N accumulated during the period divided by the number of days. NO_2^- was included in the estimates of NO_3^-. Net nitrification was calculated in the same manner (daily mean accumulation of NO_3-N) as net N mineralisation. With this method of calculation, the nitrification rate can exceed the mineralisation rate if the NH_4^+ pool decreases during the incubation. Since N mineralisation is the transfer of organic N to inorganic N, the net mineralisation rates were expressed as g^{-1} LOI or g^{-1} organic C, not g^{-1} dw.

Estimates of nitrification were based on root-free samples, in which NO_3^- accumulation was not reduced by root uptake but was possibly reduced by microbial NO_3^- assimilation. Therefore, these estimates are called net nitrification.

Inorganic N was extracted with 100 ml 1 M KCl solution for 1 h. The filtrate was photometrically analysed for NH_4^+ and NO_3^- on a FIA STAR 5010 analyser.

The material was analysed using JMP statistical software (SAS Institute Inc. 1989). The variation in net nitrification was compared using linear models including both continuous (deposition, pH, C/N ratio, net N mineralisation and C mineralisation) and nominal factors (tree species). Because tree species turned out to be an important factor for net nitrification rate, spruce and beech sites were analysed separately by multiple and simple regression analysis.

14.2.2 Extrapolation to the Field

To obtain a rough estimate of the annual net N mineralisation and net nitrification in the field, the rates (g^{-1} day^{-1}) obtained in the laboratory at 15 °C and optimal moisture were multiplied by (1) the amount of soil in the specific soil layer (g ha^{-1}) estimated from the quantitative samplings described above and corrected for stoniness, (2) a temperature correction factor according to the Ratkowsky function (Ratkowsky et al. 1982)

$$A^{1/2} = b(T - T_{min}), \tag{1}$$

where A denotes the activity (net N mineralisation rate) at temperature T, T_{min} the minimum temperature at which the activity starts, and b the slope, and (3) a moisture correction factor. Temperature and moisture relations for a humus layer at a site close to Skogaby were studied in the laboratory within this project. The results will be published separately (Seyferth 1998; U. Seyferth and T. Persson, submitted) and the details are not given here. By estimating the A value (in µg N g^{-1} LOI day^{-1}) at 15 °C and 60% WHC (A_{ref}) for each site and soil layer, the corresponding b value [expressed as (µg N g^{-1} LOI day^{-1})$^{1/2}$ °C^{-1}] could be calculated. The present authors (U. Seyferth and T. Persson, submitted) found the b value to be linearly dependent on the logarithm of the water potential between near-optimum (60% WHC) and low (15% WHC, close to the permanent wilting point) soil moisture. By applying the parameters found in our study (U. Seyferth and T. Persson, submitted) to all sites, the correction factor (c_N) for extrapolation from laboratory to the field could be calculated as:

$$c_N = (T - T_{min})^2 / (T_{ref} - T_{min})^2$$
$$*[-0.0128 \log(\Psi) + 0.081]^2 / [-0.0128 \log(\Psi_{ref}) + 0.081]^2 \quad (2)$$

where T was the soil temperature (°C) in the field, T_{min} − 9.3 °C (Seyferth 1998), T_{ref} 15 °C, ψ the soil water potential (MPa) in the field and ψ_{ref} the soil water potential (MPa) at 60 % WHC. Where the ψ_{ref} was not available for specific sites (most sites), the ψ_{ref} for Skogaby (0.0035 MPa) was used.

Daily soil temperatures and water potentials at 5–10 cm depth were recorded and/or simulated at Skogaby, Klosterhede and Thezan during a minimum of 3 years. For the other sites, monthly mean air temperatures were recorded (see Persson et al., Chap. 2, this Vol.), and these values were used for extrapolation to the field. At many sites, the water content (kg kg^{-1} or m^3 m^{-3}) and not the water potential had been determined (see CD-ROM, Persson et al., Chap. 2, this Vol.). The water potential was roughly calculated by using the relationship between actual water content and water content at saturation and converting this relationship into water potentials using the pF curve from Skogaby. The use of air temperatures instead of soil temperatures meant higher amplitudes over the year, but the correction factor (c_N) for the whole year was only marginally affected.

14.2.3 Manipulation of pH in the Laboratory

For all sites given in Table 14.1, finely ground CaCO$_3$ was mixed into subsamples of the FH materials (or 0–5 cm mineral soil) and the 10–20 cm mineral soil. Addition of 37.5 and 3 mg CaCO$_3$ g^{-1}, respectively, to these materials, increased the pH value by about 1.5–2 units. 0.05 M H$_2$SO$_4$ were added to other FH and 10–20-cm materials to decrease the pH value by about 0.3 units.

14.2.4 Manipulation of pH, N Availability and Nitrifier Density in the Laboratory

At some sites (Åheden and Klosterhede), where no net nitrification could be detected, CaCO$_3$, urea (0.5 mg urea-N g^{-1} organic matter) and nitrifying humus from limed plots in a spruce stand (1 g fresh humus to 60 g fresh humus) were added separately or in combination to humus layers from these sites. The aim was to evaluate which of the factors pH, N deficiency or lack of nitrifiers were responsible for the lack of net nitrification. For the Klosterhede humus, pH was also increased by 1.5–2 units by addition of NaOH, to compare the effects of the cations Ca^{2+} and Na$^+$ on the nitrification potential.

14.2.5 Autotrophic Versus Heterotrophic Nitrification

For all sites and all soil layers given in Table 14.1, acetylene (C$_2$H$_2$) gas was injected twice a month into containers with gas-tight lids to inhibit autotrophic nitrification (Hynes and Knowles 1982; Hyman and Wood 1985; De Boer et al. 1993). The treatment resulted in a C$_2$H$_2$ concentration of about 1 % (v/v) during 5 h, whereupon the lids were removed for 2 h to allow the C$_2$H$_2$ to disappear. Addition of C$_2$H$_2$ can act as a carbon source to heterotrophic soil microorganisms (Klemedtsson et al. 1990;

Terry and Leavitt 1992). To avoid lack of inorganic N because of N immobilisation stimulated by this carbon source, dissolved urea (0.5 mg urea-N g^{-1} organic matter) was added to some of the 10–20-cm samples to increase the ammonium availability (through urea hydrolysis to ammonia), and the same amount of urea was also added in combination with C_2H_2.

14.2.6 Net N Mineralisation and Nitrification in N-Fertilisation Experiments

To test whether N deposition has a potential to change N turnover at northern sites, samples were taken from control and fertilised plots in the long-term N-fertilisation experiments Norrliden (northern Sweden), Stråsan (central Sweden) and Skogaby (southern Sweden). At Norrliden, the fertilised plots ($n = 2$) were treated with ammonium nitrate at a mean dose of 73 kg N ha^{-1} a^{-1} for 23 years (1971–94, no fertilisation in 1990, Tamm et al. 1995), and the sampling was performed in 1994. At Stråsan, the fertilised plots ($n = 2$) were treated with ammonium nitrate at a mean dose of 74 kg N ha^{-1} a^{-1} for 23 years (1967–89, Berdén 1994). The sampling was made in 1994, i.e. 5 years after the last fertilisation. At Skogaby, the fertilised plots ($n = 4$) were treated with ammonium sulphate at a dose of 100 kg N ha^{-1} a^{-1} for 9 years (1988–96, Bergholm et al. 1995) and the sampling was made before fertilisation in early 1997. The characteristics of each site are given in Persson et al. (Chap. 2, this Vol.).

14.2.7 Comparison of N Turnover in Similar Soils at Different Climate

Samples of humus layers were taken from a *Picea abies* stand at Andersby in central Sweden and a *Pinus halepensis* stand at La Clape in southern France for comparison of net N mineralisation and net nitrification determined in the laboratory. Both stands have similar age, soil pH, soil C/N ratio and precipitation but are different as regards temperature climate (see below). Soil samples (0–5-cm depth) were also taken from a *Quercus robur* stand at Andersby and a *Quercus pubescens* stand at Thezan in southern France for the same reasons. Sampling and laboratory treatments were similar to those described for the main sites but restricted to the uppermost soil layers.

14.2.8 Comparison of N Turnover in Sieved and Intact Soil Cores

For comparison of sieved and intact soil with regard to net N mineralisation and net nitrification, intact soil cores (83 cm^2 area, 25 cm depth) were sampled by forcing PVC cylinders into the soil at the same sampling spots and sampling times as for the ordinary samplings. Intact soil cores ($n = 4$ per plot) were sampled from each of the four plots at Åheden, Gribskov and Klosterhede, all sites with sandy soil. The soil was transported within the cylinders to the laboratory, where they were placed at 15 °C. The soil moisture was adjusted by addition of distilled water, so the soil water content was the same as in the sieved soil placed in separate containers. Each cylinder with soil layers was placed in a plastic bag with a 5-mm round opening for gas exchange. The water content was adjusted periodically to keep the soil moisture constant. However, later it turned out that the litter and humus layers became drier in comparison with the sieved layers, because the water percolated into the deeper mineral soil layers. At the end of the incubation (145 days for Gribskov, up to 207 days for Åheden), the four cores (replicates) per plot were pooled plotwise for each soil layer, and the concentrations of ammonium and nitrate were determined in the same manner as for the ordinary samples (see above). To calculate the accumulation of ammonium and nitrate, the background concentrations (identical with those in the sieved samples) were subtracted from the final estimates.

Samples were also taken from the FH layers in four control and four N-fertilised plots at Skogaby at the same time as for the ordinary sampling. The intact humus-layer samples ($n = 4$ per plot) were taken side-by-side of the samples to be sieved. The former samples were adjusted in size to fit the same type of containers as used for the sieved samples. After that, the intact and sieved samples were treated identically.

14.2.9 In Situ Studies at Aubure

In situ studies were made in the three Aubure stands (AuP90, AuP45 and AuF) in the Strengbach catchment from October 1992 to October 1996. Changes in soil mineral-N contents were monitored fortnightly (monthly during wintertime) in the field inside and outside open cylinders (Lemée 1967; Raison et al. 1987) inserted in the soil in order to suppress root uptake. At the same time, throughfall was collected by pluviometers and gutters (Dambrine et al. 1995) and leaching (L_{lys}) was calculated as the product of water flux and NO_3^--N and NH_4^+-N concentrations in the soil solution measured by zero-tension lysimeters at 10-cm depth in the mineral soil. Since March 1994, ion exchange resin (IER) bags placed at the bottom of the cores were used to trap inorganic nitrogen leached (L_{cyl}) from the soil without root uptake (Hübner et al. 1991). Each bag contained 40 g of IER (DOWEX 21K, Cl$^-$-saturated or IRN 77, Na$^+$-saturated) mixed with the same weight of glass beads.

After sieving (4-mm mesh), inorganic N in the soil samples were extracted by shaking in 1 M KCl for 1 h. Inorganic N in the resin was extracted in 1 M NaCl. Inorganic N in the filtrates was measured by colorimetry or ionic chromatography. Gross contents were corrected according to (1) the effect of 24 h of contact between soil and the KCl solution, assessed in the laboratory, (2) the resin fixing efficiency of both resin types and (3) the efficiency of the extractant on both resin types.

Fluxes during the period T0 to T1 were calculated as follows:

$$\text{Net } N_{min} = A_{iT1} - A_{iT0} - A_{idep} + A_{iLcyl} + \text{net nit} \quad (3)$$

$$\text{Net nit} = N_{iT1} - N_{iT0} - N_{idep} + N_{iLcyl} \quad (4)$$

$$N_{upt} = \text{Net } N_{min} + A_{idep} + N_{idep} - A_{OLlys} - N_{OLlys}, \quad (5)$$

where Net N_{min} is net N mineralisation, A_{iT1} and A_{iT0} denote NH_4-N inside (i) the cylinder at time T1 and T0, A_{idep} is deposition of NH_4-N into the cylinders during the period T0 to T1, A_{iLcyl} is the leaching of NH_4-N from the cylinders and net nit is net nitrification. Correspondingly, N denotes NO_3-N with the same subscripts as above. N_{upt} is total N uptake by roots, which is calculated to be net N mineralisation plus deposition of inorganic N minus leaching of NH_4-N and NO_3-N in zero-tension lysimeters at 10-cm depth outside (O) the cylinders, A_{OLlys} and N_{OLlys}, respectively.

Average seasonal and annual fluxes for the 4-year period were calculated using the available seasonal mean leaching in the cylinders (2.5 years for NO_3^- and 1 year for NH_4^+). Monthly fluxes were compared by ANOVA.

14.2.10 Denitrification

Denitrification was measured in situ as nitrous oxide emission from the soil surface and as potential denitrification enzyme activity in soil samples with or without inhibition of nitrous oxide reductase activity. The measurements included the beech forest sites at Sorø, Schacht, Aubure (AuF) and Collelongo.

Nitrous oxide emission was measured at monthly intervals (twice a month at Sorø) from spring to late autumn in 1998 by a closed chamber technique. Twenty PVC tubes (19 cm long, 10-cm inner diameter) were installed at AuF, Schacht and Collelongo, while at Sorø, 12 chambers with a diameter of 30 cm were installed. Five-ml gas samples were removed from the headspace, 0, 40, 80 and 120 min after sealing the cores, and transferred to 3-ml Venoject tubes (Terumo, Leuven, Belgium). The Venoject tubes were analysed for N_2O on a Hewlett Packard 5890 Gas Chromatograph connected to an automatic headspace sampler (Mikrolab, Aarhus, Denmark) and equipped with an electron capture detector.

Potential denitrification enzyme activity was measured by adding 10 g of soil (from 0–15-cm soil depth) and 15 ml of substrate (1 mM glucose and 1 mM KNO_3) to 116-ml incubation bottles which were sealed with a butyl rubber stopper and made anaerobic by flushing with N_2. To inhibit nitrous oxide reductase activity, 11 ml acetylene (C_2H_2) was added to a final headspace concentration of 10%. In bottles without inhibition, 11 ml N_2 was added. The bottles were incubated at 22 °C at 300 rpm on a rotary shaker. After 5, 30, 60, 90 and 150 min, a 2-ml gas sample was transferred from the headspace to a 3-ml Venoject tube prefilled with 3-ml N_2. The N_2O content was analysed as described above. Denitrification activity was calculated by fitting headspace N_2O concentrations to a simple model that takes into account synthesis of enzymes during the incubation.

14.3 Net N Mineralisation Based on Laboratory Studies

Accumulation of inorganic N (net N mineralisation), expressed as g^{-1} LOI or g^{-1} total C (mostly organic C), was generally high in the litter and humus layers and decreased with increasing soil depth (Table 14.1, Fig. 14.1). The highest net N mineralisation rates (38–50 $\mu g\,N\,g^{-1}\,C\,day^{-1}$) were found in the beech litter from Schacht and Sorø and the spruce litter from Waldstein (Table 14.1). At some sites (especially so for Thezan), the litter layer had lower net N mineralisation rate than the corresponding humus layer. Åheden differed from all other sites in having no detectable net N mineralisation in any soil layer during the first 52-day period of incubation (Table 14.1, Fig 14.1). During the further incubation between 52 and 207 days, mineralised N was produced both in the humus and litter layers at Åheden, but at greater depths almost no N was mineralised during the 207-day incubation, indicating low N availability or high microbial demand for N (Fig. 14.1). Apart from Åheden and some of the Aubure stands, the accumulation pattern for most sites for the period 50–150 days did not differ substantially from that obtained during the first 50 days (Fig. 14.2). Lower rates during the initial 50-day period than during the 50–150-day period indicate a decrease in C availability during the late part of the incubation. Negative accumulation of inorganic N during the period 100–150 days, especially in the 10–20-cm soil layer (Fig. 14.2), was probably an artefact due to denitrification after watering of nitrate-rich soil samples.

Table 14.1. Daily net N mineralisation rates (above) and potential net nitrification rates (below) ($\mu g\,N\,g^{-1}\,C\,day^{-1}$) for each soil layer and site as calculated from laboratory incubations during about 50 days

Net N mineralisation

Soil layer	Åhe	Sko	Klo	Gri	Sor	Nac	Jez	Wal	Sch	AuP90	AuP45	AuF150	The	Col	MdM
L	0	8.5	18.8	23.2	43.6	14.8	19.6	38.1	50.0	Incl. FH	11.2	16.3	0	13.5	5.4
FH	0	5.1	7.6	14.9a	18.0a	7.3	12.0	9.8	15.3	13.9	17.8	27.0	9.4a	17.2a	8.7a
0–10 cm	0	5.2	1.4	6.1b	9.6b	1.4	4.4	5.1	5.6	8.8	7.6	5.3	13.2b	7.8b	5.4b
10–20 cm	0	5.6	2.3	5.0	7.5	2.1	3.6	4.8	6.4	3.9	3.0	9.2	13.6	5.2	4.8
20–30 cm	0	2.4	0.3	4.8	6.0	0.9	2.7	2.9	4.0	5.6	1.8	3.3	9.7	4.2	3.9
30–50 cm	0	−0.1	−0.3	2.8	5.1	0	2.5	1.2	n.e.	4.6	1.6	0.6	6.0	1.6	1.1

Net nitrification

Soil layer	Åhe	Sko	Klo	Gri	Sor	Nac	Jez	Wal	Sch	AuP90	AuP45	AuF150	The	Col	MdM
L	0	0	0	3.0	44	0.8	6.4	14.2	11.1	Incl FH	0	1.7	0	12.6	5.3
FH	0	0	0	7.2a	14.1a	1.9	12.2	3.4	6.4	7.1	0.2	9.7	0.3a	25.3a	18.7a
0–10 cm	0	0	0.1	3.8b	8.6b	0.8	4.5	4.1	4.7	5.2	1.8	4.6	2.3b	6.2b	5.4b
10–20 cm	0	0.2	0	2.1	5.7	1.2	3.6	5.8	6.8	3.3	2.1	6.1	0.4	2.3	4.8
20–30 cm	0	0.4	0	2.7	5.1	1.4	2.8	3.0	5.0	5.3	1.6	3.2	0.5	2.7	3.9
30–50 cm	0	0.3	0.2	1.6	4.2	0.2	1.6	1.7	n.e.	4.2	1.8	1.1	0.4	1.7	1.1

SE is approximately 10–20% of the mean value. 0 indicates a value less than 0.05. If net nitrification estimates are higher than net mineralisation estimates, ammonium present in the background samples has been nitrified. Where no FH layer was found, the 0–10-cm layer was subdivided into 0–5 (a) and 5–10 cm (b). See Persson et al. (Chap. 2, Table 2.1, this Vol.) for complete site designation. At AuP90, the L layer was included in the FH layer (incl. FH).

n.e. Not estimated.

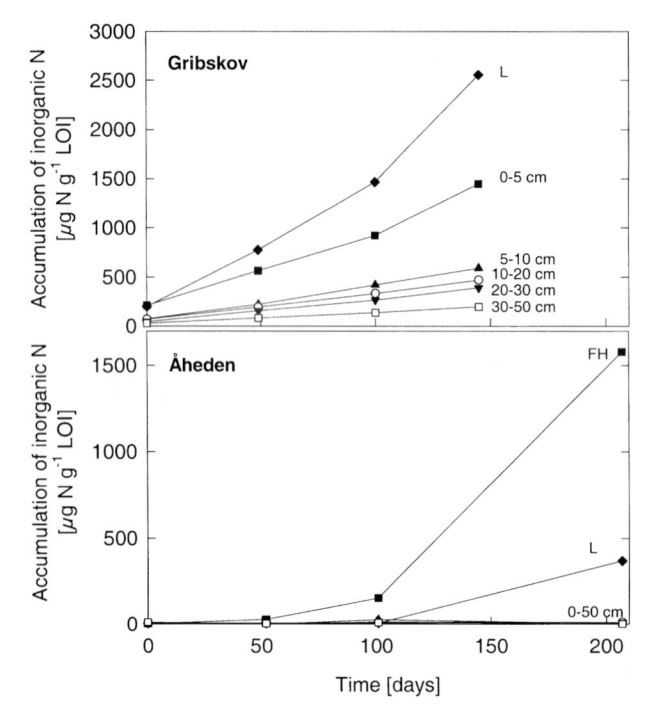

Fig. 14.1. Accumulation of inorganic N (NH_4-N + NO_3-N) in different soil layers exemplified by data from Gribskov and Åheden. *LOI* Loss-on-ignition. Note the different scales

When taking the soil layer pools (see Persson et al., Table 2.8, Chap. 2, this Vol.) into consideration, the laboratory estimates could be extrapolated to the field by using the monthly correction factors based on the temperature and moisture response functions calculated for each site (Fig. 14.3). The estimated field net N mineralisation amounted to between 0 (Åheden) and 191 (Schacht) $kg\,N\,ha^{-1}\,a^{-1}$ (Table 14.2, Fig. 14.4). High net N mineralisation (>100 $kg\,N\,ha^{-1}\,a^{-1}$) was estimated for two of the spruce sites (Waldstein and Monte di Mezzo) and all beech sites (Schacht, Collelongo, Sorø, AuF, Gribskov and Jezeri). Where a FH layer had developed, a considerable fraction (32–79%) of total net N mineralisation originated from the organic soil layers (L and FH) (Table 14.2). At the sites where a humus layer was lacking (Thezan, Collelongo, Monte di Mezzo, Gribskov and Sorø), the litter layer accounted for 0–21% of total net N mineralisation.

There was no clear relationship between latitude and net N mineralisation in the field (Fig. 14.4). There probably exists a gradient in net N mineralisation at higher latitudes, from southern Scandinavia at Skogaby (about 56°N) to northern Scandinavia at Åheden (about 64°N), indicated by the data in Table 14.3 and other studies (Persson 1996; Falkengren-Grerup et al. 1998). This gradient coincides with a tenfold decrease in N deposition (Lövblad et al. 1995). Furthermore, there was no clear relationship

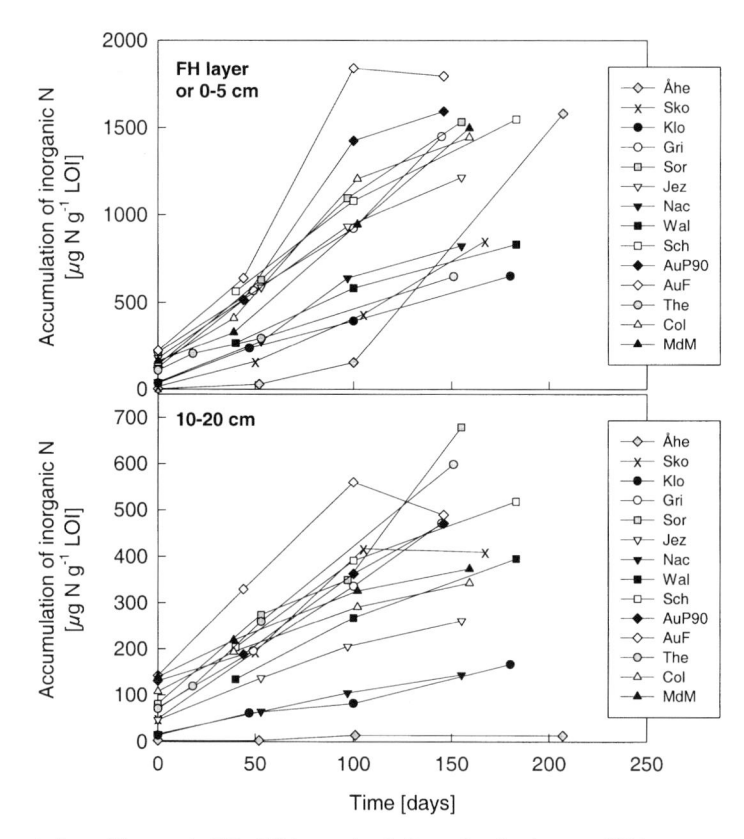

Soil Nitrogen Turnover – Mineralisation, Nitrification and Denitrification

Fig. 14.2. Accumulation of inorganic N in FH layers (or 0–5-cm depth where an FH layer was lacking) and at 10–20-cm depth at the sites in the NIPHYS/CANIF transect. Site abbreviations given in Chapter 2, Tables 2.1 and 2.3

between mean annual air temperature and net N mineralisation in the field (not shown). There was a positive correlation, $r^2 = 0.57$, $p < 0.005$, between the soil N pool and total net N mineralisation for the sites with an N pool between 1 and 9 t ha^{-1} (Fig. 14.5). The Italian sites Collelongo and Monte di Mezzo, both sites with a calcareous subsoil and with about 18 t N ha^{-1}, deviated from this pattern and had a comparatively low net N mineralisation. For comparable sites, net N mineralisation seemed to be higher in the beech than in the spruce stands (Fig. 14.6).

14.4 Net Nitrification Based on Laboratory Studies

Net nitrification differed pronouncedly between sites (Table 14.1). At Åheden, there was no net N mineralisation and, thus, no net nitrification. Also at Skogaby and Klosterhede, there was almost no net nitrification in any of the soil layers. At Aubure (P45) and Thezan, net nitrification was almost negligible in the organic (LFH) soil

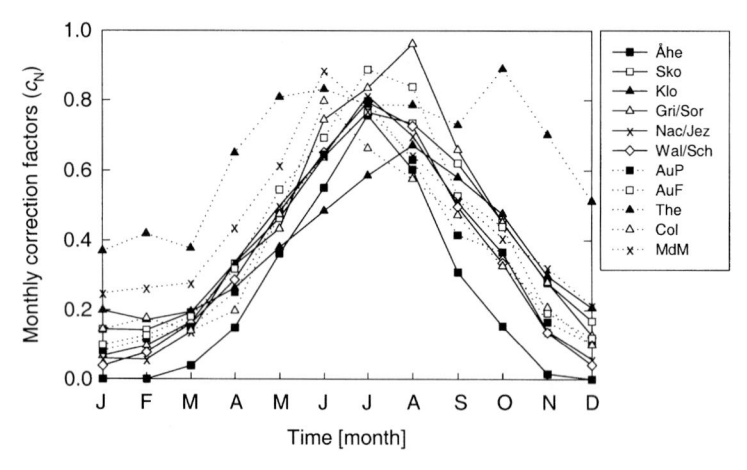

Fig. 14.3. Monthly correction factors (c_N) for different sites used to convert N mineralisation rates obtained in the laboratory at 15 °C and 60% WHC to N mineralisation rates in the field

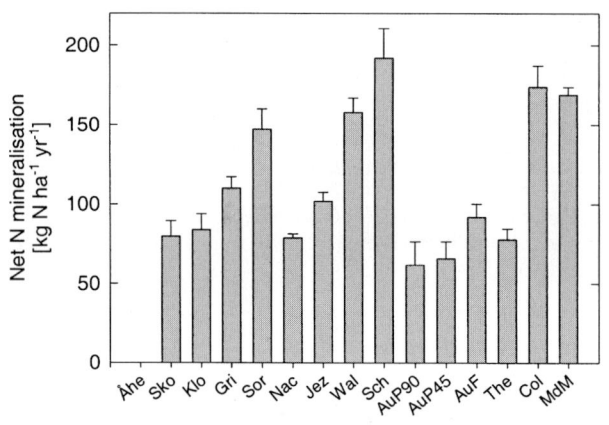

Fig. 14.4. Annual net N mineralisation in the field estimated for the whole soil profile to a depth of 50 cm (30 cm at Schacht) at different sites (from Åheden in the north to Monte di Mezzo in the south). Note that Åheden has no net N mineralisation. For details, see Table 14.2

layers, but could be detected in the mineral soil. At some sites, for example, Collelongo, Monte di Mezzo and Sorø, almost all mineralised N was converted to nitrate during the incubation period of 50 days.

Relative net nitrification (the fraction of mineralised N that was converted to nitrate) was positively correlated with pH and negatively correlated with C/N ratio in the L and FH layers (including the 0–5-cm mineral soil, where an FH layer was lacking) (Fig. 14.7). The correlation with the C/N ratio was especially evident when the non-nitrifying sites Klosterhede, Skogaby and Åheden was not included in the

Table 14.2. Annual net N mineralisation (above) and potential net nitrification (below) (kg N ha^{-1} a^{-1}) for each soil layer and site as calculated from laboratory incubations of sieved soil, soil pools in the field (see Persson et al., Chap. 2, this Vol.) and corrected for soil temperature and moisture (Fig. 14.3). Where no FH layer was found, the 0–10-cm layer was subdivided into 0–5 (a) and 5–10 cm (b). See Persson et al., Chapter 2, Table 2.1, this Volume for complete site designation. SE is approximately 10–20% of the mean values. The variation is lower for net N mineralisation than for net nitrification, and the variations are often largest in the L layer. Note that net nitrification values are overestimates, because in the field roots are present and can take up ammonium and, thereby, reduce nitrification

Net N mineralisation

Soil layer	Åhe	Sko	Klo	Gri	Sor	Nac	Jez	Wal	Sch	AuP90	AuP45	AuF	The	Col	MdM
L	0	4	13	16	31	13	14	21	45	Incl FH	6	9	−1	14	12
FH	0	22	53	45a	43a	49	62	70	57	31	29	31	16a	60a	70a
0–10 cm	0	30	9	12b	21b	9	15	29	46	18	21	40	14b	30b	19b
10–20 cm	0	18	9	15	21	7	5	21	27	3	4	21	18	36	32
20–30 cm	0	7	1	13	13	2	3	11	16	4	2	6	14	19	24
30–50 cm	0	0	−1	9	18	0	4	5	n.e.	5	4	8	16	16	10
Total	0	81	84	110	147	80	101	157	191	61	66	115	77	175	167
% in LFH	–	32	79	15	21	78	75	58	53	51	53	35	0	8	7

Net nitrification

Soil layer	Åhe	Sko	Klo	Gri	Sor	Nac	Jez	Wal	Sch	AuP90	AuP45	AuF	The	Col	MdM
L	0	0	0	2	31	2	5	8	10	Incl FH	0	1	0	13	12
FH	0	0	0	22a	34a	7	63	24	24	16	0	11	0a	88a	75a
0–10 cm	0	0	0	8b	19b	8	15	24	39	11	5	32	2b	23b	15b
10–20 cm	0	1	0	6	16	6	4	25	29	3	3	14	1	19	34
20–30 cm	0	1	0	7	11	3	3	12	20	4	2	6	1	12	25
30–50 cm	0	1	0	5	16	2	4	8	n.e.	5	4	9	1	17	11
Total	0	3	0	50	127	28	94	101	122	39	14	73	5	172	172
% in LFH	–	0	–	4	24	32	72	32	28	41	0	16	0	8	7

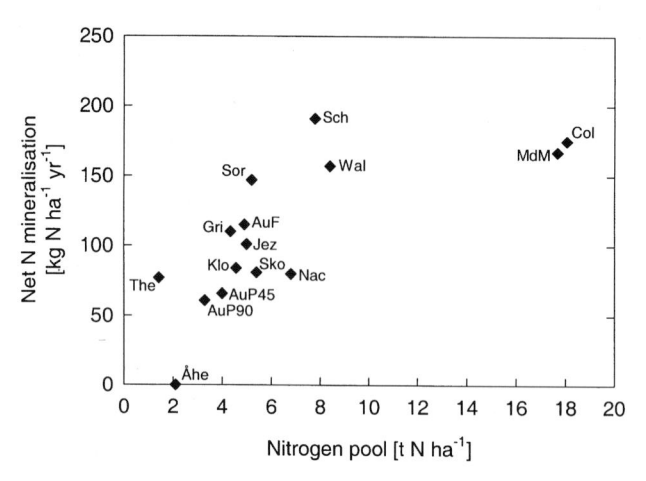

Fig. 14.5. Relationship between the total N pool and annual net N mineralisation in the field estimated for all sites. Data from Tables 2.5 (Chap. 2) and 14.2. $r^2 = 0.57$, $p < 0.005$ for the sites with N pools between 1 and 9 t ha^{-1}

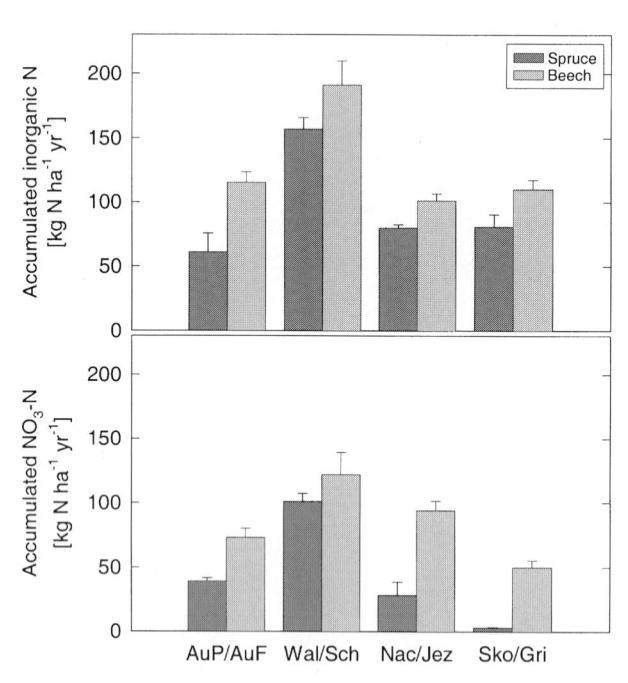

Fig. 14.6. Annual net N mineralisation (*above*) and potential net nitrification (*below*) estimated for the soil profile to a depth of 50 cm in comparable beech and spruce stands

Soil Nitrogen Turnover – Mineralisation, Nitrification and Denitrification 311

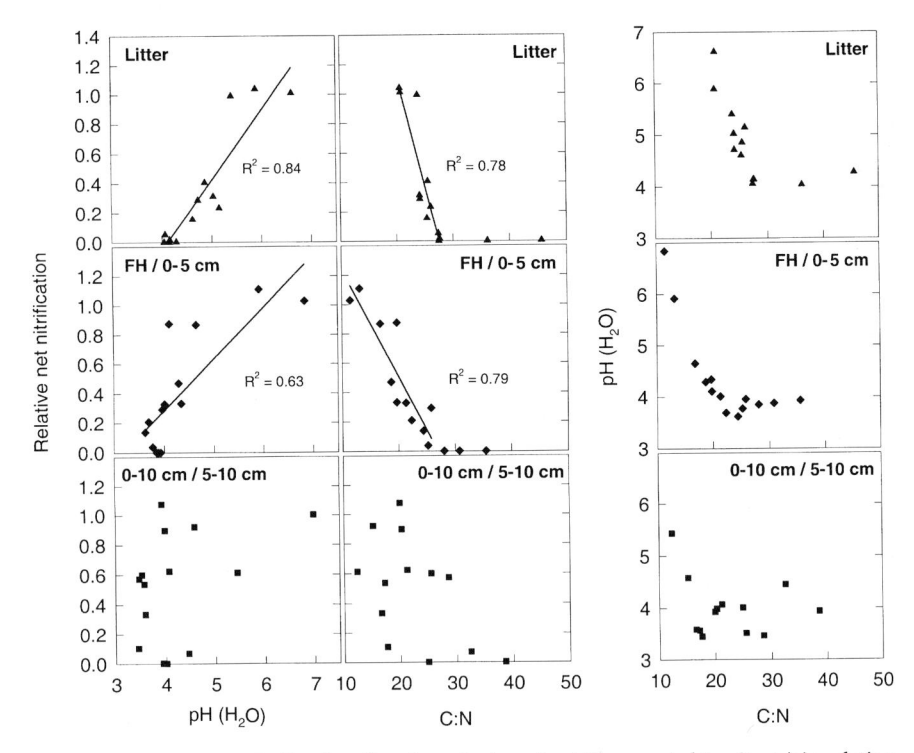

Fig. 14.7. Relative net nitrification (fraction of mineralised N converted to nitrate) in relation to pH (H$_2$O) and C:N ratio in the litter, humus and upper mineral soil layers at all beech and spruce sites. The relation between C:N ratio and pH is also given

regression. According to these data, net nitrification did not occur when pH was lower than 4 in the L layer and 3.5 in the FH/0–5 cm layer. Further, no net nitrification was detected in these layers when the C/N ratio exceeded 28. Deeper in the soil (illustrated by 0–10 or 5–10 cm depth, Fig. 14.7), there was no clear correlation with either pH or C/N ratio, and relative net nitrification was substantial also at pH 3.5 (Waldstein, Nacetin and Aubure, AuF) and at C/N ratios above 28 (Nacetin). In the L and FH layers, there was also a negative correlation between pH and C/N ratio, indicating that the (negative) correlation between C/N ratio and relative net nitrification could to a large extent be explained by pH in these layers.

Net nitrification data from the laboratory were extrapolated to the field in the same manner as for net N mineralisation (Table 14.2), remembering that these data were potential ones presupposing no ammonium uptake by roots. Calculated in this way, net nitrification was always higher in beech than in comparable spruce stands (Fig. 14.6). In the field, ammonium uptake by plants will probably reduce the possibility for ammonium oxidisers to perform the first step in the nitrification process (Myrold 1998). Thus, the data on net nitrification given in Table 14.2 and Fig. 14.6 are overestimates and are probably realistic in the field only after clearcutting.

The difference in net nitrification between beech and spruce was highest in the upper soil layers. This could partly be related to differences in C/N ratios in the FH

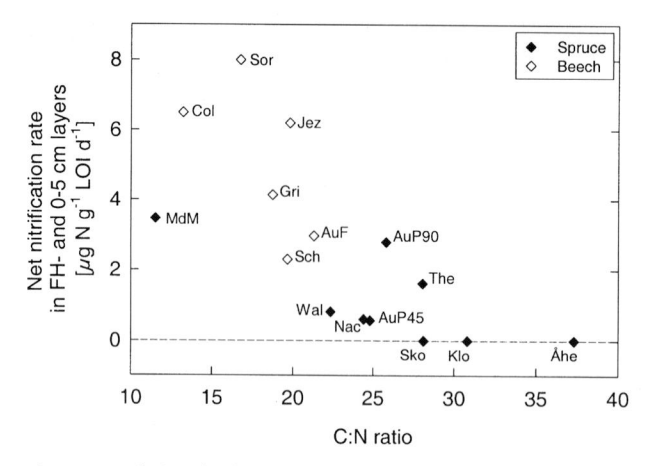

Fig. 14.8. Relationship between C:N ratio and net nitrification rate in the FH and 0–5-cm layers determined in the laboratory at 15°C and 60% WHC during the whole incubation period (150–207 days). Note that the spruce site Monte di Mezzo (MdM) was exceptional in being an afforested arable field

and 0–5 cm soil layers (Fig. 14.8). In these soil layers, all beech sites had lower C/N ratios than the spruce sites, with the exception of Monte di Mezzo, which is a spruce plantation on former arable land. With increasing soil depth, the difference in C/N ratio between tree species decreased, and at greater depths than 20 cm there was no difference (see Persson et al., Table 2.8, Chap. 2, this Vol.).

Net nitrification, which is more directly related to ammonium availability than to C/N ratio, was positively related to net mineralisation rate in the mineral soil layers and especially so in the beech stands (Fig. 14.9). In the spruce soils, the correlation was weaker in the upper mineral soil, but improved with increasing soil depth (Fig. 14.9). In the organic soil layers there was no clear correlation between net N mineralisation and net nitrification. On the other hand, there was a good correlation between pH and relative net nitrification in the organic layers (see Fig. 14.7).

14.5 Manipulation of pH, N Availability and Nitrifier Density in the Laboratory

Addition of $CaCO_3$ in the laboratory always resulted in increased pH, and addition of H_2SO_4 in decreased pH. In the FH layer (replaced by the 0–5 cm mineral soil where an FH layer was absent) an increase in pH by about 1.5 to 2 units generally resulted in increased net nitrification (Fig. 14.10). In one case (Monte di Mezzo), where the background pH was high (pH 6.8), the increase in pH (to 7.4) led to an insignificant reduction in net nitrification. For two of the sites (Skogaby and Klosterhede), where no background net nitrification could be detected, the addition of $CaCO_3$ did not result in any net nitrification. For the same soil layer, acidification, decreasing the pH value by about 0.2–0.5 units, decreased net nitrification at sites where the background samples had a detectable net nitrification.

In the 10–20 cm mineral soil layer, the increase in pH did not result in a marked effect on net nitrification. Both increases and decreases were observed (Fig. 14.10),

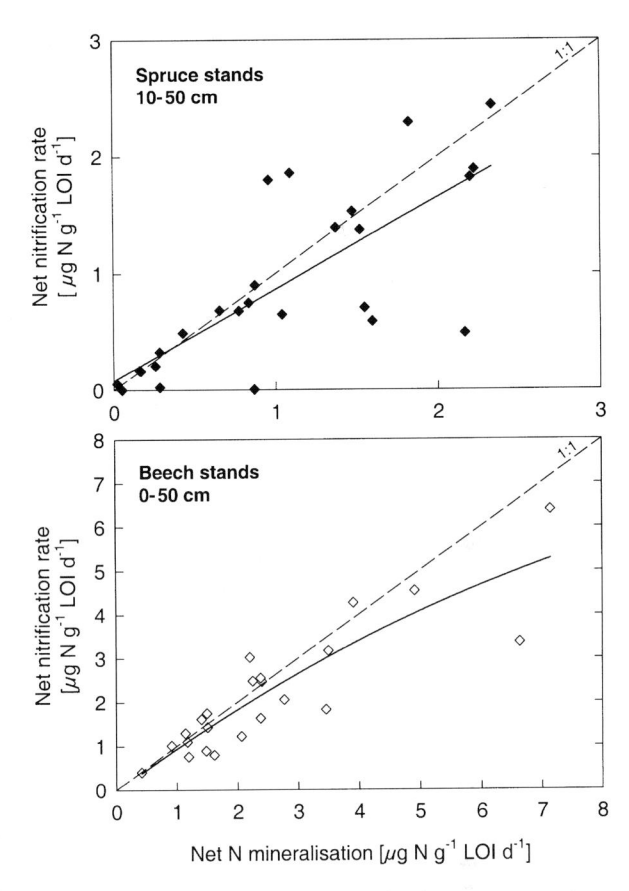

Fig. 14.9. Relation between net N mineralisation rate and net nitrification rate at 10–50 cm depth in the mineral soil in spruce stands and at 0–50 cm depth in beech stands determined in the laboratory at 15°C and 60% WHC during the whole incubation period (150–207 days)

but in most cases there were no significant changes. Addition of H_2SO_4 mostly decreased net nitrification or had no significant effects. Consequently, changes in pH had a greater impact on net nitrification in the FH layer or 0–5 cm mineral soil than in the 10–20-cm mineral soil.

Treatment of the non-nitrifying FH layers at Klosterhede and Åheden with urea (U) as a nitrogen (ammonium) source, $CaCO_3$ (Ca) to increase pH and inoculation with nitrifying humus (I) to add nitrifying bacteria showed that urea alone could not stimulate net nitrification at any of the two sites (Fig. 14.11). Addition of $CaCO_3$ (resulting in an initial pH of about 6) had no effect on net nitrification at Klosterhede during the 180-day incubation period but caused a slight increase at Åheden after 207 days. The combination of the Ca and U treatment had no effect on net nitrification during the first 100 days of incubation but led to a considerable net nitrification at both sites at the end of the incubation. Also the I and I + U treatments increased net nitrification only during the last incubation period. The Ca + I and Ca + U + I treatments resulted in an earlier start of net nitrification, and the accumula-

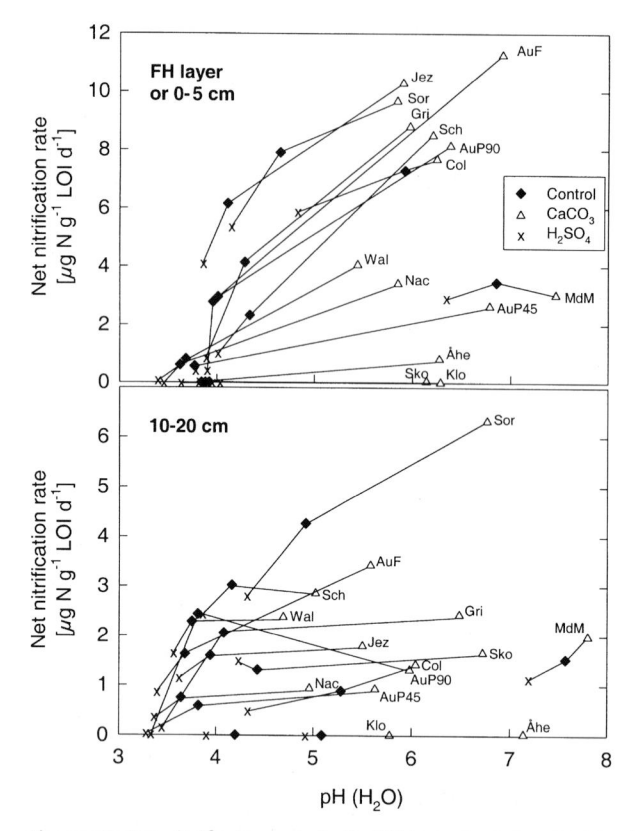

Fig. 14.10. Net nitrification rate in the FH layer (or 0–5-cm mineral soil where an FH layer was lacking) and the 10–20-cm mineral soil after treatment with $CaCO_3$ or H_2SO_4 in the laboratory and incubated for 150–207 days at 15 °C and 60% WHC in the laboratory

tion of nitrate became higher than in any of the other combinations of treatments. The study showed that low pH (both sites), low amounts of ammonium (Åheden) and low amounts of nitrifiers (both sites) were all probable explanations for the lack of net nitrification. However, the Ca + U + I treatment, that appeared to result in an almost complete net nitrification at Åheden, did not lead to a complete net nitrification at Klosterhede.

We hypothesised that high deposition of sodium (Na^+) in sea salt at Klosterhede could be a possible explanation for low nitrifier activity. However, addition of NaOH resulting in the same pH as for $CaCO_3$, led to high net nitrification (not shown here) when combined with the I and U treatments. Thus, the lack of net nitrification was probably not due to the sodium deposition.

14.6 Autotrophic Versus Heterotrophic Nitrification

Treatment with acetylene (C_2H_2) gas every second week during the entire incubation period showed that net nitrification (where it occurred) was more or less suppressed

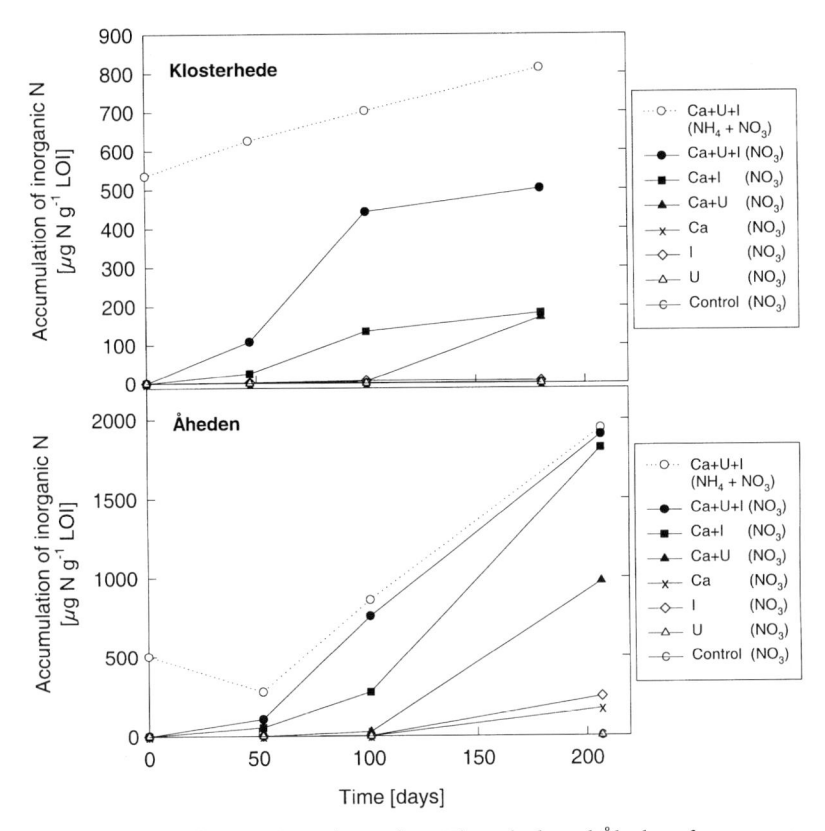

Fig. 14.11. Accumulation of nitrate in FH layers from Klosterhede and Åheden after no treatment (*Control*) or treatment in the laboratory with urea (*U*), CaCO$_3$ (*Ca*), inoculation with nitrifying humus (*I*) and combinations of these treatments. Accumulation of inorganic N (NH$_4$ + NO$_3$-N) in the Ca + U + I treatment (*dotted line*) is given for comparison

by the treatment. At the coniferous sites in Scandinavia (Skogaby, Klosterhede and Åheden), net nitrification occurred in the mineral soil only after a long time of incubation. At these sites the acetylene treatment resulted in a complete inhibition of net nitrification, suggesting that autotrophic nitrification was the dominant process (Fig. 14.12, Skogaby). A complete or an almost complete inhibition was also found at the coniferous sites at Aubure, Monte di Mezzo and Thezan despite sometimes high net nitrification (Fig. 14.12, Thezan). At the central European coniferous sites Waldstein and Nacetin, the blocking effect of acetylene was incomplete, especially in the humus layer (Fig. 14.12, Nacetin). In the six beech stands studied, some net nitrification could occur despite acetylene treatment (Fig. 14.12, Gribskov). Typically, net nitrification was reduced to 20–50% in the 0–5 and 5–10-cm soil layers as compared with the untreated soil. In the litter layer and at depths greater than 10 cm, the inhibiting effect of acetylene was more complete. In conclusion, there was a strong indication that all beech stands and the coniferous stands studied in Germany and the Czech Republic had not only autotrophic but also heterotrophic nitrification, and the latter process was most obvious in the humus layer and in the uppermost mineral soil.

14.7 Net N Mineralisation and Nitrification in N-Fertilisation Experiments

Long-term addition of ammonium nitrate corresponding to a mean dose of 73 kg N ha^{-1} a^{-1} for 23 years at Norrliden and Stråsan resulted in a marked increase in net N mineralisation rate to a depth of 30 cm in comparison with control plots (Table 14.3). The control plots at Norrliden had a net N mineralisation rate almost as low as at Åheden (Table 14.1), situated at about the same latitude (about 64 °N). The control plots at Stråsan (60–61 °N, see Persson et al., Table 2.3, Chap. 2, this Vol.) had slightly higher net N mineralisation rates but still very low ones. Addition of N fertilisers at Norrliden and Stråsan resulted in net N mineralisation rates similar to those found in spruce stands in central Europe (Table 14.1).

Addition of 100 kg N ha^{-1} a^{-1} in the form of ammonium sulphate at Skogaby during nine years resulted in a significant ($p < 0.05$) increase in net N mineralisation rate in the L layer. This increase was associated with a decrease in the C/N ratio from 36 (Persson et al., Table 2.8, Chap. 2, this Vol.) to 23. In the other soil layers there was no significant increase in net mineralisation rate (Table 14.3) and no clear decrease in

Table 14.3. Net N mineralisation and net nitrification rates (μg N g^{-1} C day^{-1}) at soil layers in control (C) and N-fertilised plots (N) in the N fertilisation experiments Norrliden, Stråsan and Skogaby during an incubation period of 50 days at 15 °C in the laboratory. Negative values indicate N immobilisation. A higher net nitrification than net mineralisation rate indicates that ammonium present at the start of the incubation was nitrified

Net N mineralisation rate

Soil layer	Norrliden C	Norrliden N	Stråsan C	Stråsan N	Skogaby C	Skogaby N
L	Incl. in FH	Incl. in FH	0.8	15.7	8.5	31.3
FH	0.2	9.9	0.1	13.6	5.1	7.4
0–10 cm	0.0	3.4	1.3	4.5	5.2	1.4
10–30 cm	−0.1	1.5	2.0	6.5	4.0	2.5
30–50 cm	−0.7	0.0	–	–	−0.1	1.1

Net nitrification rate

Soil layer	Norrliden C	Norrliden N	Stråsan C	Stråsan N	Skogaby C	Skogaby N
L	Incl. in FH	Incl. in FH	0.0	1.9	0.0	8.4
FH	0.0	−0.1	0.0	0.3	0.0	0.0
0–10 cm	0.0	−0.2	0.0	2.5	0.0	0.0
10–30 cm	0.0	1.0	0.0	5.0	0.3	3.4
30–50 cm	−0.1	0.2	–	–	0.3	3.1

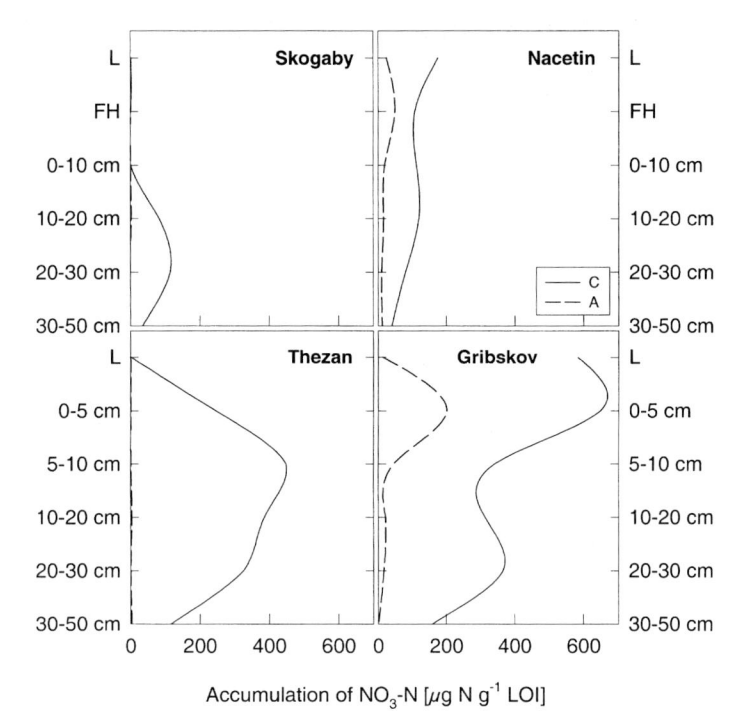

Fig. 14.12. Accumulation of NO$_3$-N in soil layers from Skogaby, Nacetin, Thezan and Gribskov without (*C Solid line*) or with (*A Broken line*) addition of acetylene

C/N ratio. There was even a significant decrease in the rate at 0–10 cm depth, possibly because of high ammonium concentrations.

Long-term N fertilisation significantly increased net nitrification rate in the L layer but not in the FH layer. Nor was there an increase at 0–10-cm depth for Norrliden and Skogaby. However, at greater soil depths, the N fertilisation resulted in increased net nitrification rates, and these rates were similar to or sometimes even higher (because of high ammonium levels at start) than net N mineralisation.

In conclusion, the study showed that long-term N fertilisation simulating high N deposition could change N mineralisation rates from low to high ones, similar to those in central Europe. This indicates that the N factor is more crucial than climatic factors in determining N mineralisation potentials at high latitudes. According to the different results obtained after 9-year fertilisation compared to those after 23 years, we conclude that narrowing of the C/N ratio of the organic matter rather than the abundance of inorganic N determines net N mineralisation. The results furthermore confirms the conclusion (see above) that increases in net nitrification rates in the organic horizons need increases in pH, whereas those in the deeper mineral soil (10–50-cm depth) need increases in ammonium levels and not pH.

14.8 Comparison of N Turnover in Similar Soils at Different Climate

The humus layer in the *Picea abies* stand at Andersby (AnP) in central Sweden (moder) and that in the *Pinus halepensis* stand at La Clape (LaC) in southern France (xeromoder) had similar pH, C/N ratio and annual precipitation but very different annual mean temperatures (Table 14.4). In the *Quercus robur* stand at Andersby (AnQ) and the *Quercus pubescens* stand at Thezan (ThQ) no humus layer had developed and the humus form was a mull with fairly similar characteristics (Table 14.4).

The pattern of net N mineralisation was very similar in the humus layers from the two coniferous forest sites AnP and LaC (Fig. 14.13). A main difference between the sites was that LaC had much higher initial concentrations of inorganic N (360 μg g^{-1} LOI) than AnP (40 μg g^{-1} LOI) at the time of sampling (late April and early May in 1993, respectively). Net nitrification followed a similar pattern. Initially, the nitrate levels were low at both sites, but the accumulation of nitrate increased dramatically with time (Fig. 14.13), indicating that the nitrification activity was probably initially suppressed by drought (LaC), low ammonium levels and/or low nitrifier populations (AnP).

For the two broadleaf sites, inorganic N accumulated at a slightly higher rate in the soil from AnQ at 0–5-cm depth than from ThQ at the same depth. The concentration of organic matter (loss-on-ignition) was higher at AnQ (25% of dry weight) than at ThQ (8% of dry weight), probably indicating that the organic matter at AnQ was less decomposed. At both sites, nitrate readily accumulated from a situation with a dominance of ammonium at the start of the incubation. Net nitrification seemed to be more complete in the ThQ stand with less organic matter.

For the two coniferous sites, treatment with acetylene resulted in an almost complete inhibition of net nitrification, indicating autotrophic nitrifiers (Fig. 14.13). In the broadleaf stands, the acetylene blocking was not complete. This might indicate

Table 14.4. Characteristics of coniferous and broadleaf stands at Andersby (AnP and AnQ) in Central Sweden and La Clape (LaC) and Thezan (ThQ) in Southern France. Estimates given as means ± SE ($n = 4$)

Site/stand	Coniferous		Broadleaf	
	AnP	LaC	AnQ	ThQ
Stand	*Picea abies*	*Pinus halepensis*	*Quercus robur*	*Q. pubescens*
Latitude	60°09′	43°09′	60°09′	43°07′
Annual mean air temp. (°C)	5.5	14.8	5.5	14.4
Annual mean precip. (mm)	570	587	570	578
Soil layer	FH	FH	0–5 cm	0–5 cm
pH (H$_2$O)	5.04 ± 0.02	4.71 ± 0.01	5.84 ± 0.30	6.47 ± 0.12
C/N ratio	24.0 ± 0.2	22.8 ± 0.1	15.0 ± 0.7	18.9 ± 0.6

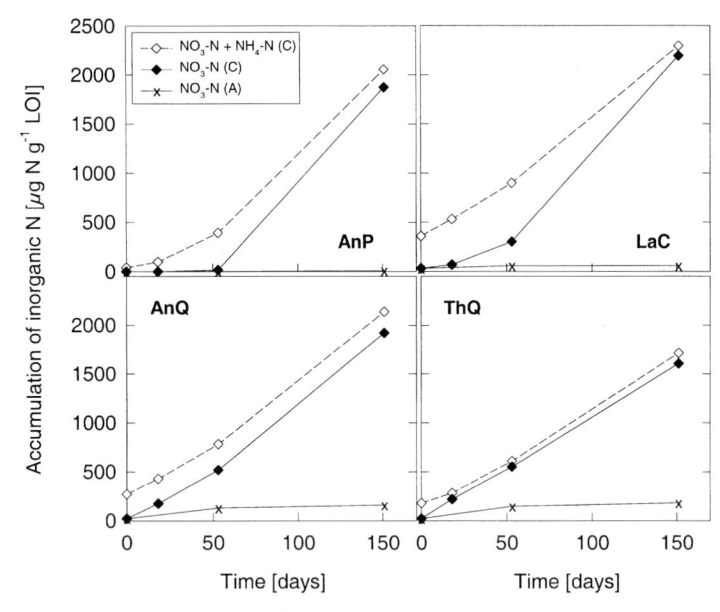

Fig. 14.13. Accumulation of NH_4-N + NO_3-N (*broken line*), NO_3-N (*upper solid line*) and NO_3-N after addition of acetylene (*lower solid line*) in the laboratory at 15 °C and near-optimal soil moisture in sieved humus layers (AnP and LaC) and 0–5-cm topsoil layers (AnQ and ThQ). *AnP* a *P. abies* stand in central Sweden; *LaC* a *P. halepensis* stand in southern France; *AnQ* a *Q. robur* stand in central Sweden; *ThQ* a *Q. pubescens* stand in southern France

presence of heterotrophic nitrifiers but can also be an effect of incomplete diffusion of acetylene into the clayish soil. The results of incomplete acetylene inhibition of net nitrification in the oak stands resemble those obtained in the beech stands (see above), which were also found in the topsoil.

In conclusion, the comparison in the laboratory of topsoil layers from two different latitudes and temperature climates showed that net N mineralisation and net nitrification potentials were astonishingly similar when plant species, C/N ratios and pH values were similar (although not identical). The similarity in these soil properties was probably due to the fact that decomposition of soil organic matter is restricted by drought at the Mediterranean sites and by cold at the Scandinavian sites.

14.9 Comparison of N Turnover in Sieved and Intact Soil Cores

Sieving of the FH layers at Skogaby had a tendency to reduce net N mineralisation, by 26% in the untreated plots and by 21% in the N-fertilised plots, in comparison with the intact FH layers (Fig. 14.14). However, no significant effect ($p > 0.05$) could be shown. A decrease in net N mineralisation after sieving could possibly be explained by the removal of roots and adherent mycorrhizal fungi that otherwise should have contributed to the release of inorganic N during decomposition.

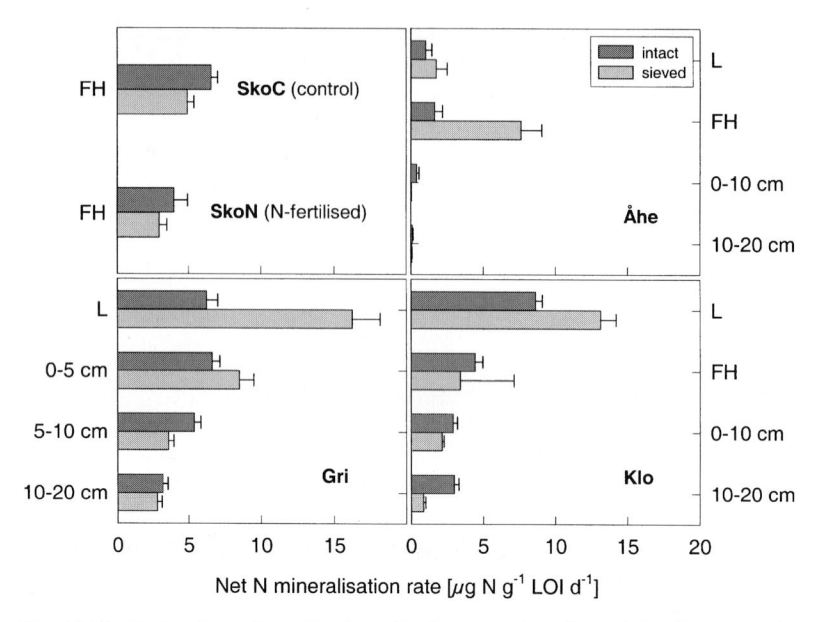

Fig. 14.14. Comparison of net N mineralisation rates (µg g^{-1} LOI^{-1} day^{-1}) in sieved and intact soil layers from Skogaby, control (*SkoC*) and N-fertilised plots (*SkoN*), Åheden (*Åhe*), Gribskov (*Gri*) and Klosterhede (*Klo*) for a period of 145 to 207 days. *Error bars* indicate one SE

For the other sites, the soil layers in intact soil columns were compared with the sieved soil layers kept in separate containers. In the intact columns, especially those from Åheden and Klosterhede with coarse sand, the soil moisture became lower than intended in the litter and humus layers and higher than intended at 10–20-cm depth because of redistribution of the soil water.

At Åheden sorting/sieving initially appeared to increase net N mineralisation in the litter and humus layers, whereas no significant effect was found in the mineral soil (Fig. 14.14). A check of the soil moisture showed, however, that the intact soil cores had much lower water content (close to the wilting point) in the litter and humus layers than the sieved litter and humus layers kept separately (near optimum moisture). Therefore, the difference was more likely to depend on effects of soil moisture [see Eq. (2) above] than on effects of sieving.

Also at Gribskov, the moisture was lower in the litter layer on the soil column than in the sorted litter, explaining the lower net N mineralisation (Fig. 14.14). In the other soil layers, there was no significant difference between sieved and intact soil layers. At Klosterhede, sorted needle litter had a tendency to have higher net N mineralisation rates than the undisturbed litter lying on top of the intact cores. Again lower moisture in the undisturbed litter can explain the difference. In the other soil layers, no difference could be assigned to the sieving treatment. At 10–20 cm depth, the higher net N mineralisation in the intact cores could be explained by higher soil moisture.

It has been hypothesised that soil architecture controls microbially mediated decomposition processes in terrestrial ecosystems (VanVeen and Kuikman 1990).

Physical disruption of soil can, therefore, increase microbial access to readily mineralisable organic matter (Cabrera and Kissel 1988). Raison et al. (1987) found that sieving can either increase net N mineralisation or cause immobilisation of N. They concluded that soil disturbance (e.g. sieving) markedly affects rates of soil N mineralisation. Other studies have shown less pronounced effects. Fine sieving (1-mm mesh) of mineral soil has been shown to cause a temporary (0–14 days) increase in net N mineralisation (Hassink 1992). The relative increase in mineralisation was larger in loams and clays than in sandy soil. However, in a longer time perspective (14–70 days), small differences in net N mineralisation were found both in fine and coarse (8-mm mesh) sieving in comparison with undisturbed soil samples (Hassink 1992). The latter findings are in agreement with ours.

The comparison of sieved and intact soil cores as regards net N mineralisation in our study did not show any marked effect of sieving after adjusting for differences in soil moisture. There was a tendency to lower net N-mineralisation rates in sieved versus intact humus layers, possibly explained by removal of roots and adherent mycorrhizal fungi in the former layers. At least the mycorrhizal fungi are often rich in N and should stimulate net N mineralisation during long-term decomposition in intact soil cores. The study indicated that net N mineralisation can be determined in freshly sieved soil (no freezing or drying) without any severe artefacts.

14.10 In Situ Mineralisation Studies at Aubure

Inorganic N concentrations varied considerably throughout the year in the 15-cm topsoil (organic layers and 10 cm of the mineral soil) in the three stands at Aubure (Jussy 1998). Concentrations of inorganic N were higher in the AuF stand than in the two spruce stands (Table 14.5). Ammonium was the dominant form of inorganic N in all three stands, but in AuP45 the nitrate concentrations were especially low (Table 14.5; see also Persson et al., Table 2.9, Chap. 2, this Vol.).

Inorganic N in throughfall (about 65% nitrate) was significantly higher in AuP90 than in AuP45 and AuF (Table 14.5). N leaching (at 15-cm depth) estimated by zero-tension lysimeters was also higher in AuP90 than in the other two stands. Nitrate dominated in the lysimeter solution in the AuP90 and AuF stands, whereas small amounts were estimated to leach in the AuP45 stand.

At the end of the in situ incubations (fortnightly during summer, monthly during winter), the concentrations of inorganic N was always higher inside than outside the

Table 14.5. Estimated throughfall (input), mean concentrations (kg N ha^{-1}) and leaching (out of the cylinders) of ammonium and nitrate (kg N ha^{-1} a^{-1}) in the uppermost soil profile to a depth of 15 cm (10 cm in the mineral soil) in the Aubure stands

	Throughfall (kg N ha^{-1} a^{-1})			Inorg. N conc. (kg N ha^{-1})			Leaching (kg N ha^{-1} a^{-1})		
	NH$_4$-N	NO$_3$-N	Inorg. N	NH$_4$-N	NO$_3$-N	Inorg. N	NH$_4$-N	NO$_3$-N	Inorg. N
AuP45	2.6	4.9	7.5	8.5	0.5	9.1	1.3	1.0	2.3
AuP90	6.4	11.0	17.4	6.5	3.2	9.7	2.8	16.4	19.2
AuF	3.3	5.9	9.2	9.2	2.9	12.1	1.7	4.2	5.9

Table 14.6. Mean annual leaching and net N mineralisation ($kg\,N\,ha^{-1}\,a^{-1}$) in the in situ cylinders in the uppermost soil profile to a depth of 15 cm in the Aubure stands

	Leaching ($kg\,N\,ha^{-1}\,a^{-1}$)			Net N mineralisation ($kg\,N\,ha^{-1}\,a^{-1}$)		
	NH_4-N	NO_3-N	Inorg. N	NH_4-N	NO_3-N	Inorg. N
AuP45	7.1	5.5	12.6	61.5	5.9	67.4
AuP90	9.1	51.7	60.8	43.6	82.2	125.9
AuF	12.9	67.9	80.8	50.8	137.7	188.4

incubation cylinders ($p < 0.05$) indicating root uptake of N outside the cylinders. Nitrate leaching from the cylinders as measured by the ion exchange resin was much higher than outside the cylinders, but the leaching from the AuP45 cylinders was much less than from the AuP90 and AuF cylinders.

Annual net N mineralisation and net nitrification as calculated from the cylinder incubations were about 1.5-fold higher in the AuF soil (almost $190\,kg\,N\,ha^{-1}\,a^{-1}$) than in the AuP90 soil (about $125\,kg\,N\,ha^{-1}\,a^{-1}$) (Table 14.6). Net N mineralisation rate in the young spruce soil (AuP45) was lower, and its nitrifying capacity was very low. Laboratory incubations at 6 °C (average field temperature) of the 0–15-cm soil layer showed that net N mineralisation and net nitrification rates were always higher in the AuF soil than in the AuP90 and AuP45 soils. No substantial net nitrification occurred in the AuP45 soil (Jussy 1998).

Deposition of mineral nitrogen increases with stand age (Emmett et al. 1993) and is higher in coniferous than in deciduous stands (Van Praag and Weissen 1984). Both effects explain the larger input in AuP90. In AuP45 and AuF, atmospheric N input and net N mineralisation were assumed to be almost consumed by root uptake, while in AuP90 root uptake was insufficient, which led to nitrate leaching and soil acidification (Van Breemen et al. 1984). The uptake of N into the trees was allocated very differently between the stands. In the AuP45 stand, N allocated into tree growth, litterfall and production of root litter was estimated to be 17, 38 and $23\,kg\,N\,ha^{-1}\,a^{-1}$, respectively. The corresponding figures for AuP90 were 6, 29 and $77\,kg\,N\,ha^{-1}\,a^{-1}$, and for AuF 6, 43 and $138\,kg\,N\,ha^{-1}\,a^{-1}$, respectively (Jussy 1998). Production of root litter was not measured independently but was calculated as the difference between the computed N uptake and the measured N accumulation in perennial parts and annual litterfall. The high estimates for AuP90 and AuF indicate either a very high root/mycorrhizal turnover, or an overestimation in the net mineralisation data.

Low net nitrification and relative high biomass increment in comparison with the low deposition explained the low leaching rate in AuP45. Inhibition of root uptake increased nitrate leaching inside the cylinders, especially in the AuF stand, where the net nitrification rate was the highest. Hence, nitrate leaching was enhanced by heavy nitrification but was reduced and modified by root uptake.

The mineralisation rates obtained from the in situ studies were in the upper range of published values using field incubation methods (Runge 1974; Strader et al. 1989; Reich et al. 1997). Factors such as pH, soil type and C/N ratio were not sufficiently different between the soils to explain the magnitude of mineralisation and nitrification. N inputs to the soil did not appear as a decisive factor. The two spruce stands had very different nitrification rates, although their location and physical and chemical soil characteristics were very similar, especially as regards pH. Possible differences may be due to the following factors:

1. Inhibition due to leachates (polyphenols, Northup et al. 1995) or to microbial N assimilation associated to roots. Inhibition would be reduced in the old stand, where the root biomass was very low (see Stober et al., Chap. 5, this Vol.). Transpiration was 50% higher in AuP45 than in AuP90 (Biron 1994), indicating a higher fine root biomass, due to the better health status of the trees.
2. Because N deposition is lower in young than in old stands, and inorganic N is readily accumulated in the tree biomass and probably also by heterotrophic microorganisms, the ammonium oxidisers are probably suffering from substrate limitation. Another possibility is that gross nitrification might occur but that the microbial biomass was assimilating all the nitrate produced, leading to a net nitrification close to zero (Stark and Hart 1997).

In conclusion, net N mineralisation and net nitrification varied very much between the soils in the three Aubure stands and seemed to be independent of soil acidity, soil type or humus type. N inputs to the soil did not appear as a decisive factor. At the three stands, the leaching was probably controlled by the root activity. In the old spruce stand, AuP90, root uptake was lower than the sum of mineralisation and throughfall, which led to nitrate leaching, soil impoverishment and acidification (Van Breemen et al. 1984).

14.11 Comparison of in Situ and Laboratory-Based Mineralisation Studies

The in situ mineralisation studies resulted in higher estimates of annual net N mineralisation (Table 14.6) than in those based on laboratory incubations and extrapolated to the field (Table 14.2). For the soil layers involved (to a depth of 10 cm in the mineral soil), the in situ and the laboratory/extrapolation methods concerning the AuP45, AuP90 and AuF sites resulted in estimates of 67 and 56, 126 and 49, and 188 and 80 kg N ha^{-1} a^{-1}, respectively. The correspondence between the two methods was fairly good for the AuP45 stand, where the dominating N form was ammonium, but for the other two stands the differences were large.

One possible explanation for the differences is that the soil moisture was higher inside than outside the cylinders, because the root uptake of water was suppressed. The extrapolation from the laboratory to the field assumed that suboptimal moisture reduced net N mineralisation by, on average, 18% at AuP45 and AuP90, whereas the moisture was near optimal for AuF situated in a north-facing slope. If the open cylinders had near-optimal soil moisture, this can explain the discrepancy between the in situ and laboratory results for AuP45 but not for the other stands.

Another difference between intact and sieved soil is that dead roots may be mineralised (Hatch et al. 1990), but cut roots within the cylinders may also immobilise nitrogen as long as they are alive (Raison et al. 1987). The combination of both effects led Arnold et al. (1994) to conclude that cut roots had a minor effect. The comparison of intact and sieved soil in the laboratory (see above) supported the latter conclusion, although there was a tendency for higher net N mineralisation in intact soil cores, probably because of decomposition of N-rich mycorrhiza.

Another possibility is that the laboratory/extrapolation method underestimated net N mineralisation at AuP90 and AuF. The estimated net mineralisation rates determined in the laboratory were, however, high in these stands (Fig. 14.2) and cannot explain the differences. On the other hand, the soil pools were estimated to be comparatively low (see Persson et al., Tables 2.5 and 2.8, Chap. 2, this Vol.). It is possible that the samplings (performed once) at AuP90 and AuF was not representative in hitting spots with high stoniness, leading to an underestimation of the organic soil pool and the field net N mineralisation.

In conclusion, the in situ method and the laboratory/extrapolation method agreed well for AuP45 but not for the other two stands. Both methods indicated the same proportion in net N mineralisation for AuP90 and AuF, but estimated different levels.

14.12 Denitrification

Emission of nitrous oxide was studied at four beech sites. The emission from the beech forest soil at Aubure was at least one order of magnitude larger than at the other sites (Table 14.7). Because the measurement period at Aubure was, furthermore, restricted to May to October, the emission represents a significant N loss from the system. The high emission rates could probably be ascribed to high soil moisture and presence of nitrate (Table 14.5, see also Persson et al., Table 2.9, Chap. 2, this Vol.).

Nitrous oxide emission rates from the different soils were influenced not only by potential denitrification activity but also by the activity of nitrous oxide reductase. Inhibition of this reductase by means of acetylene showed that there was no effect at Aubure, whereas there was a large effect for the wet site at Sorø (Table 14.8). Thus, the high emission rates from the Aubure site may in part be due to a lack of nitrous oxide reductase activity in the denitrifying community. Nitrous oxide reductase is more sensitive to low pH than the other enzymes involved in denitrification, and the low pH in the Aubure soil (about 3.6 at 0–10-cm depth) may partly explain the lack of nitrous oxide reductase activity. Alternatively, differences in the composition of the denitrifying communities among the sites may influence nitrous oxide emission rates.

Table 14.7. Nitrous oxide emission from the CANIF beech forest soils for the period May to October (for Sorø March to December)

Site	N_2O efflux rate ($\mu g\ N_2O\text{-}N\ m^{-2}h^{-1}$)	N_2O emission (kg N_2O-N ha^{-1} per period)
Sorø (wet)	7.4	0.49
Sorø (dry)	6.8	0.45
Aubure	75.0	3.30
Schacht	3.5	0.15
Collelongo	3.1	0.14

Table 14.8. Potential denitrification activity in soil samples from the CANIF beech forest sites and the relation between potential denitrification activity without or with inhibition (by 10% C_2H_2) of nitrous oxide reductase

Site	Potential denitrification activity ($\mu g\ N_2O\text{-}N\ g^{-1}\ h^{-1}$)	Relation (%) between denitrification activity at 10 and 0% C_2H_2
Sorø (wet)	1.06 ± 0.03	11
Sorø (dry)	0.17 ± 0.00	86
Aubure	0.63 ± 0.01	100
Schacht	0.76 ± 0.01	92
Collelongo	0.71 ± 0.01	74

Table 14.9. Nitrous oxide emission from temperate forest soils as reported in the literature. Number of sites investigated given within parentheses

Tree cover	Nitrous oxide emission ($kg\ N\ ha^{-1}\ a^{-1}$)
Coniferous	0.28 (21)
Deciduous	2.10 (24)
Picea abies	0.35 (9)
Fagus sylvatica	2.20 (12)

N addition may increase N loss through denitrification. However, other parameters like pH and soil water status may influence denitrification rates, and the fraction of added N lost as nitrous oxide from temperate forest soils ranges from 0.03 to 1.6% (Bowden et al. 1991; Brumme and Beese 1992; Matson et al. 1992; Klemedtsson et al. 1997). It is possible that chronic additions via deposition (Brumme and Beese 1992) may lead to increasing emission over time as the heterotrophic microbial demand for N is met (Matson et al. 1992). Soil pH may decrease concomitantly with increased N deposition and influence N loss through denitrification as denitrification rates are generally decreased at low pH (e.g. Struwe and Kjøller 1994).

A literature review of nitrous oxide emission from temperate forest soils revealed that beech forest soils emit higher amounts of nitrous oxide than soils planted with Norway spruce (Table 14.9). In contrast, nitric oxide emission may be responsible for a significant N loss from the spruce forest sites as the NO/N_2O ratio has a preference for nitric oxide in spruce forest soils and for nitrous oxide in beech forest soils (Butterbach-Bahl et al. 1997; Rennenberg et al. 1998). However, at least a part of the emitted nitric oxide may be converted inside the canopy to NO_2 that might then be absorbed by the leaves (Rennenberg et al. 1997). Overall, we believe that denitrification represents a smaller N loss at the spruce sites than at the beech sites in the CANIF transect.

14.13 Final Discussion

Both sieved and intact soil cores were studied in the laboratory to compare whether the estimates of net N mineralisation and net nitrification differed depending on the treatment. The comparison showed no marked effect of sieving when incubated for a period of 150–200 days. The study indicated that net N mineralisation could be determined in freshly sieved soil (no freezing or drying) without any severe artefacts when incubated for a long time in the laboratory. In short-term incubations (10–20 days), net N mineralisation in sieved and intact soil will probably differ, as indicated by many reports on N mineralisation (e.g. Raison et al. 1987; Hassink 1992) and our data on C mineralisation rates (Fig. 12.3b, Chap. 12, this Vol.). Therefore, we made a compromise by estimating net N mineralisation rates and net nitrification rates for a period of 50 days. We considered this period to be long enough to have got stabilised conditions after the initial disturbance (see Fig. 14.2) and short enough to avoid too high concentrations of inorganic N.

Although the main comparison of sites ($n = 15$) and soil layers (mostly $n = 6$) was based on the same methodology, whereby net N mineralisation and net nitrification rate was determined at the same constant temperature (15 °C) and the same soil moisture (50–60% WHC) in the laboratory, we could not take samples from all sites at the same time for logistic reasons. This meant that the quality of the litter and soil substrates was not only impacted by the site conditions but also by the season. Beech litter sampled in April (e.g. at Schacht and Collelongo) was probably less decomposed than litter sampled in early October (e.g. Jezeri and Sorø) just before the main litter-fall. Furthermore, we could not exclude the possibility of seasonality in the microbial community, which might have affected the comparisons. However, we could not see any consistent effect of sampling time on the data of net N mineralisation or net nitrification. Therefore, we found it justified to make the comparisons of sites and soil depths.

Net N mineralisation rate differed markedly between sites. No or very low net N mineralisation was estimated for the soil layers at Åheden and the complementary sites Norrliden and Stråsan (all boreal sites in northern or central Sweden), whereas all temperate sites in southern and central Europe had a pronounced net N mineralisation. After the initial incubation period of 50 days, inorganic N started to accumulate also at Åheden and the other boreal sites. The lag of about 50 days in the laboratory (corresponding to 200–300 days in the field at the annual mean temperature) indicated that the microorganisms were strongly N-limited also in the field. Trees and ground vegetation roughly took up 20–60 kg N ha^{-1} a^{-1} at these sites (data derived from Tamm et al. 1995; Eriksson et al. 1996; Persson 1996) and N deposition was about 2–5 kg N ha^{-1} a^{-1} (see Persson et al., Table 2.1, Chap. 2, this Vol.). The low net N mineralisation estimated (close to 0 kg N ha^{-1} a^{-1}) could not match the plant uptake of N. Because soil microorganisms are generally considered to be better competitors for inorganic N than plant roots (Zak et al. 1990), two explanations seem possible for the long-lasting lag in net N mineralisation and N-limitation for the soil microorganisms. (1) Roots with mycorrhizal fungi can assimilate amino acids and small peptides (Smith and Read 1997; Näsholm et al. 1998) whereby the uptake of organic N can partly or wholly replace the uptake of inorganic N and leave other microorganisms to decompose substrates depleted in N. (2) N mineralisation occurs mainly in hot spots in the soil, where the mycorrhizal fungi can find inorganic N sources, leaving soil microorganisms at a distance from the hot-spots to be N-limited. Studies of ectomycorrhizal fungi (see Taylor et al., Chap. 16, this Vol.) indicate that the boreal site Åheden had a higher diversity of proteolytic fungi than any of the other sites. This indicates that the first explanation is quite plausible.

Long-term N fertilisation simulating high N deposition at some of these sites showed that the mineralisation pattern changed dramatically. In contrast to the control plots with a lag in net N mineralisation, inorganic N (mainly ammonium) accumulated without any lag phase in the same manner and at the same rates as in central Europe. This indicates that the N factor is more crucial than climatic factors in determining N mineralisation potentials at high latitudes. Long-term annual N fertilisation for 23 years resulted in a narrowing of the C/N ratio of the organic matter in the litter and humus layers, whereas N fertilisation during 9 years only resulted in a narrowing of the C/N ratio in the litter but not in the humus layer. These results indicate that narrowing of the C/N ratio in the soil is more likely to occur as an effect of N-enriched litterfall rather than through N immobilisation via the soil microbial biomass into the soil organic matter, that has been found to be an important mechanisms for short-term N immobilisation (Sjöberg and Persson 1998).

The comparison of sites in the NIPHYS/CANIF transect showed that net nitrification was always higher in beech than in spruce stands in the same area. Beech stands had generally higher pH and lower C/N ratio in the organic layers than corresponding spruce stands (except for Monte di Mezzo on a former arable soil). Gundersen et al. (1998) found that nitrate leaching and nitrate concentrations were negatively correlated with forest floor C/N ratios in a large European dataset. They also found no nitrate present in the subsoil at sites with C/N ratios above 30 in the forest floor. In the NIPHYS/CANIF transect no net nitrification was found at the two sites with C/N ratios above 30 in the humus layer. However, pH turned out to be an even more determining factor for net nitrification in the organic soil layers than the C/N ratio or net N mineralisation, whereas the role of pH was less decisive in the mineral soil layers. In these layers, net N mineralisation (and factors determining this process) was more important. At some Scandinavian sites (Åheden and Klosterhede), where net nitrification could not be detected, a combination of low pH, low amounts of ammonium and low amounts of nitrifiers were probable explanations of the lack of net nitrification.

The lowest pH level at which net nitrification was observed to occur varied within the European transect. Net nitrification was relatively high at the central European sites Waldstein (pH 3.7) and Nacetin (pH 3.6) in the humus layer, whereas no net nitrification occurred at the Scandinavian sites Åheden and Klosterhede (pH 3.9) despite addition of nitrifying humus. In the mineral soil (10–20-cm depth) net nitrification could be maintained even at pH 3.3 (after acidification) at both Aubure and Nacetin. A possible explanation is that the nitrifiers in central Europe have experienced a long-term N deposition of ammonium and, therefore, have evolved a high tolerance for low pH to cope with the favourable N availability. In Scandinavia, and especially in northern Scandinavia, this tolerance has probably not been obtained because of ammonium limitation.

Recent studies of gross nitrification in North American forest soils (New Mexico and Oregon) by Stark and Hart (1997) showed that gross nitrification rates could be much greater than net nitrification rates and even be high despite no net nitrification observed. In the NIPHYS/CANIF transect, some studies were made using the $^{15}NO_3^-$

isotope-dilution technique to determine gross nitrification. This work indicated that gross nitrification (based on a 24-h study) was high where net nitrification (184-day incubation) was high (a site close to Skogaby) and gross nitrification was low where net nitrification was low (Skogaby) (A. Rudebeck and C. Ste-Marie, in prep.). A longer observation period will favour net nitrification, and low or non-detectable net nitrification during such a long period (also used for some studies in the NIPHYS/CANIF study) will strongly indicate that gross nitrification is also low. This is further supported by Klemedtsson et al. (1999), who found that the number of ammonium-oxidising bacteria determined by the MPN method, was below the detection limit (500 cells g^{-1} dw) in the humus layer from unlimed plots (about pH 4.2) at a range of forested sites from southern to northern Sweden, whereas limed soils (pH 5–6) had numbers ranging from 10^3 to $10^5 g^{-1}$. This is in agreement with our data (Fig. 14.10) showing that liming increased net nitrification rate in the humus/topsoil layer.

14.14 Conclusions

- N turnover differed pronouncedly between boreal forest soils in northern Sweden and temperate forest soils in central Europe, including southern Scandinavia and central Italy. The main difference was that no net N mineralisation could be detected after incubation in the laboratory for 50 days at 15 °C at the northern sites, whereas inorganic N was readily formed at all southern sites. A possible explanation for the northern situation is either that the decomposer microorganisms become N-limited because roots/mycorrhiza are efficient competitors of organic N or that N mineralisation occurs at hot spots in the soil, leaving most microorganisms to be N-limited in between.
- A positive correlation existed between soil N pool and net N mineralisation in the NIPHYS/CANIF transect.
- Net N mineralisation and net nitrification (per ha and year) was generally higher in beech than in comparable spruce stands (in the same region).
- Net nitrification tended to be positively correlated with pH in the litter and humus layers but not in the mineral soil. Net nitrification was positively correlated with net N mineralisation in the mineral soil but not in the litter and humus layers.
- Net nitrification in the humus/topsoil layer responded with an increase after addition of $CaCO_3$ and with a decrease after addition of H_2SO_4 at practically all sites. However, the sensitivity to low pH values was different for different sites. At the central European sites, net nitrification could occur at lower pH than at the northern sites. We hypothesise that high NH_4^+ availability at the central European sites has favoured an adaptation of nitrifiers to acid conditions, whereas such an adaptation is not developed in soils with low NH_4^+ availability.
- At the northern sites, a combination of factors (low pH, low ammonium levels and low nitrifier populations) all contribute to low levels of net nitrification. This explains why liming alone or N fertilisation alone will not increase net nitrification in the organic/topsoil layers.
- Long-term N fertilisation (>20 years) at boreal sites increased net N mineralisation to levels similar to those in central Europe. The treatment also increased net nitrification in the mineral soil layers. In the organic soil layers, N fertilisation increased net nitrification only if pH was also increased. This indicates that an

adaptation of nitrifiers to acid conditions (pH < 4.0) could need a longer time than 20 years.
- At most European beech and spruce forest sites, autotrophic nitrifiers seemed to dominate over heterotrophic nitrifiers in regulating net nitrification. Heterotrophic nitrification (indicated by insensitivity to C_2H_2) was most pronounced, although never dominating, in the topsoil layers in central Europe.
- Results from long-term N-fertilisation experiments indicate that increased N deposition will not increase soil N pools and net N mineralisation independent of plants. It seems that inorganic N added needs to be circulated through the plants and be returned to the soil as N-enriched needle, leaf or root litter to affect N pools and net N mineralisation. Consequently, the buildup of a soil N pool after N addition is more delayed in spruce stands (6–10-year classes of needles) than in deciduous stands with all leaves falling each year.
- Nitrous oxide emission rates were very variable at the beech sites (spruce sites were not studied). High rates were measured at one site (Aubure) with high soil moisture, plenty of nitrate and low pH.

Acknowledgments. The study was supported by grants from EEC (ENV5V-CT92-1043 and ENV4-CT95-0053) and the Swedish Environmental Protection Agency.

References

Ambus P (1998) Nitrous oxide production by denitrification and nitrification in temperate forest, grassland and agricultural soils. Eur J Soil Sci 49:495–502

Arnold G, Van Beusichem ML, Van Diest A (1994) Nitrogen mineralization and H^+ transfers in a Scots pine (*Pinus sylvestris* L.) forest soil as affected by liming. Plant Soil 161:209–218

Azhar El Sayed, Verher R, Proot M, Sandra P, Verstraete W (1986) Binding of nitrite-N on polyphenols during nitrification. Plant Soil 94:369–382

Berdén M (1994) Ion leaching and soil acidification in a forest haplic podzol: effects of nitrogen application and clear-cutting. PhD Thesis, Swedish University of Agricultural Sciences, Department of Ecology and Environmental Research, Uppsala, Report 73, pp 1–22

Bergholm J, Jansson P-E, Johansson U, Majdi H, Nilsson L-O, Persson H, Rosengren-Brinck U, Wiklund K (1995) Air pollution, tree vitality and forest production – the Skogaby project. General description of a field experiment with Norway spruce in south Sweden. European Commission, Luxembourg, Ecosystem Research Report 21, pp 70–87

Biron P (1994) Modélisation du bilan hydrique de deux peuplements d'épicéa du bassin versant d'Aubure. PhD Thesis, Université Louis Pasteur, Strasbourg, 200 pp

Bowden RD, Melillo JM, Steudler PA, Aber JD (1991) Effects of nitrogen additions on annual nitrous oxide fluxes from temperate forest soils in the northeastern United States. J Geophys Res 96D:9321–9328

Brumme R, Beese F (1992) Effects of liming and nitrogen fertilization on emissions of CO_2 and N_2O from a temperate forest. J Geophys Res 97D:12851–12858

Butterbach-Bahl K, Gasch R, Breuer L, Papen H (1997) Fluxes of NO and N_2O from temperate forest soils: impact of forest type, N deposition and liming on the NO and N_2O emissions. Nutrient Cycling Agroecosyst 48:79–90

Cabrera ML, Kissel DE (1988) Potentially mineralizable nitrogen in disturbed and undisturbed soil samples. Soil Sci Soc Am J 52:1010–1015

Dambrine E, Bonneau M, Ranger J, Mohamed AD, Nys C, Gras F (1995) Cycling and budgets of acidity and nutrients in Norway spruce stands in northeastern France and the Erzgebirge (Czech Republic). In: Landmann G, Bonneau M (eds) Forest decline and atmospheric deposition effects in the French mountains. Springer, Berlin Heidelberg New York, pp 233–258

Davidson EA, Kingerlee W (1997) A global inventory of nitric oxide emissions from soils. Nutrient Cycling Agroecosyst 48:37–50

De Boer W, Klein Gunnewiek PJA, Kester RA, Tietema A, Laanbroek HJ (1993) The effect of acetylene on N transformations in an acid oak-beech soil. Plant Soil 149:292–296

Emmett BA, Reynolds B, Stevens PA, Noris DA, Hughes S, Görres J, Lubrecht I (1993) Nitrate leaching from afforested Welsh catchments – interactions between stand age and nitrogen deposition. Ambio 22:386–384

Eriksson HM, Berdén M, Rosén K, Nilsson SI (1996) Nutrient distribution in a Norway spruce stand after long-term application of ammonium nitrate and superphosphate. Water Air Soil Pollut 92:451–467

Falkengren-Grerup U, Brunet J, Diekmann M (1998) Nitrogen mineralisation in deciduous forest soils in south Sweden in gradients of soil acidity and deposition. Environ Pollut 102:415–420

Gundersen P, Callesen I, de Vries W (1998) Nitrate leaching in forest ecosystems is related to forest floor C/N ratios. Environ Pollut 102:403–407

Hart SC, Stark JM, Davidson EA, Firestone MK (1994) Nitrogen mineralization, immobilization and nitrification. SSSA Book Series 5. Methods of soil analysis. Part 2. Microbiological and biochemical properties. Soil Science of America, Madison, pp 985–1017

Hassink J (1992) Effects of soil texture and structure on carbon and nitrogen mineralization in grassland soils. Biol Fertil Soils 14:126–134

Hatch DJ, Jarvis SC, Phillips L (1990) Field measurement of nitrogen mineralization using soil core incubation and acetylene inhibition of nitrification. Plant Soil 124:97–107

Hübner C, Redl G, Wurst F (1991) In situ methodology for studying N-mineralization in soils using anion exchange resins. Soil Biol Biochem 23:701–702

Hyman MR, Wood PM (1985) Suicidal inactivation and labelling of ammonia monooxygenase by acetylene. Biochem J 227:719–725

Hynes RK, Knowles R (1982) Effect of acetylene on autotrophic and heterotrophic nitrification. Can J Microbiol 28:334–340

Jenkinson DS, Fox RH, Rayner JH (1985) Interactions between fertilizer nitrogen and soil nitrogen – the so-called 'priming' effect. J Soil Sci 36:425–444

Jussy JH (1998) Minéralisation de l'azote, nitrification et prélèvement radiculaire dans différents écosystèmes forestiers sur sol acide. PhD Thesis Université Henri Poincaré, Nancy-I, 171 pp

Killham K (1986) Heterotrophic nitrification. Spec Publ Soc Gen Microbiol 20:117–126

Klemedtsson L, Hansson G, Mosier A (1990) The use of acetylene for the quantification of N_2 and N_2O production from biological processes in soil. In: Revsbech NP, Sørensen J (eds) Denitrification in soil sediments. Plenum Press, New York, pp 167–180

Klemedtsson L, Klemedtsson ÅK, Moldan F, Weslien P (1997) Nitrous oxide emission from Swedish forest soils in relation to liming and simulated increased N-deposition. Biol Fertil Soils 25:290–295

Lemée G (1967) Investigations sur la minéralisation de l'azote et son évolution annuelle dans des humus forestiers in situ. Œcol Plant 2:319–324

Lövblad G, Kindbom K, Grennfelt P, Hultberg H, Westling O (1995) Deposition of acidifying substances in Sweden. Ecol Bull 44:17–34

Matson PA, Gower ST, Volkmann C, Billow C, Grier CC (1992) Soil nitrogen cycling and nitrous oxide flux in a Rocky Mountain Douglas-fir forest: effects of fertilization, irrigation and carbon addition. Biogeochemistry 18:101–117

Myrold DD (1998) Transformations of nitrogen. In: Sylvia DM, Fuhrmann JJ, Hartel PG, Zuberer DA (eds) Principles and applications of soil microbiology. Prentice Hall, New Jersey, pp 259–294

Näsholm T, Ekblad A, Nordin A, Giesler R, Högberg M, Högberg P (1998) Boreal forest plants take up organic nitrogen. Nature 392:914–916

Nömmik H, Vahtras K (1982) Retention and fixation of ammonium and ammonia in soils. Agronomy 22:123–171

Northup RR, Yu Z, Dahlgren RA, Vogt KA (1995) Polyphenol control of nitrogen release from pine litter. Nature 377:227–229

Persson T (1996) Kan skogen ta hand om allt kväve? Fakta Skog Konf 2(1996):55-73
Prosser JI (ed) (1986) Nitrification. Spec Publ Soc Gen Microbiol 20, 217 pp
Raison RJ, Connell MJ, Khanna PK (1987) Methodology for studying fluxes of soil mineral-N in situ. Soil Biol Biochem 19:521-530
Ratkowsky DA, Olley J, McMeekin TA, Ball A (1982) Relationship between temperature and growth rate of bacterial cultures. J Bacteriol 149:1-5.
Reich PB, Grigal DF, Aber JD, Gower ST (1997) Nitrogen mineralization and productivity in 50 hardwood and conifer stands on diverse soils. Ecology 78:335-347
Rennenberg H, Kreuzer K, Papen H, Weber P (1998) Consequences of high loads of nitrogen for spruce (*Picea abies*) and beech (*Fagus sylvatica*) forests. New Phytol 139:71-86
Runge M (1974) Die Stickstoff-Mineralization im Boden eines Sauerhumus-Buchenwaldes. Œcol Plant 9:201-230
SAS Institute Inc (1989) SAS/STAT Users Guide, version 6, 4th edn, vol 2. SAS Inst, Cary, North Carolina, 846 pp
Seyferth U (1998) Effects of soil temperature and moisture on carbon and nitrogen mineralisation in coniferous forests. Dept of Ecology and Environmental Research, Swedish University of Agricultural Sciences, Uppsala. Licentiate Thesis, No 1
Sjöberg RM, Persson T (1998) Turnover of carbon and nitrogen in coniferous forest soils of different N-status and under different $^{15}NH_4$-N application rate. Environ Pollut 102:385-393
Smith SE, Read DJ (1997) Mycorrhizal symbiosis, 2nd edn. Academic Press, San Diego
Stark JM, Hart SC (1997) High rates of nitrification and nitrate turnover in undisturbed coniferous forests. Nature 385:61-64
Strader RH, Binkley D, Wells CG (1989) Nitrogen mineralization in high elevation forests of the Appalachians. I. Regional patterns in southern spruce-fir forests. Biogeochemistry 7:131-145
Struwe S, Kjøller A (1994) Potential for N_2O production from beech (*Fagus sylvatica*) forest soils with varying pH. Soil Biol Biochem 26:1003-1009
Tamm CO, Aronsson A, Popovic' B (1995) Nitrogen saturation in a long-term forest experiment with annual additions of nitrogen. Water Air Soil Pollut 85:1683-1688
Terry RE, Leavitt RW (1992) Enhanced acetylene biodegradation in soil with history of exposure to the gas. Soil Sci Soc Am J 56:1477-1481
Van Breemen N, Driscoll CT, Mulder J (1984) Acidic deposition and internal proton sources in acidification of soils and waters. Nature 307:599-604
Van Praag HJ, Weissen F (1984) Potential nitrogen transfer and regulation through brown acid soils under beech and spruce stands. Plant Soil 82:179-191
Van Veen JA, Kuikman PJ (1990) Soil structural aspects of decomposition of organic matter. Biogeochemistry 11:213-233
Zak DR, Groffman PM, Pregitzer KS, Christensen S, Tiedje JM (1990) The vernal dam: plant-microbe competition for nitrogen in northern hardwood forests. Ecology 71:651-656

15 Nitrogen and Carbon Interactions of Forest Soil Water

B.R. ANDERSEN and P. GUNDERSEN

15.1 Introduction

In terrestrial ecosystems, nitrogen in soil water has mainly been considered in its inorganic plant available forms (NH_4, NO_3). Only few studies in forests have included the dissolved organic nitrogen fraction (e.g. Koopmans et al. 1996), although DON may be the dominant N form in soil water in many ecosystems (Sollins and McCorison 1981; Qualls et al. 1991). DON accounted for 95% of N losses in streams in an undisturbed natural temperate forest at pristine conditions in Chile (Hedin et al. 1995). Nitrogen losses in deep percolating soil water amounted to $2.2\,kg\,N\,ha^{-1}\,a^{-1}$ in both spruce and beech at Solling, Germany (Matzner 1988), a region with much higher atmospheric depositions.

In order to predict effects of N deposition in forest ecosystems and to estimate critical loads of nitrogen, it is important to quantify the contribution of DON to the N balance. Losses of DON will not normally amount to large quantities but may still constitute an important N input to downstream freshwater and coastal ecosystems or groundwater reservoirs.

Against this background, this project aimed at investigating (1) the relative distribution of dissolved nitrogen in inorganic (DIN) and organic (DON) compounds, respectively, in forest soil waters, and (2) a possible correlation between DON and dissolved organic carbon (DOC). Measurements were performed in forest stands paired coniferous and deciduous at five latitudinal positions along the north-south European transect introduced in the beginning of this Volume. Differences in climate and atmospheric nitrogen deposition along the transect were expected to allow a broader interpretation of the results.

Inorganic N forms are easily measured by a number of routine methods (e.g. ion chromatography and flow injection analysis). DON concentrations cannot be measured directly, but must be estimated by subtracting DIN from a measure of the total dissolved nitrogen (TDN) concentration. Typical methods for determining TDN in soil solution (e.g. persulfate digestion, UV-oxidation and total Kjeldall nitrogen) are difficult and time-consuming. The difficulty in measuring TDN is probably the reason for the relatively few studies in forest considering DON in the N balance.

Thus, an important step of this project was to develop a method for routine measurement of TDN in small volumes of soil water samples to allow analysis of a large number of samples. This new method is briefly reported in this chapter.

15.2 Approaches to Studying the Forest Soil Waters

15.2.1 Soil Water Sampling

Soil water samples were obtained using poly(tetrafluoroethene) (PTFE) porous suction cups with a mean pore size somewhat less than 10 µm (PRENART standard, Prenart Equipment ApS, Frederiksberg, Denmark) installed in different soil depths (25, 55 and 90 cm). The cups were acid-washed prior to installation. Contact between the cup surface and the soil matrix was enhanced by applying a slurry of acid-rinsed quartz powder and demineralised water during installation. Glass bottles placed in insulated boxes served both as partial vacuum reservoirs and sample collectors. The boxes were buried in the ground to reduce heating during summer and avoid freezing during winter. Each glass bottle was connected by polyethylene tubes to two individual sampler cups installed in parallel (i.e. in the same soil depth) and thus yielded a combined sample. The inlet of the tubes in the cup samplers was at the bottom of the cups, whereby water drawn into the cups was continuously transferred to the glass bottles avoiding prolonged contact time between the cup material and the sample. Sampling was semicontinuous as a partial vacuum of ca. −60 kPa relative to the soil water matrix was established every 2–4 weeks. The sampling rate inevitably varied during each sampling period as the suction power decreased. During a sampling period the individual glass bottles were isolated from each other, preventing one pair of cups with poor or no contact to the surrounding soil water pool from influencing the sampling from other cup pairs. Samples were brought to the laboratory facilities of the project partner attending the individual sites at 2–4-week intervals and stored in a refrigerator (+4 °C) without prior addition of preserving agents. No significant changes in measured parameters during storage could be detected, even after prolonged periods. At most sites it was impossible to sample throughout the year because of either winter periods with frozen soil and/or snow cover preventing access, or dry summer conditions.

15.2.2 Chemical Analyses

All soil water samples were sent by rapid air freight to our laboratory in Hoersholm, Denmark. Analysis for dissolved carbon, DOC, content was carried out using a Shimadzu TOC5000 analyser. NH_4 and NO_3 were determined using standard colorimetrical methods. Total dissolved nitrogen, TDN, was determined by the method described below, and dissolved organic nitrogen, DON, finally calculated by difference [DON = TDN − ($NH_4 + NO_3$)]. Analytical detection limits for TDN, NH_4 and NO_3 were 0.1, 0.04 and 0.03 mg N dm^{-3}, respectively.

15.2.3 Analysis for Total Dissolved Nitrogen

As part of this work we developed a method for the determination of total dissolved nitrogen in aqueous samples using an FIA system with in-line catalytic oxidative decomposition mediated by a persulfate reagent and microwave energy. The FIA system consisted of an autosampler (Perkin Elmer AS 90), an FIA pump and controller unit (Perkin Elmer FIAS 300) and a spectrometer (Perkin Elmer UV/VIS Lambda 2). The in-line decomposition step replaced a relatively labour-intensive separate decomposition step, whereas the final determination of nitrogen as NO_3 was identical to the standard colorimetrical method. Catalytic oxidative decomposition was carried out applying a solution of a persulfate-containing reagent Oxisolv (Merck Ltd., no. 112936) to the sample stream. Following this addition, the stream was passed through the oven room of a modified conventional kitchen microwave oven providing the energy necessary to decompose organic compounds in the samples and oxidise nitrogen to nitrate. Merck Ltd. does not specify the nature of the catalytic compound in the Oxisolv reagent.

The modifications to the microwave oven were: (1) making a couple of holes allowing the tube carrying the sample stream to pass through the oven room; and (2) transfering the timer control from the microwave oven itself to the FIA apparatus, allowing the FIA software to control when and for how long power should be applied. When testing the method, it turned out that the most critical part was the exact positioning of the sample stream coil inside the oven room, indicating that the energy distribution was highly uneven. The method could most likely be improved using a small focusing microwave device developed for laboratory purposes now commercially available.

The completeness of the decomposition step was tested using solutions of some selected pure compounds, showing N recovery to be greater than 90% even in complex heterocyclic organic compounds. Calibration curves were linear for TDN concentrations in the range 0.1 to $5\,mg\,N\,dm^{-3}$, and the detection limit of the method was $0.01\,mg\,N\,dm^{-3}$. Having developed the method to this stage, which was considered adequate for the expected N contents in the forest soil waters under study, we did not pursue possible further refinements. Throughout the project we used a certified aqueous sample with known TDN concentration as an internal reference.

15.3 Soil Water Concentrations of Nitrogen and Carbon

The total period covered at the individual sites was highly different, as shown in Table 15.1. Attempts to sample were performed throughout the entire project period at the two German sites (except when snow cover prevented access) and the Danish beech site. At Klosterhede, the Danish spruce site, a large number of samples were available from a 1-year period, as we could take advantage of the installations and on-going sampling programme of two other large ecosystem projects (EXMAN and NITREX). The NITREX plot at Klosterhede gave the added opportunity to test if experimentally elevated inorganic N-input to the forest floor influenced the composition of the soil water.

We discontinued our sampling at Aaheden in northern Sweden and at Thezan near the French Mediterranean coast early in the project period, because it quickly became apparent that the majority of samples that could be obtained had very low N contents. Another problem at Aaheden was that spring snowmelt took place when the upper soil layers were still frozen, preventing our cup samplers from yielding any samples, resulting in surface runoff. Earlier studies on stream water in the area suggest that a sizeable amount of N is removed from the forest in surface runoff during snowmelt. (P. Högberg, pers. comm.)

The two Italian sites were not included until the latter part of the project period, coinciding with two consecutive summers (1997 and 1998) with drier than normal weather, resulting in a much lower number of soil water samples than planned. Only 20–25% of the sampling attempts yielded enough soil water for analyses.

Table 15.1. Summary of the sample periods at the ten stands included in our study. The sampling periods varied between 2 and 4 weeks

Site (study period)	Tree	No. sampling Periods	No. Obs.[a]	Compound	No. Analysed[b]	above DL (%)
Aaheden (7/94–10/95)	Pinus	17	102	NO_3	44	2
				NH_4	46	2
				TDN	23	9
Aaheden (7/94–10/95)	Betula	17	102	NO_3	45	4
				NH_4	46	2
				TDN	24	4
Klosterhede (1/93–10/96)	Picea	39	1152	NO_3	638	29
				NH_4	642	52
				TDN	170	71
Gadevang (3/94–12/98)	Fagus	53	318	NO_3	208	37
				NH_4	132	12
				TDN	218	89
Waldstein (7/93–9/98)	Picea	65	390	NO_3	287	97
				NH_4	227	14
				TDN	287	99
Schacht (7/93–9/98)	Fagus	71	425	NO_3	273	29
				NH_4	200	7
				TDN	271	68
Thezan (6/93–1/95)	Pinus	21	126	NO_3	84	15
				NH_4	83	16
				TDN	58	28
Thezan (6/93–1/95)	Quercus	21	126	NO_3	103	5
				NH_4	105	15
				TDN	73	25
Monte di Mezzo (5/97–10/98)	Picea	18	108	NO_3	25	100
				NH_4	25	24
				TDN	25	100
Collelongo (5/97–10/98)	Fagus	27	162	NO_3	31	55
				NH_4	31	42
				TDN	31	100

[a] Maximum number of soil water samples available for analyses from each stand if all samplers in all soil depths had yielded water in every sampling period.
[b] For each N compound the actual number of samples analysed, whereas the last column gives the percentage of these having a concentration above the analytical detection limit (DL).
DOC is not included as all samples but one had contents well above the DL.

15.3.1 Dissolved Organic Carbon

As could be expected, DOC concentrations in soil water decreased with soil depth at all sites, as illustrated in Fig. 15.1. The decrease was less pronounced at the extremes of the gradient, possibly because the soils at Aaheden and Collelongo are less developed (cf. Chap. 2, this Vol. for soil characteristics), and would have demanded soil water sampling equipment with a much smaller sampling radius to resolve a likely concentration gradient in narrow soil layers near the surface.

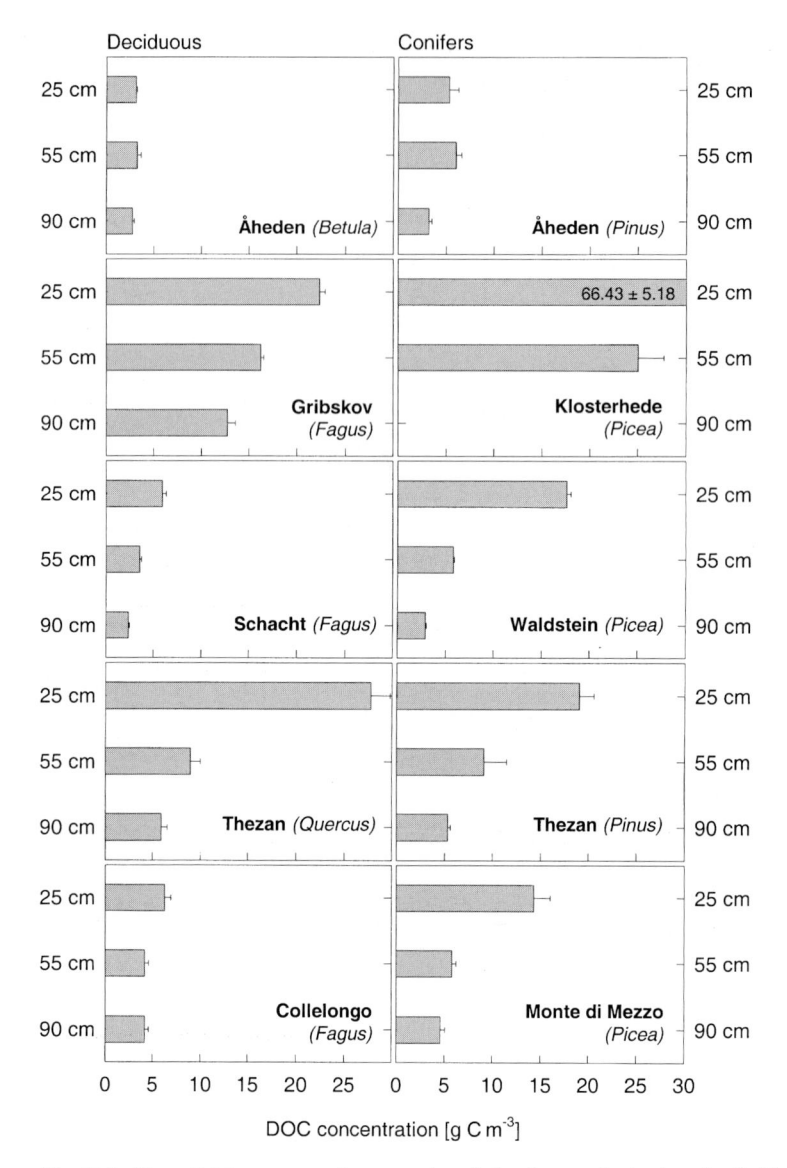

Fig. 15.1. Mean DOC concentration at each soil depth at each site (mg dm^{-3}). The subfigures are arranged according to the position of the site along the north to south gradient and with deciduous stands at *left* and coniferous at *right*. *Error bars* indicate standard error of the mean. At Klosterhede the value for the 25-cm soil depth was 66.4 ± 5.2, exceeding the range at the other sites by more than a factor of 3

At the Danish spruce site Klosterhede DOC concentrations were consistently very high, especially in 25-cm soil depth, but the average concentration in 55-cm depth even exceeded the values found in the topmost soil layer sampled at all other sites. Seasonal variations in DOC concentrations were not very marked at the sites; at most sites the seasonal high and low concentration did not differ more than between the sets of samplers operated (not depicted). The NITREX plot at Klosterhede receiving extra inorganic N to the forest floor showed low DOC concentrations compared to the average of the control plots, but the relatively large variation between the sampler sets in the control plots masked the possible difference, which was not significant in most sampling periods.

15.3.2 Total and Organic Nitrogen in Soil Solution

The mean concentration of total nitrogen in the forest soil waters varied markedly along the gradient and between tree species at comparable latitudes (Fig. 15.2). At the extreme north at Aaheden only very few samples had a TDN content above our analytical detection limit, as explained above. This is in good accordance with the notion of a pristine environment with very little freely available N, as explained in the site description in Chapter 2 (this Vol.). At Klosterhede TDN concentrations were consistently low, with a mean even lower than at the Danish beech site. In contrast, in Germany and Italy there was generally much higher TDN contents in soil waters from spruce stands than from beech stands.

The TDN concentration tended to decrease down through the soil profiles at the majority of the stands. However, the two Italian sites both showed a higher average TDN concentration in 55 cm depth than in 25 cm. At Collelongo this pattern was also seen in the deepest soil layer sampled (90 cm). This different pattern at the Italian sites may be ascribed to the differing soil types, calcareous with high pH (see Persson et al., Chap. 2, this Vol.). At Waldstein there was an almost uniform average TDN concentration profile.

The average DON concentration profiles were almost uniform at most sites (Fig. 15.2), with only a weak, non-significant tendency towards lower concentrations at greater depth. The only exceptions to this pattern were the samples from Klosterhede (detailed in Fig. 15.3) and under oak at Thezan, but too few measurements were available from this latter site to base conclusions on them. The difference in TDN and DON concentration profiles indicates that inorganic N is preferentially removed from solution at most sites, but give no insight into the processes causing it.

It is immediately evident from Fig. 15.2 that DON is a significant part of TDN in soil water at most of the sites. Exceptions are the two Italian sites, and, to some extent, Waldstein. No consistent seasonal patterns in the fraction of TDN in soil water consisting of DON were found, indicating that a strong influence of microbial activity and/or root uptake was not present.

15.4 Correlation Between Dissolved Organic Nitrogen and Carbon

We investigated a possible relationship between the soil water contents of DON and DOC using a simple linear model, restricting it by demanding the intercept term to be zero. Several combinations of site and soil depth showed some correlation, but only at the control plots at Klosterhede were there at the same time a highly significant correlation and a significant probability that the intercept term was indeed zero (see also Fig. 15.4):

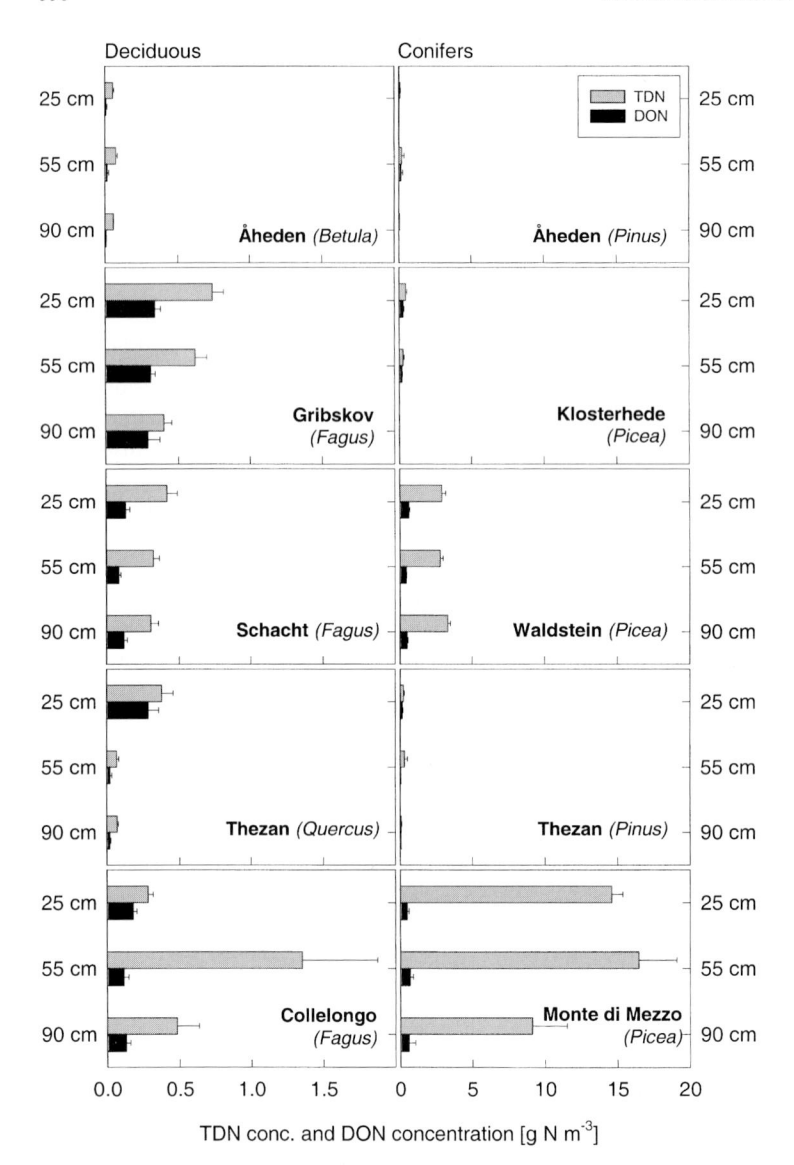

Fig. 15.2. Mean measured TDN (*upper bar* at each soil depth) and calculated DON (*lower bar*) concentrations, including standard error bars. Units: mgN dm^{-3}. The figure is constructed similarly to Fig. 15.1. Note that the scale for deciduous stands (*left part*) is much finer

25 cm depth: DON = 0.028 ± 0.003 × DOC, adjusted R^2 = 0.69, $p < 0.001$,
55 cm depth: DON = 0.051 ± 0.004 × DOC, adjusted R^2 = 0.78, $p < 0.001$.

The slopes of these linear relations corresponds to overall C:N ratios of 36 and 20, respectively (based on weight), in the dissolved organic compounds. The value for

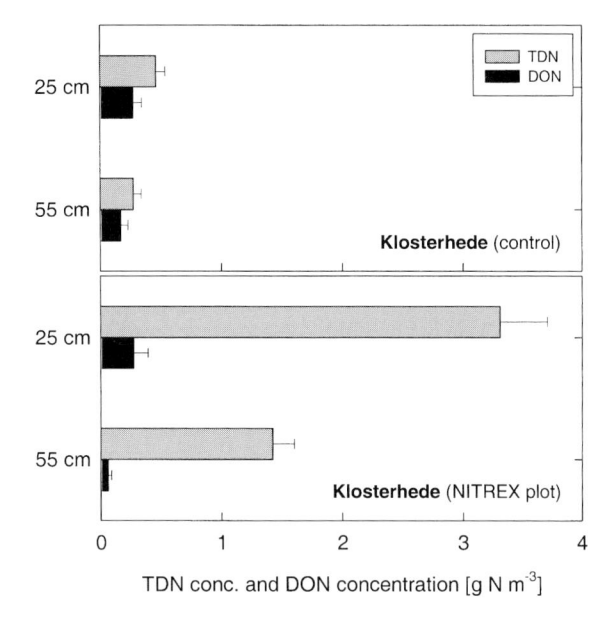

Nitrogen and Carbon Interactions of Forest Soil Water

Fig. 15.3. As Fig. 15.2, but including measurements from the NITREX plot at Klosterhede. This plot received an extra $35\,\mathrm{kg\,N\,ha^{-1}\,a^{-1}}$ as dissolved NH_4NO_3

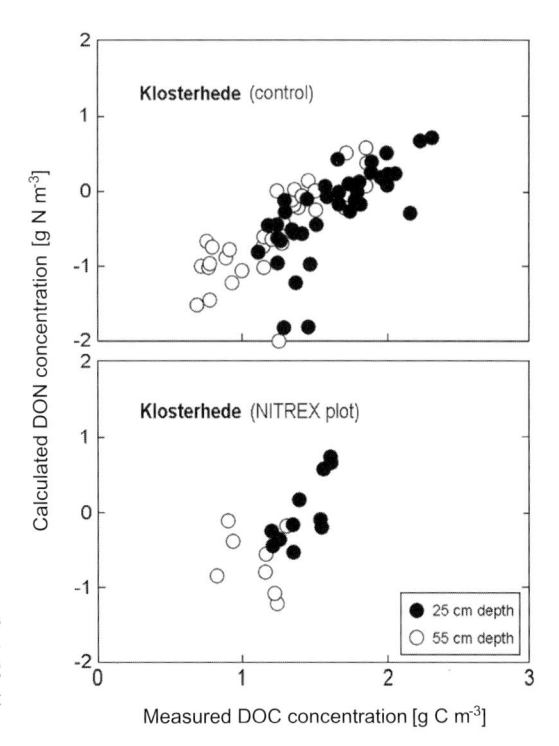

Fig. 15.4. Klosterhede control plots (*left*) and NITREX plot (*right*). Calculated DON vs. measured DOC concentrations $(\mathrm{mg\,dm^{-3}})$. Note that both axes are logarithmic

25 cm depth is very near the value (33) found for the bulk of the solid organic layer (Gundersen and Rasmussen 1995).

15.5 Conclusions

We succeeded in eliminating a labour-intensive separate decomposition step in TDN analysis by incorporating it as an in-line step in a new continuous-flow FIA method. DON was a significant fraction of TDN in the majority of samples from all sites, except at Aaheden, which had extremely low concentrations of any N compound. This supports our hypothesis that a complete assessment of potential N uptake and N leaching cannot be made without taking DON into account. The concentration profiles of TDN and DON down through the soil layers varied at most of the sites, indicating a preferential uptake of inorganic N – mainly available as NO_3 in these forest soils – relative to the fraction of TDN present as DON. The Italian sites, having calcareous soils with high pH, were singled out by showing higher NO_3 concentrations in the deeper soil layers, indicating a high nitrification activity even in the mid-soil layers. DON and DOC could not be shown to be significantly correlated in these stands, except at Klosterhede.

Acknowledgements. This study was begun at the Laboratory of Environmental Sciences and Ecology, Technical University of Denmark. The research group and laboratory staff moved to the Danish Forest and Landscape Research Institute in 1992. It was in part financed by the EC Environment Programme as part of the NIPHYS and CANIF projects (contracts EV5V-CT92-0143, EV5V-CT94-0433 and ENV4-CT95-0053), and in part by the Danish Strategical Environment Programme. We are deeply grateful to our foreign colleagues attending our equipment at the individual sites, and our laboratory staff, in particular P. Frederiksen, for thorough and engaged work during treatment and analysis of all samples included in the study.

References

Gundersen P, Rasmussen L (1995) Nitrogen mobility in a nitrogen limited forest at Klosterhede, Denmark, examined by NH_4NO_3 addition. For Ecol Manage 71:75–88

Hedin LO, Armesto JJ, Johnson AH (1995) Patterns of nutrient loss from unpolluted, old-growth temperate forests: evaluation of biogeochemical theory. Ecology 76:493–509

Koopmans CJ, Tietema A, Boxman AW (1996) The fate of ^{15}N enriched throughfall in two coniferous stands at two nitrogen deposition levels. Biogeochemistry 34:19–44

Matzner E (1988) Der Stoffumsatz zweier Waldökosysteme im Solling. Berichte des Forschungszentrum Waldökosysteme/Waldsterben, Reihe A, Bd 40. Univ Göttingen, Germany

Qualls RG, Haines BL, Swank WT (1991) Fluxes of dissolved organic nutrients and humic substances in a deciduous forest. Ecology 72:254–266

Sollins P, McCorison FM (1981) Nitrogen and carbon solution chemistry of an old-growth coniferous forest watershed before and after cutting. Water Resour Res 17:1409–1418

Part D
Diversity-Related Processes

16 Fungal Diversity in Ectomycorrhizal Communities of Norway Spruce [*Picea abies* (L.) Karst.] and Beech (*Fagus sylvatica* L.) Along North-South Transects in Europe

A.F.S. TAYLOR, F. MARTIN, and D.J. READ

16.1 Introduction

Spruce (*Picea abies* [L.] Karst.) and beech (*Fagus sylvatica* L.) which are among the most important tree species, respectively, of boreal and temperate forest ecosystems in Europe are characteristically ectomycorrhizal (Meyer 1973). While the forests dominated by these plants have a low diversity of tree species, the trees themselves typically support a very diverse community of fungal symbionts (Trappe 1962; Väre et al. 1996). In recent years, however, concern has been expressed over an apparent reduction in the number of fungal species represented in the form of carpophores in European forests (Arnolds 1991). While this is an important concern in itself, from the standpoint of tree nutrition and forest health the key issue is the structure of the fungal community on the roots rather than that observed above ground as carpophores.

It has been hypothesised that the observed loss in carpophore diversity is attributable to increases in pollutant nitrogen deposition onto forest ecosystems in which N limitation has been a characteristic feature (Baar and ter Braak 1996). In this chapter a test of the hypothesis that pollutant N deposition has contributed to a reduction in fungal diversity is presented. This was achieved by examining both mycorrhizal tips in soil cores and carpophores in spruce and beech forests along transects that represent gradients of increasing mineral N deposition. The transects extend from relatively unpolluted sites in Scandinavia towards sites in Central Europe which suffer greater N-pollution loads. Diversity is considered at both the inter- and intraspecific levels. Interspecific diversity has been considered at the within core (point diversity), within site (α diversity) and between site (β diversity) levels.

Recently, studies at the intraspecific level have shown that there can be as much functional diversity within a single species of ectomycorrhizal fungus, as between them (Wagner et al. 1988 1989; Gay et al. 1993). These observations have increased awareness of the need to characterise the extent of genotypic variation in naturally occurring populations of individual fungal taxa. Because of the inherently intensive nature of the molecular studies required to identify genotypic variability, such studies, initially at least, are most effectively focussed at a single site where attention is devoted to a species known to be an important component of the ectomycorrhizal fungal community. On this basis the basidiomycete *Laccaria amethystina* (Bolt. ex Hooker) Murr. was selected and its spatial genetic structure examined in a closed 150-year-old beech forest at Aubure, NE France.

Finally, the pattern of N utilisation by selected mycorrhizal fungi of spruce isolated from different parts of the transect was analysed, with a view to testing the hypothesis that pollutant N deposition has selected in favour of mineral nitrophilous fungi with lower ability to utilise organic forms of N.

16.2 Analysis of Ectomycorrhizal Community Structure and Diversity

16.2.1 Morphotyping

The past 15 years have witnessed a revolution in our ability to determine the structure of ECM root communities. A combination of morphological characterisation (Agerer 1986–1998) and molecular identification using PCR-RFLP analysis of the ITS region (Gardes and Bruns 1996) has enabled significant advances to be made. Morphological characterisation (here referred to as morphotyping) has the great advantage that it is relatively inexpensive, enabling large numbers of root tips to be assessed.

Root material for morphotyping was collected from the four spruce and four beech sites shown in Table 16.1. For detailed site characteristics see Persson et al. (Chap. 2, this Vol.). Root tips were collected by taking replicate (nine per site) soil cores, each of 4.5 cm diameter and consisting of all of the organic matter. Cores were returned to the laboratory and fine root material was extracted by soaking the samples in water, followed by wet sieving through a combination of 1.5- and 0.5-mm sieves. The percentage of tips lost using this procedure is very small (<1%). Root tips were then examined microscopically and, where possible, the associated mycorrhizal fungi were identified to morphotype level using published descriptions (e.g. Agerer 1986–1998;

Table 16.1. Characteristics of the ectomycorrhizal communities in four spruce and four beech forests along north-south transects in Europe

Site[a]	Total no. of tips sampled	Level of colonisation (%)	Total no. of morphotypes	Mean no. of morphotypes per sample[b]	SWD index[c] (Hs)	Evenness[d]
Spruce						
Åheden	1311	99.4	19	6.0 ± 0.4a	3.55	0.840
Klosterhede	1873	90.6	19	5.2 ± 0.8a	3.34	0.838
Waldstein	1114	99.2	14	2.8 ± 0.6b	2.57	0.509
Aubure	2610	98.8	18	4.6 ± 0.7a	3.29	0.759
Beech						
Gribskov	2123	99.2	14	4.1 ± 0.4	2.83	0.697
Schacht	2141	100.0	22	4.3 ± 0.4	3.93	0.870
Aubure	698	99.9	11	2.8 ± 0.6	2.34	0.561
Collelongo	3209	99.8	19	3.8 ± 0.6	3.28	0.718

[a] Spruce: Åheden – Northern Sweden; Klosterhede – Western Jutland, Denmark; Waldstein – Fichtelgebirge, Germany; Aubure – Vosges Mountains, NE France. Beech: Gribskov – N. Sealand, Denmark; Schacht – Fichtelgebirge, Germany; Collelongo – Appenines, Italy.
[b] Means sharing the same letter are not significantly different at $p = 0.05$.
[c] Shannon-Wiener diversity Index – Hs (Krebs 1989).
[d] Evenness – Berger-Parker index of community evenness (Magurran 1988).

Brand 1991; Ingleby et al. 1990; Haug and Pritsch 1992). The term morphotype is used here both in the case of mycorrhizas where the mycobiont is known (e.g. *Russula ochroleuca*) and where it is unknown (e.g. "Piceirhiza bicolorata"). With the exception of tips colonised by *Cortinarius* species and E-type fungi, where it is not possible to discriminate the structures at specific level, each morphotype is considered to be formed by a single fungal taxon. The number of tips of each morphotype was recorded and expressed as a percentage of the total root tips found at each site.

In addition to the assessment of community structure below ground, a single visit was made to each of the four spruce sites at the height of the fruiting season and the carpophores of the ectomycorrhizal fungi were recorded and collected over an area of approx. 100 × 100 m. Nomenclature follows that of Moser (1983) for Agarics and Boleti and Breitenbach and Kränzlin (1986) for Aphyllophorales.

16.2.2 Assessment, Characterisation and Comparison of Ectomycorrhizal Community Diversity at Spruce and Beech Sites

Several methods were selected to examine different aspects of community diversity.

1. Rank-abundance diagrams (β diversity)

The analysis of species abundance patterns within communities provides a means of comparing overall community structure between sites. Tokeshi (1993) compared the different models and plotting methods used for displaying abundance data and suggested that a plot of the arithmetic rank (of the species) versus log (abundance) was the most versatile and convenient method of presentation. We have followed this recommendation.

The abundance of each ECM morphotype on each of the two tree species at any site was first placed in rank order with the most abundant first. The total number of tips at each site colonised by the individual morphotypes was then converted to a log (base 10) value and plotted against the ranked abundance values.

The shape of the curves derived from plotting abundance data in this way may take the form of a straight line which reflects communities where a few species are dominant, to forms with more than one slope, referred to as broken-stick types, which indicate that abundance is more uniformly distributed through a number of species (Magurran 1988). May (1974) proposed that the broken-stick type of curve was produced in situations where availability of a single prevailing ecological resource was constraining all species present in a natural community equally.

2. Rarefaction (β diversity)

It is evident that, as in the present study, if the total number of individuals sampled (in this case root tips) at each site varies, then direct between-site comparisons of values such as species (or morphotype) richness will be invalid. We therefore employed the method of rarefaction (Krebs 1989), designed for use under these circumstances, which calculates the number of species that would be expected in samples of a standard size. The simplest version of this technique was used, which takes the number of individuals in the smallest sample as the standardised sample size (in this case, 1114 at Waldstein for spruce and 698 at Aubure for beech). In addition, we also calculated the predicted number of morphotypes in sample sizes smaller and larger than this to enable a clearer comparison of the differences between sites.

To achieve this, the predicted number of morphotypes (morphotype richness) was calculated in samples of increasing size from 50 tips, at intervals of between 50 and 200 tips, up to the actual number of tips sampled at a given site. The expected morphotype richness at all sites, for each host species, was then compared by plotting the predicted richness against the number of tips in a hypothetical sample.

3. Bray-Curtis index of similarity (point and β diversity)

The homogeneity of species (morphotype) composition in samples was calculated, within and between sites, using the Bray-Curtis measure of similarity (Krebs 1989).

$$\text{Bray-Curtis similarity index} = 1 - \left(\sum_{i=1}^{n} [X_{ij} - X_{ik}] \Big/ \sum_{i=1}^{n} [X_{ij} + X_{ik}] \right),$$

where X_{ij}, X_{ik} = number of individuals in morphotype i in each sample j and k, n = number of morphotypes in each sample.

It is standard practice to use its complement value (i.e. 1 minus the index) in order to obtain positive relationships between similarity and the value of the index, such that calculated values range from 0 (totally dissimilar) to 1 (identical). The analysis was based upon the numbers of mycorrhizal tips in each of the morphotype categories.

4. Shannon-Wiener diversity index (α diversity)

In order to facilitate a comparison with other studies, overall community diversity was measured using the Shannon-Wiener diversity (Hs) index (Krebs 1989). This incorporates a measure of both species richness and community evenness by determining the relative contribution of each species to the community.

$$\text{Shannon-Wiener index} = -\sum_{i=1}^{n} p_i \ln p_i,$$

where p_i = proportion of individuals found in the *i*th morphotype.

5. Berger-Parker evenness index (α diversity)

This is a simple means of evaluating community structures which expresses the proportional importance of the most abundant species (Magurran 1988). As with the Bray-Curtis index, the complement of the index was used, hence the higher the value of the index the greater the evenness of the community (i.e. the less the dominance of the most abundant species).

$$\text{Berger-Parker index} = 1 - (N_{max}/N),$$

where N_{max} = number of individuals in the most abundant morphotype, N = total number of individuals recorded

6. Point diversity

The average number of morphotypes per sample was compared between sites using one-way analysis of variance (Minitab 1998).

7. Relationships between fungal diversity and site soil characteristics

Relationships in the form of Pearson product moment correlation coefficients (Minitab 1998) were sought between Shannon Wiener diversity indices (calculated as described above), for the communities under both host tree species at each site and the following site characteristics (1) soil pH, (2) total nitrogen (N) and sulphur deposition, (3) extractable N (as $NH_4 + NO_3$, as NH_4 alone and as NO_3 alone) (4) extractable aluminium. Site data for this analysis were derived from Persson et al. (Chap. 2, this Vol.). Correlation coefficients were also calculated for relationships between morphotype richness, defined as the number of morphotypes found at a site and the same site parameters.

16.3 ECM Communities of Spruce Forests

16.3.1 Analysis Based on Ectomycorrhizal Morphotypes

A total of 43 morphotypes were distinguished on the 6908 root tips of spruce examined (Table 16.2). Colonisation at all sites, calculated as the percentage of root tips converted to mycorrhizas, was high: the lowest value, at Klosterhede, being over 90%. The number of morphotypes recorded at a site ranged from 14 to 19. The community diversity of the mycorrhizal roots, as measured by the Hs index, was highest at Åheden, with a value of 3.55 (Table 16.1). No single morphotype colonised more than 20% of the root tips at this site. Nine types each colonised between 5 to 20% of the root tips. The evenness of the community at Åheden was also the highest, although only marginally more than at Klosterhede. This situation contrasted strongly with that seen at Waldstein, which had the lowest Hs of 2.57 and where more than 76% of the root tips were colonised by only two morphotypes. This disjunct morphotype distribution also leads to a very low evenness index (Table 16.1). Intermediate Hs indices were found at Klosterhede and Aubure. The community structures at these sites resembled each other with respect to both morphotype richness and relative abundance (Table 16.2).

Rank abundance curves (Fig. 16.1a) for the four sites revealed strong similarities between community structures at Åheden, Kosterhede and Aubure, but a distinctly different pattern at Waldstein where, as shown in Table 16.2, two morphotypes, *Tylospora* sp. and *L. rufus*, dominated. This causes a rapid early drop in the curve, displacing it relative to the other sites. The difference between Waldstein and the other sites is further exacerbated by the absence of significant numbers of relatively rare morphotypes, a feature which gives rise to a truncated tail in its curve.

The disparity in morphotype richness between Waldstein and the other sites is also highlighted in the rarefaction curves (Fig. 16.2a). The results suggest that, irrespective of sample size, Waldstein has fewer morphotypes. This was supported from the analysis of the number of morphotypes per sample, which demonstrated that significantly fewer morphotypes per sample were recovered from Waldstein than from any of the other spruce sites (Table 16.1).

Morphotype composition of samples from the four spruce sites was more similar within sites than between (Table 16.3). Samples from Åheden were clearly less similar to those from other sites, reflecting the fact that 74% of the morphotypes found at Åheden were unique to that site (Table 16.2).

Table 16.2. Relative abundance of ectomycorrhizal morphotypes, expressed as the percentage of colonised roots in samples collected under Norway spruce, at four study sites along a north-south European transect. (Sites: Åheden – northern Sweden; Klosterhede – western Jutland, Denmark; Waldstein – Fichtelgebirge, Germany; Aubure – Vosges Mountains, NE France)

Species/type (Total no. = 43)	Aheden (19)	Klosterhede (19)	Waldstein (14)	Aubure (18)
(cf.) *Amphinema*	–	0.5	–	–
Cenococcum geophilum	5.9	0.2	0.8	0.9
Cortinarius spp.	16.3	–	–	–
Dermocybe cinnamomea	–	–	–	0.5
D. cinnamomeobadia	–	–	–	0.1
E-type	15.7	2.4	0.9	0.8
Elaphomyces sp.	–	0.3	–	5.6
Fagirhiza globulifera	–	0.3	–	–
F. setifera	0.3	–	–	–
Hygrophorus pustulatus	–	0.9	–	2.9
Laccaria sp.	–	–	–	9.8
Lactarius (cf.) *necator*	–	–	5.2	–
L. rufus	–	–	26.1	8.9
Lactarius sp.	–	1.9	–	–
Picierhiza bicolorata	–	8.5	–	–
P. gelatinosa	5.7	–	0.8	6.3
P. oleiferans	–	–	–	4.5
P. punctata	–	5.4	–	–
P. rosa-nigrescens	11.8	–	–	–
P. rosea	–	0.3	–	–
Piloderma croceum	0.1	–	–	–
Russula ochroleuca	–	–	3.1	1.6
R. (cf.) *xerampelina*	3.4	–	–	–
Russula 1	–	–	–	0.4
Russula 2	–	1.4	–	–
Russula 3	–	4.3	–	–
Russula 4	–	9.0	–	–
Russula 5	–	16.0	–	–
Russula 6	0.1	–	–	–
Russula 7	8.8	–	–	–
Russula 8	0.5	–	–	–
Russula 9	–	–	6.0	–
Thelephora sp.	–	0.5	–	24.1
Tylospora sp.	0.7	14.2	49.1	4.5
Xerocomus spp.	–	1.0	2.2	19.6
Unknown morphotypes	28.9	23.5	5.0	8.3
Non-mycorrhizal roots	0.6	9.4	0.8	1.2

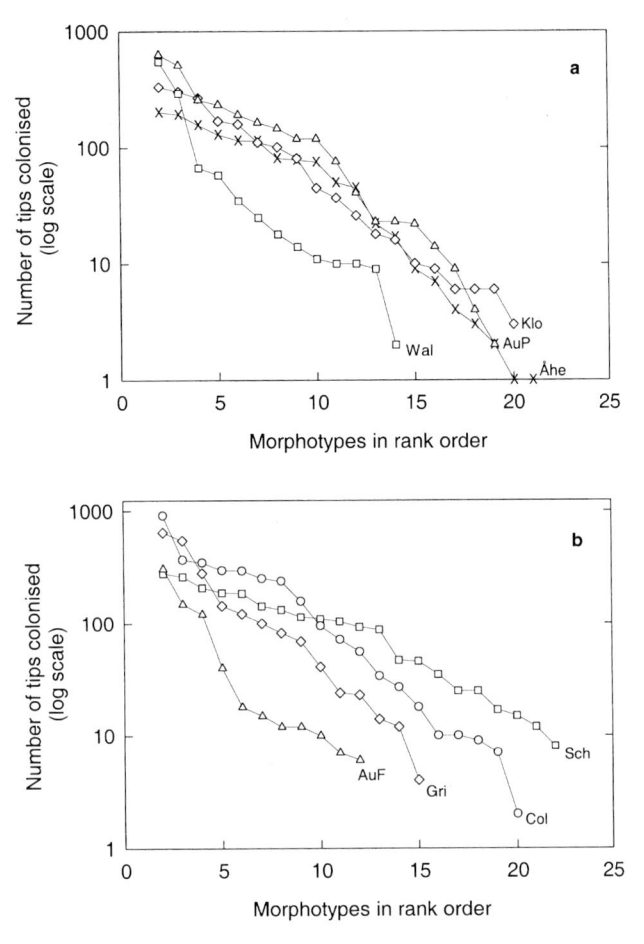

Fig. 16.1. Rank-abundance of morphotypes in a spruce and b beech ectomycorrhizal fungal communities, using log abundance ($\log_{10}$ no. of tips colonised by each mycorrhizal morphotype) plotted against ranked morphotype abundance.
a Spruce: *Åhe* Åheden, N. Sweden; *Klo* Klosterhede, Denmark; *Wal* Waldstein, Germany; *AuP* Aubure, NE France. b Beech: *Gri* Gribskov, Denmark; *Sch* Schacht, Germany; *AuF* Aubure, NE France; *Col* Collelongo, Italy

The occurrence and abundance of some taxa showed a clear trend along the transect from north to south. At Åheden, *Tylospora* sp. occupied less than 1% of the root tips, but its relative abundance increased to 4.5% at Aubure, 14.2% at Klosterhede and 49.1% at Waldstein. There was some indication that the opposite pattern was evident with ectomycorrhizas formed by members of the Cortinariaceae. At Åheden, 16.3% of the roots were colonised by fungi of this family while they accounted for 0.6% at Aubure and were not detected at either Klosterhede or Waldstein.

There were significant negative correlations between community diversity, morphotype richness and amounts of extractable inorganic soil N (Table 16.4). No

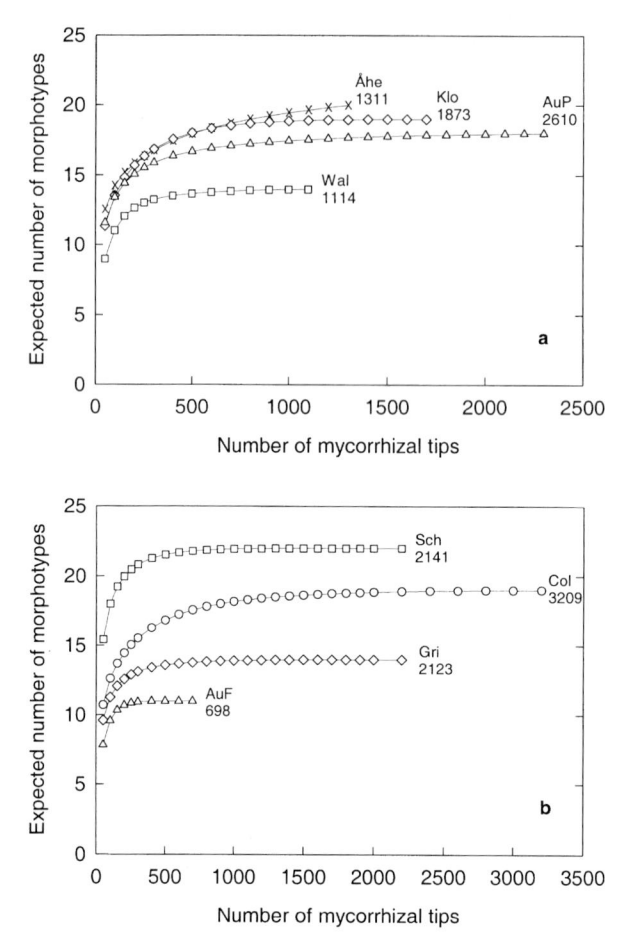

Fig. 16.2. Relationships between predicted number of morphotypes and numbers of mycorrhizal tips as revealed by rarefaction analysis in **a** spruce and **b** beech forests in Europe. In both **a** and **b**, the *number* after each line is the actual number of mycorrhizal tips examined at each site.
a Spruce: *Åhe* Åheden, N. Sweden; *Klo* Klosterhede, Denmark; *Wal* Waldstein, Germany; *AuP* Aubure, NE France. **b** Beech: *Gri* Gribskov, Denmark; *Sch* Schacht, Germany; *AuF* Aubure, NE France; *Col* Collelongo, Italy

significant relationship was found between the composition of the ectomycorrhizal community and the other site parameters tested.

16.3.2 Analysis Based on Carpophore Occurrence

A total of 40 ECM fungi were recorded as carpophores on the four sites (Table 16.5). As in the case of the mycorrhizal analysis, species richness was greatest at Åheden,

Table 16.3. Bray-Curtis similarity indices (Krebs 1989) achieved from within and between site comparisons of the ectomycorrhizal communities in four beech and four spruce forests. Indices are calculated from abundance data of the number of root tips colonised by different ectomycorrhizal morphotypes

Spruce	Åheden[a]	Klosterhede	Waldstein	Aubure
Åheden	**0.181**			
Klosterhede	0.014	**0.204**		
Waldstein	0.015	0.054	**0.149**[b]	
Aubure	0.023	0.064	0.075	**0.120**

Beech	Gribskov[a]	Schacht	Aubure	Collelongo
Gribskov	**0.136**			
Schacht	0.110	**0.068**		
Aubure	0.149	0.084	**0.121**	
Collelongo	0.004	0.009	0.003	**0.121**

[a] Sites. Spruce: Åheden – northern Sweden; Klosterhede – western Jutland, Denmark; Waldstein – Fichtelgebirge, Germany; Aubure – Vosges Mountains, NE France). Beech: Gribskov – N Sealand, Denmark; Schacht – Fichtelgebirge, Germany; Collelongo – Appenines, Italy.
[b] Figures are mean values and those marked in bold are within site comparisons.

Table 16.4. Relationships (expressed as correlation coefficients) between KCl extractable inorganic soil N in the LFH soil horizons, the Shannon-Wiener diversity indices and morphotype richness of ectomycorrhizal communities in four spruce and four beech forests

	Extractable nitrogen ($\mu g\,N\,g^{-1}$ LOI)		
	$N-NH_4 + N-NO_3$	$N-NH_4$	$N-NO_3$
SWD (spruce)	−0.998**[a]	−0.982*	−0.964*
Richness (spruce)	−0.971*	−0.937 ns	−0.986*
SWD (beech)	0.957*	0.032 ns	0.794 ns
Richness (beech)	0.952*	0.032 ns	0.834 ns

[a] ns No significant correlation.
* Weak correlation ($0.01 < p < 0.05$).
** Strong correlation ($0.001 < p < 0.01$).

where 18 species were identified. At Klosterhede and Aubure, 14 and 12 species, respectively, were found, while Waldstein yielded only 7 species. No species was common to all sites. Thirty-one species occurred only at single sites. With the exception of Waldstein, all sites supported a high proportion of species that were unique to its community. Again, the occurrence of members of the Cortinariaceae appeared

Table 16.5. Ectomycorrhizal fungi recorded as carpophores in Autumn 1993 under Norway spruce in four study plots along a north-south European transect. (Sites: Åheden – northern Sweden; Klosterhede – western Jutland, Denmark; Waldstein – Fichtelgebirge, Germany; Aubure – Vosges Mountains, NE France)

Fungal species (Total no. = 39)	Åheden (18)	Klosterhede (14)	Waldstein (7)	Aubure (12)
Amanita spissa var. *excelsa* (Fr.)	–	–	–	+
Cantharellus tubaeformis Fr.	–	+	–	–
Chroogomphus rutilus (Schff. ex Fr.) O.K. Miller	+	–	–	–
Cortinarius anomalus (Fr. ex Fr.) Fr.	+	–	–	–
C. collinitus Fr.	–	+	–	–
C. gentilis (Fr.) Fr.	+	–	–	–
C. integerrimus Kuehn.	+	–	–	–
C. laniger Fr.	+	–	–	–
C. ochrophyllus Fr.	+	–	–	–
C. cf. *paragaudis* Fr.	+	–	–	–
Dermocybe cinnamomea (L. ex Fr.)	–	–	–	+
D. cinnamomeobadia (R. Hry.) Mos.	–	–	–	+
D. crocea (Schff.) Mos.	+	–	–	–
D. semisaguineus (Fr.) Mos.	+	–	–	–
Hebeloma (cf.) *claviceps* (Fr.) Kummer	–	+	–	–
H. mesophaeum (Pers. ex Fr.) Quél	–	–	–	+
H. (cf.) *pumilum* Lge.	–	+	–	–
Hydnum repandum L.:Fr.	–	+	–	–
Hygrophorus olivaceoalbus Fr.	+	–	+	–
H. piceae Kuehn.	+	–	–	–
H. pustulatus (Pers. ex Fr.) Fr.	–	–	+	+
Laccaria amethystina (Bolt. ex Hooker) Murr.	–	+	–	–
L. (cf.) *bicolor* (R. Mre.) Orton	–	–	–	+
Laccaria laccata (Scop. ex Fr.) Bk.	–	+	+	+
Lactarius fuscus Roll.	+	–	–	–
L. necator (Bull. em Pers. ex Fr.) Karst.	–	+	–	+
L. rufus (Scop.) Fr.	+	–	+	+
L. theiogalus (Bull.) Fr.	–	+	–	–
Paxillus involutus (Batsch) Fr.	–	+	+	–
Russula aeruginea Lindbl.	–	+	–	–
R. decolorans Fr.	+	–	–	–
R. emetica var. *silvestris* Sing.	–	+	–	–
R. ochroleuca (Pers.) Fr.	–	–	+	+
R vinosa Lindbl.	+	–	–	–
Tricholoma flavovirens (Fr.) Lund	+	–	–	–
T. inamoenum (Fr.) Quél.	+	–	–	–
T. portentosum (Fr.) Quél.	–	+	–	–
Xerocomus badius (Fr.) ex Gilb.	–	+	+	+
X. chrysenteron (Bull. ex St. Amans) Quél.	–	–	–	+

to change along the transect. They were most common at Åheden (eight spp.), less well represented at Klosterhede and Aubure (each with three spp.) and absent from Waldstein.

16.4 ECM Communities of Beech Forests

The total of 39 morphotypes were distinguished on 8171 root tips of beech (Table 16.6). Non-mycorrhizal tips were very uncommon, only 0.3% of all tips examined being uncolonised. The number of morphotypes found at each site varied considerably, from 14 to 22. This may reflect the difference in the number of root tips collected from the different sites, but the rarefaction curves (Fig. 16.2b) strongly suggest that there were marked differences in morphotype richness between the sites. Overall, morphotype richness was highest at Schacht. This was reflected in the high diversity index and the high value for community evenness (Table 16.1). The lowest diversity and evenness indices were recorded at Aubure, where over 40% of the tips were colonised by *Russula ochroleuca* (Table 16.6).

The rank-abundance plots clearly separated the communities at the four beech sites (Fig. 16.1b). There was an increasing steepness of slope in the rank-abundance plots in the following sequence; Schacht: Collelongo: Gribskov: Aubure. The reduction in morphotype richness through these sites further accentuated the difference in the patterns of morphotype abundance.

The number of morphotypes common to all four sites was low, only four in total. However, the three most northerly sites shared a total of eight morphotypes. Of these, *R. ochroleuca* and *L. subdulcis* were the two most commonly encountered. The community at Collelongo was effectively separated from the other three sites by the high level of morphotypes, 11 of the 19, being unique to that site (Table 16.6). The distinctive nature of the Collelongo site was highlighted by the results from the Bray-Curtis analysis (Table 16.3). Surprisingly, within-site similarity in sample composition at Aubure, Gribskov and Schacht was often lower than that occurring between sites.

In contrast to the situation seen at the spruce sites, there were significant positive correlations between morphotype richness, community diversity and total extractable inorganic soil N (Table 16.4). However, neither $N-NH_4$ nor NO_3, when analysed separately, were significantly correlated with diversity or morphotype richness. No significant relationship was found between the composition of the ectomycorrhizal community and other site parameters tested.

16.5 Genetic Diversity Within a Population of *Laccaria amethystina*

16.5.1 Sampling Plots and Collection of Carpophores

Carpophores of *L. amethystina* were collected from three different 100-m^2 sampling plots located along a 120-m transect crossing the beech stand located between two roadways. Each sampling site corresponds to a 10×10 m square. Beech was the only tree species occurring within 20 m around the plots. Within the studied area the fruiting period of *L. amethystina* extends from early August to end-October, depending on the climatic conditions. In 1994, 1148 carpophores were collected in early September (sampling 1) and late-September (sampling 2) during the height of the fruiting period

Table 16.6. Relative abundance of ectomycorrhizal morphotypes, expressed as the percentage of roots colonised, in root samples collected from four beech (*Fagus sylvatica* L.) forests. (Gribskov – N Sealand, Denmark; Schacht – Fichtelgebirge, Germany; Aubure – Vosges Mountains, NE France; Collelongo – Appenines, Italy

Morphotype (Total no. = 39)	Gribskov (14)	Schacht (22)	Aubure (11)	Collelongo (19)
Amanita sp.	0.2	–	–	–
Boletus sp.	13.2	4.3	21.3	0.1
Byssocorticium atrovirens	–	–	–	0.3
Cenococcum geophilum	1.1	4.1	1.0	0.8
Cortinarius sp.	4.7	5.3	2.6	0.3
Dermocybe sp.	5.7	8.6	–	–
Elaphomyces muricatus	3.3	2.2	1.7	–
Fagirhiza (cf.) *fusca*	–	4.9	–	11.5
F. arachnoidea	–	–	–	7.8
F. oleifera	–	8.8	1.4	2.2
Genea sp.	0.6	0.4	–	9.1
Hebeloma sp.	–	–	1.7	–
Laccaria amethystina	3.9	6.7	2.1	–
Lactarius (cf.) *pallidus*	–	1.2	–	1.7
L. rubrocinctus	–	5.1	–	–
L. subdulcis	6.8	13.0	17.3	–
Lactarius sp. 1	1.9	–	–	–
Lactarius sp. 2	–	9.7	–	–
Lactarius sp. 3	–	0.7	–	–
Lactarius sp. 4	–	–	–	4.9
Piceirhiza bicolorata	0.7	0.2	0.9	7.4
Russula fellea	–	6.2	–	–
R. ochroleuca	30.3	12.2	43.9	–
Russula sp. 1	25.7	–	–	–
Russula sp. 2	–	1.2	–	–
Russula sp. 3	–	–	–	9.2
Russula sp. 4	–	–	–	1.0
Russula sp. 5	–	–	–	0.2
Tomentella sp.	–	–	–	10.8
Tuber sp.	–	1.6	–	–
Unknown no. 1	1.1	–	–	–
Unknown no. 2	–	2.1	–	–
Unknown no. 3	–	0.8	–	–
Unknown no. 4	–	0.6	–	–
Unknown no. 5	–	–	5.7	–
Unknown no. 6	–	–	–	28.2
Unknown no. 7	–	–	–	2.9
Unknown no. 8	–	–	–	1.1
Unknown no. 9	–	–	–	0.3
Non-mycorrhizal	0.8	–	0.1	0.2

so that it can be assumed that a low proportion of the carpophores has been missed. In 1997, only 87 carpophores were found and collected in mid-September (sampling 3) on the three sampling plots. On each occasion, all carpophores were sampled and their coordinates were noted on a sampling grid (10 × 10 m) to prepare distribution maps. Mapped carpophores were numbered, bagged individually, transported to the laboratory within 6 h following sampling and stored at −20 °C pending analysis.

16.5.2 DNA Extraction and Amplification

Total DNA was extracted directly from carpophore tissues using the hexadecyltrimethylammonium bromide (CTAB)/proteinase K protocol as described by Henrion et al. (1994). Random amplified microsatellite analysis (RAMS) or interrepeat PCR was carried out as described in Martin et al. (1998) using $(GTG)_5$ and $(CCA)_5$ primers. Carpophores having identical banding patterns were regarded as belonging to the same genetic individual (genet) and were mapped according to the distribution of collected carpophores. Amplification products were separated by electrophoresis using 2% (w/v) FMC-MetaPhor agarose/0.25% BRL-agarose gel electrophoresis in 1X tris-borate-EDTA buffer.

16.5.3 Genotyping of *L. amethystina* Using RAMS

Molecular typing was carried out using RAMS to ascertain the genetic identity of collected *L. amethystina* carpophores. For each carpophore, $(GTG)_5$-primed PCR displayed 5 to 20 amplification products within the range 0.3- to 1.5-kbp (data not shown). The banding patterns obtained using the $(GTG)_5$ and $(CCA)_5$ microsatellites were highly polymorphic, but some markers (e.g. bands at 570 and 980 bp) were shared by several carpophores. Up to 382 different banding patterns were identified among the 572 carpophores analyzed from the three sampling plots during the autumn of 1994 and 1997. The average number of carpophores per genet was 1.5 (Table 16.7). Associations between genets were assessed by principal-

Table 16.7. Genetic diversity of *a Laccaria amethystina* population in a 150-year-old beech stand at Auberge, NE France

	Sampling 1 (1994)			Sampling 2 (1994)			Sampling 3 (1997)		
	Plot I	Plot II	Plot III	Plot I	Plot II	Plot III	Plot I	Plot II	Plot III
No. of carpophores sampled	127	104	109	242	438	147	32	27	28
No. of RAMS analyses performed	81	94	96	101	67	47	31	27	28
No. of genets found	63	57	36	65	53	40	27	18	23
No. of carpophores genet^{-1}	1.7	1.7	2.7	1.7	1.4	1.3	1.2	1.5	1.2
Average size of genets (m)	0.25	0.35	0.65	0.25	0.25	0.20	0.15	0.20	0.20
Estimated number of genets ha^{-1}		5200			5266			2266	

coordinates analysis of the similarity matrix for the combined data set (see http://mycor.nancy.inra.fr/Documents/Aubure/Aubure.html); no clustering of individuals was revealed. The lack of close relationship between genets was substantiated by UPGMA clustering.

16.5.4 Spatial Distribution of *L. amethystina* Genets

Mapping of the carpophores having identical RAMS patterns revealed a high number of small-sized genets (Fig. 16.3). Samples taken from a group of carpophores were usually genetically identical. However, two carpophores less than 2–5 cm apart from each other could belong to different genets as a result of intermingling genets (e.g. coordinates J2 and H2; Fig. 16.3). Of the 270 carpophores typed by RAMS on plot I in 1994 and 1997, 162 different genotypes were identified (Fig. 16.3; grey zones and circles), whereas 128 and 98 genets were found on plots II and III (data not shown), respectively, indicating a high density of genets. On plot I, 131 different genets were observed during the fall of 1994. The largest genet covered about 1 m^2 and comprised

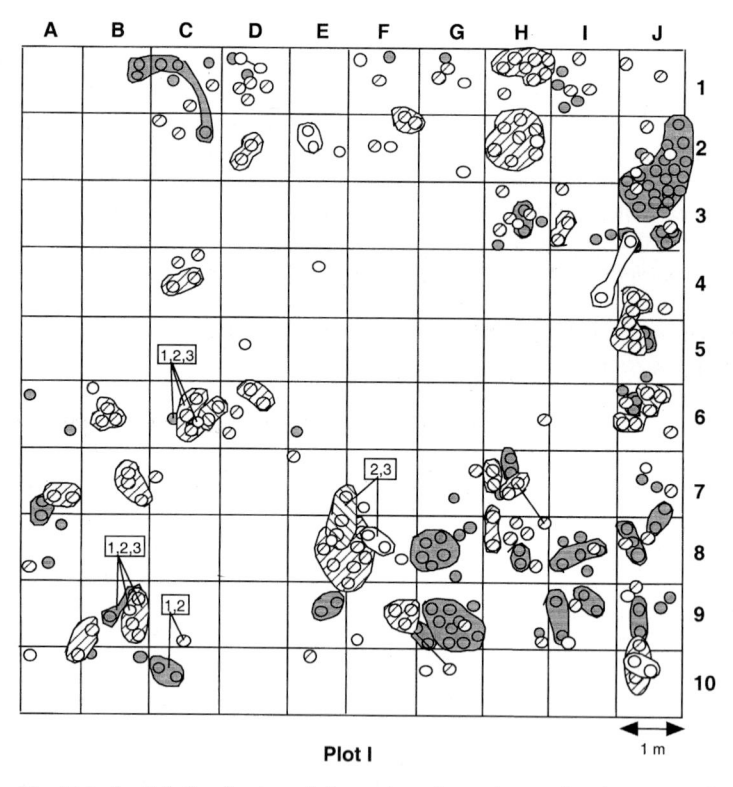

Fig. 16.3. Spatial distribution of the various *Laccaria amethystina* genets found on plot I in early September 1994. Trees are represented by *closed dark circles*. *Gray zones* encompass sporophores belonging to the same genet as determined by RAMS. Some zones are *hatched* to facilitate visualization of intermingling genets

21 carpophores (J2,3; Fig. 16.3), but most genets comprised a single carpophore. The three plots showed no common genets.

The carpophore surveys revealed obvious differences among weeks and years in the composition of genets. Although three genets (~5%) identified in early September 1994 were still producing carpophores in end-September 1994, most genets were not observed 2 weeks later (Fig. 16.3). Conversely, several genets were not fruiting in early September, whereas their carpophores were identified in end-September the same year.

The high degree of genotypic spatial diversity can be explained either by (1) frequent establishments of genets followed by their extinction after 1 to 2 years of vegetative growth or (2) initial spore establishment with persistence but not expansion and erratic fruiting. If the former hypothesis is true, the *L. amethystina* population experienced a high extinction level due to unknown factors (e.g. competition from other species, hard winters). This extinction would be compensated by yearly recolonization of plots by basidiospores, as reported for the ruderal strategist *Hebeloma cylindrosporum* (Gryta et al. 1997). However, survival of a few genets up to 3 years (Fig. 16.3) showed that the latter hypothesis cannot be ruled out. It should be noted that the ground location of these few lasting genets was similar in 1994 and 1997, confirming the low expansion of these genets.

The observed population structure (prolific carpophore production, small genets) is a preferred reproductive strategy of early-stage, ruderal ectomycorrhizal species occurring under young trees (Selosse et al. 1998, 1999) or on disturbed sites, where novel genet establishment relies on small-scale disturbances (wind-thrown trees, animal tracks, soil movements) (Gryta et al. 1997). None of these disturbances has been recorded in the old Aubure beech forest. A continuous tree layer has been present for the past 150 years, and the wood production yield is within the range of recorded values in northeastern France, ruling out any obvious stresses.

16.6 Isolation and Growth of ECM Fungal Isolates on an Organic N Source

Representative tips of each spruce morphotype were surface-sterilised by exposure for 2 min to a 30%, 100 vol solution of hydrogen peroxide. The tips were then washed before being plated onto MMN agar (Marx 1969). Where the symbiont was known to be a basidiomycete, $5\,mg\,l^{-1}$ benomyl was added to the MMN (Schild et al. 1988). In addition, attempts were made to obtain isolates from the carpophores found at each spruce site.

The ability of isolates derived from mycorrhizas and carpophores to produce mycorrhizas was first confirmed in synthesis trials with aseptically grown spruce seedlings in peat-vermiculite using the method of Duddridge (1986). Only fungi shown to form typical ectomycorrhizas were used in studies of nitrogen utilisation. These were screened for their ability to utilise organic N, using the plant protein gliadin (mw 30 000, N content 14%) as a surrogate soil organic N source. The gliadin ($1.62\,g\,l^{-1}$, i.e. $0.23\,g\,N$) was added to agar containing a basal mineral nutrient solution (Norkrans 1950) from which other N sources were excluded. Isolates were grown on the medium, with and without supplementary carbon in the form of a $2.5\,g\,l^{-1}$ addition of glucose. Radial growth and/or clearing of the opaque medium were used as indicators of the ability of the isolates to utilise the gliadin. All measurements were taken 1 month after inoculation of the culture plates.

Sixty-six isolates were obtained from surface-sterilised spruce mycorrhizas, over half of these originating from Åheden. The fungi on the roots of spruce at Åheden appeared to be more amenable to growth in culture than those from the other sites. Out of a total of 19 morphotypes found at Åheden, 13 yielded fungal isolates. Analysis of the isolation success using the χ^2 test demonstrated that the possibility of obtaining an isolate from a mycorrhiza was not independent of site and that more isolates were obtained from Åheden than would be expected if all the morphotypes had an equal chance of growing ($\chi^2 = 14.0$, $p < 0.01$). In addition to the mycorrhizal isolates, a further 22 isolates were obtained from carpophores, giving a total of 88 isolates from 28 fungal species.

In the presence of glucose, all the isolates from Åheden demonstrated some growth on the gliadin medium (Table 16.8). The isolates from Waldstein, in contrast, grew very poorly on the medium. Isolates from the other two sites produced radial growth rates which were intermediate between these two extremes. In the absence of glucose, the total number of isolates capable of growing on the medium was markedly reduced. However, isolates from Åheden were still more able to grow on the organic medium.

16.7 Comparative Evaluation of Ectomycorrhizal Diversity

The observation that more than 90% of all root tips are converted to mycorrhizas, irrespective of position on the transect or the extent of nitrogen deposition, permits two primary conclusions. First, that nitrogen deposition, while it may influence the structure of the fungal community, does not reduce the quantitative importance of the symbiosis in the systems examined. Second, that nutrient uptake by all the dominant tree species of the forests sampled must occur, for the most part, through roots ensheathed by fungal mantles. It follows that by occupying the interface between the soil and the absorptive region of the root surface, these fungi are in a position to exercise major influences upon the nutrition of the trees and their responses to environmental perturbation.

In the absence of quantitative changes in the extent of mycorrhiza formation, any such influences can only be effected through qualitative changes in the structure of the fungal community. At this level a striking contrast emerges between the inherent simplicity of the often monospecific stands of trees occupying the study sites and the considerable richness and complexity, at the genus as well as inter- and intraspecific levels, of the symbiotic fungal community to which these trees are hosts. Such structural complexity poses two major challenges, the first being to devise sampling strategies that are adequate to provide a realistic description of the community of mycorrhizal fungi occurring on the roots; the second, even more challenging, to ascribe possible functional significance to any between-site differences in composition which might be observed.

With regard first to the problem of description, many, faced with the complexity of the below-ground situation, have sought to simplify analysis by considering only that part of the community which manifests itself above ground as fruit bodies. While this reductionist approach has provided valuable information, particularly for those concerned with fungi as a component of human diet, we now know that records of

Table 16.8. The ability of ectomycorrhizal fungal isolates to utilise gliadin, an organic nitrogen source, in relation to the substrate from which the isolates originated. Isolates obtained from ectomycorrhizas and fruit bodies from four spruce sites along a north-south European transect. (Sites: Åheden – northern Sweden; Klosterhede – western Jutland, Denmark; Waldstein – Fichtelgebirge, Germany; Aubure – Vosges Mountains, NE France)

Site	Isolate/species[a]	Substrate[b]	Growth[c] Org N + glucose	Org N − glucose
Åheden	*Dermocybe semisanguinea* (F)	n/a	+	+
	Hebeloma subsaponaceum (F)	n/a	+	−
	Tricholoma flavovirens (F)	n/a	+	+
	Cenococcum geophilum (M)	Org	+	−
	Cortinarius sp. 1 (M)	Min	+	−
	Cortinarius sp. 2 (M)	Org	+	+
	Tylospora fibrillosa (M)	Org/min	(+)	−
	E-type (M)	Org/min	+++/−	++/−
	Unknown 1 (M)	Org/min	+	+
	Unknown 3 (M)	Min	+	+
	Unknown 4 (M)	Org/min	+++	+++
	Unknown 8 (M)	Org	(+)	(+)
	Unknown 12 (M)	Min	+	−
Klosterhede	*Hebeloma* sp. (F)	n/a	+++	+++
	Laccaria amethystina (F)	n/a	+	−
	Xerocomus badius (F)	n/a	+	−
	Cenococcum geophilum (M)	Min	+/−	+/−
	Picierhiza punctata (M)	Org/min	−	−
	Russula sp. (M)	Org	+	−
	Tylospora fibrillosa (M)	Org/min	(+/−)	−
Waldstein	*Lactarius rufus* (F)	n/a	(+)	−
	Paxillus involutus (F)	n/a	−	−
	Xerocomus badius (F)	n/a	(+)	−
	Tylospora fibrillosa (M)	Org/min	(+/−)	−
Aubure	*Amanita spissa* (F)	n/a	+	+
	Hebeloma mesophaeum (F)	n/a	+++	++
	Xerocomus badius (F)	n/a	−	−
	Cortinarius sp. (M)	Org	(+)	−
	E-type (M)	Org	++	++
	Piceirhiza gelatinosa (M)	Org/min	+	−
	Tylospora fibrillosa (M)	Org/min	+	−

[a] Origin of isolate: M = mycorrhizal isolate, F = fruit body isolate.
[b] Soil fraction in which the ectomycorrhizal tips were found, Org = Organic, Min = Mineral, n/a = not applicable.
[c] Diameter of fungal colony after 1 month on solid agar medium containing the insoluble plant protein gliadin as the sole nitrogen source, ± glucose as a carbon source. (+) slight growth; + 1–3 cm; ++ 3–6 cm; +++ > 6 cm; / intraspecific differences between isolates of the same species.

carpophore occurrence do not provide a realistic picture of the community structure on the roots themselves. The problems arise because some of the fungi which can be dominant components of the mycorrhizal community are represented above ground, if at all, only as inconspicuous fruiting structures, e.g. *Tylospora fibrillosa* (Taylor and Alexander 1989) *Tomentella sublilacina* (Gardes and Bruns 1996), while others may fruit prolifically but constitute only a small proportion of the mycorrhizas present, e.g. *Suillus pungens* (Gardes and Bruns 1996). The solution to these problems, adopted in the present study, is to combine recording of carpophore production with intensive analysis of the mycorrhizas themselves.

Interpretation of the functional significance of any shift in community structure is also constrained, in this case by a lack of knowledge of the functional capabilities of some of the most important taxa. This is particularly true of a key group notably represented by the two large genera *Cortinarius* and *Russula* which, because they are extremely difficult, usually impossible, to grow in culture, are not tractable as experimental organisms. What we know of their biology comes from field analysis of their distribution and host preferences (Molina et al. 1992) and from a small number of studies using excised mycorrhizal roots or slow growing cultures, all of which suggest that they have specialised nutrient requirements. The recalcitrance of this group contrasts with the relative tractability of taxa such as *Laccaria laccata, L. amethystina, Lactarius rufus, Paxillus involutus,* and *Thelephora terrestris*, which grow readily in culture, have unspecialised nutrient requirements, show little host specificity, and so have been widely used in experimental studies (Smith and Read 1997) as well as in seedling inoculation programmes. The distinction between the first and second groups is sufficiently large that some (e.g. Dighton and Mason 1985), applying ecological strategy theories developed by Gadgil and Solbrig (1972) and Grime (1979), have sought to categorise them respectively as S or stress-tolerant and R or ruderal strategists. The fast-growing fungi of low host specificity clearly have many of the attributes of ruderals, while some, at least, of the characteristics of *Cortinarius* and *Russula* spp. are those of S strategists. In particular, they are successful in environments where, for climatic and biotic reasons, mineralisation processes are extremely slow.

What emerges from the study of spruce stands on the transect is that the southward decline in fungal species diversity, demonstrated by the reducing Shannon-Wiener indices, are largely attributable to the loss of putative S strategists at the expense of species most readily categorised as being of the R type. It is possible to hypothesise that this shift arises as a result of opportunistic replacement of the S strategists by ruderals in response to increased inputs of mineral nitrogen, and there is considerable circumstantial evidence, most of it based upon records of fruit body production, to support such a view.

At the broadest level, data, for example from The Netherlands, show that over the past two decades, during which N inputs have increased markedly in north-central Europe, there has been a considerable decrease in the recorded frequency of most ectomycorrhizal basidiomycetes (Arnolds 1991; Arnolds and Jansen 1992; Arnolds 1995). More specifically some genera, amongst which *Cortinarius, Tricholoma* and *Russula* are identified in these surveys, appear to be particularly sensitive to nitrogen enrichment and to have declined severely as a consequence. The observation that six *Cortinarius* species were recorded as fruit bodies in spruce forest at Åheden, one at Klosterhede and none at Aubure or Waldstein is thus strongly suggestive of an N-based impact on the genus, a view reinforced by the observation that *Cortinarius* was not recorded as a component of the morphotype community at any sites other than Åheden. Lilleskov and Fahey (1996) reported that *Cortinarius* was particularly sensitive to a point source of N input from a fertiliser factory in otherwise pristine forests of Alaska, and further evidence to suggest sensitivity of the genus to N comes from fertiliser trials. Brandrud (1995) showed that following application of 35 kg $NH_4 NO_3$ $ha^{-1} a^{-1}$ to mature spruce forest there was a severe decline of *Cortinarius* carpophore production. Similar results were reported by Wästerlund (1982) from fertilised pine stands.

Some indication of the physiological basis for increases in the R component of the fungal community in more southerly spruce forests can also be gained from the literature. *Tylospora*, for example, which made up half the mycorrhizal root tips at the most heavily N-enriched site, Waldstein, is known from forest fertiliser trials to be insensitive to high N inputs (Taylor and Alexander 1989; Saunders et al. 1996). Other species, notably *Lactarius rufus*, an important component of the root community at Waldstein and *Paxillus involutus*, seen to fruit only at Waldstein and Klosterhede, have also been reported to respond positively to N enrichment in both unfertilised (Arnolds 1991; Arnolds and Jansen 1992) and fertilised trial plots (Hora 1959; Laiho 1970; Ohenoja 1978; Salo 1979; Wästerlund 1982; Brandrud 1995). Collectively, these studies indicate a functional divergence between those fungi that show strong negative responses to enhancement of mineral N in forest soil and those that respond positively. Such an interpretation is strengthened by the results of analyses reported here, showing that the only measured soil variable to be positively correlated with loss of diversity in spruce mycorrhizal communities is the quantity of extractable mineral N.

It is interesting to speculate as to the mechanism involved in the apparent decline of the S-type fungi. One feature which distinguishes these fungi is their relatively well-developed ability to mobilise nitrogen from polymeric organic residues, thus providing the host plant with otherwise unavailable sources of the element. Abuzinadah and Read (1986a) categorised several readily culturably representatives of the group, e.g. *Amanita muscaria* and *Suillus bovinus*, as being protein fungi on the basis of such ability. In the present work it is revealed that several *Cortinarius* species of importance in the Åheden mycorrhizal community have similar attributes. Abuzinadah and Read contrasted the proteolytic capability of such fungi with the failure of several species of the R category, such as *Lactarius rufus* and *Laccaria laccata* to use protein, and referred to these as non-protein fungi. It is a notable feature of the Waldstein community that *L. rufus* occupies a prominent proportion of mycorrhizal morphotypes. Even when some fungi of the R type such as *Paxillus involutus* do show some proteolytic activity, they appear to transfer relatively little of the organic N acquired by their vegetative mycelium to their host plants (Abuzinadah and Read 1986b). It is also striking that the isolate of *P. involutus*, which was obtained from Waldstein, showed no proteolytic capabilities when grown on the gliadin agar. This would suggest either that proteinase production has been suppressed due to high availability of mineral N or that there has been a selection for genotypes within *P. involutus* which lack the ability to produce proteinase enzymes.

Of particular functional significance may be the observation of Arnebrant (1994) that there are considerable interspecific differences in the sensitivity of the vegetative mycelial systems of mycorrhizal fungi to exposure to mineral N. She applied sprays of NH_4 and NO_3 ions to mycelia growing across natural substrates in observation chambers. An unidentified isolate showed extreme sensitivity to application of both ions, its mycelium ceasing growth altogether. A number of strains of the ruderal species *P. involutus*, in contrast, were relatively little affected. The implications of such observations are considerable and require further experimental analysis. Inhibition of the growth of extraradical mycelium, if it occurred widely as a response in sensitive species, would explain the decline in their ability to produce carpophores, since a major consequence would be reduction in the assimilate supply upon which construction of fruiting structures is dependent (Lamhamedi et al. 1994). In this connection the observation of Dutch workers that removal of surface organic layers which have become enriched with mineral N as a consequence of wet deposition can restore fruiting of a number of S strategists, including *Cortinarius* species, is important (Baar and ter Braak 1996).

The structures of the mycorrhizal communities in *Fagus* stands are less easily explicable in terms of anthropogenic N enrichment. This may be in part because these stands did not occupy such distinctive gradient of deposition. In addition, it may be that since *Fagus* occurs naturally upon soils which have a higher capacity both for N mineralisation and nitrification than those typically supporting spruce (Ellenberg 1988), it can be hypothesised that selection will have favoured nitrophilous symbionts and hence an inherently greater resistance to perturbation from increased anthropogenic inputs of NH_4 and NO_3. Such a hypothesis is supported by the observation that, in striking contrast to the spruce stands, there was significant positive correlation between quantities of extractable mineral N in soils of the *Fagus* site and morphotype richness.

Whatever the functional basis of the apparent shifts in community structure observed particularly in spruce, it is evident from the study that in general lack of specificity in the mycorrhizal association confers advantages which contribute to the resilience of forest systems when they are subjected to perturbation. When the vigour of one cohort of naturally selected symbionts is weakened, in this case by mineral N enrichment of the soil environment, compatibility of the trees with alternative fungal species or genotypes which are better able to function under the new circumstances ensure that the normal processes of nutrient capture assimilation and transfer are maintained.

16.8 Conclusions

- Over 90% of all tree root tips at all sites are ectomycorrhizal.
- The trees are therefore heavily dependent upon mycorrhizal fungi for the acquisition of nitrogen and other essential nutrients.
- The within- and between-site taxonomic diversity of the fungi involved in mycorrhizal formation is large.
- Within spruce forests, the decline in diversity towards the south of the transect is accompanied by increases in single species dominance and is paralleled by changes in the nitrogen nutrition of the fungi involved.
- Characteristic fungal species of northern sites, which are able to utilise organic N, decrease in number as availability of mineral N increases towards the south.
- The ectomycorrhizal fungal community associated with beech appears to have a greater resilience to increased levels of pollutant N.

References

Abuzinadah RA, Read DJ (1986a) The role of proteins in the nitrogen nutrition of ectomycorrhizal plants. I. Utilization of peptides and proteins by ectomycorrhizal fungi. New Phytol 103:481-493

Abuzinadah RA, Read DJ (1986b) The role of proteins in the nitrogen nutrition of ectomycorrhizal plants. III. Protein utilisation by *Betula*, *Picea* and *Pinus* in mycorrhizal association with *Hebeloma crustuliniforme*. New Phytol 103:507-514

Agerer R (1986-1998) Colour atlas of Ectomycorrhizae. Einhorn-Verlag, Schwäbisch-Gmünd

Arnebrant K (1994) Nitrogen amendments reduce the growth of extramatrical ectomycorrhizal mycelium. Mycorrhiza 5:7-15

Arnolds E (1991) Decline of ectomycorrhizal fungi in Europe. Agric Ecosyst Environ 35:209-244

Arnolds E (1995) Conservation and management of natural populations of edible fungi. Can J Bot 73:S987-S998

Arnolds E, Jansen E (1992) New evidence for changes in the macromycete flora in The Netherlands. Nova Hedwigia 55:325-351

Baar J, ter Braak CJ (1996) Ectomycorrhizal sporocarp occurrence as affected by manipulation of litter and humus layers in Scots pine stands of different age. Appl Soil Ecol 4:61-73

Brand F (1991) Ektomykorrhizen an *Fagus sylvatica*. Charakterisiering und Identifizierung, ökologische Kennzeichnung und unsterile Kultivierung. Libri Bot 2:1-229

Brandrud TE (1995) The effects of experimental nitrogen addition on the ectomycorrhizal fungus flora in an oligotrophic spruce forest at Gårdsjön, Sweden. For Ecol Manage 71:111-122

Breitenbach J, Kränzlin F (1986) Fungi of Switzerland, vol 2 Non-gilled Fungi. Verlag Mykologia, Lucerne, Switzerland

Dighton J, Mason PA (1985) Mycorrhizal dynamics during forest tree development. In: Moore D, Casselton LA, Wood DA, Frankland JC (eds) Developmental biology of Higher Fungi. Cambridge University Press, Cambridge, pp 117-139

Duddridge JA (1986) The development and ultrastructure of ectomycorrhizas III. Compatible and incompatible interactions between *Suillus grevillei* (Klotzsch) Sing. and 11 species of ectomycorrhizal hosts in vitro in the absence of exogenous carbohydrates. New Phytol 103:457-464

Ellenberg H (1988) Vegetation ecology of central Europe. Cambridge University Press, Cambridge

Gadgil M, Solbrig OT (1972) The concept of r-and K-selection: evidence from wild flowers and some theoretical considerations. Am Nat 106:14-31

Gay C, Marmeisse R, Fouillet P, Buntertreau M, Debaud JC (1993) Genotype/nutrition interactions in the ectomycorrhizal fungus *Hebeloma cylindrosporum* Romagnesi. New Phytol 123:335-343

Gardes M, Bruns TD (1996) Community structure of ectomycorrhizal fungi in a *Pinus muricata* forest: above- and below-ground views. Can J Bot 74:1572-1583

Grime JP (1979) Plant strategies and vegetation processes. John Wiley, New York

Gryta H, Debaud J-C, Effosse A, Gay G, Marmeisse R (1997) Fine scale structure of populations of the ectomycorrhizal fungus *Hebeloma cylindrosporum* in coastal sand dune forest ecosystems. Mol Ecol 6:353-364

Haug I, Pritsch K (1992) Ectomycorrhizal types of spruce [*Picea abies* (L.) Karst.] in the Black Forest. Kernforschungszentrum Karlsruhe.

Henrion B, Chevalier G, Martin F (1994) Typing truffle species by PCR amplification of the ribosomal DNA spacers. Mycol Res 98:37-43

Hora FB (1959) Quantitative experiments on toadstool production in woodlands. Trans Br Mycol Soc 42:1-14

Ingleby K, Mason PA, Last FT, Fleming LV (1990) Identification of ectomycorrhizas. Institute of Terrestrial Ecology research publication no 5. HMSO, London

Krebs CJ (1989) Ecological methodology, 2nd. Harper & Row, New York
Laiho O (1970) *Paxillus involutus* as a mycorrhizal symbiont of forest trees. Acta For Fenn 106:1-73
Lamhamedi MS, Godbout C, Fortin JA (1994) Dependence of *Laccaria bicolor* basidiome development on current photosynthesis of *Pinus strobus* seedlings. Can J For Res 24:412-415
Lilleskov EA, Fahey TJ (1996) Patterns of ectomycorrhizal diversity over an atmospheric nitrogen deposition gradient near Kenai, Alaska. In: Szaro TM, Bruns TD (eds) Abstr 1st Int Conf on Mycorrhizae. Berkeley, CA, USA. University of California, Berkeley 76
Magurran AE (1988) Ecological diversity and its measurement. Croom Helm, London
Martin F, Costa G, Delaruelle C, Diez J (1998) Genomic fingerprinting of ectomycorrhizal fungi by microsatellite-primed PCR. In: Varma A, Hock B (eds) Mycorrhiza manual. Springer Lab Manual, Berlin Heidelberg, New York, pp 463-474
Marx DH (1969) The influence of ectotrophic ectomycorrhizal fungi on the resistance of pine roots to pathogenic infections. I. Antagonism of mycorrhizal fungi to pathogenic fungi and soil bacteria. Phytopathology 59:153-163
May RM (1974) General introduction. In: Usher MB, Williamson MH (eds) Ecological stability. Chapman and Hall, London, pp 1-14
Meyer FH (1973) Distribution of ectomycorrhizae in native and man-made forests. In: Marks GC, Kozlowski TT (eds) Ectomycorrhizae. Academic Press, New York, pp 79-105
Minitab (1998) Minitab statistical software. Release 12. Minitab Inc, Pennsylvania
Molina R, Massicotte H, Trappe JM (1992) Specificity phenomena in mycorrhizal symbiosis: community ecological consequences and practical applications. In: Allen MF (ed) Mycorrhizal functioning. Chapman & Hall, London, pp 357-423
Moser M (1983) Keys to agarics and boleti (Polyporales, Boletales, Agaricales, Russulales) Roger Phillips, London
Norkrans B (1950) Studies in growth and cellulolytic enzymes in *Tricholoma*. Symb Bot Ups 11:126
Ohenoja E (1978) Mushrooms and mushroom yields in fertilized forests. Anna Bot Fenn 15: 38-46
Salo K (1979) Mushrooms and mushroom yield on transitional peatlands in central Finland. Ann Bot Fenn 16:181-192
Saunders E, Taylor AFS, Read DJ (1996) Ectomycorrhizal community response to simulated pollutant nitrogen deposition in a Sitka, spruce stand, North Wales. In: Szaro TM, Bruns TD (eds) Abstr 1st Int Conf on Mycorrhizae. Berkeley, CA, USA, University of California Berekeley 106.
Schild DE, Kennedy A, Stuart MN (1988) Isolation of symbiont and associated fungi from ectomycorrhizas of sitka spruce. Eur J For Pathol 18:51-61
Selosse M-A, Jacquot D, Bouchard D, Martin F, Le Tacon F (1998) Temporal persistence and spatial distribution of an American inoculant strain of the ectomycorrhizal basidiomycete *Laccaria bicolor* in European forest plantations. Mol Ecol 7:561-573
Selosse M-A, Martin F, Le Tacon F (1999) Structure and dynamics of experimentally introduced and naturally occurring *Laccaria* spp. genets in a Douglas fir plantation. App Environ Microbiol
Smith SE, Read DJ (1997) Mycorrhizal Symbiosis, 2nd edn. Academic Press, San Diego
Taylor AFS, Alexander IJ (1989) Demography and population dynamics of ectomycorrhizas fertilised with nitrogen. Agric Ecosyst Environ 28:493-496
Tokeshi M (1993) Species abundance patterns and community structure. Adv Ecol Res 24: 112-186
Trappe JM (1962) Fungal associates of ectotrophic mycorrhizae. Bot Rev 28:538-606
Väre H, Ohenoja E, Ohtonen R (1996) Macrofungi of oligotrophic Scots pine forests in northern Finland. Karstenia 36:1-18
Wagner F, Gay G, Debaud JC (1988) Genetic variability of glutamate dehydrogenase activity in monokaryotic and dikaryotic mycelia of the ectomycorrhizal fungus *Hebeloma cylindrosporum*. Appl Microbiol Biotechnol 28:566-576

Wagner F, Gay G, Debaud JC (1989) Genetical variation on nitrate reductase activity in mono- and dikaryotic populations of the ectomycorrhizal fungus *Hebeloma cylindrosporum* Romangnési. New Phytol 113:259-264

Wästerlund I (1982) Försvinner tallens mycorrhizasvampar vid gödsling? Sven Bot Tidskr 76:411-417

17 Diversity and Role of the Decomposer Food Web

V. Wolters, A. Pflug, A.R. Taylor, and D. Schroeter

17.1 Introduction

Only a small part of primary production is consumed by phytophagous organisms (Ellenberg et al. 1986). It is generally agreed that the food web based on detritus is more important for the flow of energy and nutrients through terrestrial ecosystems than the food web based directly on autotrophic production (Swift et al. 1979). The decomposer community that is responsible for this flow is composed of microorganisms and invertebrates. The major components of the decomposer microorganisms are bacteria and fungi. The dominant decomposer fauna groups belong to the Protozoa, Nematoda, Oligochaeta and Arthropoda (Fig. 17.1). There is evidence that the effects of soil organisms on ecosystem functioning critically depend on both the structural diversity of the decomposer community and environmental conditions (Freckman 1994; Hall 1996). However, little information is available concerning the impact of soil biodiversity on the integrity, function and sustainability of terrestrial ecosystems (Wolters 1998a).

The project reported on here focused on the decomposer community of four selected coniferous sites forming a European north-south transect. It aimed at refining the tools for understanding and predicting the effects of environmental conditions (climate, soil conditions, deposition) on C and N turnover by below-ground organisms. The following questions were addressed:

- Can the structure of the decomposer community be linked to different temperature and humidity regimes as well as to external inputs? To what extent does trophic connectivity of the decomposer food web vary between geographical regions?
- Do alterations in the structure of decomposer communities change the C and N transfer through specific channels of the below-ground food web? What is the role of key species, functional groups and trophic levels?
- What is the quantitative contribution of the decomposer community (microflora and fauna) to the C and N cycle within the forest ecosystem?

Bacteria, fungi, testate amoebae, nematodes, enchytraeids, and microarthropods (Collembola and Acari) were determined in terms of trophic connectivity, functional groups, and diversity. In addition, the C and N transfer through the soil community was estimated on the basis of a food web model (De Ruiter et al. 1994).

Diversity and Role of the Decomposer Food Web

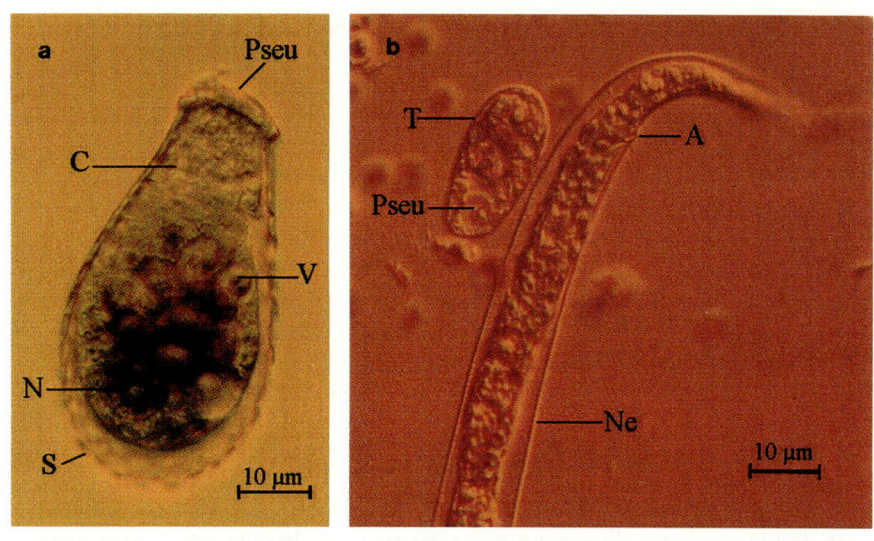

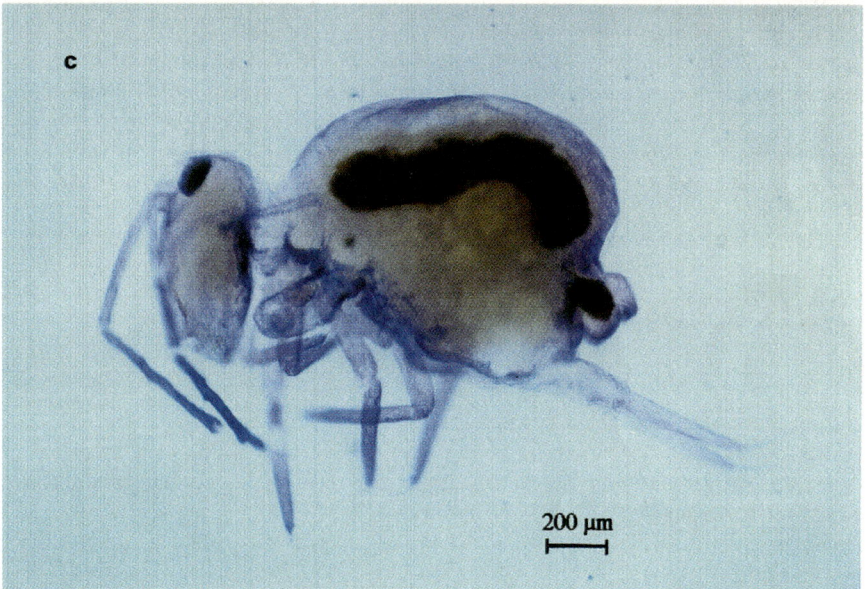

Fig. 17.1. a *Nebela militaris* PENARD 1890, a common species of testate amoebae at all sites (stained with aniline blue in watery suspension, 400×, bright field illumination). *S* Shell; *C* cytoplasm; *V* food vacuole; *N* nucleus; *Pseu* pseudostome (the shell aperture through which the pseudopodia emerge). **b** *T Trinema lineare* Penard 1890, one of the most dominant testate amoebae at all sites, and *Ne* part of a nematode (stained with anilin blue in watery suspension, 400×, DIC); *Pseu* pseudostome (see above). **c** *Dicyrtoma fusca* (Lucas 1842) a symphypleone collembolan species

17.2 Approaches to Investigating Decomposer Communities

The study sites were selected to form a north-south transect of European coniferous forests: Åheden (northern Sweden; Åhe), Skogaby (southern Sweden; Sko), Waldstein (Germany; Wal) and Aubure (France; AuP) (see Table 17.1 for a detailed description). Samples were collected at four sampling times: (1) October/November 1996, (2) May/June 1997, (3) September 1997 and (4) March/April 1998. At each time, samples were taken from the organic layer (L + F + H) of each site by means of soil corers (Ø 5 cm). Water content and pH_{H2O} were determined using standard methods. Carbon content and C:N ratio of the organic layer were measured gas chromatographically (fourth sampling time only). Soil microbial carbon (C_{mic}) and nitrogen (N_{mic}) were determined using the fumigation-extraction method (Brookes et al. 1985; Vance et al. 1987). The amount of N extracted from the unfumigated controls was used as an estimate of mineral N content (N_{min}). The metabolic potential of the microflora was estimated by measuring the CO_2 release at 10 °C and at optimal water content (300% dw). After the substrate samples had been preincubated for 6 days in the dark, CO_2 was captured in NaOH for a further 6 days and C mineralisation was then measured titrimetrically. Functional diversity of the microflora was determined using Biolog GN plates (methods and results see Kjøller et al., Chap. 18, this Vol.). The material collected at the fourth sampling time was used for determining a number of additional microbiological parameters. These include: (1) bacteria: number of living cells, biomass, frequency of dividing cells, cell length, width and volume, and (2) fungi: ergosterol content to estimate fungal biomass. The bacterial parameters were measured with automatic confocal laser-scanning microscopy picture analysis after

Table 17.1. Characteristics of the four selected coniferous sites (from data given in Persson et al., Table 2.1 Chap. 2, this Vol.)

	Åhe	Sko	Wal	AuP
Dominant tree species[a]	Ps, Pa, Bp	Pa	Pa	Pa
Type of stand	natural	planted	planted	planted
Stand age in 1995 (a)	180	31	142	92
Latitude	64°13′ N	56°33′ N	50°12′ N	48°12′ N
Altitude asl (m)	175	95–115	700	1050
Climate[b]	b	ho	hc	ho
Mean annual air temperature (°C)	1.0	7.6	5.5	5.4
Mean annual precipitation (mm)	588	1237	890	1192
Total N deposition (kg N ha^{-1} a^{-1})	1.7	16.4	20.1	14.7
pH (H$_2$O) FH layer	3.9	4.1	3.7	3.5
C content of the organic layer (%)[c]	42.6	44.4	37.6	38.9
C pool LFH layer (kg C 10^3 ha^{-1})	14.9	31.1	59.0	17.5
N pool LFH layer (kg N 10^3 ha^{-1})	0.4	1.1	2.7	0.7

[a] Pa = *Picea abies*; Bp = *Betula pubescens*; Ps = *Pinus sylvestris*.
[b] b = boreal; ho = humid oceanic; hc = humid continental.
[c] Calculated using C pool data and mean dry mass data (Persson et al., Chap. 2, this Vol.).

fluorescent staining (Bloem et al. 1997). The ergosterol content was measured using HPLC analysis (Djajakirana et al. 1996).

Microarthropods were extracted by means of the high-gradient-canister method (MacFadyen 1953; Wolters 1983), nematodes and enchytraeids were extracted by means of the O'Connor wet-funnel-technique. Testate amoebae were counted directly from aqueous suspensions after aniline blue staining (differentiation of cysts, active and empty shells) using inverse microscopy. Due to the enormous amount of material collected, not all taxa have as yet been determined to a high resolution level. Collembola (sampling 2, 3, 4) and testate amoebae (sampling 1, 2) were determined to species level, Acari (sampling 1, 2, 3, 4) were sorted into feeding groups (Luxton 1972) and nematodes (sampling 4) were identified to genus level (Yeates et al. 1993). Enchytraeids (sampling 1, 2, 3, 4), Collembola (sampling 1) and nematodes (sampling 1, 2, 3) were counted without further taxonomic processing. Invertebrate biomass was calculated from abundance values using conversion factors (Volz 1951; Luxton 1972; Schönborn 1975; Huhta 1976; Persson and Lohm 1977; Persson et al. 1980; Petersen and Luxton 1982; Huhta et al. 1986; Dunger and Fiedler 1989; Schaefer and Schauermann 1990; Wanner 1991; Teuben and Smidt 1992; Yeates et al. 1993; Berg 1997). For brevity, we often refer to the following groups: microflora (bacteria + fungi), microfauna (testate amoebae + nematodes) and mesofauna (enchytraeids + Acari + Collembola).

Statistical analysis was performed using the Statistica software package (StatSoft Inc 1995). Homogeneity of variances was tested according to Sen and Puri. The effect of the factors site and time (sampling times) was analysed with adequately transformed data using two-way Anovas. For some data sets analysis was restricted to one-way Anovas on the effect of site specific differences between means were tested with the Tukey HSD test. Pearson product-moment coefficients were calculated to test for significant relationships between environmental parameters and animal abundance. The food web approach of De Ruiter et al. (1994) was used to model C and N mineralisation by the below-ground community. In short, diagrams of the food webs were constructed by aggregating species into functional groups. Then energy flow descriptions of the food webs were computed, in which the feeding rates were calculated from the observed population sizes, death rates and energy conversion efficiencies. Annual feeding rates were calculated by assuming that the annual average production of the organisms balances the rate of loss through natural death and predation. Calculations of feeding rates started with the top predators which suffer only from natural death, and proceeded working backwards to the lowest trophic levels. Physiological parameters were taken from the literature (see De Ruiter et al. 1993 for details).

17.3 The Microflora

Significant site effects revealed by the Anovas indicate strong effects of regional conditions on microbial parameters (Table 17.2). The total microbial C storage ($kg\,C_{mic}\,ha^{-1}$) varied between 287 and $759\,kg\,C_{mic}\,ha^{-1}$, with fungi contributing by far the largest part to the microbial C pool (Table 17.3). Standing crop of microflora was highest at Wal and did not differ very much between the other sites. Microbial C concentration ($\mu g\,C_{mic}\,g^{-1}\,dw$) shows a completely different pattern than the total micro-

Table 17.2. Anova results on the effects of site and time (sampling times) on microbial parameters and animal abundance (two-way Anova: df effect (time) = 3, df effect (site) = 3, df effect (time × site) = 9; 1-way Anova: df effect (site) = 3)

	Time		Site		Time × site	
	F	p^a	F	p^a	F	p^a
Microflora						
Microbial C concentration (mg C_{mic} g^{-1} dw)[b]	37.9	**	103.4	**	21.0	**
Microbial N concentration (µg N_{mic} g^{-1} dw)	189.8	**	44.5	**	6.4	**
Metabolic potential at 10 °C (µg C g^{-1} dw h^{-1})	26.3	**	137.0	**	14.5	**
Microfauna						
Testate amoebae[c] (ind 10^6 m^{-2})	86.4	**	6.4	**	0.6	n.s.
Nematoda (ind 10^3 m^{-2})	22.8	**	1.3	n.s.	4.6	**
Mesofauna						
Enchytraeidae (ind 10^3 m^{-2})	6.4	**	100.5	**	2.1	*
Collembola (ind 10^3 m^{-2})	6.9	**	5.2	**	3.2	**
Acari (ind 10^3 m^{-2})	7.4	**	3.1	*	3.7	**

[a] p-level: not significant = n.s.; ≥0.05 = *; <0.01 = **.
[b] dw = dry weight.
[c] Cysts and active cells.

Table 17.3. Results of the microbiological measurements. Figures with identical letters are not significantly different from each other (Tukey HSD test). (Standard deviation in parentheses; for site abbreviations see Table 2.1, Chap. 2, this Vol.)

	Åhe	Sko	Wal	AuP
Microbial C concentration (mg C_{mic} g^{-1} dw)	10.8[a] (3.6)	5.9[b] (2.4)	4.8[c] (0.9)	6.4[b] (1.7)
Total C storage[a] (kg C_{mic} ha^{-1})	378	416	759	287
Microbial N concentration (µg N_{mic} g^{-1} dw)	76[a] (43)	104[b] (51)	96[a] (100)	158[c] (133)
Total N storage[a] (kg N_{mic} ha^{-1})	2.7	7.3	15.0	7.1
Microbial C:N ratio	142	57	50	40
Bacteria[b] (kg C ha^{-1})	20	31	36	37
Fungi[b] (kg C ha^{-1})	358	385	723	250
Metabolic potential at 10 °C (µg CO_2-C g^{-1} dw h^{-1})	9.4[a] (3.5)	5.5[b] (0.9)	4.3[c] (0.9)	4.1[c] (0.9)

[a] Conversion to area-based estimates was carried out using dry mass data from Persson et al., Chap. 2 (this Vol.).
[b] Calculated from the bacteria/fungi ratio determined at the fourth sampling time.

bial C storage, pointing to the important role of the thickness of the organic layer in determining the size of the microbial C pool.

The climatic gradient associated with the latitudinal transect chosen for this study was partly offset by an altitudinal gradient (Table 17.1). The two most obvious characteristics of the transect are (1) the increase in N deposition from north to south, decreasing again at the French site, and (2) the boreal climate at the site in northern Sweden. The total microbial N storage (kg N_{mic} ha^{-1}) revealed by the fumigation-extraction method was exceptionally low (Tables 17.1 and 17.3). This is probably due to methodological problems associated with this approach in organic materials. Further on in this text we will thus refer to 'extractable microbial N' to account for this potential bias. The total amount of extractable microbial N follows the pattern of atmospheric N deposition. Moreover, high N deposition coincides with low fungal biomass (Tables 17.1 and 17.3). Atmospheric N deposition has been shown to be a prime factor in reducing fungal biomass in the organic layers of coniferous forests (Berg et al. 1998). We thus conclude that external N input impacts both microbial N immobilisation and the composition of the microflora.

The concentration of extractable microbial N ($\mu g N_{mic} g^{-1}$ dw) is not correlated to N deposition, but is exceptionally high at Aubure. A possible explanation is that this site is strongly influenced by former land use (see Dambrine et al., Chap. 19, this Vol.) and probably still is in a phase of humus accumulation. As a consequence of the methodological problems mentioned above, the microbial C:N ratios measured in our study massively exceed values reported in the literature (C:N ratio: fungi 4, bacteria 10, De Ruiter et al. 1993). This parameter was nevertheless negatively correlated to the N_{min} content of the substrate ($r^2 = 0.11$, $p < 0.001$, $n = 107$), indicating that increased availability of mineral nitrogen results in higher nitrogen content of the microflora at both the local and the regional scale. Laboratory and field studies have shown that N incorporation into microorganisms is strongly affected by the amount of available nitrogen (Anderson and Domsch 1980; Joergensen et al. 1995). Variations in the microbial C:N ratios found in our study indicate that the microflora of acid forest soils responds to changes in N availability not only by growth but also by alterations in internal N concentrations.

The metabolic potential of the microflora (i.e. the CO_2 release under standardised conditions of 10 °C and optimal water content; cf. Sect. 17.2) increases from south to north (Table 17.3). The high metabolic potential at Åheden compared to the other sites is confirmed by the findings of other authors (cf. Cotrufo et al., Chap. 13, this Vol.). High metabolic activity of the soil microflora has often been attributed to a shift towards bacteria, which generally have a higher turnover rate than fungi (Paul and Clark 1989). In this case, however, no shift towards the bacterial community was observed (Table 17.3). Compared to the ambient conditions of the boreal site, an elevation of the temperature to 10 °C is a rather drastic shift. Under these conditions C sources may be metabolised, which were formerly not used due to harsh climatic conditions (cf. Table 17.1). The large amount of C incorporated per microbial biomass unit ($C_{mic} g^{-1}$ dw) at the site in northern Sweden (Table 17.3) indicates an accumulation of carbon in the microbial biomass. This is consistent with the finding that considerable amounts of nutrients are being locked into the microbial biomass at high latitudes due to adverse environmental conditions (Jonasson et al. 1999). We conclude that the microflora at Åheden is very sensitive to climate change and might respond to elevated temperature by activating a comparatively large labile C pool.

17.4 The Soil Fauna

17.4.1 Major Taxa of the Decomposer Food Web

The abundance range of the different taxa found in our study is consistent with the results reported in the literature (Table 17.4; see Petersen and Luxton 1982), with the exception that nematode abundance was surprisingly low. Protozoa have rarely been included in analyses on the structure of the soil fauna community (Foissner 1987). The dominance of testate amoebae demonstrated in our study indicates that the exclusion of this microfauna group might lead to a considerable bias in the evaluation on the role of different invertebrate taxa in terrestrial ecosystems.

Some general patterns are apparent in the abundance data. Testate amoebae were the dominant group at all sites, followed by nematodes, microarthropods and enchytraeids, with Acari being the dominant microarthropod group (Table 17.4). Significant site effects revealed by the Anovas nevertheless indicate strong effects of regional conditions on the decomposer fauna (Table 17.2). The most obvious differences between sites were (1) exceptionally high density of testate amoebae at Waldstein, (2) reversed abundance maxima and minima of Acari and Collembola at Skogaby and Aubure and (3) low density of enchytraeids at the two Swedish sites (Table 17.4). The four coniferous forest sites investigated thus host characteristic decomposer communities.

The Pearson product-moment coefficients revealed that the number of testate amoebae, nematodes, and enchytraeids were positively related to moisture content (Table 17.5). This suggests that small and soft-skinned taxa are particularly sensitive to dry conditions. The data shown in Fig. 17.2 indicate that differences in water content explained both within-site as well as between-site variation in the number of testate amoebae. The strong response of testate amoebae to alterations in water content is consistent with the results of Lousier (1974). In our study, the close relationship between these two variables seemed to be confined to conditions when the water content was below 220% (Fig. 17.2). While the water content was still important above this level, additional factors caused a strong internal variability. This result confirms the threshold hypothesis for the relationship between water content and soil

Table 17.4. Density of various decomposer fauna groups. Figures with identical letters are not significantly different from each other (Tukey HSD test). (Standard deviation in parentheses; for site abbreviations see Table 2.1, Chap. 2, this Vol.)

	Åhe	Sko	Wal	AuP
Microfauna				
Testate amoebae[a] (ind 10^6 m^{-2})	656^a (486)	679^a (623)	1765^b (1829)	776^a (774)
Nematoda (ind 10^3 m^{-2})	219^a (44)	197^a (44)	373^a (354)	309^a (250)
Mesofauna				
Enchytraeidae (ind 10^3 m^{-2})	2^a (4)	5^b (2)	38^c (14)	43^c (17)
Collembola (ind 10^3 m^{-2})	92^{a,b} (46)	60^b (19)	65^b (10)	127^a (58)
Acari (ind 10^3 m^{-2})	194^{a,b} (130)	255^a (173)	186^{a,b} (42)	135^b (37)

[a] Cysts and active cells (two sampling times only).

Table 17.5. Results of the Pearson product-moment correlation between environmental parameters and animal abundance (only significant results with $p < 0.05$ are shown, $r =$ correlation coefficient)

	Water content (% dw)		pH_{H2O}		N_{min} ($\mu g N g^{-1}$ dw)		N_{mic} ($\mu g N g^{-1}$ dw)	
	r	n	r	n	r	n	r	n
Testate amoebae[a] (ind $10^6 m^{-2}$)	0.62	26	−0.71	26	0.70	25	−	−
Nematoda (ind $10^3 m^{-2}$)	0.64	20	−0.47	20	−	−	0.55	20
Enchytraeidae (ind $10^3 m^{-2}$)	0.81	20	−0.57	16	0.77	20	0.77	20
Adult Acari (ind $10^3 m^{-2}$)	−0.47	19	−	−	−	−	−0.53	19

[a] Empty shells, cysts and active cells.

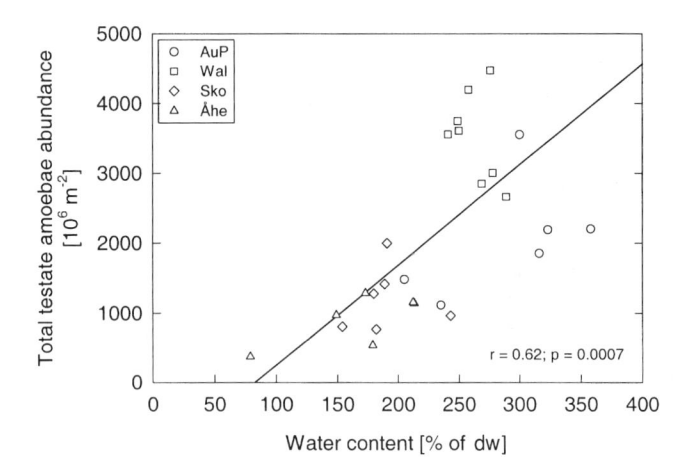

Fig. 17.2. Correlation between abundance of testate amoebae (cysts, active cells, empty shells) and water content of the substrate (for site abbreviations see Table 2.1, Chap. 2, this Vol.)

biota drawn from field experiments (Huelsmann et al. 1997). Negative correlation between water content and adult Acari supports the conclusion that microarthropods with a sclerotised cuticle are favoured by drier conditions (Table 17.5). They can be adversely affected by high water content, because mobility is reduced in a wet and compact organic layer. Several other factors such as pH, extractable microbial N and extractable N_{min} were also correlated to various decomposer fauna taxa (Table 17.5). This reflects that the composition of soil fauna is correlated to a complex set of factors including humus form, elevation, tree growth, mineral content of leaf litter and several soil parameters such as pH and C:N ratio (Ponge et al. 1997).

17.4.2 Trophic Groups Within Major Taxa

A closer look at the trophic structure within the major taxa of the decomposer food web may reveal basic differences between the food webs of the different sites. Variations in the density of trophic groups offer an opportunity to evaluate the influence of environmental conditions on the trophic connectivity of the decomposer food web (Table 17.6). Contrary to expectations, several trophic groups of nematodes and Acari showed no significant differences in abundance between the sites. Hence, various components of the decomposer food web in coniferous forest soils seem to remain remarkably constant over a wide range of environmental conditions. This holds for bacterivorous, fungivorous, predaceous and omnivorous nematodes as well as for detritivorous mites.

The dominance of bacterivorous nematodes at all sites is consistent with data reported by Huhta et al. (1986) for two Finnish coniferous forest soils. This indicates that bacteria might be more important to the microtrophic food web of the investigated sites than expected by the low bacterial biomass (Table 17.3). High numbers of rhizophagous nematodes (Table 17.6) and testate amoebae (Table 17.4) at Waldstein point to very specific conditions at this site, which is particularly distinguished from

Table 17.6. Trophic group density (ind $10^3 \, m^{-2}$) of nematodes and Acari. Figures with identical letters are not significantly different from each other (Tukey HSD test). (Standard deviation in parentheses; for site abbreviations see Table 2.1, Chap. 2, this Vol.)

	Åhe	Sko	Wal	AuP
Nematoda[a]				
Bacterivorous	147[a]	105[a]	168[a]	202[a]
	(24)	(36)	(59)	(61)
Fungivorous	41[a]	39[a]	47[a]	58[a]
	(24)	(27)	(26)	(38)
Predaceous	12[a]	0[a]	3[a]	0[a]
	(19)	(0)	(7)	(0)
Omnivorous	8[a]	20[a]	43[a]	21[a]
	(10)	(22)	(30)	(19)
Rhizophagous	12[a,b]	33[a,c]	112[c]	27[a]
	(18)	(16)	(66)	(36)
Acari				
Detritivorous	89[a]	118[a]	102[a]	86[a]
(Oribatida)	(58)	(58)	(33)	(32)
Panphytophagous	98[a,c]	125[a]	54[b,c]	31[b]
(Oppioidea, Astigmata, undetermined nymphs)	(77)	(118)	(23)	(9)
Predaceous	8[a]	11[a,b]	30[c]	18[b,c]
(Gamasina, Uropodina)	(1)	(2)	(12)	(6)

[a] Abundance estimates based on the results of all four sampling times, while the relative contribution of the individual trophic groups was estimated from determinations made at the fourth sampling time only.

the others by high N input and dense ground vegetation. The fact that alterations at the microfauna level were accompanied by a particularly high density of predaceous mites indicates that changes at the level of primary consumers feed up to higher levels of the trophic cascade (Table 17.6).

17.4.3 Diversity Within Major Taxa

Testate amoebae and Collembola were selected to analyse the relationship between species diversity and site conditions. Protozoa and microarthropods play key roles in the carbon and nitrogen cycles in many soils, but it is not known what level of diversity is essential for these processes (Coûteaux and Darbyshire 1998; Wolters 1998a). A total of 41 testate amoebae species and 61 collembolan species was determined in our study, covering a broad range of morphological and functional types. The number of species found at each site varied between 31 and 38 for testate amoebae and between 21 and 33 for Collembola. This roughly corresponds to the number of species commonly found in other forest ecosystems (Wanner 1991; Wolters 1998b).

The calculation of diversity parameters revealed remarkable variations between sites and taxa (Table 17.7). The most consistent trends are: (1) species richness of Collembola and testate amoebae was highest at the French site, (2) species richness of Collembola gradually increased from north to south and (3) the diversity index of both taxa was lowest at the northern Swedish site. Low invertebrate diversity in northern Sweden confirms that adverse climatic conditions lead to an impoverished decomposer community. Interestingly, the southward increase of collembolan species richness parallels the increase in microbial N content (microbial C:N ratio, Table 17.3). Several authors have shown that the nutritional value of fungi to Collembola increases with increasing N availability (Leonard and Anderson 1991). It can be speculated that a greater amount of high-quality food allowed more collembolan species to coexist.

Table 17.7. Diversity parameters of testate amoebae and Collembola (for site abbreviations see Table 2.1, Chap. 2, this Vol.)

	Åhe	Sko	Wal	AuP
Testate amoebae[a]				
Species richness (n)	31	31	31	38
Diversity Hs[b]	3.0	3.4	3.7	3.6
Evenness E	0.68	0.76	0.78	0.75
Collembola[c]				
Species richness (n)	21	27	32	33
Diversity Hs	2.0	2.6	2.3	2.2
Evenness E	0.67	0.78	0.66	0.63

[a] 1st and 2nd sampling time only.
[b] Hs: Shannon-Wiener Index.
[c] 2nd, 3rd and 4th sampling time.

17.5 Contributions of the Decomposer Food Web to Carbon and Nitrogen Flow

The amount of carbon stored in the biomass of the total decomposer food web varied between 329 and 818 kg C ha^{-1} (Table 17.8), corresponding to 1.4–2.7% of the total C pool of the LFH layer (cf. Table 17.1). Thus only a small portion of carbon stored in the organic layer is fixed in the biomass of edaphic biota. The micro- and mesofauna made up 7 to 13% of the biomass, while the rest is accounted for by the microflora. Compared to the results of other authors, the share of the fauna is nevertheless high (e.g. Swift et al. 1979; Persson et al. 1980). This is largely due to the inclusion of testate amoebae in our study, and emphasises again the need for considering Protozoa in the analysis of the below-ground communities.

Food web modelling has proven to be a powerful tool for estimating both the functional role of soil biota and critical links in the below-ground community (De Ruiter et al. 1998). The structure of the food web used for the modelling approach in this study is shown in Fig. 17.3. The annual amount of C mineralised by the soil community varied from 1600 to a maximum of 3300 kg C ha^{-1} at Waldstein (Fig. 17.4). C mineralisation was quite similar on the investigated sites except for a maximum value in Waldstein. The particular conditions at this site and the associated shift in the microfood web have already been mentioned above. When using the same approach to compare seven agricultural soils, De Ruiter et al. (1998) calculated annual C-mineralisation rates varying between 2000 and 5000 kg C ha^{-1}. This indicates that recalcitrant substrates accumulating on the surface of coniferous forest soils are decomposed at a much slower rate than organic residues of arable soils.

Our results can be compared to the data for the LFH layer reported in other chapters of this volume: flux calculations based on a ^{14}C technique (Harrison et al., Chap. 11, this Vol.), extrapolated mineralisation rates from laboratory incubation (Persson et al., Chaps. 12 and 14, this Vol.), and simulation of C and N cycling by means of the

Table 17.8. Biomass C (kg C ha^{-1}) of various groups of the decomposer fauna. For site abbreviations see Table 2.1 (Chap. 2, this Vol.)

	Åhe	Sko	Wal	AuP
Microfauna				
Nematoda	0.3	0.2	0.4	0.4
Testate amoebae	13.1	13.6	35.3	15.5
Total	13.4	13.8	35.7	15.9
Mesofauna				
Enchytraeidae	0.6	1.2	8.4	9.5
Collembola	6.6	4.3	4.6	9.0
Acari	11.1	14.6	10.6	7.7
Total	18.2	20.0	23.7	26.3
Total decomposer food web (microflora + fauna)	410	450	818	329

Diversity and Role of the Decomposer Food Web

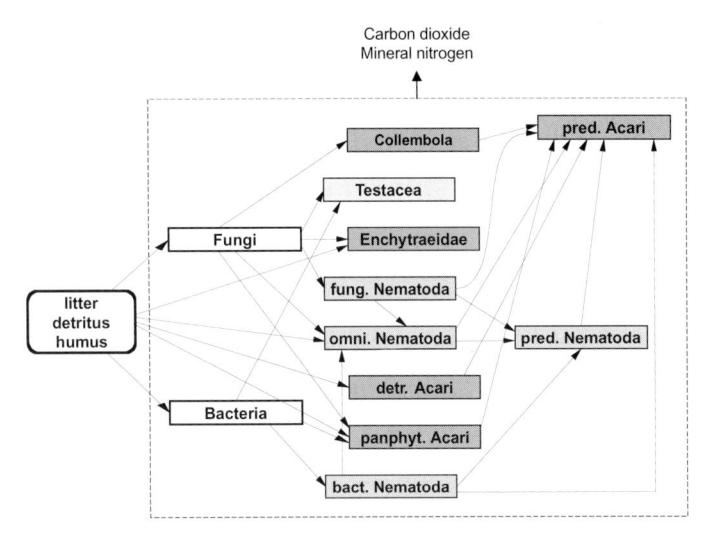

Fig. 17.3. Diagram of the trophic connectivity within the decomposer food web. *Arrows* represent feeding relationships pointing from prey to predator. *Pred.* Predaceous; *fung.* fungivorous; *omni.* omnivorous; *detr.* detritivorous; *panphyt.* panphytophagous; *bact.* bacterivorous

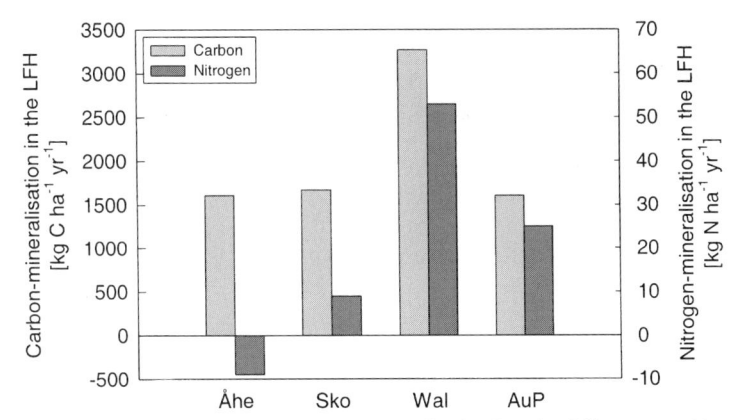

Fig. 17.4. Annual C and N mineralisation by the decomposer food web at the different sites (for site abbreviations see Table 2.1, Chap. 2, this Vol.)

NUCOM model (van Oene et al., Chap. 20, this Vol.). The C mineralisation rates derived from the food web model for the two Swedish sites are in the same order as those gained from the other approaches, whereas the modelled values for Waldstein and Aubure are considerably higher. Differences between sites closely correspond to the C-flux calculations reported in Harrison et al., Chap. 11, this Vol., with the exception of the northernmost site. Here, the food web approach results in a slightly lower C mineralisation rate than the ^{14}C technique and was very close to the value estimated by the NUCOM model (van Oene et al., Chap. 20, this Vol.). The modelled rates of C mineralisation in the LFH layer do not meet our expectation of a slow decomposition process at the boreal site. This is illustrated by the turnover times of carbon, which can be calculated by dividing the estimated annual C-mineralisation rates (Fig. 17.4) by the C pool of the organic layer (Table 17.1). The resulting turnover times are 9 (northern Sweden; Åheden), 11 (France; Aubure), 18 (Germany; Waldstein) and 19 years (southern Sweden; Skogaby) respectively. A delayed mineralisation of carbon at the boreal site seems thus not to be attributable to a slow turnover rate in the uppermost organic layer.

The annual mineralisation of N in the LFH layer varied between 0 and 53 kg N ha^{-1} (Fig. 17.4). The values were slightly lower than those gained from the other techniques (Harrison et al., Chap. 11; Persson et al., Chap. 14; van Oene et al., Chap. 20, this Vol.). However, at each site at least one of the techniques alternatively used within the CANIF project delivered very similar estimates to the food web model. Differences between sites vary in a similar fashion as the extrapolated N mineralisation rates reported in Chapter 14 (this Vol.). The fact that no net mineralisation of N occurred at the northern Swedish site leads to the conclusion that the N cycle is closed at this site (cf. Persson et al., Chap. 14, this Vol.). N limitation may thus be a major factor controlling ecosystem processes in boreal forests. Field experiments with fertilisers have shown that N addition can significantly accelerate decomposition processes and tree growth in boreal forests (Bergh et al. 1999). How this fits the fact that the metabolic potential measured by short-term incubations does not seem to be limited by N availability (cf. Sect. 17.3) has to be tested in the future.

The food web model also allows estimation of the relative contribution of the various components of the decomposer community to C and N cycling. Bacteria and fungi contributed about 90% of the C mineralised by the decomposer food web. This is consistent with the finding that the microflora is of enormous importance for ecosystem functioning and is responsible for 80–99% of non-root CO_2 losses from soil (e.g. Wolters and Schaefer 1994). The contribution of the micro- and mesofauna varies around 10%, but can be considerably higher for annual N mineralisation. This confirms the conclusions drawn from studies in a wide range of ecosystems (Anderson 1995). Testate amoebae accounted for 4–5% of total C mineralisation. The contribution of the Nematoda was 0.1–0.2%, that of Enchytraeidae 0–4% and that of microarthropods 2–4%. Preliminary sensitivity analyses revealed that the C and N mineralisation rates calculated by the food web model were strongly affected by alterations in the size of the microbial C and N pool. Despite the subordinate role of the fauna, elimination of invertebrates from the model reduces C mineralisation by more than a factor of 2, indicating the important indirect impact of invertebrate feeding on microbial mineralisation rates. However, even strong alterations in the biomass of groups such as nematodes and enchytraeids only caused minor changes in overall C mineralisation. This suggests that the consumer level is quite robust to changes in the biomass of individual taxa.

17.6 Conclusions

- Total extractable microbial N storage follows the pattern of atmospheric N deposition. Ectorganic horizons are central in retaining N inputs (Currie et al. 1999). Our results imply a certain capacity of the microflora to act as a sink for atmospheric N deposition.

- The metabolic potential of the microflora at the boreal site is high. This may be the result of easily degradable C sources accumulating in the organic layer. The stability of stored carbon under changing climatic conditions has been the subject of recent concern, because global warming is predicted to be most pronounced in northern continental regions (IPCC 1996). Potential feedback mechanisms on climatic change deserve more scientific attention (cf. Christiansen et al. 1999).
- The rapid C turnover time calculated for the organic layer of the site in northern Sweden indicates that C accumulation at boreal sites cannot be attributed to a delayed mineralisation of carbon in the uppermost organic horizons containing up to 90% of the decomposer organisms.
- The sites host characteristic decomposer communities with pronounced differences in the species pattern of some closely studied taxa. However, the relation of microflora, micro- and mesofauna remains remarkably constant and the trophic connectivity is quite similar along the transect. This may partly be explained by the strong restrictions set to the decomposer food web by poor substrate conditions in coniferous forests. The German site with heavy atmospheric N deposition and a rich herb layer provides some remarkable exceptions.
- Low invertebrate diversity in northern Sweden confirms that harsh climatic conditions lead to an impoverished decomposer community. This points to the sensitivity of the boreal site to changes in environmental conditions, because impoverished communities may lack the functional redundancy allowing the belowground biota to compensate for changes at the species level (Wolters 1998a).
- The outcome of the food web models was quite insensitive to changes at the consumer level. Even drastic alterations in the biomass of keystone groups such as nematodes and enchytraeids only caused marginal changes in the overall C and N flow. This partly reflects the functional redundancy of the soil community. However, C and N budgets only tell part of the story, because non-trophic effects and specific interactions are largely ignored (Wolters 2000). These relationships may have a strong impact on process rates and must consequently remain a central target of ecosystem analyses.
- The close similarity between the results of the food web model and the values gained by a wide array of different techniques is encouraging. It shows that mathematical and conceptual models are very useful for converting results of soil biological analyses into budgets for whole soils.

Acknowledgements. We are very grateful to P. De Ruiter and J. Moore for giving us the opportunity to use their food web model. We thank M. Leonhardt, C. Meder, S. Vesper and B. Wasmus for technical assistance. We also thank the CANIF partners for continuous support and stimulating discussions. This study was funded by the European Commission (contract no. ENV4-CT95-0053).

References

Anderson JM (1995) Soil organisms as engineers: microsite modulation of macroscale processes. In: Jones CG, Lawton JH (eds) Linking species and ecosystems. Chapman & Hall, London, pp 94–106

Anderson JPE, Domsch KH (1980) Quantities of plant nutrients in the microbial biomass of selected soils. Soil Sci 130:211–216

Berg M (1997) Decomposition, nutrient flow and food web dynamics in a stratified pine forest soil. Department of Ecology and Ecotoxicology, Section Soil Ecology. Vrije Universiteit, Amsterdam, 310 pp

Berg MP, Kniese JP, Verhoef HA (1998) Dynamics and stratification of bacteria and fungi in the organic layers of a Scots pine forest soil. Biol Fertil Soils 26:313–322

Bergh J, Linder S, Lundmark T, Elfving B (1999) The effect of water and nutrient availability on the productivity of Norway spruce in northern and southern Sweden. For Ecol Manage 119:51–62

Bloem J, Veninga M, Shepherd J (1997) Vollautomatische Messung von Bodenbakterien mit Hilfe der konfokalen Laser-Scanningmikroskopie und der Bildanalyse. Mitt Wiss Technik XI:143–148

Brookes PC, Landman A, Pruden G, Jenkinson DS (1985) Chloroform fumigation and the release of soil nitrogen: a rapid direct extraction method for measuring microbial biomass nitrogen in soil. Soil Biol Biochem 17:837–842

Christensen TR, Jonasson S, Callaghan TV, Havström M (1999) On the potential CO_2 release from tundra soils in changing climate. Appl Soil Ecol 11:127–134

Coûteaux M-M, Darbyshire JF (1998) Functional diversity amongst soil protozoa. Appl Soil Ecol 10:229–237

Currie WS, Nadelhoffer KJ, Aber JD (1999) Soil detrital processes controlling the movement of N-15 tracers to forest vegetation. Ecol Appl 9:87–102

De Ruiter PC, van Veen JA, Moore JC, Brussaard L, Hunt HW (1993) Calculation of nitrogen mineralization in soil food webs. Plant Soil 157:263–273

De Ruiter PC, Neutel A-M, Moore JC (1994) Modelling food webs and nutrient cycling in agroecosystems. TREE 9:378–383

De Ruiter PC, Neutel A-M, Moore JC (1998) Biodiversity in soil ecosystems: the role of energy flow and community stability. Appl Soil Ecol 10:217–228

Djajakirana G, Joergensen RG, Meyer B (1996) Ergosterol and microbial biomass relationship in soil. Biol Fertil Soils 22:299–304

Dunger W, Fiedler HJ (1989) Methoden der Bodenbiologie. Gustav Fischer, Stuttgart

Ellenberg H, Mayer R, Schauermann J (1986) Ökosystemforschung – Ergebnisse des Sollingprojektes 1966–1986. Ulmer, Stuttgart

Foissner W (1987) Soil Protozoa: fundamental problems, ecological significance, adaptations in ciliates and testaceans, bioindicators, and guide to the literature. Progr Protistol 2:69–212

Freckman DW (1994) Life in the soil. Soil biodiversity: its importance to ecosystem processes. Report of a Workshop held at the Natural History Museum, London, 30 August–1 September, 26 pp

Hall GS (ed) (1996) Methods for the examination of organismal diversity in soils and sediments. CAB International, Oxon, UK

Huelsmann A, Schroeter D, Pflug A, Wolters V (1997) Response of the decomposer community to experimental alterations in spruce litter humidity. In: van de Geijn SC, Kuikman PJ (eds) Prospects for co-ordinated activities in core projects of GCTE, BAHC and LUCC. CCB-Wageningen, Wageningen, pp 152–153

Huhta V (1976) Effects of clear-cutting on numbers, biomass and community respiration of soil invertebrates. Ann Zool Fenn 13:63–80

Huhta V, Hyvönen R, Kaasalainen P, Koskenniemi A, Muona J, Mäkelä I, Sulander M, Vilkamaa P (1986) Soil fauna of Finnish coniferous forests. Ann Zool Fenn 23:345–360

IPCC (1996) Climate change 1995. Impacts, adaptations and mitigation of climate change: scientific-technical analyses. Cambridge University Press, Cambridge

Joergensen RG, Anderson TH, Wolters V (1995) Carbon and nitrogen relationships in the microbial biomass of soils in beech (*Fagus sylvatica* L.) forests. Biol Fertil Soils 19:141–147

Jonasson S, Michelsen A, Schmidt IK (1999) Coupling of nutirent cycling and carbon dynamics in the Arctic, integration of soil microbial and plant processes. Appl Soil Ecol 11:135–146

Leonard MA, Anderson JM (1991) Growth dynamics of Collembola (*Folsomia candida*) and a fungus (*Mucor plumbeus*) in relation to nitrogen availability in spatially simple and complex laboratory systems. Pedobiologia 35:163-173

Lousier JD (1974) Effects of experimental soil moisture fluctuations on turnover rates of Testacea. Soil Biol Biochem 6:19-26

Luxton M (1972) Studies on the oribatid mites of a Danish beech wood Soil - I. Nutrition biology. Pedobiologia 12:434-463

Macfadyen A (1953) Notes on methods for the extraction of small soil arthropods. J Anim Ecol 22:65-78

Paul EA, Clark FE (1989) Soil microbiology and biochemistry. Academic Press, San Diego

Persson T, Lohm U (1977) Energetical significance of the annelids and arthropods in a Swedish grassland soil. Ecol Bull 23:218

Persson T, Bååth E, Clarholm M, Lundkvist H, Söderström BE, Sohlenius B (1980) Trophic structure, biomass dynamics and carbon metabolism of soil organisms in a Scots pine forest. In: Persson T (ed) Structure and function of northern coniferous forests - an ecosystem study. Swedish Natural Science Research Council (NFR), Stockholm, pp 419-459

Petersen H, Luxton M (1982) Quantitative ecology of microfungi and animals in soil and litter - a comparative analysis of soil fauna populations and their role in decomposition processes. OIKOS 39:286-388

Ponge JF, Arpin P, Sondag F, Delecour F (1997) Soil fauna and site assessment in beech stands of the Belgian Ardennes. Can J For Res 27:2053-2064

Schaefer M, Schauermann J (1990) The soil fauna of beech forests: comparison between a mull and a moder soil. Pedobiologia 34:299-314

Schönborn W (1975) Ermittlung der Jahresproduktion von Boden-Protozoen. I. Euglyphidae (Rhizopoda, Testacea). Pedobiologia 15:415-424

StatSoft Inc (1995) STATISTICA 5.0 A for windows (computer program manual). Tulsa, Oklahoma

Swift MJ, Heal OW, Anderson JM (1979) Decomposition in terrestrial ecosystems. Blackwell, Oxford

Teuben A, Smidt GRB (1992) Soil arthropod numbers and biomass in two pine forests on different soils, related to functional groups. Pedobiologia 36:79-89

Vance ED, Brookes PC, Jenkinson DS (1987) An extraction method for measuring soil microbial C. Soil Biol Biochem 19:703-707

Volz P (1951) Untersuchungen über die Mikrofauna des Waldbodens. Zool Jb Syst 79:514-566

Wanner M (1991) Zur Ökologie von Thekamöben (Protozoa: Rhizopoda) in süddeutschen Wäldern. Arch Protistenkd 140:237-288

Wolters V (1983) Ökologische Untersuchungen an Collembolen eines Buchenwaldes auf Kalk. Pedobiologia 25:73-85

Wolters V (1998a) Functional aspects of animal diversity in soil - introduction and overview. Appl Soil Ecol 10:185-190

Wolters V (1998b) Long-term dynamics of a collembolan community. Appl Soil Ecol 9:221-227

Wolters V (2000) Invertebrate control of soil organic matter stability. Biol Fertil Soils 31:1-19

Wolters V, Schaefer M (1994) Effects of acid deposition on soil organisms and decomposition processes. In: Hüttermann A, Godbold D (eds) Effects of acid rain on forest processes. John Wiley, New York, pp 83-127

Yeates GW, Bongers T, de Gode RGM, Freckman DW, Georgieva SS (1993) Feeding habits in soil nematode families and genera - an outline for soil ecologists. J Nematol 25:315-331

18 Diversity and Role of Microorganisms

A. Kjøller, M. Miller, S. Struwe, V. Wolters, and A. Pflug

18.1 Introduction and Background

The focus on biodiversity changes is often on the conspicuous elements of ecosystems, like trees and flowers, birds and mammals, while the less visible part operating below ground has often been ignored. It is well established that soil communities are among the most species-rich compartments of terrestrial systems (Hall 1996). The implications of this enormous diversity for soil function are still very little known and difficult to address (Wolters 1997). The elucidation of species richness of soils in conjunction with sustainability assessments of soil-mediated ecosystem processes must therefore have a high priority in global biodiversity efforts (Freckman 1994). This holds particularly for soil microorganisms, which have the greatest impact on soil nutrient storage and turnover, but are the most poorly investigated components of the below-ground community. The major factors affecting microbial diversity in soils are climate, substrate conditions and other organisms. In recent times, these factors have been strongly modified by anthropogenic forces.

The European transect of the CANIF project offered the opportunity for a comparative approach to soil microbial diversity. Several important topics of recent soil microbial research were addressed:

– Expression of microbial enzymes in situ.
– Functional role of microfungi in complex ecosystems.
– Substitution of genera with same function.
– Functional diversity of bacteria.
– Metabolic activity of bacteria in relation to site factors.

The studies concentrated on two case studies. The first focused on microfungi in the four southernmost beech forest sites in Italy, France, Germany and Denmark. Microfungi were isolated from beech litter in litterbags on soil extract agar six times during 2 years of incubation and identified to generic level for determination of fungal diversity. Cellulase activity in the same litter was determined using fluorescent substrate analogues of cellulose for linking fungi to functional diversity. The second case study concentrated on the functional diversity of bacteria in four coniferous sites forming a European north-south transect: northern and southern Sweden, Germany and France. This study was carried out by means of the Biolog technique. To provide a general framework, a short summary of the present understanding of soil microbial diversity is given as a background.

Different groups of microorganisms interact during the course of litter decomposition in a very complex way. Certain groups have specific roles alone or in interaction with other groups. The aim of the study presented here was to link biodiversity with the function of soil microbial populations (1) by analysing the changes in the community structure of microfungi seen in relation to the cellulase activity and (2) by studying changes in the functional structure of bacteria in relation to site conditions. The following questions were examined:

- Will the composition of the fungal flora on a standard litter incubated along a European transect change according to local climate (cf. Cotrufo et al., Chap. 13, this Vol.)?
- When different European litters are incubated in Italy, will the composition of the litter floras gradually become similar?
- Will cellulase activity depend more on local climate than on origin of litter?
- Changes in the composition of the fungal population may be reflected in cellulase activity.
- Will the functional structure of microbial communities of a European transect change according to local conditions?
- Do specific patterns in the ability of the microflora to use a complex set of C-sources reflect their role in the decomposer system?
- Do site-specific differences in the functional diversity of soil microbiota indicate the impact of environmental factors on the below-ground community?

18.2 Experimental Background

Many microorganisms are ubiquitous and local species diversity is generally high. However, microbial species diversity is not well documented, and it is suggested that everything is everywhere as a consequence of small body size and concomitant large species population size (Fenchel et al. 1997). Further, it is claimed from studies carried out in aquatic systems that microbial diversity is never so impoverished that the microbial community cannot play its full part in biogeochemical cycling, because the species complement of the microbial community quickly adapts even to critical changes in the local environment (Finlay et al. 1997). This conclusion is not necessarily true for the soils of terrestrial ecosystems, where there is no free movement as in aquatic systems.

There are various ways to enumerate the enormous range of bacteria and fungi in soils, including viable counts, direct counts and DNA-content analysis. However, the results obtained by using these methods are not directly comparable to each other, because they relate to different traits of the microbes. Comparisons between different litter types or soil profiles are thus only meaningful if the same methods are applied throughout the examination. To be able to identify the individual members of a microbial population it is necessary to isolate these on laboratory substrates. This raises the problem that only those organisms that are able to grow on the media chosen for a particular study will be detected. For example, it is claimed that only a minor fraction of the bacteria can be grown in culture. Further, identification to species level of bacteria and of Fungi Imperfecti is a very demanding task. Skilled personnel is imperative to carry out the different test programs for identifying these taxa even if automated test systems were available. However, such systems have not been developed so far and careful microscopic observation is still needed. Moreover, it is not possible to identify all the microbial components in a soil sample, despite the fact that the introduction of molecular methods seems to solve some of the problems associated with the determination of bacteria by analysing rRNA sequences. The artificially composed fungal group of Fungi Imperfecti is an assemblage of microfungi without any sexual reproduction stage and the isolated strains can so far only be identified on the basis of both morphology and vegetative reproduction structures.

An evaluation of microbial diversity in soils cannot only be based on the enumeration and identification of the organisms involved. It is also essential to investigate the functional role of the microorganisms in order to define the significance of microbial diversity. The function of certain groups of microorganisms can be estimated by measurements of processes directly in the field. One example is the production of CO_2 as an estimate of the total aerobic activity of bacteria and fungi. By applying selective inhibition with antibiotics it may be possible to discriminate between fungal and bacterial activity on the basis of CO_2 measurements. N_2O production is used as a measure of nitrate reduction or nitrification. CH_4 release is an end product of anaerobic processes occurring in moist and waterlogged soil and has also been used to measure functional aspects of soil microbiota. To study specific processes and capacities of individual populations or strains, it is essential to grow the organisms in laboratory cultures. This is a very laborious and time-consuming process and the task to identify the ecological role of microorganisms in a large project with many variables is extensive and expensive and consequently alternative approaches will be attractive.

In recent years, the functional diversity of microorganisms in soil samples has been investigated using Biolog microtitre plates allowing for testing of the ability to use different substrates. This approach has proven to be a sensitive tool for determining large-scale differences in the functional diversity of soil microorganisms (Zak et al. 1994). Studies using the Biolog system revealed significant differences between microbial communities of different soil types (Timonen et al. 1998), different management systems (Buyer and Drinkwater 1997), contaminated versus unpolluted soils (Wünsche et al. 1995) and different vegetation types such as heath and grasslands (Lee and Caporn 1998; Johnson et al. 1998). However, interpretations have to be made with care, because Biolog data neither reflect the total functional diversity of the complex microbial community in soil (Lindstrom et al. 1998), nor do they necessarily measure the functional potential of the numerically dominant taxa (Smalla et al. 1998). In general, the method is selective for bacteria actively metabolising under the conditions of the microtitre plates (Garland and Mills 1991; Bossio and Scow 1995; Staddon et al. 1998). The suitability of the Biolog approach for measuring the functional structure of microbial communities is nevertheless proved by the similarity between the analysis of fatty acids and carbon source utilisation data (Buyer and Drinkwater 1997; Ibekwe and Kennedy 1998). Methodological standardisation is essential for gaining reliable results. The influence of inoculation density can be eliminated by extending the inoculation time over 48h (Wünsche et al. 1995). Moreover, reproducibility requires samples of approximately equivalent inoculum densities (Haack et al. 1995).

An alternative strategy has recently been suggested by Degens and Harris (1997), who, instead of relying on growth of bacteria on the substrates offered in the Biolog plates, estimated substrate-induced respiration. The same substrates as in Biolog were used but added to soil samples and the pattern of in situ catabolic potential (ISPC) was determined. The conclusion was that a physiological technique could provide "a reasonably rapid and simple method to assess the catabolic diversity of microbial communities without extraction or culturing organisms from soil". This method was later used to demonstrate that a decrease in functional diversity may not necessarily result in a decline in decomposition rates (Degens 1998).

Enzyme activity is essential for the degradation of organic matter and can thus be used as an indicator of soil functionality. It has been argued that this could be the most direct way to show fungal functional diversity in decomposition studies (Miller 1995; Zak and Visser 1996). Recently methylumbelliferyl-labelled (MUF) model substrates have been implemented for determination of enzyme activities in soil (Miller et al. 1998) and litter (Møller et al. 1999), hereby providing a sensitive method for estimation of enzyme activities in decomposition studies with high spatial resolution. Depending on the specificity of the enzyme analysis, it is possible to compare microbial communities in different soils or litter types, to identify community succession and consequently also to identify effects of disturbances (Sinsabaugh et al. 1991).

Molecular techniques have been developed during recent years to provide insight into microbial – so far primarily bacterial – diversity by sequence analysis of DNA obtained independently of culture techniques. The comparison between phenotypic diversity and DNA heterogeneity of bacteria was initiated by Torsvik et al. (1990). Since then a large number of studies have been published, but it is difficult to define the relationships between non-growing and active cell numbers and biomass. It will presumably take several years for a new microbial ecology to be written on the basis of knowledge obtained from molecular techniques. A particularly promising approach is to use these methods as an extension of traditional methods (Palleroni 1997). This author has also stated that it is inevitable that strains will be isolated on appropriate substrates to study how microorganisms act in natural environments. Preparations for analysing denitrifying bacteria with nitrite reductase probes have been made in the context of the CANIF project, but we have otherwise not applied these methods in the project.

18.3 Community of Microfungi in Beech Forests

Changes in the composition, spatial pattern and relative abundance of the microflora occurring during the succession on organic substrates lead to characteristic alterations in biodiversity from fresh litter through different stages of decomposition until complete mineralisation. Only a full investigation of the enzymatic ability of the isolated fungi and a concomitant determination of the disappearance of carbon compounds (e.g. cellulose, lignin) in the litter may explain the shift in the fungal populations (see Kjøller and Struwe 1992 and Frankland 1998 for a more detailed discussion).

18.3.1 Approaches to the Study of the Diversity of Microfungi

The main ideas behind the combined transplant-transect experiment are described in Cotrufo et al. (Chap. 13, this Vol.) Beech leaves from sites along the European transect including forests in Italy (Collelongo), France (Aubure), Germany (Schacht) and Denmark (Sorø) were collected in October 1996 and placed in litterbags inserted into the litter layer at the Collelongo site (transplant experiment, Sect. 18.2.2). Similarly, Italian beech leaves were placed at the three other European sites (the transect

Table 18.1. The 11 most frequent fungal genera from all litter types and sites (overall frequency >2%)

Acremonium	Acr[a]
Alternaria	Alt
Aureobasidium	Au
Chalara	Cha
Cladosporium	Cl
Fusarium	Fus
Mortierella	Mort
Mucor	Muc
Paecilomyces	Pae
Phoma	Pho
Trichoderma	Tric

[a] These abbreviations are used in Tables 18.3 and 18.4.

experiment, Sect. 18.2.3). Sampling took place in March, July and November 1997 and in March, August and October 1998. The soil wash method was used for isolating fungi. The litter samples were blended and small particles (less than 2 mm) were washed with sterile water on a sieve to remove conidia attached to the particles. The particles were then placed on soil agar and incubated at ambient temperature for 2–4 weeks. The aim was to isolate the most frequent actively growing saprophytic decomposer microfungi. The isolated strains from the particles were identified to generic level. Simultaneously, enzyme activity (cellulase) in the litter was determined by a novel method to estimate cellulase activity in soil and litter, using a fluorogenic methyl umbelliferyl (MUF)-labelled substrate analogue, namely cellobiohydrolase (see Chap. 13, this Vol.).

The 11 most frequent fungi (occurring at a frequency higher than 2% of the total number of strains) isolated from the different litter types and age groups during the 2 years of incubation are shown in Table 18.1. Besides these common microfungi, a wide variety of genera recorded only once or a few times were identified. These genera were observed in all four litter types placed at the Italian site, but also when Italian litter was incubated in France, Germany and Denmark. A survey of these fungi is shown in Table 18.2. Besides the above-mentioned identified genera, a certain part of the fungal flora remained unidentified; in certain samples these sterile strains made up as much as approx. 50% of the total isolated fungi (Table 18.6).

The succession of fungi during decomposition of the different types of beech litter incubated either in Italy (transplant experiment) or Italian litter incubated along the transect (transect experiment) is shown in Tables 18.3 and 18.4. The fungal community structure was characterised using two different indices: Simpson's diversity index (measure of dominance) and Shannon's index (fungal community evenness) (Margurran 1988; cf. Table 18.5).

18.3.2 Transplant Experiment

The fungal flora from the various European beech forest litters was first analysed 4 months after litterfall and incubation in the litter layer at the Collelongo site. Since

Table 18.2. The less frequent fungal genera from all litter types and sites (overall frequencey <2%)

Beauveria
Botrytis
Chrysosporium
Cylindrocladium
Dactylaria
Dreschlera
Geotrichum
Menispora
Verticillium
Yeasts

Table 18.3. Occurrence of the most frequent fungal genera during decomposition of beech litter in the transplant experiment (expressed as a percentage of the total number of isolates in each sample; t_1, t_2, t_3 = March, July, November 1997, t_4, t_5 = March, August 1998; for abbreviations see Table 18.1)

	t_1		t_2		t_3		t_4		t_5	
Col I	Cha	40.9	Cha	19.7	Pen	30.2	Cl	25.0	Cl	26.4
	Au	18.2	Cl	15.2	Cl	14.3	Pen	21.8	Pen	16.7
	Cl	13.6	Pen	7.6	Acr	8.4	Acr	5.0	Au	5.6
			Au	3.0	Alt	6.4	Au	5.6	Mort	4.2
							Cha	3.3	Pae	4.2
AuF F	Cha	25.0	Cha	19.7	Fus	9.5	Pen	15.3	Pen	38.4
	Cl	7.4	Cl	8.8	Pen	7.9	Acr	15.3	Cl	21.5
	Au	4.4	Au	7.0	Acr	7.9	Cl	6.8	Acr	6.2
			Acr	5.3	Cl	4.5	Fus	4.8	Au	4.6
									Mort	3.9
Sch DE	Cha	27.4	Cha	12.1	Pen	19.7	Acr	12.9	Pen	26.4
	Muc	12.9	Cl	6.9	Cl	11.5	Cl	8.1	Cl	19.4
	Cl	11.3	Muc	5.2	Muc	8.2	Pen	4.8	Mort	12.5
	Acr	4.8	Pho	3.5	Cha	4.9	Fus	4.8	Au	5.6
	Pen	3.2			Fu	3.3	Mort	3.2	Muc	4.2
					Alt	3.3	Pho	3.2	Alt	2.8
					Pae	3.3			Pae	2.8
					Pho	3.3				
Sor DK	Cl	17.2	Cl	23.3	Cl	37.5	Cl	36.0	Cl	32.1
	Au	10.9	Au	13.7	Au	2.8	Acr	9.3	Pen	14.8
	Alt	10.9	Cha	6.9	Alt	2.8	Pen	8.0	Alt	6.2
	Tric	9.3	Alt	5.5	Tric	2.8	Au	4.0	Mort	4.9
	Cha	7.8	Pen	2.7			Au	2.5	Tric	4.9
	Fus	4.6							Acr	3.7
									Muc	3.7

Table 18.4. Occurrence of the most frequent fungal genera during decomposition of beech litter in the transect experiment (expressed as a percentage of the total number of isolates in each sample; t_1, t_2, t_3 = March, July, November 1997, t_4, t_5 = March, August 1998; for abbreviations see Table 18.1)

	t_1		t_2		t_3		t_4		t_5	
Col I	Cha	40.9	Cha	19.7	Pen	30.2	Cl	25.0	Cl	26.4
	Au	18.2	Cl	15.2	Cl	14.3	Pen	21.8	Pen	16.7
	Cl	13.6	Pen	7.6	Acr	8.4	Acr	5.0	Au	5.6
			Au	3.0	Alt	6.4	Au	5.6	Mort	4.2
							Cha	3.3	Pae	4.2
AuF F	Cha	62.9	Cha	50.9	Cl	10.4	Cl	11.5	Pen	9.0
	Au	3.2	Au	13.6	Mort	10.4	Acr	9.9	Cl	7.5
			Cl	8.5	Pho	8.3	Alt	6.6	Mort	6.0
			Mort	5.1	Cha	4.2	Pen	4.9	Acr	4.5
					Pen	4.2	Mort	4.9	Fus	4.5
					Au	4.2	Cha	3.3		
					Acr	2.1	Muc	3.3		
Sch DE	Cha	53.5	Cha	30.5	Cl	18.8	Cl	25.0	Cl	20.0
	Au	6.9	Au	10.2	Cha	7.8	Pen	15.6	Pen	15.4
	Acr	6.9	Cl	8.5	Au	7.8	Acr	7.8	Mort	15.4
	Cl	5.2	Fus	5.1	Mort	7.8	Mort	7.8	Acr	9.2
	Mort	5.2	Alt	3.4	Acr	6.3	Cha	4.7	Tric	3.1
					Pae	6.3	Tric	3.1		
					Pho	6.3	Fus	3.1		
Sor DK	Cha	43.3	Cl	18.8	Pen	14.5	Acr	23.8	Pen	20.9
	Mort	5.4	Cha	15.6	Cl	8.7	Fus	11.1	Tric	11.9
	Cl	2.7	Acr	10.9	Acr	5.8	Cl	6.4	Mort	11.9
	Au	2.7	Au	7.8	Mort	4.4	Pho	6.4	Cl	4.5
	Acr	2.7	Mort	4.7	Fus	2.9	Pen	4.8	Fus	4.5
			Pen	3.1	Alt	2.9			Acr	3.0

the Collelongo site is covered with snow during the winter months, the March sample could be valid to indicate the original flora on the different litter samples, but it seems as if the foreign litters have already been invaded by the same dominant genera found on the native Italian litter; therefore there was a major similarity in the composition of the fungal flora on all litter samples (Table 18.3). The overall dominating fungal genera on Italian, French and German litter were primarily *Chalara*, constituting 40, 25 and 27% of the total number of isolates, second *Aureobasidium* and *Cladosporium*; on German litter *Mucor* and on Danish litter *Alternaria* were initially also among the most frequent fungal genera, and increasing frequency of sterile mycelia at all sites (Table 18.6).

The cellulase activity was low in March and July 1997 and almost at the same level in the four different litters transplanted to Collelongo (Table 18.7). The (cellulase) activity was highest in the Italian litter at each sampling during the 2 years of incu-

Table 18.5. Simpson's and Shannon's indices calculated for each sampling and the two experiments, transplant and transect. t_1–t_5 as in Table 18.3. Values of Simpson index near 1 shows high diversity while high Shannon values show a more even distribution of isolates among the genera (species) present

		Transplant experiment		Transect experiment	
		Simpson index	Shannon index	Simpson index	Shannon index
Col I	t_1	0.37	1.21	0.37	1.21
	t_2	0.28	1.46	0.28	1.46
	t_3	0.30	1.52	0.30	1.52
	t_4	0.26	1.63	0.26	1.63
	t_5	0.24	1.83	0.24	1.83
AuF F	t_1	0.28	1.54	0.79	0.51
	t_2	0.23	1.86	0.45	1.08
	t_3	0.15	2.05	0.17	1.92
	t_4	0.21	1.79	0.14	2.09
	t_5	0.27	1.72	0.17	1.89
Sch DE	t_1	0.25	1.69	0.44	1.29
	t_2	0.19	1.82	0.25	1.74
	t_3	0.16	2.17	0.14	2.21
	t_4	0.11	2.47	0.16	2.10
	t_5	0.20	1.89	0.19	1.87
Sor DK	t_1	0.18	1.86	0.50	1.21
	t_2	0.26	1.61	0.16	2.05
	t_3	0.49	1.28	0.16	2.08
	t_4	0.29	1.66	0.19	2.04
	t_5	0.23	1.84	0.18	1.94

bation except in August 1998. The dominating fungal flora from these first two samplings was composed of non-cellulolytic fungi; the most frequent genus, *Chalara*, is a strong proteolytic microfungus. At the two subsequent samplings in November 1997, after 1 year of incubation, and in March 1998 the cellulolytic activity increased except in the German litter in November and in the Danish litter in March. During the same period there was a shift in the fungal flora towards a composition with more cellulolytic fungi as *Penicillium, Alternaria, Trichoderma* and *Acremonium*, *Penicillium* and *Cladosporium* were the most frequent in all four litter types, *Mortierella* appeared in the German and the Danish litter, *Trichoderma* only in the Danish.

No major differences in composition of the fungal flora (biodiversity) were observed on the beech litter from four European beech forests transplanted to the Italian site. The most frequent fungi were almost the same, shifting from non-cellulolytic strains to a dominance of cellulolytic after 1 year of incubation. However, when litterbags with leaves from France, Germany and Denmark were placed in the Italian litter layer, some specific fungal genera appeared, *Phoma* only on the German litter and *Trichoderma* only on the Danish. These new genera did not appear later

Table 18.6. The percentage of sterile mycelia at each sampling in the transplant and transect experiments. t_1–t_5 as in Table 18.3

		Transplant	Transect
Col I	t_1	18.2	18.2
	t_2	42.4	42.4
	t_3	25.4	25.4
	t_4	26.7	26.7
	t_5	26.4	26.4
AuF F	t_1	33.8	14.5
	t_2	33.3	15.3
	t_3	46.0	37.5
	t_4	42.4	44.3
	t_5	13.8	20.9
Sch DE	t_1	24.2	0.0
	t_2	48.3	15.3
	t_3	3.3	26.6
	t_4	35.5	17.2
	t_5	18.1	16.9
Sor DK	t_1	32.8	10.8
	t_2	39.7	20.3
	t_3	44.4	39.1
	t_4	28.0	23.8
	t_5	22.2	23.9

and the composition of the fungal flora was not completely similar from sample to sample. The Italian litter incubated in Italy showed slightly higher cellulase activity than litter from France, Germany and Denmark, virtually in all samplings during the 2 years of decomposition.

18.3.3 Transect Experiment

When the Italian litter was incubated in France, Germany and Denmark, the dominating fungal genus was also *Chalara* (sometimes more than 50% of the isolates) at all sites at the samplings in March and July 4 and 7 months after litterfall, Table 18.4. As already mentioned above, this fungus is not cellulolytic, and cellulase activity was correspondingly low in March, increasing during the next one and a half years more at the sites outside Italy, where the activity remained virtually at the same level as in July 1997.

After July 1997 (271 days after litterfall) *Chalara* was only rarely isolated from the litter, and was absent after August 1998; *Aureobasidium* was much less frequent than *Chalara* but also disappeared after the first two samplings. These two fungal genera were present in the initial low-cellulase period while the group of cellulolytic fungi was invading the litter after 9 months. Important genera are *Penicillium*, *Acremonium*, *Fusarium*, *Mortierella* and *Trichoderma* in the Danish litter; *Cladosporium* in the

Table 18.7. Cellulase activity (nmoles 4-MU g^{-1} leaf dw $hour^{-1}$) on beech leaves at each sampling in the transplant and transect experiment. t_1, t_2, t_3 = March, July, November 1997; t_4, t_5, t_6 = March, August, October 1998

		Transplant	SE ($n = 9$)	Transect	SE ($n = 9$)
Col I	t_1	18	3	18	3
	t_2	26	4	26	4
	t_3	62	7	62	7
	t_4	72	13	72	13
	t_5	43	8	43	8
	t_6	68	5	68	5
AuF F	t_1	9	3	15	3
	t_2	20	6	50	10
	t_3	44	13	64	13
	t_4	54	6	59	10
	t_5	79	8	87	7
	t_6	58	4	89	9
Sch DE	t_1	16	3	14	1
	t_2	25	6	62	11
	t_3	27	4	38	4
	t_4	76	15	86	12
	t_5	57	4	118	38
	t_6	56	6	83	7
Sor DK	t_1	21	3	15	3
	t_2	20	4	52	11
	t_3	49	12	86	13
	t_4	27	5	91	19
	t_5	36	2	123	39
	t_6	52	2	122	9

Italian, French and German litter in March and August 1998. The lower cellulolytic activity on the Italian site was not caused by the composition of the flora, but must be seen as a consequence of the climatic conditions, dry, cool summers and snow for 6 months during the winter. The same low activity was seen when other European litters were incubated in Collelongo, the transplant experiment.

The fungal flora on the Italian litter remained almost the same at all sites during the first 9 months of decomposition, thereafter the cellulolytic activity increased, caused by growth of *Penicillium* and *Cladosporium* in Italy and Germany, and a wide variety of dominating genera, *Penicillium, Acremonium, Trichoderma, Fusarium* and *Mortierella* in Denmark, and at all sites a high occurrence of sterile mycelia (Table 18.6).

18.3.4 Diversity Indices

In order to be able to compare the fungal communities at different stages of decomposition of the beech litter two diversity indices were used, the Simpson index indicating the fungal diversity and the Shannon index measuring fungal community evenness (Table 18.5). Both indices are calculated on the basis of genera only as the isolates were not identified to species level. In most cases only one species of each genus was observed in a separate sample. The sterile and unidentified strains were not included in the calculations. In a similar study Houston et al. (1998) found that results from calculations including or excluding unidentified strains varied only slightly and also differences between index calculations using species or genera were very small.

18.3.4.1 Transect

When the Italian litter was placed at the other European sites the Simpson diversity index initially showed a higher value – one or a few dominating species (e.g. *Chalara*) – but the index value decreased faster at these sites to lower values than at the Italian site. This is paralleled by an increase in the Shannon evenness index (except for the last sample in France and Germany). It seems that the changes in diversity were highest at the French site at Aubure. The diversity changes are larger than in the transplant experiment.

18.3.4.2 Transplant

In the transplant experiment the initial diversity values were comparatively low and were little changed during the 2 years of decomposition period. The transplanted litter from Sorø responded in a different way and had a very low value of the Simpson index after the first 4 months, increasing during the first year and remained at a rather high level even after 2 years. The changes in values of the diversity indices from sample to sample can easily be traced to the prominence of particular strains (cf. Tables 18.3 and 18.4). In Table 18.4 it can be seen that the dominating *Chalara* strain on the litter disappeared after approximately 7 months, resulting in a rapid decrease in the Simpson index at all sites.

An earlier examination of the fungal succession on beech litter from similar beech forest in Denmark is provided in Table 18.8 for comparison with results in the present study. The main part of the fungi isolated are the same as in the CANIF experiments but with some prominent exceptions. *Cladosporium* was isolated only in the first 6 months of decomposition here, while in the CANIF experiment both litter trans-

Table 18.8. Occurrence of the most frequent fungal genera during decomposition of beech litter in a Danish forest expressed as a percentage of the total number of isolates in each sample. t_1–t_5 as in Table 18.3. (Chry = *Chrysosporium*, Het = *Heteroconium*, Psf = *Psudofusarium*, for other abbreviations see Table 18.1)

t_1		t_2		t_3		t_4		t_5	
Au	15.0	Acr	15.2	Tric	25.0	Mort	18.1	Mort	28.8
Het	13.8	Au	10.1	Mort	6.5	Chry	4.3	Pen	26.4
Cl	11.3	Cl	6.1	Acr	5.6	Acr	3.2	Tric	17.7
Psf	5.0								

planted to Italy and Italian litter at all sites including Sorø had a certain incidence of *Cladosporium* in the analyses throughout the experiment; all samples of Sorø litter incubated in Italy contained *Cladosporium*. The same pattern of occurrence applies to *Aureobasidium*. These changes in diversity are included in the calculated diversity indices; but the question of interest for the discussion of diversity in relation to function is which mechanisms are causing the changes. Concerning *Cladosporium*, it was previously shown that this genus contributes especially to pectin decomposition - in Danish forests (Kjøller and Struwe 1980). The new pattern points to different properties of the Italian litter and also environmental factors may be important, e.g. the importance of temperature for enzyme activity and optimal decomposition. *Mortierella* has previously been identified as an active chitin decomposer occurring in the later part of decomposition (Kjøller and Struwe 1990). This pattern was confirmed in the transplant experiment and *Mortierella* was only isolated late after 15-19 months. The Italian litter in situ was also inhabited by *Mortierella* but only once; when placed along the transect *Mortierella* appeared in all samples. This unexpected distribution indicates a fundamental difference in conditions for decomposition, whether related to substrate quality or environmental factors.

18.4 Functional Diversity of Bacteria in the Litter of Coniferous Forests

18.4.1 Approaches to Investigating Functional Diversity of Bacteria

The investigation was carried out with material sampled at the coniferous sites in Aubure (France, AuP), Waldstein (Germany, Wal), Skogaby (southern Sweden, Sko) and Åheden (northern Sweden, Åhe; see Table 17.1 in Wolters et al. Chap. 17, this Vol., for a description of sites). The results reported here are part of the transect study described in Chapter 17 (this Vol.; sampling scheme see Chap. 17.2, this Vol.). They include data from three sampling times: May/June 1997, September 1997, March/April 1998. Effects of time are not considered, because phenological differences in the functional diversity of bacteria are beyond the scope of this chapter.

Functional diversity of the soil bacteria was determined using the Biolog method (Garland and Mills 1991). The Biolog GN microplates (Biolog Inc., Hayward, CA) used in our study contain 96 wells preloaded with a buffered nutrient medium and a tetrazolium violet redox dye (Bochner 1989; Biolog 1993). Each of the 95 wells additionally contains a different carbon source, the remaining well is left without a carbon source to function as a control. Inoculum was gained by diluting an aliquot of 5 g fresh weight from each sample in 45 ml sterile tetra natriumdiphosphatdecahydrat (0.18%). After extracting the microorganisms by shaking the solution for 30 min at 4 °C, the cell suspension was attenuated with sterile NaCl solution (0.85%) to a mean inoculum density of 10^6–10^7 cells ml^{-1}. Cell density was determined by counting the bacteria in a Thoma chamber (microscope magnification: 400×). Each well of the Biolog plates was inoculated with 100 µl cell suspension. Five to eight plates per sampling time and site were inoculated. The plates were incubated for 48 h at 28 °C. Absorption was measured with an E_{max} microplate reader at 590 nm using the Soft Max Pro (Molecular Devices Corp.) software.

Following Zak et al. (1994), the 95 different carbon sources were assorted to six substrate groups: carbohydrates (CH; $n = 28$), carboxylic acids (CA, $n = 24$), amides and amines (AM, $n = 6$), amino acids (AA, $n = 20$), polymers (PO, $n = 5$) and miscellaneous (MI, $n = 12$). The average well colour development (AWCD) was calculated as a measure of Biolog measurable metabolic activity either for individual substrate groups (AWCD$_S$) or for the mean of all substrates (AWCD$_T$). Functional diversity was measured as substrate richness (S), substrate diversity (H), and substrate evenness (E; Zak et al. 1994). S is the number of different substrates that are used by the bacterial community. H encompasses both substrate richness and the intensity of substrate use and was quantified according to information theory (Magurran 1988). Substrate evenness (E) measures the equitability of activity across all utilised substrates. In addition, several abiotic and microbiological parameters (water content, pH, C and N content of the organic layer, CO_2 evolution, microbial biomass, bacterial and fungal biomass, number of bacteria, number of dividing bacterial cells) were determined (see Wolters et al. Chap. 17, this Vol., for methods).

Statistical treatment was performed using the Statistica software package. Site specific differences in substrate diversity, substrate evenness and substrate richness were tested by means of the Kruskal-Wallis test. The effect of the factors site and substrate group on AWCD was tested by means of a two-way Anova. Homogeneity of variances was tested using Sen and Puri's non-parametric test. If necessary, data were transformed adequately. Differences between means were tested with the Tukey HSD test. Cluster analysis was performed to analyse the functional structure of the bacterial community. In addition, correlation analysis (Pearson-product-moment correlation) was performed using Biolog parameters as dependent and both abiotic and microbial parameters as independent variables.

18.4.2 Substrate-Specific Metabolic Activity

On average, 84% of all substrates offered were decomposed by the microorganisms extracted for the Biolog analyses. The significant main effect of the factor substrate group revealed by the Anova (Table 18.9) reflects the decrease of substrate specific metabolic activity (AWCD$_S$) from carbohydrates (CH) to amino acids (AM, Fig. 18.1A). The small part of total variance explained by this factor points to the strong variability of Biolog measurable substrate use. A significant difference between means could thus only be established for CA and AM (Fig. 18.1A). A slightly different picture emerges from the cluster analysis (Fig. 18.1B). Two functional clusters were identified, separating CA (carboxylic acids) and CH from all other substrates. In addition, a particularly close association was revealed for AWCD$_S$ of MI (miscellaneous) and AA (amino acids).

Kreitz and Anderson (1997) argue that the decomposition of freshly fallen litter is retarded in acidic soils, because the ability of bacteria to degrade carboxylic acids and other C sources is strongly reduced. Our investigations, in contrast, indicate a particularly high activity of functional groups metabolising carboxylic acids and carbohydrates, while the microbial use of amides and amines seems to be retarded. We do not have an explanation for this apparent contradiction, except that the investigations of Kreitz and Anderson (1997) were carried out with mineral soil, while our study only included pure organic matter. Therefore it is possible that the two studies focused on different components of the decomposer community, because the organic matter accumulating in mineral soils generally is strongly humified and often protected in

Diversity and Role of Microorganisms

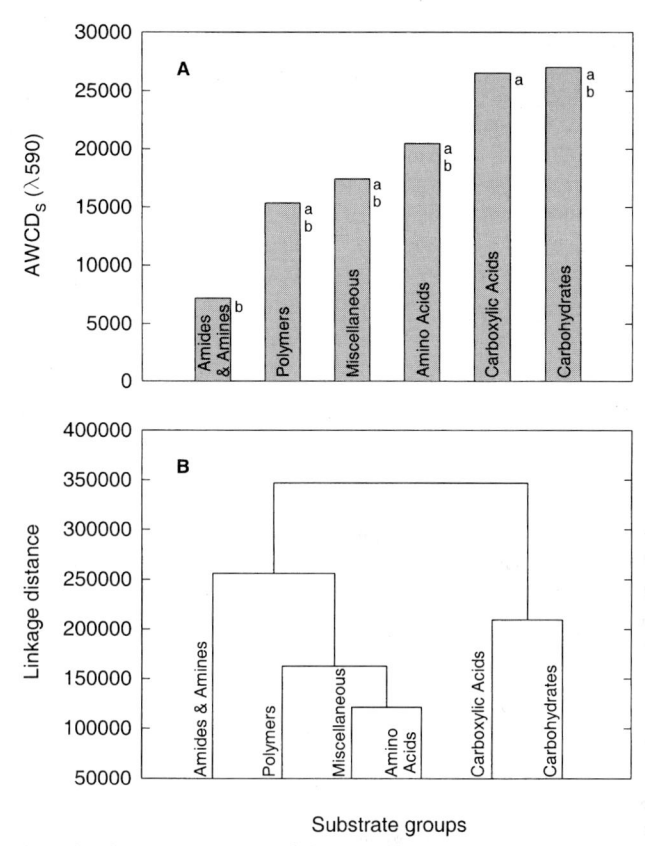

Fig. 18.1. **A** Average well colour development (AWCD$_S$) of the six substrate groups; $n = 72$. *Bars represent standard deviations of means. Values with the same letter* are not significantly different (Tukey HSD test). **B** Result of the cluster analysis on the AWCD$_S$ of the six substrate groups. *CH* Carbohydrates; *CA* carboxylic acids; *AA* amino acids; *AM* amines and amides; *PO* = polymers; *MI* miscellaneous

aggregates, while organic layers usually contain a wide array of free organic substrates at all stages of decomposition.

Most interestingly, the interaction between the two factors substrate group and site was not significant (Table 18.9), suggesting that intrasite differences in bacterial substrate use were more pronounced than intersite differences. It can thus be concluded that the activity pattern of functional groups shown in Fig. 18.1A was strikingly similar at all sites. This is consistent with the finding that microbial communities from similar habitats are similar in functional potential despite strong differences in geographic location (Goodfriend 1998). It confirms that specific substrate conditions in the organic layer of coniferous forests set such strong restrictions to certain components of the decomposer foodweb that other environmental factors are overruled (see Wolters et al. Chap. 17, this Vol.).

Table 18.9. Results of the two-way ANOVA on the effects of the factors site and substrate group on the average well colour development ($n = 72$; df: degrees of freedom, MS: mean squares)

	df effect	MS effect	F	p level[a]
Substrate group	5	3.41	2.63	*
Site	3	24.60	19.01	***
Interaction	15	0.05	0.04	n.s.

[a] p levels: n.s. > 0.05, * < 0.05, ** < 0.01, *** < 0.001.

18.4.3 Site-Specific Differences in Metabolic Activity

The main effect of the factor site on mean metabolic activity ($AWCD_T$) was highly significant and explained the largest part of total variance (Table 18.9). This confirms that the Biolog technique is a useful approach for measuring site-specific differences in the functional structure of microbial communities. Though $AWCD_T$ was low at both Swedish sites, significant differences between means could only be established for the site in southern Sweden compared to all other sites (Fig. 18.2).

According to the results of the correlation analysis, substrate humidity, pH and C content of the organic layer are the major environmental factors determining the Biolog measurable component of the microbial community at the sites investigated (Table 18.10). The water content was positively correlated to $AWCD_T$ and to $AWCD_S$ of several functional groups. This highlights the sensitivity of bacteria to changes in substrate humidity (Paul and Clark 1989). The negative correlation between pH and $AWCD_T$ and $AWCD_S$ of various functional groups seems to contradict the results of other authors reporting a positive correlation between pH and Biolog data (Kreitz and Anderson 1997; Staddon et al. 1998). It should be noted, however, that the relationship found in our study only applies to a very narrow range of low pH values provided by the acidic substrates included in our study. Microbial parameters have been shown to be highly variable in acid forest soils (Joergensen et al. 1995).

18.4.4 Diversity Parameters

Both substrate diversity and substrate evenness were positively correlated to the frequency of dividing bacterial cells (Table 18.10). This suggests that the active part of the bacterial community is adequately covered by the Biolog approach. Accompanying investigations show that this component of the soil microflora may be more important for the microtrophic foodweb at the sites investigated than is indicated by its low biomass (see Wolters et al. Chap. 17, this Vol.). However, no significant correlation between Biolog data and any other of the microbiological parameters was found. This emphasises again that the interpretation of Biolog results should be confined to bacteria.

Diversity and Role of Microorganisms 397

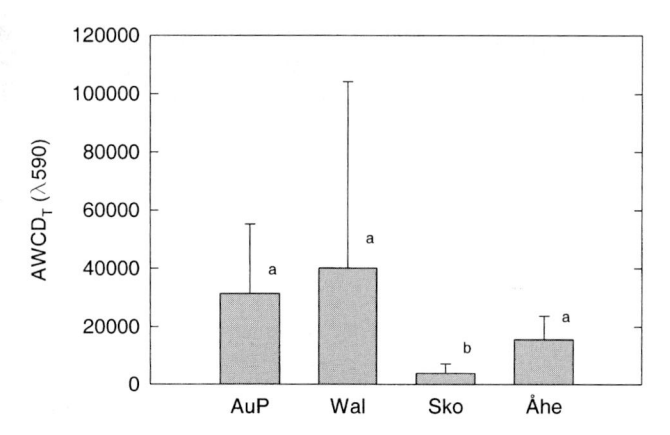

Fig. 18.2. Site effect on the average well colour development ($AWCD_T$) of all substrates; $n = 72$. *Bars* represent standard deviations of means. Values with the same letter are not significantly different (Tukey HSD test). *AuP* Aubure; *Wal* Waldstein; *Sko* Skogaby; *Åhe* = Åheden

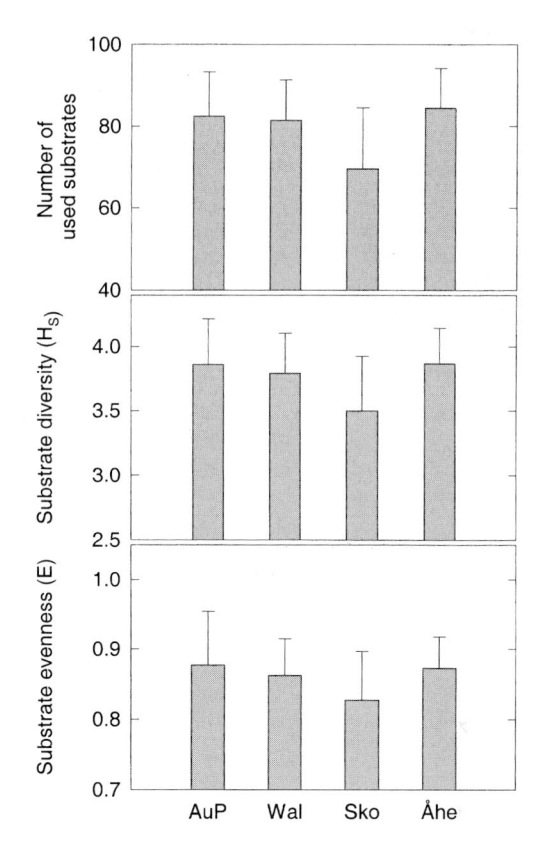

Fig. 18.3. Site effect on the parameters substrate richness, substrate diversity and substrate evenness (n = 82). *Bars* represent standard deviations of means. *AuP* Aubure; *Wal* Waldstein; *Sko* Skogaby; *Åhe* = Åheden

Table 18.10. Significant results of the correlation analysis. Variables marked with an asterisk were measured only at the last sampling time; for abbreviations see Section 18.3.2

	r	p level[a]	n
Water content (% dw)			
AWCD$_T$	0.24	*	82
CA	0.26	*	82
AA	0.25	*	82
AM	0.26	*	82
Substrate richness	0.25	*	81
Substrate diversity	0.28	*	80
Substrate evenness	0.24	*	80
pH$_{H2O}$			
AWCD$_T$	−0.38	**	82
CH	−0.38	***	82
CA	−0.36	**	82
AA	−0.37	**	82
MI	−0.31	**	82
PO	−0.38	**	82
AM	−0.37	**	82
C content of organic layer (% dw)*			
AWCD$_T$	−0.52	**	32
CH	−0.40	*	32
CA	−0.57	**	32
AA	−0.50	**	32
MI	−0.48	**	32
AM	−0.57	**	32
Substrate evenness	0.44	*	31
Frequency of dividing bacterial cells (%)*			
Substrate diversity	0.44	*	31
Substrate evenness	0.54	**	31

[a] p levels: n.s. > 0.05, * < 0.05, ** < 0.01, *** < 0.001.

The Kruskal-Wallis test revealed a significant effect of site on substrate richness, substrate diversity, and substrate evenness. This confirms that the Biolog approach is a sensible tool for comparing the functional structure of bacterial communities of different habitats (e.g. Goodfriend 1998). Staddon et al. (1998) related decreasing functional diversity of the microflora with increasing latitude to increasing harshness of soil environments from south to north (i.e. decreased productivity, nutrient limitation and higher acidity). This fits very well to the conclusions drawn from our own investigations on the decomposer foodweb (see Wolters et al. Chap. 17, this Vol.). We thus expected a similar trend in our Biolog data. However, this is not supported by our results. While the functional diversity of the bacteria at the northernmost site was very similar to the sites in Germany and France, that of the bacteria at the southern Swedish site was remarkably low (Fig. 18.3).

Diversity and Role of Microorganisms

The similarity of the sites in northern Sweden, Germany and France is probably due to the fact that the latitudinal gradient was partly offset by the altitudinal gradient (cf. Table 17.1 in Wolters et al. Chap. 17, this Vol.). However, the low functional diversity at the southern Swedish site is surprising. The results of the correlation analyses suggest that specific climatic conditions are prime factors for explaining these unexpected results (Table 18.10). The positive correlation between water content and all parameters of functional diversity points again to the dominating role of substrate humidity in determining bacterial communities in soil. While the C content of the organic layer was positively correlated to substrate evenness, it was negatively correlated to $AWCD_T$ and various $AWCD_S$s. This suggests that the depression of several functional groups in spots of comparatively high C content causes a more even distribution in substrate use. A possible explanation is that the accumulation of recalcitrant substrates leads to an impoverishment of the microbial community. This is consistent with the finding that some specialisation of function occurs among microorganisms with increasing recalcitrance of SOM during later stages of the decomposition process (Ladd et al. 1996). We conclude that low functional diversity of bacteria at the southern Swedish site might be explained by comparatively mild climatic conditions, leading to a very specific combination of substrate humidity, slightly increased pH and high C content of the organic layer. Future studies have to show if this result is partly due to the extraordinarily dry conditions in 1997. The fact that low functional diversity of bacteria at the southern Swedish site coincides with low metabolic activity supports the hypothesis that an impoverishment of soil communities could adversely affect decomposition processes.

18.5 Conclusions

18.5.1 Microfungi in Four Beech Forest Sites

- The composition of the fungal flora on the litter from the Italian site (Collelongo) was different from the litter floras from the sites in Aubure (F), Schacht (DE) and Sorø (DK). However, this is a statement with modifications, as the litter was investigated only 4 months after litterfall and the transplanted litter from France, Germany and Denmark was already invaded by the dominant fungus in Italy *Chalara*. When the Italian litter was incubated at the other sites in Europe, *Chalara* still remained, but the litter was rapidly colonized by the local fungal flora.
- The transect experiment showed that there was a higher cellulase activity when the standard Italian litter was placed on the other sites. This could be interpreted as adverse conditions in Italy and there was consequently a faster decomposition at the northern sites. The adverse conditions at the Italian site include snow cover lasting until May in addition to hot and very dry summers. Consequently, the litters from Aubure, Schacht and Sorø showed the same cellulase activity as the Italian litter when transplanted to Italy but with a lower diversity than the native litter in most cases; the environmental conditions allowed only some of the fungi to develop in Italy. On the contrary, the fungal diversity increased initially in the Italian litter deposited along the transect in Europe but also decreased more rapidly with time.
- The diversity changes can partly be related to changes in function and there are several examples of functional substitution at different stages of decomposition of the various litter at the European sites.

18.5.2 Functional Diversity of Bacteria in Four Coniferous Sites

- The Biolog technique usefully complements other approaches to the analysis of soil communities by focusing on the active bacterial component of the microflora (cf. Wolters et al. Chap. 17, this Vol.). It is a useful approach for measuring both differences in bacterial use of different C sources and site specific differences in the functional structure of bacterial communities.
- Under the conditions of the sites selected for our study, the functional structure of the bacterial component of the microflora was very similar over a wide geographical range. This is probably due to a strong selective force excerted by specific features of the organic substrates accumulating in coniferous forests and to the fact that the latitudinal gradient was partly offset by the altitudinal gradient.
- The results from the southern Swedish site clearly prove that regional conditions can strongly modulate the microbial community. Low diversity and activity of the Biolog measurable microflora at this site can be explained by the fact that macroclimatic conditions lead to a very specific combination of environmental factors affecting the performance of bacteria colonising the organic layer.
- The combination of low functional diversity and low metabolic activity of bacteria at the southern Swedish site suggests that an impoverishment of soil communities could adversely affect decomposition processes.

References

Biolog (1993) Instructions for use of Biolog GP and GN microplates. Biolog, Hayward, California

Bochner B (1989) Breathprints at the microbial level. ASM News 55:536–539

Bossio DA, Scow KM (1995) Impact of carbon and flooding on the metabolic diversity of microbial communities in soils. Appl Environ Microbiol 61:4043–4050

Buyer JS, Drinkwater LE (1997) Comparison of substrate utilisation assay and fatty acid analysis of soil microbial communities. J Microbiol Methods 30:3–11

Degens BP (1998) Decreases in microbial functional diversity do not result in corresponding changes in decomposition under different moisture conditions. Soil Biol Biochem 30:1989–2000

Degens BP, Harris JA (1997) Development of a physiological approach to measuring the catabolic diversity of soil microbial communities. Soil Biol Biochem 29:1309–1320

Fenchel T, Esteban GF, Finlay BJ (1997) Local versus global diversity of microorganisms: cryptic diversity of ciliated protozoa. Oikos 80:220–225

Finlay BJ, Maberly SC, Cooper JI (1997) Microbial diversity and ecosystem function. Oikos 80:209–213

Frankland JC (1998) Fungal succession – unravelling the unpredictable. Mycol Res 102:1–15

Freckman DW (1994) Life in the soil. Soil biodiversity: its importance to ecosystem processes. Report of a Workshop held at the Natural History Museum, London, Aug 30 – Sept 1, 1994, 26 pp

Garland JL, Mills AL (1991) Classification and characterization of heterotrophic microbial communities on the basis of patterns of community-level sole-carbon-source utilisation. Appl Environ Microbiol 57:2351–2359

Goodfriend WL (1998) Microbial community patterns of potential substrate utilisation: a comparison of salt marsh, sand dune and seawater-irrigated agronomic systems. Soil Biol Biochem 30:1169–1176

Haack SK, Garchow H, Klug MJ, Forney LJ (1995) Analysis of factors affecting the accuracy, reproducibility, and interpretation of microbial community carbon source utilisation patterns. Appl Environ Microbiol 61:1458–1468

Hall GS (1996) Preface. In: Hall GS (ed) Methods for the examination of organismal diversity in soils and sediments. CAB International, Wallingford

Houston APC, Visser S, Lautenschlager RA (1998) Microbial processes and fungal community structure in soils from clear-cut and unharvested areas of two mixedwood forests. Can J Bot 76:630–640

Ibekwe AM, Kennedy AC (1998) Phospholipid fatty acid profiles and carbon utilisation patterns for analysis of microbial community structure under field and greenhouse conditions. FEMS Microbiol Ecol 26:151–163

Joergensen RG, Anderson TH, Wolters V (1995) Carbon and nitrogen relationships in the microbial biomass of soils in beech (*Fagus sylvatica* L.) forests. Biol Fertil Soils 19:141–147

Johnson D, Leake JR, Lee JA, Campbell CD (1998) Changes in soil microbial biomass and microbial activities in response to 7 years simulated pollutant nitrogen deposition on a heathland and two grasslands. Environ Pollut 103:239–250

Kjøller A, Struwe S (1980) Microfungi of decomposing red alder leaves and their substrate utilisation. Soil Biol Biochem 12:425–431

Kjøller A, Struwe S (1990) Decomposition of beech litter. A comparison of fungi isolated on nutrient rich and nutrient poor media. Trans Mycol Soc Jpn 31:5–16

Kjøller A, Struwe S (1992) Functional groups of microfungi in decomposition. In: Carroll GC, Wicklow DT (eds) The fungal community. Marcel Dekker, New York pp 619–630

Kreitz S, Anderson T-H (1997) Substrate utilisation patterns of extractable and non-extractable bacterial fractions in neutral and acidic beech forest soils. In: Insam H, Rangger A (eds) Microbial communities. Functional versus structural approaches. Springer Berlin Heidelberg New York, pp 149–160

Ladd JN, Foster RC, Nannipieri P, Oades JM (1996) Soil structure and biological activity. In: Stotzky G, Bollag J-M (eds) Soil biochemistry vol 9. Marcel Dekker, New York, pp 23–77

Lee JA, Caporn SJM (1998) Ecological effects of atmospheric reactive nitrogen deposition on semi-natural terrestrial ecosystems. New Phytol 139:127–134

Lindstrom JE, Barry RP, Braddock JF (1998) Microbial community analysis: a kinetic approach to constructing potential C source utilisation patterns. Soil Biol Biochem 30:231–239

Magurran AE (1988) Ecological diversity and its measurement. Princeton University Press, Princeton, pp 179

Miller M, Palojärvi A, Rangger A, Reeslev M, Kjøller A (1998) The use of fluorogenic substrates to measure fungal presence and activity in soil. Appl Environ Microbiol 64:613–617

Miller SL (1995) Functional diversity in fungi. Can J Bot 73 (Suppl 1):50–57

Møller J, Miller M, Kjøller A (1999) Fungal-bacterial interaction on beech leaves; influence on decomposition and dissolved organic carbon quality. Boil Biol Biochem 31:367–374

Palleroni NJ (1997) Prokaryotic diversity and the importance of culturing. Antonie van Leeuwenhoek Int J Gen Mol Microbiol 72:3–19

Paul EA, Clark FE (1989) Soil microbiology and biochemistry. Academic Press, San Diego, pp 273

Sinsabaugh RL, Antibus RK, Linkins AE (1991) An enzyme approach to the analysis of microbial activity during plant litter decomposition. Agric Ecosyst Environ 34:43–54

Smalla K, Wachtendorf U, Heuer H, Liu W-T, Forney L (1998) Analysis of Biolog GN substrate utilisation patterns by microbial communities. Appl Environ Microbiol 64:1220–1225

Staddon WJJ, Trevors JT, Duchesne LC, Colombo CA (1998) Soil microbial diversity and community structure across a climatic gradient in western Canada. Biodivers Conserv 7:1081–1092

Timonen S, Jørgensen KS, Haahtela K, Sen R (1998) Bacterial community structure at defined locations of *Pinus sylvestris* – *Suillus bovinus* and *Pinus sylvestris* – *Paxillus involutus* mycorrhizospheres in dry pine forest humus and nursery peat. Can J Microbiol 44:499–513

Torsvik V, Salte K, Sørheim R, Goksøyr J (1990) Comparison of phenotypic diversity and DNA heterogeneity in a population of soil bateria. Appl Environ Microbiol 56:776–781
Wolters V (ed) (1997) Functional implications of biodiversity in soil. Ecosyst Res Rep 24:133
Wünsche L, Brüggemann L, Babel W (1995) Determination of substrate utilisation patterns of soil microbial communities: an approach to assess population changes after hydrocarbon pollution. FEMS Microbiol Ecol 17:295–306
Zak JC, Visser S (1996) An appraisal of soil fungal biodiversity: the crossroads betwIen taxonomic and functional biodiversity. Biodivers Conserv 5:169–183
Zak JC, Willig MR, Moorhead DL, Wildman HG (1994) Functional diversity of microbial communities: a quantitative approach. Soil Biol Biochem 26:1101–1108

Part E
Integration

19 Spatial Variability and Long-Term Trends in Mass Balance of N and S in Central European Forested Catchments

E. Dambrine, A. Probst, D. Viville, P. Biron, M.C. Belgrand,
T. Paces, M. Novak, F. Buzek, J. Cerny, and H. Groscheova

19.1 Introduction

Acid deposition linked to SO_2 emissions has provoked a large-scale acidification of soils and waters in Europe. Losses of nutrients caused by soil acidification, enhancement of forest growth by nitrogen deposition, and climatic stress were determinant factors of the large scale spruce decline in central Europe. Beginning in the mid-1970s after the petrol crisis in western Europe, and more recently following the political and economic changes in Eastern Europe, a large reduction in SO_2 emissions and acid deposition has occurred over Europe. Simultaneously, the deposition of nitrogen compounds has remained stable or increased. These changes in S and N deposition have been often reflected by nitrate and sulphate concentrations in runoff. However, internal soil processes may interact in such a way that the trends in deposition may not be seen in the stream clearly (Durka et al. 1994). The relative influence of these processes is difficult to quantify as there has been only few long-term investigations integrating both the stand spatial variability and the role of recharge areas. This chapter focuses on the mean and long-term fate of S and N inputs at the catchment scale and present comparative budgets at the forest stand and whole-catchment scales. The comparison reflects the environmental changes along the west east cross-section through central Europe.

19.2 Approaches to Studying Long-Term Changes in Watersheds

One French catchment and two Czech catchments receiving different loads of acid deposition have been monitored over 10 to 25 years. The basic characteristics of the three catchments are summarised in Table 19.1. The small (80 ha) Strengbach catchment at Aubure is located on the eastern ridge of the Vosges Massif. This region receives the highest load of acid deposition in France (see Landmann and Bonneau 1995). Acid sandy soils are developed from a base-poor granite. The catchment vegetation is composed of 45% of the first generation (>80-year-old) Norway spruce [*Picea abies* (L) Karst], 23% of the first or second generation young spruce stand (<45 years), 17% old beech (*Fagus sylvatica*) (mixed with a minor proportion of old fir).

Water and mineral fluxes and budgets were measured at three forest stands – an old declining spruce (P90) and young spruce (P45) located side by side on the south-facing slope, and an old beech stand (F150) on the north-facing slope (Fichter et al. 1998) – and at the catchment scale (Probst et al. 1995).

The catchment Jezeri (X-16) is located in the NW part of the Krusne hory (Erzgebirge) Mountains near the city of Most. It is located on the SE-oriented slope facing

Table 19.1. Major characteristics of three small forested catchments in the Czech Republic and France

No.	X-8	X-16	Strengbach
Location	Salacova Lhota	Jezeri	Aubure
Region	Rural	Industrial	Rural
	Bohemian-Moravian Upland	Krusne hory Mts.	Vosges Mts.
Study period	1976–1998	1978–1998	1986–1996
Latitude	49°31′ N	50°33′ N	48°12′ N
Longitude	14°59′ E	13°30′ E	7°11′ E
Minimum altitude (m)	557	480	883
Maximum altitude (m)	744	900	1146
Mean slope (%)	13.3	17.8	22
Area (km^2)	1.68	2.6	0.8
Annual precipitation (mm)	685	812	1357
Annual runoff (mm)	128	431	811
Prevailing bedrock	Biotite-muscovite gneiss, quartzite	Biotite-sillimanite gneiss granite, quartzite	Base-poor granite
Prevailing soil	District to eutric cambisol	District to eutric cambisol	District Cambisol to Podzol
Prevailing vegetation	100% spruce	25% Beech 80 years old	45% Spruce (90 years old)
		7% Spruce 50 years old	23% Spruce (<40 years old)
		65% Young mixed plantation	17% Old beech (>140 years old)
		3% Grassland	15% Grassland + young plantations

one of the most industrialised regions in the country. This catchment represents one of the most acidified and polluted regions in central Europe, called the black triangle (Cerny and Paces 1995). The bedrock in Jezeri is base-poor biotite-muscovite orthogneiss and a coarse-grained two-mica granite. Soils range from Eutric Cambisol (lower part) to Distric Cambisol Gleyic (upper part). The forest before dieback was composed of beech (*Fagus sylvatica*) at elevations between 480 and 750 m asl and spruce (*Picea abies*) at elevations between 750 and 900 m asl. After dieback of spruce due to acid deposition in the second half of the 1970s, the original spruce forest was reduced to small residual stands. Grass, rowan, birch and young spruce (*Picea pungens*) replaced spruce. A neighbouring catchment, X-14, with very similar properties was monitored during the period 1978-1983 (Paces 1985; Cerny and Paces 1995). The following interpretation of the long-term trends in N and S deposition, runoff and accumulation or depletion combines our monitoring of both the catchments X-14 and X-16. This connection is justified by the simultaneous monitoring of both catchments in 1983, yielding similar results. Monitoring and sampling of throughfall and soil solution started in two plots within X-16 in November 1993 representing the remaining spruce stand and the beech stand. The budget in the catchment X-16 (years 1994 to 1998) was calculated as the difference between runoff output and throughfall plus stemflow input weighted by the area of individual types of vegetation (Table 19.1). The catchment budget does not include biological fixation.

The catchment Salacova Lhota (X-8) is located in the eastern part of the Bohemian-Moravian Uplands on the southern slope of the hill Straziste facing a rural countryside without major point sources of industrial pollution near the town of Pacov. The distance between the extremely polluted catchment X-16 and the relatively clean catchment X-8 is 160 km (Fig. 19.1). The bedrock is a base-poor biotite and biotite-sillimanite paragneiss and quartzite. Soil is eutric cambisol to district cambisol, with a large proportion of residual quartzite debris. Vegetation is represented by 40-year-old spruce (*Picea abies*). Bulk precipitation and runoff have been monitored since 1976 (Paces 1985) and throughfall since 1994. Sampling and analytical procedure were the same as in the catchment X-16. Biological fixation has been calculated from growth rates in individual forest departments and our chemical analysis of wood.

19.3 Temporal Variations and Trends

19.3.1 Ambient Concentrations of SO_2 and NO_X (NO_2)

The differences in the concentrations of the acidifying oxides in the three catchments are shown in Fig. 19.2.

The average concentrations of SO_2 in air at X-16, X-8 and Aubure were 90, 20 and 10 $\mu g\,m^{-3}$ by the end of the 1980s. Extreme SO_2 concentrations at X-16 were mostly related to local brown coal combustion. In the beginning of the monitoring period, concentrations were stable at X-16, increasing at X-8, and decreasing at Aubure. During the 1990s, SO_2 concentrations decreased at all sites. The previous differences in the concentrations between the catchments have remained. The average concentrations of NO_2 were 25, 10 and 5 $\mu g\,m^3$ at X-16, X-8 and Aubure by the end of the 1980s. The nitrogen oxide concentrations remained almost stable at X-16 during the monitoring period. The concentrations decreased at Aubure and X-8 to meet similar values.

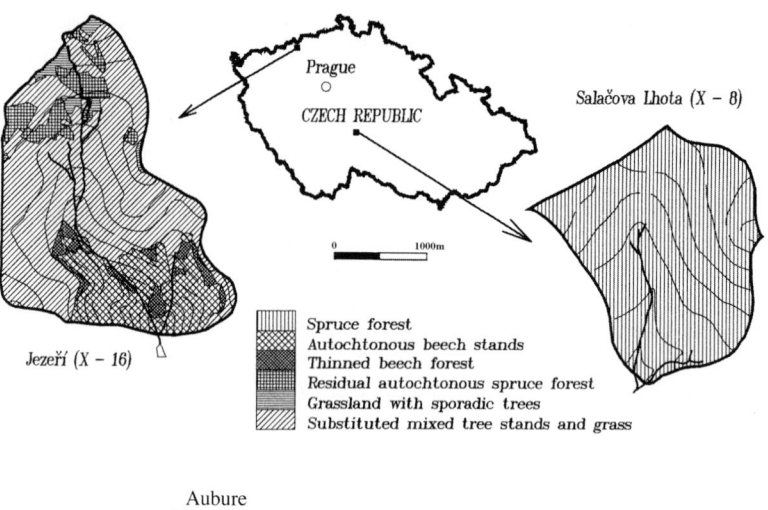

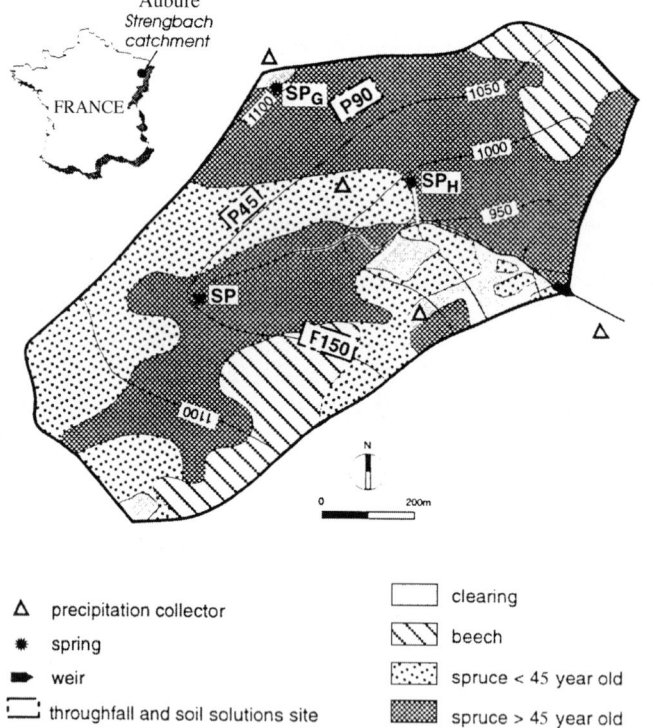

Fig. 19.1. Location and maps of the catchments

Fig. 19.2. Time variations or ambient air SO_2 and NO_2 concentrations ($\mu g\ m^{-3}$)

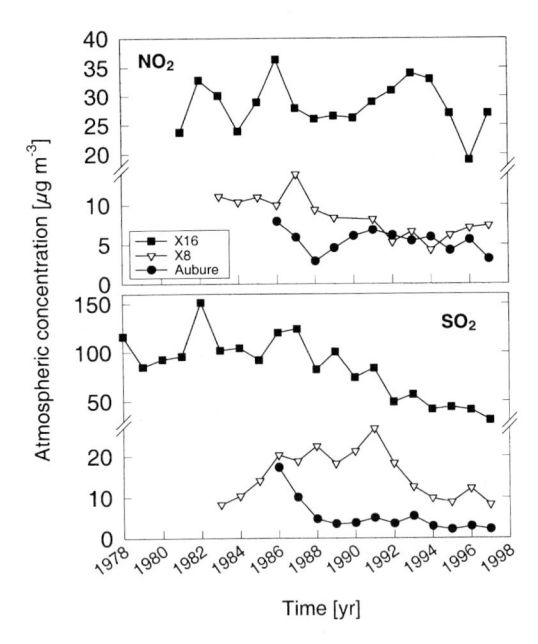

19.3.2 Precipitation and Runoff

The hydrological characteristics and their temporal variations in the three catchments are summarised in Fig. 19.4. Three dry periods can be distinguished from the precipitation amounts: 1976–1978; 1982–1985; 1990–1993. Wet periods were 1980–1981; 1986–1988; 1994–1995. Low drainage was often measured during wet years preceded and followed by dry periods, showing the importance of water storage in the catchments. Annual variations were much larger at Aubure, in relation to higher precipitation. On average, water annual budgets were slightly above 500 mm at X-16 and Aubure and below 500 mm at X-8. The budgets are close to the average mean potential evapotranspiration.

19.3.3 S and N in Bulk Precipitation

Sulphate was the dominant anion at all sites (Fig. 19.3). The average SO_4 concentration over the period 1986–96 changed from west to east from 36 $\mu Eq\ l^{-1}$ at Aubure, 168 $\mu Eq\ l^{-1}$ at X-16 and 94 $\mu Eq\ l^{-1}$ at X-8. Sulphate concentration decreased during the study period in all the catchments. The ratios between mean SO_4 concentrations before 1988 and after 1994 were 1.5, 1.5 and 1.3 for Aubure, X-16 and X-8, respectively.

Using mean annual concentrations, NH_4 and NO_3 in bulk precipitation could be ranked in the order: $NH_{4Aub} = NO_{3Aub} \ll NO_{3X-8} = NO_{3X-16} < NH_{4X-8} < NH_{4X-16}$. Variations of NH_4 and NO_3 concentrations in each site were poorly related except at X-16. The NO_3 concentrations decreased at X-16 and Aubure, while NH_4 increased at Aubure and decreased at X-16 during the period 1986–1995. The NH_4 and NO_3 concentrations were almost stable at X-8.

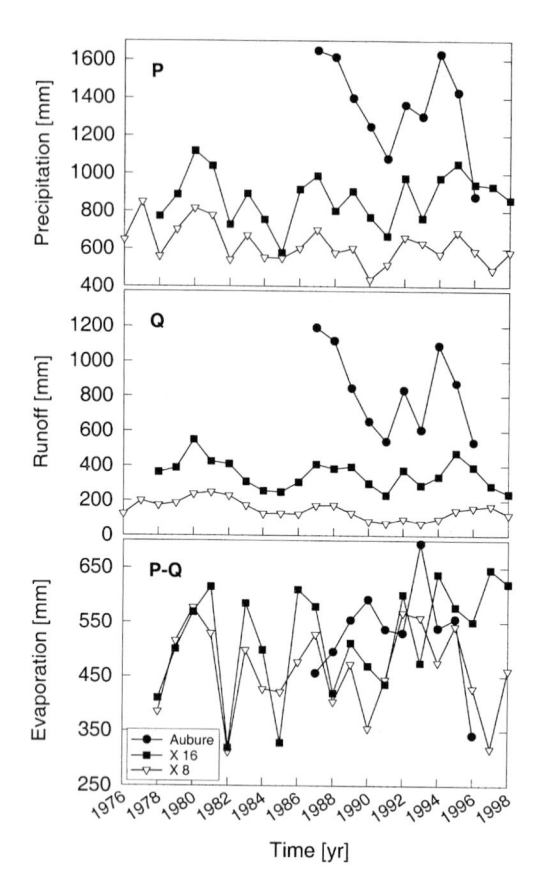

Fig. 19.3. Time variations of precipitation, runoff and hydrological budgets for the three catchments studied

More significant for the evaluation of the temporal changes in the Czech catchments is the comparison of concentrations and fluxes obtained by the monitoring in the most polluted period of 1976 to 1983, prior to the political and economic changes in central Europe (Paces 1985) and the monitoring during the NIPHYS and CANIF projects in 1994 to 1998 (Tables 19.2, 19.3). The average concentration of total nitrogen has slightly increased in the less polluted catchment X-8 and decreased in the acidified catchment X-16. The concentration of sulphur decreased from 130 to 92 µEq l^{-1} in X-8 and from 186 to 127 µEq l^{-1} in X-16. This indicates that a major impact on the forest ecosystem behaviour was due to the reduction of pollution by sulphur.

The trends in flux of S and N due to bulk deposition are illustrated in Fig. 19.4. The flux of SO_4 in bulk precipitation decreased continuously from the end of the 1970s to the present day at all sites. The deposition of SO_4 decreased from about 1.7 kEq ha^{-1} a^{-1} to 1.2 kEq ha^{-1} a^{-1} at X-16 and from 0.8 kEq ha^{-1} a^{-1} to 0.5 kEq ha^{-1} a^{-1} at X-8. Although SO_2 concentrations in air and SO_4 concentrations in precipitation were higher at X-8 than at Aubure, the SO_4 fluxes in precipitation and trends at X-8 and Aubure were very similar. The deposition of NH_4 and NO_3 was higher at X-16 and about the same at X-8 and Aubure. NH_4 deposition remained at about 0.5 kEq ha^{-1} a^{-1} at X-8, decreased from 0.7 kEq ha^{-1} a^{-1} to 0.6 kEq ha^{-1} a^{-1} at X-16

Table 19.2. Long-term change in NH_4, NO_3 and SO_4 concentrations ($\mu Eq\,l^{-1}$) in Czech catchments

		X-16 Jezeri		X-8 Salacova Lhota	
		1978–1983	1994–1998	1976–1983	1994–1998
Bulk precipitation	NH_4	78	80	66	73
	NO_3	61	55	48	52
	N total	139	134	114	125
	SO_4	200	118	130	113
Runoff	NH_4	0	3	0	0
	NO_3	200	106	25	22
	N total	200	108	25	22
	SO_4	1406	1240	366	354

Table 19.3. Long-term change in the flux of NH_4, NO_3, and SO_4 ($kEq\,ha^{-1}\,a^{-1}$) in Czech catchments

		X-16 Jezeri		X-8 Salacova Lhota	
		1978–1983	1994–1998	1976–1983	1994–1998
Bulk precipitation	NH_4	0.39	0.55	0.26	0.49
	NO_3	0.54	0.43	0.35	0.29
	N total	0.93	0.97	0.61	0.78
	SO_4	1.23	0.91	0.66	0.54
Runoff	NH_4		0.04		0
	NO_3	0.86	0.29	0.04	0.04
	N total	0.86	0.33	0.04	0.04
	SO_4	6	3.8	0.56	0.49

(Table 19.3) and increased from 0.21 to $0.35\,kEq\,ha^{-1}\,a^{-1}$ at Aubure during the study period. A slight decrease in NO_3 deposition was noticeable at all sites. The total N deposition almost did not change at either site.

These observations indicate a reduction of the level of pollution by sulphur oxide but not so in the case of nitrogen compounds.

19.3.4 Throughfall Deposition

Throughfall deposition monitoring was not done at all sites, therefore the comparison was restricted to the recent period. Throughfall deposition of S and N was lower in beech and young spruce stands and higher in old spruce (Table 19.4). Variations were mostly related to the filtering efficiencies of the canopies (Ignatova and Dambrine 2000). At X-16, the difference between spruce and beech was also partly due to the fact that spruce is located on the upper plateau, where fog deposition is

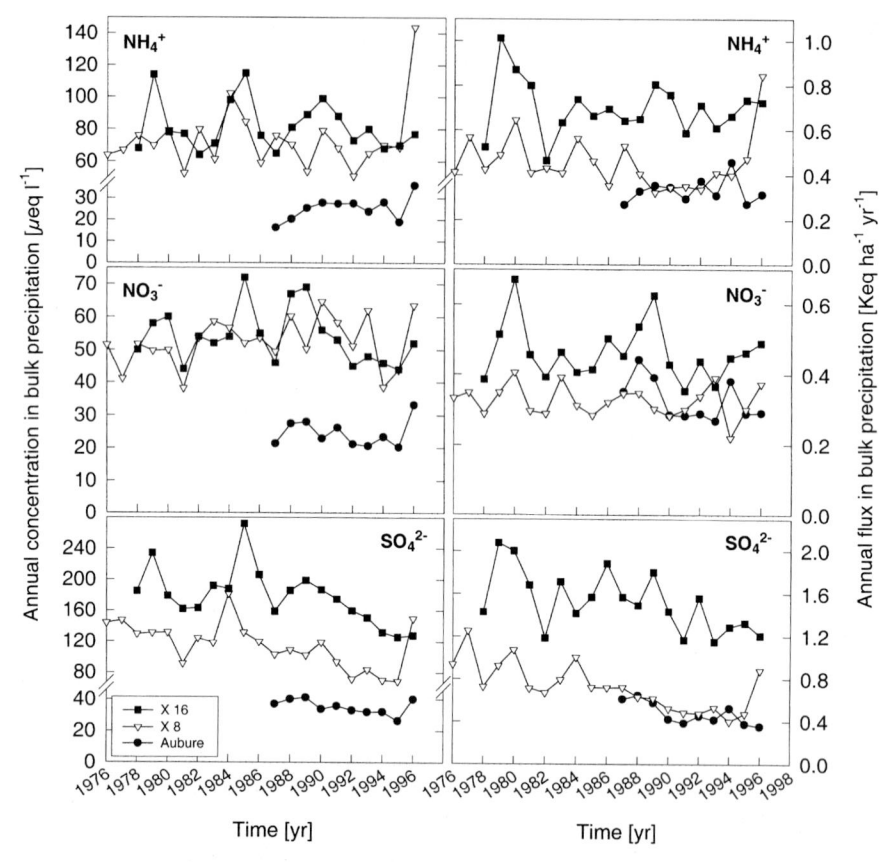

Fig. 19.4. Time variations of SO_4, NH_4 and NO_3 mean annual concentrations ($\mu Eq\,l^{-1}$) and fluxes in bulk precipitation for the three catchments studied

very high, while beech is located on the slope of X-16. Throughfall deposition of S and N below beech at X-16 and below young spruce at X-8 was twice as high as below stands of similar characteristics at Aubure. These differences between sites are in agreement with the differences in SO_2 and NO_2 concentration in ambient air (Fig. 19.2).

19.3.5 Streamwater Runoff

The trends in S and N concentrations in streams are illustrated in Fig. 19.5. At X-16, SO_4 decreased over 20 years, from about 1400 to 1100 $\mu Eq\,l^{-1}$. Since 1978, sulphate concentrations dropped rapidly at the beginning of the observation. Later, the decrease in the sulphate concentration levelled off. At Aubure, the SO_4 concentration decreased linearly down to 200 $\mu Eq\,l^{-1}$, with a slope of $-3\,\mu Eq\,yr^{-1}$. The concentration of SO_4 at X-8 showed large interannual variations from 250 to 500 $\mu Eq\,l^{-1}$, but no long-term trend. Nitrate concentrations were highest at X-16 and lowest at X-8. Nitrate concentrations at X-16 decreased from 150 to 70 $\mu Eq\,l^{-1}$ in 20 years, the decrease being rapid

Table 19.4. Budgets of N, Cl and S (kEq ha^{-1} a^{-1}) at the stand and at the catchment scale. Time period: Aubure: 1993–1995, X-8 and X-16: 1994–1998. *Picea* 40: *Picea* stand 40 years old

Catchment Scale	Aubure																	
	Picea 90			*Picea* 45			*Fagus* 150			Whole catchment		
	N	Cl	S	N	Cl	S	N	Cl	S	N	Cl	S
Bulk precipitation (1)	0.64	0.18	0.43	0.64	0.18	0.43	0.64	0.18	0.43	0.64	0.18	0.43
Throughfall (2)	1.34	0.63	1.23	0.56	0.51	0.91	0.66	0.49	0.71	0.94	0.51	0.94
Biomass immobilisation (3)	0.24			0.73			0.33			0.38		
Drainage at root depth (4)	1.07	0.61	1.26	0.15	0.37	0.99	0.11	0.45	0.91	0.63	0.51	1.12
Soil budget (2-3-4)	0.03	0.02	−0.04	−0.31	0.14	−0.08	0.21	0.04	−0.20	−0.07	0.00	−0.17
Runoff (6)										0.31	0.45	1.63
Catchment budget (2-3-6)										0.24	0.07	−0.68

Catchment Scale	X-8						X-16					
	Picea 40			Whole catchment			*Fagus* 100/*Picea* 60			Whole catchment		
	N	Cl	S	N	Cl	S	N	Cl	S	N	Cl	S
Bulk precipitation	0.78	0.11	0.54	0.78	0.11	0.54	0.97	0.13	0.91	0.97	0.13	0.91
Throughfall + stem flow beech (2a)							1.30	0.20	1.40			
Throughfall spruce (2b)	1.15	0.27	1.66				1.85	0.35	3.90			
Throughfall weighted average (2c)[a]				1.14	0.27	1.66				1.18	0.19	1.37
Biomass immobilisation (*Picea*) (3)				0.36		0.05	0.35		0.07			
Runoff (6)				0.04	0.09	0.49				0.33	0.25	3.80
Catchment budget (2c-3-6)				0.74	0.18	1.12				0.85	−0.06	−2.43

[a] Comment: throughfall input was calculated as an average weighted by area forested with beech, spruce, and covered with grass and deciduous trees represented by throughfall measurements under birch.

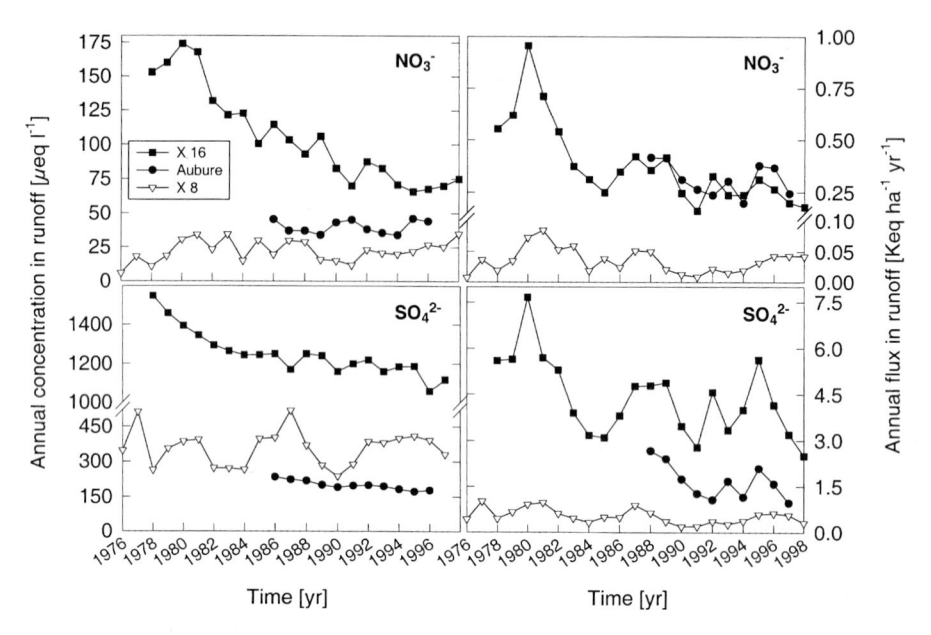

Fig. 19.5. Time variations of SO$_4$, NH$_4$ and NO$_3$ mean annual concentrations (µEq l^{-1}) and fluxes in runoff for the three catchments studied

during 1980–1985 and slowing down later. The mean annual nitrate concentrations fluctuated around the average values at Aubure (40 µEq l^{-1}) and X-8 (25 µEq l^{-1}) without obvious trend. The dry period 1990–1991 produced higher NO$_3$ concentrations at Aubure and lower concentrations in the Czech catchments.

The long-term decrease in concentrations and fluxes at X-16 (Tables 19.2 and 19.3) indicates that the reduction of sulphur pollution led to a more efficient retention of nitrogen in the catchments and probably higher accumulation of carbon. Similar significant changes have not been observed in the catchment X-8, indicating that the reduction of pollution in the Black Triangle is a relatively short-range effect that does not influence the ecological behaviour of forest which is 160 km away from the sources of pollution.

Trends in runoff are presented in Fig. 19.5. The interannual variations in the discharge of the dissolved components mirror the interannual variations in water drainage. At X-16, SO$_4$ drainage decreased abruptly from more than 6 kEq ha^{-1} a^{-1} in 1980 to less than 4 kEq ha^{-1} a^{-1} in 1983 and later fluctuated around this value. SO$_4$ drainage decreased also at Aubure and to a smaller extent at X-8. Nitrate drainage decreased abruptly from 1 to 0.3 kEq ha^{-1} a^{-1} at X-16, between 1980 and 1984. In the following period, the decrease was much smoother. The trends were almost the same at Aubure and X-16. Nitrate drainage was low and without statistically significant trend at X-8.

The rapid decrease in the discharge of NO$_3$ and SO$_4$ was not related to any change in bulk precipitation composition or amount at X-16. A large reduction of dry deposition of SO$_4$, linked to the removal of needles from damaged spruce stands at higher elevation, is the most likely cause for the decrease in SO_4 output. The NO_3 peak in drainage was related to the dieback of spruce at the end of the 1980s. The steep decrease in NO_3 output from 1980 to 1984 is related to the relatively low runoff of water during this period. The most significant decrease in the runoff was measured in the case of SO_4, from 5.6 to $3.9\,kEq\,ha^{-1}\,a^{-1}$ at X-16 (Table 19.3). The decrease in nitrate leaching is in agreement with the reforestation of the damaged part of X-16. In the less polluted and less acidified catchment X-8, the changes were much less pronounced.

19.4 Budgets

The budgets of sulphur, nitrogen and chloride in the catchments and in the monitored forest stands are presented in Table 19.4.

At X-16, the 4-year average budget is $-2.43\,kEq\,ha^{-1}\,a^{-1}$ for sulphur, $0.86\,kEq\,ha^{-1}\,a^{-1}$ for nitrogen and $-0.06\,kEq\,ha^{-1}\,a^{-1}$ for chloride ions. The budget of chloride ions near zero indicates that our estimates of the atmospheric inputs have been probably close to the real situation. The results show that sulphur is leached now from the storage pool accumulated during the period of high pollution. This release of previously accumulated sulphate occurred not only due to desorption but also because of mineralisation of organic matter in the deforested area and in the soil below the surviving spruce stands (see Sect. 19.5, below). Sulphur mean residence time in soil was estimated to be between 11 (Novak et al. 1996) and 29 years (Novak and Prechova 1995). This explains the smooth seasonal fluctuation of sulphate in runoff in a response to larger short-term variations of sulphate in atmospheric input. The situation was reversed in the period of the highest pollution (1978–1983), when an accumulation of $0.77\,kEq\,ha^{-1}\,a^{-1}$ of S in the catchment was observed (Paces 1985).

The nitrogen budget in the recent period (1994–1998) is $+0.85\,kEq\,ha^{-1}\,a^{-1}$. A large part of this accumulated nitrogen is fixed in the new growth of trees and grass. We observed much smaller accumulation of nitrogen during the period of highest pollution (1978–1983) when old damaged spruce was not yet replaced by the new growth of trees and grass. Our estimate of the biological immobilisation of nitrogen at this time was $0.18\,kEq\,ha^{-1}\,a^{-1}$ and drainage of nitrate was much higher (Paces 1992).

Recent data on soil solution at root depth from site Nacetin, a residual spruce stand near X-16, show that NO_3 concentrations in soil solution were high ($>200\,\mu Eq\,l^{-1}$) and close to NO_3 concentration in springs from the upper part of the catchment X-16 (Havel et al. 1999). On the other hand, data from lysimeters below beech at the lower part of X-16 indicate very low nitrate concentration ($<20\,\mu Eq\,l^{-1}$). This explains the decrease in nitrate concentration along the stream. N losses by denitrification in groundwater may also occur, as is indicated by the data on ^{15}N in nitrate and ammonium mass balance (Buzek et al. 1998).

The budgets of Cl, S and N in the less polluted catchment X-8 are different. The budget of chloride ions in the period 1976–1983 was closer to zero than that observed during the recent period ($0.18\,kEq\,ha^{-1}\,a^{-1}$). The accumulation of sulphur was estimated to be $0.56\,kEq\,ha^{-1}\,a^{-1}$ as compared to the present measured accumulation of $1.12\,kEq\,ha^{-1}\,a^{-1}$. The difference may be influenced by the different evaluation methods in the observation periods and not due to the changes in the atmospheric input of sulphur. The input of sulphur for the period 1994–1998 was calculated from throughfall measurements. The input of S for the period 1976–1983 was estimated from the measured bulk precipitation and the dry deposition of sulphur calculated from the concentrations of SO_2 in air and air deposition velocities (Paces 1985). Even if we consider the uncertainties in the budget evaluation, the data indicate that even in one of the least polluted rural regions in the Czech Republic there has been an accumulation of sulphur in spruce forests at least during the last 22 years. Our data suggest that sulphur is still accumulating in the forest soils in the rural region. Nitrogen budget indicates that the accumulation of nitrogen in the soil and denitrification have proceeded with a rate of about $0.7\,kEq\,ha^{-1}\,a^{-1}$ in 1994–1998. During the whole period of monitoring from 1976 to 1998, the leaching of nitrate was low, on average $0.035\,kEq\,ha^{-1}\,a^{-1}$. This can be compared to the high average leaching rate of $0.48\,kEq\,ha^{-1}\,a^{-1}$ in the acidified catchment X-16.

At Aubure, nitrate drainage was very low at P45 because nitrification was extremely poor, and at F150 because of intense root turnover. It was higher at P90, in relation to the higher deposition and the poor health of the stand (see Persson et al., Chap. 14, this Vol.). On average over the 1993–95 period, considering the flux of elements in throughfall as input, and biomass immobilisation + drainage as outputs, the soil nitrogen budget was well balanced at P90, negative at P45 and positive at F150 (Table 19.4). N accumulated mainly in the biomass at P45 and in the soil at F150. Because these N budgets do not consider the N flux directly taken up by the canopies, which was estimated to at least $5\,kg\,ha^{-1}\,a^{-1}$ in spruce stands (Ignatova et Dambrine 2000), soil N budgets were most probably positive in all stands. Depending on the stands, the soil S budget was balanced (P90) or slightly negative (P45 and F150). The relatively low total S store ($3-4\,kEq\,ha^{-1}$) in soils, and especially of inorganic S ($0.3-0.5\,kEq\,ha^{-1}$), compared to the present throughfall deposition ($0.7-1.2\,kEq\,ha^{-1}\,a^{-1}$), explains this apparent equilibrium. Cl budgets were almost balanced. At the catchment scale, input-output budgets were computed taking into account the different proportions of stands in the catchment. The soil budget, computed using lysimeter data at root depth in the different stands, was balanced for N and slightly negative for S. In comparison, the catchment budget computed using runoff output was positive for N and strongly negative for S (Probst et al. 1995). The differences between these budgets provided an indication on the S flux released from the regolith and the N flux either accumulated in the soils and taken up by the vegetation along the stream, or denitrified. The relative variation of NO_3 and Cl concentrations from soil solutions to various springs and to the stream (Probst et al. 1995) suggested that denitrification below the root zone was very likely along the slope and in the bottom of the catchment. This N loss below the root zone was essential to rise the pH from about 4.4 in soil solutions at root depth to 6.2 in the stream (Probst et al. 1990).

19.5 Biological Cycling of Sulphur

The existence of contrasting $\delta^{34}S$ signatures of individual S fluxes at Jezeri (Novak et al. 1995; Groscheova et al. 1998) makes it possible to quantify the proportion of organically cycled S in the catchment. Elevated inputs of sulphur were traditionally believed to remain largely in an inorganic form in forest soil before being flushed out (cf. Johnson 1984). Novak et al. (1996) have shown that at atmospheric depositions $>30\,kg\,S\,ha^{-1}\,a^{-1}$, typical of Central Europe, sulphur content in the O+A soil horizon is directly proportional to sulphur input. Our data from X-16 indicate that a large proportion of the accumulated sulphur is organically cycled before it is discharged from the catchment.

The $\delta^{34}S$ value of the surface discharge in 1992–1995 was 4.3‰. The $\delta^{34}S$ value of the discharge is probably a product of mixing of two isotopically different S sources. The isotopically heavier source of sulphur is atmospheric deposition (5.9‰). The isotopically lighter source is the soil sulphur (3.9‰). During the monitoring period, S discharge (5.4 kEq ha^{-1} a^{-1}) exceeded atmospheric input (1.8 kEq ha^{-1} a^{-1}). The contribution of sulphur released from the soil reservoir to the runoff equals 66% (3.6 kEq ha^{-1} a^{-1}). Since sulphur is isotopically stratified in the vertical soil profile (mean $\delta^{34}S$ becomes heavier with depth from 2.6‰ in O+A to 5.2 and 5.3 in Bv and B/C, respectively), the contribution of the organic soil horizons to sulphur in the runoff can be estimated. In a mixing model equation

$$Q_{OUT}\, \delta_{OUT} = Q_{IN}\, \delta_{IN} + Q_{SOIL}\, \delta_{SOIL},$$

where Q are sulphur fluxes (kEq ha^{-1} a^{-1}), subscripts IN, OUT and SOIL denote catchment-level atmospheric input, surface discharge and soil source, and δ stands for $\delta^{34}S$. δ_{SOIL} is the isotopic composition of the soil component that is selectively leached out from the soil pool. The δ_{SOIL} value calculated from the above equation is 3.4‰, falling between $\delta^{34}S$ of the organic and mineral soil horizons. The soil component of sulphur to runoff, whose isotopic signature is $\delta^{34}S$ equal to 3.4‰, is made up by 70% of the organic soil horizons (2.6‰) and 30% of that of the mineral soil horizons (5.3‰). The soil storage of sulphur was 46 kEq ha^{-1}. The catchment mass balance indicates that the depletion in the soil pool of sulphur over the 3 years has amounted to approximately 8% of the soil pool. Sulphur of the organic soil horizons at the beginning of the observation period (November 1992) was made up by 96% of organic sulphur, the percentage of organic S at the end of the observation period was statistically indistinquishable ($p < 0.05$). Therefore, it is possible to approximate the ratio of the masses of organic and inorganic S released form the organic soil horizons to the stream by their abundance in the soil (96:4). Consequently, organically cycled S from the humus layer contributed 45% of sulphate in X-16 discharge. This estimate is surprisingly high but agrees well with ^{34}S-labelling experiments on German forest soils (Prietzel et al. 1995).

19.6 Conclusions

At all studied sites, S input decreased linearly during the study period, while N input remained approximately stable. Time trends of S inputs and S outputs were in the same direction, but the amplitudes and shapes of the trends differed owing to variations in dry deposition, sorption-desorption-mineralisation processes and rain amounts. Variations in the balance between soil nitrification and tree uptake of nitrogen, variations in tree rooting, changes in forest health, and processes occurring in the water-saturated areas obscured the relationships between trends in nitrogen deposition and nitrogen discharge from the catchments. At both Aubure and X-16, drainage waters poor in nitrate originating from healthy beech stands have diluted the drainage waters rich in nitrate provided by declining spruce stands.

References

Buzek F, Cerny J, Paces T (1998) The behaviour of nitrogen isotopes in acidified forest soils. Water Air Soil Pollut 105:155–164

Cerny J, Paces T (eds) (1995) Acidification in the Black Triangle region. In: Acid reign, 5th Int Conf On Acidic Deposition Science and Policy, Göteborg, 1995, Excursion, Prague

Durka W, Schulze E-D, Gebauer G, Voerkelius S (1994) Effects of forest decline on uptake and leaching of deposited nitrate determined from ^{15}N and ^{18}O measurements. Nature 372:765–767

Fichter J, Dambrine E, Turpault MP, Ranger J (1998) Base cation supply in spruce and beech ecosystems of the Strengbach catchment (Vosges Mts, France). Water Air Soil Pollut 104:125–148

Groscheova H, Novak M, Havel M, Cerny J (1998) Effect of altitude and tree species on d^{34}S of deposited sulfur (Jezeri catchment, Czech Republic). Water Air Soil Pollut 105:287–295

Havel M, Peters NE, Cerny J (1999) Longitudinal patterns of stream chemistry in a catchment with forest dieback, Czech Republic. Environ Pollut 104:157–167

Ignatova N, Dambrine E (2000) Evidence for spruce canopy uptake of dry and occult deposited N from comparative throughfall measurements below plastic and living Christmas trees. Ann For Sci 57:113–120

Johnson DW (1984) Sulfur cycling in forests. Biogeochemistry 1:29–43

Landmann G, Bonneau M (1995) Forest decline and atmospheric deposition effects in the French Mountains. Springer, Berlin Heidelbery New York

Novak M, Prechova E (1995) Movement and transformation of ^{35}S-labelled sulphate in the soil of a heavily polluted site in the northern Czech Republic. Environ Geochem Health 17:83–94

Novak M, Bottrell SH, Groscheova H, Buzek F, Cerny J (1995) Sulphur isotope characteristics of two North Bohemian forest catchments. Water Air Soil Pollut 85:1641–1646

Novak M, Bottrell SH, Fottova D, Buzek F, Groscheova H, Zak K (1996) Sulfur isotope signals in forest soils of Central Europe along an air pollution gradient. Environ Sci Technol 30:3473–3476

Paces T (1985) Sources of acidification in central Europe estimated from elemental budgets in small basins. Nature 325:31–36

Paces T (1992) Acidification: biogeochemical signature of economic and political systems. In: Kharaka YK, Mast AS (eds) Water-rock interaction. Balkema, Rotterdam, pp 413–417

Prietzel J, Mayer B, Krouse HR, Rehfuess KE, Fritz P (1995) Biogeochemical transformation of simulated wet sulfate deposition in forest soils assessed by a core experiment using stable sulfur isotopes. Water Air Soil Pollut 79:243–260

Probst A, Dambrine E, Viville D, Fritz B (1990) Influence of acid atmospheric inputs on surface water chemistry and mineral fluxes in a declining spruce stand within a small granitic catchment (Vosges massif, France). J Hydrol 116:101–124

Probst A, Fritz B, Viville D (1995) Mid-term trends in acid precipitation, streamwater chemistry and element budgets in the Strengbach catchment (Vosges massif, France). Water Air Soil Pollut 79:39–59

20 Model Analysis of Carbon and Nitrogen Cycling in *Picea* and *Fagus* Forests

H. van Oene, F. Berendse, T. Persson, A.F. Harrison, E.-D. Schulze,
B.R. Andersen, G.A. Bauer, E. Dambrine, P. Högberg, G. Matteucci,
and T. Paces

20.1 Introduction

The CANIF project experimentally investigates the carbon and nitrogen flows in *Picea abies* and *Fagus sylvatica* forest stands. The experimental subprojects encompass a large diversity of research subjects ranging from root studies for uptake of nutrients, root turnover, the diversity and the role of mycorrhizae, soil fauna and microorganisms, soil organic matter dynamics, tree growth and nutrient relations to measurements of leaching fluxes. Only by putting all these different aspects together is an overview of the functioning of the ecosystem possible. Process-based models that incorporate the carbon, nutrients and water flows of the ecosystem are very appropriate to use for this integrative function. Their great advantage is that such models with site-specific input data for climate and deposition levels can highlight the major differences between sites.

Here, in the CANIF project, the NUCOM (nutrient cycling and competition) model (Van Oene et al. 1999a) is used as ecosystem model that can explore the functioning of the sites at ecosystem level. The main emphasis of this model is on the carbon and nitrogen interactions, especially the feedback relations between plant growth and decomposition and mineralisation processes.

A variety of models exist that are directed to forest growth and element cycling, e.g. Century (Parton et al. 1996; Ryan et al. 1996), Biome-BGC (Forest-BGC) (Running and Gower 1991; Ryan et al. 1996) and PnET (Aber and Federer 1992; Aber et al. 1995). Century differs from the NUCOM model in describing soil organic matter dynamics in more detail including relations to lignin content in Century. The major differences, however, lie in the N dynamics, where Century couples soil organic N fluxes in proportion to soil organic C fluxes, NUCOM includes microbial characteristics thereby allowing for variable N fluxes in relation to C fluxes. Also in the plant part, Century differs in the N dynamics by controlling N concentrations in relation to growth dynamics more strictly whereas in NUCOM, N concentrations are free to vary within the same range for any growth rate. In addition, in Century, potential net primary production is not calculated from intercepted photosynthetic radiation but given as a parameter. The Biome-BGC (Forest-BGC) model includes a more detailed description of assimilation and transpiration processes than NUCOM; however, litter and soil organic matter dynamics are more simple and simulated at an annual time step. Soil organic C decomposition and N mineralisation are, as in Century, coupled in proportion to litter decomposition that is determined by actual evapotranspiration and lignin content. PnET predicts net primary production in relation to temperature and water availability but is not an ecosystem model that includes the complete cycling of C and N. N dynamics are not included and leaf N concentration is needed as input parameter. In PnET, soil respiration is included using an empirical relationship, and no feedback exists between biomass production and litter production and decomposition. Some other models exist that, like NUCOM, are focussed on C and N interactions. These are MBL-GEM (Rastetter et al. 1991) and Q (see Ryan et al. 1996; Ågren and Bosatta 1996), but both work on an annual time step and do not include hydrological processes. The TEM model (Raich et al. 1991; Melillo et al. 1993) is applied at larger (global) scales and uses therefore combined vegetation C and N pool, and soil C and N pool. With this model net primary production is calculated for an equilibrium situation where no changes in state variables occur.

Compared to these other models, the NUCOM model is thus specific in including the complete C and N cycling as well as water flows through the system, all described in a process-oriented manner. In the dynamics of soil organic matter, microbial characteristics determine N fluxes. The model has no constraints on (leaf) N concentrations within a large range between minimum and maximum concentration. Thus, N availability in combination with water-limited assimilation determine plant concentrations fully. Characteristic for the NUCOM model, however, is that, in contrast to the models described above that do not include interactions between plant species, NUCOM is able to describe the competition between the dominant plant species and the feedback of changed plant species competition on soil organic matter dynamics and nutrient availability (Berendse 1994). This feature of NUCOM is, however, not used in this CANIF project but applications are given in Van Oene et al. (1999a,b).

In this chapter, first an overview of the NUCOM model is given with a description of the included processes, followed by its application to the CANIF sites and an analysis of the functioning of these sites. The sites included in this analysis are the main CANIF sites: the Swedish sites Åheden (mixed *Picea* and *Pinus*) and Skogaby (*Picea*), the Czech sites Nacetin (*Picea*) and Jezeri (*Fagus*), the German sites Waldstein (*Picea*) and Schacht (*Fagus*), the French sites at Aubure (*Picea*, *Fagus*) and the Italian sites Collelongo (*Fagus*) and Monte di Mezzo (*Picea*).

20.2 Model Description

20.2.1 General

The NUCOM model is a process-based model that links vegetation and soil processes. The model is especially developed to analyse plant growth and soil organic matter dynamics and the effects of environmental changes on these processes. It describes the flow of nutrients, carbon and water through the ecosystem. Growth of plants can be limited by light, nitrogen or water. Other macronutrients are included but do not affect plant growth. The model is built in modules describing plant growth, soil organic matter dynamics, soil chemistry and hydrology. Carbon and nitrogen interactions strongly determine the feedbacks between plant growth and decomposition and mineralisation through C:N ratios in both plant material and soil organic matter; these modules are the main parts of the model (see Fig. 20.1). Although these modules are conceptually most important, NUCOM is a balanced model that includes the different processes on an equivalent detailed base. In this application, the hydrological processes are calculated with a time step of 1 day. The other processes are simulated with a time step of 1 month, using monthly averaged data calculated in the hydrology module for hydrological variables.

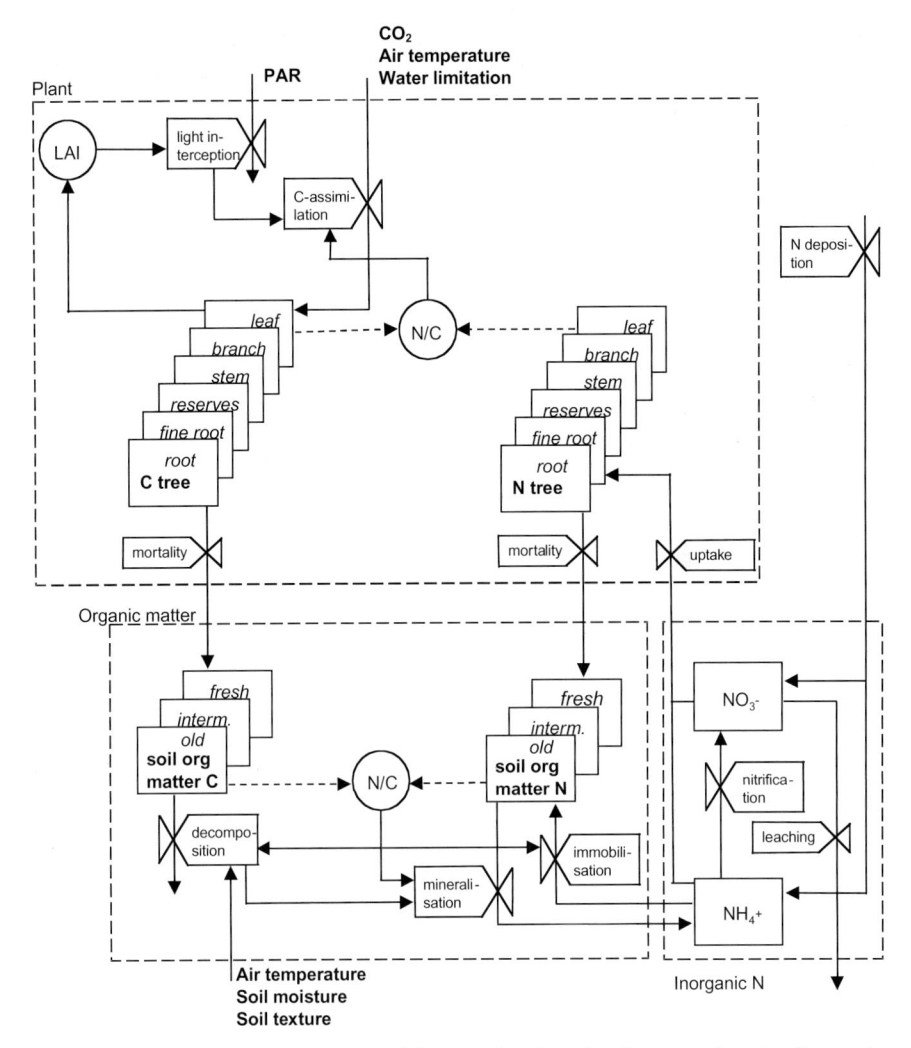

Fig. 20.1. A schematic representation of the C and N flows in plant growth and soil organic matter dynamics in the NUCOM model (after Van Oene et al. 1999a). Each soil organic matter class is further subdivided in different plant organ classes. Here, the soil organic matter classes correspond to the organic matter in the litter layer, the FH layer and the 0–30 cm soil layer. The hydrology module simulates actual and potential evapotranspiration and soil moisture. Plant C assimilation is limited by water stress through closure of stomata; this is simulated by using the simulated ratio of actual to potential transpiration as a reduction factor on assimilation. Nitrogen limitation is determined by the amount of N taken up and the amount of assimilated C. Per plant organ, the resulting N concentration varies between a species-specific minimum and maximum concentration. Nitrogen or carbon in excess over growth requirement are stored in the plant reserve pools. These reserves are used by the plants to start growth in spring. Soil organic matter decomposition is affected by simulated soil moisture and soil texture. Per soil organic matter class, N:C ratios result from simulated N:C ratios in litter input and decomposition and mineralisation

20.2.2 Plant Dynamics

The plant is divided in several organs: leaves, stems, branches, coarse and fine roots. Dry weight (or C amount) and N amount of each plant organ are state variables that are tracked in time.

20.2.2.1 Potential Plant Assimilation

Plant assimilation is driven by interception of photosynthetic radiation that is described by using the Lambert-Beer equation for exponential decay of light within the canopy [Table 20.1; Eq. (1)]. If more than one species is present, each species

Table 20.1. The equations of the model describing light interception, assimilation, nitrogen uptake, allocation, mortality, decomposition and mineralisation, and their temperature, CO_2 and moisture dependencies

Light interception
$$I_{abs_{i,n}} = \frac{k_{ext,i} \cdot LAI_{i,n}}{\sum_i k_{ext,i} \cdot LAI_{i,n}} \cdot I_n \cdot \left(1 - e^{-\sum_i k_{ext,i} \cdot LAI_{i,n}}\right). \quad (1)$$

Potential assimilation
$$A_{pot,i} = RUE_i \cdot \sum_n I_{abs_{i,n}} \cdot k_i(CO_2) \cdot k_i(T) \quad (2)$$

with
$$\begin{array}{ll} k_i(T) = 0 & \text{if } T < T_{min,i} \text{ or } T > T_{max,i} \\ k_i(T) = (T - T_{min,i})/(T_{opt1,i} - T_{min,i}) & \text{if } T_{min,i} < T < T_{opt1,i} \\ k_i(T) = 1 & \text{if } T_{opt1,i} < T < T_{opt2,i} \\ k_i(T) = (T_{max,i} - T)/(T_{max,i} - T_{opt2,i}) & \text{if } T_{opt2,i} < T < T_{max,i} \end{array} \quad (3)$$

with $k_i(CO_2) = 1 + \beta_i \cdot \ln\left(\dfrac{CO_2}{CO_{2,ref}}\right)$ and $CO_{2,ref} = 350\,ppm$. $\quad (4)$

Water-limited assimilation $A_{w,i} = Tr_i/Tr_{pot,i} \cdot A_{pot,i}$. $\quad (5)$

Actual N uptake and actual assimilation (without N redistribution and storage) $\quad (6)$

if $U_{pot,i} < n_{con,min,i} \cdot A_{w,i}$ then
$\quad A_{act,i} = U_{pot,i}/n_{con,min,i}$
$\quad U_{act,i} = U_{pot,i}$

if $U_{pot,i} < n_{con,max,i} \cdot A_{w,i}$ then
$\quad A_{act,i} = A_{w,i}$
$\quad U_{act,i} = U_{pot,i}$
$\quad n_{con,i} = \dfrac{U_{act,i} - n_{con,min,i} \cdot A_{act,i}}{(n_{con,max,i} - n_{con,min,i}) \cdot A_{act,i}}$

if $U_{pot,i} \geq n_{con,max,i} \cdot A_{w,i}$ then
$\quad A_{act,i} = A_{w,i}$
$\quad U_{act,i} = n_{con,max,i} \cdot A_{w,i}$.

Potential N uptake
$$U_{pot,i} = \sum_l \left(\frac{SRL_i \cdot B_{r,i,l}}{\sum_i SRL_i \cdot B_{r,i,l}}\right) \cdot Navail_l. \quad (7)$$

Root distribution $\quad Y_i = 1 - k_{rf,i}^d \quad (8)$

Allocation $\quad G_{j,i} = k_{all,j,i} \cdot A_{act,i} \quad (9)$

Mortality $\quad M_{j,i} = k_{m,j,i} \cdot B_{j,i} \quad (10)$

with $k_{m,j,i} = \left(\dfrac{Age_{j,i}}{Age\,max_{j,i}}\right)^4$ for leaves and roots, and $k_{m,j,i} \leq 1$.

Table 20.1. *Continued*

Decomposition	$\dfrac{dC_{org,c,i,j}}{dt} = -C_{org,c,i,j} \cdot k(T)_{dec,c,i,j} \cdot k(clay\%)_c \cdot k(moist)_c$		(11)
with	$k(T)_{dec,c,i,j} = e^{[a_{c,i,j} + b \cdot T \cdot (1 - 0.5 \cdot T/T_{opt,dec})]}$		(12)
with	$k(clay\%)_c = 1$	if clay% <= 5	(13)
	$k(clay\%)_c = -0.0133 \cdot clay\% + 1.0666$	if 5 > clay% <= 50	
	$k(clay\%)_c = 0.40$	if clay% > 50	
with	$k(moist)_c = 0.7383 \cdot \dfrac{\theta_c}{0.60 \cdot \theta_{field,c}} + 0.2657,$	if $\theta_c < 0.60 \cdot \theta_{field,c}$	(14)
	$k(moist)_c = 1,$	if $\theta_c \geq 0.60 \cdot \theta_{field,c}.$	
Mineralisation	$\dfrac{dN_{org}}{dt} = \sum_c \sum_i \sum_j \left[\left(\dfrac{N_{org,c,i,j}}{C_{org,c,i,j}} - n_{crit,c,j,i} \right) \cdot \dfrac{dC_{org,c,i,j}}{dt} \cdot \dfrac{1}{1-\epsilon} \right]$		(15)
with	$n_{crit,c,j,i} = \dfrac{f_N \cdot \epsilon}{f_C}$		

Parameters: Agemax$_{j,i}$ = maximum life span of plant organ j of species i (day), a$_{c,j,i}$ = parameter determining absolute decomposition rate of litter class c of plant organ j and species i, b = parameter determining temperature dependency decomposition rate, β_i = CO_2 growth enhancement factor of species i (–), ϵ = microbial assimilation efficiency (–), f$_C$ = microbial C concentration (g C g^{-1}), f$_N$ = microbial N concentration (g N g^{-1}), k$_{all,j,i}$ = allocation parameter to plant organ j of species i (–), k$_{ext,i}$ = light extinction coefficient of species i (–), k(T)$_{dec,c,j,i}$ = relative decomposition rate of class j and plant organ j of species i (g C g C^{-1} time step^{-1}), k(clay%)$_c$ = soil texture dependency of decomposition of litterclass c (–), k$_i$(CO$_2$) = CO$_2$ dependency of assimilation of species i (–), k$_{rf,i}$ = root distribution parameter (–), k(moist)$_c$ = soil moisture dependency of decomposition of litterclass c (–), k$_i$(T) = temperature dependency of assimilation of species i (–), k$_{m,j,i}$ = mortality parameter plant organ j of species i (time step^{-1}), n$_{con,min,i}$ = minimum N concentration g^{-1} new produced plant material of species i (g N g^{-1}), n$_{con,max,i}$ = maximum N concentration per gram new produced plant material of species i (g N g^{-1}), n$_{crit,c,j,i}$ = critical N concentration of class c and plant organ j of species i (g N g C^{-1}), RUE$_i$ = potential radiation use efficiency of species i [g MJ(PAR)$^{-1}$], SRL$_i$ = specific root length (m kg^{-1}), T$_{min,i}$ = minimum temperature for assimilation of species i (°T), T$_{opt,dec}$ = optimum temperature for decomposition (°C), T$_{opt1,i}$ = optimum temperature 1 for assimilation of species i (°C), T$_{opt2,i}$ = optimum temperature 2 for assimilation of species i (°C), T$_{max,i}$ = maximum temperature for assimilation of species i (°C), $\theta_{field,c}$ = volumetric moisture content at field capacity in litter class c (m^3 m^{-3}).

Variables: A$_{act,i}$ = actual assimilation rate of species i (kg ha^{-1} time step^{-1}), A$_{pot,i}$ = potential assimilation rate of species i (kg ha^{-1} time step^{-1}), A$_{w,i}$ = water-limited assimilation rate (kg ha^{-1} time step^{-1}), Age$_{j,i}$ = age of plant organ j of species i (day), B$_{j,i}$: biomass of plant organ j of species i (kg ha^{-1}), B$_{r,i,l}$: fine root biomass of species i in root layer l (kg ha^{-1}), C$_{org,c,i,j}$ = soil organic C of class c and plant organ j of species i (kg C ha^{-1}), d = depth (cm), G$_{j,i}$: growth rate of plant organ j of species i (kg ha^{-1} time step^{-1}), I$_{abs,i,n}$ = absorbed radiation by species i in canopy layer n (MJ(PAR) ha^{-1} time step^{-1}), LAI$_{i,n}$ = leaf area index of species i in canopy layer n (ha ha^{-1}), I$_n$ = incoming radiation in canopy layer n (MJ(PAR) ha^{-1}), M$_{j,i}$: mortality rate of plant organ j of species i (kg ha^{-1} time step^{-1}), Navail$_l$ = available N in root layer l (kg N ha^{-1}), n$_{con,i}$ = N concentration in new produced plant material of species i (g N g^{-1}), N$_{org,c,i,j}$ = soil organic N of class c and plant organ j of species i (kg N ha^{-1}), T = air temperature (°C), Tr$_i$ = actual transpiration rate of species i (mm H$_2$O time step^{-1}), Tr$_{pot,i}$ = potential transpiration rate of species i (mm H$_2$O time step^{-1}), U$_{act,i}$ = actual N uptake rate of species i (kg N ha^{-1} time step^{-1}), U$_{pot,i}$ = potential N uptake rate of species i (kg N ha^{-1} time step^{-1}), Y$_i$ = cumulative root fraction of species i (–), θ_c = volumetric moisture content in litter class c (m^3 m^{-3}).

absorbs an amount of light equal to its share in the total leaf area times the total light absorption of the vegetation. Five canopy layers per plant species are distinguished. Leaf area is calculated from leaf biomass and specific leaf area. A constant leaf area density over canopy height is assumed.

Potential assimilation is calculated using a radiation-use efficiency that describes how much dry weight is formed by interception of one unit of photosynthetic active radiation, i.e. optimal canopy photosynthesis is linearly related to intercepted radiation (Monteith 1977, 1994; Cannell et al. 1987; Sands 1996). Haxeltine and Prentice (1996) show, using the hypothesis of optimal N allocation within the canopy, that the standard non-rectangular hyperbola formulation for the instantaneous response of leaf net photosynthesis to absorbed photosynthetic active radiation leads to this linear equation for time-integrated response of canopy net photosynthesis to absorbed radiation as expressed by the radiation use efficiency. Air temperature and CO_2 dependencies [Table 20.1; Eqs. (2)–(4)] modify this potential assimilation. The CO_2 dependency is a descriptive equation for the response of crops, restricted to the range of 200–1000 ppm CO_2 (Goudriaan et al. 1985). The beta-factor is used to calculate effects of CO_2 on potential assimilation. The potential assimilation calculated with Eq. (2) is modified by nitrogen and water limitation. The actual response to increased CO_2 is thus dependent on nitrogen and water limitation on plant growth.

20.2.2.2 Actual Plant Assimilation

The potential assimilation is reduced by limitation of water and nitrogen. Water limitation is calculated in the hydrology module (see below) as the ratio between actual and potential transpiration [Table 20.1; Eq. (5)]. This is based on the idea that closure of stomata as a result of water stress also limits CO_2 uptake and thus reduces assimilation. Wong et al. (1979) experimentally demonstrated that the rate of assimilation nearly proportionally changed with stomatal conductance at equal external CO_2 concentration. This implies a constant intercellular to external CO_2 concentration as also supported by many observations (Long and Hutchin 1991). Maintenance of this constant ratio with elevation of external CO_2 concentration implies a decreased stomatal conductance and hence decreased transpiration per unit leaf area (Long and Hutchin 1991). This leads to the observed increased water use efficiency of many plant species at elevated CO_2 concentrations (Arp et al. 1998). The latter feedback of decreasing stomatal conductance at increasing CO_2 levels is not yet included in NUCOM.

Nitrogen limitation is dependent on the N uptake of the plant species and the demand of the plant species [Table 20.1; Eq. (6)], in line with the view that resource availability and replenishment control plant growth (Ingestad 1982; Ingestad and Lund 1986; Rastetter and Shaver 1992; Ingestad and Ågren 1995). The potential N uptake is a function of N availability (that is N mineralised plus N deposition) and root length [Table 20.1; Eq. (7)], where root length is calculated as specific root length times root biomass per soil layer. Root distribution with depth is described by a function of Gale and Grigal 1987 [Table 20.1; Eq. (8)] giving the cumulative root fraction from the soil surface to depth d in centimetres. Uptake of NH_4^+ and NO_3^- is proportional to their supply, which in most situations will be dominated by NH_4^+. The N demand of a species is dependent on the assimilation rate and the N requirement of the species. This requirement varies between a minimum and maximum N amount based on a minimum and maximum concentration for each plant organ. In addition, reserve pools for C and N are present.

20.2.2.3 Allocation, Reserve Pools and Mortality

The actual assimilation, i.e. amount of dry weight produced, is allocated to the different plant organs following a fixed allocation scheme [Table 20.1; Eq. (9)]. A pool for labile carbohydrates and nitrogen is included to store surpluses of carbohydrates and nitrogen. If assimilation is too low to accommodate all available nitrogen up to a maximum nitrogen concentration and to fill this plant reserve pool, then the nitrogen still in excess is not taken up (the plant species is light- or water-limited). This excess can subsequently be leached. Conversely, if N uptake is too low to sustain the assimilation, actual assimilation is lower and plant growth is N-limited. These reserve pools are also used by the plants to start growth in spring.

The mortality rate varies between plant organs and for leaves and roots is dependent on the organ's age, while for branches a fixed turnover is taken [Table 20.1; Eq. (10)]. Mortality of stems is neglected in this study. The age of leaves and roots is calculated from the age of new produced material and the age of plant material that dies off weighted by biomass. Above-ground dead tissue is transferred to a dead material pool and from there gradually into the litter layer of soil organic matter. Dead roots are immediately transferred to the litter layer. Part of the nitrogen in the dying leaves and fine roots is withdrawn, up to a minimum concentration in the litter produced, and reallocated to living tissue by adding it to the pool of nitrogen taken up.

20.2.3 Soil Organic Matter Dynamics

The soil organic matter is distinguished in three age classes, and subclasses for plant organs. The age classes correspond in this study to the soil organic matter of litter layer, the FH layer and the 0-30-cm layer. Decomposition of soil organic matter is calculated per plant species, age class and plant organ class. A fraction of the soil organic matter in the L and FH layer is yearly transferred to the succeeding age class. The decomposition rate is dependent on temperature, moisture and clay content [Table 20.1; Eq. (11)]. The temperature dependency of decomposition [Table 20.1; Eq. (12)] follows Kirschbaum (1995). In this equation, the a term determines the absolute decomposition rate at optimum temperature, and b, in conjunction with the optimum temperature, its temperature dependency. It describes an exponential curve that starts to level off above ca. 25 °C and reaches an optimum at 36.9 °C. The stabilising effect of clay on soil organic matter is expressed in Table 2.1, Eq. (15) (Bloemhof 1992). The description of the moisture dependency of the decomposition rate is based on data from Seyferth (1998) calculating the relative change in decomposition rate at different moisture levels.

Nitrogen mineralisation is dependent on decomposition rate, the simulated C:N ratio of the organic matter, the C:N ratio of the microbial biomass and the assimilation efficiency of the microbes [Bosatta and Berendse 1984; Table 20.1; Eq. (16)]. If the C:N ratio of the organic matter is lower than the critical C:N ratio for the microbes, a net release of N may be expected. A net immobilisation occurs when the C:N ratio is above this critical value.

20.2.4 Soil Chemistry

In the soil layers (LFH layer and 0–30 cm), soil solution chemistry, pH and exchangeable fractions of cations at the cation exchange complex are calculated through a number of chemical equilibria and exchange reactions. Simulated pH does not affect any other variable or parameter in the other modules. For calcareous soils, the module describes the dissolution of calcite that neutralises the acid input (cf. De Vries et al. 1989). In non-calcareous soils, the acid input is neutralised by exchange with cations and aluminium hydroxide dissolution. The process description for non-calcareous soils used in the NUCOM model is given in Van Oene et al. (1999a).

20.2.5 Hydrology

The water balance is described assuming a free-draining soil profile. Two soil layers are distinguished: the LFH layer and the 0–30-cm layer. Precipitation, interception, soil evaporation, plant transpiration, percolation and soil water retention characteristics determine the water stored in each layer. Interception is calculated as a function of precipitation with a maximum interception reached at a LAI of 5 [Ward 1975; Table 20.2; Eq. (1)]. Interception is subtracted from precipitation. Net radiation that supplies the energy for evaporation and transpiration is estimated from general equations following Monteith and Unsworth (1990) and Kropff (1993). Soil evaporation follows the description of Van Laar et al. (1992), with evaporation taking place mainly from the upper soil layer, with some carryover to the lower layer. The Penman-Monteith equation (Monteith and Unsworth 1990) is used to calculate potential transpiration of the forest [Table 20.2; Eq. (2)]. The aerodynamic resistance is calculated following Thornley and Johnson (1990) [Table 20.2; Eq. (3)]. In accordance with Monteith, the canopy resistance of the stand is taken as the stomatal conductance of all leaves in parallel with stomatal conductance as described by Lohammer et al. (1980) [Table 20.2; Eqs. (4), (5)]. Water uptake is a function of water availability and root length [analogue to potential N uptake given in Table 20.1; Eq. (7)]. Water availability is the total amount of water in the soil layer minus that retained at permanent wilting point (pF = 4.2) and minus soil evaporation in that layer. The water demand is the potential transpiration divided proportionally with root length over the soil layers. Actual plant transpiration is the lower amount of water uptake and water demand. The ratio between actual and potential transpiration is used to describe water limitation on plant growth [Table 20.1; Eq. (5)].

Soil water characteristics relating volumetric soil moisture content and soil water potential are estimated based on an empirical relationship [Table 20.2; Eq. (6)] given by Thornley and Johnson (1990). The parameter b in this equation can be determined by a single point measurement of volumetric moisture content and soil potential, here taken to be at field capacity defined as pF = 2 (or $10 J kg^{-1}$) [Thornley and Johnson 1990; Table 20.2; Eq. (7)]. Percolation is calculated as the amount of water in excess of field capacity.

Model Analysis of Carbon and Nitrogen Cycling in *Picea* and *Fagus* Forests

Table 20.2. Equations used in the model describing interception, transpiration and soil moisture content

Interception	$IC_i = P \cdot \min\left(\dfrac{IC_{max,i}}{5} \cdot LAI_i, IC_{max,i}\right).$	(1)
Penman-Monteith	$\lambda E = \dfrac{\Delta \cdot R_n + \rho c_p \delta e / r_a}{\Delta + \gamma(1 + r_c / r_a)}.$	(2)
Aerodynamic resistance	$r_a = \dfrac{\ln[(Z + \zeta - d)/\zeta] \cdot \ln[(Z + \zeta_m - d)/\zeta_m]}{k^2 u_a}.$	(3)
Canopy resistance	$r_c = (g_s LAI)^{-1}$	(4)
	with $g_s = \dfrac{R_g}{R_g + R_0} \cdot \dfrac{g_{max}}{1 + a\delta e}.$	(5)
pF curve	$\theta_l = \sqrt[b]{\dfrac{\Psi_{field,l}}{\Psi_l} \cdot \theta_{field,l}^b}$	(6)
	with $b = \ln\left(\dfrac{\Psi_{field,l}}{0.375}\right) \Big/ \ln\left(\dfrac{0.557}{\theta_{field,l}}\right).$	(7)

Parameters: a = constant in stomatal conductance equation (kPa^{-1}), b = constant in pF curve equation (−), c_p = specific heat of air at constant pressure (1005 J kg^{-1} K^{-1}), d = zero plane displacement (0.64 × height, m), g_{max} = maximal stomatal conductance (m s^{-1}), $IC_{max,i}$ = maximum intercepted fraction of rain by species i (−), k = Karman's constant (0.40), R_0 = constant in stomatal conductance equation (W m^{-2}), γ = psychrometric constant (66 Pa K^{-1}), λ = latent heat of vaporisation of water (2465 J g^{-1}), ρ = density of dry air (1.2 kg m^{-3}), $\theta_{field,l}$ = volumetric moisture content at field capacity in soil layer l (m^3 m^{-3}), ζ = roughness parameter (0.026 × height, m), ζ_m = roughness parameter (0.13 × height, m), $\Psi_{field,l}$ = water potential at field capacity in soil layer l (10 J kg^{-1}). Variables: g_s = stomatal conductance (m s^{-1}), IC_i = intercepted rain by species i (mm H$_2$O timestep^{-1}), LAI_i = leaf area index of species i (−), R_g = global radiation above stand (W m^{-2}), R_n = net radiation above stand (W m^{-2}), r_a = aerodynamic resistance (s m^{-1}), r_c = canopy resistance (s m^{-1}), u_a = windspeed at reference height (m s^{-1}), Z = reference height (m), Δ = temperature derivative of saturated vapour pressure function (Pa K^{-1}), δe = vapour pressure deficit (Pa), θ_l = volumetric moisture content in soil layer l (m^3 m^{-3}), Ψ_l = water potential in soil layer l (J kg^{-1}).

20.3 Input Data and Parameter Values

20.3.1 Climatic and Deposition Data

Site-specific climate and deposition data are used. The climate data are daily values for global radiation, average air temperature, precipitation, vapour pressure and wind speed. These are used in the hydrological module to calculate the water balance on a daily time step. Monthly averaged values for hydrological variables are used in the other modules that run with a time step of 1 month. Since the sites had different numbers of climatic years available, the available data between 1993 and 1997 were averaged on a daily basis. This averaged climatic year is used as input for the simulations. Each simulation year thus has the same climatic data. A summary of these climatic data for temperature and precipitation are given in Table 2.2, Chapter 2, this Volume.

The input data for deposition are wet deposition data that are measured at the sites (or nearby) and EMEP dry deposition data. These latter data are model calculations based on emission points and atmospheric transport (Barrett and Berge 1996; Berge 1997). The data represent averaged dry deposition values for a grid of 150 × 150 km and do not include any scavenging of deposition by the forest. The available deposition data were averaged over the years 1993 to 1997 giving an averaged yearly value for wet and dry deposition which in the model are evenly divided over the months. Forests, however, scavenge deposition from the atmosphere. To include this effect, a forest-filtering factor is included for dry deposition. These filtering factors are estimated (based on Ivens et al. 1989; De Vries et al. 1993) to be for *Picea* 1.1, 1.9 and 1.6 for respectively NO_3^-, NH_4^+ and SO_2. For *Fagus* these values are 1.1, 1.2 and 1.2. The dry cation deposition is assumed to relate to SO_2 and has that filtering factor. The yearly averaged values for S and N wet and dry deposition are given in Table 2.1, Chapter 2, this Voulume.

20.3.2 Parameter Values

The parameter values describing plant species characteristics of *Picea* and *Fagus* are kept similarly for all sites (Table 20.3). Only the parameters that give the allocation of dry weight to different plant organs (Table 20.4) are calibrated to observed data (see Sect. 20.4 calibration). The parameter values that are based on data of the sites are averaged data for the *Picea* and *Fagus* sites separately. The N concentrations for branch, stem, coarse and fine roots are kept constant while the N concentration in the foliage can vary between a maximum and a minimum concentration. This minimum N concentration is set 15% lower than the lowest measured value of the sites, while the maximum concentration is estimated to be 45% higher than the highest observed value.

The decomposition rates of soil organic matter ($gCgC^{-1}day^{-1}$) of the sites determined by Persson et al. (see Chap. 12, this Vol.) are used for decomposition of leaves and fine roots in L and FH layer, and all plant organs in the 0–30-cm layer. The decomposition rates of branches and coarse roots in L and FH layer are put to half the measured values of these layers because the measured data were based on sieved material and thus larger parts are excluded. These values impinge only on dead material of branches and coarse roots produced during the simulations, which is a smaller part of total mortality. Persson determined the decomposition rates at a temperature of 15 °C and 60% of soil moisture content at field capacity. From these values we calculated the site-specific a-values in Eq. (12) (Table 20.1) using the original value for a of –3.764, for b of 0.204 and the optimal temperature ($Topt_{dec}$) of 36.9 (Kirschbaum 1995). The effect of clay on the decomposition rate is assumed not be included in the measured data since soil moisture content was high. The clay content of the 0–30-cm layer is given in Table 20.4. The parameter values determining N mineralisation, the assimilation efficiency of the microbes and the critical N concentration, were taken similarly for eight of the ten sites. These values are based on Bosatta and Berendse

Table 20.3. Parameter values for Picea and Fagus

Parameter	Unit	Picea	Fagus	References
k_{ext}	(–)	0.50	0.65	Hoffman (1995), Bossel (1996)
SLA	($cm^2 g^{-1}$)	39.0	18.9	Data sites
RUE	($g\,dw\,MJ^{-1}$)	1.9	1.9	Derived Monteith (1977), Cannell et al. (1987), Cannell (1989)
T_{min}	(°C)	–4	–1	Hoffman (1995), Bossel (1996)
T_{opt1}	(°C)	10	15	Hoffman (1995), Bossel (1996)
T_{opt2}	(°C)	26	25	Hoffman (1995), Bossel (1996)
T_{max}	(°C)	40	40	Hoffman (1995), Bossel (1996)
β	(–)	0.70	0.70	Goudriaan and Unsworth (1990)
$n_{c,max}$-leaf	($mg\,g^{-1}$)	22.0	40.0	Estimate
$n_{c,min}$-leaf	($mg\,g^{-1}$)	6.5	9.3	Data sites
n_c-branch	($mg\,g^{-1}$)	4.6	4.0	Data sites
n_c-stem	($mg\,g^{-1}$)	1.1	1.1	Data sites
N_c-roots	($mg\,g^{-1}$)	0.8	0.2	Data sites
n_c-fine roots	($mg\,g^{-1}$)	12.5	12.5	Data sites
SRL	($m\,g^{-1}$)	14.5	12.6	Data sites
k_{rf}	(–)	0.95	0.93	Jackson et al. (1996)
Age_{max}-leaf	(day)	3880	200	Estimate
Age_{max}-fine roots	(day)	300	300	Estimate
k_m-branch	(yr^{-1})	0.0040	0.0014	Derived De Vries et al. (1990)
k_m-stem	(yr^{-1})	0	0	Assumption
Max fraction of N reallocated – leaf	(–)	0.3	0.55	Estimate Chapin and Kedrowski (1983), Del Arco et al. (1991), Killingbeck (1996)
Max fraction of N reallocated – fine roots	(–)	0.2	0.2	Estimate Chapin and Kedrowski (1983), Del Arco et al. (1991), Killingbeck (1996)
IC_{max}	(–)	0.30	0.20	Ward (1975)

(1984) and McClaugherty et al. (1985), and correspond to microbial N:C ratios of 0.08–0.115. For two sites, Wal and AuF, the critical N concentration is set lower after calibration (see Sect. 20.4).

The measured soil moisture content at field capacity for the layers is given in Table 20.4. For Åhe, MdM, Sch and Col no data were available; these were estimated based on the data of the other sites. The data for Sko are calculated from the given moisture content at saturation (pF = 0). The parameters used in the calculation of stomatal conductance [Eq. (5), Table 20.2] are normally determined by regression of measured stomatal conductance to global radiation and vapour pressure deficit for a specific site. Measured stomatal conductance data are, however, lacking. Only for Sko have these parameter values been calculated (Cienciala et al. 1992). Therefore the parameter values of Sko were used as starting point for the other sites where g_{max} is determined further by calibration (see following section).

Table 20.4. Parameter values for allocation of carbohydrates, decomposition and hydrology per site (see text and Tables 20.1, 20.2)

Code	Åheden	Skogaby	Nacetin	Waldstein	Aubure-P	Monte di Mezzo	Jezeri	Schacht	Aubure-F	Collelongo	Åheden	Waldstein	Waldstein
	Åhe	Sko	Nac	Wal	AuP	MdM	Jez	Sch	AuF	Col	Åhe2	Wal2	Wal3
Allocation													
Shoot	0.65	0.65	0.60	0.70	0.75	0.55	0.70	0.70	0.70	0.70	0.65	0.70	0.70
Leaf	0.30	0.30	0.35	0.45	0.30	0.30	0.38	0.38	0.38	0.38	0.30	0.45	0.45
Branch	0.17	0.17	0.15	0.20	0.17	0.17	0.19	0.19	0.19	0.19	0.17	0.20	0.20
Stem	0.53	0.53	0.50	0.35	0.53	0.53	0.43	0.43	0.43	0.43	0.53	0.35	0.35
k_{all}-roots[a]	0.05	0.05	0.05	0.05	0.05	0.05	0.05	0.05	0.05	0.05	0.05	0.05	0.05
k_{all}-fine roots	0.30	0.30	0.35	0.25	0.20	0.40	0.25	0.25	0.25	0.25	0.30	0.25	0.25
Decomposition													
Clay%-0–30 cm	3.0	6.4	9.7	12.6	20.2	35.0	11.3	11.5	7.7	30.5	3.0	12.6	12.6
k_{ass}	0.20	0.20	0.20	0.20	0.20	0.20	0.20	0.20	0.20	0.20	0.50	0.20	0.20
n_{crit}-L layer[a]	0.016	0.016	0.016	0.016	0.016	0.016	0.016	0.016	0.016	0.016	0.016	0.016	0.016
n_{crit}-L layer[b]	0.023	0.023	0.023	0.023	0.023	0.023	0.023	0.023	0.023	0.023	0.023	0.023	0.023
n_{crit}-FH-layer	0.016	0.016	0.016	0.010	0.016	0.016	0.016	0.016	0.010	0.016	0.016	0.010	0.010
n_{crit}-0–30 cm	0.016	0.016	0.016	0.010	0.016	0.016	0.016	0.016	0.010	0.016	0.016	0.010	0.010
Hydrology													
θ_{field}-LFH layer	0.54	0.40	0.54	0.54	0.54	0.54	0.54	0.54	0.54	0.54	0.54	0.54	0.54
θ_{field}-0–30 cm	0.30	0.30	0.40	0.51	0.25	0.37	0.34	0.51	0.31	0.37	0.30	0.51	0.51
R_0	136.8	136.8	136.8	136.8	136.8	136.8	136.8	136.8	136.8	136.8	136.8	136.8	136.8
g_{max}	0.023	0.023	0.045	0.012	0.110	0.015	0.030	0.004	0.090	0.010	0.023	0.009	0.009
a	2.79	2.79	2.79	2.79	2.79	2.79	2.79	2.79	2.79	2.79	2.79	2.79	2.79

[a] For the plant organs leaves and fine roots.
[b] For the plant organs branch, stem and coarse root.

20.4 Model Calibration and Comparison with Measured Data

20.4.1 Calibration Method

Simulations were run for 25 years with constant climate. The parameters for allocation, the parameter g_{max}, and the fractions of soil organic matter that yearly pass through to the next organic material class (Table 20.5) are used to calibrate to measured data of leaf biomass and leaf N concentration. The parameter g_{max} can be used for calibration because it is a parameter in the calculation of potential evaporation; it can thus affect the ratio of actual to potential evaporation (a measure of stomatal closure) that determines assimilation. All other parameters are kept constant for each tree species. The calibration is performed in such a way that an equilibrium situation is created with constant leaf biomass, constant leaf N concentration, nearly constant growth and nearly constant N mineralisation during these 25 years. However, soil organic C and N pools are increasing. Simulated values are given as the average over these 25 years. This equilibrium situation is representative for the actual field situation, where mature forests have closed canopies and are adjusted to climatic conditions and N availability. The calibrated parameters of the sites give information on existing differences between the sites.

The calibrated values for g_{max} (Table 20.4) correspond to the different climatic conditions of the sites compared to the reference site Sko (Table 20.6). Higher radiation and higher vapour pressure deficit increase evaporative demand. Compared to Sko, sites with higher global radiation have increased g_{max}, while sites with higher vapour pressure deficit have decreased g_{max} in order to diminish evaporative demand. Only the French sites, AuF and AuP, deviate in this respect; both needed an increased g_{max} value to fit to measured leaf biomass and leaf N concentration. Comparison of calibrated g_{max} values with literature data of maximal stomatal conductances of leaves is not appropriate, since how g_{max} is used here in calibration expresses canopy properties rather than leaf properties. Highest simulated values for stomatal conductance (Table 20.6) are lower than the parameter values for g_{max}. Highest simulated values for canopy conductance (Table 20.6) lay in the range given by Schulze et al. (1995), except for the French and Czech sites.

For Wal, MdM and Åhe, the calibration is not optimal, while for the other sites the calibrated simulation results are satisfactory (see next section). Therefore, further calibrations were performed for these three sites. For MdM, no better calibrated situa-

Table 20.5. Calibrated yearly fractions of soil organic matter in L layer and FH layer that pass through to the next soil organic matter class (see text)

	Fractions of soil organic matter passing through to next class[a] ($1 \, yr^{-1}$)												
	Åhe	Sko	Nac	Wal	AuP	MdM	Jez	Sch	AuF	Col	Åhe2	Wal2	Wal3
L layer	1/6	1/2	1/4	1/5	1/1	1/4	1/4	1/6	1/6	1/4	1/6	1/2	1/2
FH layer	1/60	1/20	1/90	1/110	1/90	1/1	1/150	1/80	1/20	1/1	1/60	1/70	1/70

[a] At AuP no L layer is present, at MdM and Col no FH layer is present. At these sites the soil organic matter is passed through to the next layer in 1 year.

Table 20.6. Annual global radiation, averaged daily vapour pressure deficit, simulated leaf area index (LAI), highest simulated stomatal conductance, and highest simulated canopy conductance

	Åhe	Sko	Nac	Wal	AuP	MdM	Jez	Sch	AuF	Col	Åhe2	Wal2	Wal3
Annual global radiation ($MJ\,m^{-2}\,yr^{-1}$)	3225	3083	5192	3291	3513	5162	5192	3291	3513	4314	3225	3291	3291
Average daily vapour pressure deficit (kPa)	0.247	0.188	0.212	0.876	0.156	0.258	0.212	0.876	0.156	0.312	0.247	0.876	0.876
Simulated LAI	1.8	5.9	5.4	6.8	3.5	6.6	7.0	6.2	4.1	5.7	2.4	7.5	7.0
Highest simulated stomatal conductance ($m\,s^{-1}$)	0.007	0.007	0.021	0.002	0.036	0.008	0.014	0.001	0.029	0.004	0.007	0.002	0.002
Highest simulated canopy conductance ($m\,s^{-1}$)	0.014	0.042	0.106	0.014	0.125	0.047	0.068	0.003	0.109	0.021	0.014	0.011	0.010

tion could be found. For the Åhe and Wal sites, the new calibrated points are denoted Åhe2 and Wal2. In addition, a third point, Wal3, is included which equals Wal2 in parameter setting but includes also an understorey vegetation (represented as *Vaccinium myrtillus*). For Wal2 and Wal3 the new calibrated parameter values are: g_{max}: 0.009, Age_{max}: 2880, RUE: 2.1, k_{ass}: 0.5 and n_{crit} in FH and 0–30 cm layer: 0. For the Åhe2 site only microbial efficiency was adjusted (kass:0.5). These additional calibrated sites are included in the tables and figures.

20.4.2 Comparison of Measured and Simulated Forest Growth Data

In Fig. 20.2, the measured and calibrated simulated data are given. Simulated leaf biomass closely corresponds to the measured data except for Åhe. Simulated leaf N concentration shows a similar picture, with the largest deviation for Jez. At this site the measured data give very low values, probably due to collection in October after a dry period (T. Paces, pers. comm.). Leaf growth is overestimated for MdM and underestimated for Wal. For Wal, stem growth is also underestimated while for all other

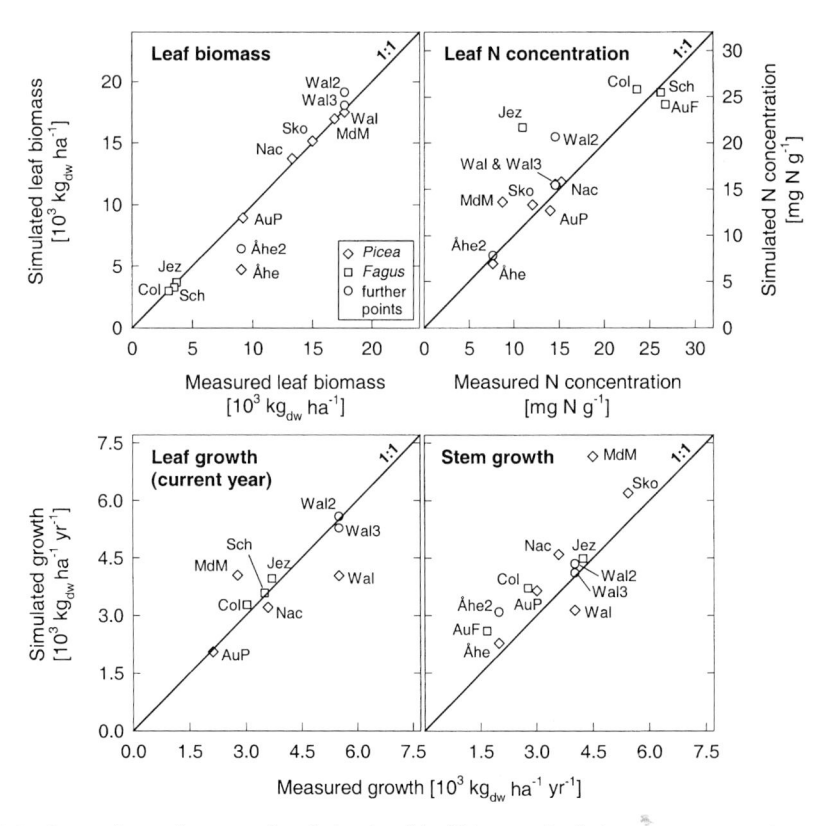

Fig. 20.2. Comparison of measured and simulated leaf biomass, leaf nitrogen concentration, leaf and stem growth for the CANIF sites that had these data available. Åhe *measured* stem growth is for *Pinus* only, while *simulated* is for *Picea* plus *Pinus*. Diamonds *Picea* forests; *squares Fagus* forests; *circles* further calibrated points for Åhe2, Wal2 and Wal3 see text

sites it is overestimated. Simulated stem growth for Åhe includes both *Picea* and *Pinus*, while measured data includes only *Pinus*. Simulated values for *Pinus* alone underestimate somewhat.

Table 20.7 gives the simulated N fluxes and simulated forest N demand. Comparison with measured N demand for stem and leaf growth agrees with the above picture. The simulated N demand of MdM is overestimated compared to the measured demand. At the Jez site, the measured demand is low due to the low measured N concentration. Taking into account also N leaching at Wal, Nac and AuP, the simulated N availability is low at these sites since only demand is met and simulated N leaching is zero. For Wal and AuF, the parameters for mineralisation (critical N concentration of soil organic matter) are already set lower to meet the demand of the forest. For Åhe, the simulated N availability is too low to keep the forest at equilibrium, even when immobilisation of N by microbes is set to zero. The underestimation of N leaching is also a consequence of model description where no horizontal heterogeneity is accounted for.

The simulation of the Sko site where no calibration, except for allocation, is performed indicates that the model is able to reproduce correct biomass and leaf N concentration when parameters for stomatal conductance are known. The mean simulated ratio of actual to potential transpiration of 0.72 agrees with the measured value of 0.68 (Cienciala et al. 1994). Also simulated potential transpiration (552 mm) corresponds to calculated potential transpiration (480 mm) (Cienciala et al. 1994). The approach used for the other sites by calibrating one of the involved parameters to observed leaf biomass and leaf N concentration therefore gives a realistic picture of the forest dynamics.

The simulation of the Åhe site indicates that not enough N becomes available through mineralisation to keep the forest in steady-state situation. High immobilisation by microbes in this strongly N-limited system is a major explanatory factor. However, the fact that the forest manages to grow indicates that other sources of N are used, probably small organic N compounds such as amino acids and protein, thereby shortcutting the normal mineralisation pathway, as demonstrated by Näsholm et al. (1998). In the Åhe2 calibrated site, N availability was artificially enlarged by increasing microbial efficiency. However, the Åhe2 site still hardly matches N demand.

The Wal site behaves differently from the other sites with respect to allocation of carbohydrates. The forest at the Wal site allocates more carbohydrates to leaf growth than to stem growth while all other sites have the reversed allocation pattern. The underestimated leaf growth at this site indicates that the foliage has a faster turnover than is simulated (too high maximum leaf life span at this site). Besides this, it also indicates that the actual leaf biomass is not able to produce the necessary carbohydrates for the measured growth and that intercepted radiation at this site is used more effectively than given in the radiation use efficiency parameter. An explanation for the different allocation pattern might be that N taken up directly into the canopy, estimated to amount up to 20% of N use at the Wal site (see Harrison et al. Chap. 8, this Vol.), is assimilated directly and not transported downwards. Simulated N deposition (including scavenging) is about 20% higher than total N deposition but is clearly underestimating canopy uptake. However, even when simulating the canopy uptake correctly, it would require a more detailed description of where the N is used in order to simulate the changed allocation pattern.

Table 20.7. Simulated and measured N fluxes for the CANIF sites

	Åhe	Sko	Nac	Wal	AuP	MdM	Jez	Sch	AuF	Col	Åhe2	Wal2	Wal3[a]
Simulated (kg N ha^{-1} yr^{-1})													
N availability, of which	40	119	115	98	51	183	104	104	60	129	56	163	166
N deposition[b]	2	18	22	24	18	12	22	21	15	11	2	24	24
N mineralisation	39	100	94	73	34	171	82	83	45	117	54	139	142
N leaching	0	0	0	0	0	0	1	0	0	7	0	0	0
N uptake	40	119	115	98	51	183	103	103	60	122	56	163	127 (39)
Total N demand, of which	41	133	131	117	59	200	144	142	90	134	57	191	151 (39)
leaf and stem	10	55	57	68	30	66	90	93	58	89	15	124	87 (30)
fine roots	26	67	67	40	23	123	47	42	27	39	35	55	52 (7)
Measured (kg N ha^{-1} yr^{-1})													
N demand leaf+stem[c]			59	84	29	33	45	92		75		84	84
N leaching[d]	–	–	+	+	+	?	?	–	–	?	–	+	+

[a] In brackets the values for the understorey.
[b] Total N deposition including forest filtering.
[c] Measured leaf growth times measured leaf N concentration plus measured stem growth times 1.1 mg N g^{-1}.
[d] Not measured but based on inorganic N concentration in soil solution at depth >50 cm.

Based on these observations, the parameter settings for the additional calibrated points Wal2 and Wal3 were chosen in combination with artificially enlarged N mineralisation by increasing microbial efficiency and partly neglecting immobilisation. The Wal2 calibrated site attains the high measured growth rates of leaves and stem, but overestimates leaf biomass and leaf N concentration. Conversely, at the Wal3 site also leaf biomass and leaf N concentration correspond to measured data due to a competitive understorey vegetation that takes part (23%) of the available N.

At the MdM site, growth is overestimated, which might be due to an underestimated effect of water limitation on growth and/or an underestimated effect of moisture reduction on decomposition and mineralisation. For MdM, no better calibrated situation could be found mainly because too few measured soil data were available and indicating the fact that soil moisture limitation on decomposition for the 0–30 cm layer may be different from the curve used that was determined for LFH layer by Seyferth (1998).

The calibration of g_{max} for the AuP and AuF sites does not correspond to the climatic conditions at these sites. However, these sites are situated close to the mountain ridge at a slope of, respectively, 30 and 26%. Therefore soil moisture content may be overestimated and less water may be available for transpiration, resulting also in a reduction of potential assimilation.

20.4.3 Soil Organic Matter Dynamics

The simulated C and N mineralisation can be compared to measured data at the same sites. Persson et al. (see Chaps. 12 and 14, this Vol.) extrapolated mineralisation rates from laboratory incubations. Harrison et al. (see Chap. 11, this Vol.) calculated C and N fluxes using a ^{14}C technique. However, the latter values are more difficult to compare with because the subdivision into soil layers has been done differently.

Simulated C mineralisation rates in L and FH layer and totally are higher than those calculated by Persson et al. (Fig. 20.3), whereas C mineralisation rates for the mineral soil 0–30 cm correspond well. The simulated C mineralisation use the optimal relative C mineralisation rates ($gCgC^{-1}day^{-1}$) determined under laboratory conditions by Persson et al. The differences therefore can only be due to the different calculation methods for temperature and moisture dependencies. Differences in calculating temperature dependency play no role – the two methods result in less than 2% difference – because these dependencies are almost identical up to 25 °C, where Seyferth's measurements stop, while Kirschbaum's data are valid up to 30 °C. Differences in calculation of soil moisture effects on mineralisation do occur, although the calculated reduction effect is based on the same study for two Swedish sites (Seyferth 1998). Seyferth determined reduction effects on decomposition at different volumetric soil water contents expressed relatively to volumetric soil water content at 60% of field capacity. Persson et al. calculated the soil moisture reduction effect related to the water potential values. In this modelling study we expressed the soil moisture reduction effect related to volumetric soil water content. Differences between the two methods arise due to varying water potential – soil water content relationships (pF curves) in different soil layers and at different sites. It appears unclear if lower decomposition rates under dryer conditions caused by malfunctioning microbes need to be related to higher water potentials or to lower soil water contents. However, in the simulations, soil moisture hardly affected C mineralisation at any of the sites. The cal-

Fig. 20.3. Comparison of measured and simulated C mineralisation in the L, FH, 0–30 cm and L-30-cm (total) soil layers. Measured data are from Harrison et al. (Chap. 11, this Vol.) and Persson et al. (Chap. 12, this Vol.). Note that Harrison et al. subdivided soil layers into LF layer (in figure L layer) and 0–5 cm including the H layer (in figure FH layer). Persson et al. and the model study subdivide into L, FH and mineral soil layers. *Diamonds Picea* forests; *squares Fagus* forests; *open points* measured data Persson et al.; *filled points* measured data Harrison et al.

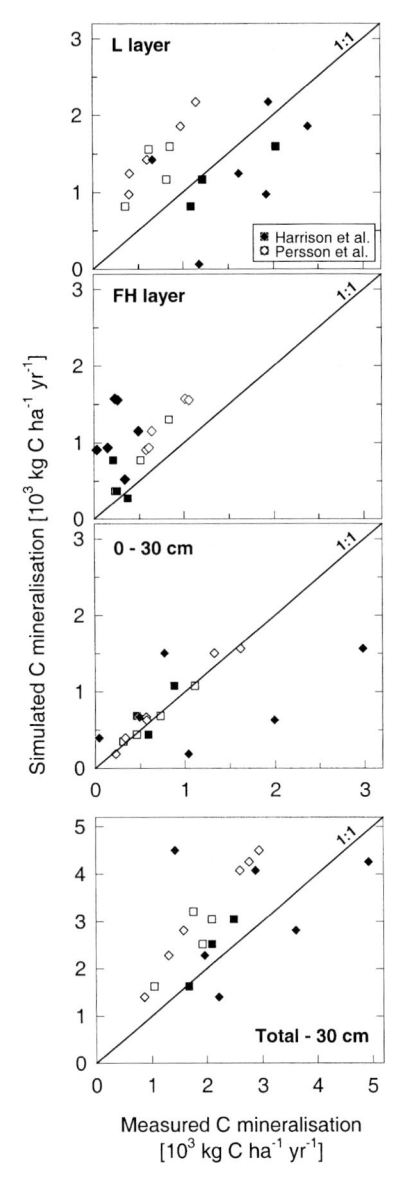

culated effect of soil moisture by Persson et al. alone is not known. There appears to be no conforming view how to describe soil moisture limitations on decomposition. As in this study, O'Connell (1990) related decomposition to soil moisture contents. Other approaches used relate decomposition to the ratio between available water and potential evapotranspiration (Parton et al. 1996), or to actual evapotranspiration (Berg et al. 1993).

More importantly in this study, the difference in C mineralisation rates seems to be related to the pool of soil organic C that increases during the simulation of 25 years and thus also gives higher C mineralisation values. However, also the first year's simulation results are higher than those of Persson due to also an input through litterfall and root decay in that year, while in Persson's measurements no input of new litter occurs.

Comparison of simulated data with measured data of Harrison et al. show for the 0–30-cm layer and totally a fluctuating pattern; some sites are underestimated and some overestimated. The measured data in the L layer are higher than the simulated ones, probably caused by the inclusion of the F layer in the measured data. For the same reason, the measured data in the FH layer are lower due to the fact that measured data consist of the 0–5-cm layer that includes the H layer.

Comparing mean residence time for soil organic C, calculated as C pool divided by C mineralisation for Persson et al. and the model results and calculated by Harrison et al. based on the ^{14}C-bomb approach, correspond to the observed differences in C mineralisation rates (Fig. 20.4). The simulated mean residence times agree reasonably well with the values of Persson et al. and since they are derived from the same basic data, it indicates that differences in calculation method for soil moisture reduction did not play such a large role and that the higher simulated C mineralisation rates are similarly related to the higher simulated soil organic C pools. Where Harrison et al. find a positive correlation of mean residence time with latitude for the 0–5-cm layer (here taken as FH layer) and a negative correlation with temperature, both the simulated and Persson's mean residence time depict highest values at the mid-continental sites where N deposition levels are highest. However, also Harrison et al. find a higher mean residence in relation to N deposition, with Åhe being an outlier. In the simulated data, temperature is not a major determining factor for mean residence time. However, the calculation of mean residence time as C pools divided by C mineralisation neglects other C fluxes such as DOC leaching, although in the simulated values vertical fluxes of C are included by transfer of one soil organic matter class to another (Table 20.5). The discrepancies might become smaller when knowledge on the most northern and most N-limited site, Åhe, would be better.

Comparing the simulated with the measured N mineralisation data (Fig. 20.5) of Persson et al. (see Chap. 14, this Vol.), the simulated data show a tendency to overestimate at lower measured N mineralisation rates and to underestimate at higher N mineralisation rates in the L and FH layer, whereas in the 0–30-cm layer simulated values are mostly lower than the measured ones. Related to the calculated data of Harrison et al., an almost similar picture arises as for C mineralisation with the MdM site as an extreme outlier in the 0–30-cm layer. In the simulations the N mineralisation rates are dependent on decomposition rates and microbial characteristics [Table 20.1, Eq. (15)]. Due to immobilisation the mean residence times based on the N pools and N fluxes are thus always larger than the C-based mean residence times (Table 20.8). Only for the Wal2 and Wal3 sites, where in the FH and 0–30-cm layer microbial immobilisation is set to zero and assimilation efficiency is enlarged (the latter at the Åhe2 site also) (Table 20.4), are N-based mean residence times larger than the C-based. The measured data of Persson et al., however, depict in the L layer for the German sites, and in the FH and 0–30-cm layer for all but two sites, higher N than C turnover rates. The MdM site and the Åhe site deviate from this picture with a strong immobilisation at the Åhe site and no net N mineralisation. The N mineralisation data of Harrison et al., however, are precisely calculated on the presumption that C and N mean residence times are equal. The description of N mineralisation in the model is based

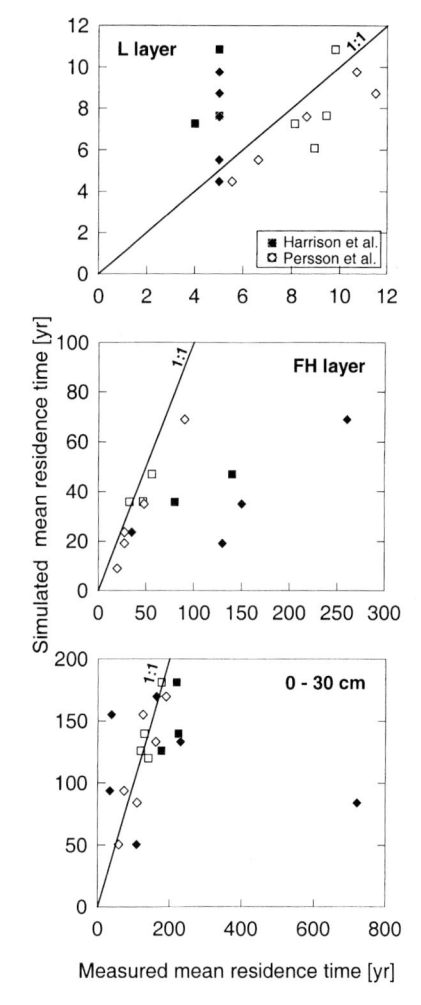

Fig. 20.4. Comparison of measured and simulated mean residence times of soil organic C at the CANIF sites. Measured data are from Harrison et al. (Chap. 11, this Vol.) and Persson et al. (Chap. 12, this Vol.). *Notes* and *points* as in Fig. 20.3

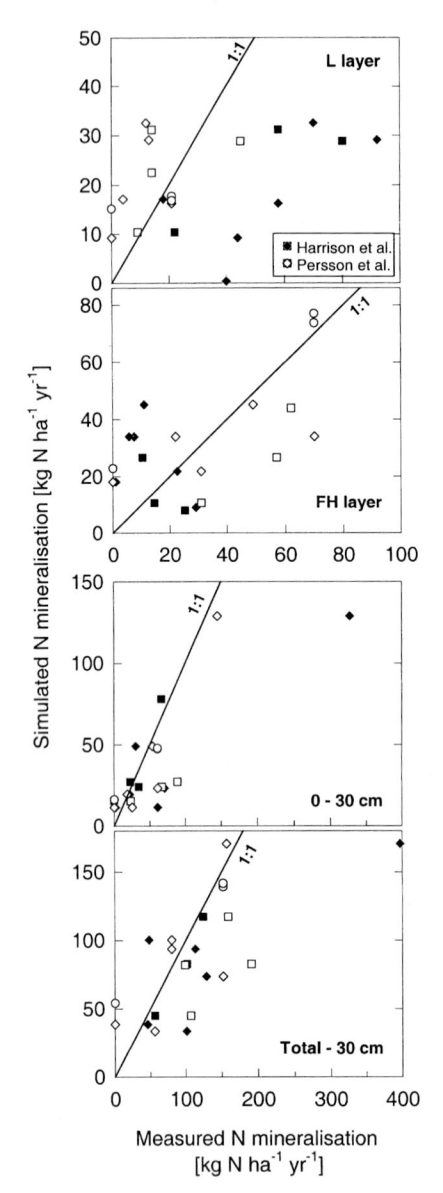

Fig. 20.5. Comparison of measured and simulated N mineralisation in the L, FH, 0–30-cm and L-30-cm (total) soil layers. Measured data are from Harrison et al. (Chap. 11, this Vol.) and Persson et al. (Chap. 12, this Vol.). *Notes* and *points* as in Fig. 20.3

on C limitation of microbial growth of decomposers and that N is released when it is in excess of microbial needs (Bosatta and Berendse 1984; Ågren and Bosatta 1987; Bosatta and Ågren 1991). The measured data of Persson et al. for the Åhe, Sko and MdM are consistent with this view. Also the L layer of most sites corresponds to this view, except for the German sites. In addition to the different soil layer division, neglect of immobilisation by setting N turnover equal to C turnover in the calculations of Harrison et al. leads to their high N mineralisation data in the L layer. The relative decomposition rates (g C g C^{-1} day^{-1}) of Persson et al. depict a tendency to

Table 20.8. Mean residence times of soil organic matter based on N (i.e. N pools divided by N fluxes) divided by mean residence times of soil organic matter based on C (i.e. C pools divided by C fluxes) for the measured data of Harrison et al. (Chap. 11, this Vol.), Persson et al. (Chaps. 12, 14, this Vol.), and the simulated data

		Mean residence times – N-based/mean residence times – C-based												
		Åhe	Sko	Nac	Wal	AuP	MdM	Jez	Sch	AuF	Col	Åhe2	Wal2	Wal3
Persson et al.	L	0.0	4.2	2.8	0.8		4.6	1.8	0.8	1.5	2.8			
	FH	0.0	1.6	0.9	0.4	0.8		0.7	0.5	0.4				
	0–30	0.0	1.0	1.2	0.4	0.6	1.2	0.7	0.4	0.4	0.6			
Simulated	L	2.6	2.2	1.9	2.0		1.9	2.0	1.7	2.1	1.4	1.9	1.2	1.4
	FH	2.4	1.5	1.3	1.1	1.6		1.3	1.2	1.1		1.9	0.5	0.6
	0–30	1.4	1.3	1.4	1.1	1.1	1.0	1.2	1.2	1.0	1.0	0.9	0.5	0.5
Harrison et al.[a]	all layers	1.0	1.0	1.0	1.0	1.0	1.0	1.0	1.0	1.0	1.0	1.0	1.0	1.0

[a] Presumption that N turnover is equal to C turnover in order to calculate N fluxes from N pools.

decrease at lower C:N ratio of the soil organic matter in the L and FH layer (Persson et al. Chap. 12, this Vol.) consistent with the view that at higher N availability (deposition rates) decomposition is hampered probably due to a slower decomposition of lignin compounds at higher N concentrations (Berg and Ekbohm 1991). Contrarily, in the 0–30-cm layer an opposite tendency is present, but at most sites the soil organic material in this layer is probably too old to be affected by the high N deposition levels of the recent decades. Thus, in the measured data of L and FH layer of Persson et al., lower relative decomposition rates go together with higher relative N mineralisation rates (g N g C^{-1} day^{-1}) for the sites that have higher N than C turnover rates. It remains unclear if there is an overestimation of measured relative N mineralisation rates due to presence of inorganic N or already decomposed small organic N compounds in the starting material since, e.g. at the Sch site, the measured N mineralisation seems overestimated compared to growth demand (even when an understorey vegetation is included) and the fact that there is probably no (or hardly) leaching at this site (Table 20.7). It also might indicate that microbial efficiency is larger (at similar decomposition rates lower C mineralisation rates but relatively higher N mineralisation rates) or that the chemistry of the soil organic matter is different. The N mineralisation rates at the AuF, AuP and Wal sites, simulated using the above approach, are too low to meet demand and expected N leaching rates. Only by setting microbial immobilisation lower and increasing microbial efficiency could the demand for tree growth be met. The mean decomposer community properties as simulated by Wolters et al. (Chap. 17, this Vol.) using a foodweb approach, however, do not indicate different microbial efficiencies or microbial C:N ratios for the Åhe, AuP, Sko and Wal sites.

Conclusively, the interactions between C and N in decomposition and mineralisation are not yet fully understood. In addition, the uptake of organic N at the Åhe sites requires another modelling approach, but knowledge for this is still lacking.

20.5 Model Analyses

In the first part of this section, sites will be compared in their performance for forest growth, mineralisation and net ecosystem productivity. These components will be related to different ecosystem variables. How the different sites will react to changes in environmental conditions is tested in the second part to this section. MdM is excluded from these sensitivity analyses since simulated growth did not correspond to measured data due to a lack of soil data that made calibration difficult and probably also due to the land-use history. In addition, an example is given of the effect of logging on net ecosystem productivity.

20.5.1 Site Comparisons and Relationships Between Variables

20.5.1.1 Forest Growth

Forest growth is strongly determined by N availability (Fig. 20.6A). Increased N availability leads to higher net primary productivity (NPP), defined as assimilation minus respiration, thus the new C growth of all plant organs. The net increment in the C pool of the vegetation (Fig. 20.6B) is, however, also determined by losses by litterfall and root decay. Trees at sites with higher allocation to long-living plant organs as stems, branches and coarse roots have a relatively higher increment in C pool over

Fig. 20.6A,B. Simulated (**A**) net primary productivity (NPP) and (**B**) net increment of C in vegetation related to N availability. *Points* as in Fig. 20.2

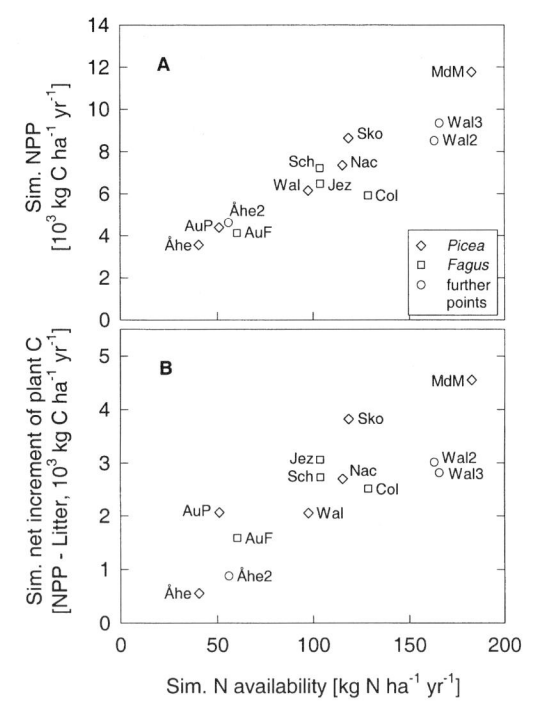

time. N availability is the sum of N mineralisation and N deposition (see Table 20.7). Where N deposition is mainly determined by external factors, N mineralisation is determined by site-specific factors. The N mineralisation rate is controlled by the amount of soil organic N present as well as by the decomposition and mineralisation characteristics of the microbial community at a site in combination with quality aspects of soil organic matter. Where N mineralisation is positively correlated with the amount of soil organic N (Fig. 20.7A), the quality aspect expressed in the C:N ratio of SOM hardly shows a relationship (Fig. 20.7B). Linking N mineralisation directly to litter input and root decay, the source of soil organic matter, a strong relationship exists with the total amount of litterfall and root decay (Fig. 20.8A) and a weak relationship with the C:N ratio of this dead material (Fig. 20.8B). Splitting the total amount of litterfall and root decay into above-ground and below-ground parts of decaying material, a similar relationship exists for each separately (Fig. 20.8C,D). Decaying leaves and fine roots make up >70% (except Åhe 40%) and >95%, respectively, of above- and below-ground, the remainder is branches and coarse roots. This picture also shows that differences in allocation pattern between plant organs with a fast turnover (leaves and fine roots) is negated in the effect of total litterfall and root decay on growth. Forest growth is thus strongly controlled by N availability, and, in turn, N mineralisation is strongly controlled by litterfall and root decay that is higher at higher growth rates and strongly dependent on leaf and fine root turnover. Noteworthy is that there are no differences between *Fagus* and *Picea* forests (except for lower C:N ratios in litterfall in *Fagus*), each site taking its own point along the curves.

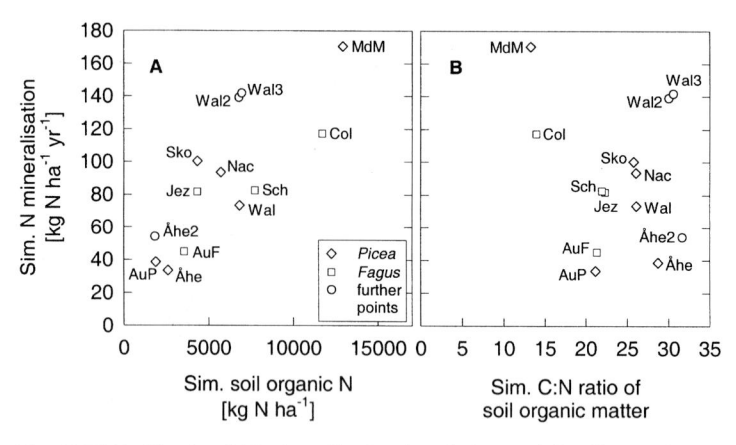

Fig. 20.7A,B. Simulated N mineralisation in relation to (**A**) soil organic N pool and (**B**) C:N ratio of soil organic matter. *Points* as in Fig. 20.2

The final N concentration in the leaves, affecting also the quality of the decaying material, is a result of both N availability and growth (carbohydrates). The latter amount is determined not only by N availability but also by other growth-limiting factors such as reduction of assimilation by closure of stomata by water stress [Table 20.1, Eq. (5)]. Figure 20.9 depicts the relative importance of N and water limitation at the different sites. Åhe clearly is strongly N-limited. Most other sites are rather balanced between the two limitations, whereas AuP and Nac show largest growth reduction by water stress. Even under conditions of high water supply, the ratio of actual to potential transpiration may be reduced, as indicated by the irrigated plot at the Skogaby experimental site that had a ratio of actual to potential transpiration of 0.79 whereas the drought plot had a ratio of 0.44 (Cienciala et al. 1994).

20.5.1.2 C Mineralisation

The dynamics of soil organic matter are determined by the rate at which it is decomposed and the rate at which it is replenished by litterfall and root decay. The C mineralisation rate, as measure for the decomposition rate, is dependent on temperature [Table 20.1, Eq. (12)]. The simulated C mineralisation, however, does not show a relationship with annual temperature (Fig. 20.10A) or with higher soil organic C pools (Fig. 20.10B). The input of litterfall and root decay to the soil organic pool, however, appears to be a controlling factor for C mineralisation (Fig. 20.10C). The measured relative mineralisation rates ($gCgC^{-1}day^{-1}$) at optimum temperature for the litter layer are on average 15 times (but vary between sites from 7 to 23 times) higher than in the 0–30-cm layer and on average 5 times higher (but vary between sites from 2 to 7 times) than in the FH layer (Persson et al. Chap. 12, this Vol.). Although average mean residence time of soil organic matter in the litter layer is only a few years (Fig. 20.5), simulated C mineralisation in the litter layer amounted on average to 39% of total C mineralisation. The contribution of the FH layer was about equal in size whereas the deeper 0–30-cm layer made up the remaining part. For the Åhe site it might be noted that optimal relative mineralisation rates of litter and FH layer were

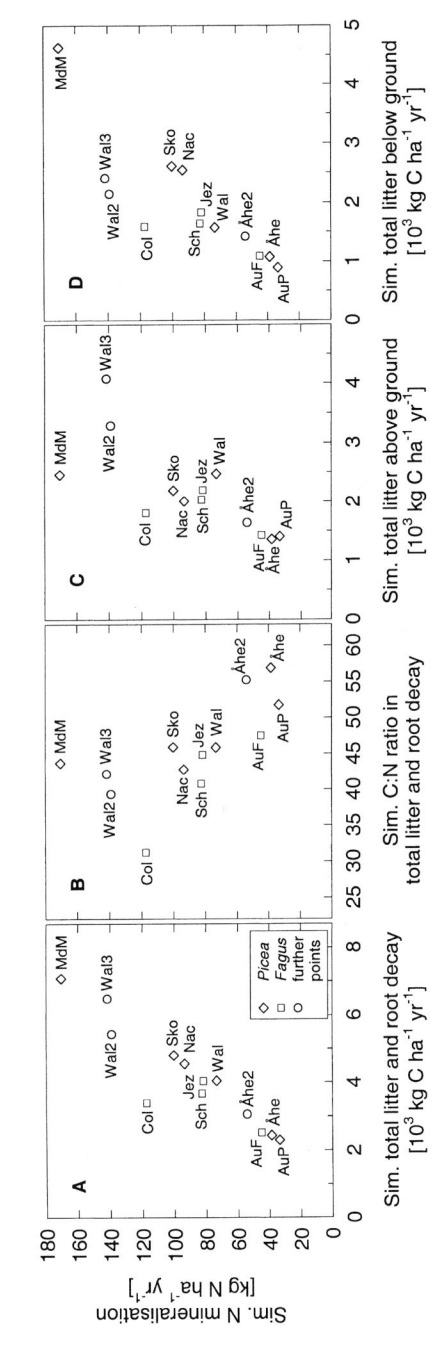

Fig. 20.8A–D. Simulated N mineralisation in relation to (**A**) total litterfall and decay and (**B**) C:N ratio in total litter and root decay, (**C**) above-ground and (**D**) below-ground litterfall and decay. *Points* as in Fig. 20.2

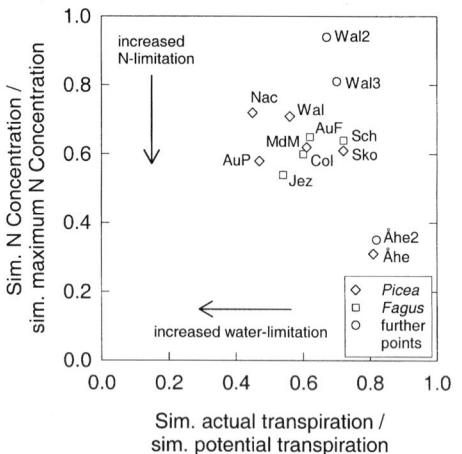

Fig. 20.9. Simulated leaf N concentration relative to maximum leaf N concentration, expressing nitrogen limitation, against actual to potential transpiration, expressing water limitation. *Points* as in Fig. 20.2

absolutely highest of all sites. Soil moisture content also affects decomposition but at none of the sites did this play a relevant role (although for the MdM site this is probably incorrectly simulated). The effect of different microbial properties at Wal and Wal2,3 become apparent. The latter have higher N availability and higher growth rates, for Wal3 even more enhanced by an understorey vegetation, resulting in higher litterfall rates but similar C mineralisation rates.

Figure 20.10C also shows that at all sites the C input to the soil organic C is larger than the C mineralisation and that only the Åhe site is closest to equilibrium with the smallest increase of the soil C pool.

20.5.1.2 Net Ecosystem Productivity

Net ecosystem productivity (NEP), the net change in C within an ecosystem, can be calculated as the difference between C assimilation of the plants and the C losses by autotrophic and heterotrophic respiration. In the NUCOM model the radiation use efficiency is used to simulate C assimilation. This concept is defined as the produced amount of dry matter per unit intercepted radiation. Thus, in this concept, respiration losses of the plants (growth respiration and maintenance respiration) are already accounted for, therefore net primary production (NPP) is directly calculated. Net ecosystem productivity can then be calculated as the difference between net primary production and the heterotrophic respiration (C mineralisation) from the soil. Both these components have been discussed in the two sections above. Forest growth appeared strongly controlled by N availability (Fig. 20.7A). N mineralisation that makes up a larger part of N availability appeared to increase with soil organic N pool, and was strongly related to the total input into the soil organic matter by litterfall and root decay (Figs. 20.7A, 20.8A). Likewise, C mineralisation showed a tendency to increase with soil organic C pool and was also strongly connected to the total input into the soil organic matter (Fig. 20.10B,C). Therefore, both NPP and NEP exhibit a strong relation to soil organic matter input (Fig. 20.11). The effect of temperature, that appeared not to be a major contributing factor to C mineralisation (Fig. 20.10A), may be of greater importance for both NPP and NEP (Fig. 20.12), although it should

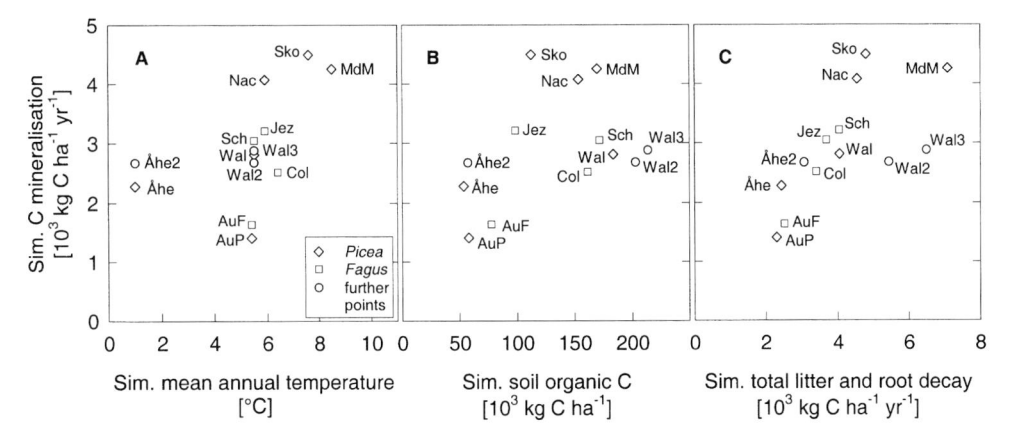

Fig. 20.10A–C. Simulated C mineralisation in relation to (**A**) mean annual temperature, (**B**) soil organic C pools and (**C**) total (above-ground and below-ground) litterfall. *Points* as in Fig. 20.2

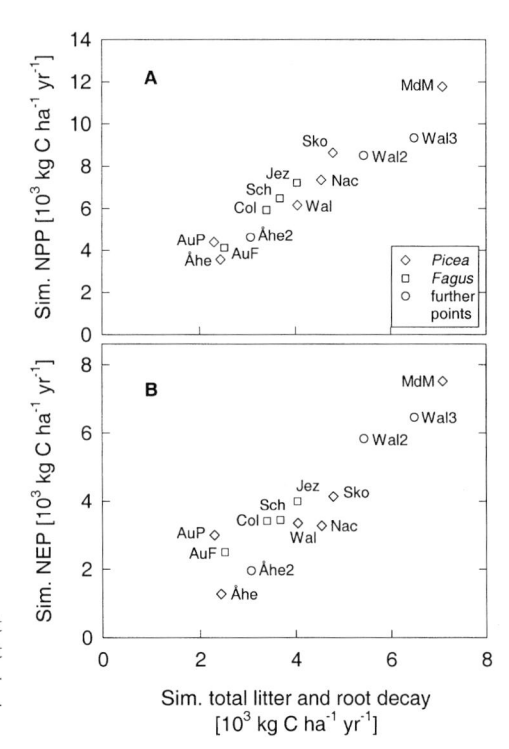

Fig. 20.11A,B. Simulated (**A**) net primary productivity (NPP) and (**B**) net ecosystem productivity (NEP) in relation to total (above-ground and below-ground) litterfall. *Points* as in Fig. 20.2

be noted that for the MdM site NPP is overestimated and thus also NEP. The variation in annual temperature between most of the sites is, however, too small to give a convincing relationship. Plotting the two components of NEP, C mineralisation and NPP, against each other (Fig. 20.13A) shows that the sites with slow-growing forests

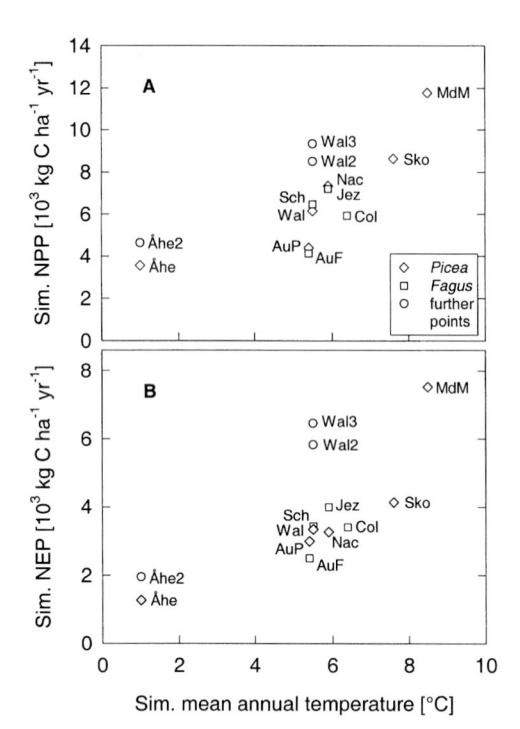

Fig. 20.12A,B. Simulated (**A**) net primary productivity (NPP) and (**B**) net ecosystem productivity (NEP) in relation to mean annual temperature. *Points as in Fig. 20.2*

lie closer, in absolute values, to the point of switching from a positive NEP (C fixation) towards a negative NEP (C emission). There is a positive relation between NPP and NEP, as suggested by Schulze and Heimann (1998), but this has not always been observed (Schulze et al. 1999). The linearity emerging in the simulated variables is mainly a consequence of the closeness in simulated values for the sites (six sites with simulated N mineralisation between 98 and 129 kg N ha^{-1} a^{-1}, four of which with almost similar climate). The sites with underestimated N mineralisation therefore fall in the linear part of the curves and not in the higher ranges with higher N mineralisation rates where NPP (and NEP) starts to level off and the relations become non-linear. The simulations show (Fig. 20.13B) that proportionally NEP can vary considerably for ecosystems with rather similar NPP. Negative values for NEP arise when C mineralisation is larger than NPP. Here in these simulations, C mineralisation may be underestimated by using the air temperature to calculate C mineralisation in the 0–30-cm soil layer where it can be expected that in winter soil temperature is higher than air temperature. In addition, C mineralisation from deeper (>30-cm) soil layers is not included.

20.5.1.3 Discussion

From the above simulations, it becomes clear that climatic factors determine to a large extent the positioning of the sites along the curves. Partner sites, lying close to each other and experiencing almost similar climatic conditions (see Persson et al. Chap. 2, this Vol.), often can be found close to each other along a curve. An exception to this might be MdM and Col, although MdM is not satisfactorily simulated; it also has an

Fig. 20.13A,B. Simulated (**A**) C mineralisation and (**B**) ratio between net ecosystem productivity and net primary productivity (NEP/NPP) in relation to net primary productivity (NPP). *Points* as in Fig. 20.2

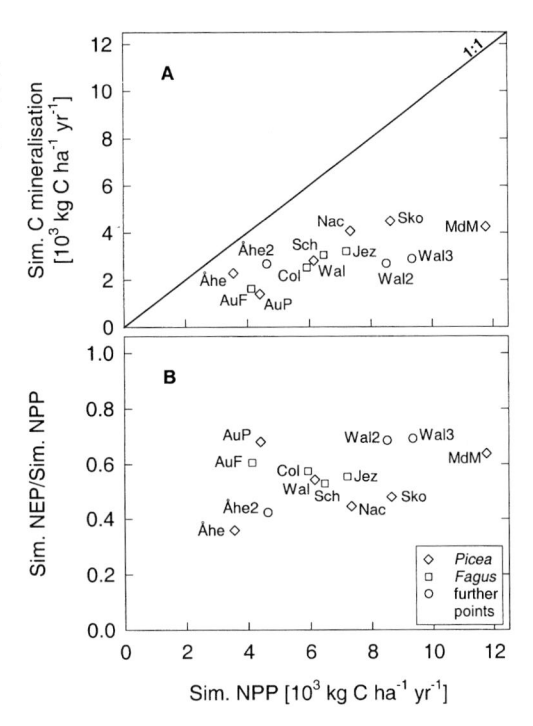

agricultural site history and lower elevation that make it different from its partner site. The midcontinental sites Wal, Sch, Nac and Jez are often near to each other. Although the AuF and AuP sites climatically do not differ that much from these midcontinental sites, the former behaviour corresponds more to the north Swedish site, Åhe. The growth-limiting conditions, at AuF and AuP reduction of assimilation by water stress, at Åhe N limitation, however, make these sites alike in growth rate, in soil organic C and N pools, NPP and NEP. Whether water stress is the limiting factor at the AuF and AuP sites is not certain; the calibration and simulations indicate, however, that a reduction in assimilate production is necessary to explain the actual behaviour. However, water stress may be a means for other factors that are not included in the model, one of which is the site history of litter raking. Additionally, it should be mentioned that in the early 1980s, AuP suffered successive droughts from which the stand never recovered, probably due to poor soils and old stand age (E. Dambrine, pers. comm.).

A key point emerging is that sites may be limited to different extents by water stress (cf. Fig. 20.9), but over the sites the relationships among different ecosystem variables all exhibit a strong N-controlled character. NPP, and thus NEP, is strongly related to N availability (Fig. 20.6); neglecting MdM that is not well simulated, the increase in NPP decreases at higher N availabilities. Site-specific conditions put each site at its position along the curves. Any differences between *Picea* or *Fagus* stands do not emerge, as also noted by Reich et al. 1997. This argues for the idea that deciduous and coniferous forests should not be seen as contrasting life forms but only as a continuum in gradient of leaf life span with associated other traits (Reich et al. 1995). Different stand age also has no straightforward importance, as Wal, as one of the oldest *Picea* stands, has one of the highest growth rates. Effects of stand age are, however, also difficult to clarify in this project because stand age effects are entangled with site-specific climatic conditions. However, in successional stages of chronosequences, that by definition have similar climatic conditions, major effects of stand age are demonstrated with accumulating soil organic matter and increasing N mineralisation rates (Berendse 1990, 1998; Berendse et al. 1998).

Controlling factor for N mineralisation, C mineralisation and thus also NEP appears the amount of above-ground litterfall and below-ground root decay. The ratio between above-ground litterfall and below-ground root decay varies between 0.5 (MdM) to 1.6 (Wal, see also Stober et al. Chap. 5, this Vol.). Root turnover is on average 1.1 times the root biomass where root biomass varies between 800 (AuP) to 3500 kg C ha^{-1} (MdM). This root turnover is on the conservative side where values mentioned in literature range between 1 and 7 (Persson 1983; Van Praag et al. 1988). Leaf turnover for the *Picea* stands is on average 0.25 times leaf biomass and may be on the high side due to the assumption of steady state in leaf biomass. Leaf life span, taken as equal for all *Picea* stands (set lower for Wal2,3) is also however, dependent on site-specific conditions. Leaf life span tends to increase at lower temperatures, lower radiation levels, and lower nutrient and water availability (Reich et al. 1995). A better understanding of litterfall, but especially of root turnover, will allow better estimates of C mineralisation and NEP. In addition, the strong control of litterfall and root turnover on C mineralisation, NPP and NEP also implies that year-to-year and within-year variations in litter input will lead to considerable variations in CO_2 fluxes of the system. Measurements of these fluxes, e.g. by eddy correlation technique, may therefore give very varying pictures from year to year but also effects of temperature on C mineralisation may change NEP between years. Thus, the relation between NEP and NPP needs further investigation.

A latitudinal trend with lower C sequestration at higher latitude is often inferred (Valentini et al. 2000) and related to temperature levels. Although the CANIF sites are located along a latitudinal gradient, mean annual temperatures were very similar for most of the sites (see Persson et al., Chap. 2, this Vol.). In accord with the latitudinal trend, the most northern CANIF site, Åhe, has lowest NEP but net ecosystem productivity varied in a range from ca. 2000 to 6500 kg C ha^{-1} a^{-1} for CANIF sites that varied less than 1 °C in annual temperature. Clearly, factors other than temperature alone determine NEP. Major explanatory factors are N availability that strongly enhances NPP, and soil organic C pools that codetermine C mineralisation rates together with site-specific decomposition rates. At the CANIF sites large soil organic C pools, a result of favourable growing conditions, are located within a high N deposition region and have reduced decomposition rates. These conditions result in relatively smaller C mineralisation rates and higher NEP than sites in less polluted areas that have higher decomposition rates and lower NEP.

20.5.2 Model Experiments

20.5.2.1 Sensitivity of NPP, C Mineralisation and NEP to Changes in N Availability

An important point in the interactions between C and N is how sensitive NPP and NEP are to changes in N availability. Both increasing N mineralisation and N deposition lead to an increased N availability. Here, the sensitivity to changes in N availability is analysed in additional simulations where per site N deposition varies between zero and two times the actual N deposition. Thus, the relative changes for all sites are equal, whereas the absolute changes differ per site depending on the amount of N deposition.

For most of the sites NPP was lower when N deposition ranged from zero to half the actual N deposition (Fig. 20.14). The increment of NPP at some of the sites at lower than actual deposition is due to the larger amounts of carbohydrates that are present as reserve pool at lower N supply rates. At certain N supply levels, the obtained leaf area led to a totally higher assimilate production (dry mass and reserves) than at actual N deposition level. Higher N deposition did not increase NPP. Only the N-limited site Åhe deviates by increasing NPP at higher N deposition ranges. For the *Fagus* sites an N deposition higher than at present led during the first years to an increased leaf area and assimilate production, but after some years a smaller carbohydrate reserve pool negatively affected the buildup of leaf area during the start of growth in spring, resulting in a totally decreased NPP. This illustrates the higher dependency on reserve pools of deciduous species compared to coniferous species. However, here it is mainly indicative of situations where N supply is ample, leading to high leaf N concentrations, but where assimilate production is constrained by water stress set rigid in calibration. At all sites, except Åhe, increased N deposition levels led to increased leaf N concentrations and increased N leaching. As a consequence of the lower NPP, amounts of litterfall (cf. Fig. 20.11) and thus C mineralisation rates (cf. Fig. 20.10) were smaller (Fig. 20.14), although these changes were relatively smaller than those in NPP. Also, similarly, N mineralisation rates were smaller; thus decreased N deposition resulted in an even stronger decrease in N availability. NEP follows a similar behaviour as NPP with hardly any sensitivity for change increases in N availability at those sites that already have a high N mineralisation rate and or where limitations other than N are dominating (cf. Fig. 20.9). These simulated events are, to a large extent, a result of the used fixed climatic conditions that impose equal water stress from year to year. In reality, water stress (and thus assimilate production) will vary, leading in situations with ample water supply and smaller vapour pressure deficits to an enhancement of growth and NPP at higher N availabilities (see Sect. 20.5.2.3, below).

20.5.2.2 Sensitivity of NPP, C Mineralisation and NEP to Changes in Temperature

Expected environmental changes include increases in temperature due to increased atmospheric concentration of greenhouse gases. Temperature affects NPP both directly through assimilation and indirectly by effects on transpiration and water flows. Also, C mineralisation is affected by temperature. Thus, changes in NEP are a result of the combined effects on NPP and C mineralisation.

The sensitivity to changes in temperature was analysed in simulation runs where temperature was increased by 1° or 2° during the whole year.

The direct effect of temperature increased assimilation between 0 and 6% at 1 °C increase in temperature and between 4 and 11% at 2 °C increase in temperature. The variability is caused by an interaction with intercepted radiation and thus leaf area. The indirect effect of temperature through changes in transpiration, however, works

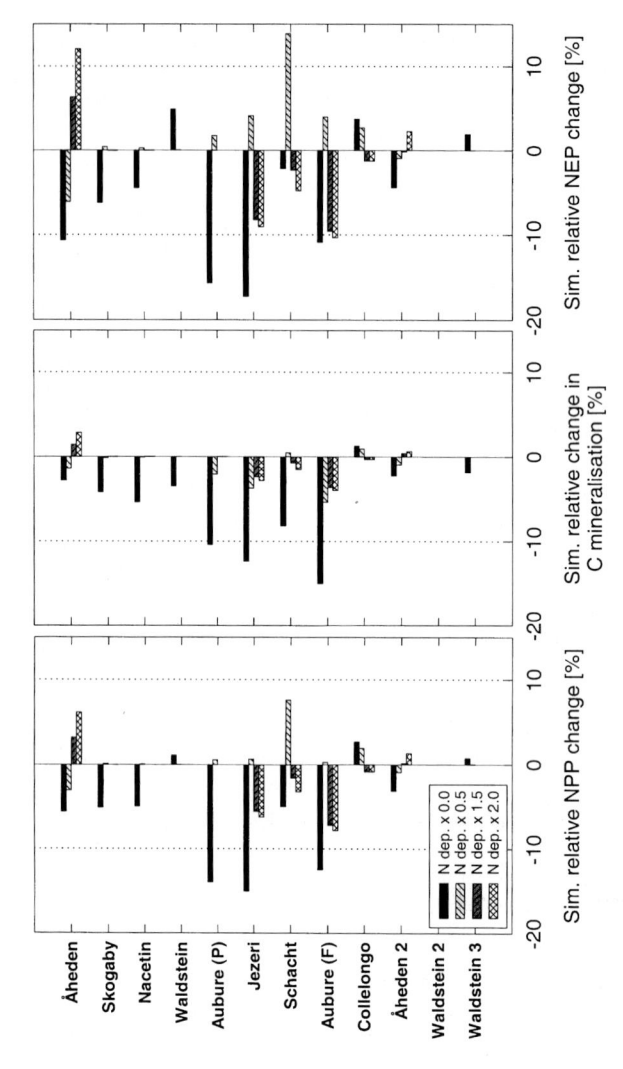

Fig. 20.14. Simulated relative changes in net primary productivity (NPP), C mineralisation and net ecosystem productivity (NEP) at changed N deposition levels. N deposition is varied by 0.0, 0.5, 1.5 and 2.0 times the actual N deposition at each sites. Changes are expressed relatively to 1.0 times the actual N deposition

in the opposite direction. Increased temperature increased potential transpiration, but at most sites soil water availability did not allow actual transpiration to follow to the same extent, thereby increasing water stress and stomatal closure, which led to decreased assimilation. The ultimate effect on most sites therefore is decreased NPP (Fig. 20.15). Note that the indirect effect of temperature here probably is overestimated by including only a change in temperature and not an accompanying change in vapour pressure and thus probably overestimating vapour pressure deficit. The increase in temperature also increased N mineralisation at all sites, leading to higher leaf N concentrations and increased N leaching. Only the strongly N-limited site Åhe takes advantage of the increased N availability, resulting in an increase in NPP (Fig. 20.15) that is larger than that caused by the direct temperature effect on assimilation alone. At the Sch, Wal, Wal2 and Wal3 sites, potential transpiration is hardly affected probably because the vapour pressure deficit at these sites is already high (Table 20.6). Here, the direct effect of temperature, increase of assimilation, can profit and NPP is increased.

C mineralisation is enhanced by a temperature increase (Fig. 20.15) and, as a consequence of increased NPP, with an associated increase in litterfall and root decay, an even strengthened effect for the Åhe, Sch, Wal, Wal2 and Wal3 sites.

The combined effect of a temperature increase on NPP and C mineralisation results for NEP in a general decline for most sites (Fig. 20.15). Only at the Åhe site, where NPP compensates for the increased C mineralisation rates does NEP increase.

Additional simulation runs (not shown) where temperature was increased only during the growing season (months 4–9) or only during the winter period (months 1–3 and 10–12), showed that the above results are mainly caused by temperature changes during the growing season.

20.5.2.3 Sensitivity of NPP, C Mineralisation and NEP to Changes in Precipitation

In addition to changes in temperature, also changes in precipitation amount and pattern are expected to occur as a consequence of global warming. The expected changes in precipitation are, however, much more uncertain. Changes in precipitation affect soil moisture conditions, and thus decomposition and transpiration rates. Here, sensitivity to changes in precipitation were analysed by increasing annual precipitation with 15 and 30% divided in proportion with actual precipitation over the year.

Increased precipitation improved soil water conditions and actual transpiration increased thereby reducing the effect of water stress on assimilation although potential transpiration also increased as a result of increased leaf area. Therefore NPP increased for most sites (Fig. 20.16). The increased NPP, however, is associated with lower leaf N concentrations since N availability is not directly affected but, on the contrary, becomes smaller through a feedback by higher C:N ratios in the produced litter. The increase in NPP therefore is largest for those sites that were most water-stressed but simultaneously had a large N availability.

Linked to the increase in NPP and the associated increase in litter production, also C mineralisation increases (Fig. 20.16). Direct effects of precipitation on C mineralisation via soil moisture effects on decomposition do not play a role since soil moisture contents were already optimal.

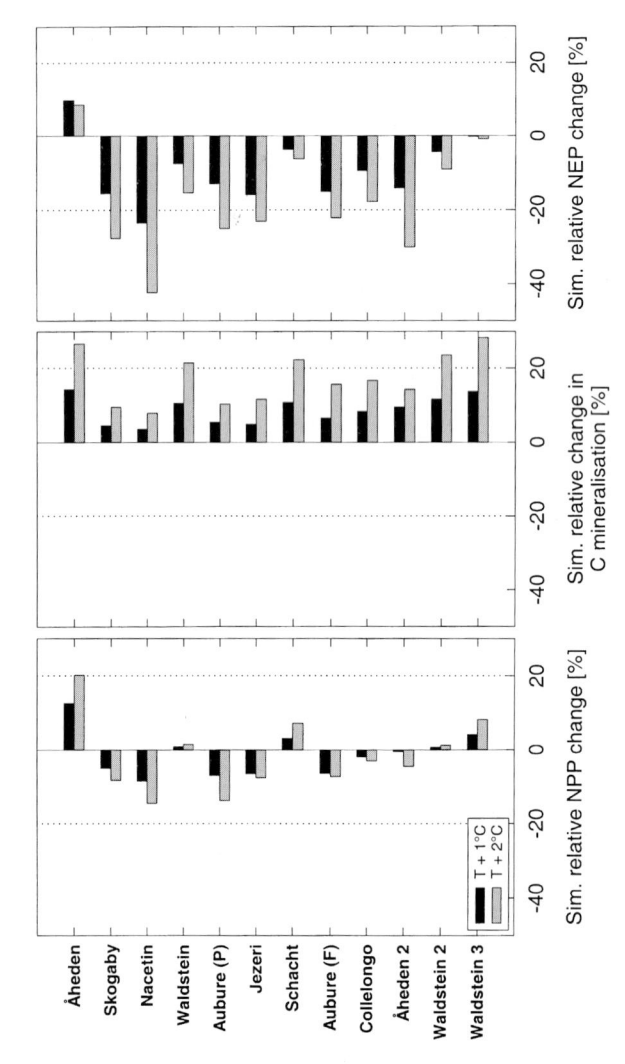

Fig. 20.15. Simulated relative changes in net primary productivity (NPP), C mineralisation and net ecosystem productivity (NEP) at increased temperature levels. Temperature is increased by 1 or 2 °C during the whole year. Changes are expressed relatively to the actual temperature situation

Fig. 20.16. Simulated relative changes in net primary productivity (NPP), C mineralisation and net ecosystem productivity (NEP) at increased precipitation levels. Precipitation is increased with 15 or 30% divided over the year in proportion with actual precipitation. Changes are expressed relatively to the actual precipitation situation

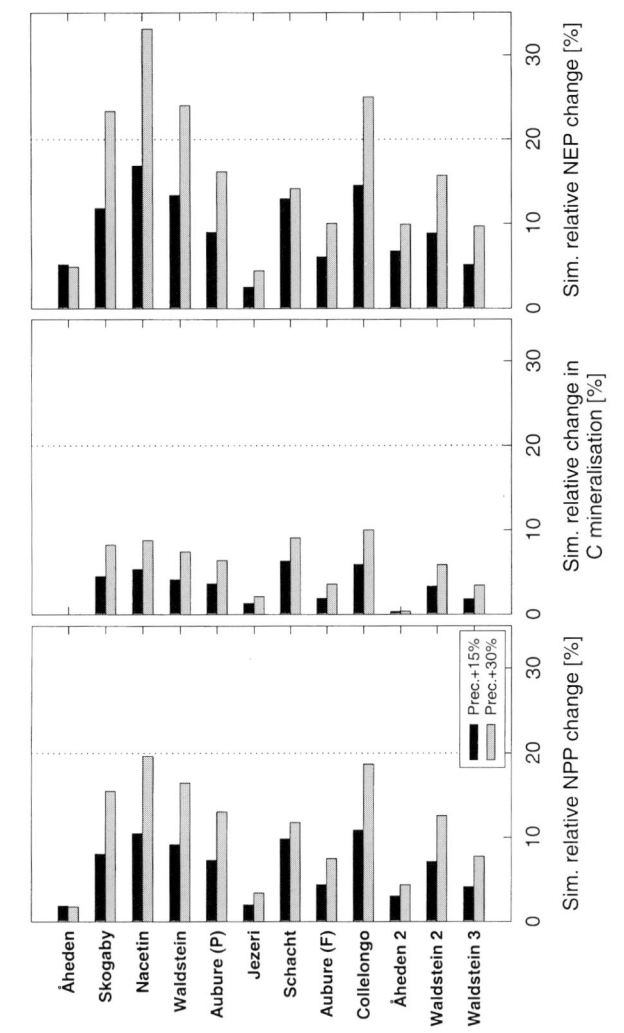

As a net result of the changes in NPP and C mineralisation, NEP increases at all sites (Fig. 20.16).

Additional simulation runs (not shown) where precipitation was increased only during the growing season (months 4–9) or only during the winter period (months 1–3 and 10–12), showed that the above results are mainly caused by changes in precipitation during the growing season. Here, only increases in precipitation were considered, since expected changes foresee an increase in annual precipitation (Alcamo and Kreileman 1996). However, within-year variations in precipitation pattern may be large and may also include a decrease in precipitation during the growing season. Based on the above simulations, one can, however, deduce that opposite effects to an increase in precipitation may occur.

20.5.2.4 Interactions Between Temperature and Precipitation Changes

The combined effects of changes in temperature and in precipitation were analysed in simulation runs where temperature was increased with 2 °C and precipitation with either 15 or 30% in a manner similar to the above simulations.

Since temperature and precipitation work oppositely on transpiration, and thus water stress and NPP, as described above, their combined effect is the result of each one's relative effect. Increases in precipitation compensate the negative effect on water stress for a majority of the sites, leading to increased NPP instead of decreased NPP at only a change in temperature, although at smaller changes in precipitation the compensating effect may not be large enough (Fig. 20.17). Since higher temperatures increased N mineralisation, here the precipitation-stimulated growth goes together with an increase in leaf N concentration. Smaller changes in precipitation led to less increase in growth but highest leaf N concentrations and highest N leaching.

Higher temperatures increased C mineralisation rates (Fig. 20.15) as did also increased growth at higher precipitation levels (Fig. 20.16). The combined effect of temperature and precipitation increase thus results in even stronger increase in C mineralisation rates (Fig. 20.17).

Their combined effect on NEP is that at sites where temperature alone caused a decrease in NEP, the compensating effect of increased precipitation led to smaller changes in NEP or even compensated for the amount that NEP increased (Fig. 20.17). For Åhe, the positive effect of increased N availability at higher temperature is strengthened also by a precipitation-stimulated growth, resulting in a strongly increased NPP and NEP.

20.5.2.5 Sensitivity of NPP, C Mineralisation and NEP to Changes in CO_2 Concentration

Increased atmospheric CO_2 concentration is a clear environmental change component. Scenario studies (Alcamo and Kreileman 1996) predict that CO_2 levels will increase to about 510 ppm in 2050 and 720 ppm in 2100 if no adequate CO_2 emission reductions are made. Increased atmospheric CO_2 concentration affects NPP directly through increased assimilation. The sensitivity to increased CO_2 levels was analysed in simulation runs where CO_2 was set to 540 and 720 ppm, respectively, a 50 and 100% increment compared to present levels (360 ppm).

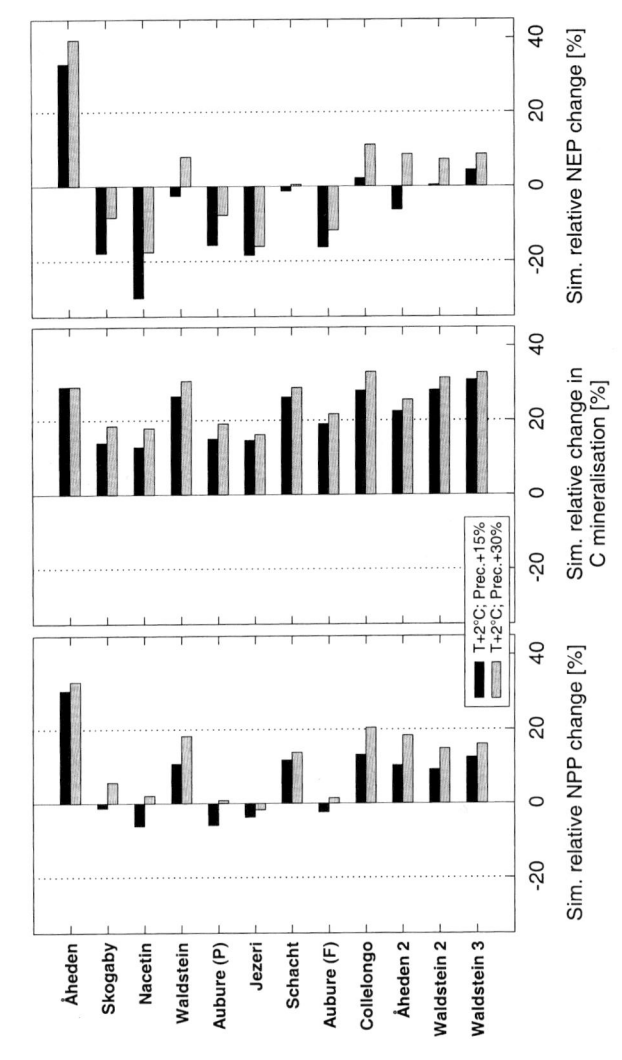

Model Analysis of Carbon and Nitrogen Cycling in *Picea* and *Fagus* Forests

Fig. 20.17. Simulated relative changes in net primary productivity (NPP), C mineralisation and net ecosystem productivity (NEP) at increased temperature and precipitation levels. Temperature is increased by 2 °C during the whole year simultaneously precipitation is increased with 15 or 30% divided over the year in proportion with actual precipitation. Changes are expressed relatively to the actual temperature-precipitation situation

Higher atmospheric CO_2 levels increased potential assimilation rates between 28 and 44% at 540 ppm and between 47 and 66% at 720 ppm. Dependent on the other growth-limiting factors, these higher potential assimilation rates were utilised to increase NPP (Fig. 20.18). At the strongly N-limited Åhe site, NPP hardly changed, whereas at the Col site with ample N supply an increase of 55% occurred at 720 ppm. The increase in NPP was associated with decreasing leaf N concentrations. As a consequence of increased NPP, litterfall increased (cf. Fig. 20.11), but due to the lower leaf N concentration the C:N ratio in litterfall was increased. These higher C:N ratios negatively affected N mineralisation rates, resulting in decreases up to 9% (Col), thus giving a feedback through N availability on plant growth. With higher litterfall, also C mineralisation (cf. Fig. 20.10) increased (Fig. 20.18). The combined effect of NPP and C mineralisation on NEP was an increase in NEP (Fig. 20.18). Effects of CO_2 on stomatal conductance leading to decreased transpiration rates per unit leaf area and higher water use efficiencies (Long and Hutchin 1991; Arp et al. 1998) were not simulated. It may be expected that increased water use efficiencies will increase NPP most at sites where water limitation is more dominating than N limitation.

20.5.2.6 Effects of Logging on NPP, C Mineralisation and NEP

Logging is expected to have major effects on C mineralisation and NEP, since with removal of biomass, actual plant growth and NPP are strongly reduced but C mineralisation continues and becomes the dominating C flux. Here, simulations were done for the Wal, Åhe and AuP sites, the oldest *Picea* sites within the CANIF project. Logging was performed in the 10th year of simulation by removing all above-ground stem and branch biomass but leaving all needles behind at the site. These needles and the below-ground biomass were added to the dead material pools and from there gradually transferred to soil organic matter, thus being handled exactly as dead material from normal mortality. In the third year after logging, the sites were reforested with 3000 *Picea* plants ha^{-1} of 0.70 m height. After reforestation, no other management practices were performed. The Wal2 and Wal3 sites were included to compare with the Wal site for differences in N availability (comparison Wal and Wal2) and the effect of understorey vegetation (comparison Wal2 and Wal3) on NEP.

After logging, C stocks (above- and below-ground) are reduced with only the dead material pools and part of the understorey vegetation remaining (Fig. 20.19). NPP successively falls to zero, except at the Wal3 site. Here, the understorey vegetation continues growing and NPP is considerable (Fig. 20.19). After reforestation, NPP increases again (with the hump mainly a model consequence) and C stocks increase. Sites with higher N availability gain higher NPP and higher C stocks than sites with lower N availability. At Wal3 the understorey vegetation increases NPP but C stocks are not higher than at Wal2 as plant turnover is fast for this ground-layer vegetation.

After logging, C mineralisation continues and increases at Åhe, AuP and Wal3 (at the latter site also by damage and dying off of part of the understorey). At Wal and Wal2 C mineralisation decreases due to a drier topsoil that hampers decomposition. In time, C mineralisation continues to increase as soil organic C pools become larger. Only at Åhe does the C mineralisation not attain levels similar to those before, probably due to a lower NPP by changing from a mixed *Pinus-Picea* forest to a monoculture *Picea* forest. At sites with lower N availability the litter production is smaller and therefore also C mineralisation is lower. The short life span of the ground-layer

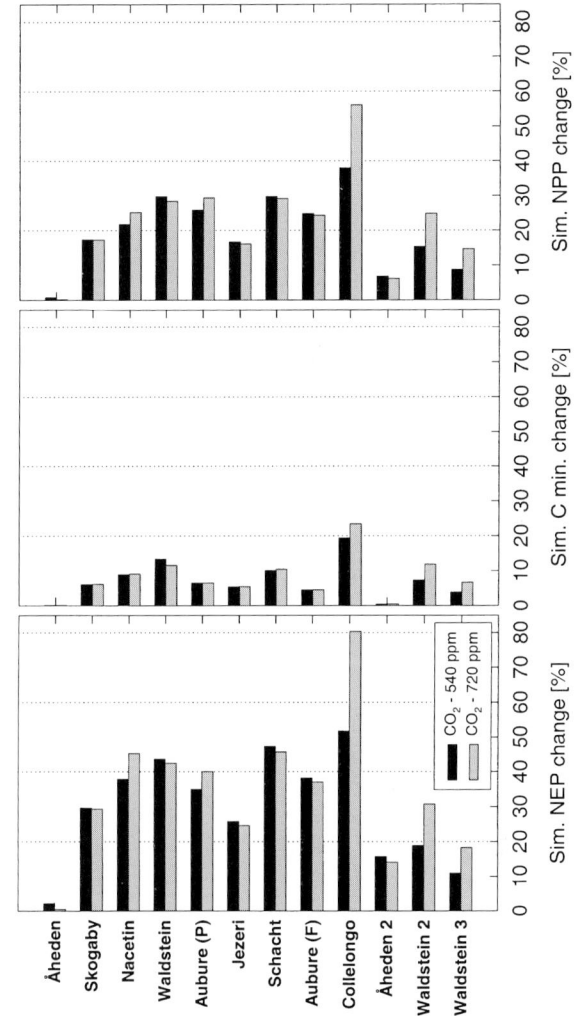

Fig. 20.18. Simulated relative changes in net primary productivity (NPP), C mineralisation and net ecosystem productivity (NEP) at increased atmospheric CO_2 concentrations. CO_2 levels are increased to 540 and 720 ppm. Changes are expressed relatively to the actual CO_2 levels (360 ppm)

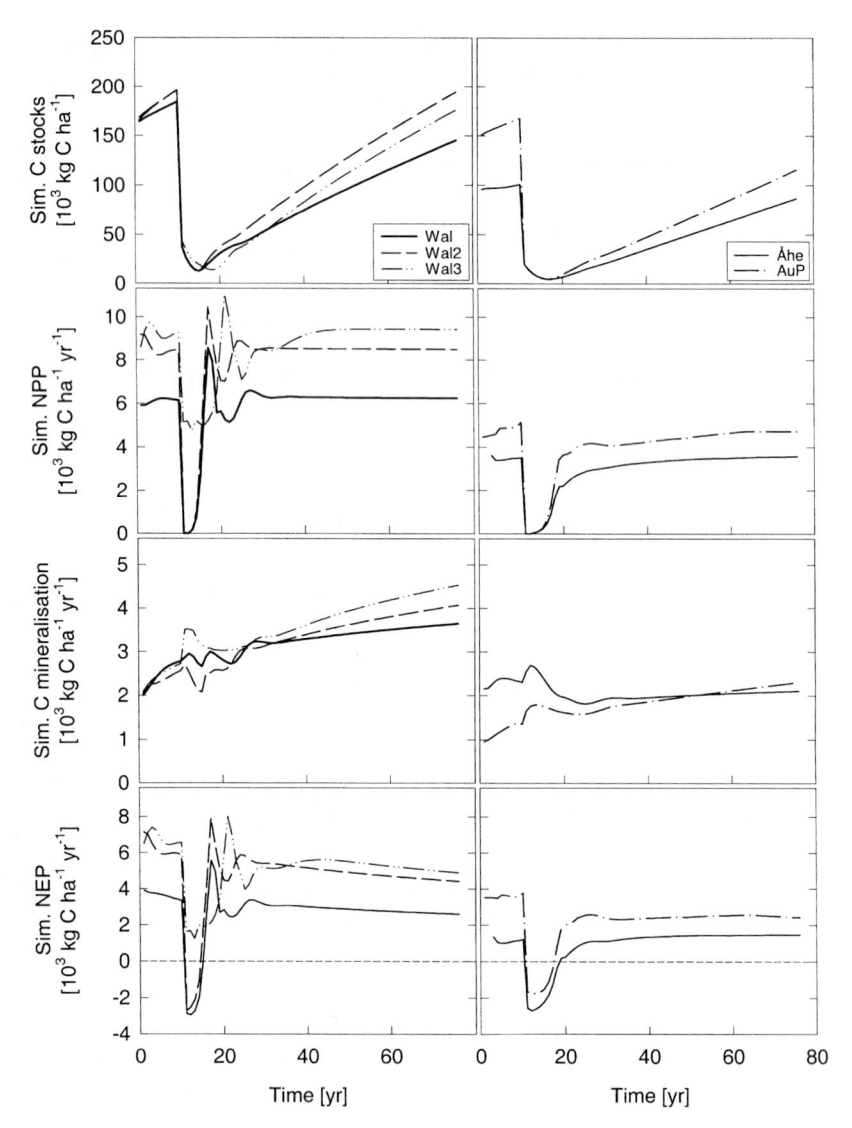

Fig. 20.19. Time course of simulated vegetation C stocks, NPP, C mineralisation and NEP after logging at the Wal, Wal2, Wal3, Åhe and AuP sites. Logging was performed in the 10th year of simulation and the sites were replanted 3 years after logging. At logging, stem and branch material was removed and needles were left

vegetation (Wal3) leads to a higher C mineralisation than without a ground-layer vegetation. Here, another property of the ground-layer vegetation, higher decomposability, is neglected.

As a result of the changes in NPP and C mineralisation, NEP becomes negative at the sites where NPP is zero. At the Wal3 site, NEP is reduced but still positive because of the NPP by the understorey vegetation. In the long term, however, the NEP at the sites (except Åhe) is lower than before logging because NPP reaches a level equal to that before logging whereas C mineralisation is larger.

NEP thus is greatly affected directly after logging and the years following reforestation. C emission (a negative NEP) occurs when the site has no plant growth. To what extent an understorey vegetation is capable of attaining growth rates and NPP high enough to prevent net ecosystem C emissions depends on factors such as presence and damage at logging. Larger soil evaporation can dry out the topsoil layer and reduce decomposition and C mineralisation rates. The time period needed for the forest to attain NPP similar to before depends on planting schemes (what year, how many plants etc.) at reforestation and growing conditions afterwards. The simulated NPPs during the first decades after planting are uncertain and probably overestimated because the model is parameterised for mature closed forests and less suited for open replantations with different growth patterns and mortality rates of young plants.

20.5.2.7 Discussion

The above simulations analysing the sensitivity of changes in environmental conditions on NPP, C mineralisation and NEP demonstrate the interlinkages of C, N and water cycles in ecosystems. Where higher N availability tends to increase NPP (up to non-N-limiting conditions) and, indirectly, C mineralisation, the net result will be an increased net ecosystem productivity. Higher precipitation levels act similarly to higher N availability, only the effects are smaller due to a feedback of lower N availability at higher C:N ratios in produced litter. Higher temperatures not only directly increase primary production and decomposition but also indirectly affect these processes through changes in N availability and transpiration rates. Higher atmospheric CO_2 concentrations act in a direction comparable with increasing N and precipitation levels, as more CO_2 can be utilised when these factors are less growth-limiting. In Fig. 20.20 a generalised scheme gives the directional changes of each factor but split for temperature in a direct component and an indirect component where negative effects on NPP arise due to increased stomatal closure at higher potential transpiration. The outcomes of temperature changes therefore are site-specific

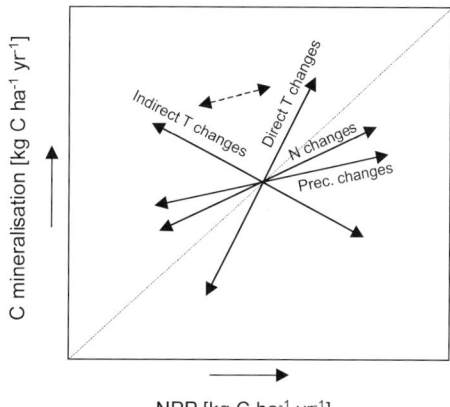

Fig. 20.20. A generalised scheme of the directional effects of changes in N availability, precipitation and temperature on net primary productivity (NPP) and C mineralisation. Temperature changes are divided into a direct component (on assimilation and C mineralisation) and an indirect component where a negative effect on NPP arises through increased potential transpiration causing closure of stomata. Higher atmospheric CO_2 concentrations act in a direction comparable with increasing N and precipitation levels. These directions are based on shown simulations

and dependent on the relative importance of the growth-limiting factors at that site. Changes in both temperature and precipitation therefore can act as compensating factors on net ecosystem productivity when precipitation levels increase, but also as amplifying factors when precipitation levels decrease in conjunction with a temperature increase. Only considering the effects of temperature change when discussing possible consequences of global warming, e.g. Kirschbaum (1995), neglects these interactions and may not give a complete picture of expected events. Here, net ecosystem productivity at the most northern site, Åhe, reacts positively through increased N availability that reduces the N-limiting conditions for primary production. In such a way these types of sites may contribute to the C sink in the northern latitudes mentioned by Myeni et al. (1997) and Schimel (1995). The results are presented here as annual values; within-year variation, however, can occur, as noted by Goulden et al. (1996) and Lindroth et al. (1998). The simulation results underlined the importance of changes that occur during the growing season. The earlier start of growth in spring is not accounted for in this study but can have pronounced effects on primary production (Kramer 1996; Bergh et al. 1998). Of course, the ultimate effect of temperature change on primary production and decomposition is dependent on the magnitude of the change, that may differ with the season.

20.6 Conclusions

The simulation results show that the CANIF sites differ in many respects. The modelling approach used seems to be able to produce reliable growth rates, biomass and leaf N concentrations for most of the sites. Major uncertainties lay in the water balance and transpiration rates. Lack of data imposed uncertainties for the systems' hydrological boundaries. The applied calibration method, however, appeared suitable to analyse the sites under nearly steady-state conditions, and give a reliable picture of the forest dynamics.

Climatic factors determine to a large extent the characteristics of each site. No clear differences between *Picea* and *Fagus* forests emerged, although differences in humus form and soil dynamics are known for both vegetation types. N availability is a strong controlling factor for net primary production and net ecosystem productivity. The controlling variable for N mineralisation, C mineralisation and thus net primary production and net ecosystem productivity appears the amount of above-ground litterfall and below-ground root decay. This variable forms an intermediate in a feedback process where forest growth is strongly controlled by N availability, and, in turn, N mineralisation is strongly controlled by litterfall and root decay that is higher at higher growth rates. C:N ratios in litter affect N mineralisation but to a lesser extent than amounts of litter.

Net ecosystem productivity is dependent on temperature, but also other factors have a clear impact. Major explanatory factors for net ecosystem productivity are N availability, that strongly enhances net primary production, and soil organic C pools, that codetermine C mineralisation rates together with site-specific decomposition rates. At the CANIF sites, large soil organic C pools, a result of favourable growing conditions, are located within a high N deposition region and have reduced decomposition rates that may be explained by high N concentrations that presumably suppress decomposition of lignin compounds. These conditions result in relatively smaller C mineralisation rates and higher NEP than sites in less polluted areas, that have higher decomposition rates and lower NEP. These different decomposition rates along the latitudinal gradient associated with N deposition levels represent a major clue to heterotrophic respiration and net ecosystem productivity. At high N deposition sites, low decomposition rates are associated with high N turnover; the mechanism behind this is not yet clear and limits modeling exercises. In contrast, at strongly N-limited sites, microbial competition prevents net N mineralisation. At such sites uptake of organic N takes place, but the understanding of this process also is still very limited.

Logging has a major impact on net ecosystem productivity. Many uncertainties remain as to how understorey vegetation, growth of young replantations and changed soil moisture conditions at clearcuttings affect C emission and C fixation.

C sequestration in forests still has many uncertainties. The sensitivity analyses to the effects of changes in environmental conditions on net ecosystem productivity highlight some of the possible interactions that can occur. The outcome of global change appears hard to predict and might be site-specific, determined by the relative importance of the growth-limiting factors at that site.

Thus, although many of the feedback and feedforward processes in C, N and water interactions are understood, many questions remain when interacting factors are joined together.

Acknowledgments. Without the help of many persons in gathering the required field data, this study could not have been performed. We thank Ghasem Alavi, Johann Bergholm, Jiri Cerny, Pavel Cudlin, Hana Groscheova, Michaela Hein, Patrick Karlsson, Martina Mund, Lars-Owe Nilsson, Martin Novak, Anna Rudebeck and Maud Quist for their great help.

References

Aber JD, Federer CA (1992) A generalised, lumped-parameter model of photosynthesis, evatranspiration and net primary production in temperate and boreal forest ecosystems. Oecologia 92:463–474

Aber JD, Ollinger SV, Federer CA, Reich PB, Goulden ML, Kicklighter DW, Melillo JM, Lathrop RG Jr (1995) Predicting the effects of climate change on water yield and forest production in the northeastern United States. Clim Res 5:207–222

Ågren GI, Bosatta E (1987) Theoretical analysis of the long-term dynamics of carbon and nitrogen in soils. Ecology 68:1181–1189

Ågren GI, Bosatta E (1996) Theoretical ecosystem ecology – understanding element cycles. Cambridge University Press, Cambridge

Alcamo J, Kreileman E (1996) Emission scenarios and global climate protection. Global Environ Change 6:305–334

Arp WJ, Van Mierlo JEM, Berendse F, Snijders W (1998) Interactions between elevated CO_2, nitrogen and water. Effects on growth and water use of six perennial plant species. Plant Cell Environ 21:1–11

Barrett K, Berge E (eds) (1996) Transboundary air pollution in Europe. MSC-W Status Report 1996. EMEP/MSC-W, Report 1/96, Norwegian Meteorological Institute, Oslo

Berendse F (1990) Organic matter accumulation and nitrogen mineralisation during secondary succession in heathland ecosystems. J Ecol 78:413–427

Berendse F (1994) Litter decomposability – a neglected component of plant fitness. J Ecol 82:187–190

Berendse F (1998) Effects of dominant plant species on soils during succession in nutrient-poor ecosystems. Biogeochemistry 42:73–88

Berendse F, Lammerts EJ, Olff H (1998) Soil organic matter accumulation and its implications for nitrogen mineralisation and plant species composition during succession in coastal dune slacks. Plant Ecol 137:71–78

Berg B, Ekbohm G (1991) Litter mass-loss rates and decomposition patterns in some needle and leaf litter types. Long-term decomposition in a Scots pine forest. VII. Can J Bot 69: 1449–1456

Berg B, Berg MP, Bottner P, Box E, Breymeyer AI, Calvo de Anta R, Couteaux M-M, Escudero A, Gallardo A, Kratz W, Madeira M, Mälkönen E, McClaugherty CA, Meentemeyer V, Virzo de Santo A, Piussi P, Remacle J (1993) Litter mass loss rates in pine forests of Europe and Eastern United States: SOM relationships with climate and litter quality. Biogeochemistry 20:127–159

Berge E (ed) (1997) Transboundary air pollution in Europe. MSC-W Status Report 1997. EMEP/MSC-W, Report 1/97, Norwegian Meteorological Institute, Oslo

Bergh J, McMurthie R, Linder S (1998) Climatic factors controlling the productivity of Norway spruce: a model-based analysis. For Ecol Manag 110(1–3):127–139

Bloemhof HS (1992) Simulatie van de productie en nutrientenhuishouding van wegbermvegetaties. Centre for Agro-Biological Research, Agricultural Research Department, Report, Wageningen, The Netherlands

Bosatta E, Ågren GI (1991) Theoretical analysis of carbon and nutrient interactions in soils under energy-limited conditions. Soil Sci Soc Am J 55:728–733

Bosatta E, Berendse F (1984) Energy or nutrient regulation of decomposition: implications for the mineralisation-immobilisation response to perturbations. Soil Biol Biochem 16:63–67

Bossel H (1996) TREEDYN3 forest simulation model. Ecol Model 90:187–227

Cannell MGR (1989) Physiological basis of wood production: a review. Scand J For Res 4:459–490

Cannell MGR, Milne R, Sheppard LJ, Unsworth MH (1987) Radiation interception and productivity of willow. J Appl Ecol 24:261–278

Chapin FS, Kedrowski RA (1983) Seasonal changes in nitrogen and phosphorus fractions and autumn retranslocation in evergreen and deciduous taiga trees. Ecology 64:376–391

Cienciala E, Lindroth A, Cermák J, Hällgren, Kucera J (1992) Assessment of transpiration estimates for *Picea abies* trees during a growing season. Trees 6:121–127

Cienciala E, Lindroth A Cermák J, Hällgren JE, Kucera J (1994) The effects of water availability on transpiration, water potential and growth of *Picea abies* during a growing season. J Hydrol 155:57–71

Del Arco JM, Escudero A, Garrido MV (1991) Effects of site characteristics on nitrogen retranslocation from senescing leaves. Ecology 72:701–708

De Vries W, Posch M, Kämäri J (1989) Simulation of the long-term soil response to acid deposition in various buffer ranges. Water Air Soil Pollut 48:349–390

De Vries W, Hol A, Tjalma S, Voogd JCH (1990) Literatuurstudie naar voorraden en verblijftijden van elementen in bosecosystemen. Staring Centrum Report 94, Wageningen, The Netherlands

De Vries W, Posch M, Reinds GJ, Kämäri J (1993) Critical loads and their exceedance on forest soils in Europe. The Winand Staring Centre for Integrated Land, Soil and Water Research, Agricultural Research Department Report 58, Wageningen, The Netherlands

Gale MR, Grigal DF (1987) Vertical root distributions of northern tree species in relation to successional stage. Can J For Res 17:829–834

Goudriaan J, Unsworth MH (1990) Implications of increasing carbon dioxide and climate change for agricultural productivity and water resources. In: Kimball BA, Rosenberg NJ, Allen LH (eds) Impact of carbon dioxide, trace gases, and climate change on global agriculture. American Society of Agronomy, Spec Edn No 53, Madison, Wisconsin, pp 111–130

Goudriaan J, Van Laar HH, Van Keulen H, Louwerse W (1985) Photosynthesis, CO2 and plant production. In: Day W, Atkin RK (eds) Wheat growth and modeling. Plenum Publishing, New York, pp 107–122

Goulden ML, Munger JW, Fan S-M, Daube BC, Wofsy SC (1996) Exchange of carbon dioxide by a deciduous forest: response to interannual climate variability. Science 271:1576–1578

Haxeltine A, Prentice IC (1996) A general model for the light-use efficiency of primary production. Funct Ecol 10:551-561

Hoffman F (1995) FAGUS, a model for growth and development of beech. Ecol Model 83: 327-348

Ingestad T (1982) Relative addition rate and external concentration; driving variables used in plant nutrition research. Plant Cell Environ 5:443-453

Ingestad T, Ågren GI (1995) Plant nutrition and growth: basic principles. Plant Soil 168-169: 15-20

Ingestad T, Lund AB (1986) Theory and techniques for steady state mineral nutrition and growth of plants. Scand J For Res 1:439-453

Ivens W, Klein Tank A, Kauppi P, Alcamo J (1989) Atmospheric deposition of sulphur, nitrogen and basic cations onto European forests: observations and model calculations. In: Kämäri J, Brakke DF, Jenkins A, Norton SA, Wright RF (eds) Regional acidification models. Springer, Berlin Heidelberg New York, pp 103-112

Jackson RB, Canadell J, Ehrlinger JR, Mooney HA, Sala OE, Schulze E-D (1996) A global analysis of root distribution for terrestrial biomes. Oecologia 108:389-411

Killingbeck KT (1996) Nutrients in senesced leaves: keys to the search for potential resorption and resorption proficiency. Ecology 77:1716-1727

Kirschbaum MUF (1995) The temperature dependence of soil organic matter decomposition, and the effect of global warming on soil organic C storage. Soil Biol Biochem 27:753-760

Kramer K (1996) Phenology and growth of European trees in relation to climate change. PhD Thesis Agricultural University, Wageningen, The Netherlands

Kropff MJ (1993) Mechanisms of competition for water. In: Kropff MJ, Van Laar HH (eds) Modelling crop-weed interactions. CAB International, Wallingford, UK, pp 63-76

Lindroth A, Grelle A, Morén A-S (1998) Long-term measurements of boreal forest carbon balance reveal large temperature sensitivity. Global Change Biol 4:443-450

Lohammer T, Larsson S, Linder S, Falk SO (1980) FAST - simulation models of gaseous exchange in Scots pine. In: Person T (ed) Structure and function of Northern coniferous forests - an ecosystem study. Ecol Bull 32:505-524

Long SP, Hutchin PR (1991) Primary production in grasslands and coniferous forests with climate change: an overview. Ecol Appl 1:139-156

McClaugherty CA, Pastor J, Aber JD, Melillo JM (1985) Forest litter decomposition in relation to soil nitrogen dynamics and litter quality. Ecology 66:266-275

Melillo JM, McGuire AD, Kicklighter DW, Moore B III, Vorosmarty CJ, Schloss AL (1993) Global climate change and terrestrial net primary production. Nature 363:234-240

Monteith JL (1977) Climate and the efficiency of crop production in Britain. Philos Trans R Soc Lond B 281:277-294

Monteith JL (1994) Validity of the correlation between intercepted radiation and biomass. Agric For Meteorol 68:213-220

Monteith JL, Unsworth MH (1990) Principles of environmental physics, 2nd edn. Edward Arnold, London

Myeni RB, Keeling CD, Tucker CJ, Asrar G, Nemani RR (1997) Increased plant growth in the northern high latitudes from 1981 to 1991. Nature 386:698-702

Näsholm T, Ekblad A, Nordin A, Giesler R, Högberg M, Högberg P (1998) Boreal forest plants take up organic nitrogen. Nature 392:914-916

O'Connell AM (1990) Microbial decomposition (respiration) of litter in eucalypt forests of South-western Australia: an empirical model based on laboratory incubations. Soil Biol Biochem 22:153-160

Parton WJ, Coughenour MB, Scurlock JMO, Ojima DS, Gilmanov TG, Scholes RJ, Schimel DS, Kirchner TB, Menaut J-C, Seastedt TR, Garcia Moya E, Kamnalrut A, Kinyamario JI, Hall DO (1996) Global grassland ecosystem modelling: development and test of ecosystem models for grassland systems. In: Breymeyer AI, Hall DO, Melillo JM, Ågren GI (eds) Global change: effects on coniferous forests and grasslands. John Wiley, Chichester, pp 229-269

Persson H (1983) The distribution and productivity of fine roots in boreal forests. Plant Soil 71:87–101

Raich JW, Rastetter EB, Melillo JM, Kicklighter DW, Steudler PA, Peterson BJ, Grace AL, Moore B III, Vörösmarty CJ (1991) Potential net primary productivity in South America: application of a global model. Ecol Appl 1:399–429

Rastetter EB, Shaver GR (1992) A model of multiple-element limitation for acclimating vegetation. Ecology 73:1157–1174

Rastetter EB, Ryan MG, Shaver GR, Melillo JM, Nadelhoffer KJ, Hobbie JE, Aber JD (1991) A general biogeochemical model describing the responses of the C and N cycles in terrestrial ecosystems to changes in CO2, climate, and N deposition. Tree Physiol 9:101–126

Reich PB, Koike T, Gower ST, Schoettle AW (1995) Causes and consequences of variation in conifer leaf life-span. In: Smith WK, Hinckley TM (eds) Ecophysiology of coniferous forests. Academic Press, San Diego, pp 225–254

Reich PB, Grigal DF, Aber JD, Gower ST (1997) Nitrogen mineralisation and productivity in 50 hardwood and conifer stands on diverse soils. Ecology 78:335–347

Running SW, Gower ST (1991) FOREST-BGC, A general model of forest ecosystem processes for regional applications. II Dynamic carbon allocation and nitrogen budgets. Tree Physiol 9:147–160

Ryan MG, Hunt ER Jr, McMurtrie RE, Ågren GI, Aber JD, Friend AD, Rastetter EB, Pulliam WM, Raison RJ, Linder S (1996) Comparing models of ecosystem function for temperate conifer forests. I. Model description and validation. In: Breymeyer AI, Hall DO, Melillo JM, Ågren GI (eds) Global change: effects on coniferous forests and grasslands. John Wiley, Chichester, pp 313–362

Sands PJ (1996) Modelling canopy production. III. Canopy light-utilisation efficiency and its sensitivity to physiological and environmental variables. Aust J Plant Physiol 23:103–114

Schimel DS (1995) Terrestrial ecosystems and the carbon cycle. Global Change Biol 1:77–91

Schulze E-D, Heimann M (1998) Carbon and water exchange of terrestrial systems. In: Galloway J, Melillo JM (eds) Asian change in the context of global climate change. Cambridge University Press, Cambridge, pp 145–161

Schulze E-D, Leuning R, Kelliher FM (1995) Environmental regulation of surface conductance for evaporation from vegetation. Vegetatio 121:79–87

Schulze E-D, Lloyd J, Kelliher FM, Wirth C, Rebmann C, Lühker B, Mund M, Knohl A, Milyukova IM, Schulze W, Ziegler W, Varlagin AB, Sogachev AF, Valentini R, Dore S, Grigoriev S, Kolle O, Panfyorov MI, Tchebakova N, Vygodskaya NN (1999) Productivity of forests in the Eurosiberian boreal region and their potential to act as a carbon sink – a synthesis. Global Change Biol (in press)

Seyferth U (1998) Effects of soil temperature and moisture on carbon and nitrogen mineralisation in coniferous forests. Licentiate thesis no. 1. Department of Ecology and Environmental Research, Swedish University of Agricultural Sciences, Uppsala

Thornley JHM, Johnson IR (1990) Plant and crop modelling – a mathematical approach to plant and crop physiology. Clarendon Press, Oxford

Valentini R, Matteucci G, Dolman AJ, Schulze E-D, Rebmann C, Moors EJ, Granier A, Gross P, Jensen NO, Pilegaard K, Lindroth A, Grelle A, Bernhofer Ch, Grünwald T, Aubinet M, Ceulemans R, Kowalski AS, Vesala T, Rannik Ü, Berbigier P, Lousteau D, Gudmundsson J, Thorgeirsson H, Ibrom A, Morgenstern K, Clement R, Moncrieff J, Montagnani L, Minerbi S, Jarvis PG (2000) The carbon sink strength of forests in Europe: novel results from the flux observation network. Nature 404:861–865

Van Laar HH, Goudriaan J, Van Keulen H (1992) Simulation of crop growth for potential and water-limited production situations: as applied to spring wheat. CABO-DLO Wageningen, The Netherlands

Van Oene H, Berendse F, De Kovel CGF (1999a) Model analysis of the effects of historic CO$_2$ levels and nitrogen inputs on vegetation succession. Ecol Appl 9(3):920–935

Van Oene H, Van Deursen EJM, Berendse F (1999b) Plant-herbivore interaction and its consequences for succession in wetland ecosystems: a modeling approach. Ecosystems 2:122–138

Van Praag HJ, Sougnez-Remy S, Weissen F, Carlette G (1988) Root turnover in a beech and a spruce stand of the Belgian Ardennes. Plant Soil 105:87–103
Ward RC (1975) Principles of hydrology. McGraw-Hill, London
Wong SC, Cowan IR, Farquhar GD (1979) Stomatal conductance correlates with photosynthetic capacity. Nature 282:424–426

21 Interactions Between the Carbon and Nitrogen Cycles and the Role of Biodiversity: A Synopsis of a Study Along a North-South Transect Through Europe

E.-D. Schulze, P. Högberg, H. van Oene, T. Persson, A.F. Harrison, D. Read, A. Kjøller, and G. Matteucci

21.1 Introduction

The NIPHYS/CANIF project of the EEC has provided the unique opportunity to examine forest ecosystem processes and diversity along a transect through Europe ranging from north Sweden to central Italy. The main objectives of this study were (1) to identify and quantify effects of N deposition on ecosystem processes, particularly the C cycle, by extending the range of observations across a deposition maximum in central Europe, and (2) to study feedback effects between ecosystem processes and biodiversity. The study resulted in a very comprehensive and consistent set of data on ecosystem processes over a 5-year period. However, in contrast to earlier large-scale ecosystem studies (IBP: Reichle 1981; Acid Rain Programme: Last and Watling 1991), the samples were collected and data generated by the same scientists at all sites. This assured comparisons of results on a broad geographic scale. In addition, key parameters were assessed by different methods, and integrating parameters were collected for different processes, in order to test and verify predictions made at higher and lower scales, ranging from physiological responses to ecosystem level processes.

In the following we attempt to reassess the data presented in detail in the preceding chapters and seek for an integrated analysis that addresses the following general questions:

- What regulates the C and N fluxes in forest ecosystems?
- Are there thresholds and non linear responses?
- What are net ecosystem productivity (NEP) and net biome productivity (NBP) in managed systems and can they be inferred from process studies?
- What role does biodiversity play in ecosystem functioning?

21.2 Change of Ecosystem Processes Along the European Transect

The study of a transect through Europe was confronted with a range of intrinsic problems, one of which relates to the weather pattern. If a high-pressure cell resulted in dry and hot weather in Scandinavia, very wet conditions are found in southern Europe, and vice versa. Thus, it was difficult to collect data under conditions of comparable water status along such a transect in a single year. Even though the project lasted 5 years, we were not able to study the effects of the interannual variability of key parameters. The changing weather pattern, however, gave confidence that we are dealing with general (long-term) patterns in this analysis. Another problem arises from the fact that some key parameters, such as canopy N uptake, still remain unmeasured in a comparative way over a broad geographic scale. Since it was only possible to study a limited number of sites, this project selected sites with *Picea abies* and *Fagus sylvatica* as dominant species, and with emphasis on acid soils, because these represent a major proportion of European forests. Due to local variations that are associated with changes in soil conditions, the local variation of some key parameters (e.g. leaf N concentration) was just as large as the variation along the continental transect (see Bauer et al., Chap. 4, this Vol.), but this local variation remained unstudied.

The environmental conditions along the transect (as summarised by Persson et al., Chap. 2, this Vol.) show a decrease in precipitation (Fig. 21.1F) from the south to the north, indicating that the Mediterranean climate of Italy was offset by the high elevation at the two Italian sites (Collelongo, Col, and Monte di Mezzo, MdM). Annual mean temperatures of the selected sites were similar at the south and central European sites but appreciably lower in north Sweden. N deposition in rainfall showed the expected maximum in central Europe with higher rates than expected in Italy, which could be due to local emissions from the Rome metropolitan area (see Schulze, Chap. 1, this Vol.).

Ammonium and nitrate concentrations in the organic layer (Fig. 21.1E) exhibited a characteristic change with latitude. Concentrations generally decreased from south to north, with soils under *Fagus* always having higher concentrations than soils under *Picea*. Nitrate was below the detection limit north of Denmark, and in the boreal forest also ammonium reached very low concentrations. Because the samplings were performed at different seasons at different sites, the conclusions about the latitudinal changes of ammonium and nitrate concentrations in soil, however, remain tentative.

The plant and soil C pools (Fig. 21.1D) did not follow the same trend. Soil C and the C content of the organic layer (O horizon) generally decreased with latitude. To our knowledge, the remarkably low level of soil C at the French site at Aubure was due to cattle grazing and litter raking in the past. Along the transect, there was no consistent difference between *Fagus* and *Picea* sites. Total plant C showed the highest values in the old Danish beech forest. The variation in C pools was due to stand age and management rather than latitude, except for slightly lower plant C in north Sweden.

The latitudinal trend of Net Primary Productivity expressed in C units (NPP_C) showed a trend similar to the soil C content (Fig. 21.1.C). The main uncertainty of NPP_C at all sites was with the estimates of fine roots and their turnover (see Scarascia et al., Chap. 3; George et al., Chap. 5, this Vol.). The present NPP_C estimate is based on the assumption that fine roots up to 2 mm thick will turn over once a year (see Stober et al., Chap. 5, this Vol.). Because of lower root growth in Scandinavia, 1-mm fine roots were taken as basis to estimate fine root growth in north Sweden (Åheden site, Åhe). The low soil C of the French site was mirrored by a distinctly lower NPP_C. Despite this, there was no consistent difference between NPP_C for *Fagus* and *Picea* along the transect. Leaf area index (LAI) varied fairly little in central Europe, but was lower in the boreal forest.

C mineralisation in the whole soil profile and litter inputs to the soil (foliage plus fine root material) were fairly similar along the European transect (Fig. 21.1B) with the exception of low values at Aubure (with low amounts of organic matter and NPP_C). The litter estimates do not include losses of bark and twigs and may thus be

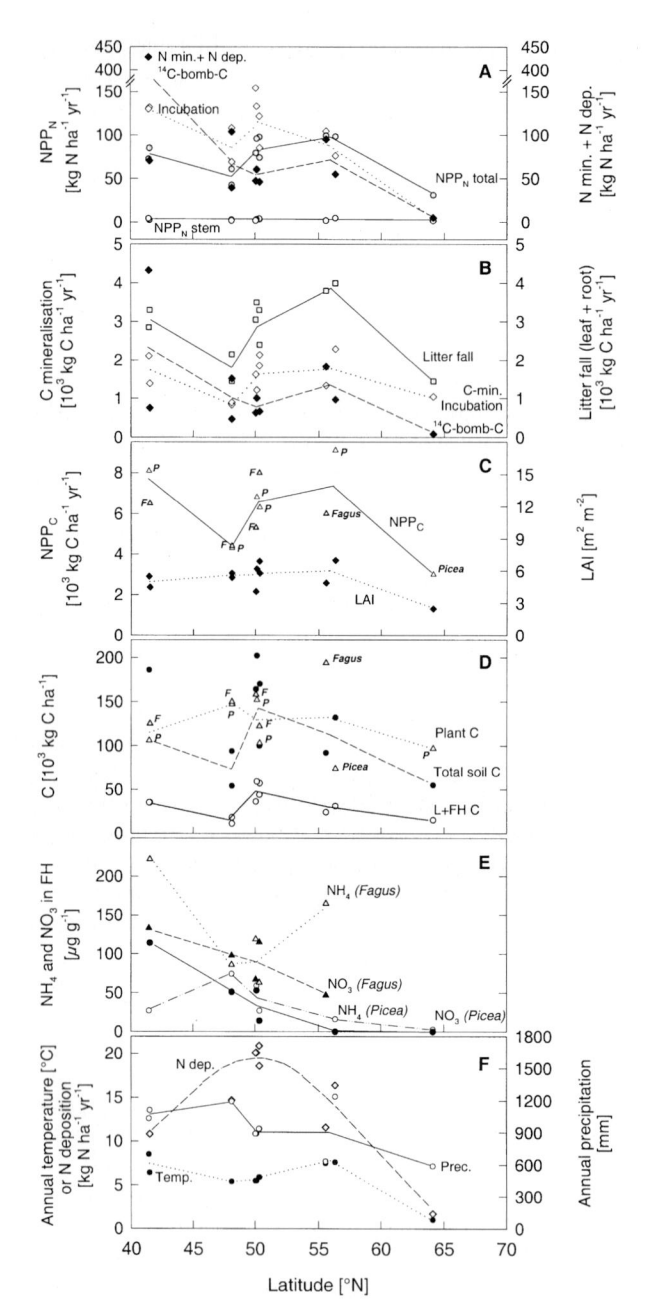

Fig. 21.A–F. Trends of ecosystem parameters along a north-south transect through Europe

underestimated. Different methods were used to determine C mineralisation: the ^{14}C-bomb carbon method, which is based on C age and litterfall (see Harrison et al., Chap. 11, this Vol.), and the incubation method, which incubates soils under a range of constant conditions and uses these data to estimate rates under field conditions (see Persson et al., Chap. 12, this Vol.). When based on the same amount of carbon in the profile, the ^{14}C-bomb carbon method appeared to yield lower rates (1.23 tC ha^{-1} a^{-1} ± 1.20, av ± S.D.) of C mineralisation than the incubation method (1.52 ± 0.52 tC ha^{-1} a^{-1}). The main difference between the two methods was observed at the MdM site (4.33 and 2.11 tC ha^{-1} a^{-1} for the ^{14}C and the incubation method, respectively). Excluding the MdM site, the incubation methods resulted in an estimate that was 0.54 ± 0.75 tC ha^{-1} a^{-1} higher than by the ^{14}C method ($p < 0.05$). The data obtained for the $O_f + O_h$ horizon ($O_e + O_a$ horizon according to FAO nomenclature) by food web studies were generally higher than the values determined by incubation (see Wolters et al., Chap. 17, this Vol.), but only few sites were studied for food webs.

N mineralisation was also determined by two methods, namely by incubation under constant conditions in the laboratory and extrapolation to field conditions, and by the ^{14}C turnover rate of C mineralisation assuming a constant linkage of C and N mineralisation. In order to compare both estimates, the rates were based on an average N content in each horizon of the soil profile. Thus, N weighted data are different from the data presented in Harrison et al. (Chap. 11, this Vol.) and Persson et al. (Chap. 14, this Vol.).

Average N mineralisation based on ^{14}C bomb carbon measurements was 81.1 ± 121 kgN ha^{-1} a^{-1}, but this average included a very high estimate of N mineralisation at the MdM site (417 kgN ha^{-1} a^{-1}). Excluding this agricultural afforestation site, N mineralisation based on ^{14}C was 43.8 ± 28 kgN ha^{-1} a^{-1}. In contrast, N mineralisation as estimated by incubation was on average 86.9 ± 40 kgN ha^{-1} a^{-1}. Excluding the MdM site had little effect on the average N mineralisation as measured by incubation (83.7 ± 41 kgN ha^{-1} a^{-1}). The difference between the two methods was significant ($p < 0.05$).

The N requirement for growth, NPP$_N$, was estimated by taking into account 20% reallocation of N from foliage and roots before foliage is shed as litter (Chapin and Kedrovski 1983; Chapin et al. 1990; see van Oene et al., Chap. 20, this Vol.). There is an uncertainty in assuming a constant reallocation across the transect which cannot be resolved at this stage. The N requirement for growth, NPP$_N$, showed no significant differences between *Fagus* and *Picea* (av. NPP$_N$: 75.8 ± 23 kgN ha^{-1} a^{-1}). Again, site-specific conditions overrode a general trend of decreasing NPP$_N$ from the south to the north. Only 2.89 ± 1.1 kgN ha^{-1} a^{-1} were required for stem growth (3.8% of NPP$_N$). This contrasts to the N demand for foliage and root growth, which require 60.9 ± 15.9 kgN ha^{-1} a^{-1} or 80% of NPP$_N$. It is obvious that this fraction is essential for C assimilation including nutrient uptake and transpiration. In fact, it is the basis for NPP$_C$ including stem growth. This interaction between N requirement for root and foliage growth versus stem growth has not been fully appreciated in studies evaluating the effects of N deposition on stem growth (Nadelhoffer et al. 1999).

Nitrate concentration in soil-water was measured only at selected sites (see Anderson, Chap. 15, this Vol.), and in combination with simulated water fluxes (see van Oene et al., Chap. 20, this Vol.) the nitrate loss to ground-water was calculated. About ten times higher rates of NO$_3^-$ leaching were estimated to occur under *Picea* than under *Fagus*. Estimated leaching reached a maximum of almost 29 kg ha^{-1} a^{-1} at the *Picea* site of Aubure. At that site, leaching was 84% of the estimated potential nitrification rate (incubation in absence of root uptake). Nitrate leaching was lower at Waldstein and Nacetin (9 and 11 kgN ha^{-1} a^{-1}). Although the export of nitrate was only 10 to 30% of the potential nitrification rate at these sites, leaching matched nitrate input by wet deposition (Durka et al. 1994). No nitrate leaching could be detected at Åheden or Skogaby, the two northernmost sites.

The export of DOC decreased with soil depth, indicating that a large fraction of DOC was precipitated or mineralised on its path through the soil profile. The total amount that was exported to the groundwater was <1% of the mineralisation rate and 3% of the net biome productivity.

Summarising the geographic trends, we find
- no parallel response of all parameters,
- that site-specific conditions override the latitudinal trends for some parameters,
- that there is an obvious effect of past land use at the litter-raked site, a maximum of variation for many processes in Denmark and south Scandinavia, and generally higher rates in Italy, and lower rates in north Scandinavia,
- no major differences between *Fagus* and *Picea* sites, except for higher concentrations of ammonium and nitrate in the O-horizon, higher net nitrification, but lower nitrate leaching under *Fagus*.

21.3 What Limits the C and N Fluxes in These Forest Ecosystems?

We may assume (see Schulze, Chap. 1, this Vol.) that under steady-state conditions in old-growth forests C and N mineralisation as measured over the whole soil profile should balance litter production (foliage and root material), and N requirement for growth should balance N mineralisation plus inputs by deposition. Figure 21.2A shows that C mineralisation generally did not match litterfall. Averaged over all sites and age classes, C mineralisation (1.23 to 1.52 t C ha^{-1} a^{-1} based on the ^{14}C and incubation determinations respectively) was considerably lower than litter (foliage plus fine root material) production (2.75 ± 0.85 t C ha^{-1} a^{-1}). The difference increased with increasing litter input. Since litter inputs do not include losses of bark, cones and twigs, the gap between resource supply and use by mineralisation was probably underestimated. C mineralisation balanced litterfall only at the MdM site where a high microbial activity at that site is related to the high pH value of this calcareous soil (see Persson et al., Chap. 2, this Vol.). Also at the litter-raked site of Aubure the *Picea* forest was close to an equilibrium of litterfall and mineralisation.

The results raise two questions: (1) which factors determine litter production, and (2) why does C mineralisation in soil not generally balance litter input? We are aware that there will always be short-term differences between litter production and C mineralisation, but it will have far-reaching implications if both processes do not match in the long term. The litter decomposition studies exposed different types of litter across the transect (see Cotrufo et al., Chap. 13, this Vol.) and demonstrated that an adjustment of decomposition should be possible over a 1- to 3-year period.

In order to investigate the first question about the factors causing high litter production, Fig. 21.2B relates the forest N requirement for growth, NPP$_N$, to the N availability as expressed by the N mineralisation rate plus wet deposition of N. The deposition rates increase by an additional 20 kg N ha^{-1} a^{-1} for Central Europe if dry deposition is included (see Harrison et al., Chap. 8, this Vol.). Neglecting the MdM site as a special case, NPP$_N$ was close to the average rate of N supply for most sites. The ^{14}C method generally estimated mineralisation rates that were lower than the N use by the forest, while the incubation methods generally yielded N availabilities that

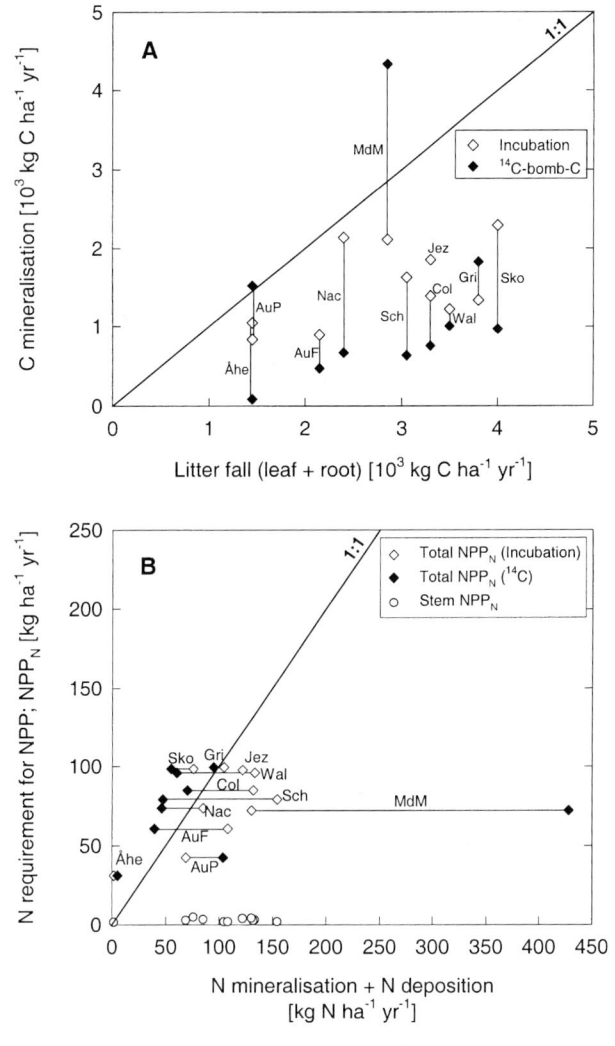

Fig. 21.2. A Carbon mineralisation as related to litterfall. B N requirement for growth (NPP$_N$) as related to net N mineralisation plus N deposition. Sweden: *Åhe* Åheden *Picea* and *Pinus*; *Sko* Skogaby *Picea*; Denmark: *Gri* Gribskov *Fagus*; Germany: *Sch* Schacht *Fagus*; *Wal* Waldstein *Picea*; Czech Republic: *Nac* Nacetín *Picea*; *Jez* Jezerí; France: *AuF* Aubure *Fagus*; *AuP* Aubure *Picea*; Italy: *Col* Collelongo *Fagus*; *MdM* Monte di Mezzo *Picea*

were larger than the N requirement for growth. There was no clear difference between *Fagus* and *Picea* forests. The boreal site (Åhe) is a special case, where both methods underestimated the N availability. In this case, NPP$_N$ indicates uptake of organic N by the mycorrhiza (see Wallenda et al., Chap. 6, this Vol.). Also at Skogaby, both methods indicate a higher uptake of N by the forest than when being estimated as supply of inorganic N. This may be taken as an indication that uptake of organic N remains an important fraction also in south Scandinavia. The fact that NPP_N exceeded the ^{14}C estimate of N availability also at other sites further south might indicate that uptake of organic N may continue also in the presence of inorganic N, but this needs further investigation. In contrast, the *Picea* plantation at the MdM site was not able to make full use of the higher rate of N supply. The MdM site had reached full canopy closure and showed signs of self-thinning due to competition for light between tree individuals. Also the *Picea* forest at Aubure showed signs of excess N supply despite its low NPP_C. Highest NPP_N was not observed at the highest N supply, but in the intermediate range of N availability, namely at the Skogaby (Sko) and the Gribskov (Gri) site.

For the majority of sites, an increase in NPP_N with N availability is apparent. It has been suggested that N deposition was largely responsible for the recent increase in forest growth in Europe (Spieker et al. 1998). Broadly speaking, our data confirm a relation between N supply and tree growth, but there seems to be an upper limit of N uptake which is close to about $100\,kgN\,ha^{-1}\,a^{-1}$. Dry deposition will add up to $20\,kgN\,ha^{-1}\,a^{-1}$, and nitrate leaching will export up to $20\,kgN\,ha^{-1}\,a^{-1}$. Thus, we think that Fig. 21.2B represents the right order of magnitude of forest N availability and N use. Independent of soil type and stand age, none of the forests used more than $100\,kgN\,ha^{-1}\,a^{-1}$, but also the average N supply was generally below this value. We do not know and therefore cannot take into consideration the N demand by either microbial biomass or ground vegetation. Thus, NPP_N for the whole ecosystem will be higher than NPP_N for the tree cover alone (Aber et al. 1989). There are indications that NPP_N seems to decline when N availability exceeds $100\,kg\,ha^{-1}\,a^{-1}$. We think that this level of N availability is close to a threshold that describes N saturation of these ecosystems which may not necessarily correlate with nitrate leaching, which has been the "hydrological" definition of N saturation. Apparently, most of the central European forests are close to N saturation conditions (see also the discussion on decomposition) independent of age and substrate, but there are only few sites where N availability clearly exceeds this threshold.

Long-term fertilisation experiments can add some insight into the discussion on N saturation. At the CANIF site Skogaby, a number of plots received extra input of nitrogen when the stand was 24 years old. From 1988 onwards, $100\,kgN\,ha^{-1}\,a^{-1}$ in the form of ammonium sulphate or as ammonium nitrate plus macro- and micronutrients was added to certain plots (n = 4 for each treatment). The inputs thus increased from 73 to about $170\,kgN\,ha^{-1}\,a^{-1}$. For the ammonium sulphate treatment the 24-year-old stand reacted with an uptake of $145\,kgN\,ha^{-1}\,a^{-1}$ allocated into growth or storage of stems, bark, foliage, branches and cones (52), needle litter (29), coarse roots (10) and root litter (54, not measured but calculated as a difference), whereas the remaining input was leached out as ammonium (6) and nitrate (9) or increased the inorganic N pool ($10\,kg\,ha^{-1}\,a^{-1}$) as a mean for the first 3 years (Nilsson and Wiklund 1994; Persson 1996). The corresponding uptake for the ammonium nitrate treatment was lower than in the ammonium sulphate treatment ($131\,kgN\,ha^{-1}\,a^{-1}$), which was explained by higher leaching of the mobile nitrate. The experiments showed for young stands with initially low N concentrations in the foliage ($12\,mg\,N\,g^{-1}$) that these have the capability to increase both the needle N concentration (15 to $16\,mg\,N\,g^{-1}$) and growth beyond an N use of $100\,kgN\,ha^{-1}\,a^{-1}$. These data do not invalidate the results obtained along the European transect, but they demonstrate that a sudden increase in N availability does not necessarily mean a simultaneous increase in N leaching. The Skogaby experiment also showed that the growth response was transient even though the N addition was constant. Eight to 10 years after the onset of the experiment, the growth rate in the ammonium sulphate treatment declined and the needles showed signs of magnesium deficiency. (L.-O. Nilsson, pers. comm.).

The growth response to added N may also be related to forest management. For economic reasons, most study sites are thinned to function at a stand density below the self-thinning level set by competition for light because the silvicultural aim was the production of a few big trees rather than many small trees. In addition, growth at full stocking can exceed the availability of cations, especially of Mg in central Europe, which has resulted in forest decline (Schulze 1989). Thus, stands were thinned below canopy closure in the 1980s and 1990s, which is indicated by the profuse ground cover of grasses in these stands (see Figs. 2.8, 2.10, Chap. 2, this Vol.). At low stand density a response to added N of the remaining trees is still possible without reaching saturation.

Canopy uptake may become a decisive factor in the process of eutrophication. N deposition enters into the ecosystem as inorganic N, some of which inputs straight into the soil (via canopy throughfall) and may be leached to groundwater. Another fraction of dry deposition which yet remains difficult to quantify (15 to 40 kg N ha^{-1} a^{-1}) are assimilated by the twigs and foliage into organic N. This added N would reach the ground as litter and increase the N mineralisation rate in the long-term perspective. Therefore, chronic input of N from the atmosphere would enhance eutrophication and contribute to the discrepancy between N availability and N use.

In the context of N use and supply it is important to note that the chemical nature of the N supply is not constant across the transect. A comparison of the N cycle between north and south European forest shows (Fig. 21.3) that in the boreal forest, no ammonium or nitrate could be detected in the organic layer, and there was no net mineralisation as measured by incubation (see Persson et al., Chap. 14, this Vol.) and N deposition is very low. In the boreal forest, the mycorrhizal fungi appear to short-cut the N cycle by taking up organic sources, e.g. amino acids from fresh litter, leaving

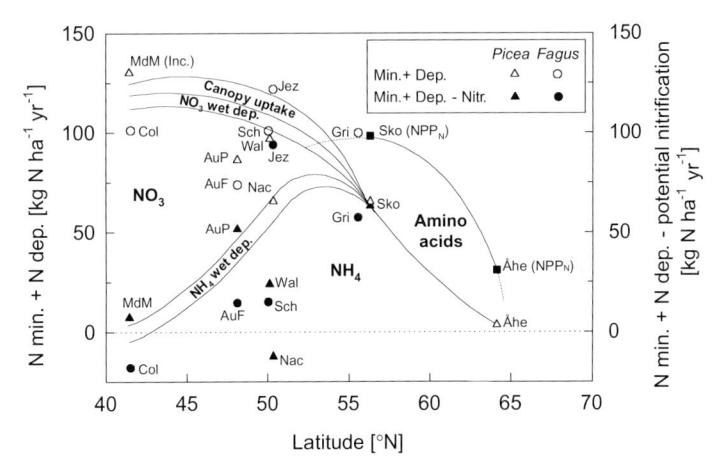

Fig. 21.3. Latitudinal change of net N mineralisation plus deposition, and potential nitrification. For abbreviations of sites see Fig. 21.2

little N for mineralisation by other microorganisms (Read 1991; Näsholm et al. 1998). Thus, the N availability at Åheden can only be estimated from NPP_N.

Moving south, from northern Sweden, N becomes detectable in the organic layer as ammonium (under *Picea*) and ammonium plus nitrate (under *Fagus*) in south Scandinavia and Denmark (Fig. 21.1E). NPP_N remained higher than the supply at the Skogaby *Picea* site, which might indicate uptake of organic N. Ammonium production by mineralisation and potential nitrate production by nitrification (nitrification in absence of roots) increase moving further south and reach equal shares in central Europe. Potential nitrification takes an increasing share of mineralisation on the calcareous soils of Italy, but high nitrification was also observed at Nacetin, a study site with a dominant grass layer as groundcover. We may interpret these shifts in N availability from north to south as a result of a decrease in the ratio between N demand by the trees and the N supply resulting from decomposition. Microorganisms and plants have increasing access to N when moving further south. This results from generally better growing conditions as well as from anthropogenic N deposition. The process could be enhanced by N-dependent changes in mycorrhizal communities (see Taylor et al., Chap. 16, this Vol.). Under conditions of excess N, autotrophic nitrification by bacteria transferred an increasing fraction of the mineralised N into nitrate. Obviously, further studies are needed to quantify the actual nitrification rate in soils, especially since nitrate plays such a pivotal role in the ecosystem's hydrochemistry.

A full understanding of the fate of nitrate is further complicated by translocation processes in the soil profile. Nitrate is not only displaced from upper to lower soil horizons but also produced and utilised below the rooting zone, partially in association with the consumption of DOC (see Andersen, Chap. 15, this Vol.). Obviously, the transport of nitrate from upper to lower root horizons will have negative effects on the cation balance in the rooting zone. It will uncouple the N cycle within the soil profile and change the edaphic conditions for growth in the top soil in the long term (see below). Independent of these changes in N availability, there seem to be other physiological and structural limitations to C assimilation by trees. Such limitations may determine the growth of *Picea* at the MdM site, where self-thinning took place.

The increase of production with N availability does not resolve the parallel problem at the decomposer side of the C and N cycle. It remains uncertain why mineralisation does not balance litter supply. Figure 21.2A indicates an increasing gap between litter supply and decomposition, and without any apparent difference between *Fagus* and *Picea*.

There are several reasons for this gap between resource supply and demand at the decomposer side of the C cycle. One possible cause is the change in the fungal community between north Sweden and central Europe. Apparently, the lignin-decomposing fungi and lignin decomposition may decrease with increasing availability of ammonium (see Taylor et al., Chap. 16, this Vol.). This would cause an accumulation of organic matter at the central European sites. The increase in MRT of organic matter (see Harrison et al., Chap. 11; Persson et al., Chap. 12, this Vol.) gives independent support to this idea (Fig. 21.4). Besides effects of community structure, abiotic parameters set the conditions for C mineralisation. The low levels of mineralisation and nitrification in the boreal forest may be explained by the low pH value of the acid soils. The experiments by Persson et al. (Chap. 14, this Vol.) in the boreal

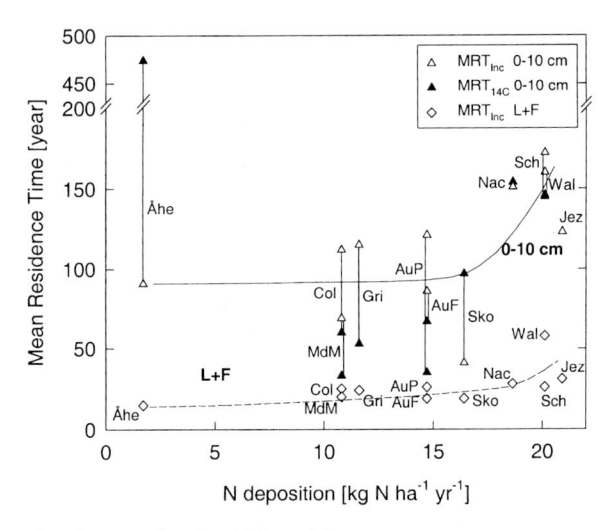

Fig. 21.4. Mean residence time of carbon as related to N deposition

forest suggest that the activity of nitrifying microorganisms is limited by the low levels of pH. The study of a small-scale gradient (the Betsele transect) gives an outstanding example of the significance of cation transport in ground water, the accumulation of Ca in a groundwater discharge area and a threshold-type initiation of microbial ammonium and nitrate production (Giesler et al. 1998).

It has also been suggested that the thickness of the litter layer determines the rate of mineralisation (see Matteucci et al., Chap. 10, this Vol.). Figure 21.5 shows that this is only the case at low C contents in the soil. Mineralisation increased initially with the C content of the organic layer, but a saturation of mineralisation is reached at 20 to 30 t C ha^{-1} a^{-1}. Most European sites operate above that saturation level. It is obvious that an accumulation of organic matter can only take place if mineralisation is not proportional to the thickness of the organic layer.

Temperature and water status in the organic layer may limit the activity of the decomposers (see Matteucci et al., Chap. 10; Wolters et al., Chap. 17, this Vol.), but the hydrological balance was positive at all sites. However, differences are expected between a litter layer of needles and of deciduous leaves, and there are indications from the Skogaby irrigation plot that mineralisation was water-limited in *Picea* also at this fairly humid site, where precipitation exceeds 1000 mm a^{-1} (see van Oene et al., Chap. 20, this Vol.).

21.4 What Are Net Ecosystem Productivity (NEP) and Net Biome Productivity (NBP), and How Do They Relate to Ecosystem Parameters?

Net ecosystem productivity, NEP, is defined as the difference between C assimilation and all respiratory processes (Melillo et al. 1993). Since NPP is the balance of plant assimilation and plant respiration (including roots) the difference between NPP$_C$ (as

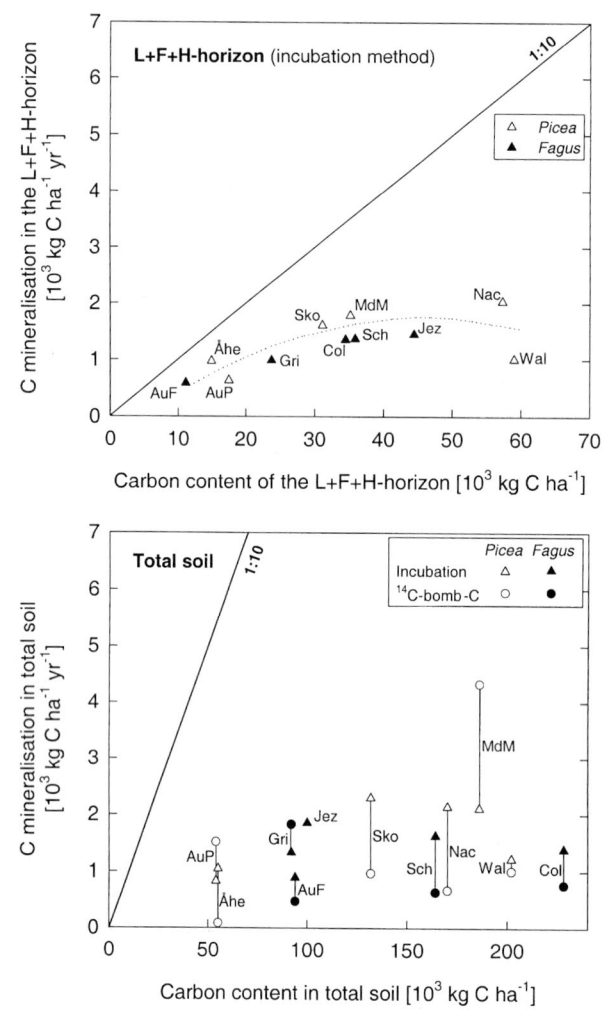

Fig. 21.5. C mineralisation as related to the carbon content (*top*) of the L + F + H horizon, and (*bottom*) as related to total soil carbon. For abbreviations of sites see Fig. 21.2

ecosystem C input) and C mineralisation is a measure of NEP (Fig. 21.6). NEP is zero, if NPP_C is equal to C mineralisation (1:1 line in Fig. 21.6), and NEP increases with NPP_C above this line.

The NEP rates calculated by this approach were of the same order of magnitude (Table 21.1), but generally higher than those obtained from direct flux measurements (Valentini et al. 1999). A direct comparison can only be made for the Collelongo site, where both approaches to determine NEP were used on the same plots. At Waldstein the eddy covariance system measured a neighbouring stand of younger age (Weidenbrunnen) which showed acute symptoms of Mg deficiency due to special soil

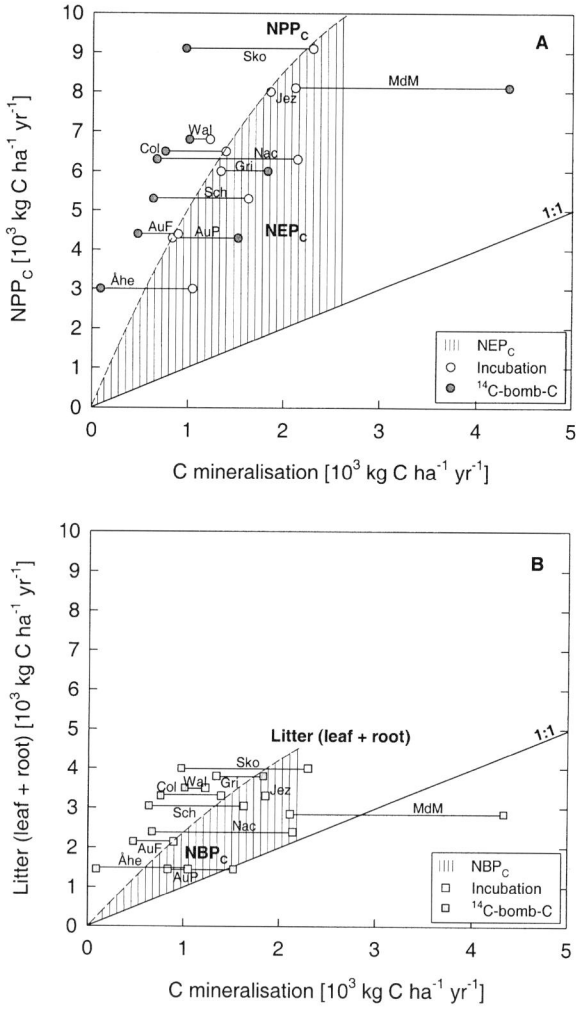

Interactions between the Carbon and Nitrogen Cycles and the Role of Biodiversity 479

Fig. 21.6. A NPP in carbon units (NPP$_C$) as related to C mineralisation. B Litter fall as related to C mineralisation. The *hatched areas* in **A** equal net ecosystem productivity (NEP) and in **B** the *hatched area* equals net biome productivity (NBP) For abbreviations of sites see Fig. 21.2

conditions. In north Sweden the eddy covariance system measured a younger stand in the same region. Thus, for the German and Swedish sites the comparisons can only be approximated (Table 21.1). It is encouraging to see that the bottom-up approach of process studies and the top-down approach of flux measurements resulted in data that are very close; but the data also indicate that the local variation may be large, and therefore it is essential that ecosystem and flux measurements are performed on the same plot.

NPP$_C$ increased with C mineralisation (Fig. 21.6A). NEP is illustrated in Fig. 21.6A by the difference between the regression line of NPP$_C$ versus C mineralisation and the

Table 21.1. Comparison of NEP from process studies and inventories and from direct measurements. The Waldstein measurements are not made on the same plot. The present work refers to the Coulissenhieb (Cou) while the eddy covariance measurements were made at Weidenbrunnen (Wei) stand nearby. At Åheden the eddy covariance data refer to the Flakaliden (Fla) site which is a younger stand in the same region

Site	NEP from NPP and mineralisation $tC\,ha^{-1}a^{-1}$ Incub. ^{14}C		NEE from eddy covariance flux measurements $tC\,ha^{-1}a^{-1}$	NBP from litter fall and mineralisation $tC\,ha^{-1}a^{-1}$ Incub.[a] ^{14}C	
Colleongo	5.1	5.7	4.7 (1993/94) to 6.6 (1996/97)	1.91	2.54
Waldstein	5.6	5.8 (Cou)	0.77 (1997/98) (Wie)	2.28	2.49 (Cou)
Åheden	2.0	2.9 (Åhe)	1.93 (1997) (Fla)	0.40	1.36 (Åhe)

[a] Incub.: refers to the incubation method and ^{14}C to the ^{14}C-bomb method to determine C mineralisation.

1:1 line (Fig. 21.6A, hatched area). Thus, NEP, which is to a large degree determined by the increment of woody biomass, increased with increasing C mineralisation and NPP$_C$. Since we are dealing with managed forests, part of the above-ground woody biomass (stems) would be harvested and will not support heterotrophic respiration. Therefore, at the ecosystem level, we think that NEP (as measured by process studies and inventories as well as by flux measurements) overestimate the long-term storage of C in forest ecosystems. Therefore Fig. 21.6B shows the C input to the soil by litter (foliage plus root material) as related to C mineralisation. The difference between litter input and C mineralisation (as represented by the 1:1 line in Fig. 21.6) approximates the net biome productivity (NBP) as defined by Schulze and Heimann (1998). NBP increases with litter fall and mineralisation (as a result of stimulated NPP). The approach for estimating NBP presented in this study does not include the mineralisation processes following logging of commercial forest. Logging produces slash which is added to the soil C pool, which would increase NBP, but logging also accelerates decomposition of existing soil organic matter due to mechanical disturbance of the soil and changes in microclimate as well as species composition of the groundflora that would decrease NBP. Model simulations by van Oene (Chap. 20, this Vol.) show a significant carbon loss in the first decade after logging. Thus, the rate of NBP as it is derived in Fig. 21.6 is probably higher than the long-term C immobilisation, depending on the management system and stand age.

On average, NBP is $1.44 \pm 0.92\,tC\,ha^{-1}a^{-1}$, which is 30% of NEP ($4.84 \pm 1.54\,tC\,ha^{-1}a^{-1}$). NEP appeared to increase ($p = 0.06$) with the C store in the organic layer (O horizon), while NBP was independent of the C content in the organic layer (Fig. 21.7). This would suggest that there is á priori no maximum level for C storage in soils, and that other factors, such as disturbances, will disrupt the transfer of C through the soil layers. The MRT shows that this C comprises not only labile but also recalcitrant C. NEP declined with stand age ($p < 0.05$) as would be expected from NPP$_C$ and stem growth. In contrast, there is a tendency for NBP to increase with

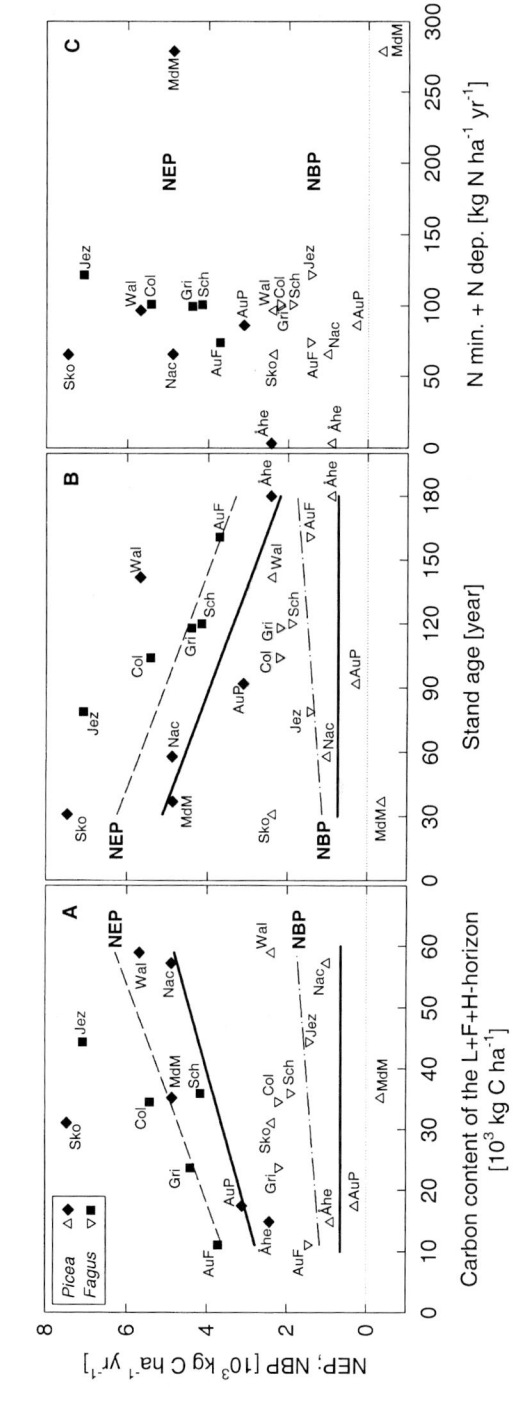

Interactions between the Carbon and Nitrogen Cycles and the Role of Biodiversity

Fig. 21.7A–C. Net ecosystem productivity (NEP), and net biome productivity (NBP) as related to (A) the carbon content of the L+F+H-horizon, **B** stand age, **C** N availability. *Bold lines* indicate the same relation as predicted by the NUCOM model (see Van Oene et al., Chap. 20, this Vol.). For abbreviations of sites see Fig. 21.2

stand age (slope not significant). This is confirmed by other studies where negative NBP was found in young stands depending on site conditions (Buchmann and Schulze 1999).

It is a misconception to believe that C storage would increase by cutting and replacing old growth stands (for further discussion see WBGU 1998), because NBP is mainly determined by litterfall, and this quantity is larger in old than in young stands. It is important to note that the site which would most closely meet the requirements of the Kyoto protocol, the MdM site, is a net C source. In contrast, the oldest stands, that are unimportant in the context of the Kyoto protocol (because they were established before 1990), and which are most likely to be harvested, have the highest C immobilisation rates. There is no effect of N availability or N deposition on NEP, but there appears to be a decline in NBP at high N availability.

The average rate of NEP (all sites: 4.84 ± 1.54) was higher than the rate estimated for Europe from flux measurements (3.03 $tC\,ha^{-1}\,a^{-1}$, Valentini et al. 1999). NBP, representing the C immobilisation in soils only (average 1.44 ± 0.92 $tC\,ha^{-1}\,a^{-1}$) would be half of the rate measured by eddy flux covariance (without taking into account C losses after logging). Since measurements were carried out across a broad range of climatic and environmental conditions, we may be in a position to extrapolate these rates across Western Europe (146 10^6 ha). The result suggests that the West European forest are a net carbon sink of 0.71 ± 0.22 $Gt\,a^{-1}$ including woody biomass growth, and 0.21 ± 0.14 $Gt\,a^{-1}$ if only carbon accumulation into soils is considered. NEP and NBP could probably decrease by 20% if the expected C loss after logging were taken into account (see van Oene et al., Chap. 20, this Vol.). NBP may be even higher if the longevity of wood products were considered.

21.5 Are There Thresholds and Non-Linearities?

There are several relations between ecosystem processes that are not linear, showing saturation characteristics which are important when considering predictions for the future. Most important is the response of NPP_N to N availability. NPP_N appears not to increase beyond 100 $kgN\,ha^{-1}\,a^{-1}$ available N. We think that forests in Europe have reached or are close to this threshold. The ecosystem effects of anthropogenic N deposition and canopy uptake depend on the chemical environment where the deposition takes place. The effect will be different in habitats with low or high rates of mineralisation. Since canopy uptake of N would be assimilated into amino acids, this fraction will enter the soil as organic N, and it will thus be measured as part of the mineralisation rate. Chronic N deposition would push the system into a range of higher N mineralisation rates and thus enhance the discrepancy between N supply and N use.

There are other response functions that show a saturation characteristic. C mineralisation saturated in relation to litterfall or to C storage in the organic layer. We also think that the response of NBP with N availability and stand age are ecologically and environmentally of importance and concern, respectively. NBP did not change with the amount of C stored in the organic layer. This suggests that there is no maximum of C storage in the organic layer, unless C is being mobilised by other processes, such as disturbances.

A destabilisation of the ecosystems by N deposition is not obvious from the C and N data, but there are hidden thresholds, such as the increasing nitrate production and leaching at high N availability. This will cause a cation imbalance in the long term (Ulrich 1987). P and K limitations have also been implicated as part of an unbalanced nutrient cycle (Stevens et al. 1993; Harrison et al. 1995, 1999), causing nitrate N to be leached from the soil profile. Along the CANIF transect, however, there was no consistent P or K deficiency, although there are several indications of site-specific nutrient limitations (see Bauer et al., Chap. 4, this Vol.). A plantation of forest that cannot utilise nitrate on habitats exhibiting strong rates of nitrification will lead to leaching and eutrophication and, therefore, imply a long-term ecological risk.

21.6 What Role Does Biodiversity Play in Ecosystem Processes?

After having investigated the clear-cut "roadmap" of the C and N cycle of forest ecosystems, we enter now into the demise of biodiversity. Having discussed the different C and N processes and analysed the magnitude of pools and fluxes, we will now attempt to see how far the diversity, above and below ground, will account for these differences. It is self-evident that all fluxes (except leaching of nitrate) in ecosystems are mediated by organisms. However, it remains uncertain if few or many organisms are needed to carry out this task.

21.6.1 Effects of Tree Species

Starting with the plant cover, we have been dealing with tree monocultures (except for north Sweden), but the question arises if it matters if the tree cover is deciduous or coniferous. There are many known differences between *Fagus* and *Picea* forests with respect to stand climate, water relations and productivity (Schulze 1982). However, if the edaphic conditions are similar, as in this study, then the differences between the two species with respect to the overall ecosystem fluxes are surprisingly small. In most cases, the overall variability between sites is larger than the difference due to the tree species. It is possible that our observations have been too short-term to uncover trends that would lead to long-term differences; but the 25-year model runs (see van Oene et al., Chap. 21, this Vol.) also support the conclusion that there are no major differences between tree species in ecosystem fluxes.

There are, however, significant differences at the process level, mainly concerning the adaptation to the use of nitrate. *Fagus* utilised nitrate at a higher rate than *Picea*. Ammonium and nitrate extractions showed higher concentrations in *Fagus* than in *Picea* litter, but this was reversed in the mineral soil, where DOC and DON concentrations in soil water were higher under conifers. Thus, the total N use by *Fagus* was higher than by *Picea*, whereas leaching of nitrogen was tenfold higher under *Picea* than under *Fagus*. This will affect the cation balance in the long term (Ulrich 1987; Schulze and Ulrich 1991), and the present study was too short to detect this type of change. Forest management has changed deciduous to coniferous forests at a large scale in Europe, and it is expected that this will affect the ecosystem cation balance and thus have consequences also for the C and N cycle in the long run. The saturation responses of NPP and of mineralisation could be a consequence of this indirect effect. In the case of central and eastern Europe, this effect was strongly enhanced by air pollution (see Dambrine et al., Chap. 19, this Vol.). In order to ameliorate this adverse effect of conifers, liming has become a routine practice in managed coniferous stands in parts of central Europe. Liming will also affect the P cycle (Harrison 1982, 1989).

21.6.2 Mycorrhizas

Mycorrhizal fungi are essential root symbionts for most forest trees and the study of Taylor et al. (Chap. 16, this Vol.) demonstrates not only a large diversity at the species level, but also a tremendous genetic diversity within a single mycorrhizal fungus. The key question concerns the extent to which this inter- and intraspecific diversity contributes to ecosystem functioning.

At the interspecific level it appears that both qualitative and quantitative aspects of the nitrogen supply exert strong influences upon selection of mycorrhizal fungi. Thus, sites which have large stocks of organic N and low rates of mineralisation support fungi which can utilise organic N without addition of glucose, while those with greater N availability or increased nitrification rates are dominated by fungi which lack these attributes and are minerotrophic (Fig. 21.8). We can therefore recognise distinctive functional groups. Since trees occurring in environments with low rates of N mineralisation are dependent upon the fungi for mobilisation of N from organic compounds, the functional group responsible for maintenance of this ecosystem flux can clearly be seen to have keystone properties. We do not know at this stage how many of these keystone species are required to maintain the fluxes, though in view of the physicochemical complexity of the soil organic residues in which N is sequestered, it can be hypothesised that greater genetic and biochemical diversity will be required in these environments than in those requiring expression only of single enzymes such as nitrate reductase or glutamate dehydrogenase.

More pertinent to the issue of the role of mycorrhizal biodiversity in overall ecosystem function is the observation that even in sites such as Åheden, where fungi capable of mobilising organic N are favoured, some of the minerotrophic species are retained. These representatives of a distinct functional group, e.g. *Tylospora fibrillosa*, which assume dominance when the systems are disturbed by N inputs, are far from being redundant. They provide the ecosystem with the potential for rapid response to circumstances such as mineral N enrichment which threaten its equilibrium, and thus will contribute to ecosystem resilience. Amongst the essential functions in addition to mineral N capture which such fungi might perform in the perturbed situation would be capture of essential anions, particularly phosphate, as well as cations, whose decreasing availability limits productivity in N-saturated environments. It is more difficult to determine whether the converse situation applies in the minerotrophic environments, but some of the *Russula* species observed to be present at low levels here can be judged, on the basis of their known ecological characteristics, to be likely members of the organic N-functional group and hence to provide buffering if mineral N inputs should decline.

Changes in the physicochemical quality of substrates down the soil profile represent, in microcosm, some of the shifts seen at the latitudinal scale, and for the same reasons, they can be predicted to select in favour of diversity. At this spatial scale the potential advantage of both intra- and interspecific variability for optimisation of habitat exploitation can be appreciated. There is good evidence from earlier studies of genetic structure of the ectomycorrhizal species *Suillus bovinus* (Dahlberg and Stenlid 1990) that selection operates over time within a population of siblings to

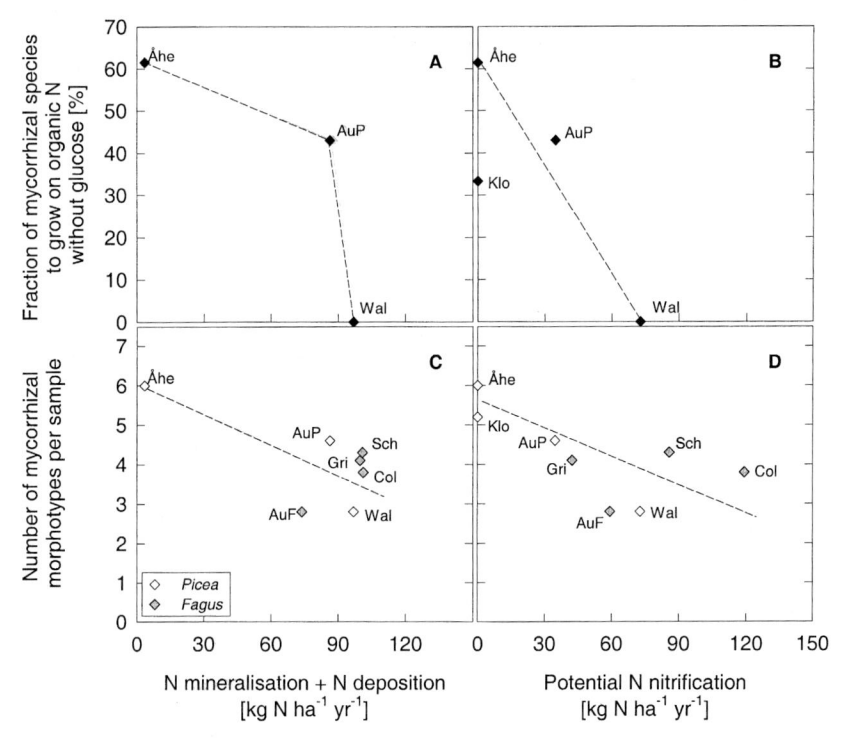

Fig. 21.8A–D. Fraction of mycorrhizal species (**A,B**) and number of mycorrhizal morphotypes per sample (**C,D**) as related to N mineralisation and deposition (**A,C**) and to potential N nitrification (**B,D**). For abbreviations of sites see Fig. 21.2

reduce to a smaller number the initially vast number of individual genets produced by a fruit body. Not only can it be predicted that those genets surviving under equilibrium conditions have attributes appropriate for their particular niche, but also that the diversity found in the initial population would enable rapid response to perturbation by providing alternative suites of genets appropriate to the conditions such as those associated with N enrichment.

The importance of mycorrhizal fungal activities in the broader process of decomposition should not be ignored. Fungi are capable of lignin degradation and their role in facilitating the breakdown of these and related lignocellulose compounds is therefore fundamental to an understanding of the biogeochemical process discussed in this Volume. While there have been few studies of the abilities of mycorrhizal fungi to degrade these compounds, those that have been made indicate considerable potential for lignophylic (Trojanowski et al. 1984; Haselwandter et al. 1990) and cellulolytic (Cao and Crawford 1993a,b) activities. Laboratory analyses (Griffiths and Caldwell 1992) confirm greater amounts of cellulolytic activity in those fungi which form mycorrhizal mats (i.e. k-strategists) than in non-mat forming r-strategists. Furthermore, rates of lignin and cellulose decomposition in the field are significantly greater in these mats than in adjacent non-mat soils (Entry et al. 1991).

Because mycorrhizal fungi, particularly those of the k-type, are often unculturable or slow-growing, most of what we know of the biochemistry and control of lignin decomposition comes from studies of a model organism, the white-rot fungus *Phanerochaete chrysosporium*. Two features of direct relevance to the present study emerge from analysis of this fungus. The first is that expression of the lignase enzyme requires growth cessation and a switch from primary to secondary metabolism. Second, and perhaps most important, lignolytic activity is inhibited in the presence of excess mineral N and triggered primarily by N limitation (Keyser et al. 1978). If, as appears likely, the biochemical mechanisms and the controls upon them are the same in mycorrhizal fungi as in *P. chrysosporium*, we would expect greater lignolytic activity and hence more rapid decomposition rates in the slow-growing mineral N-limited environments of the boreal forest than in the more N-enriched systems of the south.

At this moment we do not know if N in soil solution affects mycorrhizal activity and survival directly or indirectly. Berg and Matzner (1997) proposed that (1) soil acidification (as a consequence of acid rain or N deposition) may result in B deficiency in microorganisms including fungi, (2) Mn deficiency (cofactor of peroxidase), or (3) NH_4 binds in a stable form to humus precursors. None of these alternatives holds for the entire transect. The Italian sites were not acidified and were driven by nitrate rather than ammonium but showed similar unbalance between resource supply and use as the other sites.

Although a clear role of mycorrhizas becomes apparent, the fungal community responds (adjusts) to the chemical environment of the habitat possibly with significant effects on decomposition, but they are not able to alter the chemical environment in a way that the N status of the system remains favourable for the k-strategists.

21.6.3 Microfungal, Bacterial and Invertebrate Communities

The difference in ammonium and nitrate concentrations in the litter layer of *Fagus* and *Picea* already indicates differences in biological activity between forest types even though total rate of mineralisation may not be different (Table 21.2). Wolters et al. (Chap. 17, this Vol.) and Kjøller et al. (Chap. 18, this Vol.) quantify a large diversity in species that can decompose a very large assortment of organic compounds ranging from simple organic acids to carbohydrates to some polymeric compounds. No significant changes in the potential functions of the bacterial community could be detected in the coniferous forests along the European transect except for the site in southern Sweden, where the diversity was low. The accumulated litter may exert a strong selection and the low functional diversity in southern Sweden may indicate impoverished communities affecting decomposition and enhancing site conditions in a feedback (see Kjøller et al., Chap. 19, this Vol.).

The microfungal community examined in the beech forest along the transect showed small but distinct differences in composition of the flora over the time scale involved. During decomposition, the frequency of cellulolytic microfungi increased, replacing the initial group of non-cellulolytic fungi, concurrently with increasing cellulase activity. One may consider describing this as functional diversity rather than species diversity. The use of standard beech litter showed higher cellulase activity at the northern beech sites – leading to a faster decomposition. When litters from Europe were transplanted to Italy, virtually the same cellulase activity could be main-

Table 21.2. Comparative conditions in soils of *Fagus* and *Picea* forests at the German sites Schacht and Waldstein

	Fagus (Schacht)		*Picea* (Waldstein)	
Base saturation in 0–10 cm depth (%)	9.7 ± 2.2		4.0 ± 1.1	
Concentrations in LFH-horizon ($\mu g\,g^{-1}$)				
NH_4^+	108.2 ± 44.2		44.0 ± 27.1	
NO_3^-	81.8 ± 30.5		29.5 ± 26.6	
Concentrations in soil water ($mg\,l^{-1}$)	25 cm	90 cm	25 cm	90 cm
DOC-C 25 cm	5.9 ± 1.3	2.7 ± 0.6	17.6 ± 4.1	2.9 ± 0.9
DON-N 25 cm	0.1 ± 0.2	0.1 ± 0.2	0.6 ± 0.6	0.5 ± 0.5
NO_3^--N 25 cm	0.2 ± 0.5	0.1 ± 0.3	2.3 ± 1.9	2.8 ± 1.8
NO_3^- flux at 25 cm depth ($kgN\,ha^{-1}\,a^{-1}$)	1.2		9.2	
DON flux at 25 cm depth ($kgN\,ha^{-1}\,a^{-1}$)	0.7		2.5	
DOC flux at 25 cm depth ($kgC\,ha^{-1}\,a^{-1}$)	31.7		69.7	
Mineralisation rate in LFH horizon				
$kgC\,ha^{-1}\,a^{-1}$	1100 ± 397		1387 ± 538	
$kgNO3$-$N\,ha^{-1}\,a^{-1}$	21.7 ± 10.2		28.7 ± 28.6	
Change in stable isotopes	$\delta^{15}N$	$\delta^{13}C$	$\delta^{15}N$	$\delta^{13}C$
between L and 5 cm soil ‰	+4.5 ± 0.1	0 ± 0.4	+5.6 ± 0.8	−0.34 ± 0.5

tained by a lower fungal diversity in the transplanted litters. Also it was found that the key functions could be occupied by different organisms at different sites, showing functional substitution.

This may be different for the invertebrate community, where model calculations of invertebrate food webs show that these communities are well-buffered against species losses, i.e. modelling suggests that the flux changes very little, even if major groups are lost. The total food web of invertebrates contributed only 7 to 9% to the total C mineralisation. Thus, it appears that these groups contribute very little and contain so much excess functional redundancy that the environmental change along the transect did not affect the overall function.

We may conclude that the high number of individuals and taxa among the soil organisms make it a very difficult task to identify the key regulating organism for species function. Instead, it seems feasible to identify ecological thresholds of microbial occurrence and activity in quantitative terms based on functional groups or specific activities. In the complex matrix created by litter material, soil organisms interact and regulate the processes that take place. The redundancy appears to be large, and overlapping functions are common and these traits are important elements for litter decomposition; but despite this, diversity decomposition did not match litterfall.

21.7 Conclusions

- The C and N cycles of European forests, except in the boreal region, are not balanced, i.e. resource supply exceeds resource use especially at the decomposition

level. It remains unclear why organisms cannot use these extra resources. It appears that biodiversity is not limiting these processes, and that the soil communities adapt to the soil environment readily, with a change in species composition at similar levels of diversity. We could not detect a situation where the biodiversity of soil organisms regulates the C and N flux, and must conclude that other soil chemical factors, perhaps base cation supply, Al or P and pH regulate and limit the biological activity such that resource use and supply do not match.
- Anthropogenic N deposition does not alter the short-term C and N fluxes significantly, because the internal fluxes within the ecosystem are much larger than the extra supply and plant production can cope with the supply under most conditions. However, chronic atmospheric deposition of N from air pollutions can trigger processes in the soil, which will lead in the long run to changes in the soil chemical environment, altering the resource use beyond a species-specific threshold. This threshold appears to be at about $100\,kgN\,ha^{-1}\,a^{-1}$ with regard to inorganic N. It does not matter if this supply originates from ecosystem internal or external processes. If N mineralisation rates are low, e.g. due to past land use, external supply may not be harmful, while if N mineralisation rates are high, the extra effect of N deposition will lower the site productivity. Canopy uptake of N will contribute to a high N mineralisation rate and thus eventually lead to, or enhance the unbalance between supply and demand.
- The study of the ecosystem internal transfers of C provide estimates of the ecosystem net C balance which is important in view of the Kyoto protocol. C sequestration can be managed and increased by maintaining diversity in terms of species and tree ages on the same area, and by reducing N availability, since NBP was negatively related to the C content in the organic layer. It seems that there is no upper limit to which C can be stored in the forest soil unless other environmental conditions change and support C immobilisation, such as disturbances.
- The average rate of NEP (including woody biomass increments) was about $5\,tC\,ha^{-1}\,a^{-1}$ and the rate of NBP (excluding woody biomass increments and neglecting decomposition after logging) was aout $1.4\,tC\,ha^{-1}\,a^{-1}$. These rates are higher than estimates based on the increase in stable humus and black carbon, mainly because of unknown mineralisation after logging. However, since measurements were carried out across a broad range of climatic and environmental conditions we may be in a situation to extrapolate these rates across Europe. Results suggest the sequestration of $0.7\,Gt\,a^{-1}$ of C (measured as NEP) including stem-wood increments. When excluding stem-wood growth because it will be harvested, NBP of European forest as measured by the increase of soil C (including possible effects of logging) is in the order of $0.2\,Gt\,a^{-1}$, which is almost 20% of the fossil fuel consumption of the EEC ($1\,Gt\,a^{-1}$: Oak Ridge Natl. Laboratory Web Site, Marland and Boden 1996) and thus exceed the 8% reduction commitment by the Kyoto protocol, but carbon gains in managed forests cannot be accounted for.

References

Aber JD, Nadelhoffer KJ, Steudler P, Melillo JM (1989) Nitrogen saturation in northern forest ecosystems. Bioscience 39:378–386

Berg B, Matzner E (1997) Effect of N deposition on decomposition of plant litter and soil organic matter in forest systems. Environ Rev 5:1–25

Buchmann N, Schulze E-D (1999) Net CO_2 and H_2O fluxes of terrestrial ecosystems. Global biogeochem cycles 13:751–760
Buchmann N, Schulze E-D, Gebauer G (1995) ^{15}N-nitrogen and ^{15}N-nitrate uptake of a 15-year-old *Picea abies* plantation. Oecologia 102:361–370
Cao W, Crawford DL (1993a) Carbon nutrition and hydrolytic and cellulolytic activities in the ectomycorrhizal fungus *Pisolithus tinctorius*. Can J Microbiol 39:529–535
Cao W, Crawford DL (1993b) Purification and some properties of β-glucosidase from the ectomycorrhizal fungus *Psilothus tinctorius* strain SMF. Can J Microbiol 39:125–129
Chapin FS III, Kedrovski RA (1983) Seasonal changes in nitrogen and phosphorous fractions and autumn translocation in evergreen and deciduous taiga trees. Ecology 64:373–391
Chapin FS III, Schulze E-D, Mooney HA (1990) The ecology and economics of storage in plants. Annu Rev Ecol Syst 21:423–427
Dahlberg A, Stenlid J (1990) Population structure and dynamics in *Suillus bovinus* as indicated by spatial distribution of fungal clones. New Plytol 115:487–493
Durka W, Schulze E-D, Gebauer G, Voerkelius S (1994) Effects of forest decline on uptake and leaching of deposited nitrate determined from ^{15}N and ^{18}O measurements. Nature 372:765–767
Entry JA, Donnelly PK, Cromack K (1991) Influence of ectomycorrhizal soils on lignin and cellulose degradation. Biol Fertil Soils 11:75–78
Giesler R, Högberg M, Högberg P (1998) Soil chemistry and plants in Fennoscandian boreal forests as exemplified by a local gradient. Ecology 79:119–137
Griffiths RP, Caldwell BA (1992) Mycorrhizal mat communities in forest soils. In: Read DJ, Lewis GH, Fitter AH, Alexander IJ (eds) Mycorrhizas in ecosystems. CAB International, Wallingford, UK, pp 98–105
Harrison AF (1982) Labile organic phosphorous mineralisation in relation to soil properties. Soil Biol Biochem 14:343–351
Harrison AF (1989) Phosphorous distribution and cycling in European forest ecosystems. In: Tiessen H (ed) Phosphorous cycles in terrestrial and aqautic ecosystems. Regional Workshop 1: Europe. SCOPE/UNEP Proceedings, University of Saskatchewan, Saskatoon, Canada, pp 42–76
Harrison AF, Stevens PA, Dighton J, Quarmby C, Dickinson AL, Jones HE, Howard DM (1995) The critical load of nitrogen for Sitka spruce forests on stagnopodsols in Wales: role of nutrient limitations. For Ecol Manage 76:139–148
Harrison AF, Carreira J, Poskitt JM, Robertson SMC, Smith R, Hall J, Hornung M, Lindley DK (1999) Impacts of pollutant inputs on forest canopy conditions in the UK: possible role of P limitations. Forestry 72:367–377
Haselwandter K, Bobleter O, Read DJ (1990) Degradation of ^{14}C-labelled lignin and dehydropolymer of coniferyl alcohol by ericoid and ectomyccorhizal fungi. Arch Microbiol 153:352–354
Kajimoto T, Matsuura Y, Sofronov MA, Volokitina AV, Mori S, Osawa A, Abaimov AP (1999) Above- and below-ground biomass and net primary productivity of a *Larix gmelinii* stand near Tura, central Siberia. Tree Physiol 19:815–822
Keyser P, Kirk TK, Zeikus JG (1978) Lignolytic enzyme system of *Phanerochaete chrysosporium*: synthesised in the absence of lignin in response to nitrogen starvation. J Bacteriol 135:790–797
Last FT, Watling R (1991) Acid deposition: its nature and impacts. R Soc Edinb Sect B 97B:273–324
Machonat G, Matzner E (1995) Quantification of ammonium sorption in acid forest soils by sorption isotherms. Plant Soil 168:95–101
Majdi H, Persson H (1993) Spatial distribution of fine roots, rhizosphere and bulk-soil chemistry in an acidified *Picea abies* stand. Scand J For Res 8:147–155
Melillo JM, Mcguric AD, Kicklighter DW, Moon III B, Vör Ösmarty CJ, Schloss AL (1993) Global climate change ed terrestrial met primary production, Nature 363:234–240
Melillo J, Prentice C, Schulze E-D, Farquhar G, Sala O (1995) Terrestrial ecosystems: responses to global environmental change and feedbacks to climate. IPCC-WGI 9:445–482

Meyer O (1993) Functional groups of microorganisms. In: Schulze E-D, Mooney HA (eds) Biodiversity and ecosystem function. Ecological studies 99. Springer, Berlin Heidelberg New York, pp 67–96

Nadelhoffer KJ, Emmett BA, Gundersen P, Kjonaas OJ, Koopmans CJ, Schleppi P, Tietema A, Wright RF (1999) Nitrogen deposition makes minor contribution to carbon sequestration in temperate forests. Nature 398:145–148

Näsholm T, Ekblad A, Nordin A, Giesler R, Högberg M, Högberg P (1998) Boreal forest plants take up organic nitrogen. Nature 392:914–916

Nilsson L-O, Wiklund K (1994) Nitrogen uptake in a Norway spruce stand following ammonium sulphate application, fertigation, irrigation, drought and nitrogen-free fertilisation. Plant Soil 164:221–229

Persson T (1996) Kann skogen ta hand om allt kväve? (Can the forest retain all nitrogen?) Fakta Skog Konf 2:55–73 (in Swedish)

Read DJ (1991) Mycorrhizas in ecosystems. Experimention 47:376–391

Reichle ED (ed) (1981) Dynamic properties of forest ecosystems. International Biological Programme 23, Cambridge University Press, Cambridge, 683 pp

Schulze E-D (1982) Plant life forms as related to plant carbon, water and nutrient relations. In: OL Lange, PS Nobel, CB Osmond, H Ziegler (eds) Encyclopedia of plant physiology. Physiological plant ecology, vol 12B. Water relations and photosynthetic productivity. Springer, Berlin Heidelberg New York, pp 615–676

Schulze E-D (1982) Plant life forms as related to plant carbon, water and nutrient relations. In: Lange OL, Nobel PS, Osmond CB, Ziegler H (eds) Encyclopedia of plant physiology, vol 12B. Springer, Berlin Heidelberg New York, pp 615–676

Schulze E-D (1989) Air pollution and forest decline in a spruce (*Picea abies*) forest. Science 244:776–783

Schulze E-D, Heimann H (1998) Carbon and water exchange of terrestrial systems. In: Galloway JN, Melillo J (eds) Asian change in the context of global change. Cambridge University Press IGBP-Series, vol 3, Cambridge, pp 145–161

Schulze ED, Ulrich B (1991) Acid rain – a large-scale, unwanted experiment in forest ecosystems. SCOPE 45:89–106

Schulze E-D, Oren R, Lange OL (1989) Nutrient relations of trees in healthy and declining Norway spruce stands. In: Schulze E-D, Oren R, Lange OL (eds) Air pollution and forest decline: a study of spruce (*Picea abies*) on acid soils. Ecological Studies 77. Springer, Berlin Heidelberg New York, pp 392–417

Schulze E-D, Lloyd J, Kelliher FM, Wirth C, Rebmann C, L••hker B, Mund M, Knohl A, Milykova I, Schulze W, Ziegler W, Varlagin A, Sogachov A, Valentini R, Dore S, Grigoriev S, Kolle O, Tchebakova N, Vygodskaya NN (1999) Productivity of forests in the Eurosiberian boreal region and their potential to act as a carbon sink – a synthesis. Global Change Biol 5:703–722

Smith SE, Read DJ (1997) Mycorrhizal symbiosis 2nd edn. Academic Press, London, 605 pp

Spieker H, Mielikäinen K, Köhl M, Skovsgaard JP (eds) (1996) Growth trends in European forests. In: European Forest Institute Research Report 5. Springer, Berlin Heidelberg New York, 372 pp

Stevens PA, Harrison AF, Jones HE, Williams TG, Hughes S (1993) Nitrate leaching from a Sitka spruce plantation and the effect of fertilisation with phorphorous and potassium. For Ecol Manage 58:233–247

Trojanowski J, Haider K, Hüttermann A (1994) Decomposition of ^{14}C-labelled lignin, holocellulose and lignocellulose by mycorrhizal fungi. Arch Microbiol 139:202–206

Ulrich B (1987) Stability, elasticity, and resilience of terrestrial ecosystems with respect to matter balance. In: ••. Ecological Studies 61. Springer, Berlin Heidelberg New York, pp 11–49

Valentini R, Matteucci G, Dolman AJ, Schulze E-D, Rebmann C, Moors EJ, Granier A, Gross P, Jensen NO, Pilgaard K, Lindroth A, Grelle A, Bernhofer CH, Grünwald T, Aubinet M, Ceulemans R, Kowalski AS, Vesala T, Rannik Ü, Berbigier P, Lousteau D, Gudmundsson J, Thorgeirsson H, Ibrom A, Morgenstern K, Clement R, Montcrieff J, Montagnani L, Minerbi

S, Jarvis PG (1999) Respiration in the main determinant of European forest carbon balance. Nature (in press)

WBGU (1998) The accounting of biological sinks and sources under the Kyoto protocol: a step forwards or backwards for global environmental protection? German Advisory Council on Global Change, Spec Rep 1998, ISBN 3-9806309-0-9, Alford-Urgener-Institut, Bremerhaven, 75 pp

Westoby M (1984) The self-thinning rule. Adv Ecol Res 14:167–225

Subject Index

age of stands   16p, 24p
Al   see soil
amino acids
  arginin   84pp
  aspartate   84, 87
  concentrations   83pp
  glutamate   84, 87
  glutamine   84, 87, 129, 134, 178
  glycine   134
  methods   65, 123
  uptake   123, 128p, 137, 357p
ammonia   173pp, 297pp
ammonium
  ammonium/nitrate   136
  soil   157pp, 35, 38, 297pp
  transect   470, 475
  uptake   127, 135, 137pp, 149, 162
ANPP   see NPP
arginin   see amino acids
aspartate   see amino acids

basal area of stands   18
base saturation   see soil
biomass   52pp
  allometric ratios   53
  branches, twigs   see branches, twigs
  decomposer   see decomposer
  foliage   see foliage
  method   51pp
  microorganisms   see microorganisms
  nutrient pools   57
  roots   see roots
  stem/total biomass   53, 90
  summary   60
  understorey   see understorey
BIOME BGC   see modelling
branches, twigs
  biomass   53
  growth   55
  nutrient pools   58, 66
bulk density   see soil

C
  age (^{14}C)   238, 246
  budgets   248
  climate effects   245
  concentrations   65
  cycle   3pp, 431pp
  investment   see roots
  mineralisation   see mineralisation
  partitioning in plant   53, 140
  pools   21, 31, 36pp, 40, 56, 89, 90, 242, 264, 469
  root   see roots
  soil   see soil C
  uptake   see mycorrhizae
Ca
  plant concentrations   66, 70
  plant pools   58
  soil   32
CANIF project   8, 468, 487
carbon   see C
canopy nitrogen uptake   see also N
  gas phase   174, 177
  liquid phase   174
  methods   175pp
  policy   184
  summary   183
  total   179
  wet deposition   177
catchment studies   405pp, 417
  ecosystem budget   413, 415
  long term trends   405pp
  streamwater runoff   412
cation exchange capacity   see soil CEC
CD-ROM   39
CENTURY   see modelling
climate along transect   20, 24, 470
critical loads of nitrogen   182

data bank   see CD-ROM
decomposer   378
  biomass   376

Xp = page X and following page; Xpp = page X and following pages

climate effects   373
community   368pp
diversity indices   374
food web   366pp, 374, 377
microflora   369pp, 382pp
soil fauna abundance   372
trophic structure   374
decomposition (litter)   156, 246, 278pp, 436, 473
age of stand (correlation)   268
budget   436, 444
carbon pool   264
climate effects   248, 250, 278
C/N ratio   277, 283
enzyme activity   278, 285pp
mass loss   281pp, 289pp
methods   238, 258, 275
N   156, 279, 290p
rates   260pp, 284, 377pp, 436
transect   281, 469
transplants   283
denitrification   7, 307p, 316, 324
density of stands   18, 59
deposition   21, 42
N-dry   21, 180
N-throughfall   411, 164, 197, 321
N-wet   15pp, 21, 190, 191pp, 409
S-dry   15, 21pp, 29, 42
S-wet   15, 21pp, 29, 42
diameter of stands   18
diversity
decomposer   374p, 486
fungi   345pp, 484
invertebrates   366pp
microfungi   389, 391pp
microorganisms   382pp, 398, 486
mycorrhizae   see mycorrhizae
plant   483
diversity related processes   343pp, 483pp
fungi   243pp, 344pp, 484
soil invertebrates   486
trees   483
DOC (dissolved organic carbon)   see soil water
DON (dissolved organic nitrogen)   see soil water
dry deposition   see deposition

ecosystem budgets   413, 415
C pools   36pp
N pools   36pp
thresholds   482
European transect   see transect

fertilizer trials   see mineralisation
foliage
biomass   53, 433
^{13}C   see isotopes
growth   55, 433
leaf area index   see LAI
^{15}N   see isotopes
litter fall   470, 473
specific leaf area   57
nutrient pools   58
concentrations   66, 70
contents   74, 78, 80pp
N   65pp
weight per organ   77, 80pp
fungi
community structure   247
general   243
method   344
mycorrhizae   see mycorrhizae
saprophytic fungi   207
diversity   see diversity

glutamate   see amino acids
glutamine   see amino acids
glycine   see amino acids
growth   see NPP
branches, twigs   see branches, twigs
foliage   see foliage
roots   see roots

harvest index   53
height of stands   18

isotopes
^{13}C
soil depth   197pp
foliage   202
uptake   140
^{14}C
age   see C age
bomb   238
turnover   237
methodology   239pp
^{15}N
ammonium   195pp
foliage   203
labelling   146pp, 154, 288
life forms   205pp
nitrate   195pp
N-pollution   194, see also pollution
N-uptake   124pp, 144pp
soil depth   197pp
throughfall   197

Subject Index

wet deposition   190pp
^{32}S
  soil depth   197pp, 200

K
  plant concentrations   66, 70
  plant pools   58
  soils   31
Kyoto protocol   3

LAI   18, 56p, 469p
leaf   see foliage
leaf area index   see LAI
litter fall   see foliage
logging   458pp

magnesium   see Mg
methods
  biomass   51pp
  canopy uptake   175pp
  decomposition   238, 258, 275
  DOC   233
  fungi   344
  isotopes   189, 239pp, 299
  microorganisms   368, 383, 393
  mineralisation   see mineralisation
  NBP, NEP   478
  NPP   51
  nutrition   64
  respiration   218, 230
  roots   100
  soil water   see soil
mean residence time   see MRT
Mg
  plant concentrations   66, 70
  plant pools   58
  soils   32
microflora   see decomposer
microfungi   382pp, 385pp
  abundance   387
  BIOLOG   368, 393
  climate   396
  community   386
  diversity   see diversity
  enzymes   394pp
  transect   390
  transplants   386
microorganisms
  diversity   see diversity
  methodology   368, 383, 393
  microbial biomass   370
  microbial C and N   370
  microfungi   see microfungi

mineralisation   461
  C   257pp
    budgets   248, 440
    fertilizer trials   269
    methods   238pp, 258pp
    N-interaction   248, 268pp
    rates   247, 260pp
    MRT, see MRT
  climate effects   248, 250, 451, 453
  CO_2 456
  litter   see decomposition
  N   94, 147, 156, 246, 287pp, 291, 298, 377,
      436, 473, 469
    budgets   308, 440
    C/N effect   311pp
    fertilizer effect   316
    methods   299pp
    net rates   305pp, 309
    N pool effect   311, 315
    pH effect   314
    turnover   318pp
    summary   326pp
  transect   469
modelling   419pp
  BIOME BGC   419
  CENTURY   419
  NUCOM model   420pp
    allocation   425
    assimilation   422pp
    calibration   431
    data input   427
    hydrology   426
    mortality   425
    soil chemistry   426
    soil C   425
  PnET   419
  simulations
    age of stands   460
    leaf biomass   433
    logging   459pp
    MRT   438
    mineralisation   437, 440, 444pp, 452
    N-cycle   434
    NBP   479
    NEP   447pp, 452pp, 479
    NPP   443, 447pp, 452pp
    N-availability   443
    transpiration   446
    SOM   436
    summary   461
MRT
  C   237pp, 246pp, 271pp, 378, 441
  N   249pp, 246pp, 441, 469, 477

summary   253
transect   469
mycorrhizae
  amino acid uptake   128, 357, 358
  ammonium uptake   137pp
  carbon uptake   138
  communities   347pp
  colonization of roots   349
  diversity indices   345pp
  genetic diversity   353pp
  N concentration   115
  nitrate uptake   137pp
  proteolytic activity   361
  regional distribution   351, 359
  r-s strategies   360pp
  species number   348pp, 353
  summary   362

N   140, 166, see also nutrition
  amino acids   see amino acids
  ammonium   see ammonium
  availability   450, 475
  budget   248
  canopy N   91, see also canopy nitrogen uptake
  climate effects   246
  concentrations   65, 70, 115
  cycle   3pp, 7, 172, 435pp
  deposition   see deposition
  foliage   see foliage
  isotopes   144pp, see also isotopes
  leaching   321
  mineralisation   see mineralisation
  mycorrhizae   see mycorrhizae
  nitrate   see nitrate
  partitioning in trees   82, 162
  pools   27, 31, 40, 56, 89, 90, 242pp
  pollution
    NH_3 173, 407
    NO   173, 297pp, 407
    NO_2 173, 407pp
  roots   124pp, 150, 161
  saturation   208
  soil   see soil
  throughfall   see deposition
  uptake total   52, 55, 94, 149, 172pp, 434pp
Na
  soil   32
NBP
  budget   478pp
  definition   6
  logging   460
  methodology   478
  modelling   479

rates   458
site conditions   481
needle   see foliage
NEE   6, 217pp
NEP
  budget   232, 250, 444, 478pp
  climate   448pp
  definition   6
  logging   458, 460
  methodology   478
  modelling   447pp, 452pp, 479
  site conditions   481
net biome productivity   see NBP
net ecosystem exchange   see NEE
net ecosystem productivity   see NEP
net primary productivity   see NPP
NIPHYS project   8, 468, 487
nitrate
  ammonium/nitrate   136
  ^{15}N see isotopes
  partitioning in trees   162
  reductase activity   123, 125pp, 178
  soil   35, 38, 83, 157, 297pp
  tissue   83
  transect   470, 475
  uptake   127, 135, 137pp, 149, 162
nitrite   297pp
nitrification   7, 307p, 316
nitrogen   see N
NO   173, 297pp
N_2O 324p
NPP   461
  above ground NPP   see ANPP
  ANPP   54, 66, 90, 92
  climate   448pp, 451, 453
  CO_2 456
  definition, general   6, 49, 60pp
  methods   51
  modelling   447pp
  roots   112, 443
  stem volume   18
  total NPP   54pp, 442
  transect   469p
NUCOM model   see modelling
nutrient fluxes   405
nutrient pools   57pp, 93
  branches, twigs   see branches, twigs
  foliage   see foliage
  stem   see stem
  roots   see roots
nutrient uptake   see roots
nutrition   63
  calcium   see Ca
  concentrations   65pp, 70pp

Subject Index

magnesium   see Mg
methods   64
pools   see nutrient pools
potassium   see K
phosphorus   see P
sulphur   see S

old-growth forests   69

P
plant concentrations   66, 70
pools   58
PnET   see modelling
pollution   see N, see S
potassium   see K
precipitation   see climate

respiration
  annual budget   231, 248
  carbon store   229
  daily trends   221
  ecosystem   232
  method   218, 230
  roots   see roots
  seasonal trends   223
  soil C   see soil
  soil water   224pp
  temperature   224pp
roots
  biomass   99pp, 114
  coarse roots
    biomass   53
    growth   55
  fine roots
    biomass   53, 105, 114
    C investment   111
    dead/live   106, 117
    growth   55, 108
    nutrients   68, 73, 125
    turnover   111, 113, 118
  growth   55, 99pp, 118
    in-growth core   104
    length   108, 109, 116
    root windows   100pp
    soil coring   100
  method   100
  N uptake   122pp
  nutrient pools   58, 66
  nutrient uptake   123pp
  respiration   233
  root/shoot ratio   53

S
concentrations   66, 70

deposition   see deposition
pollution
  SO_2 407
pools   58
cycle   416
simulations   see modelling
sites   see transect
sodium   see Na
soil
  Al   30
  base saturation   30, 34
  bulk density   29, 30pp
  C   68, 237pp
    concentration   28, 31, 36
    fluxes   248, 436
    model   241
    C/N ratio   31pp, 35pp
    pools   27pp, 31pp, 35pp, 40pp, 242pp
    respiration   217pp, 237pp
    turnover time   237pp
  cations   32pp
  CEC   30pp, 34
  coring   see roots
  fauna   see decomposer
  invertebrates   366pp
  loss-on-ignition   32pp
  N   31pp, 68, 159, 469
    concentrations   36
    fluxes   248
    pools   36, 242
    turnover time   249, 440pp
  nutrient profiles   76
  soil organic matter   32pp
  pH   16, 30pp
  stoniness   16, 29
  water   232pp
    DOC   232, 236, 436
    DOC/DON   238pp
    DON   232, 237
    methods   233
SOM   see soil organic matter
specific leaf area   see foliage
stand age   see age of stands
stand characteristics   16pp
stem
  biomass   53
  growth   55
  nutrient pools   58, 66
stoniness   see soil
sulphur   see S

temperature   see climate
throughfall deposition   see deposition
transect   15pp, 42pp, 468pp, 487

C and N pools   469
climate   16, 20, 469
LAI   18, 469
MRT   469
mineralisation   469
N in soil   469
NPP   469
processes   470
site aspect   42pp
site history   18

site locations   10, 15pp, 24pp
soils   16pp, 26pp
twigs   see branches, twigs

understorey
  biomass   52
  nitrogen uptake   165pp

wet deposition   see deposition, see canopy nitrogen uptake

Species Index

fungi
  Amanita spissa   352, 359
  Amphinema sp   348
  Cantharellus tubaeformis   352
  Cenococcum geophilum   348, 359
  Chroogomphus rutilus   352
  Continarius sp   348, 352, 359pp
  Dermocybe cinnamoma   348, 352
  Dermocybe cinnamomeobadia   348, 352
  Dermocybe semisanguinea   359
  Elaphomyces sp   348
  Fagirhiza globulifera   348
  Fagirhiza setifera   348
  Hebeloma various sp   352, 359
  Hydnum repandum   352
  Hygrophorus pustulatus   348
  Laccaria amethystina   207, 343, 359
  Laccaria rufus   348, 359
  Laccaria various sp   348, 352, 359pp
  Lactarius subdulcis   129, 207
  Lactarius various sp   348, 352, 359pp
  Lactarius vellereus   134
  Paxillus involutus   130, 352, 359pp
  Picierhiza bicolorata   348
  Picierhiza gelatinosa   348
  Picierhiza oleiferans   348
  Picierhiza punctata   348, 359
  Picierhiza rosa-nigricans   348
  Picierhiza rosea   348
  Piloderma croceum   134, 348
  Russula ocholeuca   130
  Russula various sp   348, 352, 359pp
  Suillus pungens   360
  Telephora sp   348, 359pp
  Tomentella sublilacina   360
  Tricholema various sp   352, 359
  Tylospora sp   348, 359pp
  Xerocomus various sp   348, 352, 359

invertebrates
  Acari   370
  Colembola   370
  Dicyrtoma fusca   367
  Enchytraeidae   370
  Nebela militaris   367
  Nematoda   366pp, 370pp
  Testate amoeba   370
  Trinema linearis   367

microorganisms and microfungi
  Acremomium   386pp
  Alternaria   386pp
  Aureobasidium   386pp
  Beauveria   387
  Botrytis   387
  Chalara   386pp
  Chrysosporium   387
  Cladosporium   386pp
  Cylindrocladium   387
  Dactylaria   387
  Dreschlera   387
  Fusarium   386pp
  Geotrichum   387
  Menispora   387
  Mortierella   386pp
  Mucor   386pp
  Paecilomyces   386pp
  Phoma   386pp
  Trichoderma   386pp
  Verticillium   387

plants
  Athyrium filix-femina   22
  Betula pubescens   15, 18, 22pp, 40, 71, 79
  Calamagrostis villosa   18, 22, 52
  Calluna vulgaris   19, 22pp, 146
  Deschampsia flexuosa   18pp, 22pp, 43pp, 52, 63, 69, 75, 92pp, 146, 205
  Dryopterix filix-mas   18
  Erica scoparia   18, 22pp, 45
  Fagus sylvatica   all chapters
  Galium odoratum   18, 22, 43pp

Xp = page X and following page; Xpp = page X and following pages

Geranium robertianum 18
Juniperus oxycedris 18, 22pp, 45
Oxalis acetosella 18, 52
Picea abies all chapters
Pinus pinaster 15, 22pp, 30, 45, 69, 71, 79
Pinus halepensis 15, 25
Pinus sylvestris 18, 23pp, 40, 43, 69, 71, 79
Polystichum spinulosum 18, 22pp

Quercus coccifera 22
Quercus ilex 18
Quercus petraea 147
Quercus pubescens 15, 22pp
Rubus sp 18
Ulex parvifolia 18, 45
Vaccinium myrtillus 18p, 22pp, 44pp, 52, 63, 146
Vaccinium vitis-idaea 18p, 24

Conditions of use and terms of warranty
(This is an unauthorized translation of the original text in German language. Only the original German text is legally binding.)

§ 1 Concluding the contract
Opening the sealed plastic cover binds the end user to the conditions of use and the terms of the warranty. If the end user does not wish to be bound by these conditions, he should return the unopened package to his supplier or to Springer-Verlag and the selling price will be refunded. For the return of goods, § 7. is valid.

§ 2 Copyright and conditions of use
1. All rights pertaining to the data are owned by the author. The information is protected by copyright.
2. Springer-Verlag grants the end user, subject to legal liability, the non-exclusive right to use the data as described by the terms of this contract. Under this contract use of the data is restricted to the instructions described..
3. The databank may, at any one time, only be used on one computer at a single workplace.
4. The databank may be copied once for backup purposes.

§ 3 Transfer of the data
1. Any transfer (e.g. sale) of the data to a third party and with it the transfer of the right and the possibility of its use may only occur with the written permission of Springer-Verlag or the author.
2. Springer-Verlag will give this permission when the end user up to this point makes a written application and the subsequent end user makes a declaration that he will remain bound by the terms of this contract. Receipt of permission terminates the right of the first end user and the transfer to the second end user may take place.

§ 4 Unauthorized use
1. The complete dataset is protected by
   - the laws of copyright
   - the laws governing the use of trademarks
   - the laws of trade and commerce
   - this contract.

Violations may lead to action being taken under civil and criminal law.
2. The buyer is liable to Springer-Verlag for any damages or detriment accruing from any infringement of these regulations.

§ 5 Functional limitations
1. Even with the lates state of technological development and with meticulous care being taken during production, errors in the databank cannot be excluded.

§ 6 Warranty
1. In response to justified claims, Springer-Verlag has, as first possibility, the option of supplying the user with another copy of the data. If the claim is still not remedied, the end user can demand the return of the selling price from his supplier when he returns the databank in compliance with the terms set out in § 7.
2. A prerequisite to making a claim under the warranty is that the end user supplies an exact description of the defect in writing.
3. The end user has no claim to a reduction in the selling price or to correction of defects. In other respects the German Code of Civil Law (BGB) concerning the warranty of goods shall apply (§§ 459 to 480 BGB).

§ 7 Returning the databank
1. The customer can only return the databank (e.g. according to § 1 or § 6 Sect. 1.) in its entirety together with the original sales receipt/invoice. In addition he has to hand over the declaration stating that no copies remain in his possession.

§ 8 Help
1. Questions should be mailed or sent via mailbox to Springer-Verlag. The answers from the author are merely forwarded by Springer-Verlag without being checked. The questions are normally answered in the order they are received. It will not be possible to answer every question.

§ 9 Liability
1. Springer-Verlag and the author are only liable for willful intent, gross negligence, and when the databank fails to fulfill its assured purpose and function. The assured purpose and functions are those which are explicitly declared in writing. There is no liability for information described in § 8.
2. The liability under German law for product liability is unaffected.
3. The plea that the end user is also at fault remains an option for Springer-Verlag.

§ 10 Conclusion
1. The location of the competent court for all legal action in connection with the databank and this contract is D-69115 Heidelberg if the contract partner is a registered trader or equivalent, or if he has no legal domicile in Germany.
2. This contract is exclusively governed by the laws of the Federal Republic of Germany with the exception of the UNCITRAL laws of trade and commerce.
3. Should any provision of the contract prove unenforceable or if the contract is incomplete, the remaining provisions will remain unaffected. The invalid provision shall be deemed replaced by the provision which in a legally binding manner comes nearest in its meaning and purpose to the unenforceable provision. This shall apply to any omission in the contract that may occur.

Nutzungs- und Garantiebedingungen
(Original German version of the text "Conditions of use and terms of warranty")

§ 1 Vertragsabschluss
Durch Öffnen der CD-ROM-Hülle vereinbart der Endnutzer mit dem Springer-Verlag die Nutzungs- und Garantiebedingungen, die auf der gegenüberliegenden Seite abgedruckt sind. Falls der Endnutzer dies nicht anerkennen will, kann er die ungeöffnete Packung mit dem Original-Kaufbeleg binnen zwei Wochen gegen volle Erstattung des Kaufpreises seinem Lieferanten oder dem Springer-Verlag zurückgeben. Für die Rückgabe gilt § 7.

§ 2 Urheber- und Nutzungsrechte
1. Alle Nutzungsrechte an der Datenbank stehen dem Autor zu. Die Datenbank ist urheberrechtlich geschützt.
2. Der Springer-Verlag überlässt dem Endnutzer die nicht ausschließliche schuldrechtliche Befugnis, die Datenbank vertragsgemäß zu nutzen. Vertragsgemäß ist nur eine Nutzung, bei der die Datenbank mit Hilfe der beschriebenen Anweisungen ausgeführt wird.
3. Die Datenbank darf zur selben Zeit nur auf einem Rechner und auf einem Arbeitsplatz benutzt werden.
4. Die Datenbank darf nur einmal zu Sicherungszwecken vervielfältigt werden.

§ 3 Weitergabe der Datenbank
1. Jede Weitergabe (z.B. Verkauf) der Datenbank an Dritte und damit jede Übertragung der Nutzungsbefugnis und -möglichkeit bedarf der schriftlichen Erlaubnis des Springer-Verlages oder des Autors.
2. Der Springer-Verlag wird die Erlaubnis geben, wenn der bisherige Endnutzer dies schriftlich beantragt und eine Erklärung des nachfolgenden Endnutzers vorliegt, dass dieser sich an die Regelungen dieses Vertrages gebunden hält. Ab dem Zugang der Erlaubnis erlischt das Nutzungsrecht des bisherigen Nutzers und wird die Weitergabe zulässig.

§ 4 Unerlaubte Nutzung
1. Die gesamte Datenbank ist durch
   - Urheberrecht,
   - Warenzeichenrecht,
   - Wettbewerbsrecht und
   - diesen Vertrag

geschützt. Verstöße hiergegen können zivilrechtlich und strafrechtlich verfolgt werden.
2. Der Käufer haftet gegenüber dem Springer-Verlag für alle Schäden und Nachteile aufgrund von Verletzungen dieser Regelung.

§ 5 Funktionsbeschränkungen
1. Nach dem Stand der Technik können Fehler der Datenbank auch bei sorgfältiger Erstellung nicht ausgeschlossen werden.

§ 6 Garantie
1. Bei berechtigten Beanstandungen hat der Springer-Verlag zunächst die Möglichkeit, dem Endnutzer ein anderes Exemplar zu überlassen. Wenn damit die Beanstandung nicht behoben ist, kann der Endnutzer von seinem Lieferanten den Kaufpreis zurückverlangen, wenn er die Datenbank entsprechend § 7 zurückgibt.
2. Die Inanspruchnahme der Garantie setzt voraus, dass der Endnutzer den Mangel schriftlich genau beschreibt.
3. Auf Minderung und Nachbesserung hat der Endnutzer keinen Anspruch. Im übrigen gelten die Regeln der kaufrechtlichen Gewährleistung (§§ 459–480 BGB) entsprechend.

§ 7 Rückgabe
1. Der Kunde kann die Datenbank (z.B. nach § 1 oder § 6 Abs. 1) nur komplett und mit dem Original-Kaufbeleg zurückgeben. Zusätzlich hat er die Erklärung abzugeben, dass keine Kopien existieren.

§ 8 Beratung
1. Anfragen sind schriftlich oder über Mailbox an den Springer-Verlag zu richten. Der Springer-Verlag vermittelt lediglich ungeprüft die Beantwortung durch den Autor. Die Antworten erfolgen üblicherweise in der Reihenfolge des Eingangs. Nicht jede Frage wird beantwortet werden können.

§ 9 Haftung
1. Der Springer-Verlag und der Autor haften nur bei Vorsatz, bei grober Fahrlässigkeit und bei Eigenschaftszusicherungen. Die Zusicherung von Eigenschaften bedarf der ausdrücklichen schriftlichen Erklärung. Für Auskünfte nach § 8 wird nicht gehaftet.
2. Die Haftung aus dem Produkthaftungsgesetz bleibt unberührt.
3. Der Einwand des Mitverschuldens des Endnutzers bleibt dem Springer-Verlag offen.

§ 10 Schluss
1. Gerichtsstand für alle Klagen im Zusammenhang mit der Datenbank und dieser Vereinbarung ist D-69115 Heidelberg, wenn der Vertragspartner Vollkaufmann oder gleichgestellt ist oder keinen allgemeinen Gerichtsstand in Deutschland hat.

2. Es gilt ausschließlich das Recht der Bundesrepublik Deutschland mit Ausnahme der NCITRAL Kaufgesetze.
3. Sollte eine Bestimmung dieses Vertrages unwirksam sein oder werden oder sollte der Vertrag unvollständig sein, so wird der Vertrag im übrigen Inhalt nicht berührt. Die unwirksame Bestimmung gilt als durch eine solche Bestimmung ersetzt, welche dem Sinn und Zweck der unwirksamen Bestimmung in rechtswirksamer Weise wirtschaftlich am nächsten kommt. Gleiches gilt für etwaige Vertragslücken.

Ecological Studies
Volumes published since 1994

Volume 103
Rocky Shores: Exploitation in Chile
and South Africa (1994)
W.R. Siegfried (Ed.)

Volume 104
Long-Term Experiments With Acid Rain
in Norwegian Forest Ecosystems (1993)
G. Abrahamsen et al. (Eds.)

Volume 105
Microbial Ecology of Lake Plußsee (1994)
J. Overbeck and R.J. Chrost (Eds.)

Volume 106
Minimum Animal Populations (1994)
H. Remmert (Ed.)

Volume 107
The Role of Fire in Mediterranean-Type
Ecosystems (1994)
J.M. Moreno and W.C. Oechel (Eds.)

Volume 108
Ecology and Biogeography
of Mediterranean Ecosystems in Chile,
California and Australia (1995)
M.T.K. Arroyo, P.H. Zedler, and M.D. Fox
(Eds.)

Volume 109
Mediterranean-Type Ecosystems.
The Function of Biodiversity (1995)
G.W. Davis and D.M. Richardson (Eds.)

Volume 110
Tropical Montane Cloud Forests (1995)
L.S. Hamilton, J.O. Juvik, and F.N. Scatena
(Eds.)

Volume 111
Peatland Forestry. Ecology and Principles
(1995)
E. Paavilainen and J. Päivänen

Volume 112
Tropical Forests: Management and Ecology
(1995)
A.E. Lugo and C. Lowe (Eds.)

Volume 113
Arctic and Alpine Biodiversity. Patterns,
Causes and Ecosystem Consequences (1995)
F.S. Chapin III and C. Körner (Eds.)

Volume 114
Crassulacean Acid Metabolism.
Biochemistry, Ecophysiology and Evolution
(1996)
K. Winter and J.A.C. Smith (Eds.)

Volume 115
Islands. Biological Diversity
and Ecosystem Function (1995)
P.M. Vitousek, L.L. Loope, and H. Adsersen
(Eds.)

Volume 116
High Latitude Rainforests and Associated
Ecosystems of the West Coast
of the Americas: Climate, Hydrology,
Ecology and Conservation (1996)
R.G. Lawford, P. Alaback, and E. Fuentes
(Eds.)

Volume 117
Global Change and Mediterranean-Type
Ecosystems (1995)
J. Moreno and W.C. Oechel (Eds.)

Volume 118
Impact of Air Pollutants
on Southern Pine Forests (1996)
S. Fox and R.A. Mickler (Eds.)

Volume 119
Freshwater Ecosystems of Alaska.
Ecological Syntheses (1997)
A.M. Milner and M.W. Oswood (Eds.)

Volume 120
Landscape Function and Disturbance
in Arctic Tundra (1996)
J.F. Reynolds and J.D. Tenhunen (Eds.)

Volume 121
Biodiversity and Savanna Ecosystem
Processes. A Global Perspective (1996)
O.T. Solbrig, E. Medina, and J.F. Silva (Eds.)

Volume 122
Biodiversity and Ecosystem Processes
in Tropical Forests (1996)
G.H. Orians, R. Dirzo, and J.H. Cushman
(Eds.)

Volume 123
Marine Benthic Vegetation. Recent Changes
and the Effects of Eutrophication (1996)
W. Schramm and P.H. Nienhuis (Eds.)

Volume 124
Global Change and Arctic Terrestrial Ecosystems (1997)
W.C. Oechel et al. (Eds.)

Volume 125
Ecology and Conservation of Great Plains Vertebrates (1997)
F.L. Knopf and F.B. Samson (Eds.)

Volume 126
The Central Amazon Floodplain: Ecology of a Pulsing System (1997)
W.J. Junk (Ed.)

Volume 127
Forest Decline and Ozone: A Comparison of Controlled Chamber and Field Experiments (1997)
H. Sandermann, A.R. Wellburn, and R.L. Heath (Eds.)

Volume 128
The Productivity and Sustainability of Southern Forest Ecosystems in a Changing Environment (1998)
R.A. Mickler and S. Fox (Eds.)

Volume 129
Pelagic Nutrient Cycles: Herbivores as Sources and Sinks (1997)
T. Andersen

Volume 130
Vertical Food Web Interactions: Evolutionary Patterns and Driving Forces (1997)
K. Dettner, G. Bauer, and W. Völkl (Eds.)

Volume 131
The Structuring Role of Submerged Macrophytes in Lakes (1998)
E. Jeppesen et al. (Eds.)

Volume 132
Vegetation of the Tropical Pacific Islands (1998)
D. Mueller-Dombois and F.R. Fosberg

Volume 133
Aquatic Humic Substances: Ecology and Biogeochemistry (1998)
D.O. Hessen and L.J. Tranvik (Eds.)

Volume 134
Oxidant Air Pollution Impacts in the Montane Forests of Southern California (1999)
P.R. Miller and J.R. McBride (Eds.)

Volume 135
Predation in Vertebrate Communities: The Białowieża Primeval Forest as a Case Study (1998)
B. Jedrzejewska and W. Jedrzejewski

Volume 136
Landscape Disturbance and Biodiversity in Mediterranean-Type Ecosystems (1998)
P.W. Rundel, G. Montenegro, and F.M. Jaksic (Eds.)

Volume 137
Ecology of Mediterranean Evergreen Oak Forests (1999)
F. Rodà et al. (Eds.)

Volume 138
Fire, Climate Change and Carbon Cycling in the North American Boreal Forest (2000)
E.S. Kasischke and B. Stocks (Eds.)

Volume 139
Responses of Northern U.S. Forests to Environmental Change (2000)
R. Mickler, R.A. Birdsey, and J. Hom (Eds.)

Volume 140
Rainforest Ecosystems of East Kalimantan: El Niño, Drought, Fire and Human Impacts (2000)
E. Guhardja et al. (Eds.)

Volume 141
Activity Patterns in Small Mammals: An Ecological Approach (2000)
S. Halle and N.-C. Stenseth (Eds.)

Volume 142
Carbon and Nitrogen Cycling in European Forest Ecosystems (2000)
E.-D. Schulze (Ed.)

CD-ROM

Contents

The attached CD-ROM contains the database of this book. It is organized in three parts:

- *Part 1:* figures, figure data, and tables of the chapters:
  The figures, figure data and tables presented in the book are included in separate files for each chapter. The figures are available as pdf files. The figure data and the tables are available as ASCII files (separation of columns by tabs).
- *Part 2:* field data for each site:
  The field data that are the basis for the assessment are presented in separate files for each study site. The collected data include data on the tree species, field-layer vegetation, soil organic matter, soil data, soil chemistry data, hydrology as well as climatic conditions and deposition data. The databank covers mainly the years 1993–1998, although information from earlier years is also included when available. The data include primary data from the sites. The availability of data varies between the sites depending on the depth of the study. For instance, the site Salacova Lhota was studied only with respect to long-term trends in whole catchments (see Chap. 19).
- *Part 3:* data on abundance and diversity of soil fauna, soil microfungi, and mycorrhizae. These data are given in separate files containing lists of species present and abundances of these species. Data are ordered by groups of organisms; each file contains data for several sites.

Additionally, jpg files of slides illustrating the sites or techniques are included.

The data files on the sites on the CD-ROM were compiled by Harnke von Oene (see address list) and Barbara Lühker, editorial assistant of E.-D. Schulze (see address list).

Rules for Citation of Data Use

- The data in this databank are free for anyone to use when crediting data products and data holders by correct citation. We also consider it important that the persons that collected the data are cited properly if these data are used for any further analyses.
- Citation of figures, figure data, and tables:
  In the files of the figures and tables of the book chapters a reference of the first author of the book chapter is given. The figures, figure data, and tables should be cited by chapters: NAME (2000) CHAPTER TITLE. In: Schulze E.-D. (ed.) Carbon and nitrogen cycling in European forest ecosystems. Ecological studies 142. Springer, Berlin Heidelberg New York, PAGES
- Citation of field data:
  This database should be treated as a publication. For all data, youl will find information on who collected this data. The format of the citation should be: NAME (2000) SUBJECT (e.g., tree biomass data), CD-ROM Database. In: Schulze E.-D. (ed.) Carbon and nitrogen cycling in European forest ecosystems. Ecological studies 142. Springer, Berlin Heidelberg New York

Conditions of Use and Terms of Warranty

see previous pages